ROUTLEDGE HANDBOOK OF FOREST ECOLOGY

This comprehensive handbook provides a unique resource covering all aspects of forest ecology from a global perspective. It covers both natural and managed forests, from boreal, temperate, subtropical and tropical regions of the world. The book is divided into seven parts addressing the following themes:

- forest types
- forest dynamics
- forest flora and fauna
- energy and nutrients
- forest conservation and management
- forests and climate change
- human impacts on forest ecology.

While each chapter can stand alone as a suitable resource for a lecture or seminar, the complete book provides an essential reference text for a wide range of students of ecology, environmental science, forestry, geography and natural resource management. Contributors include leading authorities from all parts of the world.

Kelvin S.-H. Peh is Lecturer in the Faculty of Natural and Environmental Sciences, University of Southampton, and also Visiting Fellow in the Department of Zoology, University of Cambridge, UK.

Richard T. Corlett is Professor and Director of the Centre for Integrative Conservation, Xishuangbanna Tropical Botanical Garden, Chinese Academy of Sciences, Yunnan, China. He was previously a professor at the National University of Singapore and the University of Hong Kong.

Yves Bergeron is Professor of Forest Ecology and Management at Université du Québec en Abitibi-Témiscamingue and Université du Québec à Montréal, Canada.

ROUTLEDGE HANDBOOK OF FOREST ECOLOGY

Kelvin S.-H. Peh, Richard T. Corlett and Yves Bergeron

Routledge
Taylor & Francis Group
LONDON AND NEW YORK

earthscan
from Routledge

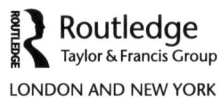

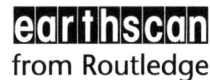

First published 2015
by Routledge
2 Park Square, Milton Park, Abingdon, Oxon OX14 4RN

and by Routledge
711 Third Avenue, New York, NY 10017

First issued in paperback 2017

Routledge is an imprint of the Taylor & Francis Group, an informa business

© 2015 Kelvin S.-H. Peh, Richard T. Corlett and Yves Bergeron,
selection and editorial material; individual chapters, the contributors

The right of the editors to be identified as the authors of the editorial material, and of the authors for their individual chapters, has been asserted in accordance with sections 77 and 78 of the Copyright, Designs and Patents Act 1988.

All rights reserved. No part of this book may be reprinted or reproduced or utilised in any form or by any electronic, mechanical, or other means, now known or hereafter invented, including photocopying and recording, or in any information storage or retrieval system, without permission in writing from the publishers.

Trademark notice: Product or corporate names may be trademarks or registered trademarks, and are used only for identification and explanation without intent to infringe.

British Library Cataloguing-in-Publication Data
A catalogue record for this book is available from the British Library

Library of Congress Cataloging in Publication Data
Routledge handbook of forest ecology / [edited by] Kelvin S.-H. Peh, Richard T. Corlett, and Yves Bergeron.
pages cm
Includes bibliographical references and index.
ISBN 978-0-415-73545-2 (hbk) -- ISBN 978-1-315-81829-0 (ebk) 1. Forest ecology. 2. Forests and forestry. I. Peh, Kelvin S.-H. II. Corlett, Richard. III. Bergeron, Yves, 1956-
QH541.5.F6R68 2015
577.3--dc23
2015011684

ISBN 13: 978-1-138-49531-9 (pbk)
ISBN 13: 978-0-415-73545-2 (hbk)

Typeset in Bembo
by Saxon Graphics Ltd, Derby

CONTENTS

Contributors   *xi*

1   Introduction   1
    Kelvin S.-H. Peh, Yves Bergeron and Richard T. Corlett

PART I
The forest   5

2   Boreal forests   7
    Jean-Pierre Saucier, Ken Baldwin, Pavel Krestov and Torre Jorgenson

3   Northern temperate forests   30
    Lee E. Frelich, Rebecca A. Montgomery and Jacek Oleksyn

4   Subtropical forests   46
    Richard T. Corlett and Alice C. Hughes

5   Tropical forests   56
    Lindsay F. Banin, Oliver L. Phillips and Simon L. Lewis

6   Managed forests   75
    Jürgen Bauhus and Patrick Pyttel

PART II
Forest dynamics — 91

7 Insect disturbances in forest ecosystems — 93
   Daniel Kneeshaw, Brian R. Sturtevant, Barry Cooke, Timothy Work,
   Deepa Pureswaran, Louis DeGrandpre and David A. MacLean

8 The role of fire in forest ecosystems — 114
   David F. Greene and Sean T. Michaletz

9 Ecological effects of strong winds on forests — 127
   Stephen M. Turton and Mohammed Alamgir

10 Forest succession and gap dynamics — 141
   Rebecca A. Montgomery and Lee E. Frelich

11 Tree genetic diversity and gene flow in forest ecosystems — 154
   Francine Tremblay

12 Changing forest dynamics: plot-based evidence — 172
   Simon Willcock and Nikée E. Groot

PART III
Forest flora and fauna — 183

13 Lianas in forest ecosystems — 185
   Stefan A. Schnitzer

14 Vascular epiphytes in forest ecosystems — 198
   David H. Benzing

15 Insects in forest ecosystems — 215
   Andrea Battisti

16 Pathogens and pests in North American forest ecosystems — 226
   Louis Bernier and Sandy M. Smith

17 Bryophytes in forest ecosystems — 239
   Nicole J. Fenton, Kristoffer Hylander and Emma J. Pharo

18 Lichens in forest ecosystems — 250
   Per-Anders Esseen and Darwyn Coxson

19	Mammals in forest ecosystems *Richard T. Corlett and Alice C. Hughes*	264
20	Birds in forest ecosystems *Jeffrey A. Stratford and Çağan H. Şekercioğlu*	279
21	Global patterns of biodiversity in forests *Christine B. Schmitt*	295

PART IV
Energy and nutrients — 307

22	Mycorrhizal symbiosis in forest ecosystems *Leho Tedersoo*	309
23	Biogeochemical cycling *David Paré, Daniel Markewitz and Håkan Wallander*	325
24	Hydrological cycling *Michael Bredemeier and Shabtai Cohen*	339
25	Primary production and allocation *Frank Berninger, Kelvin S.-H. Peh and Hazel K. Smith*	352

PART V
Forest conservation and management — 369

26	Natural regeneration after harvesting *Nelson Thiffault, Lluís Coll and Douglass F. Jacobs*	371
27	Tropical deforestation, forest degradation and REDD+ *John A. Parrotta*	385
28	Restoration of forest ecosystems *David Lamb*	397
29	Forest fragmentation *Edgar C. Turner and Jake L. Snaddon*	411
30	Ecological effects of logging and approaches to mitigating impacts *Paul Woodcock, Panu Halme and David P. Edwards*	422

31	Pollution in forests Mikhail V. Kozlov and Elena L. Zvereva	436
32	Biological invasions in forests and forest plantations Marcel Rejmánek	452

PART VI
Forest and climate change — **471**

33	Fire and climate: using the past to predict the future Justin Waito, Martin P. Girardin, Jacques C. Tardif, Christelle Hély, Olivier Blarquez and Adam A. Ali	473
34	Ecological consequences of droughts in boreal forests Changhui Peng	488
35	Assessing responses of tree growth to climate change at inter- and intra-annual temporal scale Sergio Rossi, Jian-Guo Huang and Hubert Morin	499
36	Plant movements in response to rapid climate change Richard T. Corlett	517
37	Forest carbon budgets and climate change Yadvinder Malhi, Sam Moore and Terhi Riutta	527
38	Modelling climate impacts on forest ecosystems David R. Galbraith and Bradley O. Christoffersen	544

PART VII
Human ecology — **557**

39	Multiple roles of non-timber forest products in ecologies, economies and livelihoods Charlie M. Shackleton	559
40	Agriculture in the forest: ecology and rationale of shifting cultivation Olivier Ducourtieux	571
41	Indigenous forest knowledge Hugo Asselin	586

42	Recreation in forests *Bruce Prideaux*	597
43	Impacts of hunting in forests *Rhett D. Harrison*	610
44	The ecology of urban forests *Mark J. McDonnell and Dave Kendal*	623

Index *634*

CONTRIBUTORS

Mohammed Alamgir; Doctoral Researcher, Centre for Tropical Environmental and Sustainability Science, James Cook University, Australia

Adam A. Ali; Assistant Professor, Université de Montpellier, Institut des Sciences de l'Évolution – Montpellier, UMR 5554, France

Hugo Asselin; Professor, Department of Humanities and Social Development, Université du Québec en Abitibi-Témiscamingue, Canada

Ken Baldwin; Forest Ecologist, Natural Resources Canada, Canadian Forest Service, Canada

Lindsay F. Banin; Statistical Ecologist, Centre for Ecology and Hydrology, UK

Andrea Battisti; Professor, Department of Agronomy, Food, Natural Resources, Animals and the Environment, University of Padova, Italy

Jürgen Bauhus; Professor, Chair of Silviculture, University of Freiburg, Germany

David H. Benzing; Emeritus Professor, Department of Biology, Oberlin College, USA

Yves Bergeron; Professor, Department of Applied Sciences, Université du Québec en Abitibi-Témiscamingue, Canada

Louis Bernier; Professor, Forest Research Institute, Université du Québec en Abitibi-Témiscamingue, Canada

Frank Berninger; Lecturer, Department of Forest Sciences, University of Helsinki, Finland

Olivier Blarquez; Assistant Professor, Department of Geography, Université de Montréal, Canada

Contributors

Michael Bredemeier; Professor, University of Gottingen, Germany

Bradley O. Christoffersen; Postdoctoral Researcher, School of Geosciences, University of Edinburgh, UK

Shabtai Cohen; Senior Research Scientist, Institute of Soil, Water and Environmental Sciences, Agricultural Research Organization (ARO), Volcani Center, Israel

Lluís Coll; Research Scientist, Forest Sciences Centre of Catalonia (CTFC), Spain

Barry Cooke; Research Scientist, Northern Forestry Centre, Natural Resources Canada, Canadian Forest Service, Canada

Richard T. Corlett; Professor, Centre for Integrative Conservation, Xishuangbanna Tropical Botanical Garden, Chinese Academy of Sciences, China

Darwyn Coxson; Professor, Ecosystem Science and Management Program, University of Northern British Colombia, Canada

Louis DeGrandpre; Research Scientist, Laurentian Forestry Centre, Natural Resources Canada, Canadian Forest Service, Canada

Olivier Ducourtieux; Lecturer, UFR Comparative Agriculture, UMR Prodig, AgroParisTech, France

David P. Edwards; Lecturer, Department of Animal and Plant Sciences, University of Sheffield, UK

Per-Anders Esseen; Professor, Department of Ecology and Environmental Science, Umeå University, Sweden

Nicole J. Fenton; Professor, Forest Research Institute, Université du Québec en Abitibi-Témiscamingue, Canada

Lee E. Frelich; Director, Center for Forest Ecology, University of Minnesota, USA

David R. Galbraith; Lecturer, School of Geography, University of Leeds, UK

Martin P. Girardin; Research Scientist, Laurentian Forestry Centre, Natural Resources Canada, Canadian Forest Service, Canada

David F. Greene; Professor, Chair, Department of Forestry and Wildland Resources, Humboldt State University, USA

Nikée E. Groot; Doctoral Researcher, School of Geography, University of Leeds, UK

Panu Halme; Postdoctoral Researcher, Department of Biological and Environmental Sciences, University of Jyväskylä, Finland

Contributors

Rhett D. Harrison; Professor, World Agroforestry Centre (ICRAF), Kunming, China

Christelle Hély; Director of Studies, École Pratique des Hautes Études, Institut des Sciences de l'Évolution - Montpellier, UMR 5554, France

Jian-Guo Huang; Professor, Key Laboratory of Vegetation Restoration and Management of Degraded Ecosystems, Provincial Key Laboratory of Applied Botany South China Botanical Garden, Chinese Academy of Sciences, China

Alice C. Hughes; Associate Professor, Centre for Integrative Conservation, Xishuangbanna Tropical Botanical Garden, Chinese Academy of Sciences, China

Kristoffer Hylander; Professor, Department of Ecology, Environment and Plant Sciences, Stockholm University, Sweden

Douglass F. Jacobs; Professor, Department of Forestry and Natural Resources, Purdue University, USA

Torre Jorgenson; Landscape Ecologist, Alaska Ecoscience, USA

Dave Kendal; Ecologist, Australian Research Centre for Urban Ecology, Royal Botanic Gardens Victoria and University of Melbourne, Australia

Daniel Kneeshaw; Professor, Centre d'étude de la forêt et Département des sciences biologiques, Université du Québec à Montréal, Canada

Mikhail V. Kozlov; Adjunct Professor, Section of Ecology, Department of Biology, University of Turku, Finland

Pavel Krestov; Director, Biogeographer and Vegetation Ecologist, Botanical Garden-Institute of the Far Eastern Branch of the Russian Academy of Sciences in Vladivostok, Russia

David Lamb; Honorary Research Fellow, School of Agriculture and Food Sciences and Centre for Mined Land Rehabilitation, University of Queensland, Australia

Simon L. Lewis; Reader, Department of Geography, University College London, UK

Mark J. McDonnell; Director, Australian Research Centre for Urban Ecology, Royal Botanic Gardens Victoria and Associate Professor, University of Melbourne, Australia

David (Dave) MacLean; Professor, Faculty of Forestry and Environmental Management, University of New Brunswick, Canada

Yadvinder Malhi; Professor, Environmental Change Institute, School of Geography and the Environment, University of Oxford, UK

Daniel Markewitz; Professor, Warnell School of Forestry and Natural Resources, University of Georgia, USA

Contributors

Sean T. Michaletz; Postdoctoral Research Associate, Department of Ecology and Evolutionary Biology, University of Arizona, USA

Rebecca A. Montgomery; Associate Professor, Department of Forest Resources, University of Minnesota, USA

Sam Moore; Postdoctoral Researcher, Environmental Change Institute, School of Geography and the Environment, University of Oxford, UK

Hubert Morin; Professor, Department of Fundamental Sciences, Université du Québec à Chicoutimi, Canada

Jacek Oleksyn; Professor, Polish Academy of Sciences, Institute of Dendrology, Kornik, Poland

David Paré; Research Scientist, Laurentian Forestry Centre, Natural Resources Canada, Canadian Forest Service, Canada

John A. Parrotta; Program Leader for International Science Issues, U.S. Forest Service Research and Development, USA

Kelvin S.-H. Peh; Lecturer, Centre for Biological Sciences, University of Southampton and Visiting Fellow, Department of Zoology, University of Cambridge, UK

Changhui Peng; Professor, Department of Biology, Institute of Environment Sciences, Université du Québec à Montréal, Canada

Emma J. Pharo; Senior Lecturer, School of Land and Food, University of Tasmania, Australia

Oliver L. Phillips; Professor, School of Geography, University of Leeds, UK

Bruce Prideaux; Professor, School of Business and Law, CQUniversity, Australia

Deepa Pureswaran; Research Scientist, Laurentian Forestry Centre, Natural Resources Canada, Canadian Forest Service, Canada

Patrick Pyttel; Researcher and Lecturer, Chair of Silviculture, University of Freiburg, Germany

Marcel Rejmánek; Professor, Department of Evolution and Ecology, University of California Davis, USA

Terhi Riutta; Postdoctoral Researcher, Environmental Change Institute, School of Geography and the Environment, University of Oxford, UK

Sergio Rossi; Professor, Department of Fundamental Sciences, Université du Québec à Chicoutimi, Canada

Contributors

Jean-Pierre Saucier; Directeur p. i. et Chef du Service de la sylviculture et rendement des forêts, Direction de la recherche forestière, Ministère des Forêts, de la Faune et des Parcs du Québec, Canada

Christine B. Schmitt; Assistant Professor, Chair for Landscape Management, University of Freiburg, Germany

Stefan A. Schnitzer; Professor, Department of Biological Sciences, Marquette University, USA

Çağan H. Şekercioğlu; Assistant Professor, Department of Biology, University of Utah, USA

Charlie M. Shackleton; Research Professor, Department of Environmental Science, Rhodes University, South Africa

Hazel K. Smith; Research Fellow, Centre for Biological Sciences, University of Southampton, UK

Sandy M. Smith; Professor, Faculty of Forestry, University of Toronto, Canada

Jake L. Snaddon; Lecturer, Centre for Biological Sciences, University of Southampton, UK

Jeffrey A. Stratford; Associate Professor, Department of Biology and Health Sciences, Wilkes University, USA

Brian R. Sturtevant; Research Ecologist, Northern Research Station, United States Department of Agriculture Forest Service, USA

Jacques C. Tardif; Professor, Centre for Forest Interdisciplinary Research (C-FIR), Department of Biology, The University of Winnipeg, Canada

Leho Tedersoo; Senior Researcher, Natural History Museum and Botanical Garden, Tartu University, Estonia

Nelson Thiffault; Research Scientist, Direction de la recherche forestière, Ministère des Forêts, de la Faune et des Parcs du Québec, Canada

Francine Tremblay; Research Professor, Forest Research Institute (IRF), Université du Québec en Abitibi-Témiscamingue, Canada

Edgar C. (Ed) Turner; Academic Director and Teaching Officer in Biological Sciences, Institute of Continuing Education and Post-doctoral Researcher, Department of Zoology, University of Cambridge, UK

Stephen M. Turton; Professor, Centre for Tropical Environmental and Sustainability Science, James Cook University, Australia

Contributors

Justin Waito; Researcher, Centre for Forest Interdisciplinary Research (C-FIR), Department of Biology, The University of Winnipeg, Canada

Håkan Wallander; Professor, Department of Biology, Lund University, Sweden

Simon Willcock; Research Fellow, Centre for Biological Sciences, University of Southampton, UK

Paul Woodcock; Postdoctoral Researcher, School of Biological Sciences, University of Leeds, UK

Timothy Work; Entomologist, Département des sciences biologiques, Université du Québec à Montréal, Canada

Elena L. Zvereva; Adjunct Professor, Section of Ecology, Department of Biology, University of Turku, Finland

1
INTRODUCTION

Kelvin S.-H. Peh, Yves Bergeron and Richard T. Corlett

Forests are stupendous systems. Since the first trees appeared on Earth in the Late Devonian, 390 million years ago, the complex three-dimensional structure of forests has supported the majority of terrestrial species on Earth and this is still true today. We ourselves evolved from a predominantly forest lineage, but the first humans occupied more open habitats and this non-forest origin is reflected in our ambiguous relationship with forests. On the one hand, this relationship until now has been largely destructive, with forests valued most as a source of land for cultivation, and for timber and bushmeat. On the other hand, on a crowded planet we increasingly value the services that forests can provide: clean water, erosion control, and the amelioration of local, regional and global climates. Forests are also important for recreation, and as sources of artistic and spiritual inspiration. Reconciling these incompatible objectives will require both a better understanding of forest ecology and a wider awareness of the multiple values of forests. We hope this book will contribute to both these aims.

Our forest systems today – unfortunately – are facing a suite of global challenges. Deforestation and forest degradation, biological invasions, excessive and often illegal harvesting of forest products, atmospheric pollution and climate change – to name only the biggest challenges – are impacting our forests in a profound way, and we are still learning how these systems are actually coping. Deforestation has emerged as one of the most damaging problems, as large areas are cleared and converted into agricultural land and livestock ranches to feed an ever-growing population, particularly in the tropics. Large-scale production of biofuels poses a potential additional threat. Climate change, which results in increasingly extreme weather, has also emerged as equally detrimental, leaving its mark on our forests in diverse ways – from erosive floods and deadly droughts to vanishing coastal mangrove habitats. Worse still, these environmental challenges seldom act alone, but also occur together, simultaneously and interact to further aggravate the problems, thereby putting an unprecedented pressure on the forest biodiversity and function. Policy-makers involved in ongoing international agreements, such as the Convention on Biological Diversity, and practitioners in international programmes, such as the United Nations' collaborative initiative in reducing emissions from deforestation and forest degradation (REDD), are seeking practicable solutions. Again, there is a need to provide to a wider audience reliable information on forests, their dynamics, biodiversity and responses to human disturbance and climate change, as well as the applications of ecology in the management and mitigation of these global challenges.

This handbook aims to act as a state-of-the-art summary of our current knowledge of forest ecology. It draws on the expertise of a varied international team of authors, many of whom are experts in their respective field, or practitioners with rich experience in forestry. It aims to be an informative, up-to-date resource on the literature on forests and their ecology under environmental change. The handbook does not promote any particular viewpoint, management practice or conservation approach. Instead, it covers a broad range of subjects subsumed under the realm of forest ecology, and offers a comprehensive overview for each of these topics. Most chapters aim for a global coverage as much as practicably possible, but others focus on the region where the subject is most relevant or has received most attention. By presenting each topic across different geographical areas – or at least those biomes whose processes are significantly different – and cross-comparing them as appropriate, we aim to give readers a unique perspective.

The work is grouped around seven parts. The first section of the handbook, 'The forest', is intended to help define different major forest biomes loosely according to latitudinal belts. The forest types covered are: boreal forests, northern temperate forests, subtropical forests, tropical forests and managed forests. These chapters provide an introductory overview setting out definitions, scopes and different forest types within each major forest biome. Thus, Part I provides a primer on the forest systems from five different forest biomes which provide the framework for understanding the other themes. The list of forest biomes covered is not exhaustive, in that we do not cover some important topics such as the southern temperate forests and the mountain forests. Nevertheless, the vast majority of the forest biomes are discussed in this section.

Part II, 'Forest dynamics', includes a series of chapters that explore the impacts of different disturbances – namely insects, fire and strong winds – on forests. The common theme in this section is that the forest systems often show resistance and resilience to perturbation and apparently are able to recover from disturbances. However, disturbances may exceed a critical level whereby systems rapidly and irreversibly collapse, and it is important to understand the basis for such tipping points, the circumstances under which such phenomena may be reached and how to prevent reaching them. This section also includes chapters that discuss how the biological factors, scaled from genes up to a community of hundreds of interacting species, shape and change the forests.

Part III is intended to showcase the rich biodiversity of the forests. Natural history is needed to support the understanding of major ecological processes. This section therefore concerns the ecology of important taxa such as lianas, vascular epiphytes, bryophytes, lichens, insects, mammals and birds. It also includes a brief introduction to the microbial pathogens and insect pests – the major agents of biological disturbances in both natural and managed forests. As forests harbour a huge variety of organisms, we acknowledge that the taxa discussed in this section cannot be comprehensive; for example, reptiles and amphibians are not covered. Nevertheless, this section ends with a concluding chapter that provides a global perspective on forest biodiversity, evaluating our current knowledge of species diversity in our global forests.

In Part IV, the chapters focus on the complexity of some important ecological functions that ensure energy and nutrients are acquired, utilised and recovered in a clockwork fashion. Any missing components – for instance, mycorrhizal associations – will dramatically jeopardise the net primary productivity. Likewise, any processes that are slowed down will be further diminished by a negative feedback mechanism. For example, a slower rate of nutrient cycling will substantially reduce the net primary productivity, in turn lowering plant nutrient availability and exacerbating the effects of nutrient limitations. These chapters invite the reader to appreciate the fragility of our forest systems. This section therefore builds an ecological perspective of sheer complexity in the engineering of the forests as whole systems that provide an essential foundation supporting all life.

The chapters in Part V examine the ecology behind some of the cross-cutting anthropogenic threats and some conservation approaches that inform policy and management practices. All subjects covered in these chapters either play an important ecological role in restoring the forest systems or are significant challenges that place obstacles in the path of preserving their biodiversity and function. We admittedly consider these issues in separate chapters, rather than simultaneously, although our ability to manage the forests will increasingly require an understanding of the synergistic effects of multiple challenges.

Climate change is the focus of Part VI. This part includes a series of chapters that explore the impact of different climate change phenomena, such as fire and drought. But the list of major global threats due to climate change is not exhaustive, for we do not cover topics such as the rising sea level and its impact on coastal and insular forests, which can be profound, or the impact of increased heavy precipitation events over forested areas. As in the previous section, we consider these threats caused by climate change one at a time, but the last chapter of this section shows how large-scale modelling approaches can help provide an integrated understanding of multiple global threats.

The final part of the handbook examines the direct use of forests by people and its impact in a wider context of maintaining or sustaining the forests' capacity to provide ecosystem services. The chapters within this section discuss some major services which the forests contribute – such as the provision of timber and non-timber forest products, and areas for productive cultivation and nature-based recreation, which includes urban tree-dominated 'green spaces' and forests for recreational hunting. Together these chapters summarise the connections between human well-being and the health of our global forests.

Like all works of this kind, the handbook represents a balance between the need to disseminate further knowledge, and the imperative of not missing the narrative in a welter of details. The intent for this volume is to be an authoritative text, yet at the same time appealing to the layperson interested in an introduction to forest ecology, as well as providing graduate students with a comprehensive collection of current research for further examination and discourse on the subject. Our aim is also to narrate the ecology behind some of the current approaches and trends in forest management to conservation practitioners. We are thankful to every contributor to this volume for sharing their cutting-edge knowledge. Lastly, we are also grateful to a team of outstanding anonymous reviewers for their time and hard work. For their concerns, criticisms, feedback and suggestions have also helped us bring this handbook to a higher level.

PART I

The forest

2
BOREAL FORESTS

Jean-Pierre Saucier, Ken Baldwin, Pavel Krestov and Torre Jorgenson

The boreal biome is one of the largest forested biomes on Earth and forms a circumpolar belt of forests and woodlands between the treeless arctic zone and the temperate zone. It is typically characterized by a cold continental climate with relatively short, mild summers and long, cold winters. About one-third of the biome occurs in the zone of permafrost (Brown *et al*. 1997). Boreal forests and woodlands represent approximately 33 per cent of the world's forested area (FAO 2001). Boreal forests are dominated by a relatively few, primarily coniferous, genera (*Picea, Abies, Larix, Pinus*) that are adapted to cold temperatures, low nutrient conditions and recurrent stand-replacing disturbance. Short growing seasons and cold, acidic mineral soils under conifer canopies result in extensive feathermoss carpets on upland sites. Sphagnum mosses occupy landscape positions with permanently high water tables, resulting in acidic organic soils that often develop into peatlands. On both upland and lowland sites, decomposition rates are slow and nutrient cycling is typically restricted to the upper soil layers where oxygen and increased temperatures support microorganism and fungal metabolism. Understory vegetation of boreal forests is also dominated by a relatively few botanical families, especially the Ericaceae (the heath or blueberry family) which is adapted to cold, nutrient impoverished habitat conditions. With proximity to the oceans, winters become milder and summers cooler; snow covers the ground for longer periods and the growing season is shortened. In these oceanic boreal climates, conifers are generally absent and are replaced by birch (*Betula* spp.), alder (*Alnus* spp.) or ericaceous shrublands that can tolerate such harsh conditions.

Extent of the boreal zone

The boreal zone is the northernmost forest zone and forms a large belt of forests and woodlands between latitudes 42°N and 72°N (Rivas-Martínez *et al*. 2011) (Figure 2.1). It covers approximately 12.1 million km^2 (Kuusela 1992) and represents 8.4 per cent of the surface of the earth. Boreal forests are also called taiga.

The boreal zone covers large areas of North America, Europe and Russia. In North America, boreal forests extend from Alaska (USA) through Canada to Labrador and the island of Newfoundland. In Europe, boreal forests occupy most of the areas of Sweden and Finland, a lesser part of Norway, Russian Karelia and east to the Ural Mountains. In Asia, the boreal zone stretches from the Ural Mountains eastward through Siberia and the Kamchatka Peninsula

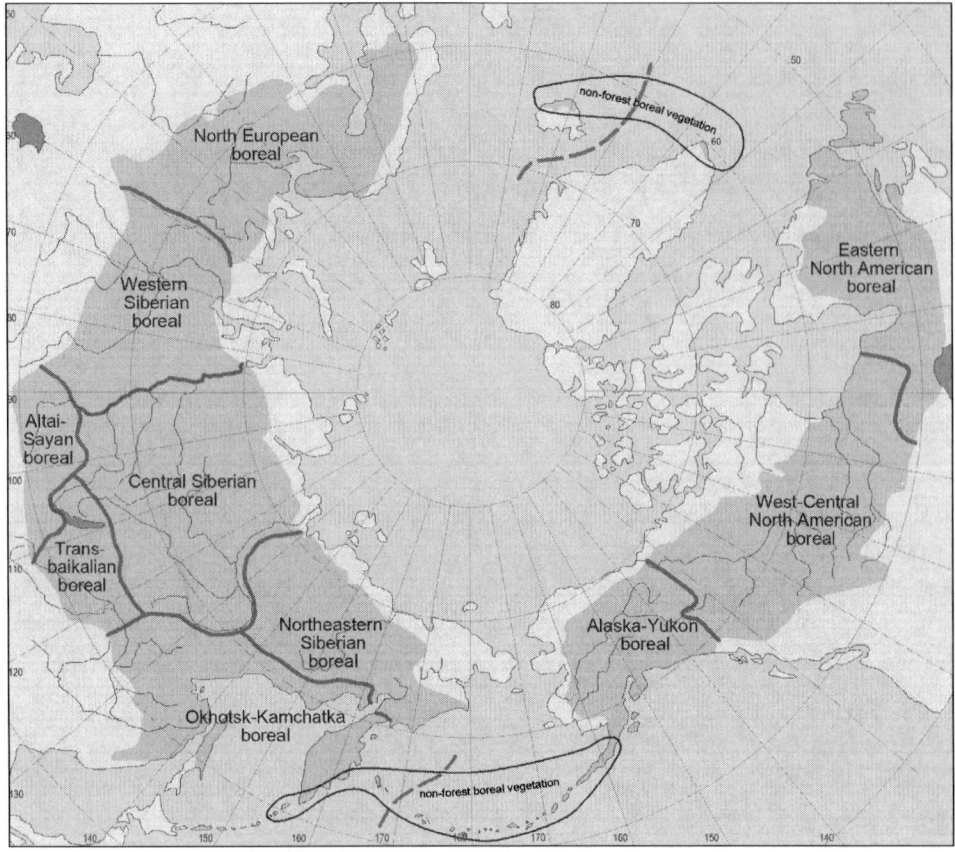

Figure 2.1 Extent of boreal forests and woodlands around the globe, with floristic subdivisions and non-forest boreal vegetation zones

to the Pacific coast. Boreal vegetation also covers some islands in the Atlantic (Iceland, Faroe Islands and the southernmost part of Greenland) and in the Pacific (Kuril and Aleutian island chains). Closed forest covers 76 per cent of the boreal zone, while commercially exploitable closed forest occupies about 53 per cent of the total area (Table 2.1).

The northern limit between the boreal zone and the arctic zone is the continental treeline (i.e., the edge of the habitat in which trees are capable of growing). Moving northward from closed forests through open forests and woodlands, climatic and/or site conditions eventually become too harsh (cold, dry, windy, infertile) to support tree growth. Trees are replaced by low or prostrate shrubs, especially dwarf ericaceous, birch and willow species, together with herbs, mosses and lichens as forest and woodland gradually changes to arctic tundra.

The southern boundary of the boreal zone is less obvious as it is based on vegetation physiognomy, species composition and ecosystem dominance on the landscape. This boundary represents a gradual shift of forest composition where, moving southward, thermophilous species replace the boreal species (Brandt 2009). It is usually marked by a change from a coniferous dominated landscape, associated with a cold climate, to a mixed forest landscape with a milder climate (see next section). In certain areas, with soils of variable fertility and very dry climate, the boreal forest is replaced by steppe at its southern margin.

Table 2.1 Area of boreal forests and woodlands by country and proportion of the country in closed forest or exploitable closed forest (adapted from Kuusela 1992)

Area of forest and other wooded land within the boreal zone (million km²)

	Russia	Alaska	Canada	Norway	Sweden	Finland	Total
Area (million km²)	7.90	0.46	3.27	0.07	0.21	0.23	12.14
Proportion of the total	65.1%	3.8%	26.9%	0.6%	1.8%	1.9%	100.0%

Proportion of the boreal zone covered by closed forest or exploitable closed forest

	Russia	Alaska	Canada	Norway	Sweden	Finland	Total
Closed forest	85.2%	10.9%	60.6%	84.3%	86.0%	85.9%	75.8%
Exploitable closed forest	57.0%	10.9%	44.0%	72.9%	75.2%	83.7%	52.6%

Transition zones

Between the boreal zone and the adjacent vegetation zones, many authors recognize transition zones.

Between the boreal zone and the arctic zone, the transition is called hemiarctic, or subarctic. This ecotone is characterized by a landscape matrix of woodlands and treeless barrens where rare stands of trees grow in sheltered locations. Accumulated snow usually protects the trees during the coldest months. These tree stands are often embedded in krummholz vegetation. Authors agree that the hemiarctic is part of the boreal zone, since its flora retains species that are at the northern limit of their ranges, such as the tree species.

Between the boreal zone and the temperate zone, the transition is called hemiboreal. In the hemiboreal, boreal species form mixed forest types with species that are less cold tolerant and are at the northern limit of their ranges. For example, in eastern North America, balsam fir (*Abies balsamea*[1]) and white spruce (*Picea glauca*), characteristic of the closed boreal forest, form mixed stands with yellow birch (*Betula alleghaniensis*) and sugar maple (*Acer saccharum*). In eastern Asia, a mix of broadleaved tree species (*Acer mono, Betula costata, Fraxinus mandshurica, Quercus mongolica*) and conifers (*Pinus koraiensis*) marks the ecotone between the boreal and temperate zones. Some authors classify the hemiboreal as a part of the boreal zone. Considering that temperate species found in this transition zone are at the northern limit of their ranges and extend far south into the temperate zone, hemiboreal vegetation fits better into the temperate zone (Saucier 2008). This treatment is logically consistent with classification of the hemiarctic vegetation into the boreal zone.

Conifer forests outside the boreal zone

In the temperate zone, certain climatically or edaphically extreme habitats support vegetation communities that share characteristics similar to those of the boreal zone.

The colder climate at higher elevations, associated with more severe exposure to wind and weather events, often with shallow or eluviated soils, favours species that show similar adaptations to those of boreal species. In some areas, high elevation mountains are refugia for forests that covered lower elevations at these latitudes during the past glaciation. Therefore, forest types similar to boreal ones are often found at high elevations of mountainous regions in the temperate zone.

---

1 For vascular plants, we follow the nomenclature of Flora of North America (1993+) for North America and Cherepanov (2007) for Europe and Asia.

In eastern North America, some high elevation forests, embedded into the northern temperate zone, share species such as *Abies balsamea, Picea glauca* and *Betula papyrifera* with the boreal zone further north, but also harbour *Picea rubens*, a more temperate species. This is the case over 1000 m elevation in New England. On a few of the highest mountain peaks, over 1300 m elevation, numerous arctic species occur as disjunct populations.

In the Rocky Mountains of western North America, a similar situation can be observed. Boreal species such as *Pinus contorta, Picea glauca* and *Abies lasiocarpa* are replaced at higher elevations southwards by temperate species of pine (e.g., *Pinus edulis*), fir (e.g., *Abies concolor*), spruce (e.g., *Picea engelmannii*) and juniper (e.g., *Juniperus osteosperma*).

In Asia, boreal-like forests occur in the mountains of the Japanese archipelago, where they are dominated by *Abies veitchii, Abies mariesii* and *Abies homolepis*. Likewise, forests of *Abies koreana* occupy the highest elevations of the mountains in the southern Korean peninsula (Nakamura and Krestov 2005). Further south, *Abies kawakamii* occurs above 3000 m in the mountains of northern Taiwan. All of these forests include some boreal species in the understory, but most of their flora is composed of temperate species.

Bogs and poor fens occurring in temperate regions contain boreal species that are characteristic of cold, acidic, nutrient-poor peatlands. In North America, *Picea mariana, Larix laricina* and *Sphagnum* spp. are typical of these oligotrophic wetlands while in Asia, *Picea glehnii* and several *Larix* species are found in these habitats.

Climate of the boreal zone

The boreal zone is generally characterized by a cold climate with relatively short summers and long winters. To distinguish boreal from other ecological zones, FAO (2001) retained the criteria proposed by Köppen and modified by Trewartha based on the number of months with a mean monthly temperature over 10°C. The boreal zone has up to three months with a mean monthly temperature over 10°C, while the temperate zone has four to eight, and the arctic zone (or polar) has none. Winter (mean monthly temperature < 0°C) usually lasts from five to seven months. The length of the vegetation growing season (mean daily temperature > 5°C) ranges from 80 to 150 days, but can be as low as 50 to 70 days at the continental treeline and along oceanic coasts, marking the transition between forests and non-forest vegetation. The boreal zone cannot be easily defined by mean annual temperature because temperature and precipitation act as compensating climatic factors. In continental climates, winters are so cold that low temperatures affect the survival of tree species, while summers are quite hot, sometimes with a water deficit for plants. Conversely, in oceanic climates, winters are not so cold but summer temperatures stay quite low, usually with very high humidity; thus, the resulting growing season is very short. Rivas-Martínez *et al.* (2011) use three climatic indices to define six bioclimates in the boreal macrobioclimate (Table 2.2). In this system, depending on continentality, the mean annual temperature of boreal sub-zones can range from less than 0°C up to 6°C (Rivas-Martínez *et al.* 2011). Conifer tree species characterize all boreal bioclimates except boreal hyperoceanic. In this bioclimate, vegetation is dominated by broadleaved woodlands, krummholz or ericaceous dwarf shrubs.

Table 2.2 Climatic values for bioclimates within the boreal macrobioclimate (from Rivas-Martínez et al. 2011)

Macrobioclimate	Bioclimate	Mean annual temperature (°C)	Continentality index[1] (°C)	Annual ombrothermic index[2]
Boreal	Boreal hyperoceanic	< 6.0	< 11	> 3.6
	Boreal oceanic	≤ 5.3	11–21	> 3.6
	Boreal subcontinental	≤ 4.8	21–28	> 3.6
	Boreal continental	≤ 3.8	28–46	> 3.6
	Boreal hypercontinental	< 0.0	> 46	–
	Boreal xeric	≤ 3.8	< 46	> 3.6

[1] The continentality index represents the annual thermic interval (mean monthly temperature of the hottest month minus the mean monthly temperature of the coldest month).
[2] The annual ombrothermic index provides an indication of the amount of water available for evapotranspiration (total precipitation of the months with mean monthly temperature > 0°C divided by the sum of mean monthly temperatures > 0°C times 10).

Vegetation of boreal forests

Main floristic subdivisions of the boreal zone

Boreal forests cover northern latitudes around the globe and overall have very similar floristic composition in their different regions. Boreal forests are characterized by coniferous tree species in the genera *Picea*, *Abies*, *Larix* and *Pinus*, often in association with broadleaved species in the genera *Betula*, *Populus*, *Alnus*, *Sorbus* and *Salix*. Understory floristic composition, dominated by mosses, lichens and ericaceous shrubs, also shows global similarity due to the shared biogeographic history of this biome. Because of these general similarities, Takhtajan (1986) combined the entire boreal zone into a single Circumboreal Floristic Region subdivided into 15 Floristic Provinces. The differences in floristic composition between these Provinces represent differences in regional climate and other environmental factors within the boreal zone, as well as differences in the history of geological and climatic events, and related evolutionary pathways for boreal species.

For this chapter, we propose ten broad floristic subdivisions of the boreal zone (Figure 2.1): North America is divided into three floristic subdivisions (Alaska–Yukon, West–Central and Eastern); there is only one European floristic subdivision (North European); and Asia is divided into six floristic subdivisions (Western Siberian, Altai–Sayan, Central Siberian, Transbaikalian, Northeastern Siberian and Okhotsk–Kamchatka). These are based on Takhtajan's Floristic Provinces in Eurasia and on floristic subdivisions developed for the Canadian National Vegetation Classification project (CNVC 2014) in North America. Each of these floristic subdivisions has its own set of characteristic tree species and forest types, although some species are present in more than one floristic subdivision. Table 2.3 presents the dominant tree species by floristic subdivisions of the boreal zone.

Table 2.3 Dominant boreal tree species by continent and floristic subdivisions of the boreal zone

North American floristic subdivisions		
Alaska–Yukon	West–Central North American	Eastern North American
Picea glauca (White spruce)	*Picea glauca* (White spruce)	*Picea glauca* (White spruce)
Picea mariana (Black spruce)	*Picea mariana* (Black spruce)	*Picea mariana* (Black spruce)
Larix laricina (Tamarack)	*Larix laricina* (Tamarack)	*Larix laricina* (Tamarack)
Betula neoalaskana (Alaska paper birch)	*Pinus contorta* (Lodgepole pine)	*Pinus banksiana* (Jack pine)
Populus tremuloides (Quaking aspen)	*Pinus banksiana* (Jack pine)	*Abies balsamea* (Balsam fir)
Populus balsamifera (Balsam poplar)	*Abies lasiocarpa* (Subalpine fir)	*Betula papyrifera* (Paper birch)
	Abies balsamea (Balsam fir)	*Betula pubescens* (Downy birch)
	Betula papyrifera (Paper birch)	*Populus tremuloides* (Quaking aspen)
	Betula nana (Dwarf birch)	*Populus balsamifera* (Balsam poplar)
	Populus tremuloides (Quaking aspen)	
	Populus balsamifera (Balsam poplar)	

European floristic subdivision
North European
Picea abies (Norway spruce)
Pinus sylvestris (Scots pine)
Betula pendula (Silver birch)
Betula pubescens (Downy birch)
Betula nana (Dwarf birch)
Populus tremula (Eurasian aspen)

Asian floristic subdivisions		
Western Siberian	Altai–Sayan	Central Siberian
Picea obovata (Siberian spruce)	*Picea obovata* (Siberian spruce)	*Larix gmelinii* (Dahurian larch)
Abies sibirica (Siberian fir)	*Abies sibirica* (Siberian fir)	*Picea obovata* (Siberian spruce)
Larix sibirica (Siberian larch)	*Larix sibirica* (Siberian larch)	*Abies sibirica* (Siberian fir)
Pinus sibirica (Siberian pine)	*Pinus sibirica* (Siberian pine)	*Larix sibirica* (Siberian larch)
Pinus sylvestris (Scots pine)	*Pinus sylvestris* (Scots pine)	*Larix cajanderi* (Cajander's larch)
Betula pendula (Silver birch)	*Betula pendula* (Silver birch)	*Pinus sibirica* (Siberian pine)
Populus tremula (Eurasian aspen)	*Betula platyphylla* (Asian white birch)	*Pinus sylvestris* (Scots pine)
	Populus tremula (Eurasian aspen)	*Betula platyphylla* (Asian white birch)
		Populus tremula (Eurasian aspen)
Transbaikalian	Northeastern Siberian	Okhotsk–Kamchatka
Larix gmelinii (Dahurian larch)	*Larix cajanderi* (Cajander's larch)	*Picea jezoensis* (Jezo spruce)
Pinus sylvestris (Scots pine)	*Pinus pumila* (Siberian dwarf pine)	*Abies nephrolepis* (Manchurian fir)
Pinus sibirica (Siberian pine)	*Pinus sylvestris* (Scots pine)	*Abies sachalinensis* (Sakhalin fir)
Picea obovata (Siberian spruce)	*Betula platyphylla* (Asian white birch)	*Larix cajanderi* (Cajander's larch)
Abies sibirica (Siberian fir)	*Populus tremula* (Eurasian aspen)	*Betula ermanii* (Erman's birch)
Larix sibirica (Siberian larch)		*Betula platyphylla* (Asian white birch)
Betula pendula (Silver birch)		*Pinus pumila* (Siberian dwarf pine)
Populus tremula (Eurasian aspen)		*Populus tremula* (Eurasian aspen)

North American floristic subdivisions

Alaska–Yukon boreal

The Alaska–Yukon boreal is dominated by mixed forests that range from early to late successional stages due to the prevalence of fire disturbance. While open to closed *Picea glauca* forests represent the potential natural vegetation on mesic slopes, *Betula neoalaskana–Populus tremuloides* forests, and spruce–birch–aspen mixed forests are intermixed in a patchy mosaic of differing aged stands, with the three dominant tree species often present in all successional stages (Viereck *et al.* 1992). The Alaska–Yukon boreal extends from western Alaska, where the forest grades into arctic tundra and boreal shrublands and heaths, to the western edge of the North American Cordillera in Yukon Territory. Latitudinally, it extends from the southern Brooks Range to the northern side of the Alaskan coastal mountains. It is differentiated from the West–Central North American boreal mainly by the presence of *Betula neoalaskana* and the absence of *Pinus contorta* var. *latifolia* and *Abies lasiocarpa*. Generally, the understory contains species that are characteristic of alpine areas or northern boreal latitudes (e.g., *Vaccinium uliginosum, Empetrum nigrum, Betula nana, Arctous rubra* and *Nephroma arcticum*). Forests are typically found up to 900 m in elevation.

The Alaska-Yukon boreal comprises a diversity of forest types related to topographic position, soil moisture, nutrients and permafrost, and disturbance by fire and thermokarst activity (Van Cleve *et al.* 1983, Chapin *et al.* 2006). In upland permafrost-free areas, vegetation succession after fire has several stages, including herb stage, shrub and sapling stage, deciduous forest (*Betula neoalaskana, Populus tremuloides*) after 30 to 130 years, and coniferous forest (*Picea glauca*) after 100 to 250 years. On permafrost-affected soils on north-facing slopes and lowlands, coniferous woodlands (*Picea mariana, Larix laricina*) predominate and are highly susceptible to fire. Thermokarst is prevalent in permafrost-rich lowlands and creates a variety of non-forested ecosystems. In the northern portion of the region, coniferous woodlands (*Picea glauca, Betula nana*) predominate, while in the southern portion ferns and herbs (*Gymnocarpium dryopteris, Dryopteris dilatata*) are abundant in the understory and other shrub species (*Salix barclayii, Alnus viridis* ssp. *sinuata*) predominate.

The non-forested region of the North Pacific, including southwest Alaska, Aleutian Islands, southeastern Kamchatka, and northern Kuril Islands have a hyperoceanic to oceanic climate and a distinctive vegetation resulting from species of both Asian and North American origin. The region supports shrublands, dwarf shrub heaths, and herbaceous meadows, with a flora dominated by circumpolar Arctic and Asian and North American boreal species and a strong component of more restricted amphi-Beringian species. But there are few species that are endemic to this region or have only a North American distribution. While the bioclimate and distinctive vegetation form the basis for a separate subdivision for this area, we included this non-forested region in the Alaska–Yukon subdivision because the focus of this chapter is on boreal forests.

West–Central North American boreal

The West–Central floristic subdivision of North America includes boreal forests of central Canada (northwestern Ontario, Manitoba, Saskatchewan, Northwest Territories and eastern Alberta), of northern British Columbia and of southern Yukon Territory, as well as the Cordilleran foothill forests of western Alberta. Sub-humid, continental climates prevail throughout the region, which extends from approximately longitude 87°W in northwestern

Ontario to approximately longitude 136°W in northwestern British Columbia/southwestern Yukon, where it extends northwards to approximately 62°N of latitude. Characteristic forest cover comprises closed-crown conifer forests, although more open forest or even woodlands are common in colder, northern areas or when site-level ecological factors become limiting. In southern portions of the region, broadleaved and mixedwood forests dominate the landscape.

The eastern portion of this floristic subdivision is underlain by the Canadian Shield, from which nutrient-poor, acidic sediments were derived by glaciation. The lithology of the rest of the region, including the boreal plains of the south-central portion, as well as the Rocky Mountains and northern British Columbia/southern Yukon, is of sedimentary and metamorphic origins and produces soils with richer nutrient content. North of approximately latitude 60°N permafrost becomes prevalent and affects soil productivity.

Conifer forests of this subdivision are characterized by spruce/feathermoss types. *Picea mariana* dominates on poorer upland soils and on forested wetlands; *Picea glauca* is the dominant species on more fertile upland and riparian sites. Both species range northwards to the continental treeline, with stand physiognomy changing into open spruce–lichen woodlands in northern areas.

With the short fire return interval in this dry, continental climate, the early seral broadleaved species *Populus tremuloides* and *Betula papyrifera* are common and abundant on the landscape. They form pure hardwood stands and also mixed stands with the spruces. *Populus tremuloides* is very common in southern portions of the region, where it often forms mixed stands with *Picea glauca*. At the southern margin of the boreal forest in Alberta, Saskatchewan and Manitoba, these species form parkland conditions as the boreal formation transitions to grasslands on the Great Plains of North America. *Betula papyrifera* ranges further north than aspen, intergrading with *Betula neoalaskana* in the northwestern part of this region. Approaching elevational and continental treeline, shrub birch species (e.g., *Betula nana*) become prevalent as the abundance of trees on the landscape declines.

In the Cordilleran foothills of western Alberta, and in northern British Columbia and southern Yukon, *Pinus contorta* var. *latifolia* is very abundant on the landscape, responding to the intense forest fire regime. It forms extensive stands following stand-replacing fire, often in mixture with either spruce species, as well as with aspen or birch. East of the foothills, *Pinus banksiana* occupies the same ecological niche as *Pinus contorta*. Its abundance increases eastwards on the Shield and is at its maximum in Manitoba and northwestern Ontario.

Abies spp. are rare on the landscape because of the intense fire regime. In the Cordillera and in British Columbia and southern Yukon, *Abies lasiocarpa* occurs occasionally on fire-protected sites. Its abundance increases with elevation, where it is an important subalpine species. Similarly, east of the Cordillera, *Abies balsamea* occurs occasionally where fire return intervals are longer. *Populus balsamifera* is characteristic of river floodplains and other rich, moist sites in the West–Central boreal region where it often occurs in mixture with *Picea glauca*.

The understory floristics of the West–Central boreal region are characterized by two factors: 1) presence and/or dominance of Cordilleran species like *Viburnum edule*, *Rosa acicularis*, *Lonicera involucrata*, *Shepherdia canadensis* and *Leymus innovatus*; and 2) absence and/or poor representation of eastern species.

Similarly, the eastern boundary of the West–Central boreal region corresponds approximately with the western range limits of understory species such as *Vaccinium angustifolium*, *Eurybia macrophylla*, *Diervilla lonicera* and *Clintonia borealis*. Other eastern species such as *Acer spicatum* and *Abies balsamea* are uncommon and infrequent in the region.

In feathermoss forests, *Pleurozium schreberi* dominates on poor sites (especially on the Canadian Shield), while *Hylocomium splendens* is dominant on the richer soils of the Cordillera and boreal plains. Associated ericaceous species are dominated by *Ledum groenlandicum, Vaccinium vitis-idaea* and *V. myrtilloides*. *Arctostaphylos uva-ursi*, together with reindeer lichens (esp. *Cladonia*[2] *mitis*), is common on dry, poor sites. *Alnus incana* is largely associated with floodplains in the West-Central region.

Distinction of the West–Central boreal from the Alaska–Yukon boreal corresponds to the range limits of *Pinus contorta* var. *latifolia* and *Abies lasiocarpa*, as well as the increased prevalence of northern species such as *Vaccinium uliginosum, Empetrum nigrum, Betula nana, Arctous rubra* and *Nephroma arcticum*.

Eastern North American boreal

The Eastern boreal floristic region of North America extends from northwestern Ontario (approximate longitude 87°W), through Québec to the Atlantic Ocean in Newfoundland and Labrador (longitude 53°W). Characteristic forest cover consists of coniferous closed-crown forests, but more open forest or even woodlands are frequent when the climate is colder or when ecological factors become more limiting. In southern portions of this region, broadleaved and mixedwood forests dominate the landscape.

The most characteristic and widely distributed forest type is the black spruce–feathermoss forest. In this forest type, *Picea mariana* dominates the canopy, with *Abies balsamea, Pinus banksiana* and sometimes *Picea glauca* as companion species in variable amounts. Closer to the northern limit of the boreal, the canopy opens to form a *Picea mariana*–lichen woodland.

Pinus banksiana is more frequent in the more continental parts of this region where the fire cycle is relatively short (around 100 years). In fact, both *Pinus banksiana* and *Picea mariana* are well adapted to fire disturbance because they have serotinous or semi-serotinous cones that can withstand fire and release seeds when mineral soil seedbeds have been exposed following fire.

Abies balsamea is associated with maritime influences and milder climates; it is most common in the humid areas of eastern Québec and Newfoundland and Labrador. Here, it mixes with *Picea mariana* after long periods without fire. In these cases, the shade-tolerant fir can regenerate under canopy cover and survive with low light availability, waiting for the canopy to open after a catastrophic event or gap-phase replacement of senescent trees. It is helped periodically by insect epidemics (e.g., spruce budworm [*Choristoneura fumiferana*] or hemlock looper [*Lambdina fiscellaria*]) that create these gaps and release the understory regeneration. Under these conditions, black spruce adopts another strategy to maintain itself — vegetative regeneration by layering. White spruce is much more frequent in hyperoceanic and oceanic bioclimates.

In southern areas of the eastern boreal, *Abies balsamea* mixes with the shade-intolerant broadleaved species, *Betula papyrifera* and *Populus tremuloides*, mostly as boreal mixedwood forests. The *Abies balsamea–Betula papyrifera* forest type is characteristic of the southern latitudes of the boreal zone in eastern Québec and Newfoundland and Labrador. In the more continental parts of this subdivision (Ontario and western Québec), mixed forests are dominated by *Populus tremuloides*.

The understory of conifer forests is typically dominated by mosses such as *Pleurozium schreberi* and *Hylocomium splendens* and, in oceanic areas, liverworts like *Bazzania trilobata*. The shrub layer is represented by heath species, in particular *Kalmia angustifolia* and *Vaccinum angustifolium*, as well as *V. myrtilloides* and *Ledum groenlandicum*. On colder or less fertile sites, often associated with coarse-textured and dry soils, the tree canopy is more open and lichens such as *Cladonia*

---

2   A recent decision about this genus put all *Cladina* species into *Cladonia*. We decided to use *Cladonia* even if Flora of North America (1993+) still uses *Cladina*.

stellaris and *C. mitis* cover the ground. On richer sites or warmer parts of the eastern boreal, the herb layer is well developed to the detriment of the moss layer, while *Acer spicatum*, *Corylus cornuta* and other broadleaved shrubs replace the heaths in the shrub layer. On wetter sites, *Sphagnum* species, as well as shrub species such as *Alnus incana* ssp. *rugosa*, dominate the understory.

European floristic subdivision

North European boreal

Takhtajan's North European floristic subdivision lies within the true boreal vegetation belt marked by forests of *Picea abies* and extends from Norway in the west (approximately longitude 5°E) to the Ural Mountains in the east (approximately longitude 60°E) between the treeless vegetation of the arctic zone to the north and mixed broadleaved–conifer forests to the south. The Northern European forests have a simple structure and species-poor floristics.

The typical and most widespread forest condition of the North European floristic subdivision on zonal sites is composed of *Picea abies* forming a tall overstory with closed-canopy cover. *Betula pendula* and *Pinus sylvestris* may occasionally be present in the tree layer. These species rapidly increase their presence after disturbance and the formation of gaps in the spruce canopy. The shrub layer is not abundant in these forests. The herb and dwarf shrub layer is usually well developed, containing *Vaccinium myrtillus*, *V. vitis-idaea*, *Deschampsia flexuosa*, *Luzula pilosa*, *Linnaea borealis*, *Melampyrum pratense* and *Trientalis europaea*. The ground layer is dominated by *Hylocomium splendens*, *Pleurozium schreberi*, *Ptilium crista-castrensis* and by a number of *Dicranum* spp. On nutrient-rich sites, productivity of forests increases and *Oxalis acetosella*, *Anemone nemorosa*, *Rubus saxatilis* and *Maianthemum bifolium* dominate in the herb layer.

Pinus sylvestris is more abundant in the canopy on nutrient-poor sites and becomes a fully dominant species on well-drained sites with sandy nutrient-poor soils with a dense lichen cover of *Cladonia* spp. and scattered ericaceous species such as *Calluna vulgaris*, *Arctostaphylos uva-ursi* and *Vaccinium vitis-idaea*. Pine also occupies paludified habitats with thick organic soils and water table at the ground surface for several months per year, indicated by the presence of *Ledum palustre*, *Vaccinium uliginosum*, *Chamaedaphne calyculata* and by a well-developed ground cover of *Sphagnum*.

Due to the strong climatic gradient between the Atlantic coast and interior regions, forest vegetation changes with distance from the ocean (Hytteborn *et al.* 2005). With proximity to the Atlantic coast, dominance of spruce decreases and the proportion of *Betula pubescens* increases. In the northeastern part of Norway, spruce disappears altogether and there is a narrow belt of pure *Betula pubescens* forests with crooked trunks and a species-rich herb layer composed of tall forbs. Where a hyperoceanic boreal climate with a short growing season prevails, in Iceland, Faroe Islands and southernmost Greenland, boreal vegetation is represented by treeless communities (Daniëls and de Molenaar 2011). In the eastern part of the province, the climate becomes more continental and *Larix sukaczewii* appears in the forest overstory.

Asian floristic subdivisions

The boreal zone of northern Asia extends from the Ural Mountains in the west (approximately longitude 60°E) to the Bering Strait (approximately longitude 170°W) in the east. It covers an enormous climatic gradient from the ultracontinental regions in eastern Yakutia, where winter temperatures fall well below −60°C and summer temperatures reach +35°C, to the Pacific coast of Kamchatka, in an oceanic climate of mild winters, cool summers and a very short growing season. About half of this area is in the zone of continuous permafrost. Compared to other parts of the boreal zone, the Asian portion is the most diverse.

Western Siberian boreal

The Western Siberian floristic subdivision extends from the Ural Mountains in the west (approximately longitude 60°E) to the Yenisei River in the east and borders the arctic zone to the north and steppe zone in the south. Floristically, this subdivision is very similar to the North European subdivision, except for replacement of the dominant conifer species by *Picea obovata*, *Abies sibirica*, *Pinus sibirica* and *Larix sibirica*. *Pinus sylvestris* occurs with increasing presence towards the south. *Betula pendula*, *Betula pubescens* and *Populus tremula* are common broadleaved species (Ermakov and Morozova 2011).

Understory composition in a typical stand of *Picea obovata* dominated forests is similar to that of a spruce stand in the North European boreal subdivision. Spruce forests with *Vaccinium myrtillus* are characteristic of sites with medium soil nutrient and moisture regimes. Greater representation of herb species in the understory, (e.g., *Oxalis acetosella*, *Maianthemum bifolium*, *Rubus saxatilis* and fern species) indicates increasing soil richness. A well-developed moss layer is characteristic for nearly all forest types, but on nutrient-poor soils the role of bryophytes in ground vegetation is increased. On sandy, dry and nutrient-poor sites *Vaccinium vitis-idaea* and *Cladonia* spp. become widespread. On moist sites with thick organic soil horizons, *Sphagnum* cover develops and species characteristic of bog vegetation may be present.

Western Siberian boreal forests vary at a landscape level. *Picea obovata* dominated forests occupy habitats with well-developed soils and with moderate characteristics of soil moisture and air humidity. *Abies sibirica* is a humidity sensitive species that occurs in habitats that receive more precipitation and have long periods of foggy weather. *Pinus sibirica* occupies large areas with poor and well-drained soils, where coarse materials from glaciers were deposited or on ridges with shallow, stony soils around rock outcrops. *Larix sibirica*, a fire-resistant and gap-dependent species, increases its presence on sites affected by ground fires.

Altai–Sayan boreal

Southern Siberia is a region where the boreal zone transitions with temperate and steppe regions. The Altai–Sayan floristic subdivision is influenced by the vegetation of adjacent mountainous areas, including the major mountain systems from the Altai in the west to the Hamar-Daban in the east.

Zonal vegetation on the low elevation arid plains is grass steppe vegetation, however in the mountains, mixedwood (*Betula pendula*, *Populus tremula*, *Pinus sylvestris*) and conifer forests (*Pinus sibirica, Abies sibirica*) prevail. These mountain systems are affected by humid Atlantic air masses and receive over 1500 mm of annual precipitation. The conditions of relatively high temperatures and high humidity that correspond to the temperate zone in central Eurasia have led to the formation of unique forests characterized by boreal dominants and well-developed layers of shrubs and tall herbs. These forests have a complete set of boreal species but differ from other boreal forests by containing a very high proportion of temperate species in the understory (Nazimova *et al.* 2014).

Central Siberian boreal

The Central Siberian floristic subdivision occupies the area between the Yenisei River in the west and the Verkhoyansk Range and the eastern border of the Aldan River Basin in the east. The climate is ultracontinental, with severe winters with no or shallow snow cover and hot summers; the entire area lies within the zone of permafrost. The combination of cold soils and harsh climatic conditions does not favour the shade-tolerant conifer species of the *Pinus*, *Picea* and *Abies* genera. Instead, *Larix* forests are prevalent.

Most of the area is occupied by *Larix gmelinii*, a tree species adapted to the most extreme growing conditions in the Northern Hemisphere. It can tolerate winter temperatures lower than −70°C, strong paludification, very low air humidity and summer precipitation, and very cold soils with shallow active layers above the permafrost. In conditions of low precipitation and summer-long droughts, permafrost is the only source of moisture for trees. *Larix gmelinii* is a long-lived species that can live for 500 years in northern regions. Because of strict intolerance to shade, larch has very low competitive ability, which limits its distribution.

Larix normally forms pure even-aged stands with a single-stratum canopy, varying in cover, without an admixture of any other tree species. Across the whole province, *Larix* stands alternate with *Pinus sylvestris* stands on sandy soils. Both species are shade intolerant and adapted to a wide range of ecological conditions. *Pinus sylvestris* saplings appear to be more fire tolerant than *Larix* saplings, and pine replaces larch when ground fires are frequent.

The shrub layer in *Larix* stands is normally well developed because canopy closure is usually low. It is composed of both circumboreal shrubs and species restricted to the eastern part of boreal Asia. On moist flats, the shrub layer may contain *Alnus fruticosa*, *Vaccinium uliginosum*, *Betula middendorffii* and *Ledum palustre*. In wet lowlands, *Carex schmidtii*, *C. appendiculata*, *C. globularis*, *Eriophorum vaginatum*, *E. gracile* and *Calamagrostis langsdorffii* can be main components of the herb layer. Some shrubs and dwarf shrubs, such as *Ledum* spp., *Myrica tomentosa*, *Chamaedaphne calyculata*, may also be present in the herb layer. On uplands, the most abundant species are *Vaccinium vitis-idaea*, *Arctostaphylos uva-ursi*, *Solidago spiraeifolia*, *Carex vanheurckii*, and *Pteridium aquilinum*. *Sphagnum* spp. occur on wet and very wet sites while *Pleurozium schreberi* and *Hylocomium splendens* form moss layers on mountain slopes. *Cladonia* spp. and *Stereocaulon* spp. are common on slopes in earlier stages of post-fire succession.

In the eastern part of the Central Siberian floristic subdivision, sites that are protected from the prevailing harsh environment are suitable for conifer forests composed of shade-tolerant *Picea obovata*, but their area is minor and they are considered to be modern refugia.

Transbaikalian boreal

This floristic subdivision lies south of Lake Baikal and spreads from the Northern Baikal Plateau in the west to the upper part of the Amur River in the east. The southern boundary of the Transbaikalian boreal is in northeastern Mongolia. The climate is ultracontinental and, although continuous permafrost is absent, soils are characterized by severe freezing. The landscape is semi-forested, moisture being the major limiting factor for forest distribution. Forests mainly occupy the northern aspects of mountain slopes, while southern aspects and valleys are covered by steppe vegetation. Most Transbaikalian boreal forests are composed of *Larix gmelinii*, except on sandy sites, which are occupied by *Pinus sylvestris*. At higher elevations, *Pinus sibirica* can form sparse stands.

Typical *Larix gmelinii* forests occur within the forest–steppe ecotone of the Transbaikalian boreal. Their open canopy is dominated by *Larix gmelinii* and *Betula platyphylla* in different proportions, with occasional occurrence of *Populus tremula*. A well-developed shrub layer usually contains *Rhododendron dauricum*, *Lonicera caerulea*, *Rosa acicularis*, *Sorbaria sorbifolia* and *Spiraea media*. Xerophilous mosses (*Abietinella abietina*, *Rhytidium rugosum*) together with boreal species (*Pleurozium schreberi*, *Aulacomnium turgidum*, *A. palustre*) predominate in the ground layer.

Northeastern Siberian boreal

The Northeastern Siberian boreal subdivision extends from the Verkhoyansk Range in the west to the coasts of the Bering and Okhotsk seas in the east, excluding the Kamchatka Peninsula. The main vegetation condition of this floristic subdivision is open *Larix cajanderi* forest and woodland, gradually changing to *Pinus pumila* thickets towards the east.

Larix cajanderi forests and woodlands are humidity dependent and occur in conditions of moderately oceanic climates along the coasts of the Arctic and northern Pacific oceans as well as in the massif mountain systems of northeast Asia (Krestov *et al.* 2009). Forest stands are characterized by: 1) a very open canopy of larch; 2) lack of species characteristic of shade-tolerant coniferous forests; 3) presence of species characteristic of larch-dominated mires; and 4) high constancy of species that are widely distributed in the subarctic/subalpine zones of northern Asia. These forests/woodlands occur across a wide range of habitats, the ecological regimes of which are influenced by shallow seasonally thawed soil over permafrost and by a very short growing season (less than three months).

With proximity to the coast of the East Siberian Sea, the role of *Pinus pumila* in the *Larix* forests increases. North of latitude 60°N, within 1000 km of the coast, it becomes the dominant species on the landscape. The main features of these forests include: 1) dominance by *Pinus pumila*, that forms a closed canopy; 2) presence of a complex of subarctic/subalpine species (e.g., *Rhododendron aureum*, *Ledum decumbens*, *Carex rigidioides*, *Sorbus sambucifolia*); and 3) presence of other tree species, usually *Larix cajanderi*.

Okhotsk–Kamchatka boreal

This subdivision includes the Kamchatka Peninsula, with the adjacent Commander Islands and Koraginskii Island, the Kuril Archipelago, Sakhalin Island, and the mainland area adjacent to the Okhotsk Sea and the Sea of Japan. The vegetation of this subdivision maintains features typical for boreal vegetation, but with forests dominated by Far Eastern tree species that are different from the rest of the Eurasian boreal zone. The major dominants are *Abies sachalinensis* var. *gracilis*, *A. nephrolepis*, *A. sachalinensis* and *Picea jezoensis*. Increasing oceanicity in the coastal areas causes dramatic changes in forest structure as shade-tolerant conifer species disappear and *Betula ermanii* becomes the dominant canopy species (Krestov 2003). This forest type changes to *Alnus fruticosa* communities in the coastal areas of Kamchatka and on the northern islands of the Kuril Archipelago, and then to treeless tall forb vegetation on the Commander Islands.

Picea jezoensis forests are characterized by the presence of species with circumboreal or trans-continental Eurasian boreal distributions, such as *Hylocomium splendens*, *Pleurozium schreberi*, *Peltigera aphthosa*, *Oxalis acetosella*, *Diplazium sibiricum* and *Rosa acicularis*. Three major forest conditions related to climatic gradients of heat and continentality can be distinguished over the range of *Picea jezoensis*: 1) pure spruce forests in the northern part of the range and in Kamchatka; 2) mixed spruce–fir forests with *Abies nephrolepis* on the mainland south of latitude 54°N, plus the Shantar Islands; and 3) mixed spruce–fir forests with *Abies sachalinensis* on the islands of Sakhalin, Iturup and Hokkaido (Krestov and Nakamura 2002).

Pure spruce forests in the northern part of the range have the simplest structure and lowest species diversity. The tree layer contains only *Picea jezoensis*, occasionally mixed with *Betula ermanii* at high elevations. Solitary individual trees of *Sorbus sibirica* occur under the main canopy. Some of the most important understory species are *Maianthemum bifolium*, *Chamaepericlymenum suecicum*, *Dryopteris expansa*, *Phegopteris connectilis*, *Streptopus streptopoides*, *Veratrum alpestre*, *Vaccinium vitis-idaea*, *Rosa acicularis*, *Juniperus sibirica* and *Ribes triste*. The moss cover consists of species similar to those found in North European and Canadian dense conifer forests.

Abies nephrolepis forests are characterized by higher diversity and more complex structure than the pure spruce forests. In addition to shade-tolerant conifer species, stands can also include *Pinus koraiensis* and broadleaved species, such as *Acer mono*, *Betula costata*, *Fraxinus mandshurica* or *Tilia amurensis*.

The spruce–fir forests on Sakhalin Island and in the southern Kurils are formed under maritime and oceanic climates and *Abies nephrolepis* is replaced with *A. sachalinensis* in the tree layer. In these forests *Sorbus commixta* or *Acer ukurunduense* also occur. The understory flora includes many species endemic to Japan and Sakhalin Island, such as *Hydrangea paniculata*, *Euonymus miniata*, *Viburnum furcatum*, *Vaccinium axillare*, *Ilex rugosa*, *I. sugeroki*, *Skimmia repens*, *Vaccinium praestans*, *Aralia cordata*, *Petasites amplus* and *Lysichiton camtschatcense*.

Forests of *Betula ermanii* characterize the maritime and suboceanic regions of northeast Asia. This species forms vegetation belts in mountains influenced by oceanic air masses and is most developed in Kamchatka, where it occurs over a wide range of ecologically different sites, except in wetlands or on permafrost. *Betula ermanii* distribution depends, among other factors, on snow depth and timing of snowmelt (length of snow-free period). The recorded maximum age is 500 years. *Alnus fruticosa*, *Pinus pumila* and *Sorbus sambucifolia* may form the tall shrub layer in *Betula ermanii* stands. The herb layer varies widely depending on local site ecology. The most common species are *Angelica ursina*, *Heracleum lanatum*, *Senecio cannabifolius*, *Cacalia hastata*, *Cirsium kamtschaticum*, *Veratrum alpestre*, *Aconitum maximum*, *Urtica platyphylla* and *Filipendula camtschatica*. Because of the well-developed herb layers, there is usually no moss–lichen layer.

In conditions of hyperoceanic boreal climate of North Kuril Islands, southeastern Kamchatka and Commander Islands (west of the Aleutian chain) boreal vegetation is dominated by shrubs with crooked stems (*Alnus fruticosa*, *Sorbus sambucifolia*) or represented by heathlands or grasslands (Krestov 2003).

Survival mechanisms of boreal species and ecosystems

Boreal species exhibit several anatomical, physiological and reproductive strategies that facilitate survival in the harsh environmental conditions of the boreal zone.

Species adaptation to cold temperatures

In order to survive boreal winters, vegetation is adapted to cold temperatures and short growing seasons. Adaptation takes different forms: physiological ability to survive extremely low temperatures; protection of buds, seeds and evergreen leaves against desiccation and freezing; and short reproductive life cycles to successfully mature seeds in an annual growing season.

Some plants, including many coniferous species, show tolerance to extreme cold temperatures during the dormant period (i.e., hardiness). Plants use three strategies to survive freezing (Sutinen et al. 2001; Brandt 2009). The first, called supercooling, is a process by which cellular water can remain liquid at temperatures as low as $-40°C$. Species that limit their adaptation to supercooling are usually more frequent in the transition between boreal and temperate zones.

A second strategy is to tolerate extracellular freezing by moving liquid outside of cell walls, thus avoiding severe damage to the internal cell structures. The third mechanism is extra-organ freezing, where water is moved out of fragile organs like buds and shoots to create a form of icy insulation when it freezes around the organs (Sutinen *et al.* 2001). Tree species that are endemic to the boreal, including those within the *Abies, Larix, Pinus, Picea, Betula* and *Populus* genera use these strategies to withstand temperatures as low as −80°C (Brandt 2009). Plant hardiness includes an ability to recover from frost damage when the warmer season returns.

Species adaptation to natural disturbance

Boreal species are well adapted to recurring natural disturbances such as fire, insect infestation and extreme weather events (windthrow, ice storms, heavy snow load) that trigger the renewal of the boreal forest.

Fire

Adaptation to fire can take several forms. Some species develop thick bark that can resist light to moderate ground fires by protecting the cambium and keeping the individual tree alive. Some temperate pine species that occur in the hemiboreal transition use this strategy. Another strategy is protection of seeds against fire. *Pinus banksiana* and *P. contorta*, for instance, have serotinous cones (cones that are protected by a waxy coating) that require intense heat, such as that provided by fire, to release their seeds (Rudolph and Laidly 1990). After a forest fire, even if the individual tree does not survive, the scales of the cone open and seeds are dispersed to sites that have been cleared of competing understory species. Fire also has the effect of reducing, if not removing altogether, the soil organic layer, thus preparing ideal seedbeds for germination and seedling establishment. Some spruce species, such as *Picea mariana*, have semi-serotinous cones (Viereck and Johnston 1990) that confer a post-fire advantage over species such as *Abies balsamea*, *Picea glauca* and *Thuja occidentalis* that do not have this adaptation.

The seeds of *Populus* spp. and *Salix* spp. are extremely light and can be dispersed over large distances by wind. The slightly heavier seeds of *Betula* spp. and *Alnus* spp., as well as those of conifers, are also winged and can be wind-dispersed over short to moderate distances. Seeds released in winter can travel long distances over snow crusts and ice. These shade-intolerant species also have the ability to outgrow competition with fast juvenile height growth in full light conditions. The hardwood species can also regenerate from stump sprouting or root suckering after fire or other catastrophic disturbance has killed the parent tree.

Insect infestation

Boreal forests are periodically subject to insect outbreaks that can have tremendous impacts on the forested landscape. When an outbreak lasts several years in forests with a high component of species that are targeted by the insect, severe mortality can result in a stand replacement or conversion event. Some shade-tolerant boreal species survive such events by establishing a bank of seedlings and saplings that persist in the understory, ready to replace dead trees as canopy gaps open.

For example, in eastern North America, *Abies balsamea* is sensitive to spruce budworm (*Choristoneura fumiferana*) outbreaks. The larvae of this insect feed on new foliage of the fir trees, killing the annual growth and leaving the old foliage, which is less efficient for photosynthesis. Since fir needles are retained on the tree for less than six years, trees become severely weakened if the infestation lasts more than three years. With a severe outbreak, high mortality rates occur after three to five years of damage by the insect, gradually opening the canopy. This shade-tolerant species produces yearly seed crops that can germinate and establish a seedling bank in the understory (Morin et al. 2009). The canopy gaps release the seedlings, effectively re-establishing a fir forest. With stands containing more than one species of different susceptibility, the result can be a shift in forest composition induced by the insect outbreak. *Picea glauca* and *P. mariana*, which are common co-associates of *Abies balsamea*, are less sensitive to this insect and may survive more years of defoliation. When hardwoods such as *Betula papyrifera* or *Populus tremuloides* are mixed with the fir, they can assume dominance over the coniferous species in a protracted outbreak. In eastern North America a cycle of 30 to 40 years between spruce budworm outbreaks is observed (Blais 1983).

Species adaptation to soil conditions

Paludification

Paludification is the accumulation of organic matter over time caused either by rising water tables or increasing soil moisture accompanied by Sphagnum colonization (Lavoie *et al.* 2005). Forest decline related to paludification is prevalent in Siberia (Russia) (Peregon *et al.* 2007), Canada (Kuhry *et al.* 1993; Lavoie *et al.* 2005), and Alaska (Jorgenson *et al.* 2013). Edaphic paludification occurs in wet topographic positions, where a high water table promotes the growth of peat-forming plants (e.g., *Carex* and *Sphagnum* species). Successional paludification can occur on well-drained soils, where peat-forming mosses accumulate at a faster rate than humus decomposition during forest succession between fire events, independent of site topography or drainage (Simard *et al.* 2007).

Paludification reduces soil temperature, decomposition rates, microbial activity and nutrient availability, which together lead to lower site productivity with time after disturbance (Swanson *et al.* 2000). Sphagnum growth and expansion, in particular, is a major contributor to acidification, peat accumulation, and a decrease in productivity (Klinger 1990). As a consequence, tree species with lower nutrient demands colonize paludified sites, which can be detrimental to timber production.

Paludification, however, can also provide a strong soil sink for carbon. For example, the extensive Siberian peatlands exhibit a general trend towards being a carbon sink rather than a source even at or near the southern limit of their distribution (Peregon *et al.* 2007). In permafrost-affected areas, the impoundment of water when drainage is impeded by permafrost or by the collapse of thawing permafrost contributes to colonization by Sphagnum and to soil carbon accumulation (Jorgenson *et al.* 2013).

Permafrost

Perennially frozen ground (permafrost) is a characteristic of arctic and subarctic regions and high mountains that is fundamental to geomorphic processes and ecological development in tundra and boreal forests. In northeast Yakutia (Eastern Siberia) the layer of permafrost can be 500 m deep, but the effect of permafrost on contemporary vegetation depends on the depth of an active soil horizon (i.e., thawed soils in the summer). Permafrost-affected regions occupy about 23 per cent of exposed lands (i.e., not under permanent ice caps) in the Northern Hemisphere; sporadic and discontinuous permafrost are common in the northern portions of the boreal biome (Figure 2.2, Brown *et al.* 1997). In addition to its short-term ecological

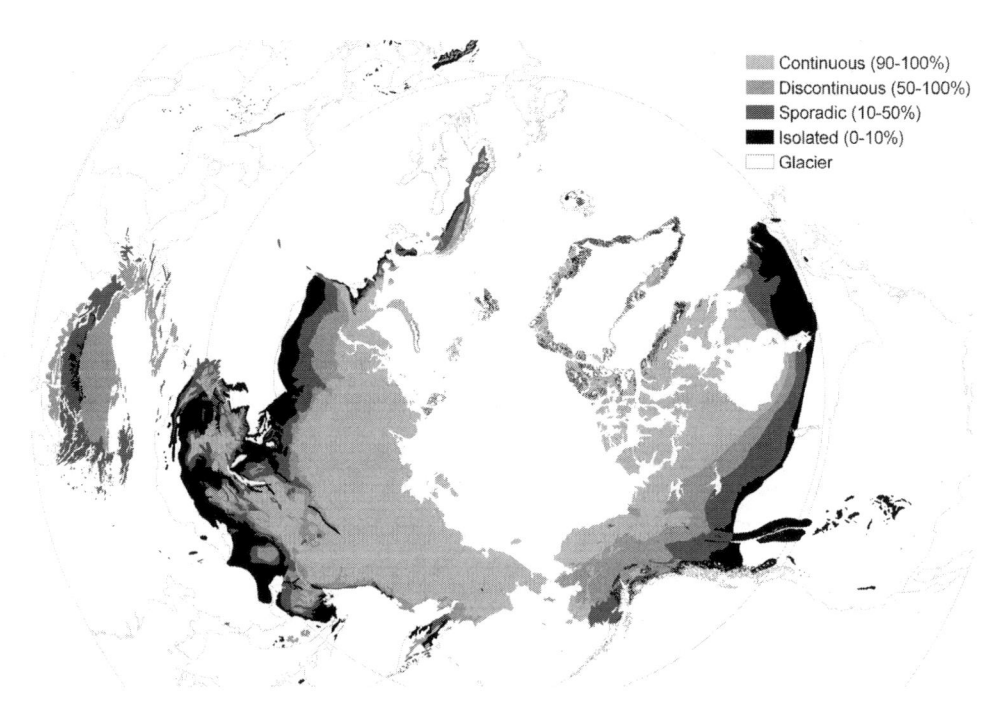

Figure 2.2 Distribution of permafrost zones in the Northern Hemisphere (adapted from Brown *et al.* 1997)

function, permafrost serves as one of the largest reservoirs of carbon and can be considered a factor in long-term boreal ecosystem dynamics (Zimov *et al.* 2006).

Generally, permafrost is considered a limiting factor for trees: cold soils and delayed thawing of the active horizon in spring cannot be tolerated by many species. However, despite a harsh climate with very low winter temperatures, very low precipitation (<300 mm annually) and very high summer temperatures, large areas of Central to Northeastern Siberia support deciduous coniferous *Larix gmelinii* and *L. cajanderi* forests. The bioclimate of this region would normally correspond to that of steppe or even desert regions. Larch species, such as *Larix gmelinii* and *L. cajanderi* have the uncommon ability to utilize moisture from the thawing permafrost surface. This adaptation explains the presence of extensive larch forests in Central to Northeastern Siberia.

Because permafrost properties are temperature dependent, degradation of permafrost in response to climate change will have large consequences for natural ecosystems and human infrastructure (Shur and Jorgenson 2007). Upon thawing everything about an ecosystem changes, including surface hydrology (by impounding of water in sinking areas and draining of upland areas), habitat changes for vegetation and wildlife, and emissions of greenhouse gases (Jorgenson *et al.* 2013).

Dynamic patterns of boreal forests

Multiple dynamic patterns can be observed in boreal forests. Bergeron *et al.* (2014) give a detailed description of the stand dynamics of North American boreal mixedwoods. Here, we provide an overview of boreal forest dynamics according to three potential pathways of stand response to disturbance: 1) auto-cyclic stand replacement after major disturbance; 2) succession of species in the absence of major disturbance; and 3) mixture maintenance through cyclic, but less severe, disturbance.

Auto-cyclic stand replacement

Major disturbances, such as fire, insect outbreak, windthrow or river channel changes, trigger the replacement of entire forest stands. As discussed in the previous section, many boreal tree species are adapted to such events and are able to replace themselves in the new stand. Some species are uniquely adapted to regenerate following specific types of stand-replacing disturbance. *Pinus banksiana* and *Picea mariana*, with their serotinous or semi-serotinous cones, protect viable seed sources from fire permitting release of their seeds at an optimal time for seedling establishment and early growth. *Abies balsamea* establishes a bank of shade-tolerant seedlings in the understory that are available to grow into the canopy of stands where high overstory mortality has occurred from insect outbreaks. *Populus tremuloides* is capable of generating numerous suckers from the surviving roots of trees that have died, especially if the soil is warmed by incident sunlight. Thus, following fire, the stand is re-colonized by vigorously growing aspen stems before other species have an opportunity to become established. This pattern of stand replacement is common in boreal forests. However, it can fail under certain circumstances, e.g., when fire occurs in a young stand before the trees have reached sexual maturity. This can result in a very open stand of the same species (i.e., a woodland) or a shift in species composition, usually towards shade-intolerant species with the ability to regenerate vegetatively or to disperse seeds from distant seed sources.

Succession of species

The gradual transition over a long period of time from a stand dominated by shade-intolerant species to one dominated by shade-tolerant species is called succession. This process takes place when no major disturbance occurs for a period exceeding the lifespan of the shade-intolerant species that initiated the stand after the last major disturbance. Over time, shade-tolerant species establish in the understory. These individuals can take advantage of openings created by the death of a single or a small group of trees to grow into the canopy (Bouchard *et al.* 2006). This process is known as gap-phase replacement. In the continuing absence of major disturbance, the species mix of the stand will evolve until no shade-intolerant trees persist. An example of this dynamic pathway in the Eastern North American boreal is the progressive transition from stands dominated by *Betula papyrifera* (a shade-intolerant species) to mixedwoods of *Betula* and *Abies balsamea* (a highly shade-tolerant species) and finally to *Abies balsamea* dominated stands that are able to auto-cycle as described above. Similarly, in the West–Central North American boreal, transition from *Populus tremuloides* stands to *Picea glauca* dominated stands is another example. Severe (i.e., stand-replacing) disturbance during the succession process returns the system to an early seral stage dominated by shade-intolerant species (e.g., *Betula* or *Populus* in our examples) from which the stand could evolve into a different pathway.

Mixture maintenance

In the presence of frequent, low intensity disturbances, a mixture of shade-intolerant and shade-tolerant tree species can establish and maintain itself. Depending on the size of the gaps created in the canopy, different species can be favoured. Large openings with more light reaching the forest floor are suitable for species that are less shade tolerant. The relative proportions of the species in the mixture can change in response to disturbance events. The mixture of *Populus tremuloides* and *Picea glauca* found in Western–Central North American boreal is an example of this process (Bergeron *et al.* 2014).

Threats to the boreal

The boreal zone has historically been sparsely populated because of climatic and soil limitations to agriculture, however indigenous peoples developed cultures that utilized seasonally abundant fauna and flora for subsistence. In recent history, the boreal has been widely exploited for industrial forestry, agriculture, mining and hydroelectric activities, and undergone increasing urban expansion. However, global boreal forests are still largely ecologically intact. Boreal forests are important for providing timber and non-timber products, habitat and forage for animals, as well as supporting ecosystem services, such as water purification, climate regulation and carbon storage (Chapin *et al.* 2006). Forest resilience in the boreal zone generally ensures regeneration of the forest after natural or anthropogenic disturbance, but the conditions of secondary forests are not similar to those of the primeval forest. Because logging usually targets dense and mature stands, it gradually replaces those stands with younger ones, with less dead wood and a different proportion of the main tree species. There are cumulative impacts of forest harvesting and other human activities over time and space, and those are augmenting the impacts of natural disturbances, not replacing them. One effect is the fragmentation and reduction of old-growth forests on the landscape (Leduc *et al.* 2009).

In a study based on remote sensing data, Potapov *et al.* (2008) mapped the world's intact forest landscapes. They estimated that two biomes were hosting the largest remaining parts of intact forest: dense tropical/subtropical forests and boreal forests accounted for 45.3 per cent and 43.8 per cent respectively of the world's intact forests (Potapov *et al.* 2008). As there is growing support in society for the protection of these intact forests, triggered by information campaigns by some environmental groups, protected areas are being extended in many boreal countries.

Ecosystems of the boreal zone, despite their floristic and structural simplicity, provide important habitat for many species of animals. The largest animals – wolves, moose, caribou, bears, beavers, porcupine, lynx and foxes – are widespread but can be represented in different regions by closely related taxa. Boreal ecosystems provide summer breeding range to many species of birds, most of which migrate south for the winter. Grazing ungulates can have an effect on forest regeneration, as they selectively choose their summer food and because they rely on bark, young twigs and buds in the wintertime (Kielland *et al.* 2006). Moose, particularly, affect the composition of early successional stands, with consequences for later forest structure. The same is true for rodents that feed on seeds of certain species, like the pines. Insects also consume large quantities of foliage and seeds of boreal vegetation. The extensive hemiarctic areas of Northern Eurasia, and North America, vegetated by woodlands of spruce, larch and lichen accommodate large populations of caribou and reindeer. In Eurasia reindeer are extensively raised by local people for multiple uses, including transportation and food. Wolves are the most important large predator of boreal forests. However, in eastern Asia, the wolf population is affected by the tiger population, because the ecological niches of these species are very similar. Black and brown bears serve as both important herbivores (mainly berries) and predators (mainly moose and caribou calves).

Human use of boreal forests, in the form of tree harvesting or land use change for mining, farming or urban development, has effects on the resident animal species. These effects could be detrimental or beneficial depending on the species. For example in eastern Canada, fragmentation of the forested landscape by access roads has been identified as a threat to the woodland caribou (*Rangifer tarandus*). This particular species inhabits the northern part of the boreal closed forest and relies on a mosaic of open stands (with a lichen understory) and more dense coniferous stands with arboreal lichens (*Usnea* spp., *Alectoria* spp.). Caribou tend to avoid areas that have been recently logged and with frequent road traffic. An alternative management option is to concentrate logging activities in certain areas while preserving large tracts of forest for a long time (Bouchard and Garet 2014). On the other hand, studies have shown that moose populations in northwestern Québec have been favoured by logging in mosaic patterns, with recently logged patches providing browse while adjacent patches of mature forest provide shelter (Potvin *et al.* 2005).

In eastern Asia, the Amur tiger (*Panthera tigris altaica*) and Amur leopard (*Panthera pardus orientalis*) represent the top of the food chain in Korean pine forest ecosystems. Their populations appear to be most influenced by contraction of forest area that reduces feeding territory and abundance of a number of key prey species. For example, as its forest habitat is modified by human use, the Amur tiger, which once lived throughout the Russian Far East, the Korean Peninsula and Northeast China, is now confined to a fragment of its former range. The major conservation task now is to avoid further fragmentation of its habitat and to achieve protection of intact Ussuri taiga tracts in the Okhotsk–Kamchatka subdivision.

Efforts have been made in many northern countries to create and maintain protected areas, in order to keep benchmarks of the natural dynamic processes and of the habitat they provide. These protected areas may harbour examples of old-growth boreal forest and may serve as refuges for some endangered species. Complementary to forest protection, are efforts to develop forest management practices that emulate patterns, at the landscape scale, that are created by natural disturbances. An important goal of ecosystem based management is to mitigate the impact of current management practices on natural ecosystems (Gauthier *et al.* 2009). With this management approach, the focus is not only on what to harvest but also on wood and forest patches to be left in place in order to ensure that essential ecosystem services continue to be provided (Leduc *et al.* 2009).

In addition to increasing pressure from logging and mining, the boreal biome is expected to be sensitive to climate change. This could result in modifications to forest composition and to the dynamic equilibrium of global carbon cycling (Chapin *et al.* 2006). The changing climate is expected to alter the natural disturbance regimes that have shaped the boreal forest. The effects of global warming will differ regionally but, in general, temperatures are expected to rise. The magnitude of the change will probably be greater in the northern boreal regions than in the southern parts, resulting in a thawing of the permafrost (Shur and Jorgenson 2007). Changes in precipitation patterns can also be expected. Many studies suggest that, given the ecological niches of boreal species, the ranges of these species will likely shift (Périé *et al.* 2014) and the known limits of the boreal zone will change. The boundary with the temperate zone is expected to move northward, as species from southern latitudes expand into the current boreal zone. The boundary with the arctic zone is also subject to change as the continental tree line could shift gradually, if soils and climate allow better survival, growth and regeneration of tree species.

Changes in boreal forests are likely to have large consequences on the global climate system through water and energy exchange, exchange of greenhouse gases, and delivery of water to the Arctic Ocean, but there is uncertainty whether the changes will enhance or mitigate global warming (Chapin *et al.* 2006). Tree canopy coverage and snow cover strongly affect albedo and energy exchange. Reduced snow cover from warmer temperatures and a longer growing season in boreal forests could provide a positive feedback to the climate system, while migration of forests into tundra areas could provide a negative feedback. The enormous accumulation of organic carbon in permafrost soils in boreal and tundra biomes potentially can have large effects on the global climate system because of their potential to thaw, decompose organic matter that has accumulated over long time periods, and release radiatively active gases to the atmosphere. Because permafrost soils account for 70 per cent of the soil organic carbon in the northern circumpolar regions (Tarnocai *et al.* 2009), there are increasing concerns about the potential release of significant amounts of this carbon as greenhouse gases, especially methane with its greater warming potential compared to carbon dioxide (Schuur *et al.* 2008). But the net effects of permafrost thaw on soil carbon dynamics are uncertain (Jorgenson *et al.* 2013). Increasing disturbance from fires, insect outbreaks, and altered land cover contributes to these changes by affecting the energy exchange of forest canopies and the combustion of vegetation and surface organic layers (Chapin *et al.* 2006). Finally, the boreal forest plays a significant role in the delivery of water to the Arctic Ocean because it dominates the landmasses of arctic watersheds (Serreze *et al.* 2006). Changes in runoff from permafrost thaw and reduced snow cover within the boreal forest could alter the water budget of the Arctic Ocean and change sea ice dynamics, thus affecting feedbacks to the climate system.

Conclusion

The boreal biome is one of the largest global ecosystems. Plant species of the circumboreal zone share several mechanisms of adaptation to the cold climate and to natural disturbance regimes. Boreal forests have evolved over long time periods, responding to alternating ice ages and warmer climate periods. One can be optimistic that this ecosystem will adapt to future climate changes. Anthropogenic pressure on the boreal zone is sometimes a matter of concern, but as societies develop awareness of their responsibilities to protect and conserve globally healthy ecosystems, boreal forests are also capable of being resilient to the impacts of human use and local environmental changes. Due to the prevalence of permafrost-affected soils, large soil carbon reservoirs, and forest structure that affects the global energy balance, changes in forest patterns and processes are likely to have large consequences for the global climate system.

Acknowledgements

We thank two anonymous reviewers whose comments contributed to the improvement of the manuscript.

References

Bergeron, Y., H.Y. Chen, N.C. Kenkel, A.L. Leduc and S.E. Macdonald, 2014. Boreal mixedwood stand dynamics: ecological processes underlying multiple pathways. *The Forestry Chronicle*, 90(02): 202–213.
Blais, J.R., 1983. Trends in the frequency, extent, and severity of spruce budworm outbreaks in eastern Canada. *Canadian Journal of Forest Research* 13: 539–547.
Bouchard, M. and J. Garet, 2014. A framework to optimize the restoration and retention of large mature forest tracts in managed boreal landscapes. *Ecological Applications*, 24(7): 1689–1704.
Bouchard, M., D. Kneeshaw and Y. Bergeron, 2006. Forest dynamics after successive spruce budworm outbreaks in mixedwood forests. *Ecology* 87: 2319–2329.
Brandt, J.P., 2009. The extent of the North American boreal zone. *Environmental Reviews* 17: 101–161.
Brown, J., O.J. Ferrians, Jr., J.A. Heginbottom and E.S. Melnikov, eds., 1997. *Circum-Arctic Map of Permafrost and Ground-Ice Conditions*. Washington, DC: U.S. Geological Survey in Cooperation with the Circum-Pacific Council for Energy and Mineral Resources. Circum-Pacific Map Series CP-45, scale 1:10,000,000, 1 sheet.

Canadian National Vegetation Classification project (CNVC), 2014. Accessed on 10 January 2014 at: http://cnvc-cnvc.ca/, and personal communication from Ken Baldwin.
Chapin III, F.S., M.W. Oswood, K. Van Cleve, L.A. Viereck and D.L. Verbyla, 2006. *Alaska's Changing Boreal Forest*. Oxford University Press, New York.
Cherepanov, S.K., 2007. *Vascular Plants of Russia and Adjacent States*. Cambridge University Press, West Nyack, New York, 528 pp.
Daniëls, F.J.A. and J.G. de Molenaar, 2011. Flora and vegetation of Tasiilaq, formerly Angmagssalik, Southeast Greenland – a comparison of data between around 1900 and 2007. *Ambio* 40(6): 650–659.
Ermakov, N. and O. Morozova, 2011. Syntaxonomical survey of boreal oligotrophic pine forests in northern Europe and Western Siberia. *Applied Vegetation Science* 14(4): 524–536.
FAO, 2001. *FRA 2000 Main Report*, FAO Forestry Paper 140. Rome.
Flora of North America Editorial Committee, eds. 1993+. *Flora of North America North of Mexico*. 18+ vols. New York and Oxford, UK.
Gauthier, S., M.A. Vaillancourt, A. Leduc, L. De Grandpré, D. Kneeshaw and others, eds., 2009. *Ecosystem Management in the Boreal Forest*. Presses de l'Université du Québec. Québec, Canada. 539 p.
Hytteborn, H., A.A. Maslov, D.I. Nazimova and L.P. Rysin, 2005. Boreal forests of Eurasia. In: Andersson F., ed., *Coniferous Forests* (Ecosystems of the world 6). Elsevier, Amsterdam, Boston, US, London *et al.*, pp. 23–100.
Jorgenson, M.T., J. Harden, M. Kanevskiy, J. O'Donnel, K. Wickland and others, 2013. Reorganization of vegetation, hydrology and soil carbon after permafrost degradation across heterogeneous boreal landscapes. *Environmental Research Letters* 8 035017. 13 pp.
Kielland, K., J.P. Bryant and R.W. Ruess, 2006. Mammalian herbivory, ecosystem engineering, and ecological cascades in Alaska boreal forests. In Chapin *et al.*, eds., *Alaska's Changing Boreal Forests*. Oxford University Press, pp. 211–226.
Klinger, L.F., 1990. Global patterns in community succession: 1. Bryophytes and forest decline. *Memoirs of the Torrey Botanical Club* 24: 1–50.
Krestov, P.V., 2003. Forest vegetation of Easternmost Russia (Russian Far East). In: Kolbek, J., M. Šrůtek and E.O. Box, eds. *Forest Vegetation of Northeast Asia*. Kluwer Academic publishers, Dordrecht, pp. 93–180.
Krestov, P.V. and Y. Nakamura, 2002. Phytosociological study of the *Picea jezoensis* forests of the Far East. *Folia Geobotanica* 37(4): 441–474.
Krestov, P.V., N.B. Ermakov, S.V. Osipov and Y. Nakamura, 2009. Classification and phytogeography of larch forests of northeast Asia. *Folia Geobotanica* 44(4): 323–363.
Kuhry, P., B. Nicholson, L.D. Gignac, D.H. Vitt and S.E. Bayley, 1993. Development of Sphagnum-dominated peatlands in boreal continental Canada. *Canadian Journal of Botany* 71: 10–22.
Kuusela, K., 1992. The boreal forests: an overview. *Unasylva* 43: 3–13.
Lavoie, M., D. Paré, N. Fenton, A. Groot and K. Taylor, 2005. Paludification and management of forested peatlands in Canada: a literature review. *Environmental Reviews* 13 (2): 21–50.
Leduc, A., S. Gauthier, M.A. Vaillancourt, Y. Bergeron, L. De Grandpré and others, 2009. Perpectives, Chapter 21. In: Gauthier, S., M.A. Vaillancourt, A. Leduc, L. De Grandpré, D. Kneeshaw and others, eds.: *Ecosystem Management in the Boreal Forest*. Presses de l'Université du Québec. Québec, Canada. pp. 519–526.
Morin, H., D. Laprise, A.-A. Simard and S. Amouch, 2009. Spruce budworm outbreak regimes in eastern North America – Chapter 7. In: Gauthier, S., M.-A. Vaillancourt, A. Leduc, L. De Grandpré, D. Kneeshaw and others, eds.: *Ecosystem Management in the Boreal Forest*. Presses de l'Université du Québec, Québec, Canada, pp. 155–182.
Nakamura, Y. and P.V. Krestov, 2005. Coniferous forests of the temperate zone of Asia. In: F. Andersson (ed.): *Coniferous Forests* (Ecosystems of the World, 6). Elsevier Academic Press, New York, Paris, London, Brussels *et al*. pp. 165–220.
Nazimova, D.I., D.M. Danilina and N.V. Stepanov, 2014. Rain-barrier forest ecosystems of the Sayan Mountains. *Botanica Pacifica* 3(1): 39–47.
Peregon, A., M. Uchida and Y. Shibata, 2007. Sphagnum peatland development at their southern climatic range in West Siberia: trends and peat accumulation patterns. *Environmental Research Letters* 2: 045014.
Périé, C., S. De Blois, M.-C. Lambert and N. Casajus, 2014. Effets anticipés des changements climatiques sur l'habitat des espèces arborescentes au Québec. *Mémoire de recherche forestière* n° 173. Gouvernement du Québec, Ministère des Ressources naturelles, Direction de la recherche forestière. 46 pp.

Potapov, P., A. Yaroshenko, S. Turubanova, M. Dubinin, L. Laestadius and others, 2008. Mapping the world's intact forest landscapes by remote sensing. *Ecology and Society*, 13 (2): 51.

Potvin, F., L. Breton and R. Courtois, 2005. Response of beaver, moose, and snowshoe hare to clear-cutting in a Quebec boreal forest: a reassessment 10 years after cut. *Canadian Journal of Forest Research*, 35(1): 151–160.

Rivas-Martínez, S., S. Rivas Sáenz and A. Penas, 2011. Worldwide bioclimatic classification system. *Global Geobotany* 1: 1–634 + maps.

Rudolph, T.D. and P.R. Laidly, 1990. Jack pine (*Pinus banksiana* Lamb.). In: Burns, R.M. and B.H. Honkala, eds.: *Silvics of North America, vol. 1. Conifers. U.S.D.A. Forest Service Agriculture Handbook 654*, 280–290.

Saucier, J.-P., 2008. Defining the boreal in the ecological land classification for Québec. In: Talbot, S.S., ed.: *Proceedings of the Fourth International Conservation of Arctic Flora and Fauna (CAFF) Flora Group Workshop*, 15–18 May 2007, Tórshavn, Faroe Islands. CAFF Technical Report No. 15. Akureyri, Iceland. 53–56.

Schuur, E.A.G., J. Bockheim, J. G. Canadell, E. Euskirchen, C. B. Field and others, 2008. Vulnerability of permafrost carbon to climate change: Implications for the global carbon cycle. *BioScience* 58: 701–714.

Serreze, M. C., A. P. Barrett, A.G. Slater, R.A. Woodgate, K. Aagaard and others, 2006. The large-scale freshwater cycle of the Arctic. *Journal of Geophysical Research* 111: C11010.

Shur, Y.L. and M.T. Jorgenson, 2007. Patterns of permafrost formation and degradation in relation to climate and ecosystems. *Permafrost and Periglacial Processes* 18: 7–19.

Simard, M., N. Lecomte, Y. Bergeron, P.Y. Bernier and D. Paré, 2007. Forest productivity decline caused by successional paludification of boreal soils. *Ecological Applications* 17 (6): 1619–1637.

Sutinen, M.-L., R. Arora, M. Wisniewski, E. Ashworth, R. Strimbeck and J. Palta, 2001. Mechanisms of frost survival and freeze-damage in nature. In: Bigras, F.J. and S.J. Colombo, eds., *Conifer Cold Hardiness*, Kluwer Academic Publishers, Tree Physiology Volume 1: 89–120.

Swanson, D.K., B. Lacelle and C. Tarnocai, 2000. Temperature and the boreal-subarctic maximum in soil organic carbon. *Géographie Physique et Quaternaire* 54: 157–167.

Takhtajan, A., 1986. *Floristic Regions of the World*. University of California Press, Berkeley, California, USA. 522 pp.

Tarnocai, C., J.G. Canadell, E.A.G. Schuur, P. Kuhry, G. Mazhitova and others, 2009. Soil organic carbon pools in the northern circumpolar permafrost region. *Global Biogeochemical Cycles* 23: GB2023.

Van Cleve, K., C.T. Dyrness, L.A. Viereck, J. Fox, F.S. Chapin III and others, 1983. Taiga ecosystems in Interior Alaska. *BioScience* 33 (1): 39–44.

Viereck, L.A. and W.F. Johnston, 1990. *Picea mariana* (Mill.) B.S.P. black spruce. In: Burns, Russell M. and Barbara H. Honkala, eds., *Silvics of North America*. Volume 1. Conifers. Agric. Handbk. 654. Washington, DC: U.S. Department of Agriculture, Forest Service: 227–237.

Viereck, L.A., C.T. Dyrness, A.R. Batten and K.J. Wenzlick, 1992. *The Alaska Vegetation Classification*. Pacific Northwest Research Station, U.S. Forest Service, Portland, OR. Gen. Tech. Rep. PNW-GTR-286., 278 pp.

Zimov, S.A., E.A.G. Schuur and F.S. Chapin III, 2006. Permafrost and the global carbon budget. *Science* 312: 1612–1613.

3
NORTHERN TEMPERATE FORESTS

Lee E. Frelich, Rebecca A. Montgomery and Jacek Oleksyn

Introduction

Biogeography

Three areas of northern temperate forest with large spatial extents occur in eastern North America, Europe, and eastern Asia; a conifer-dominated temperate rainforest also occurs in western North America (Figure 3.1, Whittaker 1975). These forests are found between boreal forest to the north, and subtropical forests, grasslands and savannas, or Mediterranean vegetation to the south. Many of the tree genera in temperate forests occur across North America, Europe and Asia: *Acer, Alnus, Betula, Carpinus, Carya, Castanea, Celtis, Fagus, Fraxinus, Larix, Pinus, Populus, Prunus, Quercus, Picea, Tilia,* and *Ulmus*. Other genera—*Liriodendron, Pseudotsuga* and *Tsuga*—occur in North America and Asia. Some tree genera are widespread across temperate and boreal biomes (e.g. some *Betula, Pinus, Picea,* and *Populus* species), others are almost totally restricted to the temperate zone (e.g. *Quercus, Tilia,* and some *Acer* species), while some occupy the transition zone plus limited portions of the southern boreal and northern temperate zones, and cannot be conveniently classified as temperate or boreal species (e.g. *Fraxinus nigra, Pinus resinosa,* and *Picea rubens* in North America). The species richness of trees and understory plants is highest in Asia, moderately high in North America, and lowest in European temperate forests, although within each continent species richness also varies from relatively high towards the southern edge of the temperate biome to relatively low at the northern edge.

Some tree communities have similar composition at the generic level across the three main occurrences of northern temperate forest. For example, *Fagus* (beech), *Acer* (maple) and *Quercus* (oak) occur together on mesic sites: *A. mono, F. crenata* and *Q. mongolica* in Japan, *A. saccharum, F. grandifolia* and *Q. rubra* in eastern North America, and *A. platanoides, F. sylvatica* and *Q. robur* in northern Europe (Runkle 1982, Yamamoto *et al.* 1995, Svenning and Skov 2005).

Climate

Northern temperate forests have strong seasonality, with a cold-induced dormant season and a warm growing season that produces distinct annual rings in trees. Growing season lengths range

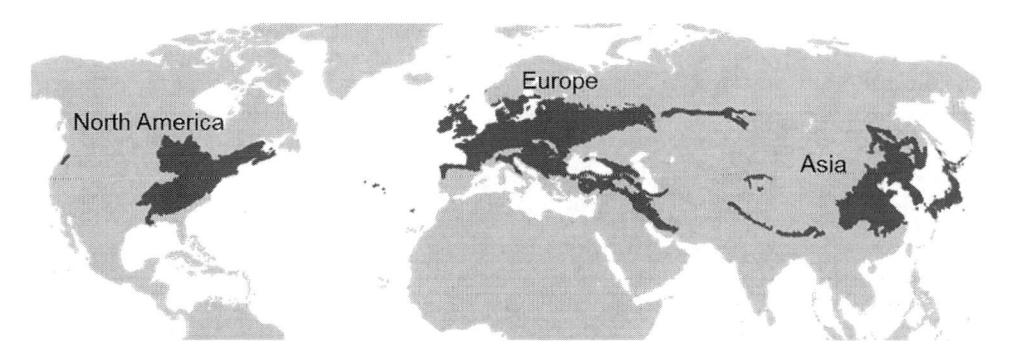

Figure 3.1 Location of the world's temperate forests
Source: Adapted from: Terpsichores (Own work) [CC BY-SA 3.0 (http://creativecommons.org/licenses/by-sa/3.0)], via Wikimedia Commons

from about four months at the temperate–boreal ecotone in the north (long, cold winters and cool summers) to about eight months at the temperate–subtropical forest ecotone (short cool winters and long hot summers) in the south (Peel *et al.* 2007). Precipitation regimes vary from wet winter/dry summer in coniferous temperate rainforests such as the Pacific Northwest USA, to rainforests wet throughout the year (*Tsuga* forests in the Southeastern USA), to wet summer regimes with snow during the dormant season in most of the cold parts of the temperate forest, to droughty summers with erratic rainfall in places where temperate forest and savanna/grassland ecotones occur, such as in the continental interiors of North America and Eurasia (Whittaker 1975). In the colder temperate climates, snow plays an important role by limiting the depth to which the soil freezes, which would otherwise cause extensive root damage in mature trees and mortality to seedlings and herbaceous plants. Some colder regions, such as Upper Michigan, USA, and northern Japan, receive extreme amounts of snow due to lake-effect or ocean-effect snowfall generated by cold winds blowing over large water bodies before hitting land. These 'snow forests' have unique species composition—for example sugar maple (*Acer saccharum*) is the dominant tree species in such forests in the Midwestern USA, since its main competitor, hemlock (*Tsuga canadensis*)—is damaged by the load of snow landing on the branches, and restricted to areas with lower snowfall.

The transition from temperate to boreal forest depends on either low winter minimum temperatures or cool summers. The deep supercooling point of the cells in the cambium under the bark of most temperate tree species ranges from −40 to −45 degrees C, and if winter temperatures are commonly that cold then many of the dominant temperate tree species in genera such as *Quercus* and *Acer* cannot survive (George *et al.* 1974). In other areas near large bodies of water (e.g. Lake Superior in North America), winter temperatures are nowhere near cold enough to limit temperate tree species but short, cool summers throw the competitive balance to boreal species, creating shoreline belts of boreal forest that transition to temperate forest inland where summers are warmer. The ecotone between northern temperate and boreal forests is broad, with species from the two biomes commonly found growing together over a 500-km-wide region (Fisichelli *et al.* 2014). In North America, temperate *Betula alleghaniensis*, *Acer saccharum* and *Fagus grandifolia* are mixed with boreal *Abies balsamea* and *Picea glauca*, in Asia, temperate *Quercus mongolica* and *A. mono* mix with boreal *Pinus koraiensis* and *P. sylvestris* var. *mongolica*, while in Europe temperate *A. platanoides*, *F. sylvatica* and *Q. robur* mix with boreal *B. pendula* and *Picea abies*.

Soil and site conditions

The northern portions of the temperate forest in North America and Europe were glaciated and have relatively young soils (about 9,000–14,000 years old) on top of glacial deposits or rocky surfaces sculpted by the glaciers, whereas older soils with a wide range of substrates, fertility, and topographical relief occur elsewhere. Mesic sites with loamy soils are most common, and usually dominated by *Acer*, *Fagus*, *Tilia*, and *Tsuga* species (Figure 3.2a). Dry-mesic to dry sites with soils that are sandy or shallow to bedrock are usually characterized by *Pinus* and *Quercus* species (Figure 3.2b). Swamps, peatlands and riparian areas are dominated by *Acer*, *Alnus*, *Betula*, *Fraxinus*, *Populus*, and *Salix* species. Gradients from wet to mesic and dry soils in complex topography are common. Temperate forest ecosystems cover an area of 10.4 million km^2 (about 6 percent of all Earth ecosystems) and store about 21 percent of carbon (139 PgC) in plants and 11.2 percent of carbon (262 PgC) in soils (Sabine *et al.* 2004).

Figure 3.2 (a) Temperate *Acer*-dominated forest on a mesic site, Michigan, USA. Photo: Jiri Schlaghamersky. (b) Prescribed fire in *Quercus* forest on a dry sandy site. Photo: Kalev Jogiste

Importance to society

High-density human settlement and land clearing, for timber and growing space for agriculture, have taken place in much of the northern temperate forest over the last several centuries in Asia, Europe, and eastern North America. This clearance was associated with the development of the Industrial Revolution in Europe during the late 1800s and its expansion to the rest of the world. For several centuries temperate forests were also essential wood sources for naval construction. Thus, the temperate forests have played a major role in shaping human society as it now exists. The need for ship wood also stimulated the development of the fields of forest science and forest management. The now recovering temperate forests constitute a large and growing source of carbon storage, wildlife habitat and watershed protection. Many important products come from the temperate forest, including sources of non-timber forest products (wild plants and fungi used as food, hunting animals for food, maple syrup [N. America] from *Acer saccharum,* and birch sap for traditional beverages and in folk medicine as an ingredient for different antiseptic and anti-inflammatory treatments [Central and NE Europe]), wood products (hardwood used for furniture and flooring, *Pinus* and *Quercus* used for construction lumber), and paper products (many species used for pulp). The forests also contribute greatly to the tourism industry through attractions such as remnant old-growth forests, large trees, spring wildflowers, brilliant fall colors, and habitat for migrant bird species, song birds, and other wildlife species commonly viewed by tourists.

Ecological processes

Nutrient cycling and decay processes

Soil structure in temperate forests usually has an organic horizon at the top of the soil profile composed of leaf litter in various stages of decay, from fresh leaves at the top to fragmented leaves in the middle and black humus at the bottom. Rate of decay is important for determining how fast nutrients are released from the organic horizon, and is determined by temperature and soil moisture, which in turn is determined by climate and soil texture. Organic horizon thickness is greater in colder climates and drier climates within the temperate zone, where decomposition of leaf litter is slower, in forests dominated by species with high C:N ratios and low nitrogen content of leaf litter (*Fagus, Picea, Pinus, Quercus*), in forests on sandy soils, and forests with low earthworm abundance. Cold temperatures in northern temperate and boreal forests depress organic matter decomposition, and may suppress nutrient movements in soil and their uptake by roots. Young, postglacial soils tend to be more N- than P-limited (Reich and Oleksyn 2004).

Changes in forestry practices that alter tree species composition, along with urbanization and other influences of human activity such as atmospheric deposition may alter cycles of ecologically important nutrients, including nitrogen, sulfur, hydrogen ions and base cations (Ca^{2+}, Mn^{2+}, Na^+, and K^+; Hedin et al. 1995). Temperate old-growth forest ecosystems are particularly vulnerable to increases in atmospheric N deposition. In response to increased N supply (both from N deposition and faster litter mineralization enhanced by increasing temperatures), the pattern of N cycling may change and be accompanied by increased selection favoring nitrophilous plant species and a decrease in ground species richness (Tamm 1991). An example of large-scale changes in nutrient cycling and soil function caused by human activity has occurred in Europe during the late nineteenth and twentieth centuries. Coniferous trees including native (*Picea abies, Pinus sylvestris*), and North American (*Picea sitchensis, Pseudotsuga menziesii*) species were widely planted to reverse eighteenth and nineteenth century deforestation and increase productivity, often replacing native deciduous species. Creation of large-scale coniferous monocultures contributed to further acidification of already acidic soils (65 percent of European topsoils have pH ≤4.5, Augusto *et al.* 2002).

Disturbance and succession

Major disturbances that kill or level the forest canopy include logging, wind, fire, ice storm, landslides, and insect infestation. Note that all of these may also create partial disturbance with scattered tree mortality contributing to the gap dynamics of multi-aged stands described below. Whether stand-leveling disturbance initiates an episode of succession—defined as a directional change in species composition over time, usually establishment of a dominant cohort of early-successional pioneer trees species followed by its gradual replacement by a suite of later-successional species—is context dependent (Heinselman 1981, Finegan 1984). For example, canopy-leveling wind storms or logging of the large trees may leave a carpet of late-successional advanced regeneration intact on the forest floor, leading to initiation of an even-aged stand followed by a stand development sequence from even-aged to multi-aged, but not necessarily change in species composition. This has been shown to occur in North American *Acer saccharum* and *Tsuga canadensis* forests. On the other hand, high-severity disturbances such as windthrow followed by high-intensity fire, or landslide could easily wipe out the advanced regeneration, leading to establishment of shade-intolerant, early-successional species succeeding to shade-tolerant, late-successional species over the subsequent one to two centuries. Illustrative case studies include *Betula papyrifera* and *Populus tremuloides* succeeding to *Acer saccharum* and *Tsuga canadensis* in North America and *Betula grossa* succeeding to *Tsuga sieboldii* and *Fagus crenata* in Japan (Yoshida and Ohsawa 1996, Frelich 2002).

Multiple disturbance regimes occur in the northern temperate forest biome. The least severe regime occurs on good quality sites with moist climates, and is dominated by single to multiple treefalls, usually wind-dominated regimes with some insect, disease, and drought mortality, where gap dynamics are prominent (described in more detail below). Such forests are dominated by mid-to-late-successional species, and under natural conditions, episodes of succession depend on rare episodes of severe disturbance mentioned above. However, in the last one to two centuries, clear cutting followed by slash burning has mimicked the rare natural wind–fire combination and created much larger expanses of early-successional forests than would have occurred naturally, in regions undergoing human settlement. A second disturbance regime is that of frequent low-intensity fire, commonly occurring in *Quercus* and *Pinus* savannas/woodlands. These forests are also multi-aged and have little succession over time unless the natural disturbance regime is disrupted by humans, usually by fire exclusion. The tree species are adapted to frequent fire due to thick bark that insulates the base of the trees from being girdled (*Quercus* and *Pinus*), and/or the ability to sprout if top-killed by fire (*Betula*, *Populus* and *Quercus*), and ability to germinate in post-fire conditions with full sun and lack of leaf litter. Once established, these trees commonly live a century or more and survive many subsequent fires. These forests occupy relatively dry climates in the transition from temperate forests to grasslands or Mediterranean vegetation, as well as excessively well drained sites on sand plains and south-facing rocky hillsides within wetter climates. A third disturbance regime is that of mixed-intensity fire, whereby fires of variable intensity occur at irregular intervals. Fires may kill some, most or all trees in a given stand, and may come at short intervals so that there is no time for succession, maintaining the current successional status, or long intervals, allowing succession to take place between fires. Although successional pathways for these forests are understood, i.e. they progress from shade-intolerant fire adapted species to shade-tolerant species, and these patterns are known for many regions, the irregular nature of the fire regime makes the actual successional trajectory in any one stand unpredictable over time (Frelich 2002). Forests with mixed-intensity fire regimes occupy sites with climates and/or soils between the first two regimes mentioned above. They include mixed *Betula*, *Pinus*, *Populus* and *Quercus* forests, sometimes succeeding to *Acer* and *Tsuga* in eastern North America, and *Pseudotsuga* potentially succeeding to *Abies*, *Picea* and *Tsuga* in western North America.

Canopy structure and gap dynamics

Large tracts of temperate forest occur on mesic sites where stand-leveling disturbances are rare, and gaps from single to several trees falling constitute the main type of disturbance (e.g. Payette *et al.* 1990). The dynamics of these treefall gaps, canopy turnover rates, canopy tree residence times, canopy structure, and recruitment of new trees into gaps have been studied at numerous sites. Such forests are usually multi-aged, and dominated by late-successional, shade-tolerant species, with some mid-tolerant species, and an occasional specimen of early-successional species. Fluctuations in relative abundance of the late-successional species occur over time due to influences on seedling success such as preferential deer grazing on seedlings, and neighborhood effects (structure and chemical properties of the leaf litter, sprouting; Brisson *et al.* 1994). The late-successional species generally exist as suppressed seedlings (or advanced regeneration) that will record a release from suppression if the tree above dies; this fact has been used to reconstruct the treefall history of forests going back as much as several centuries (Lorimer 1980, Payette *et al.* 1990). In contrast, the seedling bank of mid-tolerant tree species is rather small due to rapid turnover as seedlings die due to low light levels, but these species are adapted to enter new gaps by seed dispersal followed by rapid height growth that may surpass that of shade-tolerant species. Average tree residence times in the canopy vary from ≈100–200 years in temperate forests, with a few trees living 2–3 times the average lifespan (Runkle 1982).

Gap formation creates microsites for seedling germination including relatively dry and wet mineral soil on mounds and in pits created by uprooted trees, above-ground light gaps and below ground 'gaps' in root occupancy of the soil created by fallen trees (Beatty and Stone 1986). These gaps can have not only more light, but higher soil and air temperatures, and more water (due to lack of use by the fallen tree), leading to more nutrient availability in the first few years as the fine roots of the fallen tree decay in the relatively warm, wet environment. Tree species mid-tolerant of shade (e.g. *Fraxinus* and *Tilia*, along with some species of *Acer*, *Betula*, *Carya*, *Pinus*, and *Quercus*) often take advantage of these gaps, which therefore help to maintain local species richness over time. In addition, treefalls create coarse woody debris, which, when it reaches an advanced state of decay, can also be a microsite for germination of some tree species that cannot tolerate thick leaf litter or the dense herbaceous vegetation that may cover the forest floor. These decaying logs often last 20–40 years for angiosperm species in genera such as *Acer, Quercus* and *Tilia*, but may last 1–2 centuries for gymnosperms such as *Picea, Pinus* and *Tsuga*. Decaying logs also are small-scale hot spots of biodiversity due to the large number of species of mosses, fungi, insects, and amphibians that use them as habitat due to their relatively high water content (Harmon *et al.* 1986).

Plant species interactions

Interactions of note in the northern temperate forest include the occurrence of mosaics formed by patches of evergreen conifer and broadleaf deciduous species, shade-tolerant and intolerant species interactions, and the presence of numerous understory species that take advantage of a brief period of combined warmth and light in spring, and heterotrophic species that do not require sunlight to fuel carbon gain (e.g. saprophytic and parasitic plants).

Evergreen–deciduous mosaics of *Tsuga* or *Picea* with *Acer*, *Betula* and *Tilia* (mesic sites), or with *Pinus* and *Quercus* (drier sites) occur in the parts of the temperate forests with cooler climates, including the northern USA, northern Europe, and Asia. These mosaics (Figure 3.3) may be created by slight differences in the soil environment magnified by neighborhood effects of the trees themselves, through shading, leaf-litter chemistry, seed rain, interactions with disturbance and, in some cases, sprouting, so that the deciduous and evergreen species each favor their own reproduction and/or disfavor seedlings of the opposite group under their own canopies (Frelich *et al.* 1993).

Shade tolerance or avoiding shade is necessary for any plant species to reproduce in these forests given dense canopies and forest floor light levels ranging from 1–10 percent of full sunlight. Plant species have evolved a variety of strategies. Spring ephemerals are species that grow rapidly as soon as winter ends, taking advantage of the sunlight available on the forest floor prior to canopy leaf out. Included are species in the genera *Allium* (wild leek), *Cardamine* (toothwort), *Corydalis*, *Dicentra* (Dutchman's breetches), *Erythronium* (trout lily), and *Mertensia* (bluebell), which complete their above-ground life cycle of growth, flowering, seed set and die back to a dormant phase within 4–6 weeks (Figure 3.4). Some of these genera have vicarious species (related species that live in similar environments with similar ecological niche) among the major occurrences of temperate forest, for example, *Erythronium americanum* in the eastern USA, *E. dens-canis* in central and southern Europe, and *E. japonicum* in eastern Asia.

Other early blooming species are commonly misclassified as spring ephemerals, but they keep their leaves for much of the summer—these include geophytes such as *Trillium*, and herbs

Figure 3.3 Old-growth forest mosaic of deciduous (*Acer, Tilia, Betula*), and coniferous (*Tsuga, Thuja, Picea*) trees. Porcupine Mountains, Michigan, USA. Photo: Jiri Schlaghamersky

Figure 3.4 Spring ephemeral yellow trout lily (*Erythronium americanum*) in an *Acer* and *Betula* forest
Source: Jiri Schlaghamersky

such as *Arisaema* (jack-in-the-pulpit), *Sanguinaria* (bloodroot), and *Viola* (violets), which can use energy stored in a bulb or rhizomes to produce leaves early on in full sun, but continue to photosynthesize in deep shade after canopy leaf out. Saprophytic plants such as *Corallorhiza* (coral root orchids), *Monotropa uniflora* (Indian pipe) and parasitic plants such as *Epifagus virginiana* (beech drops) and *Conopholis americana* (squaw root) get their energy from decomposing organic matter or attachments to tree roots, and do not require sunlight.

A number of plant species take advantage of treefall gaps, where higher than average light levels last a few decades, to grow up into the canopy (e.g. *Betula, Quercus*) or to complete their life cycle and create long-lived buried seeds that await the next gap a century or more later—e.g. *Prunus pensylvanica* (pin cherry), *Rubus* (raspberry), and *Geranium bicknellii*. Another group of species has short life cycles within gaps but produce plumed seeds that float long distances through the air to 'find' new gaps (e.g. *Epilobium angustifolium*, fireweed). Shade-tolerant shrubs in forest understories can create complex spatial dynamics for tree regeneration. Understory dwarf bamboo (*Sasa* spp.) in Asian forests can exclude seedling establishment, even for shade-tolerant tree species. For example, beech (*Fagus crenata*) and maple (*Acer mono*) seedlings in Japanese forests are restricted to small patches where bamboo is absent (Yamamoto *et al*. 1995).

Plant–animal interactions

Many symbiotic and antagonistic plant–animal interactions occur in northern temperate forests, involving seed dispersal, pollination, defoliation, insect herbivory, and preferential grazing by large mammals (deer, moose) that can alter successional trajectories. Although many species of temperate trees have wind dispersed seeds, animal dispersal (zoochory) also occurs. One well-known case is for acorns of northern red oak (*Quercus rubra*) that are cached by various animal species in the fall for later use during winter. Gray squirrels (*Sciurus carolinensis*) bury acorns in the soil and are known to remember the locations of several thousand acorns, but during mast years, they often bury more than they retrieve, leaving many to germinate, and at the same time hiding those acorns from consumption by deer (*Odocoileus virginianus*), black bear (*Ursus americanus*), and turkey (*Meleagris gallopavo*). Blue jays (*Cyanocitta cristata*) can carry acorns, and commonly fly several hundred meters to two kilometers from the parent tree, scatter hoarding thousands of acorns across the landscape (Johnson and Webb 1989). In Europe, acorns of *Quercus coccifera*, *Q. robur*, *Q. petraea*, *Q. ilex*, *Q. suber*, and *Q. faginea* and *Fagus* nuts are dispersed by the Eurasian jay (*Garrrulus glandarius*). Analysis of the seed dispersal service performed by the Eurasian jay in the Stockholm National Urban Park, Sweden has shown that depending upon seeding or planting technique chosen, the cost of replacement per pair of jays through human means in the park is 2,100 US$ (seeding) to 9,400 US$ (planting) per hectare, respectively (Hougner et al. 2006). Such estimates provide a good example of the value of management strategies securing critical breeding and foraging habitats of seed dispersal animals.

Despite the fact that the Eurasian jay prefers acorns of native *Quercus robur* it is also efficient in dispersing invasive North American *Q. rubra* (Myczko et al. 2014). Increase of red oak leads to significant reduction of native species abundance and diversity (Woziwoda et al. 2014).

Seed dispersal by ants is known as myrmecochory, and is common among temperate forest understory plant species (Gomez and Espadaler 1998). The spring wildflower genera *Trillium* (North America and Asia) and *Viola* (North America, Europe and Asia) are examples, with a number of species dispersed by ants, although other vectors also exist. The seeds of these species have a fat-rich elaiosome attached to each seed, which the ant can detach for consumption, commonly after moving the seed one or two meters.

A large number of bee species occur in temperate forests (often 50 or more species in one stand), which pollinate most of the forest understory species with brightly colored flowers and limited pollen production, and a few of the tree genera such as *Acer* (in part, wind dispersal also occurs), *Prunus*, and *Tilia*. Folivorous insects are common, with many species that cause ongoing low-to-moderate levels of defoliation, that interact with plant defense compounds in leaves such as phenols and terpenoids. Other insect species have periodic outbreaks every decade or longer, during which the forest may be almost totally defoliated. For example, deciduous forests in eastern North America can be defoliated by saddled prominent moth (*Heterocampa guttivitta*) caterpillars during droughts that prevent trees from defending themselves against insects, and on a more regular 10–12 year cycle by the forest tent caterpillar (*Malacosoma americanum*) (Horsley et al. 2002). These defoliation events usually cause scattered tree mortality, especially to older trees with other health problems, but seldom cause extensive tree death at the landscape scale. Much more serious are growth losses and tree mortality after outbreaks of the common pine sawfly (*Diprion pini*). Defoliation by this late season feeder causes higher losses than herbivory by early season feeders, e.g. the European pine sawfly (*Neodiprion sertifer*). During severe outbreaks in Finland, approximately 500,000 ha were defoliated by *D. pini* resulting in high tree mortality and growth losses in the following year and later. Mortality rates in Europe after an outbreak are typically 4–24 percent and it can take 10 to 15 years for radial growth to recover (Lyytikäinen-Saarenmaa et al. 2003).

Grazing animals, especially ungulates like deer, elk and moose can regulate plant community composition and direct succession through their plant species preferences. Many ungulates in the temperate zone consume woody plants during winter and herbaceous plants during summer. If the deer to plant ratio is high, they can regulate the balance between relatively palatable and unpalatable plant species. The strength of these influences on composition is in turn embedded in a trophic cascade. For example, wolves in Wisconsin, USA create a patchy distribution in deer density across the landscape, due to predation and deer avoidance of wolf pack territories. This in turn influences lushness, composition, and species richness of herbaceous plants (Callan *et al.* 2013).

Soil animals such as earthworms, beetles, and many other taxonomic groups (whether native or invasive), also influence plant community composition by altering the structure of the organic horizon, moving seeds to different layers within the soils, consuming seeds, and regulating water and nutrient cycles within the soils. There are considerable ecological cascades within the soil (Eisenhauer *et al.* 2009); for example, earthworms can change leaf-litter structure to favor or disfavor certain fungal species that in turn may be involved in symbiotic relationships with tree roots as mycorrhizas that help trees absorb nutrients, or with symbiotic relationships with seed germination, for example orchid seeds.

Conservation issues

Land use conversion, fragmentation, and harvesting

The temperate zone has historically had high human populations, especially in Europe and Asia, and for the last century in North America as well. Therefore, much of the forest has been converted to croplands and cities, and almost all that remains forested has been logged at least once. In Europe and Asia most forests have been logged multiple times and planted regeneration is used on a widespread basis to supplement natural regeneration after harvesting, while in North America most forests have been logged once or twice, and natural regeneration is much more common. In recent decades, the wildland–urban interface, an area with widely scattered isolated houses, has encroached on large tracts of forested land, creating numerous foci for introduction of invasive species, and fragmenting the landscape, favoring certain species that do well along forest edges. A very small amount of primary temperate forest remains as compared to tropical and boreal forest biomes, commonly less than 1 percent of the original forest (Frelich 2002). Most of these primary remnants have been identified by conservationists and set aside in well-known preserves. Some examples include Great Smoky Mountains National Park (Tennessee and North Carolina), Adirondack State Park (New York), Porcupine Mountains Wilderness State Park (Michigan), and Olympic National Park (Washington) in the USA, and Bialowieza Forest in Poland and Belarus. These primary forest remnants serve as templates for restoration of secondary forests, and as a baseline for the occurrence of ecosystem processes as compared to forests that are harvested. However, in many regions with thousands of years of human influence, no natural templates for restoration exist. In such cases a multi-disciplinary synthesis of historic records, silvicultural and paleoecological evidence may be needed to develop management techniques that mimic disturbances and other conditions needed to maintain the diversity of native tree species and smaller species dependent on them, and maintenance of certain cultural features of the landscape may also become a priority (Lindbladh *et al.* 2007).

Fragmentation manifested as small woodlands of a few to a hundred hectares surrounded by agricultural lands or cities, is common in the temperate forest biome throughout the eastern USA, Europe and Asia (Wilcove *et al.* 1986). This has led to conservation problems including: (1) facilitation of invasive plant species; (2) over populations of edge-loving native wildlife species such as deer and certain birds like the cowbirds (*Molothrus* spp.), which then parasitize nests of songbirds; and (3) potential inbreeding over time and loss of populations of many species. Island biogeography theory predicts that it is not possible to maintain as many species on a small fraction of the extent of forest that existed prior to deforestation and fragmentation. The term 'extinction debt' has been used to describe the situation where many species with long lifespans still exist within fragmented landscapes, but are predicted to eventually go extinct (Tilman *et al.* 1994).

Overgrazing by ungulates and domestic livestock

Overgrazing is a common problem for maintenance of biodiversity and regeneration of trees (Côté *et al.* 2004). For example, in North America, white-tailed deer (*Odocoileus virginianus*) prefer seedlings of northern white cedar (*Thuja occidentalis*), yellow birch (*Betula alleghaniensis*), red oak (*Quercus rubra*), and hemlock (*Tsuga canadensis*) during winter, leading to widespread difficulties in regeneration of those species (Figure 3.5). This limits the ability of forest managers to direct succession towards desired tree species and to maintain species richness. During the growing season, deer prefer many herbaceous species in the families *Liliaceae* and *Orchidaceae*. Deer can drive succession to unpalatable trees such as beech (*Fagus grandifolia*) and understory species including ferns and sedges (*Carex* spp.) that are not favored or that can tolerate heavy grazing. Effects of deer overabundance in different parts of the temperate forest can be idiosyncratic. Sika deer (*Cervus nippon*) in Japan can prevent *Fagus* recruitment (rather than favoring it as in North America), but can also have similar effects on herbaceous vegetation as in North America, by favoring graminoids and ferns in the herbaceous vegetation layer (Takatsuki 2009). Lowering deer densities through hunting has been controversial since members of the public often equate more deer with a healthy ecosystem, and the effectiveness of hunting may become limiting in an increasingly urbanized world where relatively few people learn to hunt. Livestock such as cows can also cause similar problems if allowed to graze freely within woodlands without sufficient rotation to different areas throughout the growing season. Overgrazing may be a problem for national parks where plants as well as grazer populations are protected from human exploitation. For example, in the Bialowieza Forest National Park in Poland where the European bison (*Bison bonasus*) has been successfully restored after extinction in the wild at the beginning of the twentieth century, recent study has shown that consumption of trees and shrubs by bison increased with decreasing access to supplementary fodder, ranging from 16 percent in intensively fed bison to 65 percent in non-fed bison using forest habitats. Bison browsed mainly on *Carpinus*, *Corylus* and *Betula*, tree species of relatively low economic importance in the region, so that impacts on forestry may be relatively small. However, more investigation of bison impacts is needed to develop management plans to meet other objectives, such as maintaining plant diversity within the park and reducing damage to agriculture surrounding the park (Kowalczyk *et al.* 2011).

Invasive species

Intercontinental movements of temperate plant and animal species have occurred over the last few centuries, with accelerating rates of new introductions in recent decades corresponding to the larger human population and commercial trade in plants and animals (Kalusova *et al.* 2013). Some examples for plants invading mesic forests include: common buckthorn (*Rhamnus cathartica*), tatarian honeysuckle (*Lonicera tatarica*) and garlic mustard (*Alliaria petiolata*) have moved from Eurasia to North America, and black cherry (*Prunus serotina*), honey locust (*Gleditsia triacanthos*), black locust (*Robinia pseudoacacia*), and red oak (*Quercus rubra*) have moved from North America to Europe. Freed from competing species adapted to their presence, from leaf and seed-eating insects, from large herbivores that prefer them over other species, and from diseases in their native habitat, these invasive plants often become much more abundant on their new continent than their home continent, filling forest understories, reducing the

(a)

(b)

Figure 3.5 Forests dominated by *Tsuga* and *Acer* with low (a) and high (b) levels of deer browsing; note the large difference in abundance of *Tsuga* saplings. Photos: Jiri Schlaghamersky

abundance of native tree seedlings, and necessitating expensive management actions to remove them and/or to restore native species.

Invasive animals include the North American gray squirrel (*Sciurus carolinensis*) invading European forests where they displace the native red squirrel (*Sciurus vulgaris*), and the raccoon dog (*Nyctereutes procyonoides*) from Asia invading European forests (Genovesi *et al.* 2012). Earthworms as invasive species can have profound impacts on ecosystems around the world (Hendrix *et al.* 2008). A recent body of research shows that European earthworms in North America change soil structure, eliminating the organic horizon and increasing bulk density, making soils drier, depleting nutrients, and changing seedbed conditions (Hale *et al.* 2005, Resner *et al.* 2015). Ecological cascade effects include reduced tree growth, reduced native plant species richness, favoring a different suite of plant species, and facilitating invasive plant species that coevolved with the worms on their home continent (Nuzzo *et al.* 2009, Frelich *et al.* 2006, Larson *et al.* 2010).

Diseases and insect pests of trees moving from one continent to another have caused huge aesthetic, economic and habitat losses, wiping out tree species that are foundational to ecosystem function (Parry and Teale 2011). North America has been the recipient of many devastating tree pandemics including chestnut blight (*Cryphonectria parasitica*), Dutch elm disease (*Ophiostoma ulmi*), emerald ash borer (*Agrilus planipennis*), gypsy moth (*Lymantria dispar dispar*), balsam woolly adelgid (*Adelges piceae*), and hemlock woolly adelgid (*Adelges tsugae*). In combination, these pests and diseases have the capacity to greatly reduce the species richness of the tree canopy in eastern North American forests, with cascading ecological impacts on nutrient and light regimes, as well as physical structure of the forest habitat.

Climate change

The temperate forest is projected to shift to the north 200–700 km by the end of the twenty-first century, depending on local and global magnitude of warming (Galatowitsch *et al.* 2009). Two types of progression are likely to occur: warm–dry scenarios, and warm–wet scenarios. Warm–dry scenarios will likely lead to increases of oak, pine, and grass species at the expense of mesic species as the environmental niche of mesic species shrinks. This is likely to occur in mid-continental temperate forests adjacent to grassland biomes. Significant drought stress and mortality of temperate trees are expected. The second, warm–wet scenarios, would be much less stressful for most tree species, and would allow for gradual replacement of existing species with species from further south. Invasions of savanna and grasslands into mesic forest, and subtropical forest into temperate forest, are likely within several hundred kilometers of the southern margin of the northern temperate forest biome. At the same time, at the northern margin, temperate forest is likely to invade the southern boreal forest, at first forming a wider mixed temperate–boreal ecotone, as temperate species are freed from limitation of extreme winter cold and/or short summers, and later on decline of boreal tree species as warm temperature thresholds for those species are crossed.

A number of factors are likely to interact with climate change, including invasive species, deer grazing, insects, wind storms, fires, and fragmentation (Frelich and Reich 2010). Invasive species, which as a group are generally tolerant of a wide variety of climates and disturbances, have abundant seed production and long-distance dispersal, therefore giving them an advantage in a rapidly changing forest community. Deer preference for grazing on certain species of tree seedlings could either oppose climate-induced change (e.g. by consuming temperate maple and oak seedlings, but not spruce and fir seedlings at the temperate–boreal ecotone, retarding temperate invasion into boreal stands), or exacerbate climate-induced change (e.g. consuming oak seedlings along the prairie–forest border, reinforcing the movement of grasslands into temperate forest). In the first decade of the twenty-first century, several consecutive years of severe drought and heat stress that occurred in most European countries caused deterioration of *P. abies* stands, predisposing them to catastrophic bark beetle infestation (Przybył *et al.* 2008). Reconstructions of surface temperature for Europe in the last 1,500 years has shown that the late twentieth and early twenty-first century European climate is very likely warmer than that of any time during the past 500 years (Luterbacher *et al.* 2004). Therefore, this episode of drought and insect damage is a harbinger of future impacts of a warming climate. A warmer climate with higher evaporation and more erratic precipitation is likely to lead to more fires over much of the temperate forest, which will potentially increase the proportion of early-successional (e.g. *Betula, Populus, Quercus, Pinus*) forests on the landscape, and alter the dynamics of tree species migration. In large swaths of the northern temperate forest that are highly fragmented, movement of species to the north to respond to climate change will be hindered; there are a large number of species with very limited dispersal distances and long establishment times, for example the previously mentioned myrmecochorus plant species. Forest managers and the public will have to decide whether to employ the controversial practice of assisted migration for such cases (Buma and Wessman 2013).

Forest scientists and managers of forests in commercial and natural area settings face many challenges in the temperate forest biome. In a biome where climate change, overgrazing, invasive species, and fragmentation are pervasive, creative research, comparison of remnant natural forests with commercial forests, and development of adaptive management techniques, will be paramount in order to allow continued existence of productive forests capable of maintaining ecological function and native species diversity.

References

Augusto, L., Ranger, J., Binkley, D., and Rothe, A. (2002) 'Impact of several common tree species of European temperate forests on soil fertility' *Annals of Forest Science*, vol. 59, pp. 233–253

Beatty, S.W., and Stone, E.L. (1986) 'The variety of soil microsites created by tree falls' *Canadian Journal of Forest Research*, vol. 16, pp. 539–548

Brisson, J., Bergeron, Y., Bouchard, A. and Leduc, A. (1994) 'Beech-maple dynamics in an old-growth forest in southern Quebec, Canada' *Ecoscience*, vol. 1, pp. 40–46

Buma, B., and Wessman, C.A. (2013) 'Forest resilience, climate change, and opportunities for adaptation: A specific case of a general problem' *Forest Ecology and Management*, vol. 306, pp. 216–225

Callan, R., Nebbelink, N.P., Rooney, T.P., Wiedenhoeft, J.E., and Wydeven, A.P. (2013) 'Recolonizing wolves trigger a trophic cascade in Wisconsin (USA)' *Journal of Ecology*, vol. 101, pp. 837–845

Côté, S.D., Rooney, T.P., Tremblay, J-P, Dussault, C.D., and Waller, D.M. (2004) 'Ecological impacts of deer overabundance' *Annual Review of Ecology and Systematics*, vol. 35, pp. 113–147

Eisenhauer, N., Schuy, M., Butenschoen, O., and Scheu, S. (2009) 'Direct and indirect effects of endogeic earthworms on plant seeds' *Pedobiologia*, vol. 52, pp. 151–162

Finegan, B. (1984) 'Forest succession' *Nature*, vol. 312, pp. 109–114

Fisichelli, N.A., Frelich, L.E. and Reich, P.B. (2014) 'Temperate tree expansion into adjacent boreal forest patches facilitated by warmer temperatures' *Ecography*, vol. 37, pp. 152–161

Frelich, L.E. (2002) *Forest dynamics and disturbance regimes*, Cambridge University Press, Cambridge, England

Frelich, L.E. and Reich, P.B. (2010) 'Will environmental changes reinforce the impact of global warming on the prairie-forest border of central North America?' *Frontiers in Ecology and Environment*, vol. 8, pp. 371–378

Frelich, L.E., Calcote, R.R., Davis, M.B. and Pastor, J. (1993) 'Patch formation and maintenance in an old growth hemlock-hardwood forest' *Ecology*, vol. 74, pp. 513–527

Frelich, L.E., Hale, C.M., Scheu, S., Holdsworth, A., Heneghan, L. and others (2006) 'Earthworm invasion into previously earthworm-free temperate and boreal forests' *Biological Invasions*, vol. 8, pp. 1235–1245

Galatowitsch, S., Frelich, L.E. and Phillips-Mao, L. (2009) 'Regional climate change adaptation strategies for biodiversity conservation in a midcontinental region of North America' *Biological Conservation*, vol. 142, pp. 2012–2022

Genovesi, P., Carnevali, L., Alonzi, A., and Scalera, R. (2012) 'Alien animals in Europe: updated numbers and trends, and assessment of the effects on biodiversity' *Integrative Zoology*, vol. 7, pp. 247–253

George, M.F., Burke, M.J., Pellett, H.M. and Johnson, A.G. (1974) 'Low temperature exotherms and woody plant distribution' *Hortscience*, vol. 9, pp. 519–522

Gomez, C. and Espadaler, X. (1998) 'Myrmecochorous dispersal distances: a world survey' *Journal of Biogeography*, vol. 25, pp. 573–580

Hale, C.M., Frelich, L.E., Reich, P.B., and Pastor, J. (2005) 'Effects of European earthworm invasion on soil characteristics in northern hardwood forests of Minnesota, USA' *Ecosystems*, vol. 8, pp. 911–927

Harmon, M.E., Franklin, J.F., Swanson, F.J., Sollins, P., Gregory, S.V. and others (1986) 'Ecology of coarse woody debris in temperate ecosystems' *Advances in Ecological Research*, vol. 15, pp. 133–302

Hedin, L.O., Armesto, J.J. and Johnson, A.H. (1995) 'Patterns of nutrient loss from unpolluted, old-growth temperate forests: Evaluation of biogeochemical theory' *Ecology*, vol. 76, pp. 493–509

Heinselman, M.L. (1981) 'Fire and succession in the conifer forests of northern North America', in D.C. West, H.H. Shugart, and D.B. Botkin (eds), *Forest Succession, Concepts and Application*, Springer-Verlag, New York

Hendrix, P.F., Callaham, M.A., Jr., Drake, J.M., Huang, C-Y., James, S.W. and others (2008) 'Pandora's box contained bait: the global problem of introduced earthworms' *Annual Reviews of Ecology and Systematics*, vol. 39, pp. 593–613

Horsley, S.B., Long, R.P., Bailey, S.W., Hallett, R.A. and Wargo, P.M. (2002) 'Health of eastern North American sugar maple forests and factors affecting decline' *Northern Journal of Applied Forestry*, vol. 19, pp. 34–44

Hougner, C., Colding, J. and Söderqvist, T. (2006) 'Economic valuation of a seed dispersal service in the Stockholm National Urban Park, Sweden' *Ecological Economics*, vol. 59, pp. 364–374

Johnson, W.C. and Webb III, T. (1989) 'The role of blue jays (*Cyanocitta cristata* L.) in the post-glacial dispersal of fagaceous trees in eastern North America' *Journal of Biogeography*, vol. 16, pp. 561–571

Kalusova, V., Chytry, M., Kartesz, J.T., Nishino, M. and Pysek, P. (2013) 'Where do they come from and where do they go? European natural habitats as donors of invasive alien plants globally' *Diversity and Distributions*, vol. 19, pp. 199–214

Kowalczyk, R., Taberlet, P., Coissac, E., Valentini, A., Miquel, C. and others (2011) 'Influence of management practices on large herbivore diet—Case of European bison in Bialowieza Primeval Forest (Poland)' *Forest Ecology and Management*, vol. 261, pp. 821–828

Larson, E.R., Kipfmueller, K.F., Hale, C.M., Frelich, L.E., and Reich, P.B. (2010) 'Tree rings detect earthworm invasions and their effects in northern hardwood forests' *Biological Invasions*, vol. 12, pp. 1053–1066

Lindbladh, M., Brunet, J., Hannon, G., Niklasson, M., Eliasson, P. and others (2007) 'Forest history as a basis for ecosystem restoration. A multidisciplinary case study in a south Swedish temperate landscape' *Restoration Ecology*, vol. 15, pp. 284–295

Lorimer, C.G. (1980) 'Age structure and disturbance history of a southern Appalachian virgin forest' *Ecology*, vol. 61, pp. 1169–1184

Luterbacher, J., Dietrich, D., Xoplaki, E., Grosjean, M. and Wanner, H. (2004) 'European seasonal and annual temperature variability, trends, and extremes since 1500' *Science*, vol. 303, pp. 1499–1503

Lyytikäinen-Saarenmaa, P., Niemelä, P., and Annila, E. (2003) 'Growth responses and mortality of Scots Pine (*Pinus sylvestris* L.) after a Pine Sawfly outbreak' *Proceedings: IUFRO Kanazawa 2003—Forest Insect Population Dynamics and Host Influences*, pp. 81–85

Myczko, Ł., Dylewski, Ł., Zduniak, P., Sparks, T.H., Tryjanowski, P. (2014) 'Predation and dispersal of acorns by European Jay (*Garrulus glandarius*) differs between a native penduculate oak (*Quercus robur*) and an introduced northern red oak (*Quercus rubra*) in Europe' *Forest Ecology and Management*, vol. 331, pp. 35–39

Nuzzo, V.A., Maerz, J.C., and Blossey, B. (2009) 'Earthworm invasion as the driving force behind plant invasion and community change in northeastern North American forests' *Conservation Biology*, vol. 23, pp. 966–974

Parry, D. and Teale, S.A. (2011) 'Alien invasions: the effects of introduced species on forest structure and function', in J.D. Castello and S.A. Teale (eds), *Forest Health, an integrated perspective*, Cambridge University Press, Cambridge, England

Payette, S., Filion, L. and Delawaide, A. (1990) 'Disturbance regime of a cold temperate forest as deduced from tree-ring patterns: the Tantaré Ecological Reserve, Quebec' *Canadian Journal of Forest Research*, vol. 20, pp. 1228–1241

Peel, M.C., Finlayson, B.L., and McMahon, T.A. (2007) 'Updated world map of the Köppen-Geiger climate classification' *Hydrology and Earth System Sciences*, vol. 11, pp. 1633–1644

Przybył, K., Karolewski, P., Oleksyn, J., Łabędzki, A. and Reich, P.B. (2008) 'Fungal diversity of Norway spruce litter: effects of site conditions and premature leaf fall caused by bark beetle outbreak' *Microbial Ecology*, vol. 56, pp. 332–340

Reich, P.B. and Oleksyn, J. (2004) 'Global patterns of plant leaf N and P in relation to temperature and latitude' *Proceedings of the National Academy of Sciences of the United States of America*, vol. 101, pp. 11001–11006

Resner, K., Yoo, K., Sebestyen, S.D., Aufdenkampe, A., Hale, C. and others (2015) 'Invasive earthworms deplete key soil inorganic nutrients (Ca, Mg, K, and P) in a northern hardwood forest' *Ecosystems*, vol. 18, pp. 89–102

Runkle, J.R. (1982) 'Patterns of disturbance on some old-growth mesic forests of eastern North America' *Ecology*, vol. 63, pp. 1533–1546

Sabine, C.L., Heimann, M., Artaxo, P., Bakker, D.C.E., Chen, C.T.A., Field, C.B., Gruber, N. and LeQuéré, C. (2004) 'Current status and past trends of the global carbon cycle', in C.B. Field and M.R. Raupach (eds), *The Global Carbon Cycle: Integrating Humans, Climate, and the Natural World*, SCOPE 62, Island Press, Washington DC

Svenning, J-C. and Skov, F. (2005) 'The relative roles of environment and history as controls of tree species composition and richness in Europe' *Journal of Biogeography*, vol. 32, pp. 1019–1033

Takatsuki, S. (2009) 'Effects of sika deer on vegetation in Japan: A review' *Biological Conservation*, vol. 142, pp. 1922–1929

Tamm, C.O. (1991) '*Nitrogen in Terrestrial Ecosystems, Questions of Productivity, Vegetational Changes, and Ecosystem Stability*, Springer-Verlag, Berlin

Tilman, D., May, R.M., Lehman, C.L. and Nowak, M.A. (1994) 'Habitat destruction and the extinction debt' *Nature*, vol. 371, pp. 65–66

Whittaker, R.H. (1975) *Communities and Ecosystems. 2nd Edition*, MacMillan, New York

Wilcove, D.S., McLellan, C.H. and Dobson, A.P. (1986) 'Habitat fragmentation in the temperate zone', in M. Soulé (ed.) *Conservation Biology, the Science of Scarcity and Diversity*. Sinauer Associates, Sutherland, Massachusetts, USA

Woziwoda, B., Kopeć, D., Witkowski, J. (2014) 'The negative impact of intentionally introduced *Quercus rubra* L. on a forest community' *Acta Societatis Botanicorum Poloniae*, vol. 83, pp. 39–49

Yamamoto, S., Nishimura, N., and Matsui, K. (1995) 'Natural disturbance and tree species coexistence in an old-growth beech – Dwarf bamboo forest, southwestern Japan' *Journal of Vegetation Science*, vol. 6, pp. 875–886

Yoshida, N., and Ohsawa, M. (1996) 'Differentiation and maintenance of topo-community patterns with reference to regeneration dynamics in mixed cool temperate forests in the Chichibu Mountains, central Japan' *Ecological Research*, vol. 11, pp. 351–362

4
SUBTROPICAL FORESTS

Richard T. Corlett and Alice C. Hughes

Introduction

There is no widely agreed definition of what constitutes the 'subtropics', but in recent ecological literature the term has been most often applied to the two belts between the tropics (+/− 23.4°) and approximately 30° north and south of the equator (Corlett, 2013; Figure 4.1). These boundaries will be used in this chapter. Although this choice is fairly arbitrary, particularly at the outer (30°) limit, applying a uniform standard makes it possible to compare forests in the same latitudinal belt in different parts of the world, as well to make comparisons within regions between subtropical forests and the forests of the tropical (i.e. < 23.4° from the equator) and temperate (i.e. > 30°) zones. Moreover, defining the subtropics by latitude alone avoids the confusion between latitudinal and altitudinal gradients that occurs in some of the literature.

Much of the subtropical belt, as defined above, is too dry (mean annual rainfall < 600–700 mm) for forest because of the descending branches of the Hadley circulation (the subtropical high), but there are, or were, extensive forests on the southeastern side of all the continents, particularly in China and adjacent countries (71 per cent of the total current forest area in the subtropics according to the MODIS landcover layer for 2012 [NASA Land Processes Distributed Active Archive Center (LP DAAC) (2012) Mcd12q1. USGS/Earth Resources Observation and Science (EROS) Center, Sioux Falls, South Dakota, USA]), and in northern Argentina and southeastern Brazil (23 per cent), with smaller areas in northern Mexico and the southeast USA (4 per cent), eastern Australia (2 per cent), and southern Africa (1 per cent) and a very small area in southeastern Madagascar. These extant areas of subtropical forest represent only part of their potential extent; around 40 per cent if forest is assumed to have covered all land areas < 4,000 m a.s.l. and with a mean annual rainfall > 700 mm (Figure 4.1). Moreover, much of the extant forest is secondary, degraded, or consists of plantations. The mismatch between actual and 'potential' forest is greatest in southern Africa, where climate predicts a much larger area than the scattered patches that exist.

Common features of the climates of these forested regions, apart from adequate rainfall, include lowland temperatures suitable for plant growth over most or all of the year, despite the occurrence of at least mild winter frosts in most areas in most years. In coastal regions, except in South America, subtropical forests are vulnerable to cyclones, with repeated occurrences reducing stature and biomass (e.g. McEwan et al., 2011). There have also been reports of

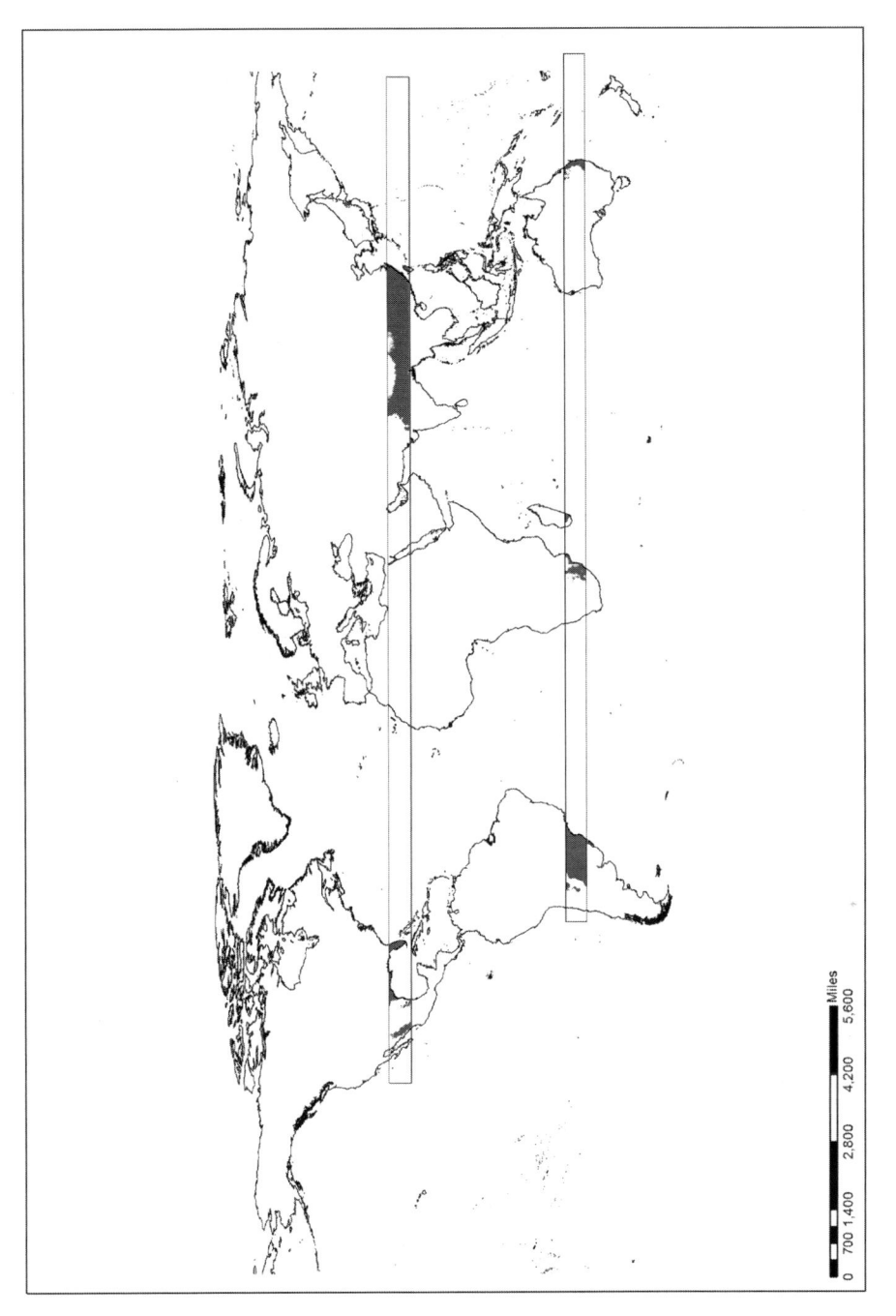

Figure 4.1 Potential forest cover between 23.4° and 30.0°, under the assumption that all areas with > 700 mm annual rainfall and < 4,000 m elevation are capable of supporting forest. This map was produced in ArcView 10.1 using climate data from WorldClim

damage by ice-storms. In early 2008, the longest cold spell for more than 40 years brought snow and ice-storms to a broad band of subtropical China, north of around 25°N, causing massive mechanical damage to native broad-leaved forests and devastation to many plantations (Zhou et al., 2011).

Subtropical forests have been intensively studied in China and have received considerable attention in South America, but studies elsewhere have been largely descriptive. The results of these studies are scattered across multiple journals, in multiple languages, since there is no subtropical equivalent of the international journals that focus on the tropics. Moreover, inconsistent nomenclature makes it hard to retrieve these papers electronically. The overview presented here is thus inevitably incomplete, but we hope it will encourage more 'pan-subtropical' comparisons.

Biogeography of the subtropics

The subtropics of the northern and southern hemisphere are separated by the 5,185-km width of the tropics. The forests of the northern subtropics, in Asia and North America, are floristically quite similar to each other, reflecting the availability of land routes across the North Atlantic during the early Tertiary, until global cooling broke this connection, probably at the end of the Eocene, c. 34 million years ago (Givnish and Renner, 2004). In contrast, the subtropical forests in the southern hemisphere have less in common with each other than they do with their nearest northern counterparts. Although these southern forests are all on fragments of the ancient southern supercontinent of Gondwana, the last terrestrial connections were in the early Cretaceous (145–100 million years ago), before most modern groups of plants originated, while tropical montane habitats have provided more recent stepping-stones from the north, across the tropical lowlands.

In southwest China, subtropical forest floras similar to those of today have existed since the early Miocene (Jacques et al., 2013), although some tropical elements in the Miocene flora have retreated to the south in response to global cooling and regional uplift (Jacques et al., 2014). In North Africa, in contrast, the shrinkage of the Tethys Sea during the Late Miocene drastically weakened the African summer monsoon, creating conditions at subtropical latitudes that were too arid for forest (Zhang et al., 2014). A relic of the North African Miocene forests persisted in the subtropical Canary Islands, which were buffered against excessive drying, but these isolated forests have been impoverished by their small total area, the difficulties of dispersal over marine barriers and changes in climate (del Arco Aguilar et al., 2010; Fernández-Palacios et al., 2011).

In eastern Australia, the rapid northward movement of the continent since the Eocene has both progressively reduced the extent of subtropical rainforests and brought the region closer to the extensive forests and floras of Southeast Asia. As a result, there is a north–south gradient in the relative contribution of recent immigrants from the Asian tropics and ancient Gondwanan lineages to humid forest floras, with immigrants enriching but not replacing the indigenous clades in subtropical forests (Sniderman and Jordan, 2011). The South American subtropics has a poorer paleoecological record, but this shows the replacement of subtropical forests by arid vegetation west of the Andes from the Middle Miocene onwards as the Andes rose and the cold Humboldt Current was established (Le Roux, 2012).

Northern subtropics

In the northern subtropics, forests are most extensive by far in East Asia. There is a much smaller area in North America and none in North Africa, although a relic of the Miocene forests of North

Africa and southern Europe persists in the Canary Islands. The most characteristic common feature of these forests is the abundance and diversity of Fagaceae (oaks and other genera) and, particularly following disturbance and at higher altitudes, the abundance of pines (*Pinus*). Fagaceae are most diverse at both the species and genus level in subtropical Asia. Pines attain their greatest diversity in the subtropical forests of the Sierra Madre Occidental of Mexico (Cord *et al.*, 2014), which is also a secondary center of diversity for oaks (*Quercus*). Other genera shared between East Asia and North America include *Abies*, *Alnus*, *Ilex*, *Magnolia* and *Picea*.

Asia

Subtropical forests were once very extensive in eastern China and the southern islands of Japan. They used to cover 25 per cent of China's total land area, but this zone is highly suitable for agriculture and has a high human population density. Most of the original forest has now been cleared and the remnants are mostly badly degraded, except at high altitude (Corlett, 2014). Some of the best-preserved examples are on Taiwan and the Ryukyu Islands. The total forest area is now increasing in the Chinese subtropics, but most of this is young secondary forests and monoculture plantations, often of exotic species.

Under the influence of the East Asian monsoon, the subtropical forests of East Asia receive more rainfall (900–2,000 mm) than most forests at these latitudes and the winter dry season is ameliorated by cooler weather that reduces evapotranspiration. As a result, these forests are composed largely of broad-leaved evergreen trees, with the Fagaceae and Lauraceae the dominant tree families. Common genera including *Ilex* (Aquifoliaceae), *Elaeocarpus* (Elaeocarpaceae), *Castanopsis*, *Cyclobalanopsis* (*Quercus*) and *Lithocarpus* (Fagaceae), *Distylium* (Hamamelidaceae), *Beilschmiedia*, *Cinnamomum*, *Cryptocarya*, *Lindera*, *Machilus*, *Neolitsea*, and *Phoebe* (Lauraceae), *Magnolia* and *Michelia* (Magnoliaceae), *Symplocos* (Symplocaceae) and *Schima* (Theaceae) (Corlett, 2014). Young secondary forests, in contrast, are often dominated by pines (e.g. *Pinus massoniana*, *P. yunnanensis*) or sometimes by winter-deciduous species (e.g. *Alnus nepalensis*) (Tang *et al.*, 2010). In cold, dry areas of northern Yunnan, there is a distinctive evergreen sclerophyllous forest, dominated by hard-leaved *Quercus* species (Tang, 2006).

Similar broad-leaved evergreen forests extend west across northern Myanmar and along the foothills of the Himalayas (Davis, 1960; Singh and Singh, 1987), although pine forests are more extensive here, probably reflecting the long history of human impacts. At higher altitudes, the broad-leaved evergreen forests give way to forests dominated by conifers (*Abies*, *Picea*, *Pinus*, *Tsuga*, etc.) and broad-leaved deciduous species. Slope aspect appears to have a large impact on forest composition in many parts of subtropical Asia but there has been little systematic study of this.

In eastern China, the subtropical lowlands have been almost entirely deforested, but in Myanmar, northern India, Bhutan and Nepal, they support mostly dry-season deciduous or semi-deciduous forests with a distinctly tropical flora. The semi-deciduous dipterocarp, *Shorea robusta* (sal), dominates large areas, and in both northeast India and northern Myanmar, lowland evergreen rainforest with dipterocarps occurs at 27–28°N (Proctor *et al.*, 1998). These forests experience annual minimum temperatures below 10°C and most suffer at least occasional frosts, but the impacts of winter cold on their structure and species composition are not obvious. It seems likely that chilling-sensitive tropical taxa have been excluded, but testing this will require comparative studies on a north–south transect across the Tropic of Cancer.

Eddy covariance measurements of carbon fluxes show that East Asian subtropical forests are currently major carbon sinks: more important on a per-area basis than tropical and temperate forests (Yu *et al.*, 2014). Net Ecosystem Productivity (NEP) between 20° and 40°N shows a strong negative relationship with forest age and a strong positive relationship with wet nitrogen deposition (from industry and agriculture), suggesting that both recovery from past disturbance and nitrogen fertilization contribute to the high carbon uptake. Year-round photosynthesis is possible in at least some sites (Yan *et al.*, 2013). Moreover, subtropical broad-leaved evergreen forests dominated by Fagaceae can attain surprisingly high above-ground biomasses (c. 70–220 Mg C ha^{-1} in Taiwan; McEwan *et al.*, 2011), making them among the most carbon-dense forests in the world.

North America

The largest forest area in the North American subtropics is on the Sierra Madre Occidental, a mountain range in northwestern Mexico that runs parallel to the Pacific coast and attains a maximum altitude of 3,300 m. The forests here are largely dominated by oaks (*Quercus* spp.), pines (*Pinus* spp.), or both, with pine dominance increasing with altitude (González-Elizondo *et al.*, 2012). This mountain range (including the southern, tropical, parts) supports 54 species of *Quercus* and 24 species of *Pinus*. Relatively minor forest types include a mixed coniferous forest at higher altitudes, dominated by *Abies*, *Pseudotsuga* and/or *Picea*, along with oaks and pines. There are also small areas of montane cloud forest with *Magnolia*, *Styrax*, *Cedrela*, *Ilex*, *Tilia*, oaks and various Lauraceae. The Sierra Madre Oriental, in eastern Mexico, also supports diverse pine–oak forests in the more humid areas. At lower altitudes on the dry northeastern coastal plains, in areas with accessible ground water, there are forests with a tree layer dominated by genera of Nearctic affinities (*Alnus*, *Carya*, *Platanus*, *Quercus*, *Salix*, *Ulmus*) (Encina-Domínguez *et al.*, 2011). Similar oak and pine-dominated forests occur in the southeastern United States, where most have been severely disturbed, and on some of the islands of the Bahamas.

Canary Islands

Remnants of subtropical broad-leaved forests persist on several of the Canary Islands. Similar forests occur north of the subtropics on the islands of Madeira and the Azores, where oceanic buffering has produced climates resembling those of the continental subtropics (Fernández-Palacios *et al.*, 2011). The Fagaceae were represented in the Canary Islands by an unidentified species of *Quercus* in the Holocene forests of Tenerife, but this declined to extinction after human settlement. The surviving broad-leaved evergreen forests are dominated by the family Lauraceae (del Arco Aguilar *et al.*, 2010). There are also extensive pine forests at higher altitudes and in drier areas, and small areas of sclerophyllous woodland at lower altitudes.

Southern subtropics

The absence of Fagaceae and native pines distinguishes the forests of the southern subtropics from those of the north, although pines are now widely planted and have become invasive in some areas (Simberloff *et al.*, 2010). Gondwanan elements (e.g. Cunoniaceae, Proteaceae, Podocarpaceae) are usually present, but most other species in these forests appear to have tropical (or northern subtropical) rather than 'southern' origins.

Madagascar

The subtropical southeastern section of Madagascar has steep altitudinal and rainfall gradients, resulting in a wide range of forest habitats, including coastal forests on sandy soils, lowland and montane humid forests and dry spiny forests, as well as a transition between rainforest and spiny forest (Goodman, 1999; Helme and Rakotomalaza, 1999; Rakotomalaza and Messmer, 1999). The flora of these forests is essentially tropical, which may reflect the absence of land to the south, but elevational zones are lower than is typical in the tropics to the north.

Southern Africa

Southern Africa is mostly too dry for forests, but there are a few thousand square kilometres of forest patches scattered through the wetter areas inland from the east coast (Midgley et al., 1997). These forests have been divided into two basic types, the interior Afromontane forests, mostly above 1,000 m elevation, and the coastal lowland Indian Ocean forests, which can then be divided further on the basis of floristics and habitats. However, there is considerable overlap in the flora of the various types, except at the extremes. Giant emergent Podocarpaceae with shade-tolerant seedlings are characteristic of old-growth Afromontane forests, but angiosperms dominate after human disturbance (Adie *et al.*, 2013). In the northern subtropics of South Africa and Mozambique, a distinctive 'sand forest' with a more tropical flora grows on deep sands (Gaugris and Van Rooyen, 2008).

South America

In northwest Argentina, subtropical forests occur across a wide range of environments, from the dry (< 1,000 mm), warm Chaco lowlands, to the cool, moist (< 3,000 mm) Yungas montane forests (Ferrero *et al.*, 2013), although the complex topography of this region produces many different combinations of rainfall and temperature, so these forests are not easily arranged along a continuum. Most of the surviving natural vegetation in the dry Chaco region would not meet most definitions of forest, but there are also areas where soil, topography and rainfall (> 500 mm) allows the formation of a more or less closed woody canopy. In comparison with tropical forests with similar dry-season length, these subtropical forests are lower, structurally simpler, smaller-leaved and floristically less diverse (Sarmiento, 1972). In comparison with the subtropical forests of China, those in South America generally have a larger deciduous component, presumably reflecting the harsher dry seasons.

In northeast Argentina, where there is no marked dry season, the subtropical forests represent an extension of the Atlantic forests of Brazil, including the largest remaining continuous area. As in subtropical China, satellite measures of greenness (NDVI) suggest that humid subtropical forests maintain high photosynthetic capacity throughout the year and could thus be large carbon sinks (Cristiano *et al.*, 2014). At the same latitude in the extreme south of Brazil, the subtropical Atlantic forest can be classified into rainforests, semi-deciduous, deciduous and mixed conifer–hardwood forests, with the southern conifer, *Araucaria angustifolia*, forming the upper canopy over a diverse broad-leaved middle and lower storey (Souza, 2007).

In the southern subtropics of Chile, to the west of the Andes, small areas of evergreen forest occur in patches on coastal mountains facing the Pacific, where they depend on fog subsidies in areas receiving only 150 mm of rainfall (van Zonneveld *et al.*, 2012). Easter Island, 3,700 km west of South America, apparently supported a palm-dominated forest until human settlement in recent millennia (Cañellas-Boltà *et al.*, 2013).

Australia

Most of Australia is too dry for forest, particularly in the subtropical zone, but a broad belt along the east coast receives enough rainfall (> c. 400 mm) to support eucalypt (*Eucalyptus* spp.) woodland and the narrow strip nearest the coast (> 800–1,000 mm rainfall) supports largely evergreen rainforest patches in a matrix of tall open eucalypt forest, with a more or less dense understorey. Rainforest was much more extensive at subtropical latitudes in Australia in the early Tertiary until the late Eocene, but has become severely restricted subsequently. The richest rainforest floras are found in areas where there is evidence for continuity of a moist climate (Weber *et al.*, 2014). These areas shelter both ancient Gondwanic lineages and species derived from Asian lineages that invaded Australia from the Miocene onwards.

Discussion

The subtropics have usually been viewed by biologists as merely a transition between the tropics and the temperate zone, although there has been little agreement on where exactly this transition occurs (Corlett, 2013). Are subtropical forests sufficiently distinct from tropical and temperate forests to be worth distinguishing as a separate ecological entity? Although the precise latitudes of the tropics of Cancer and Capricorn—the equatorward boundaries of the subtropics—are ecologically arbitrary, they are generally fairly close to the 'frost line', which seems to be a general barrier to tropical plants because of the devastating impact of ice-crystal formation inside plant cells (Corlett, 2014). In China, the poleward limit of the subtropics also coincides, more or less, with an apparent eco-physiological boundary, where the dominant broad-leaved evergreen trees give way to winter-deciduous species. This limit seems to coincide with an absolute minimum temperature of around −15°C degrees (corresponding to a January mean of around 0°C) (Corlett, 2014). However, it is not clear from the available literature if a similar boundary occurs in other parts of the world. Subtropical faunas, in contrast, are generally subsets of those nearer the equator. This presumably reflects the ability of many animals to avoid short periods of extreme cold behaviorally or, for mammals and birds, by thermoregulation, so that their poleward limits are more likely to be set by a seasonal gap in food supply than by climate directly (Corlett, 2014).

In China, where there is a good paleoecological record, forests that would be considered subtropical today have expanded north in warmer periods and retreated south in colder ones, to be replaced by broad-leaved deciduous forests or more open vegetation types (Corlett, 2014; Ni *et al.*, 2014). This suggests that it is the climate rather than the latitude that makes a 'subtropical forest'. However, while mean temperature isotherms have shifted north and south with global climate change, temperature seasonality is largely a function of latitude, so thermal regimes have not simply shifted backwards and forwards.

Subtropical forests, as defined here, are probably less extensive today than at any time in the last 20–30 million years. Long-term drying has limited their extent, except in monsoon Asia, and human impacts have further reduced their area in those regions where suitable climates still exist. These forests have received much less attention from conservationists than those in the tropics, despite their importance for both biodiversity and carbon sequestration, but much of the remaining forest area is included within Conservation International's 35 global biodiversity hotspots (Sloan *et al.*, 2014). In East Asia, breeding-bird diversity reaches a maximum at around 25°N, rather than near the equator, with the larger land area available at subtropical latitudes than on the tropical peninsula and islands further south a plausible explanation for this pattern (Ding *et al.*, 2006). Their relatively small extent reduces the global significance of subtropical forests for carbon sequestration, but as pointed out above, some have both a high biomass and a high sink capacity on a per-area basis.

Simple models of future vegetation distribution under anthropogenic climate change suggest that subtropical forests will shift polewards as the temperature rises. Even in the absence of the now ubiquitous forest fragmentation, however, tree species migration rates would probably be too slow to track temperature change (Corlett and Westcott, 2013). Real-world responses are likely to be more complex, reflecting both the complexities of climate change and the many additional anthropogenic impacts that these forests are subject to. Long-term monitoring of large, permanent, sample plots is our best hope of understanding these changes. Recent observation of rapid structural and compositional changes in subtropical forests in both China and Argentina have been interpreted as reflecting regional drying associated with warming in the former case (Zhou et al., 2014) and the possible impacts of CO_2 fertilization in the latter (Malizia et al., 2013).

References

Adie, H., Rushworth, I. and Lawes, M. J. (2013) 'Pervasive, long-lasting impact of historical logging on composition, diversity and above ground carbon stocks in Afrotemperate forest', *Forest Ecology and Management*, vol. 310, pp. 887–895.

Cañellas-Boltà, N., Rull, V., Sáez, A., Margalef, O., Bao, R. and others (2013) 'Vegetation changes and human settlement of Easter Island during the last millennia: a multiproxy study of the Lake Raraku sediments', *Quaternary Science Reviews*, vol. 72, pp. 36–48.

Cord, A. F., Klein, D., Gernandt, D. S., Pérez de la Rosa, J. A. and Dech, S. (2014) 'Remote sensing data can improve predictions of species richness by stacked species distribution models: a case study for Mexican pines', *Journal of Biogeography*, vol. 41, pp. 736–748.

Corlett, R. T. (2013) 'Where are the subtropics?', *Biotropica*, vol. 45, pp. 273–275.

Corlett, R. T. (2014) *The Ecology of Tropical East Asia, 2nd Edition*. Oxford University Press, Oxford, UK.

Corlett, R. T. and Westcott, D. A. (2013) 'Will plant movements keep up with climate change?', *Trends in Ecology and Evolution*, vol. 28, pp. 482–488.

Cristiano, P. M., Madanes, N., Campanello, P. I., di Francescantonio, D., Rodríguez, S. A. and others (2014) 'High NDVI and potential canopy photosynthesis of South American subtropical forests despite seasonal changes in leaf area index and air temperature', *Forests*, vol. 5, pp. 287–308.

Davis, J. H. (1960) *The Forests of Burma*. Department of Botany, University of Florida.

del Arco Aguilar, M., González-González, R., Garzón-Machado, V. and Pizarro-Hernández, B. (2010) 'Actual and potential natural vegetation on the Canary Islands and its conservation status', *Biodiversity and Conservation*, vol. 19, pp. 3089–3140.

Ding, T. S., Yuan, H. W., Geng, S., Koh, C. N. and Lee, P. F. (2006) 'Macro-scale bird species richness patterns of the East Asian mainland and islands: energy, area and isolation', *Journal of Biogeography*, vol. 33, pp. 683–693.

Encina-Domínguez, J. A., Rocha, E. M., Meave, J. A. and Zárate-Lupercio, A. (2011) 'Community structure and floristic composition of *Quercus fusiformis* and *Carya illinoinensis* forests of the Northeastern Coastal Plain, Coahuila, Mexico', *Revista Mexicana de Biodiversidad*, vol. 82, pp. 607–622.

Fernández-Palacios, J. M., de Nascimento, L., Otto, R., Delgado, J. D., García-del-Ray, E. and others (2011) 'A reconstruction of Palaeo-Macaronesia, with particular reference to the long-term biogeography of the Atlantic island laurel forests', *Journal of Biogeography*, vol. 38, pp. 226–246.

Ferrero, M. E., Villalba, R., De Membiela, M., Ripalta, A., Delgado, S. and Paolini, L. (2013) 'Tree-growth responses across environmental gradients in subtropical Argentinean forests', *Plant Ecology*, vol. 214, pp. 1321–1334.

Gaugris, J. Y. and Van Rooyen, M. W. (2008) 'A spatial and temporal analysis of sand forest tree assemblages in Maputaland, South Africa', *South African Journal of Wildlife Research*, vol. 38, pp. 171–184.

Givnish, T. J. and Renner, S. S. (2004) 'Tropical intercontinental disjunctions: Gondwana breakup, immigration from the boreotropics, and transoceanic dispersal', *International Journal of Plant Sciences*, vol. 165, pp. S1–S6.

González-Elizondo, M. S., González-Elizondo, M., Tena-Flores, J. A., Ruacho-González, L., and López-Enríquez, I. L. (2012) 'Vegetación de la Sierra Madre Occidental, México: una síntesis', *Acta Botánica Mexicana*, vol. 100, pp. 351–403.

Goodman, S. M. (1999) 'Description of the Réserve Naturelle Intégrale d'Andohahela, Madagascar, and the 1995 biological inventory of the reserve', *Fieldiana Zoology N.S.*, vol. 94, pp. 1–7.

Helme, N. A. and Rakotomalaza, P. J. (1999) 'An overview of the botanical communities of the Réserve Naturelle Intégrale d'Andohahela, Madagascar', *Fieldiana Zoology N.S.*, vol. 94, pp. 11–24.

Jacques, F. M. B., Shi, G. and Wang, W. (2013) 'Neogene zonal vegetation of China and the evolution of the winter monsoon', *Bulletin of Geosciences*, vol. 88, pp. 175–193.

Jacques, F. M. B., Su, T., Spicer, R. A., Xing, Y. W., Huang, Y. J. and Zhou, Z. K. (2014) 'Late Miocene southwestern Chinese floristic diversity shaped by the southeastern uplift of the Tibetan Plateau', *Palaeogeography, Palaeoclimatology, Palaeoecology*, vol. 411, pp. 208–215.

Le Roux, J. P. (2012) 'A review of Tertiary climate changes in southern South America and the Antarctic Peninsula. Part 2: continental conditions', *Sedimentary Geology*, vols. 247–248, pp. 21–38.

McEwan, R. W., Lin, Y.-C., Sun, I. F., Hsieh, C. F., Su, S. H. and others (2011) 'Topographic and biotic regulation of aboveground carbon storage in subtropical broad-leaved forests of Taiwan', *Forest Ecology and Management*, vol. 262, pp. 1817–1825.

Malizia, A., Easdale, T. A. and Grau, H. R. (2013) 'Rapid structural and compositional change in an old-growth subtropical forest: using plant traits to identify probable drivers', *PLoS ONE*, vol. 8, e73546.

Midgley, J. J., Cowling, R. M., Seydack, A. and van Wyk, G. F. (1997) 'Forests', In: *Vegetation of Southern Africa* (eds. R. M. Cowling, D. M. Richardson and S. M. Pierce). Cambridge University Press, Cambridge, UK. pp. 278–299.

Ni, J., Cao, X., Jeltsch, F. and Herzschuh, U. (2014) 'Biome distribution over the last 22,000 yr in China', *Palaeogeography, Palaeoclimatology, Palaeoecology*, vol. 409, pp. 33–47.

Proctor, J., Haridasan, K. and Smith, G. W. (1998) 'How far north does lowland evergreen tropical rain forest go?', *Global Ecology and Biogeography Letters*, vol. 7, pp. 141–146.

Rakotomalaza, P. J. and Messmer, N. (1999) 'Structure and floristic composition of the vegetation in the Réserve Naturelle Intégrale d'Andohahela, Madagascar', *Fieldiana Zoology N.S.*, vol. 94, pp. 51–96.

Sarmiento, G. (1972) 'Ecological and floristic convergences between seasonal plant formations of tropical and subtropical South America', *Journal of Ecology*, vol. 60, pp. 367–410.

Simberloff, D., Nuñez, M. A., Ledgard, N. J., Pauchard, A., Richardson, D. M. and others (2010) 'Spread and impact of introduced conifers in South America: Lessons from other southern hemisphere regions', *Austral Ecology*, vol. 35, pp. 489–504.

Singh, J. S. and Singh, S. P. (1987) 'Forest vegetation of the Himalaya', *Botanical Review*, vol. 53, pp. 80–192.

Sloan, S., Jenkins, C. N., Joppa, L. N, Gaveau, D. L. A. and Laurance, W. F. (2014) 'Remaining natural vegetation in the global biodiversity hotspots', *Biological Conservation*, vol. 177, pp. 12–24.

Sniderman, J. M. K. and Jordan, G. J. (2011) 'Extent and timing of floristic exchange between Australian and Asian rain forests', *Journal of Biogeography*, vol. 38, pp. 1445–1455.

Souza, A. F. (2007) 'Ecological interpretation of multiple population size structures in trees: the case of *Araucaria angustifolia* in South America', *Austral Ecology*, vol. 3, pp. 524–533.

Tang, C. Q. (2006) 'Evergreen sclerophyllous *Quercus* forests in northwestern Yunnan, China as compared to the Mediterranean evergreen *Quercus* forests in California, USA and northeastern Spain', *Web Ecology*, vol. 6, pp. 88–101.

Tang, C. Q., Zhao, M.-H., Li, X.-S., Ohsawa, M. and Ou, X.-K. (2010) 'Secondary succession of plant communities in a subtropical mountainous region of SW China', *Ecological Research*, vol. 25, pp. 149–161.

van Zonneveld, M. J., Gutiérrez, J. R. and Holmgren, M. (2012) 'Shrub facilitation increases plant diversity along an arid scrubland–temperate rain forest boundary in South America', *Journal of Vegetation Science*, vol. 23, pp. 541–551.

Weber, L. C., VanDerWal, J., Schmidt, S., McDonald, W. J. F. and Shoo, L. P. (2014) 'Patterns of rain forest plant endemism in subtropical Australia relate to stable mesic refugia and species dispersal limitations', *Journal of Biogeography*, vol. 41, pp. 222–238.

Yan, J., Zhang, Y., Yu, G., Zhou, G., Zhang, L. and others (2013) 'Seasonal and inter-annual variations in net ecosystem exchange of two old-growth forests in southern China', *Agricultural and Forest Meteorology*, vol. 182–183, pp. 257–265.

Yu, G., Chen, Z., Piao, S., Peng, C., Ciais, P. and others (2014) 'High carbon dioxide uptake by subtropical forest ecosystems in the East Asian monsoon region', *Proceedings of the National Academy of Sciences of the United States of America*, vol. 111, pp. 4910–4915.

Zhang, Z., Ramstein, G., Schuster, M., Li, C., Contoux, C. and Yan, Q. (2014) 'Aridification of the Sahara desert caused by Tethys Sea shrinkage during the Late Miocene', *Nature*, vol. 513, pp. 401–404.

Zhou, B., Gu, L., Ding, Y., Shao, Z., Wu, Z. and others (2011) 'The great 2008 Chinese ice storm: its socioeconomic-ecological impact and sustainability lessons learned', *Bulletin of the American Meteorological Society*, vol. 92, pp. 47–60.

Zhou, G., Houlton, B. Z., Wang, W., Huang, W. J., Xiao, Y. and others (2014) 'Substantial reorganization of China's tropical and subtropical forests: based on the permanent plots', *Global Change Biology*, vol. 20, pp. 240–250.

5
TROPICAL FORESTS

Lindsay F. Banin, Oliver L. Phillips and Simon L. Lewis

Introduction to tropical forests

Tropical forest ecosystems occur between the tropics of Cancer and Capricorn (23.5° N and S of the equator, respectively) and are characterized by dense, relatively tall, closed canopies with trees making up the dominant life form in structurally complex arrangements. Tropical forests are amongst the most speciose systems on the planet, in terms of both flora and fauna, and are estimated to contain more than half the world's recorded species. They deliver important ecosystem services, such as large-scale cycling of carbon and water, and more locally, through the provision of clean water, shelter, food and fuel to local populations. At the global scale, their limits are determined by climate and at smaller spatial scales, their presence and form is determined by local or regional topographic, geological and edaphic features, as well as anthropogenic impacts.

There is no consistent, universal definition of 'tropical forest' (see Torello-Raventos *et al.* 2013), and this is due largely to the fact that classification boundaries are drawn somewhat arbitrarily along ecological gradients; vegetation types grade from one to another along environmental continua (Woodward *et al.* 2004). Yet, it is clearly useful to define ecosystems for robust temporal and spatial comparisons and decision-making (Putz and Redford 2010). There are numerous global-scale vegetation classification schemes (e.g. Köppen 1923; Holdridge 1947; Whittaker 1975). Vegetation is classified into biomes which indicate the major ecological communities present over extensive regions of the world. Biomes correlate well with climatic regimes but are independent of floristic assemblage. The structural and functional attributes of the vegetation relate to physiological constraints on the resident communities, giving rise to similar physiognomy amongst continents.

Length and intensity of dry season are strong determinants of forest occurrence and form (Zelazowski *et al.* 2011). Tropical rain forest occurs where rainfall is abundant (approx. > 1,800 mm annually) and well-distributed throughout the year with no notable, or at most a short, dry season. Vegetation is evergreen and dominated by trees, with other life forms including lianas, epiphytes and ferns. As seasonality increases, tropical dry forests (which are also commonly referred to as monsoon forests, seasonally dry forests, deciduous or semi-deciduous forests) occur where there is a pronounced, regular dry season, during which some, many or all of the trees shed their leaves. Dry months are often defined as when precipitation falls below 100 mm – the estimated potential evapotranspiration rate of most lowland tropical forests (Shuttleworth 1988).

Tropical forests occur in a band around the equator; Central and South America constitute the neotropics and Central, West and East Africa, Madagascar, South and Southeast Asia, New Guinea, Melanesian islands and Australia make up the palaeotropical region. Currently, the largest contiguous tracts occur in South America (the Amazon basin), Central Africa (the Congo basin) and New Guinea.

Altitude has also been shown to be an important determinant of forest structure and physiognomy, mostly because of the decline in temperature (lapse rates average ca. 6°C per 1,000 m elevational difference) and hence tropical montane forest is regarded a distinct and separate formation. Locally, other distinctive forest formations arise from underlying geology, soils and proximity to water, such as coastal mangroves, peat swamp and gallery forests.

Vegetation classifications are determined by climatic envelopes and represent *potential* cover rather than *actual* cover. More recently, developments in remote sensing and geographical information systems have meant that large-scale vegetation cover can be observed using information from satellites. Estimates based on SPOT-4 VEGETATION satellite data (Global Land Cover 2000 product, see Mayaux *et al.* 2005; Bartholomé and Belward 2005) indicated that tropical moist forests (delimited following the FAO definition of 0–3 dry months, when rainfall is less than twice the mean temperature) occupy approximately 1,100 million hectares, or roughly 10 per cent of vegetated land cover globally. South American forests constitute more than half (58 per cent) of the total tropical moist forest cover, African and Asian forests contribute ~20 per cent each whilst forests in India and Australia occupy relatively small, isolated tracts where the local climate and soils permit (Mayaux *et al.* 2005). Remotely sensed data can be used to estimate changes in areal cover, but still rely on consistent definitions of what constitutes 'a forest' (see Box 5.1).

Box 5.1 Disturbance and management history of tropical forests

Forests can take many forms depending on their disturbance and management history. Putz and Redford (2010) propose the following classification scheme. Old-growth forests (sometimes referred to as primary, pristine, virgin or intact) are largely unmodified by human activities and can be determined by tree age and size distributions, number of canopy layers and species composition. Degraded forests have been modified by logging, overhunting, fires or species invasion and have lost some structural or compositional elements of old-growth forests. Secondary forests are those which have regenerated following clearance. Degraded and secondary forests represent a large proportion of tropical forest cover and there is an increasing awareness of their ecological value. Plantations include intensively managed tree farms (e.g. oil palm, rubber). By contrast, managed forests include planted stands of native trees which may permit native understories. Natural or semi-natural disturbances (including hurricanes and fires) can also 're-set' the successional status of a forest.

Physical geography of tropical moist forests

In a global context, the tropical forests are often treated as a uniform entity yet there is distinct variation in important biophysical variables (precipitation, temperature, incoming solar

Precipitation and its intra- and inter-annual seasonality

Precipitation varies over regional scales, both in terms of total annual rainfall and the seasonality of rainfall, with the lowest mean annual precipitation occurring in regions with strong intra-annual seasonality. Precipitation is largely determined by latitudinal distance from the equator; dry seasons are strongest and longest with increasing distance due to seasonal shifting of the inter-tropical convergence zone (ITCZ) (McGregor and Nieuwolt 1998). These patterns are modulated by atmospheric circulation patterns and local geography, including distance from the sea, coastline orientation and orography. On average, mean annual precipitation is highest in northwest Amazonia, Southeast Asia and New Guinea (~2,500 mm per year) where there is no regular, annual dry season and lower in Africa, South Asia and Australia (Figure 5.1a).

In the neotropics, a large gradient in rainfall is observed (Figure 5.1a). The wettest regions are close to the equator, receiving rainfall throughout the year from the ITCZ air masses, and regions on the east-facing slopes of the Andes due to orographic uplift of easterly winds (McGregor and Nieuwolt 1998). Dry seasons occur in Central America and both northern and southern fringes of the Amazon basin and these tend to be most intense in south-east Amazonia.

Mean annual precipitation is generally lower in African tropical forests where almost all regions experience a dry season (Figure 5.1a; ~ 1,800 mm per year). Some areas are subject to one longer dry season (~10°N and S) but most tropical forests in Africa receive two less intense dry periods, one when the ITCZ is furthest north (July–August) and another longer one when it is furthest south (December–March) (McGregor and Nieuwolt 1998). Geography also has an effect in West Africa; Ghana and Côte d'Ivoire receive less rainfall since their coastlines are parallel to onshore winds, and are in the rain shadow of the Liberia and Sierra Leone coastlines. The effects of lower rainfall in African forests are however, moderated by the relatively higher elevation and lower temperatures (Malhi and Wright 2004).

The climate in tropical Asia is driven by the Indo-Australian monsoon systems and modified by local orography (Walsh 1996b). Precipitation is high in insular Southeast Asia, the Malay Peninsular and New Guinea (Figure 5.1a) where monsoons deliver rain throughout the year, although drier periods occur in different regions at different times in association with surface features (Walsh 1996b). In South Asia, a short dry season occurs in the forested southern Sri Lanka, whilst the Indian and Thai forests receive much of their rain between July and September (McGregor and Nieuwolt 1998). Conversely, the wet season south of the equator in south-eastern Indonesia, southern New Guinea and northern Australia occurs from December to March, during which Australian forests receive 60–90 per cent of their annual total (Figure 5.1b). Orography influences local rainfall patterns; the eastern facing slopes of the Great Dividing Range capture most of the rains, and the mountains forming a central ridge across New Guinea produce a rain shadow to the south of the landmass (McGregor and Nieuwolt 1998).

Anomalies in sea-surface temperature can create inter-annual variability in rainfall where some regions of the tropics are subject to periodic, sometimes very severe, drought. The El Niño Southern Oscillation (ENSO) particularly affects South America and Asia. The 'Southern Oscillation' refers to the coupled shift in sea-surface temperature and atmospheric pressure that occurs in the Pacific Ocean roughly every three to five years. Under 'normal' conditions, high sea-surface temperatures in the western Pacific bring rainfall to Southeast Asia. During El Niño

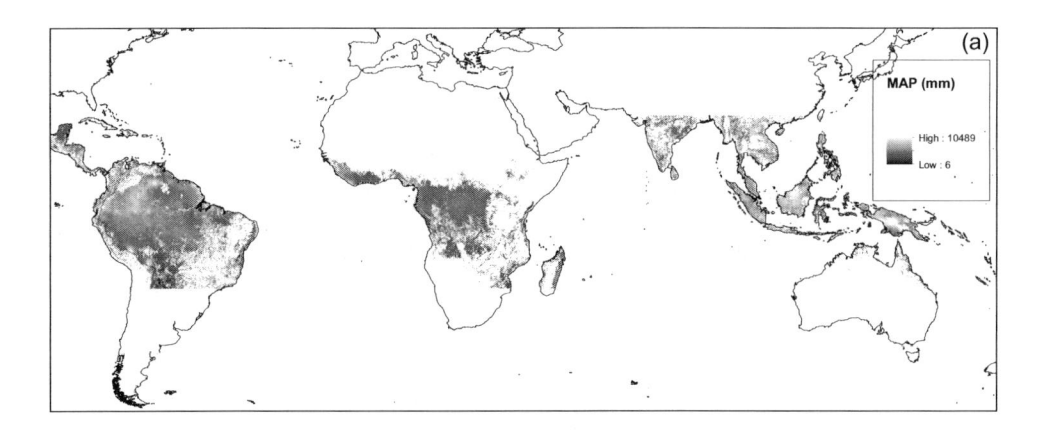

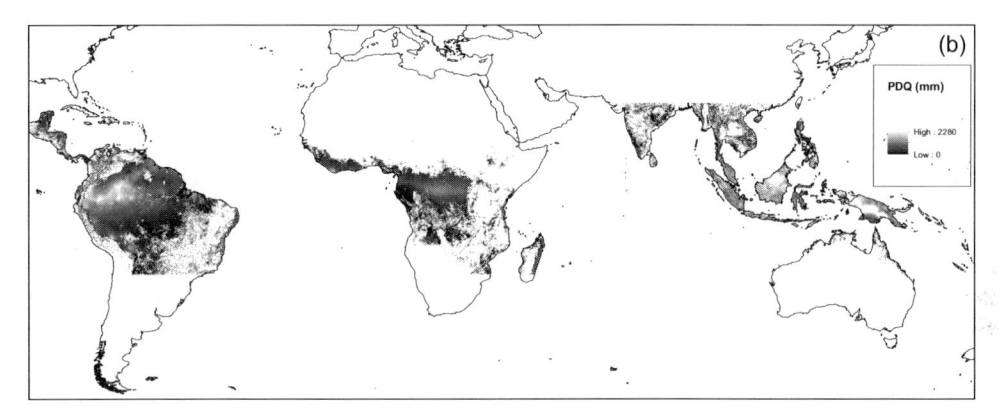

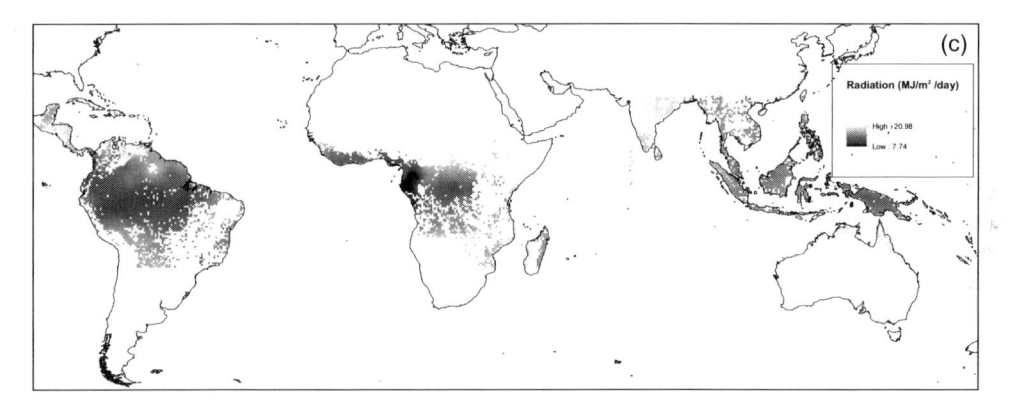

Figure 5.1 Map of: (a) mean annual precipitation (mm); (b) precipitation in driest consecutive three months (mm); and (c) solar radiation (megajoules per metre squared per day) for the tropical forest regions. Shading represents satellite-derived Global Land Cover 2000 classes: (1) broadleaved evergreen closed to open trees; (2) broadleaved deciduous closed to open trees; (7) broadleaved evergreen closed to open trees on flooded land and; (17) mosaic cultivated/managed terrestrial areas and closed–open trees (Mayaux *et al.* 2005; Bartholomé and Belward 2005) within 23.5°N and S of the equator. Rainfall data are from the WorldClim datasets; interpolations of climate station data to 30 seconds spatial resolution (a, b; Hijmans *et al.* 2005) and radiation data are based on interpolations of sunshine hours and/or cloud cover at 0.5° resolution from Climate Research Unit (c; New *et al.* 1999)

conditions, easterly trade winds slacken, allowing warmer water to propagate eastwards to the central and eastern Pacific. As a result, the large-scale distribution of atmospheric uplift and descent changes around the globe. Drier than usual weather (associated with descending air) prevails in Australia, New Guinea, Malaysia and north Brazil and the Guyanas, and sometimes East Africa. The western coast of South America is one of the areas receiving greater rainfall during El Niño events. During the opposite phase (La Niña) the trade winds strengthen and warm waters move west, changing rainfall patterns. The Atlantic Multidecadal Oscillation (AMO) refers to sea-surface temperature anomalies in the Atlantic and is associated with reduced rainfall, this time in the south and west regions of Amazonia. These inter-annual droughts are known to reduce plant growth and increase tree mortality through fatal embolism in affected forests (Phillips *et al.* 2010). They also increase risk of large forest fires which have occurred in Amazonia and Indonesia in recent years.

Temperature and altitude

Lowland tropical forests are highly productive compared with other terrestrial ecosystems because the warm environment sustains year-round growth. In the lowland (< 1,000 m above sea level) tropical rain forest regions, mean annual temperature is high (> 18°C) and there is little seasonality since the sun is almost directly overhead throughout the year. Diurnal variations in temperature tend to be stronger than seasonal ones, although temperatures can rise during dry seasons due to the lack of cloud cover (Walsh 1996a). Variation in altitude is the primary driver of differences in temperature across the tropics. Africa and Australia's lowland forested regions are at higher altitudes on average, resulting in lower regional mean annual temperatures than in tropical America and Asia, allowing forests to persist where lower rainfall would otherwise be prohibitive.

Radiation and cloud cover

Solar radiation in the tropics is largely determined by cloud cover and the diurnal timing of rainfall, which itself depends on proximity to oceans, location in relation to wind direction, coastal geometry and orography (Walsh 1996a; Churkina and Running 1998). Maritime settings tend to receive rainfall at night as a result of night-time temperature differences between surface air warmed by the oceans and the radiative cooling of cloud tops whilst continental regions have a greater amount of rainfall during the day, in association with the peak in land heating and convection (Walsh 1996a). Coastal regions with steep elevation also tend to receive rainfall in the daytime as a result of orographic uplift of sea breezes; this is the case in some coastal regions of West Africa and the mountain range on the east coast of Australia (McGregor and Nieuwolt 1998). Annual averages of solar radiation may also be affected by seasonality in cloud cover; regions experiencing strong dry seasons will have periods of high incoming solar radiation.

Mean solar radiation, calculated from standardized sunshine hours and/or cloud cover data (Figure 5.1c) is notably higher in tropical Asia and Australia compared with America and Africa. Insular Southeast Asia receives much of its rainfall at night, whereas America and Africa on the whole are subject to a continental regime in diurnal rainfall.

Climate change in the tropics

In the tropics, surface air temperatures have been rising by 0.25°C per decade for the last 30 years (Malhi and Wright 2004). Model projections indicate that the temperature of tropical forest regions will increase by at least 2–4°C by 2100, dependent on future greenhouse gas scenarios (Williams et al. 2007). As the tropics encompass the warmest climate on Earth, this change will generate novel climates and thus it is difficult to predict how biota will respond – the physiological tolerance and adaptability of tropical species to warmer temperatures is largely unknown, particularly as they have evolved in regions where seasonal variation is limited (Williams et al. 2007; Wright et al. 2009; Corlett 2011). Plants are likely to be able to tolerate small changes in temperature alone through adjustments to stomatal conductance and due to the simultaneous increased concentration of CO_2 (Lloyd and Farquhar 2008). Heat damage to leaves can occur for temperatures in excess of 40°C which are plausible at the top of the canopy, particularly if solar radiation is high, and thus seasonally dry forests could be particularly vulnerable (Corlett 2011).

Whilst there is high confidence that warming will occur in all tropical regions, there is greater uncertainty around projections of changes in precipitation. Precipitation trends in recent decades have been less clear than those of temperature, the natural variability is less well defined and the modelling of monsoons and inter-annual climatic events such as ENSO presents significant challenges (Christensen et al. 2013; Li et al. 2011). The IPCC AR5 report (Christensen et al. 2013) outlines the projected regional differences in precipitation trends at the end of the century under a given greenhouse gas emission scenario. In South America, changes in annual precipitation are not anticipated but increases in the variability and extremes of precipitation are projected. Dry-season rainfall is expected to decrease in east Amazonia, and north-east Brazil, whilst positive extreme precipitation trends may occur on the north-west coast, Peru and Ecuador. A reduction in warm-season precipitation is also projected for Central America and the Caribbean. There has been some evidence of drying in West Africa in recent decades (Fauset et al. 2012), but model-based projections of future precipitation trends are highly uncertain (Christensen et al. 2013). Little change in annual precipitation is projected for East Africa, though the annual distribution may alter. The complexity of geography and terrain in South and Southeast Asia complicate projections for this region; moderate increases in rainfall are projected for most parts of Southeast Asia and summer monsoon precipitation may increase in South Asia. Changes associated with ENSO remain uncertain in both Southeast Asia and Australia. Increased seasonality in rainfall may also occur in the Australian tropics, due to decreases in the strength of westerly winds to the north (Christensen et al. 2013). The projections focus on changes at one snapshot of time, yet changes may be dynamic over time and not unidirectional – some regions may first get wetter and then drier, according to changes in the position of the ITCZ (Hawkins et al. 2014).

Whilst the impacts of climate change on tropical vegetation are uncertain, warmer temperatures may lead to upslope migration of some species (Feeley et al. 2011) – this could, of course, be problematic for species where this option is not available. Reduced rainfall may lead to the expansion of tropical dry forests or savannas into regions that are currently occupied by tropical rain forests, for example in eastern Amazonia (Malhi et al. 2009). Higher temperatures, coupled with reduced rainfall and greater human influence may mean that these forests are more susceptible to fires. Changes in forest cover may also induce reinforcing feedbacks – tropical forests generate clouds via transpiration, and thus are a major regulator of regional weather. As forest cover is reduced, this hydrological cycling may break down leading to further drying and forest loss.

Soil and geology

Soils of the tropics are typified by old, deeply weathered and nutrient poor profiles, due to the hot, humid climate which promotes erosion and leaching. However, soil characteristics are determined by five key factors: substrate geology, topography and drainage, climate, age and biota (Jenny 1941), leading to heterogeneity at local, regional and continental scales. Commonly-occurring soils in the tropics are presented in Table 5.1. Various soil classification schemes exist; here we use the World Reference Base for Soil Resources (WRB) nomenclature (IUSS Working Group WRB 2006). Geological development of the major tropical forest regions ultimately determines the age and composition of soils. Three principal geological landforms exist across the wet tropics: i) the ancient Precambrian shield areas or 'cratons', making up large parts of eastern South America, equatorial Africa, and central and southern India which yield deeply weathered Ferralsols characteristic of the tropics; ii) the alpine fold belts comprising much of Southeast Asia, the Andes and Central America, eastern Australia and northern New Guinea; and iii) the alluvial plains of the Amazon and Congo basins and the coastal plains of the Guyanas and Mekong and Niger deltas (Beckmann 1983; IUSS Working Group WRB 2006). These landforms drive large-scale variability in soil characteristics which are modulated locally by variations in geology, topography, drainage and organisms.

There is a strong east-to-west gradient in geology and soil in South America. Central and eastern Amazonia is underlain by Precambrian shield and well-weathered, nutrient poor Ferralsols predominate. The strong topography of the Andes promotes erosion so that new, unweathered surfaces of parent material are continuously exposed, replenishing nutrients to the surface soil. Thus, in general, the soils are more fertile in the west, comprising Leptosols and Regosols at the upper-most elevations and a mixture of Andosols, Cambisols, Luvisols, Acrisols and Alisols at lower elevations (Irion 1978; Quesada et al. 2012a). Floodplain soils (Fluvisols and Gleysols) flank the Amazon River (Quesada et al. 2012a).

Precambrian shield regions predominate in West Africa and form a rim around the Congo basin. The central basin itself and some parts of West Africa are made up of sediments with restricted areas of basalt from volcanic intrusions and extrusions (Juo and Wilding 1996). Ferralsols, Nitisols and Acrisols are the most extensive soils on both shield and sedimentary substrates. Cambisols and Gleysols are also found in accordance with local topography, and more nutrient rich Lixisols are represented in regions of Côte d'Ivoire, Ghana and Nigeria (Juo and Wilding 1996).

Ancient shield geology only occurs in small, isolated patches in tropical Asia. Much more common are sedimentary and volcanic substrates (Dudal 2005; Hammond 2005). As a result Acrisols comprise ~50 per cent of the region (Dudal 2005), though Ferralsols and Nitisols on Precambrian shield can be found in India. Vulcanism at the boundaries of the Sunda shelf means that Andosols and Leptosols are more common in Asia (Dudal 2005). Arenosols, formed on ancient coastal and river terraces, are also reasonably well represented. Gleysols, Fluvisols and Cambisols occur according to local conditions.

The soils in north-eastern Australia are heterogeneous due to the variable materials which make up the Great Dividing Range and the patchy distribution of basalt (Hubble and Isbell 1983; Hubble et al. 1983). Fluvisols, Ferralsols, Acrisols, Podzols, Cambisols, Lixisols and Vertisols are commonly found in tropical Australia, depending on local geology and site conditions (Table 5.1; Hubble and Isbell 1983; Hubble et al. 1983). The northern part of New Guinea is made up of folded sedimentary rocks whilst Precambrian crust underlies southern New Guinea (Hammond 2005). Cambisols, Regosols, Leptosols and Histosols (soils with high organic matter associated with cool environments) are more commonly found in association with the strong topography. Acidic Umbrisols and Histols, both dark with high organic matter content, are also more common. More deeply weathered Acrisols represent regions in the south and Fluvisols, Lixisols and Gleysols are also represented (IUSS Working Group WRB 2006).

Table 5.1 Dominant tropical soil groups by WRB classification, organized by primary forming factor. Descriptions are based on IUSS Working Group WRB (2006), Quesada et al. (2012a) and Dudal (2005)

Soil group	Distinguishing features	Typical nutrient status
Limited age		
Leptosols	Shallow soil over hard rock/gravel, usually at medium to high altitude	Dependent on parent material. Accumulation of carbon
Cambisols	Horizon definition only beginning, with slight to moderate weathering of parent material. Found on erosional surfaces	Dependent on parent material
Climate		
Acrisols	Sub-surface accumulation of low activity clays and low base saturation	Low
Alisols	Sub-surface accumulation of high activity clays (smectite and vermiculite)	Moderate, but with Al concentrations
Ferralsols	Deep, strongly weathered, physically stable but chemically depleted	Low
Lixisols	Soil with high activity clay but high in exchangeable bases (in contrast to Alisols)	Moderate–High
Nitisols	Deep, clayey soil with well-defined nut-shaped peds with shiny surfaces. Developed on intermediate to basic substrates	High
Plinthosols	Notable presence of plinthite (Fe-quartz-kaolinite concretions) associated with a fluctuating water table	Moderate
Podzols	Illuviation of Fe, Al and organics, commonly on alluvial/coluvial/aeolian siliceous materials	Low, with high acidity and Al concentrations
Topography and drainage		
Fluvisols	Soils on recent alluvium, often shallow	Variable with parent material
Gleysols	Temporary or permanent wetness near soil surface; gleyic colours and evidence of reduction	Comparatively fertile (but anoxic conditions)
Umbrisols	Occurring at high altitudes. Accumulation of organic matter due to low temperatures and acidic	Bases dominated by high Al concentrations
Parent material		
Andosols	Dark, organic-rich soils on volcanic materials	Variable with pH
Arenosols	Sandy soil with limited soil development	Low
Luvisol	Unconsolidated alluvial/colluvial/aeolian deposits in cooler environments and young surfaces	Variable with parent material
Vertisols	Clay soils which expand and contract (smectite) to cause cracking	High (but may be physically inhibiting due to dense clay)

In sum, Ferralsols associated with Precambrian shield geology are common in large areas of tropical South America and Africa whereas Asia and eastern Australia are geologically much younger, with Acrisols particularly common in Asia. Data from the Harmonized Global Soil Database (FAO 2008) indicate that the most nutrient poor soils, in terms of total exchangeable bases (TEB: sum of Ca^{2+}, Mg^{2+}, K^+ and Na^+, a common measure of soil fertility) are absent in Asia and volcanic soils are more common meaning that, on average, TEB is higher in Asia than South America and Africa (Huston 2012). Nonetheless, the topsoil of Acrisols in Asian forests can also be very low in exchangeable bases (Banin et al. 2014). It is also noteworthy that nutrient availability to plants is not solely determined by the soils they grow in – atmospheric deposition of nutrients from marine sources, dust and anthropogenic activities also contribute and are spatially variable (Townsend et al. 2008).

Whilst soils of varying fertility are found in the tropics, in a global context they are overwhelmingly nutrient poor and yet in many localities they sustain the highest productivity of any terrestrial vegetation. This has been attributed to efficient nutrient cycling within the system to avoid losses to groundwater and the atmosphere. Mechanisms for nutrient capture and retention include dense root mats at the soil surface, symbioses with mycorrhizal fungi, algae and lichen, rapid growth of feeder roots, nutrient translocation before leaf abscission and long leaf lifespans and plant defences to prevent loss by herbivory (Stark and Jordan 1978; Jordan and Herrera 1981).

In addition to nutrients, physical soil conditions may also be limiting to plants, and these also vary with soil type. Young soils (e.g. Leptosols, Cambisols) and the formation of hardpans and concretions (e.g. in Plinthosols, Podzols and Acrisols) can prohibit deep rooting systems (Quesada et al. 2012a). Inundated soils and sandy soils, such as Fluvisols, Luvisols, Gleysols and Arenosols, have poor structure and can be unstable. In all these cases, trees are more susceptible to uprooting. Ferralsols conversely, are typically deep, well-structured and store water effectively, which can be beneficial during periods of drought.

Biogeography and diversity of tropical trees

Tropical moist forests are highly diverse compared with temperate forests, yet there is still a great variety in species richness within the tropics. For instance, African forests are markedly less diverse on average. Present day warm and wet environmental conditions may promote high diversity, but historical biotic exchange and evolution also explain some patterns in species distributions and diversity we see today.

The angiosperm flora that dominates the modern tropics is thought to have evolved in the mid-Cretaceous, coinciding with the break-up of Gondwana, approximately 100 million years ago [Mya] (Lidgard and Crane 1988; Primack and Corlett 2005). The plant species existing within tropical forests on each continent are thought to represent individualized evolutionary responses to the physical environment on each continent, though dispersal routes are also likely to have contributed to the current floristic assemblages (Primack and Corlett 2005; Corlett 2007; Donoghue 2008).

A combination of altered dispersal routes, past climatic changes and modern processes shape the distributions we see today, and several key events are thought to have had a particularly profound effect.

There was a significant drying of the tropical regions ~ 20 Mya and an associated reduction in tropical forest cover (Cox and Moore 2000). The relatively low diversity in African forests has been attributed, in part, to this period of drying and the contraction of tropical forest area. Conversely, relative climate stability coupled with regional-scale water cycling over the Amazon basin is thought to have resulted in high local diversity and low extinction rates in parts of Amazonia (Bush et al. 2004). The diversity of South American forests was also hugely influenced by the orogeny of the Andean mountains in the Tertiary and the connection formed with North America via the Isthmus of Panama ~3–4 Mya, initiating the Great American Interchange (Primack and Corlett 2005).

During the last glacial maximum colder conditions prevailed in the tropics; a drop in sea-surface temperatures of ~ 5°C reduced convective rainfall leading to dry episodes (Guilderson et al. 1994). It remains contentious whether this resulted in widespread deforestation. The glacial drying was more pronounced in Africa and Australia and led to rapid changes in forest composition. At this time, sea levels were also lower and the islands making up Southeast Asia became land-linked, facilitating movement of species around the region (Giresse et al. 1994; Cannon et al. 2009). In the neotropics, montane assemblages existed in lowland areas, and thus temperature drove migrations of Andean and Amazon floral elements up and down slope (Bush et al. 2004; Kellman and Tackaberry 1997).

Despite long periods of separation, striking similarities occur throughout the modern tropics, particularly in terms of tree familial composition. Using a pan-tropical collation of data from 35 forest plots (see Box 5.2), Gentry (1988) showed that there is a discernible overlap of the

---

Box 5.2 Measurement and monitoring of forest inventory plots

Permanent forest plots can be used to describe species composition, quantify structure and biomass, growth, carbon and population dynamics, and to monitor changes in these measures over time. An area is marked out and all trees below a minimum diameter (e.g. 1 cm or 10 cm diameter at breast height, DBH) are identified taxonomically, tagged with a unique identifier and the diameter recorded (Figure 5.2). In subsequent inventories or 'censuses' diameters of living trees are measured again, trees that died are noted, and new trees reaching the minimum threshold are identified and added to the record. Climbers, palms and understory plants may also be recorded and measured. Additional measurements, including tree height, crown dimensions and canopy cover, may be taken and samples of soil or plant material collected. Standardized methods facilitate robust cross-site comparisons. Some large plot networks have used these techniques in many sites, e.g. RAINFOR in Amazonia (Peacock et al. 2007), AfriTron in Africa (Lewis et al. 2013), CSIRO permanent plots in Australia (Bradford et al. 2014) and the pan-tropical CTFS (Centre for Tropical Forest Science) network (Losos and Leigh 2004). In extension to the forest inventory, the GEM (Global Ecosystem Monitoring) network also examines other components of the carbon budget including gas fluxes, root biomass and production and canopy production (e.g. Kho et al. 2013). These datasets are collated with the invaluable support and contribution of local institutions and communities.

Figure 5.2 Measuring tree diameters in forest inventory plots. Photo credit: LFB and Jugah Tagi

most speciose families across continents. The families commonly found in all tropical regions (except for Australia) were Leguminosae, Rubiaceae, Annonaceae, Euphorbiaceae, Lauraceae, Moraceae, Sapotaceae, Myristicaceae and Meliaceae. Lauraceae, Myrtaceae and Rubiaceae were also common in Australia. One of the most prominent differences is the dominance of the Dipterocarpaceae in Asian forests, which are largely absent elsewhere, and the comparative dominance of Leguminosae in American and African forests; these two families have been regarded as potential 'ecological equivalents' (Gentry 1988; LaFrankie et al. 2006).

Other notable differences in Gentry's (1988) plots between the regions are that: i) Bignoniaceae and Arecaceae are more speciose in the neotropics; ii) the Ebenaceae are more speciose in Africa and Asia; iii) Olacaceae and Apocynaceae are more speciose in Africa; iv) Myrtaceae are more speciose in Asia, and the Atlantic Forest of Brazil; and v) Proteaceae, Elaeocarpaceae and Monimiaceae were particularly speciose in Australia. Gentry's (1988) study also demonstrated surprisingly high overlap at the generic level, with 30 per cent of genera in African sites also found in the neotropics, and 23 per cent and 25 per cent of genera found in Asian and Australian forests, respectively, also occurring in the neotropics.

Whilst this chapter focuses on vegetation, there are also marked differences in the faunal communities of tropical forest regions (see Primack and Corlett 2005). These differences could have profound impacts on the floristic communities through key plant–animal interactions such as pollination and seed dispersal.

Tropical forest structure and dynamics

There are discernible differences in the structure and dynamics of forests across the tropics. Here, we particularly focus on findings from ground-based data (Box 5.2) and from old-growth tropical forests which better reveal 'natural' controls, since the structure and functioning of human-modified forests are substantially impacted by their disturbance histories (see Box 5.1).

Structure

Forest structure can be categorized into horizontal and vertical components. Important measures of horizontal structure include stand-level stem density, basal area and stem size distributions. Vertical structure refers to the third dimension and includes canopy characteristics such as tree height, height–diameter relations and crown dimensions and arrangement. Above-ground biomass (AGB) includes both horizontal and vertical components and is the total dry mass of living organic matter, made up of stems and branches, leaves and reproductive organs, whilst below-ground biomass is composed of mostly roots (Malhi et al. 2006). To actually measure biomass, trees must be felled so biomass is usually estimated for larger forest areas using allometric equations developed from harvested trees (e.g. Chave et al. 2005).

Tropical rain forests are characterized as structurally complex systems, storing and cycling large amounts of carbon in woody tissue, however, structure and biomass are not uniform across the tropics. On average, African forests are made up of fewer, large stems, when compared to South American and Asian forests, whilst stem density is particularly high in Australian forests (Banin et al. 2012; Lewis et al. 2013). Higher basal area and AGB values have been recorded for forests in Asia when compared with neotropical and African forests and variation in AGB is closely linked with the abundance of large-diameter trees (Lewis et al. 2013; Slik et al. 2013). Correspondingly, AGB is greater in central and eastern Amazon, where trees are long-lived and large, than in the western parts of the basin (Quesada et al. 2012b). Average maximum tree height and height–diameter relationships also vary: trees attain greater heights and are taller for a given diameter in Asia, Africa and central–east Amazon and are shorter in western Amazon and Australia (Banin *et al.* 2012). These differences are closely related to differences in basal area, mean tree size and tree lifespans. Nonetheless, maximum tree heights operate within mechanical and hydraulic limits (McMahon 1973; Ryan and Yoder 1997). Tropical dry forests are structurally distinct and tend to have lower basal area, AGB and are shorter for a given diameter (Chave *et al.* 2008; Feldpausch *et al.* 2011).

Stock parameters, such as basal area and AGB can be envisaged as the net result of tree growth inputs and mortality outputs. Whilst the relationship between above-ground net primary productivity (NPP; defined below) and AGB is strong and positive across some biomes where growth is seasonally limited, this relationship does not appear to hold for tropical forests (Keeling and Phillips 2007). Here, basal area and AGB are often determined by mortality rather than growth. Across Amazonia, structure follows a general east-to-west gradient with basal area, AGB and maximum height decreasing towards the west (Baker *et al.* 2004; Banin *et al.* 2012). Coincidently, stem dynamics increase towards the west, with higher incidence of tree falls (Chao *et al.* 2009). The varying levels of disturbance-driven mortality is perhaps why large-scale analyses of the determinants of forest structure have yielded different, and sometimes conflicting, results (e.g. Laurance *et al.* 1999; Saatchi *et al.* 2007; Slik *et al.* 2013). In addition, NPP and mortality are correlated and whilst a number of hypotheses exist, the causal relationship is unknown (Stephenson *et al.* 2011).

Growth and production

At the individual plant level, growth can be measured in terms of change in size (frequently diameter or girth in trees; see Box 5.2) or biomass change (production). Differences in growth and production rates can be due to differences in the genetic composition of a plant, its expression within a given environment and the environmental conditions and interactions with other organisms that it encounters. Plant growth is primarily limited by availability of the key resources water, light, temperature, nutrient availability and atmospheric carbon dioxide concentrations.

Water is essential for plant cell maintenance, biochemical reactions and transport of organic molecules and inorganic nutrients. As we have seen, it is a determinant of tropical forest distribution. Yet, large-scale studies have failed to identify annual rainfall as a primary determinant of above-ground wood and litter production within the wet tropics (e.g. Clark *et al.* 2001; Malhi *et al.* 2004; Chave *et al.* 2010; Schuur 2003). This may be due to two factors: first, it is rainfall in the dry season rather than annual totals that is critical for growth and second, soil water availability exploited by plant roots (determined by soil water holding capacity, texture and depth) is more important than rainfall *per se*. Overall, the net effect of water availability on stand-level growth and production in tropical moist forests is likely to be greater in regions where dry seasons are long enough for soil water reserves to be depleted. Further, dry periods may actually promote tree growth when soil water is sufficient to buffer the reduction in rainfall, because photosynthetically active radiation (PAR) is a limiting resource in tropical rain forests due to the thick cloud cover (Graham *et al.* 2003; Nemani *et al.* 2003).

Regional studies have identified large-scale correlations between soil fertility and above-ground wood production rates (Kitayama and Aiba 2002; Malhi *et al.* 2004; Paoli and Curran 2007; Quesada *et al.* 2012b; Banin *et al.* 2014). In Amazonia, above-ground wood production (AGWP) varies from ~ 3–11 tonnes per hectare per year (Malhi *et al.* 2004). Whilst climatic factors did not appear to explain this variation, the observed east-to-west gradient of increasing AGWP was consistent with differences in soil fertility.

It is most widely accepted that tropical lowland forests are P limited, and that N is in abundant supply, except in cooler, montane environments and the most nutrient poor sites (e.g. Vitousek 1984; McGroddy et al. 2004). A leading explanation for this is based on the observation that plant-available P must be obtained from weathering parent material, and in the old oligotrophic soils which dominate the tropics, P is only present in small quantities. AGWP has been found to be particularly well correlated with soil total P in forests in Amazonia and Borneo (Paoli and Curran 2007; Quesada et al. 2012b).

Fertilization experiments have also been used to determine the nature of nutrient limitation on plant growth and production. Responses to the experiments have been variable but increases in sapling and tree diameter growth, litterfall mass and litter nutrient content from a number of studies suggest that plant productivity is limited by nutrient availability in the tropics, but that nutrients other than P may be co-limiting – the base cations, calcium, potassium and magnesium and other micronutrients can also be important for plant growth, functioning and species distributions (Mirmanto et al. 1999; Kaspari et al. 2008; Wright et al. 2011).

Differences in above-ground growth and wood production rates across tropical forest regions are not well known, but rates of AGWP appear to be higher in Borneo than Amazonia (Banin et al. 2014).

At the ecosystem level, NPP is the amount of carbon assimilated via photosynthesis (gross primary productivity, GPP) less that lost via autotrophic respiration, and is of key interest in studies pertaining to carbon balance (Del Grosso 2008). NPP is composed of woody structures (stems and branches), foliage, roots, reproductive tissues (flowers, seeds, fruits), production lost to herbivory, emissions as volatile organic compounds (VOCs), root exudate and supply of carbohydrate to mycorrhiza (Luyssaert et al. 2007). Primary production may be estimated via 'top-down' methods including eddy covariance flux towers which measure canopy-level gas exchange, or 'bottom-up' via measuring the individual components (Malhi 2012).

The GPP of tropical forests ranges from ~30 to 40 tonnes C per hectare per year and is demonstrably lower at dry or montane sites – these differences are directly attributable to the reduced growing season (Malhi 2012). A synthesis of data from 35 sites across the tropics revealed that allocation between canopy, fine root and woody NPP was relatively invariant and split roughly equally across the three components, though allocation to woody NPP appears higher and roots lower in many of the Asian sites than in the American sites (Malhi et al. 2011). Stem production was estimated to account for approximately 10 per cent of GPP, indicating that stem-based measures of growth may not be solely suitable proxies for carbon fluxes or their determinants (Malhi et al. 2011).

Climate, nutrients and perhaps region therefore have effect on both above-ground growth and productivity. The role of species composition in determining rates of production is not well understood, but functional traits may affect the efficiency of carbon acquisition. A study across Amazonia determined that the effect of environment on AGWP was much stronger than that of functional composition (Baker et al. 2009).

Tropical forests in the global cycling of carbon

Plants take up atmospheric carbon as they grow and release it through respiration and senescence. Tropical forests store an estimated 460 billion tonnes of carbon in biomass and soil, comprising ~45 per cent of carbon in the terrestrial biosphere (Pan et al. 2011) and process 40 billion tonnes of carbon per year through GPP (Beer et al. 2010). Tropical forests are therefore critical to the global carbon cycle in terms of mopping up carbon from the atmosphere. Forests are also major sources of carbon to the atmosphere via deforestation and degradation and loss of forest cover reduces the ability for the terrestrial biosphere to further absorb carbon dioxide (CO_2). An estimated 2 billion tonnes of carbon was released per year during the 1990s as a result of tropical deforestation, and approximately one-third of human CO_2 emissions in the last decade have been from forest loss in the tropics (Malhi and Grace 2000; Richter and Houghton 2011). Rates of forest loss may also increase if incidence of fire and large-scale drought mortality interact with human-modified systems (Cochrane and Barber 2009; Aragão et al. 2014). There is, however, noted uncertainty and variability in estimates of CO_2 fluxes. Accurate estimation of both areal change in forest cover and carbon density (the amount of carbon stored in a given area) are needed to assess the contribution of landcover change in the tropics to global CO_2 emissions (Gibbs et al. 2007).

In addition to change in areal forest cover, the forests themselves are not in equilibrium. Data from large networks of forest inventory plots in Amazonia and Africa have, respectively, shown mean annual sinks of 0.45 (n plots = 123) and 0.63 tonnes C per hectare (n = 79) (Phillips et al. 2009; Lewis et al. 2009). There have been coincident changes in stem dynamics, showing higher rates of recruitment and mortality and increasing stand density (e.g. Phillips and Gentry 1994). The causes of these temporal trends have been hotly debated, but evidence suggests that they are at least partly attributable to increasing resource availability, such as CO_2 fertilization (Phillips and Lewis 2014). The shift to increasing biomass may, however, be transient and biomass may ultimately be limited by the constraints on canopy space, and eventually offset by increased rates of mortality (Brienen et al. 2015). Since tropical forests can potentially be major sources or sinks of CO_2, there has been impetus to incorporate forest management within policy directed at mitigating climate change. This has culminated in the development of Reduced Emissions from Deforestation and Degradation (REDD) policy (Miles and Kapos 2008). In essence, REDD provides economic incentives for countries, and particularly those in the developing world, to reduce deforestation rates and associated CO_2 emissions below a baseline (Miles and Kapos 2008). The extension, REDD+, additionally incorporates the role of sustainable management and enhancement of forest carbon stocks (Parker et al. 2009). Whilst simple in principle, a number of significant challenges must still be overcome before REDD+ policy can be successfully implemented, such as designating and distributing funds, setting baselines and ensuring permanence. Policies must also consider the needs of local populations and their need for sustainable development. Just as with the nature of these forests, the ways in which they are used by humans and their threats vary between regions, and so must strategies for their conservation.

Synthesis

Tropical forests are restricted to relatively warm and wet conditions, but strict definitions are numerous and this can be problematic for studying changes over space and time. Rather than a single, uniform entity, the tropical forest biome encompasses a great diversity of environments both in terms of climate and soil. Tropical forests have an equally rich and varied history, and the evolution of the angiosperm flora in isolated regions with different climatic and geological histories has led to unique species assemblages on different continents, and often even within regions. The diversity, structure and dynamics of these forests also vary greatly, although notably, when compared with other biomes, tropical forests store and cycle large amounts of carbon. Protection of tropical forests may provide a buffer to accelerated environmental change, but future climate change may also challenge long-term persistence of tropical forest ecosystems.

Acknowledgements

The authors thank Alan Grainger, Stephen Sitch, Patrick Meir and Richard Corlett for their helpful comments. This work was supported by a NERC studentship provided to LFB.

References

Aragão, L. E. O. C., Poulter, B., Barlow, J. B., Anderson, L. O., Malhi, Y. and others (2014) 'Environmental change and the carbon balance of Amazonian forests', *Biological Reviews*, vol. 89, pp. 913–931.
Baker, T. R., Phillips, O. L., Malhi, Y., Almeida, S., Arroyo, L. and others (2004) 'Variation in wood density determines spatial patterns in Amazonian forest biomass', *Global Change Biology*, vol. 10, pp. 545–562.
Baker, T. R., Phillips, O. L., Laurance, W. F., Pitman, N. C. A., Almeida, S. and others (2009) 'Do species traits determine patterns of wood production in Amazonian forests?' *Biogeosciences*, vol. 6, pp. 297–307.
Banin, L., Feldpausch, T. R., Phillips, O. L., Baker, T. R., Lloyd, J. and others (2012) 'What controls tropical forest architecture? Testing environmental, structural and floristic drivers', *Global Ecology and Biogeography*, vol. 21, pp. 1179–1190.
Banin, L., Lewis, S. L., Lopez-Gonzalez, G., Baker, T. R., Quesada, C. A. and others (2014) 'Tropical forest wood production: a cross-continental comparison', *Journal of Ecology*, vol. 102, pp. 1025–1037.
Bartholomé, E. and Belward, A. S. (2005) 'GLC2000: a new approach to global land cover mapping from Earth observation data', *International Journal of Remote Sensing*, vol. 26, no. 9, pp. 1959–1977.
Beer, C., Reichstein, M., Tomelleri, E., Ciais, P., Jung, M. and others (2010) 'Terrestrial gross carbon dioxide uptake: global distribution and covariation with climate', *Science*, vol. 329, no. 5993, pp. 834–838.
Beckmann, G. G. (1983) 'Development of old landscapes and soils', in Division of Soils C (eds) *Soils: An Australian Viewpoint*, CSIRO, Melbourne, Australia.
Bradford, M. G., Murphy, H. T., Ford, A. J., Hogan, D. L., and Metcalfe, D. J. (2014) 'Long-term stem inventory data from tropical rain forest plots in Australia', *Ecology*, vol. 95, no 8, p. 2362.
Brienen, R. J. W., Phillips, O. L., Feldpausch, T. R., Gloor, E., Baker, T.R., Lloyd, J. and others (2015) 'Long-term decline of the Amazon carbon sink', *Nature*, vol. 519, pp. 344–348. doi: 10.1038/nature14283
Bush, M. B., Silman, M. R. and Urrego, D. H. (2004) '48,000 years of climate and forest change in a biodiversity hot spot' *Science*, vol. 303, pp. 827–829.
Cannon, C. H., Morley, R. J., and Bush, A. B. G. (2009) 'The current refugial rainforests of Sundaland are unrepresentative of their biogeographic past and highly vulnerable to disturbance', *Proceedings of the National Academy of Sciences of the United States of America*, vol. 106, no. 27, pp. 11188–11193.
Chao, K. J., Phillips, O. L., Monteagudo, A., Torres-Lezama, A. and Martinez, R. V. (2009) 'How do trees die? Mode of death in northern Amazonia', *Journal of Vegetation Science*, vol. 20, pp. 260–268.
Chave, J., Andalo, C., Brown, S., Cairns, M. A., Chambers, J. Q. and others (2005) 'Tree allometry and improved estimation of carbon stocks and balance in tropical forests', *Oecologia*, vol. 145, pp. 87–99.
Chave, J., Condit, R., Muller-Landau, H. C., Thomas, S. C., Ashton, P. S. and others (2008) 'Assessing evidence for a pervasive alteration in tropical tree communities', *Plos Biology*, vol. 6, pp. 455–462.
Chave, J., Navarrete, D., Almeida, S., Alvarez, E., Aragão, L. and others (2010) 'Regional and seasonal patterns of litterfall in tropical South America', *Biogeosciences*, vol. 7, pp. 43–55.
Christensen, J. H., Krishna Kumar, K., Aldrian, E., An, S.-I., Cavalcanti, I. F. A. and others (2013) 'Climate phenomena and their relevance for future regional climate change' in T. F. Stocker, D. Qin, G.-K. Plattner, M. Tignor, S. K. Allen, J. Boschung, A. Nauels, Y. Xia, V. Bex and P. M. Midgley (eds.) *Climate Change 2013: The Physical Science Basis. Contribution of Working Group I to the Fifth Assessment Report of the Intergovernmental Panel on Climate Change*, Cambridge University Press, Cambridge, UK.
Churkina, G. and Running, S. W. (1998) 'Contrasting climatic controls on the estimated productivity of global terrestrial biomes', *Ecosystems*, vol. 1, pp. 206–215.

Clark, D. A., Brown, S., Kicklighter, D. W., Chambers, J. Q., Thomlinson, J. R. and others (2001) 'Net primary production in tropical forests: an evaluation and synthesis of existing field data', *Ecological Applications*, vol. 11, pp. 371–384.

Cochrane, M. A. and Barber, C. P. (2009) 'Climate change, human land use and future fires in the Amazon', *Global Change Biology*, vol. 15, pp. 601–612.

Corlett, R. T. (2007) 'What's so special about Asian tropical forests?', *Current Science*, 93, 1551–1557.

Corlett, R. T. (2011) 'Impacts of warming on tropical lowland rainforests', *Trends in Ecology and Evolution*, vol. 26, no 11, pp. 606–613.

Cox, C. B. and Moore, P. D. (2000) *Biogeography: An Ecological and Evolutionary Approach*, Sixth Edition, Blackwell Science Ltd, Oxford, UK.

Del Grosso, S. (2008) 'Global potential net primary production predicted from vegetation class, precipitation, and temperature', *Ecology*, vol. 89, pp. 2117–2126.

Donoghue, M. J. (2008) 'A phylogenetic perspective on the distribution of plant diversity', *Proceedings of the National Academy of Sciences of the United States of America*, vol. 105, pp. 11549–11555.

Dudal, R. (2005) 'Soils of Southeast Asia' in A. Gupta (ed.) *The Physical Geography of Southeast Asia*, Oxford University Press, Oxford, UK.

FAO [Food and Agriculture Organization of the United Nations] (2008) Harmonized world soil database, version 1.0. FAO Land and Water Digital Media Series 34. FAO/IIASA/ISRIC/JRC-EU/Academia Sinica, Rome, Italy. www.fao.org/nr/land/soils/harmonized-world-soil-database/en

Fauset, S., Baker, T. R., Lewis, S. L., Feldpausch, T. R., Affum-Baffoe, K. and others (2012) 'Drought-induced shifts in the floristic and functional composition of tropical forests in Ghana', *Ecology Letters*, vol. 15, pp. 1120–1129.

Feeley, K. J., Davies, S. J., Perez, R., Hubbell, S. P. and Foster, R. B. (2011) 'Directional changes in the species composition of a tropical forest', *Ecology*, vol. 92, pp. 871–882.

Feldpausch, T. R., Banin, L., Phillips, O. L., Baker, T. R., Lewis, S. L. and others (2011) 'Height-diameter allometry of tropical forest trees', *Biogeosciences*, vol. 8, pp. 1081–1106.

Gibbs, H. K., Brown, S., Niles, J. O. and Foley, J. A. (2007) 'Monitoring and estimating tropical forest carbon stocks: making REDD a reality', *Environmental Research Letters*, vol. 2, no 4, 045023.

Giresse, P., Maley, J. and Brenac, P. (1994) 'Late quaternary paleoenvironments in the Lake Barombi Mbo (West Cameroon) deduced from pollen and carbon isotopes of organic-matter', *Palaeogeography Palaeoclimatology Palaeoecology*, vol. 107, pp. 65–78.

Gentry, A. H. (1988) 'Changes in plant community diversity and floristic composition on environmental and geographical gradients', *Annals of the Missouri Botanical Garden*, vol. 75, pp. 1–34.

Graham, E. A., Mulkey, S. S., Kitajima, K., Phillips, N. G. and Wright, S. J. (2003) 'Cloud cover limits net CO_2 uptake and growth of a rainforest tree during tropical rainy seasons', *Proceedings of the National Academy of Sciences of the United States of America*, vol. 100, pp. 572–576.

Guilderson, T. P., Fairbanks, R. G. and Rubenstone, J. L. (1994) 'Tropical temperature-variations since 20,000 years ago – Modulating interhemispheric climate-change', *Science*, vol. 263, pp. 663–665.

Hammond, D. S. (2005) 'Biophysical features of the Guiana Shield' in D. S. Hammond (ed.) *Tropical Forests of the Guiana Shield*, CABI Publishing, Wallingford, Oxfordshire, UK.

Hawkins, E., Joshi, M., and Frame, D. (2014) 'Wetter then drier in some tropical areas', *Nature Climate Change*, vol. 4, no 8, pp. 646–647.

Hijmans, R. J., Cameron, S. E., Parra, J. L., Jones, P. G. and Jarvis, A. (2005) 'Very high resolution interpolated climate surfaces for global land areas', *International Journal of Climatology*, vol. 25, pp. 1965–1978.

Holdridge, L. R. (1947) 'Determination of world plant formations from simple climatic data', *Science*, vol. 105, pp. 367–368.

Hubble, G. D. and Isbell, R. F. (1983) 'Eastern Highlands (VI)', in Division of Soils C (eds) *Soils: An Australian Viewpoint*, CSIRO, Melbourne, Australia.

Hubble, G. D., Isbell, R. F. and Northcote, K. H. (1983) 'Features of Australian soils', in Division of Soils C (eds) *Soils: An Australian Viewpoint*, CSIRO, Melbourne, Australia.

Huston, M. A. (2012) 'Precipitation, soils, NPP, and biodiversity: resurrection of Albrecht's curve', *Ecological Monographs*, vol. 82, no 3, pp. 277–296.

Irion, G. (1978) 'Soil infertility in Amazonian rain-forest', *Naturwissenschaften*, vol. 65, pp. 515–519.

IUSS Working Group WRB (2006) 'World reference base for soil resources', in *World Soil Resources Reports No. 103*, FAO Rome.

Jenny, H. (1941) *Factors of Soil Formation: A System of Quantitative Pedology*. McGraw-Hill, New York.

Jordan, C. F. and Herrera, R. (1981) 'Tropical rain forests – are nutrients really critical', *American Naturalist*, vol. 117, pp. 167–180.

Juo, A. S. R. and Wilding, L. P. (1996) 'Soils of the lowland forests of West and Central Africa', *Proceedings of the Royal Society of Edinburgh Section B (Biological Sciences)*, vol. 104, pp. 15–29.

Kaspari, M., Garcia, M. N., Harms, K. E., Santana, M., Wright, S. J. and Yavitt, J. B. (2008) 'Multiple nutrients limit litterfall and decomposition in a tropical forest', *Ecology Letters*, vol. 11, pp. 35–43.

Keeling, H. C. and Phillips, O. L. (2007) 'The global relationship between forest productivity and biomass'. *Global Ecology and Biogeography*, 16, pp. 618–631.

Kellman, M. and Tackaberry, R. (1997) *Tropical Environments: The Functioning and Management of Tropical Ecosystems,* Routledge, London.

Kitayama, K. and Aiba, S. I. (2002) 'Ecosystem structure and productivity of tropical rain forests along altitudinal gradients with contrasting soil phosphorus pools on Mount Kinabalu, Borneo', *Journal of Ecology*, vol. 90, pp. 37–51.

Kho, L. K., Malhi, Y., and Tan, S. K. S. (2013) 'Annual budget and seasonal variation of aboveground and belowground net primary productivity in a lowland dipterocarp forest in Borneo', *Journal of Geophysical Research: Biogeosciences*, vol. 118, no 3, pp. 1282–1296.

Köppen, W. (1923) *Die climate de erde: grundiss der klimakunde*. De Greyte, Berlin.

LaFrankie, J. V., Ashton, P. S., Chuyong, G. B., Co, L., Condit, R. and others (2006) 'Contrasting structure and composition of the understory in species-rich tropical rain forests', *Ecology*, vol. 87, pp. 2298–2305.

Laurance, W. F., Fearnside, P. M., Laurance, S. G., Delamonica, P., Lovejoy, T. E. and others (1999) 'Relationship between soils and Amazon forest biomass: a landscape-scale study', *Forest Ecology and Management*, vol. 118, pp. 127–138.

Lewis, S. L., Lopez-Gonzalez, G., Sonke, B., Affum-Baffoe, K., Baker, T. R. and others (2009) 'Increasing carbon storage in intact African tropical forests', *Nature*, vol. 457, pp. 1003-U3.

Lewis, S. L., Sonké, B., Sunderland, T., Begne, S. K., Lopez-Gonzalez, G. and others (2013) 'Aboveground biomass and structure of 260 African tropical forests', *Philosophical Transactions of the Royal Society B: Biological Sciences*, vol. 368, 20120295.

Li, J., Xie, S. P., Cook, E. R., Huang, G., D'Arrigo, R. and others (2011) 'Interdecadal modulation of El Niño amplitude during the past millennium', *Nature Climate Change*, vol. 1, no 2, pp. 114–118.

Lidgard, S. and Crane, P. R. (1988) 'Quantitative-analyses of the early angiosperm radiation', *Nature*, vol. 331, pp. 344–346.

Lloyd, J. and Farquhar, G. D. (2008) 'Effects of rising temperatures and CO_2 on the physiology of tropical forest trees', *Philosophical Transactions of the Royal Society B-Biological Sciences*, vol. 363, 1811–1817.

Losos, E. and Leigh, E. G. (2004) *Tropical Forest Diversity and Dynamism: Findings from a Large-Scale Plot Network*, University of Chicago Press, Chicago, USA.

Luyssaert, S., Inglima, I., Jung, M., Richardson, A. D., Reichsteins, M. and others (2007) 'CO_2 balance of boreal, temperate, and tropical forests derived from a global database', *Global Change Biology*, vol. 13, pp. 2509–2537.

Malhi, Y. (2012) 'The productivity, metabolism and carbon cycle of tropical forest vegetation', *Journal of Ecology*, vol. 100, no 1, pp. 65–75.

Malhi, Y. and Grace, J. (2000) 'Tropical forests and atmospheric carbon dioxide', *Trends in Ecology and Evolution*, vol. 15, pp. 332–337.

Malhi, Y. and Wright, J. (2004) 'Spatial patterns and recent trends in the climate of tropical rainforest regions', *Philosophical Transactions of the Royal Society of London Series B-Biological Sciences*, vol. 359, pp. 311–329.

Malhi, Y., Baker, T. R., Phillips, O. L., Almeida, S., Alvarez, E. and others (2004) 'The above-ground coarse wood productivity of 104 Neotropical forest plots', *Global Change Biology*, vol. 10, pp. 563–591.

Malhi, Y., Wood, D., Baker, T. R., Wright, J., Phillips, O. L. and others (2006) 'The regional variation of aboveground live biomass in old-growth Amazonian forests', *Global Change Biology*, vol. 12, pp. 1107–1138.

Malhi, Y., Aragão, L. E., Galbraith, D., Huntingford, C., Fisher, R. and others (2009) 'Exploring the likelihood and mechanism of a climate-change-induced dieback of the Amazon rainforest', *Proceedings of the National Academy of Sciences*, vol. 106, no 49, 20610–20615.

Malhi, Y., Doughty, C. and Galbraith, D. (2011) 'The allocation of ecosystem net primary productivity in tropical forests', *Philosophical Transactions of the Royal Society B*, vol. 366, pp. 3225–3245.

Mayaux, P., Holmgren, P., Achard, F., Eva, H., Stibig, H. and Branthomme, A. (2005) 'Tropical forest cover change in the 1990s and options for future monitoring', *Philosophical Transactions of the Royal Society B-Biological Sciences*, vol. 360, pp. 373–384.

McGregor, G. R. and Nieuwolt, S. (1998) *Tropical Climatology: An Introduction to the Climates of the Low Latitudes*, Second edition, John Wiley and Sons, Chichester, UK.

McGroddy, M. E., Daufresne, T. and Hedin, L. O. (2004) 'Scaling of C:N:P: stoichiometry in forest ecosystems worldwide: implications of terrestrial Redfield-type ratios', *Ecology*, vol. 85, pp. 2390–2401.

McMahon, T. (1973) 'Size and shape in biology', *Science*, vol. 179, pp. 1201–1204.

Miles, L. and Kapos, V. (2008) 'Reducing greenhouse gas emissions from deforestation and forest degradation: global land-use implications', *Science*, vol. 320, pp. 1454–1455.

Mirmanto, E., Proctor, J., Green, J., Nagy, L. and Suriantata (1999) 'Effects of nitrogen and phosphorus fertilization in a lowland evergreen rainforest', *Philosophical Transactions of the Royal Society of London Series B-Biological Sciences*, vol. 354, pp. 1825–1829.

Nemani, R. R., Keeling, C. D., Hashimoto, H., Jolly, W. M., Piper, S. C. and others (2003) 'Climate-driven increases in global terrestrial net primary production from 1982 to 1999', *Science*, vol. 300, pp. 1560–1563.

New, M., Hulme, M. and Jones, P. (1999) 'Representing twentieth-century space-time climate variability. Part I: Development of a 1961–90 mean monthly terrestrial climatology', *Journal of Climate*, vol. 12, pp. 829–856.

Pan, Y., Birdsey, R., Fang, J., Houghton, R., Kauppi, P. and others (2011) 'A large and persistent carbon sink in the world's forests', *Science*, vol. 333, pp. 988–992.

Paoli, G. D. and Curran, L. M. (2007) 'Soil nutrients limit fine litter production and tree growth in mature lowland forest of Southwestern Borneo', *Ecosystems*, vol. 10, pp. 503–518.

Parker, C., Mitchell, A., Trivedi, M., Mardas, N. and Sosis, K. (2009) *The Little REDD+ Book*, Global Canopy Programme, Oxford, UK.

Peacock, J., Baker, T. R., Lewis, S. L., Lopez-Gonzalez, G. and Phillips, O. L. (2007) 'The RAINFOR database: monitoring forest biomass and dynamics', *Journal of Vegetation Science*, vol. 18, pp. 535–542.

Phillips, O. L. and Gentry, A. H. (1994) 'Increasing turnover through time in tropical forests', *Science*, vol. 263, pp. 954–958.

Phillips, O. L. and Lewis, S. L. (2014) 'Evaluating the tropical forest carbon sink', *Global Change Biology*, vol. 20, 2039–2041.

Phillips, O. L., Aragão, L., Lewis, S. L., Fisher, J. B., Lloyd, J. and others (2009) 'Drought sensitivity of the Amazon rainforest', *Science*, vol. 323, pp. 1344–1347.

Phillips, O. L., van der Heijden, G., Lewis, S. L., López-González, G., Aragão, L. E. O. C. and others (2010) 'Drought-mortality relationships for tropical forests', *New Phytologist*, vol. 187, pp. 631–646.

Primack, R. B. and Corlett, M. T. (2005) *Tropical Rainforests: An Ecological and Biogeographical Comparison*, Blackwell Publishing, Oxford, UK.

Putz, F. E. and Redford, K. H. (2010) 'The importance of defining "forest": tropical forest degradation, deforestation, long-term phase shifts, and further transitions', *Biotropica*, vol. 42, pp. 10–20.

Quesada, C. A., Lloyd, J., Schwarz, M., Patiño, S., Baker, T. R. and others (2012a) 'Variations in chemical and physical properties of Amazon forest soils in relation to their genesis', *Biogeosciences*, vol. 7, pp. 1515–1541.

Quesada, C. A., Phillips, O. L., Schwarz, M., Czimczik, C. I., Baker, T. R. and others (2012b) 'Basin-wide variations in Amazon forest structure and function are mediated by both soils and climate', *Biogeosciences*, vol. 9, pp. 2203–2246.

Richter, D. de B. and Houghton, R. A. (2011) 'Gross CO_2 fluxes from landuse change: Implications for reducing global emissions and increasing sinks', *Carbon Management*, 2, pp. 41–47.

Ryan, M. G. and Yoder, B. J. (1997) 'Hydraulic limits to tree height and tree growth', *Bioscience*, vol. 47, pp. 235–242.

Saatchi, S., Houghton, R. A., Dos Santos Alvala, R. C., Soares, J. V. and Yu, Y. (2007) 'Distribution of aboveground live biomass in the Amazon basin', *Global Change Biology*, vol. 13, pp. 819–837.

Schuur, E. A. G. (2003) 'Productivity and global climate revisited: The sensitivity of tropical forest growth to precipitation', *Ecology*, vol. 84, pp. 1165–1170.

Shuttleworth, W. J. (1988) 'Evaporation from Amazonian rainforest', *Proceedings of the Royal Society of London Series B Biological Sciences*, vol. 233, pp. 321–346.

Slik, J. W. F., Paoli, G., McGuire, K., Amaral, I., Barroso, J. and others (2013) 'Large trees drive forest aboveground biomass variation in moist lowland forests across the tropics', *Global Ecology and Biogeography*, vol. 22, pp. 1261–1271.

Stark, N. M. and Jordan, C. F. (1978) 'Nutrient retention by root mat of an Amazonian rain-forest', *Ecology*, vol. 59, pp. 434–437.

Stephenson, N. L., van Mantgem, P. J., Bunn, A. G., Bruner, H., Harmon, M. E. and others (2011) 'Causes and implications of the correlation between forest productivity and tree mortality rates', *Ecological Monographs*, vol. 81, pp. 527–555.

Torello-Raventos, M., Feldpausch, T. R., Veenendaal, E., Schrodt, F., Saiz, G. and others (2013) 'On the delineation of tropical vegetation types with an emphasis on forest/savanna transitions', *Plant Ecology and Diversity*, vol. 6, pp. 101–137.

Townsend, A. R., Asner, G. P. and Cleveland, C. C. (2008) 'The biogeochemical heterogeneity of tropical forests', *Trends in Ecology and Evolution*, vol. 23, pp. 424–431.

Vitousek, P. M. (1984) 'Litterfall, nutrient cycling, and nutrient limitation in tropical forests', *Ecology*, vol. 65, pp. 285–298.

Walsh, R. P. D. (1996a) 'Climate', in P. W. Richards (ed.) *The Tropical Rain Forest: An Ecological Study*, Cambridge University Press, Cambridge, UK.

Walsh, R. P. D. (1996b) 'Drought frequency changes in Sabah and adjacent parts of northern Borneo since the late nineteenth century and possible implications for tropical rain forest dynamics', *Journal of Tropical Ecology*, vol. 12, pp. 385–407.

Whittaker, R. H. (1975) *Communities and Ecosystems,* 2nd edition, McMillan, New York.

Williams, J. W., Jackson, S. T. and Kutzbach, J. E. (2007) 'Projected distributions of novel and disappearing climates by 2100 AD', *Proceedings of the National Academy of Sciences of the United States of America*, vol. 104, no 14, pp. 5738–5742.

Woodward, F. I., Lomas, M. R. and Kelly, C. K. (2004) 'Global climate and the distribution of plant biomes', *Philosophical Transactions of the Royal Society of London Series B-Biological Sciences*, vol. 359, pp. 1465–1476.

Wright, S. J., Muller-Landau, H. C. and Schipper, J. (2009) 'The future of tropical species on a warmer planet', *Conservation Biology*, vol. 23, pp. 1418–1426.

Wright, S. J., Yavitt, J. B., Wurzburger, N., Turner, B. L., Tanner, E. V. J. and others (2011) 'Potassium, phosphorus, or nitrogen limit root allocation, tree growth, or litter production in a lowland tropical forest', *Ecology*, vol. 92, pp.1616–1625.

Zelazowski, P., Malhi, Y., Huntingford, C., Sitch, S. and Fisher, J. B. (2011) 'Changes in the potential distribution of humid tropical forests on a warmer planet', *Philosophical Transactions of the Royal Society A*, vol. 369, pp. 137–160.

6
MANAGED FORESTS

Jürgen Bauhus and Patrick Pyttel

Introduction

Most forests of the world have been altered by humans for long periods of time. They have always been used to fulfil our needs for fuel, construction timber, food, fibre, and medicine. However, traditional forest management in the sense of regulating and manipulating the structure and composition of forests to meet the demands for chiefly timber and woody biomass is a relatively recent phenomenon that started ca. 300 years ago in central Europe (von Carlowitz 1713). More recently, only a few decades ago, this was expanded to supply, explicitly, also other ecosystem goods and services. In Europe, modern forest management approaches developed as consequence of previously unregulated forest exploitation for the industrial use of timber. Back in times when wood was the only widely available source of energy, large quantities of it were used for charcoal, to melt glass and metals, and for heating. Because the early industries depended on wood as source of energy, vast stretches of forest were felled to fuel these industries (Sands 2005). The realization that economic and social development could not continue with this unsustainable exploitation led to the first attempts to re-establish and manage forests (von Carlowitz 1713). The foundation of the first institutions teaching forestry and forest science also date back to this time (Sands 2005). Owing to the widespread devastation of forests in central Europe in the seventeenth and eighteenth century, early forest management mainly focused on the regeneration and re-establishment of forests. The harsh environment that young trees face when growing in open conditions, characterized by extreme temperatures, degraded soils, competition through herbs, grass, and shrubs, as well as browsing pressure through domestic and wild animals, only allowed tough tree species to be successfully regenerated. This triggered the wide scale establishment of conifer plantations mainly comprised of Norway spruce (*Picea abies*) and Scots pine (*Pinus sylvestris*). In addition, these species were easy to propagate in nurseries; they were fast growing and promised early yields. The restoration of landscapes with these plantations was so successful, mainly economically but also with regard to other ecosystem functions, that this model of forest management was adopted in many parts of the world. This explains why, even today, forests in Europe are still dominated by secondary conifer plantations, where natural forests of broadleaved tree species (e.g. European beech [*Fagus sylvatica*], oak [*Quercus robur*, *Quercus petraea*]) would occur. This approach to forest management that developed when forestry originated as a discipline was modelled on agricultural cultivation methods and therefore represented the complete opposite of the unregulated exploitation that prevailed before then (Puettmann et al. 2009). The similarity of the terms agriculture and silviculture shows this close relationship; forest land was being regarded as a field cultivated with trees. The concept of forest management followed at that time was the so-called 'regular forest' (*Normalwald* in German) (Hundeshagen 1826; see Figure 6.1). The regular forest represents an idealized forest management unit, with an even distribution of age classes, established as high forests, which provides the same, sustainable yield of timber every year. If the production cycle extended to X years to grow all the timber of the desired dimensions, then a fraction of the total production area Y that is equivalent to Y/X would be harvested and regenerated every year.

It is obvious, that this 'ideal', conceptual model is most easily implemented when only one tree species is grown, harvested through clear-cut, and replanted. As mentioned, this model depends on many assumptions, is solely focused on timber production, and ignores production risks through ecosystem disturbances. While there is, practically, not a single forest enterprise that represents this normal forest, this model has profoundly shaped forest management approaches throughout the world, leading to a dominance of even-aged high forests in many

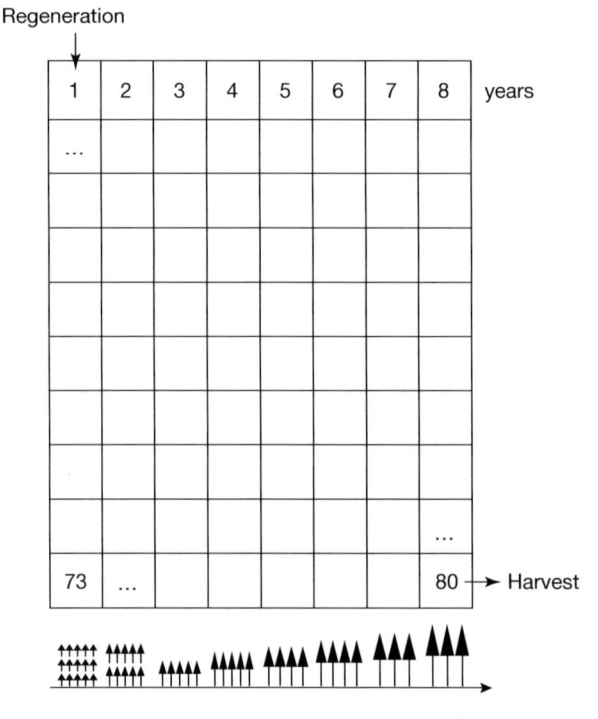

Figure 6.1 The regular forest model intends to create or maintain an even distribution of age classes and thus to ensure that the same area of forest can be harvested yearly. This very simple model for a sustainable forest resource supply is based on many assumptions such as homogeneity of site conditions and thus productivity, constancy in growth conditions over time, absence of biotic and abiotic disturbance risks and others. The sum of all squares represents the total production area (Y). Single squares represent the fraction of the yearly harvested forests (age classes) within the total production cycle (X). Stand development phases in ten year intervals are depicted below. Some basic concepts of this model are still relevant in current approaches to sustainable forest management. In more irregular and uneven-aged forests, the production of timber has to be monitored using sophisticated inventory approaches

regions. However, there have been different historical starting points and development paths of forest management in different parts of the world. In contrast to central Europe, where scientifically based forest management commenced when native forests had been largely devastated, largely intact native forests, that were to different degrees manipulated by indigenous people, provided the starting point in other parts of the world. Forest management has also evolved to encompass many more values, for which forests are being managed. Today, it is widely accepted that forests should be managed to fulfil the following goals (e.g. MCPFE 2002):

- Conservation of biological diversity
- Maintenance of the productive capacity
- Maintenance of ecosystem health and vitality
- Conservation and maintenance of soil and water resources
- Maintenance of forest contribution to global carbon cycles
- Maintenance and enhancement of long-term multiple socio-economic benefits.

These goals have been agreed upon in a number of international processes to define and promote sustainable forest management. Before we discuss how these different goals can be translated into forest management, it is helpful to depict classical forest management approaches.

Silviculture – stand level management

Forest management takes place at different spatial scales (Table 6.1). In this chapter we will focus on forest management at the stand scale because forest ecosystems are most immediately influenced by silvicultural management activities at this level. Management effects at larger spatial scales can often be regarded as the sum of management activities at the stand scale, although there are of course interactions and feedback such as the influence on disturbance patterns (Franklin and Forman 1987). For example, harvesting of adjacent stands may increase the risk of wind throw for other stands located downwind, or cutting patterns and the proportion of recently harvested areas may influence other forest patches, e.g. through increased populations of browsing animals.

Forest management at the stand scale comprises mainly silvicultural activities. Silviculture is the deliberate manipulation of forest structure, composition, and dynamics to provide ecosystem goods and services on a sustainable basis (Figure 6.2). To what extent and priority the various goods and services are pursued depends on the management objectives of the landowner (Nyland, 2002).

Table 6.1 The different scales of forest management. Lower spatial scales contribute to and affect goals at higher spatial scales. Higher spatial scales set the goals for lower spatial scales. The level of integration and harmonization of management approaches and goals across the different scales depends on the distribution of land ownership across scales. If the landscape is dominated by many small landowners with differing goals, it will be difficult to set and pursue achievable goals at a higher spatial scale (after Nyland 2002).

Spatial scale	*Type of management*	*Examples for achievable goals*
Landscape/catchment	Ecosystem management	Maintenance of plant and animal populations; catchment hydrology
Forest enterprise/district	Resource management	Sustained yield and income
Stand	Silviculture	Maintenance of productive capacity, soils and nutrient cycles

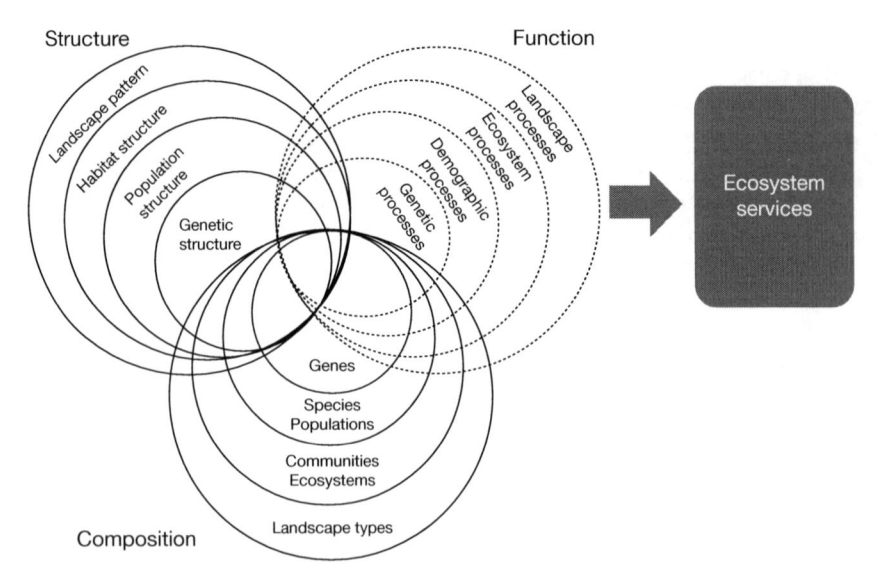

Figure 6.2 The interaction of composition and structure from the level of genetics to the level of landscapes influences ecological functions at these respective levels (adapted from Noss 1990). In forest management, composition and structure are being manipulated (circles with solid lines) at all these levels to direct ecological functions in order to provide the desired ecosystem services. Tree species composition may be changed to include more productive or valuable ones, or tree breeding may change the genetic composition. This may influence ecosystem productivity, carbon and nutrient cycling. Not all changes in forest composition and structure have only intended effects on ecosystem functioning (circles with dotted lines). For example, introducing tree species with slowly-decomposing litter may lead to soil acidification

The sum of all activities that take place at the stand level from the regeneration to the harvesting of trees is called a silvicultural system (Figure 6.3). Commonly, silvicultural systems are named after the way the forest is harvested or regeneration is initiated such as clear felling, shelterwood, seed tree system, single tree selection, coppice,[1] or coppice with standards, because the type and intensity of harvesting and the residual stand structure that remains are important determinants for all the other management activities that form a silvicultural system. Silvicultural systems intend to improve the forest stand conditions to optimize the yield of goods and services. Historically, the focus has been on the delivery of desired raw materials at predictable qualities and quantities at regular intervals and over long periods (Nyland 2002). This is still the focus for forest with an emphasis on the production function. To ensure this, desired species have to be regenerated in the preferred composition and spatial arrangement. Subsequently, growing stands are being tended to maintain or change that species composition, to ensure the desired development of tree quality and dimension, to reduce management risks such as wind throw, to maintain productivity, and to facilitate income through thinning of trees. Finally, trees are being harvested either individually (single tree selection), or in the form of groups (group selection, 'Femelschlag'), gaps, strips, or clear cuts. The removal of mature trees creates space

---

1 Coppice shoots arise from adventitious buds at the base of stems after these have been cut. In coppice systems, the new tree crop develops vegetatively from coppice shoots after cutting of all stems. Traditionally this system has been applied to produce small dimension timber for poles, fuelwood, tanning bark, and charcoal.

Regeneration
- natural
- artificial – seeding
              – planting

SILVICULTURAL SYSTEM

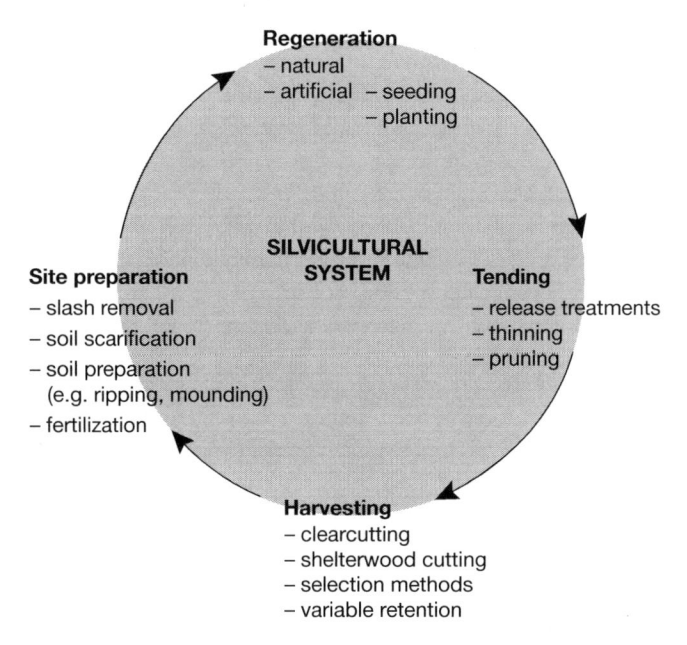

Site preparation
- slash removal
- soil scarification
- soil preparation
  (e.g. ripping, mounding)
- fertilization

Tending
- release treatments
- thinning
- pruning

Harvesting
- clearcutting
- shelterwood cutting
- selection methods
- variable retention

Figure 6.3 Silvicultural systems comprise all activities from the regeneration to the harvesting of trees at the stand level. The different activities at different points in time of stand development influence subsequent options for forest management. The sum of all these manipulations of forest structure and composition is intended to meet the forest owner's expectations for the delivery of ecosystem goods and services, and associated economic and social benefits

and provides resources for the establishment, or coppicing and growth of the next cohort of trees. The establishment and subsequent growth of trees may be further promoted through specific site preparation practices. All silvicultural decisions reflect the goals and aspirations of forest owners. In the case of community, municipal or state forests, these goals typically reflect a negotiation process. In the following we will look at these phases of silvicultural systems in more detail to provide an understanding of the rationale for these practices.

Regeneration

The desired tree species and their genetic structure may range from fast growing, possibly exotic species that have undergone intensive genetic improvement programmes to species that are native to the site-adapted forest community and have not undergone a deliberate genetic selection. Trees and forest stands may be regenerated naturally, through seedlings that establish from natural seed fall, or vegetatively through coppice shoots or suckers that emerge from roots (e.g. in *Populus sp.* or *Robinia*). Natural regeneration from seeds is the most prominent of these forms, although in some regions regeneration of forests through coppicing is still widely practiced. Natural regeneration is likely the preferred process, when the species composition and the genetic make-up of mature trees conform with the production and other stand management goals. It is often the preferred choice, when natural forest communities are to be maintained, as is the case in managed reserves such as in the European-wide Natura 2000 network of protected areas (Larsen and Nielsen 2007), or when the expected yield and financial return do not justify high investments at the time of regeneration. The expenses incurred at the time of regeneration naturally take a long time to be recovered in commercial thinning or harvesting operations. Natural regeneration is typically also employed, when dense regeneration layers with more than 5,000 or 10,000 seedlings ha^{-1} are to be established and the costs of planting so many trees would be prohibitive. Such high seedling numbers are typically preferred for many rather slow-growing hardwood species, where dense stand conditions at the sapling and pole stage are required for the self-pruning process and stem quality development in terms of straight boles (Kühne *et al.* 2013). The initial density of a seedling cohort of European beech established from natural regeneration may reach numbers in excess of 200,000 ha^{-1} (Burschel and Huss 2003).

In contrast, artificial regeneration of forest stands through planting or sowing is applied in situations where:

- land that was recently not forest is afforested or re-afforested;
- the current species or species combination does not match targets for the future stand composition;
- the genetic structure of the current stand is unsatisfactory, for example through wrong selection of provenances in the past;
- a tight control over the spatial arrangement of trees and species is desired (for example to avoid problems with overly dense stands);
- seed production to facilitate natural regeneration is unreliable;
- establishment conditions are so adverse, that they can only be overcome by tall seedlings (e.g. competition from understory vegetation, ground frost, browsing).

The density of planted seedlings depends on their growth rates, which determines the time during which they are exposed to harsh growing conditions near the ground and until they close the canopy, their planting height and on the quality of the planting material. Seedlings may be planted as bare-rooted or container stock.

Sowing is a less common form of artificial regeneration and typically restricted to situations where:

- seed availability is not a problem;
- the species to be regenerated develops a tap root, which is to be maintained (the development of tap roots is distorted or impossible in seedlings that have been lifted from nurseries);
- the initial competition from understory vegetation is not so strong as to impede the establishment of seedlings from seed;
- sowing is a much cheaper option than planting, considering also the follow-up costs for tending.

In addition, there are situations where combinations of natural and artificial regeneration are suitable. For example, low-density planting of widely spaced individuals or clusters of seedlings in a matrix of natural regeneration has been employed successfully as a cheap alternative to reforest large areas following wind throw or other disturbances (e.g. Saha *et al.* 2013). This ensures both the establishment of a minimum number of trees from the desired species as well as the maintenance of natural successional processes.

In uneven-aged silvicultural systems (e.g. single tree or group selection systems), the regeneration phase commonly overlaps with the stand tending phase and with a long harvesting phase, which may extend to several decades. Also, in even-aged shelterwood systems, the regeneration can overlap with an extended harvesting phase, in particular for shade-tolerant species. This so-called advance regeneration most commonly consists of natural regeneration, but may also be artificial for the reasons outlined above. Advance regeneration is a common approach for the establishment of shade-tolerant species, in particular those that need the protection of the canopy, for example against frost. This advance regeneration is also an insurance against disturbances in the later stages of stand development.

Tending

Tending of forest stands includes all silvicultural activities after the successful establishment of the regeneration and the final harvest of mature trees. The most important tending practices are thinning and pruning (i.e. mechanical removal of branches in order to increase branch free bole length and thereby timber quality) and, to some extent, fertilization. The latter is typically applied in intensively managed forests, such as fast growing plantations, where the replenishment of nutrients that are being removed with biomass is essential to maintain site productivity. In most other forests, it is of little relevance and therefore shall not be discussed further here.

Depending on management goals and the tree species present, thinning and pruning may also be achieved through natural processes. Self-thinning is the result of competition-induced mortality that occurs when stand density approaches the site carrying capacity for biomass (Westoby 1984). In this process, larger and more dominant trees expand their growing space at the expense of smaller, suppressed or intermediate-sized trees. In even-aged stands, natural mortality occurs at very high rates when stands and trees are in their most dynamic phase of growth since the concomitant crown expansion is the driving force of this process. Self-thinning is highest following canopy closure in the subsequent pole-size and early-mature growth stages (Oliver and Larson 1990). For a number of reasons, forest managers rarely rely on self-thinning, because stand management goals are more easily, and to a higher degree, accomplished through active thinning. Thinning is carried out for the following reasons:

- To concentrate site productivity on desired stems and species. Without thinning, trees that may eventually die, trees of poor shape, or trees of commercially undesirable species take up growing space and consume site resources such as water and nutrients that can be directed towards vigorous, high-quality trees of desired species. Thus thinning can dramatically increase the economic result of forest management. However, thinning operations, in particular when carried out with heavy machinery and in difficult terrain, can also damage the retained future crop trees.
- To regulate tree species composition. In this early dynamic phase of tree growth, differences in the competitiveness of species, owing to their height growth rates, shade-tolerance or other attributes, can lead quickly to the loss of inferior species from the stand. Thinning can increase the growing space of individuals of less competitive species and thus help to maintain tree diversity.
- To reduce time to reach desired target dimensions of trees. Increasing the growing space of trees accelerates their crown expansion and consequently their stem diameter growth.
- To increase tree health and vigour and thus reduce susceptibility to pests and pathogens. When trees have better access to site resources and can fix more carbon, they are less stressed and have more resources such as stored starch to respond to stress or disturbance. Consequently, well-thinned stands are less susceptible to secondary pests and pathogens that attack weakened or stressed individuals (e.g. Neumann and Minko 1981).

- To increase individual tree stability. The physical stability of trees depends, among other things, on the extent and depth of their root system, the length of their crown, and the thickness of the stem. These tree attributes are promoted by thinning in young stands (Cameron 2002). However, in old stands, when trees are taller and have lost much morphological plasticity, thinning destabilizes stands, at least for a period of several years after thinning.
- To utilize trees that would otherwise die and thus to provide an income to the forest owner. Alternatively, trees that die as a result of the self-thinning process would enter the pool of dead wood, which is important for the maintenance of forest biodiversity.
- To provide stand access. Young, un-thinned stands can be impenetrable and are usually not accessible to vehicles. Thinning, which is typically accompanied by the establishment of an extraction system with skidding tracks, opens up stands for all subsequent operations.
- And, in some cases, to increase water yield from the forest or to promote understory. In young un-thinned stands, the competition for light, water, and nutrients may be so intense that understory vegetation and groundwater recharge is virtually eliminated.

A thinning regime for forest stands can be described by the type of thinning, its frequency, intensity, and commencement. Thinning from above, or crown thinning, concentrates on the removal of the strongest competitors to future crop trees. In contrast, thinning from below starts at the other end of the diameter range with the removal of suppressed and intermediate trees (West 2006).

In highly uneven-aged forests, where we do find large mature trees intermingled with seedlings (young trees with a height below 1.3 m), saplings (height > 1.3 m, or when the crown has started to lift), pole-sized and early-mature trees, it becomes difficult to distinguish thinning from harvesting and initiation or promoting of regeneration.

Branches in the lower part of stems are the most important wood defect in relation to the value of timber, where dead branches are worse than live branches owing to their negative visible and technical properties, and the negative effect on quality increases with branch dimensions. Owing to their crown architecture (i.e. dimensions and insertion angles of branches) and decay resistance of wood, tree species differ in their ability to shed branches (self-pruning). Generally, self-pruning is more effective in hardwood than in softwood species. In tree species and stand conditions in which branches on the lower, most valuable part of the stem are not naturally shed after death, pruning may be a suitable or necessary practice for the production of high-quality timber (Montagu *et al.* 2003). In this case, pruning is typically concentrated on those future crop trees, which are being promoted through thinning.

Harvesting and site preparation

Harvesting commercially mature trees is not only the most important source of income for forest owners, it is also the most critical phase of stand management with potentially the strongest environmental impacts. Therefore, certain harvesting approaches like clear cutting have been strongly criticized in the past, and much research has taken place in recent decades to develop harvesting systems in which the impact on biodiversity, soils, water quality, visual aspects, etc. is reduced (Gustafsson *et al.* 2012). In contrast to thinning, the harvesting of mature trees also follows the objective to regenerate the forest stand or patches within. In general terms, harvesting systems, traditionally called regeneration methods, can be distinguished by the amount of canopy removal, the size of canopy openings, the production cycle length and consequently the residual stand structure following harvesting. These three attributes of harvesting systems match elements of ecological disturbance regimes: intensity, disturbance area, and frequency. Traditionally, harvesting systems have been grouped into those that regenerate a new tree cohort through coppicing and those that regenerate them from seed. The latter are high forest systems that are further distinguished into those that create or maintain uneven-aged or even-aged stand structures. Uneven-aged regeneration methods comprise single tree and group selection systems. Even-aged regeneration methods comprise shelterwood, seed tree, and clear cutting systems (Table 6.2 and Figure 6.4).

Single tree and group selection systems are typically applied in temperate broadleaved forests, mixed-species mountain forests, lowland tropical rainforests, dry coniferous forests, and dry eucalypt forests. These systems have in common that the economically important tree species typically do not require intensive disturbances for their regeneration. Often, these forests are characterized by high structural and compositional heterogeneity and hence there is always only a small proportion of trees that meet the current economic targets regarding dimension and wood properties (Figure 6.4). Under these conditions, seedlings of the desired, moderately to very shade-tolerant species can be regenerated beneath the existing canopy and in advance of the removal of canopy trees. Some scarification of the forest floor or surface soil may be required to facilitate establishment of tree seedlings from natural seed fall.

Table 6.2 Overview of traditional regeneration methods in coppice and high forests and their advantages and disadvantages

Regeneration method / Description	Advantages	Disadvantages
Single tree and group selection (Figure 6.4) Regeneration and maintenance of uneven-aged stand structures through removal of some trees, individually or in groups, in all size classes. Groups no larger in diameter than twice the stand height (Helms 1988).	• harvesting is focused on valuable and large trees, no waste of immature growing stock • low visual impact and contrast, high social acceptance • maintenance of most of the forest structure and thus continuity of habitat conditions, nutrient and hydrological cycles • enables natural regeneration and incorporation of advance growth	• only suitable for regeneration of shade-tolerant tree species • requires a dense road and extraction network and highly skilled forest workers • reduced efficiency of harvesting and logistics • frequent return intervals • physical damage to the retained stand and established regeneration can be high
Shelterwood methods (Figure 6.4) Cutting of a large proportion of commercially mature trees, leaving enough to provide seeds and a desired degree of shade or frost protection for the regeneration. The number, intensity, and frequency of cuts during shelterwood regeneration differ greatly. Cuts can be uniform, in groups, or strips. Reserve trees may be left to establish a two-aged forest stand (Helms 1988).	• harvesting can be focused initially on valuable and large trees • stem increment of retained canopy trees may be accelerated and stems can be harvested at optimal diameter • initially reduced visual impact and high social acceptance • high flexibility to regenerate tree species with different ecological/ physiological niches • maintenance of forest structure and thus continuity of habitat conditions, nutrient and hydrological cycles as long as mature trees remain • enables natural regeneration and incorporation of advance growth	• dense shelter not suitable for regeneration of shade-intolerant tree species • physical damage through harvesting or wind throw to the retained stand and established regeneration can be high • requires a dense road and extraction network and highly skilled forest workers • frequent return of harvesting to gradually reduce canopy

Regeneration method / Description	Advantages	Disadvantages
Clear cutting The removal of essentially all trees of all dimensions in a single harvesting operation to create a fully exposed microclimate and regenerate a new even-aged cohort of trees (Helms 1988).	• very efficient harvesting of trees and biomass • relatively low planning requirements • requires slash removal operations and permits intensive, follow-up site preparation (e.g. burning, ripping) and mechanized planting • disturbance effect concentrated in space and time	• high visual impact, depending on size and location • complete loss of forest structure and microclimate • strong disruption of many ecosystem processes and functions, e.g. water and nutrient cycles, large nutrient losses possible • large quantities of harvesting slash may need to be removed for subsequent operations • maximum development of competing vegetation may impede regeneration • requires in most cases artificial regeneration (no seed sources) • likely strong effects on adjacent landscape patches (escape fires from slash burns, wind throw)
Coppicing systems (also low forest) The removal of all trees in one harvesting operation and regeneration of the next cohort from stump sprouts or root suckers.	• the same advantages as for clear cutting except for site preparation, since the stumps need to be protected • no costs for seeds or seedlings (except for some filling in, where stumps have died) • rapid re-establishment of forest through fast growing coppice shoots • highly suitable for species that require frequent disturbance to persist in forest landscapes	• the same disadvantages as for clear cutting with regard to visual impact, loss of structure, disruption of ecosystem functioning; however, harvesting slash is usually less owing to smaller dimensions of harvested trees • no opportunity for genetic improvement through artificial regeneration • only suitable for species that coppice and for producing small dimension timber • problems of physical instability of stems at junction with stumps and stem rot spreading from stumps • commonly high export of biomass and nutrients (owing to small tree dimensions and high level of utilization) • high disturbance frequency is unsuitable for species requiring stable forest conditions

Table 6.2 continued

Regeneration method/ Description	Advantages	Disadvantages
Coppice with standards (or reserves) (Figure 6.4) The same as coppicing but standards (reserve trees) are retained to attain goals other than regeneration (Helms 1988). The method creates two-layered stands, similar in structure to shelterwoods. It can be regarded as an intermediate system between coppice (low) and high forests.	• the same advantages as for coppicing, which apply, however, only to the coppice layer • retention of standards maintains some structure, often also valuable habitat, and higher level of ecosystem functioning • permits production of large dimension and high-quality timber in standards, when compared to coppicing • facilitates silvo-pastoral uses, e.g. where standards produce acorns for pigs	• the same disadvantages as in coppice systems for the regeneration (coppice) layer • the stem quality of reserve trees may suffer from sudden exposure (i.e. epicormic branches) following removal of the coppice layer

Shelterwood methods are commonly used and therefore of high importance. The most important characteristic of these methods is the establishment of a new tree generation before completion of the preceding rotation (Figure 6.4). They are especially applied where seedlings and saplings either need protection from climatic extremes such as severe frosts or high insolation, or where some shade is needed to control the competition between other vegetation (herbs, grasses, and shrubs) and young trees. Classical examples for the application of this system comprise temperate broadleaved forests such as those with European beech, boreal and mountain forests, dry coniferous forests, and Patagonian *Nothofagus* forests, to name a few. The canopy shelter may be applied uniformly or in groups and strips. In cases where the sole purpose of the canopy shelter is to secure regeneration establishment, it may be removed in a single cut upon successful completion of the regeneration phase. In cases where canopy trees are also retained to grow into larger dimensions that increase the value of timber, the two-tiered stand phase may last several decades during which target diameter harvesting continues. When the sole purpose of retained mature trees is the provision of seeds for the regeneration, this is called a seed tree system. The latter is more common in forests with light demanding tree species with wind-dispersed seeds. The level of retained basal area is considerably lower than in shelterwood systems.

Clear cutting is the most common regeneration method globally and is applied to all types of forests. The complete removal of all trees provides for very efficient harvesting and has, owing to the immediate loss of most above ground structure, a strong impact on ecosystem functioning. Clearfelling has been likened to catastrophic ecosystem disturbances such as intensive wildfires and has therefore been used preferentially in those types of forests, where economically important tree species depend on or thrive in these conditions of full light and soil disturbance. Typical examples are moist conifer, pine and eucalypt forests as well as boreal forests. However, harvesting disturbances differ from natural disturbances in a number of ways and not just in clearfelling systems (e.g. Lindenmayer and McCarthy 2002). In the case of natural disturbances, little biomass is removed, except in the case of severe fires, soils are not disturbed, except when trees are uprooted, soils are not compacted, and the disturbance leaves highly irregular structures behind including large quantities of dead wood, living trees including advance regeneration and propagules.

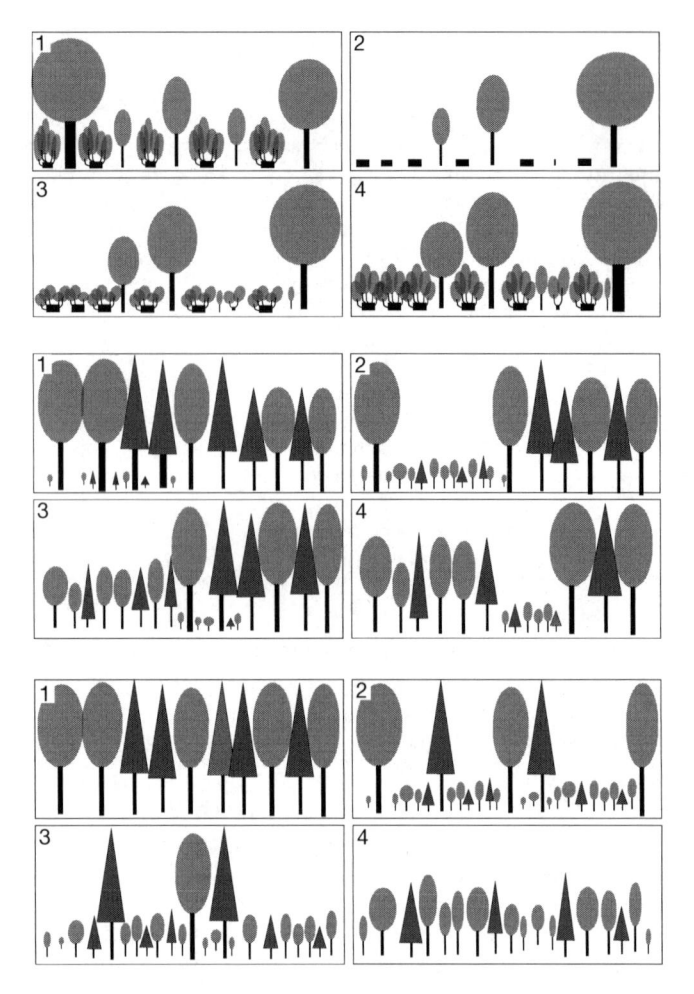

Figure 6.4 Chronosequences of stand structures to depict the dynamic development (from plate 1 to 4) in selected silvicultural systems

Top: Coppice with standards. The coppice layer is managed on short rotations and regenerated through re-sprouting of stumps. The standards are retained over several coppice regenerations to produce large dimension timber and possibly acorns or other fruits or nuts. At each harvest, new standards are recruited from natural regeneration or planting. The coppice system would be similar, but without the large standards.

Centre: Group selection system. Groups of economically mature trees are removed to initiate regeneration in gaps or to release groups of already established advance regeneration. Subsequently, these groups are enlarged and new groups of mature trees harvested in different patches. This creates a permanently uneven-aged stand structure. Group selection systems are suitable to shade-tolerant and moderately shade-tolerant tree species. In single tree selection systems, gaps would be substantially smaller, the mixture of different tree dimensions more intimate and the structure less variable over time.

Bottom: Shelterwood System. In relatively even-aged stands the canopy is opened, often following a mast year of the main tree species, to establish the regeneration. Following successful establishment, increasing light demand of the regeneration is met through successive removal of overstory trees until all mature trees are removed. However, some mature individuals may be left as habitat trees. The removal of trees ideally follows their dimensional and economic development. Thus shelterwood systems may be combined with target-diameter harvesting, where trees that were initially the smallest are given time to develop into valuable dimensions

Since it has been found that post-harvesting residual structures and organisms are very important for the continuity of habitat and ecosystem functioning, silvicultural systems have been modified in many regions of the world (Gustafsson et al. 2012). These modifications are characterized by the deliberate retention of structural elements such as life habitat trees, lying and standing dead wood, or islands of undisturbed vegetation. Particular attention is often paid to minimize harvesting-related disturbances to vegetation along water courses or around rocky outcrops. One important aspect of this approach to planning, managing, and harvesting forests is the shift in focus to the type and quantity of forest structures that are left behind rather than the conventional focus on what is being harvested (Franklin et al. 1997). Often the distribution pattern of residual trees and vegetation patches is not uniform and dispersed but patchy and aggregated or a combination of dispersed and aggregated. Therefore, this approach is often called 'variable retention'. The retention of these structural elements can be combined with any silvicultural system and regeneration method discussed above.

In addition to the type and frequency of harvesting, the intensity of silvicultural systems depends to a large degree on silvicultural activities that are being carried out between harvesting and the establishment of regeneration. The relevance of this inter-rotational management, which involves different site preparation practices, is mostly restricted to silvicultural systems that create or maintain even-aged forests. In uneven-aged forests, where mature trees with their extensive root systems remain on site and need to be protected, there are few opportunities for site preparation, except for the removal of harvesting slash, for example through burning. In addition, when only a relatively small proportion of trees is being removed, there may be little need to remove the harvesting slash to facilitate subsequent access and operations. However, in even-aged systems these needs and opportunities for intensive site preparation arise. These may include the harvesting or removal of slash material, scarification of soil surfaces to facilitate germination of seeds, the ripping of soils to break up soil layers that impede root growth, the ploughing of soils to reduce weed competition and redistribute nutrients, the mounding of soils to create favourable micro-sites for seedling growth, the extraction of tree stumps to avoid problems with pest species that breed in them (e.g. large pine weevil *Hylobius abietis*) or to use them for energy, and the application of fertilizers and herbicides (cf. Nyland 2002). Intensive site preparation practices may represent also substantial economic investments and are therefore typically restricted to intensively managed plantation forests, where tree species are used that can respond vigorously to these treatments and where these investments can be recouped after relatively short periods. This inter-rotational management can have a large impact on soils, carbon storage and water and nutrient cycles. Prominent examples of non-sustainable forest management in the past have been related to mistakes that have been made in this phase of management. See for example the growth decline in second rotation *Pinus radiata* plantations in South Australia that was subsequently reverted through improved soil organic matter management between tree rotations (Powers 1998).

Global development of forest types and management systems

In broad terms, forests may be classified into three categories: natural, semi-natural, and plantation forests. These different types of forests differ greatly in their structural complexity and compositional diversity, silvicultural systems and management intensity, and their spatial extent.

Natural, primary forests are those where industrial harvesting of wood and silvicultural activity does not occur. Once harvesting has occurred in these forests, they are classified as secondary forests or 'naturally regenerated forest' (FAO 2015). More than one-third of the global forest area (36 per cent) is, according to the last census (FAO 2010), still primary forest, that consists of native species and where ecological processes have not been substantially disturbed. This area decreased from 2000 to 2010 by 40 million ha, which equates to an annual loss of 0.4 per cent.

The area of forest designated for conservation of biodiversity comprises 11.5 per cent of the global forest area and this increased by 6.3 million ha per annum between 2000 and 2010. Also, planted forests, which cover more than 264 million ha, representing 6.6 per cent of the world's forest area, increased over the first decade of this century at an average rate of close to 5 million ha per annum (FAO 2010). At the same time, there has been a declining trend in area designated primarily for production functions of forests. The higher management intensity of these planted forests is reflected in the large share of 25 per cent of introduced tree species that are used in these systems (FAO 2010).

Given that the net loss in global forest area has been continuing at a rate of more than 5 million ha yr^{-1}, that the proportion of the remaining forest area available for multiple functions or the production of timber is also shrinking, and that the demand for timber and wood products will continue to increase in the next decades (Carle and Holmgren 2009), it is clear that some intensification of forest management needs to occur on the remaining area, unless wood is substituted by other products. The increasing demand is driven by an increasing population, a growing per capita consumption of an increasingly urbanized population, and the political will to replace fossil fuels and energy intensive products such as concrete and steel with renewable materials. However, the potential to meet an increasing demand through intensification of forest management is enormous. The total global removals of industrial round wood and of fuel wood from forests in 2005, which amounted to 3.36 billion m^3, could have been met by an area of intensively managed forests of less than 224 million ha with an average productivity of 15 m^3 ha^{-1} a^{-1}. This is less than the current area of planted forests and this level of average productivity can be achieved by many types of plantations (Cossalter and Pye-Smith 2003) and also some semi-natural forests. This simple calculation also points to the enormous potential of intensively managed forests to reduce harvesting pressure on other native or semi-natural forests and thus to contribute to biodiversity conservation (Bauhus et al. 2010). Thus when considering sustainability of managed forests, it may not be helpful to compare different forest management and silvicultural systems that vary greatly in their intensity at the scale of stands or forest management units. More important is the question of how forest reserves, semi-natural and multifunctional forests, and intensively managed, planted forests can be combined at the landscape scale to meet society's demand for all forest products and services. It has been recognized that sustainable forest landscapes rely on a combination of these three different approaches to forest management at the landscape scale. The landscape-level combination of these approaches, which provides different forest functions in different intensities in different parts of the landscape, has been formulated and trialled in the so-called TRIAD concept (Messier et al. 2009). On the one side, the TRIAD approach offers adequate space for forest reserves to provide biodiversity conservation functions, which cannot be fulfilled in actively managed forests. On the other side, plantations and intensively managed semi-natural forests offer opportunities to increase the area in forest reserves and to extend forest management in other natural and semi-natural forests, for example through retention forestry approaches (Gustafsson et al. 2012). Extensively managed, multifunctional forests can then form the landscape matrix that connects reserves to facilitate the exchange among populations and the movement of animals. The latter type of managed forests, represent a functional integration, whereas the dedication of forest areas to production or conservation functions represents a functional separation. However, the resulting forests are not mono-functional since both reserves as well as plantations still provide many other ecosystem goods and services (e.g. Bauhus *et al.* 2010).

References

Bauhus, J., Pokorny, B., van der Meer, P., Kanowski, P. and Kanninen, M. (2010) 'Ecosystem good and services – the key for sustainable plantations' in Bauhus, J., van der Meer, P. and Kanninen, M. (eds) *Ecosystem Goods and Services from Plantation Forests*. Earthscan, London, pp. 205–227

Burschel, P. and Huss, J. (2003) *Grundriss des Waldbaus*, Third edn. Ulmer Verlag, Stuttgart, 487 pp.

Cameron, A. D. (2002) 'Importance of early selective thinning in the development of long-term stand stability and improved log quality: a review', *Forestry*, vol. 75, no. 1, pp. 25–35

Carle, J. B. and Holmgren, L. P. B. (2009) 'Wood from planted forests: global outlook to 2030', in Evans, J. (ed.) *Planted Forests: Uses, Impacts and Sustainability*, CAB International and FAO, pp. 47–59

Cossalter, C. and Pye-Smith, C. (2003) *Fast-Wood Forestry: Myths and Realities*. Center for International Forestry Research, Bogor Barat, Indonesia

FAO, Food and Agricultural Organization of the United Nations (2010) *Global Forest Resources Assessment 2010: Main Report*, FAO Forestry Paper 163, FAO, Rome

FAO (2015) *FRA 2015: Terms and Definitions*. Forest Resources Assessment Working Paper 180, 31 pp.

Franklin, J. F. and Forman, R. T. T. (1987) 'Creating landscape patterns by forest cutting: ecological consequences and principles', *Landscape Ecology*, vol. 1, no. 1, pp. 5–18

Franklin, J. F., Berg, D. R., Thornburgh, D. A., Tappeiner, J. C. (1997) 'Alternative silvicultural approaches to timber harvesting: variable retention systems', in Kohm, K. A. and Franklin, J. F. (eds) *Creating a Forestry for the 21st Century: The Science of Ecosytem Management*, Island Press, pp. 111–139

Gustafsson, L., Baker, S. C., Bauhus, J., Beese, W. J., Brodie, A. and others (2012) 'Retention forestry to maintain multifunctional forests: a world perspective', *Bioscience*, vol. 62, no. 7, pp. 633–645

Helms, J. A. (1988) *The Dictionary of Forestry*. Society of American Foresters, Bethesda, US

Hundeshagen, J. C. (1826) *Die Forstabschätzung auf neuen wissenschaftlichen Grundlagen*, Laupp, Tübingen, Germany, 428 pp.

Kühne, C., Kublin, E., Pyttel, P., Bauhus, J. (2013) 'Growth and form of *Quercus robur* and *Fraxinus excelsior* respond distinctly different to initial growing space: results from 24-year-old Nelder experiments', *Journal of Forestry Research*, vol. 24, no. 1, pp. 1–14

Larsen, J. B. and Nielsen, A. B. (2007) 'Nature-based forest management – Where are we going? Elaborating forest development types in and with practice', *Forest Ecology and Management*, no. 238, pp. 107–117

Lindenmayer, D. and McCarthy, M. A. (2002) 'Congruence between natural and human forest disturbance: a case study from Australian montane ash forests', *Forest Ecology and Management*, no. 155, pp. 319–335

MCPFE, Ministerial Council on the Protection of Forests in Europe (2002) 'Improved Pan-European Indicators for sustainable forest management', Liaison Unit Vienna, 5 pp., http://www.foresteurope.org/docs/ELM/2002/Vienna_Improved_Indicators.pdf, accessed 28 February 2014

Messier, C., Tittler, R., Kneeshaw, D.D., Gelinas, N., Paquette, A. and others (2009) 'TRIAD zoning in Quebec: Experiences and results after 5 years', *Forestry Chronicle*, no. 85, pp. 885–896

Montagu, K. D., Kearney, D. E., and Smith, R. G. B. (2003) 'The biology and silviculture of pruning planted eucalypts for clear wood production – a review', *Forest Ecology and Management*, no. 179, pp. 1–13

Neumann, F. G., and Minko, G. (1981) 'The sirex wood wasp in Australian radiata pine plantations', *Australian Forestry*, vol. 44, no. 1, pp. 46–63

Noss, R. F. (1990) 'Indicators for monitoring biodiversity: a hierarchical approach', *Conservation Biology* no. 4, pp. 355–364

Nyland, R. D. (2002) *Silviculture: Concepts and Applications'*, Second edn. McGraw-Hill, New York

Oliver, C. D. and Larson, B. C. (1990) *Forest Stand Dynamics*, McGraw-Hill, New York

Powers, R. F. (1998) 'On the sustainable productivity of planted forests', *New Forests*, no. 17, pp. 263–306

Puettmann, K. J., Coates, K. D., and Messier, C. (2009) *A Critique of Silviculture: Managing for Complexity*, Island Press, Washington, 189 pp.

Saha, S., Kuehne, C., and Bauhus, J. (2013) 'Tree species richness and stand productivity in low-density cluster plantings with oaks (*Quercus robur* L. and *Q. petraea* (Mattuschka) Liebl.)', *Forests*, vol. 4, no. 3, pp. 650–665

Sands, R. (2005) *Forestry in a Global Context'* CABI Publishing, Wallingford, UK
von Carlowitz, H. C. (1713) *Sylvicultura Oeconomica – Hausswirthliche Nachricht und Naturmäßige Anweisung zur Wilden Baum-Zucht.* Braun, Leipzig, Germany 456 pp.
West, P. W. (2006) *Growing Plantation Forests.* Springer, Berlin, Heidelberg
Westoby, M. (1984) 'The self-thinning rule'. *Advances in ecological research*, no. 14, pp. 167–226

PART II

Forest dynamics

7
INSECT DISTURBANCES IN FOREST ECOSYSTEMS

Daniel Kneeshaw, Brian R. Sturtevant, Barry Cooke, Timothy Work, Deepa Pureswaran, Louis DeGrandpre and David A. MacLean

Tree-feeding insects are ubiquitous in forest ecosystems. While relatively few species cause widespread mortality, insect outbreaks are important ecological disturbances that, in some cases, can have devastating economic effects. Insect outbreaks have been considered in the context of forest disturbances for decades (Mattson and Addy 1975). However, the interest of forest managers and policy makers often waxes and wanes in cycles that follow the population dynamics of the insects themselves, despite the fact that the area affected by insect outbreaks in Canada, the United States and Europe is greater than that disturbed by fire or harvesting (FAO 2014; Kneeshaw *et al.* 2011) (Figure 7.1).

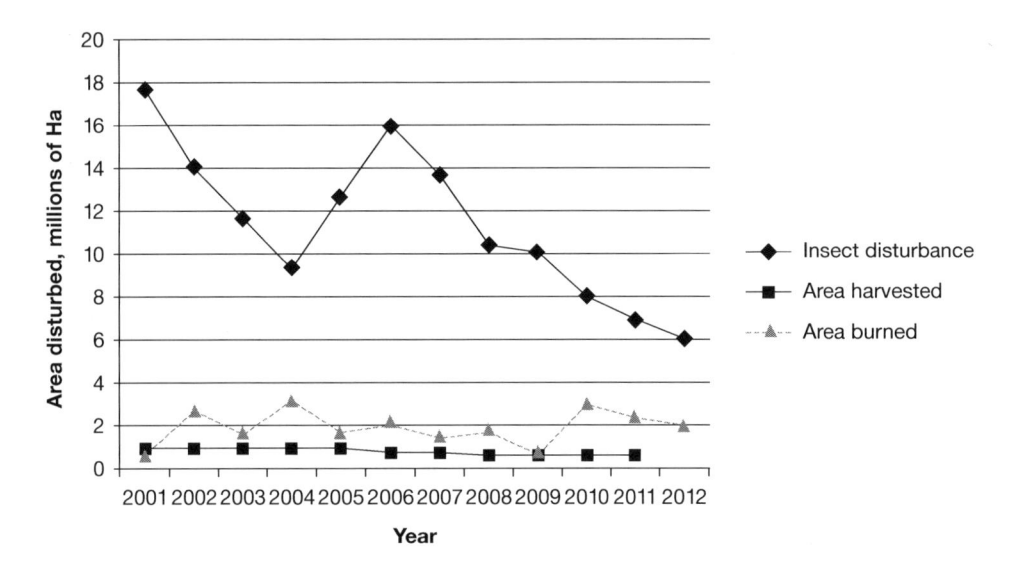

Figure 7.1 Area disturbed by insects (mountain pine beetle, forest tent caterpillar and spruce budworm combined), by fire and by harvesting from 2001 to 2012 in Canada. Data from NRCanada (http://www.nrcan.gc.ca/forests/canada/sustainable-forest-management/criteria-indicators/13259)

Species with extreme population fluctuations that vary over several orders of magnitude are capable of disrupting forest ecosystem functions (Wallner 1987) and generally fall into two main taxonomic groups: aggressive tree-killing scolytid bark beetles that feed within the inner bark and exhibit eruptive population dynamics (Raffa *et al.* 1993); and Lepidopteran and Hymenopteran species whose larvae defoliate tree leaves and needles and whose populations follow high-amplitude oscillations (Cooke *et al.* 2007). The spatial extent of outbreaks (i.e., spatial synchrony) combined with the degree of damage caused by the outbreak (i.e., severity) determines the degree to which humans define insects as "pests" (Liebhold *et al.* 2012). Depending on jurisdictional policies, large-scale tree mortality can temporarily flood regional and even continental wood markets as forest industries attempt to salvage trees before they deteriorate and lose their value (e.g., Walton 2012). Long-term legacies of severe insect outbreaks may have even greater economic impacts as extensive mortality leads to gaps in forest age–class structure and reductions in annual allowable cuts (Abbott *et al.* 2009).

There is also evidence that outbreaks in North America and Europe: 1) have increased in severity and duration; 2) have become more synchronous; 3) have expanded their geographic range; and 4) are affecting new host species as outbreak ranges expand due to changes in climate and forest structure (see sections below). For example, outbreaks of bark beetles (e.g., *Dendroctonus* spp., *Ips* spp. and *Scolytus* spp.) in western North America and Eurasia have or are predicted to reach historically unprecedented scales due to interactions between weather and changes in forest structure (Raffa *et al.* 2008; Jonsson *et al.* 2009). Outbreaks of conifer-feeding budworms (*Choristoneura* spp.) may have increased in extent, length and intensity over the past century compared to preceding centuries (Blais 1983; Swetnam and Lynch 1993). Latitudinal shifts in outbreak behaviour have been observed in Europe for species such as the pine processionary moth (*Thaumetopoea pityocampa*), an important pest of pine forests (Battisti *et al.* 2006; Buffo *et al.* 2007) and the great spruce bark beetle (*Dendroctonus micans*) (Gilbert *et al.* 2003). Population cycles of the larch budmoth (*Zeiraphera diniana* Gn.) cycled approximately every 9.3 years for almost 1,200 years before abruptly ending in the 1980s (Esper *et al.* 2007). Outbreak range shift in the spruce budworm has also been linked to climate change (Régnière *et al.* 2012).

Differences among biomes

Insect disturbances differ from other natural disturbances such as wind or fire in that dietary and other biological constraints limit the taxonomic range of tree species they can damage. Monophagous forest insects are restricted to a single tree species or genus by their specific adaptations to particular secondary metabolites, polyphagous species are more tolerant of a wider range of secondary metabolites permitting feeding on a wider (but not unlimited) taxonomic range, and oligophagous species fall between these two extremes (Jactel and Brockerhoff 2007).

Tropical forests, with high tree diversity and reduced concentration of individual tree species, are thought to inhibit widespread insect outbreaks (Janzen 1970). While many tropical insects are polyphagous (Novotny *et al.* 2002), insect outbreaks causing widespread damage and mortality in tropical regions are most commonly observed in stands dominated by one or a few tree species. Extensive mortality due to insect outbreak is not commonly observed in unmanaged forests, but more common within simple plantations consisting of few or a single tree species. For example, outbreaks of the coffee berry borer, *Hypothenemus hampei* (Ferrari), in coffee plantations are well known (Damon 2000). Indeed, the ecological dogma that "there are no outbreaks in tropical forests" (Elton 1958) has been challenged (Wolda 1987; Dyer *et al.* 2012;

Nair 2012). Research shows that the population cycles of herbivorous insects in natural tropical forests vary from low to high population densities (Nair 2012) even if widespread mortality is not observed. Dyer *et al.* (2012) summarized examples of insect damage within natural tropical forests, but acknowledged challenges limiting our understanding of tropical outbreaks. These include a general lack of understanding of tropical insect natural history, difficulties in monitoring outbreak activity (for example, monitoring insect populations with multiple generations per year), and the absence of annual tree rings for traditional historical outbreak reconstructions using tree-ring analysis.

On a global scale North American forests experience some of the largest and most severe insect outbreaks (Shorohova *et al.* 2012), and these outbreaks tend to occur in boreal and montane regions dominated by natural monocultures such as spruces, firs, pines and poplar species (Kneeshaw *et al.* 2011). Two extreme examples for North America are the mountain pine beetle, *Dendroctonus ponderosae* (Hopkins) and spruce budworm, *Choristoneura fumiferana* (Clemens). Each species has destroyed, within a single decade, millions of hectares of cordilleran and boreal forest of relatively low tree-species diversity. Such extreme examples have not been observed in comparable forests of Eurasia. For example, in 2005, 17.3 million hectares (Mha) were affected by forest insects in Canada and 5.6 Mha in the USA whereas elsewhere in the world insects affected 3.2 Mha in China, 1.7 Mha in Russia, and 1.3 Mha in Romania (FAO 2014). Although it is tempting to speculate that higher levels of insect depredation in high-latitude forests are causally associated with lower tree-species diversity, the Eurasian boreal serves as a cautionary example of the risk of over-generalization. Tree-species diversity is not the only variable that tends to be correlated with latitude and elevation. Other factors to consider include faunal diversity, especially natural enemy diversity, and climate. It is an open question why twentieth century post-glacial boreal and cordilleran forests in the eastern and western hemisphere seem to differ so greatly in the extent and intensity of insect disturbance.

Forest–insect pest patterns and processes – from the regulation of population fluctuations to the rise and fall of outbreaks

Forest pests are most often found in endemic or low population densities. Outbreaks occur infrequently as populations transition synchronously from endemic to epidemic levels across large spatial extents, from stands of trees to entire regions. According to Berryman (1987), outbreak species of forest insects fall into six classes based on the pattern of outbreak development in space and time. These six classes are defined by two axes relating to the fine-scale pattern of occurrence (either "eruptive" or "gradient") and the larger-scale pattern of recurrence (either "pulse", or "periodic", or "sustained"), in both time and space (Figure 7.2). "Eruptions" are defined as populations that occur abruptly in space and time, with sharp, clear transitions between endemic and epidemic states. In contrast, "gradients" refer to populations that vary more continuously in density through time and space; fluctuations are not abrupt and there are no clear transitions between discrete endemic and epidemic states. Having described these basic patterns of occurrence, Berryman described the patterns of recurrence in time and space as either "pulse", "cyclical" or "sustained", based on the regularity and duration of episodes (or patches) of high abundance (1987).

Thirty years later we find this pattern classification scheme still has value, although range-wide analysis of many decades of new data from many pest species leads us to observe that the spatial and temporal scale of analysis, and the type of data used in that analysis (e.g., point samples of insect population density versus polygons of defoliation in discrete damage classes) will influence one's perceptions and conclusions. In short, scale matters (Figure 7.3a). Moreover,

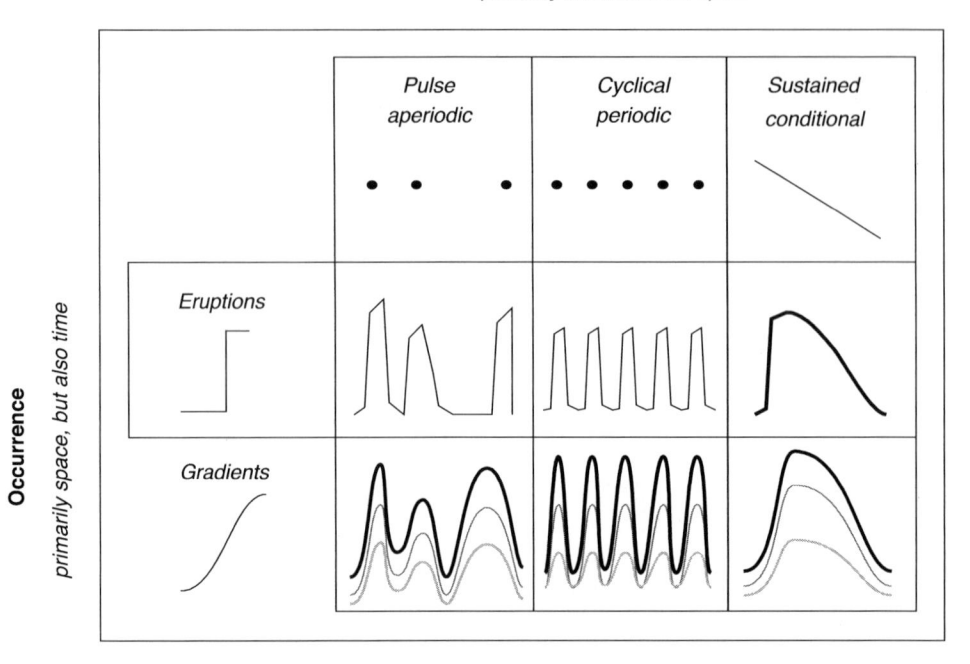

Figure 7.2 A schematic representation of different outbreak patterns in time. In the upper panel, occurrence represents the two main occurring outbreak patterns (gradient vs eruptive) which are then further broken down by their recurrence into pulsed, cyclical or sustained patterns. In the lower panel we show how all of these patterns can be observed during an outbreak in time and space for a given outbreaking insect

the ability to infer deterministic patterns of dynamics in space and time depends to what degree those dynamics are subject to random variations in ecological conditions (Figure 7.3b). The more stochastic the population process, the longer the system needs to be observed under a wider range of conditions in order to apply Berryman's classification scheme (1987) to a given data set (Figure 7.3).

Although different forest–insect pest systems exhibit different outbreak patterns, there are some universal principles at play that serve as a common basis to understand the wide array of behaviour. All outbreaking insects are influenced by changes in forest conditions, natural enemies and climatic conditions (dF/dt, dP/dt and dC/dt; Figure 7.4). The understanding of the influence of these factors on rates of herbivore population change (dH/dt) has evolved to our current understanding of reciprocal feedbacks between natural enemies, pest herbivores and forests, where all three components are simultaneously influenced by climate. Our ability to understand and predict the outcomes of these interactions is further complicated by nonlinear population processes and feedbacks that interact at different spatio-temporal scales (Holling 1973). For example, insect population turnover operates on timescales ranging from weeks to years and is therefore fast relative to tree biomass accumulation. Generally, complex dynamics are the result of the fast processes layered on top of the slow processes (Holling 2001). The same scale-crossing phenomena can also occur in space (Peters *et al.* 2004), for example, if the spread of outbreaks is enabled by landscape abundance of host species (Raffa *et al.* 2008). Cross-scale

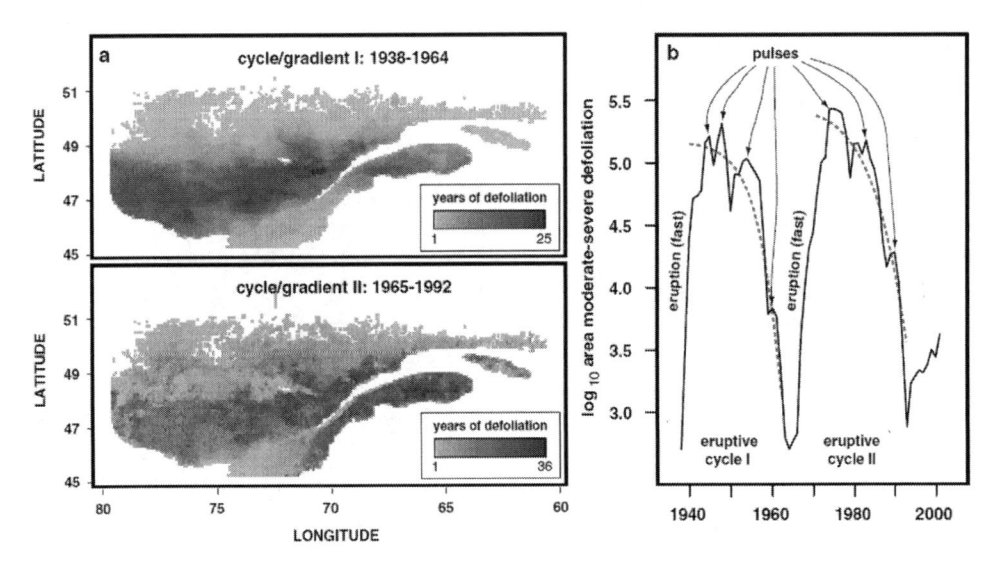

Figure 7.3 (a) Spatially, outbreak intensity is distributed according to a latitudinal belt-shaped gradient pattern, shown for spruce budworm defoliation in Quebec and Ontario, Canada from 1938–1992. (b) Cyclic eruptions occur every few decades, with pulses of defoliation occurring every 3–7 years, which are sustained through a slow decline in host forest condition. The dashed line indicates the pattern of decay associated with slow decline in forest condition, followed by a more rapid collapse due to faster predator–prey cycling. This example illustrates how a single insect species, if it is observed long enough over a large enough spatial extent, may exhibit all five behaviours (eruptive, gradient, pulse, cyclic, sustained) conceptualized in Berryman's scheme (1987; Figure 7.2)

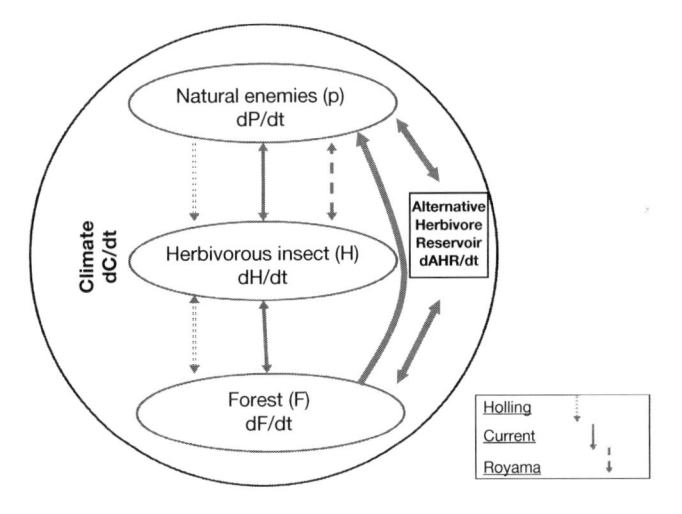

Figure 7.4 Factors influencing insect populations. Arrows indicate factors considered by different authors and the current viewpoint. The dynamic nature of the system is represented by changes in climate (dC/dt), changes in forest conditions (dF/dt), changes in natural enemies (dP/dt), changes in alternative herbivore resource (dAHR/dt) and their combined effect on outbreaking herbivorous insects (dH/dt). The current vision considers interactions between the target outbreaking insect, climate, natural enemies and the fact that these natural enemies often need alternate insect food sources which can also act as competitors with the outbreaking insect, and forest composition and structure which affects insect herbivores and thus natural enemies

interactions can include both amplifying interactions via positive feedbacks that intensify the process (such as the Allee effect, which leads to rapid population release), and attenuating interactions via negative feedbacks that diffuse the process (e.g., local plant diversity effect on herbivore population growth) (Allen 2007). Feedbacks can occur within or across spatial scales, but when they occur across scales our ability to forecast the future dynamics of the system is complicated by strong interactions among the fast and slow variables.

Despite these difficulties, there are also emergent properties of insect outbreak dynamics that are surprisingly reliable. Population dynamics of defoliators versus bark beetles behave somewhat differently due to the relative strength of host plant defences that must be overcome within each respective system. Defoliators consume a relatively renewable plant resource (foliage) while bark beetles consume an essentially non-renewable plant resource (phloem) with far more serious consequences for plant fitness on a biomass equivalent basis that are reflected by weak versus strong plant defences, respectively (Mattson *et al.* 1991). The mass-attack strategy that is the hallmark of tree-killing bark beetle systems (described below) is therefore unnecessary for the exponential growth of defoliator populations. The population dynamics of each respective system is therefore considered separately here.

Dynamics of defoliators

Defoliating larvae consume foliage and reduce the photosynthetic capacity of the tree resulting in growth loss and eventually tree mortality. Plant secondary metabolites such as terpenes and phenols function as feeding deterrents or inhibit digestion of leaf material to influence feeding rate and development of defoliator larvae and thus the level of defoliation experienced by trees, but may not have much influence during outbreak conditions (Bauce *et al.* 2006; Kumbasli *et al.* 2011). Adaptation to specific secondary metabolites and other species or genus-specific traits such as foliar phenology underlies host-specificity within defoliating species. Stand age, tree size and other characteristics of individual trees influence severity of outbreaks. For example, spruce budworm damage is greater in mature rather than immature stands (Hennigar *et al.* 2008). The closely-related jack pine budworm is dependent on pollen cones for its earliest growth stages, where severe defoliation can restrict pollen cone production the following year. Similarly, there is evidence that in some deciduous forests, trees can increase production of secondary metabolites to defend against elevated herbivory (Palo and Thomas 1984).

Among the most damaging defoliators of North American forests are spruce budworm, forest tent caterpillar (*Malacosoma disstria* Hbn.), jack pine budworm (*Choristoneura pinus pinus* Free.), gypsy moth (*Lymantria dispar* L) and hemlock looper (*Lambdina fiscellaria fiscellaria* Gn.). Gypsy moth is an introduced species from Europe where it also causes large-scale damage along with the nun moth (*Lymantria monacha* L.), Siberian silk moth (*Dendrolimus sibiricus* [Chetverikov]), larch budmoth (*Zeiraphera diniana*), autumnal moth (*Epirrita autumnata*), pine processionary moth (*Thaumetopoea pityocampa*), and others. For example, an outbreak of the nun moth in the mid-1800s defoliated spruce (*Picea abies*) forests for a decade in western Russia and eastern Prussia (modern-day Poland) (Bejer 1988). Many of these species share a common trait in that they are "early season" defoliators that time their primary feeding stage with the spring flush of new foliage that is both poorly-defended and nutrient-rich. However, as with all attempts at generalizing there are also outbreaking species with late season phenologies e.g., winter moth (*Operophtera brumata*) and autumnal moth.

Defoliator outbreaks are characteristically periodic or quasi-periodic in time and spatially-synchronized (Myers and Cory 2013). It has been suggested that the primary oscillation in populations is generally caused by a delayed density-dependent reaction of natural enemies to their defoliator prey/host populations, leading to a classical predator–prey cycle (Royama 2005). Vertebrate predators such as songbirds and rodents can functionally respond to increasing defoliator populations, but have limited numerical response and are consequently most influential during endemic population phases. For example, birds consumed 84 per cent of spruce budworm larvae and pupae at low population levels, 22 per cent during intermediate levels, but had negligible impacts on epidemic populations (Jennings and Crawford 1985). Parry et al. (1998) also found avian predation eliminated small populations of forest tent caterpillar. By contrast, insect natural enemies, in particular parasitoids, have more rapid numerical response and capacity to control outbreak populations (Roland 2005). Parasitoids (primarily wasps [Hymenoptera] and flies [Diptera]) deposit eggs on, in, or near specific immature life stages of the defoliator (i.e., eggs, larvae, or pupae) that later hatch and consume the defoliator host. The parasitoid community affecting the spruce budworm is dominated by generalist species with comparatively slow numerical response (Eveleigh et al. 2007) resulting in a characteristically long cycle with long duration (Régnière and Nealis 2007). In contrast, hemlock looper and forest tent caterpillar parasitoid communities are dominated by specialists which have rapid numerical response, resulting in more frequent outbreaks of shorter duration (Hébert et al. 2001). Lepidoptera are also attacked by viral diseases. For example, forest tent caterpillar larvae are colonial and thus the rates of lateral transmission are very high in larval aggregations (Roland and Kaupp 1995). Cycle length has additional consequences for the impact of a given outbreak, as it is the combination of defoliation intensity (i.e., the percent foliage damaged) and duration (i.e., consecutive years of defoliation) that determines tree damage (MacLean 1980; Cooke et al. 2012). For those pest systems that cause widespread mortality of tree hosts, forest collapse may contribute to the outbreak decline (Régnière and Nealis 2007; Cooke et al. 2007).

Locally oscillating populations subjected to small, random but regionally correlated perturbations, typically by weather, can lead to the synchronization of those oscillations across space – a phenomenon known as the "Moran effect" (Royama 1984). Random weather correlates with variations in survival, these random perturbations accumulate over time, and if the random perturbations are spatially correlated, the cycle phases will necessarily converge (Cooke et al. 2007). The Moran effect has been suggested as an important factor contributing to the regional outbreak synchrony for a wide range of Lepidopteran species (Ranta et al. 1997; Bjørnstad et al. 1999; Myers and Cory 2013), including species such as the Douglas-fir tussock moth (Orgyia pseudotsugata [McDunnough]) with wingless females incapable of flight. Nonetheless, adult dispersal is another process by which independently oscillating populations can become synchronized (Régnière and Lysyk 1995; Kaitala and Ranta 1998; Williams and Liebhold 2000). Imperfect synchronization can manifest as "epicentres" spreading to adjacent areas (Berryman 1987), or as "travelling waves" where outbreaks spread across large geographic areas (e.g., Bjørnstad et al. 2002; Tenow et al. 2013).

Defoliator pest species have been broadly characterized as either grazers which attack the current year's foliage or wasteful feeders which attack either foliage from previous years or destroy a disproportionate quantity of foliage which is not consumed (Cooke et al. 2007). Grazers such as spruce budworm can repeatedly remove new-year foliage over five to seven years before a tree dies (MacLean and Ostaff 1989). On hardwood species such as aspen (Populus spp.), trees produce a second flush of leaves following early season defoliation. However, continued defoliation for more than three years can deplete metabolic reserves enough to cause widespread tree mortality (Cooke et al. 2012). Wasteful feeding by species such as hemlock looper can cause tree death more rapidly than grazers as it attacks both current and previous years' foliage during a single year (MacLean and Ebert 1999; Iqbal et al. 2011). Differences in feeding behaviour can lead to characteristic spatial patterns of tree mortality on the landscape (Cooke et al. 2007). Wasteful-feeding defoliators generally cause moderate to severe mortality that is aggregated locally as outbreaks collapse before they can spread across landscapes. In contrast, moderate defoliation by grazing defoliators is generally more diffuse in pattern and extensive in spatial scale. Some insect species exhibit hybrid epidemiological behaviour that defies ready classification. For example, early stages of spruce budworm outbreaks often manifest in small emergent pockets of damage (i.e., epicentres; Bouchard and Auger 2014) that later coalesce into more regionally-continuous damage patterns.

Dynamics of bark beetles

Several species of economically important bark beetles have been recently active in western North America (Raffa et al. 2008) while the southern pine beetle (*Dendroctonus frontalis* Zimmerman) has been a long-term issue in pine-dominated systems of the south-eastern US (Coulson and Klepzig 2011). Across Europe bark beetles affected 8.2 Mha or 3.0 per cent of the forest area in 2005 which was second in Europe only to wind and storm damage (FAO 2014).

Bark beetles develop from eggs laid in galleries within the phloem of trees under the bark (i.e., subcortical environment). Larvae hatching from these eggs feed on nutrient-rich phloem resulting in tunnels under the bark. The subcortical environment offers some protection from thermal extremes, desiccation (whereas to survive in external environments defoliators had to adapt through feeding shelters, motility, colonial behaviour, and increased water use efficiencies), and natural enemies, relative to the exposed conditions experienced by defoliator larvae feeding on foliage. Extensive feeding can girdle the tree thereby preventing transport of nutrients (Wood 1982). Consequently, the subcortical environment is among the most intensely defended tissues within a tree.

Genus-specific defences of conifers attacked by bark beetles include the production of resins that physically eject or isolate burrowing insects, the ability of wound periderm to suberize and isolate damaged phloem, and the production of toxic phenolic compounds (Franceschi et al. 2005). To overcome host defences, bark beetles rely on a mass attack where large numbers of attacking beetles combined with symbiotic fungal associates overwhelm host defences. Bark beetles can detoxify host terpenes through oxidation and convert them to aggregation pheromones which further encourage the mass attack (Borden 1974; Wood 1982; Seybold and Tittiger 2003). This adaptation to resin-based defences in conifers is thought to account for the affiliation of tree-killing bark beetles with coniferous tree species (Seybold et al. 2006).

During the endemic stage, tree-killing bark beetles persist by attacking weakened trees with poor defences. A certain threshold population density is required to successfully mass attack and kill healthy trees. However, once that threshold is reached, the reproductive capacity of tree-killing bark beetles increases exponentially before levelling or dropping off at extreme population densities (Boone et al. 2011). Outbreaks occur as key population thresholds are surpassed due to triggering events such as widespread wind disturbance or drought that weaken large populations of otherwise healthy host trees (Raffa et al. 2008). Threshold temperatures may also heavily influence both generation time and overwintering success – each of which may have dramatic effects on bark beetle outbreak dynamics (Powell and Bentz 2009; see "Response to climate change" section below).

The natural enemy complex affecting bark beetles includes predatory beetles and woodpeckers, and parasitic flies and wasps (Wermelinger 2004; Safranyik and Carroll 2006; Coulson and Klepzig 2011). Predation and parasitism impacts are far more challenging to measure within the subcortical environment by comparison with free-ranging defoliators. Clerid beetles have been shown to have a significant influence on southern pine beetle population dynamics (Reeve 1997). There is an absence of quantitative evidence for analogous effects in other bark beetle systems, including mountain pine beetle, where the bottom-up effects of host plant defences are thought to outweigh the top-down influence of predation and parasitism. For example, Safranyik et al. (1999), in a model of mountain pine beetle population dynamics, include the effects of predation as a weak fixed effect. A key limitation in all studies of bark beetle population dynamics is the context-dependent nature of predation and parasitism. Unfortunately, there is a heavy bias toward the study of post-eruptive populations and a paucity of studies targeted outside the focal epicentre, on the endemic populations in the surrounding landscape.

Transitions between discrete endemic and epidemic states are abrupt within tree-killing bark beetle systems. These state transitions are accompanied by both behavioural shifts in host preferences and attack behaviour (Safranyik and Carroll 2006) and a loss of spatial patterning, as focal epicentres coalesce into relatively homogeneous, landscape-scale patterns of mortality. The rate of transition from endemic to eruptive behaviour depends on thermal parameters governing the number of generations per year. Southern pine beetle, occupying the temperate southern United States (and also areas of Mexico and Central America), may have several generations per year (i.e., multi-voltine), so spot growth proceeds relatively quickly. Mountain pine beetle, in contrast, usually has one generation per year (i.e., univoltine), and sometimes has fewer than one generation per year (i.e., partial voltinism), so spot growth occurs across years rather than within the year.

Severe outbreaks of bark beetle tend to collapse as a result of exhaustion of suitable host trees, although they may be truncated prematurely by extremely unfavourable weather (Raffa et al. 2008). The spatial pattern of bark beetle outbreak – either clustered or continuous – depends on whether the outbreak is allowed to develop to maturity. Mountain pine beetle outbreaks often terminate before host trees are completely exhausted. As the forest is thinned, the outbreak tends to fizzle, and it is even thought that there is a critical host patch size below which outbreaks are not possible (Heavilin and Powell 2008). This illustrates that bark beetles both influence, and are influenced by, host forest structure – the definition of reciprocal feedback.

Invasive species

The previous two sections illustrate how the interplay between bottom-up (i.e., tree defence) and top-down (i.e., natural enemies) forces result in consequent spatio-temporal patterns of insect damage within native forest–pest systems. Introduction of nonindigenous insects via global trade represents an increasingly prevalent threat to forests because of the potential for unrestricted population growth due to the lack of natural enemies and/or host resistance (Work et al. 2005). It is important to understand that accidental introductions are quite common, but only a select few species establish in new areas (Gandhi and Herms 2010). Yet the arrival of species such as the emerald ash borer (*Agrilus planipennis*) in North America has had and will continue to have catastrophic effects on entire genera of trees. Understanding factors that permit establishment and spread of newly arrived forest insects has therefore become a critical issue in both forest health and biodiversity worldwide (Gandhi and Herms 2010).

Two brief case studies illustrate the processes and issues underlying outbreak behaviour as species invade new systems. The emerald ash borer is a wood-boring beetle of the family *Buprestidae*. While little is known about its life history within its native range in China, it causes only minor damage to associated ash (*Fraxinus* spp.) trees there. Emerald ash borer was first detected in North America in the interior port city of Detroit, where it became apparent that North American ash species have virtually no resistance to its attack. The species readily attacks and kills ash trees of any size down to 2 cm diameter. While the species has limited dispersal capability, its spread has been massively exacerbated by human transport of colonized wood. While a few introduced predators hold promise for biocontrol, the emerald ash borer exemplifies the worst case scenario of unrestricted population growth that threatens the existence of entire tree genera at biome to continental scales.

European gypsy moth (*Lymantria dispar*) was introduced to North America as a potential silk producer, but subsequently escaped captivity and became established in hardwood forests of Massachusetts at the turn of the nineteenth century. This species is a polyphagous defoliator with a particular affinity for oak species (*Quercus* spp). Many natural enemies have been introduced for biocontrol of gypsy moth, with unfortunate unintended consequences for some native Lepidopteran species. Defoliation damage by the gypsy moth, while extensive, is far less virulent in its impacts. Following invasion its behaviour is similar to that in its native range, i.e., periodic, regionally synchronized outbreaks causing moderate damage. Despite the widespread damage caused by the introduction of gypsy moth to North America, its activities have not fundamentally transformed the ecosystem function of the forests it has invaded.

The impacts of introduced insect species therefore range from benign to catastrophic. The very real risk of the worst case scenario requires extreme vigilance to avoid accidental introductions, requiring close cooperation across jurisdictional boundaries around the globe. If exotic species become established and invasive, quick response strategies can limit large-scale tree loss including effective and early identification of infested trees, the eradication of isolated populations, slowing the spread (for example, by reducing or eliminating human transport) and diversifying forests (i.e., urban forests).

Insect disturbance effects on forest dynamics

As forest characteristics affect outbreak severity, resultant tree mortality from outbreaks also alters forest structure and ecosystem processes. Tree death creates canopy gaps of various sizes and increases the volume of snags and coarse woody debris (Kneeshaw and Bergeron 1998; Taylor and MacLean 2007; Coulson and Klepzig 2011). As stressed throughout this chapter, insect disturbance differs from most other types of disturbances in that it is host-specific – and this host-specificity has implications for resulting patterns of growth reduction, tree mortality, and forest response to those processes through time. In some cases succession is accelerated as shade intolerant overstory tree species are selectively killed during outbreaks and composition shifts toward more shade tolerant species. For example, multiple years of trembling aspen defoliation by the forest tent caterpillar has led to a shift from shade intolerant aspen to more shade tolerant balsam fir (Moulinier *et al.* 2013). Shade tolerant species can persist following overstory mortality when advance regeneration is abundant. However, non-host species may recruit if advance regeneration is limited (Reyes *et al.* 2010).

The balsam fir–spruce budworm system is a classic example, in which the insect was labelled a "super silviculturist" by Baskerville (1975a) as overstory mortality resulted in the release of an understory seedling bank of the host species balsam fir (Spence and MacLean 2012) that would then be ready to be "harvested" by the insect in a subsequent outbreak. Similar re-establishment of shade tolerant species have been observed in central Europe after severe bark beetle (*Ips typographus*) outbreaks where Norway spruce forest regenerates well after the overstory mortality of spruce (Jonasova and Prach 2004; Svoboda *et al.* 2010). In real forests, however, these dynamics are a function both of forest structure and outbreak severity (Figure 7.5). In mixedwood stands, mortality of mature balsam fir trees caused by the spruce budworm has been identified as providing opportunities for companion species to recruit (Bouchard et al. 2006) or for equally shade tolerant species such as eastern white cedar to increase in dominance (Kneeshaw and Bergeron 1998) (Figure 7.5). Similar dynamics are also observed following mountain pine beetle outbreaks where mortality of pine in mixed stands can accelerate succession to more shade tolerant companion species.

In contrast, in the spruce budworm system balsam fir is a shade tolerant species and deciduous species, such as paper birch (*Betula papyrifera*) and aspen (*Populus tremuloides*), where present, may also increase in abundance in response to outbreaks opening the canopy (Taylor and Chen 2011). This could lead to an increase in site productivity given their positive effect on soil nutrient availability (Légaré et al. 2005). The presence of hardwood species can reduce the level of defoliation of host species during spruce budworm outbreaks (Su et al. 1996) possibly due to greater herbivore and natural enemy diversity.

Following insect outbreaks, spikes in nitrogen due to the accumulation of insect frass and insect mortality have been observed (Paré and Bergeron 1996) creating a fertilizing effect with the greatest effect occurring in nitrogen limited ecosystems. Forests defoliated by gypsy moth have been shown to release nitrogen loads into the watershed (Townsend et al. 2004). Other macro and micro nutrients did not increase following spruce budworm outbreaks, for example, nutrients such as phosphorous and potassium continue to decline with insect outbreaks and are, in contrast, naturally rejuvenated by disturbances such as fire. On poor-quality sites canopy opening resulting from insect outbreak could lead to further losses in productivity by permitting ericaceous species (e.g., *Kalmia* spp.) to increase; such ericaceous species subsequently inhibit regeneration of spruce and other tree species and are associated with reduced nutrient availability and slow tree growth (Mallik 2003).

Forest management effects – positive and negative

Insect populations are constrained by food availability, which is affected by forest management through changes to tree-species composition (i.e., hosts to non-hosts), to size or age–class structure (size and age/vigour affect insect growth and damage) and configuration (fragmentation and isolation affect dispersal). In the following we will use examples from the spruce budworm and mountain pine beetle systems as case study examples of the potential effects of forest management.

Forest management practices that increase host species will lead to more severe (greater mortality) outbreaks and these may be occurring over larger areas than in the past (Blais 1983; Raffa et al. 2008). Among the most pervasive effects of forest management on insect outbreaks is fire suppression, which can favour host species and older stands, both of which increase forest vulnerability (Sturtevant et al. 2012). Forest management and fire suppression have promoted the availability of large contiguous stands of lodgepole pine, a major host species for the mountain pine beetle, and this was identified as a cause of the extent of the current outbreak (Raffa et al. 2008). Stand-level treatments by themselves have little if any influence on defoliator dynamics but changes in forest composition affect damage severity (Muzika and Liebhold 2000; Wesołowski and Rowiński 2006). Pure stands of mature host trees lead to greater and widespread mortality while forests mixed with hosts and phylogenetically distant tree species tend to have reduced herbivory (Jactel and Brockerhoff 2007). For example, higher content of non-host hardwoods can reduce spruce budworm caused defoliation and mortality in host fir and spruce stands (Bergeron et al. 1995; Su et al. 1996).

moderate

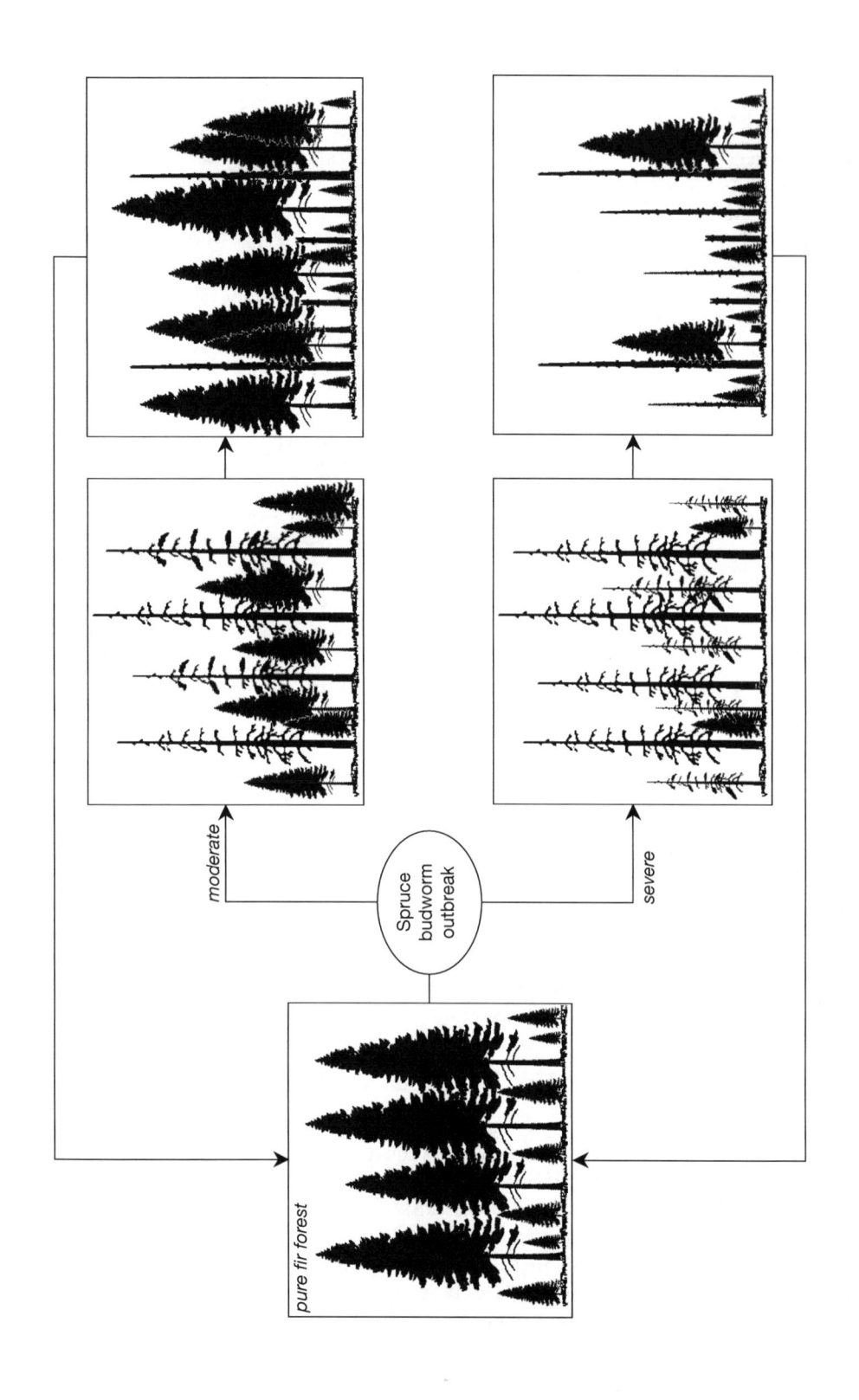

severe

Spruce budworm outbreak

pure fir forest

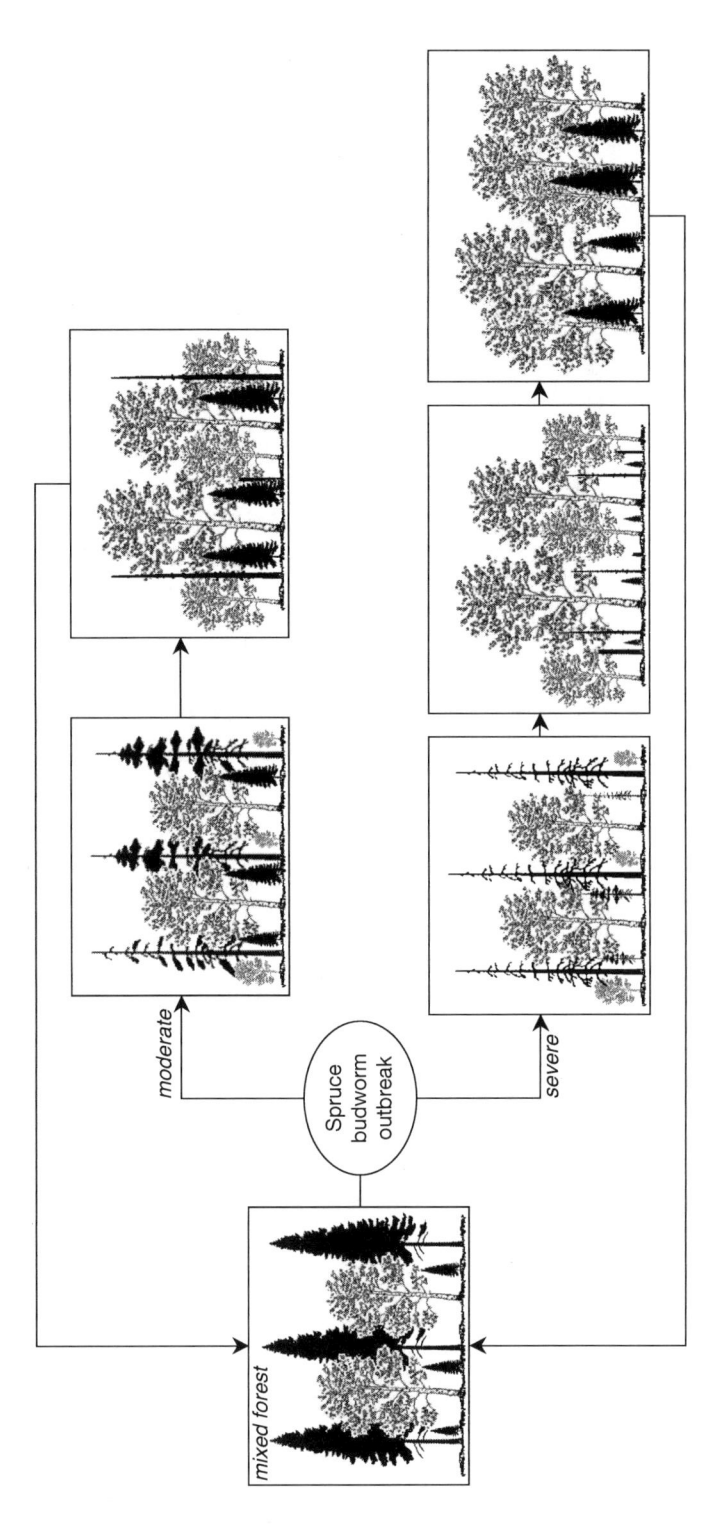

Figure 7.5 Forest dynamics following different outbreak severities. In the panel on the preceeding page a moderate outbreak in a host dominated forest releases the understory which then recruits to the overstory, following a severe outbreak the advance regeneration is also reduced and stand development to a fully stocked host stand takes much longer. In the panel on this page the initial stand conditions are a mixedwood stand. A moderate outbreak can lead to the maintenance of a mixed stand although species proportions may change. After a severe outbreak, stand composition can be shifted to dominance of companion species. Over time shade tolerant conifer host species may re-establish

Age and size class are directly modified by forest management. For example, Hennigar *et al.* (2008) report that mature stands of host species experience greater mortality than immature stands of the same species. The mountain pine beetle on the other hand attacks only trees greater than 20 to 25 cm in diameter. Harvesting and the establishment of plantations can lead to younger forests that are typically less vulnerable, but not invulnerable to species such as the spruce budworm.

Landscape studies have demonstrated that human-caused modifications to forested landscapes affect outbreak characteristics. For example, spruce budworm outbreaks within unmanaged forest landscapes were found: 1) to be more synchronous; 2) to have more trees per site affected; and 3) to be less frequent than within commercially managed forests (Robert *et al.* 2012). In aspen dominated forests, outbreaks of the forest tent caterpillar were more severe and lasted longer in fragmented landscapes whereas in Europe the winter moth was found to cause less defoliation during outbreaks in management-fragmented landscapes (Wesołowski and Rowiński 2006). Host connectivity may influence outbreak synchrony and mitigate forest damage by affecting defoliator dispersal (Ims *et al.* 2008), movement of natural enemies from adjacent habitats (Roland and Taylor 1997), or both. Although guidelines have been proposed to reduce risk and loss of timber in severe outbreaks (e.g. reducing the connectivity of the most vulnerable host species and size classes), the mechanisms and the efficacy of the proposed treatments can still be questioned. Some authors suggest that greater tree diversity offers a protective effect (Su *et al.* 1996) but there is still controversy on the strength of any effect (Miller and Rusnock 1993; Koricheva *et al.* 2006). An explanation for the lack of a consistent relationship is that diversity effects are strongest when mixtures are composed of phylogenetically distant trees (Jactel and Brockerhoff 2007). Even in the absence of protective effects, tree mixtures of non-host tree species will reduce the potential for widespread mortality compared to contiguous stands of host species (Baskerville 1975b). However, increases in non-host species may not always be desirable if these are not commercially valuable. In such a case foresters are losing potentially valuable timber fibre to protect against uncertain losses. In other cases increasing the proportion of non-host species for one insect may simply mean increasing the proportion of a species that is vulnerable to another insect pest (Robert 2014). For example, increasing the proportion of aspen and birch in a forest as non-host species to the spruce budworm may increase vulnerability to forest tent caterpillar outbreak. Tradeoffs that occur due to manipulating forest composition need to be carefully evaluated, which can be done using stand growth models and decision support systems for forest insects (e.g., MacLean *et al.* 2001; Hennigar *et al.* 2011; Iqbal *et al.* 2012).

Response to climate change

Weather and climate affect survival and development of individual life stages and thus can contribute to incipient outbreaks or to their collapse – in both defoliator and bark beetle systems. In general, warm conditions favour rapid development of larvae whereas cold temperatures can limit the northern distribution of insects if thermal units are insufficient to complete a generation. Increased access to alternative susceptible hosts at higher altitudes and milder winter temperatures have contributed to the recent unprecedented mountain pine beetle outbreak in western North America (Bentz *et al.* 2010). In particular, temperature-related transitions from bivoltine to univoltine life-cycles have enabled widespread eruptions of both mountain pine beetle and northern spruce beetle (Werner *et al.* 2006). For species such as spruce budworm warm autumn temperatures delay diapause and increase mortality as larvae exhaust metabolic resources needed to overwinter (Régnière *et al.* 2012). Climate governs at least the broadest outlines of species distributions (Sexton *et al.* 2009). Our greatest certainty is that following climate change current patterns in disturbances will not remain the same (Dale *et al.* 2001; Haynes *et al.* 2014). In terms of insect outbreaks, climate change will influence outbreak dynamics by modifying insect reproduction, survival and ranges (Candau and Fleming 2011). As described below, changes can also be expected between hosts, insect pests and controlling parasitoids and predators.

Some authors have suggested that insect outbreaks will increase in frequency and severity as climate warms (Fleming *et al.* 2002; Schelhaas *et al.* 2003). Range shifts are also predicted with outbreaks extending beyond traditional limits as climate becomes more or less hospitable (Régnière *et al.* 2012). These predictions are already supported by observations of outbreak range expansions and phenological matching of pests and tree hosts in some northern forest ecosystems (Jepsen *et al.* 2011). These range shifts can expose previously unattacked (or rarely attacked) tree species in the same family or genus to novel herbivory which may be exacerbated by a decoupling of forest pests and their parasitoids (Stireman *et al.* 2005) thus leading to greater severity outbreaks in the new range. A recent study in Europe has shown individualist responses of different species to climate change with some species' outbreaks increasing in severity and others having outbreak population cycles collapse (Haynes *et al.* 2014)

Tree hosts are also affected by climate and may be more vulnerable to insect attack if stressed by non-optimal conditions (Mattson and Haack 1987; Zhang *et al.* 2014). For example, drought creates stress and lowers tree defences, making them less resistant to herbivorous insect attacks (the plant-stress hypothesis) (Larsson 1989). Drought stress has been implicated as a major driver underlying widespread bark beetle outbreaks by multiple species in western North American forests (Raffa *et al.* 2008).

In contrast, higher mean temperatures may be of greater benefit to the natural enemy complex by increasing the development rate, fecundity and search rate of multiple parasitoid species (Gray 2008). However, the control response of parasitoids is equivocal as Stireman *et al.* (2005) suggested that parasitoids may be disadvantaged by slower dispersal during range expansion. Greater knowledge is also needed about how temperature and precipitation at different times of the year affect insect pests and their parasites. Although the effects of climate on major insect pests has been studied to some extent, very little is known about climatic controls on parasitoid species.

Conclusion

Although insects occur throughout the world, those that cause economically damaging large-scale outbreaks occur primarily in temperate and boreal forests with relatively few tree species. The outbreaking insects can be separated into two large groups: defoliators and bark beetles. For both groups, it is the larval stages that cause damage and mortality and outbreaking species are characterized by populations with large oscillations. Combinations of weather, forest condition (i.e., host abundance, age and size class distribution and landscape configuration) and natural enemies control the population dynamics of outbreaking insects. The factors all work together to influence the rise and fall of outbreaks. In the case of invasive exotic species, insect outbreaks may occur in the new ranges due to the absence of natural enemies and reduced host resistance.

It is our contention that to reduce the impacts of forest outbreaks greater focus should be placed on these forest pests not only during epidemic phases during which management responses are primarily reactive, but also during endemic phases when landscape-scale forest management decisions will have a large effect on subsequent outbreak severity. Modelling approaches based on simple rules that link outbreak severity with forest conditions are being used more frequently (MacLean et al. 2001; Hennigar et al. 2011). These decision support systems are useful during outbreaks to target interventions (pesticide application, salvage, and/or harvest rescheduling) so as to minimize both short-term and long-term losses and for preventive planning (forest restructuring) during endemic stages. During these stages landscapes and stands can be managed to reduce host connectivity, reduce host composition, and maximize stand and age–class diversity as pest management strategies to reduce the impact of future outbreaks.

Changes in forest landscapes and climate change are currently leading to, or are projected to lead to, changes in the extent and distribution of forest types and insect populations. Changes in host, predator and parasite relationships may either aggravate or attenuate outbreaks such that some insect pests will be of greater concern and others will cause less damage, and locations of outbreaks may differ. Combined, these nonlinear and cross-scale interactions (Holling 1992; Peters et al. 2004) conspire to increase long-term uncertainty in future forest–insect dynamics. Indeed, there is already evidence from around the world of changes in outbreak characteristics that have been linked to global changes (Dale et al 2001; Schelhaas et al. 2003).

Although insect outbreaks can cause widespread economic impact by causing millions of hectares of damage, outbreaks are also a natural process. As such they can be creative forces of re-organization leading to nutrient pulses, shifts in tree composition and forest structure from stand to landscape scales.

References

Abbott, B., Stennes, B. and van Kooten, G.C. 2009. Mountain pine beetle, global markets, and the British Columbia forest economy. *Canadian Journal of Forest Research* 39: 1313–1321.
Allen, C.D. 2007. Interactions across spatial scales among forest dieback, fire, and erosion in northern New Mexico landscapes. *Ecosystems* 10: 797–808.
Baskerville, G. 1975a. Spruce budworm: super silviculturist. *The Forestry Chronicle* 51: 138–140.
Baskerville, G. 1975b. Spruce budworm: the answer is forest management: or is it? *The Forestry Chronicle* 51: 157–160.
Battisti, A., Stastny, M., Buffo, E. and Larsson, S. 2006. A rapid altitudinal range expansion in the pine processionary moth produced by the 2003 climatic anomaly. *Global Change Biology* 12: 662–671.
Bauce, E., Kumbasli, M., Van Fankenhuyzen, K. and Carisey, N. 2006. Interactions among white spruce tannins, *Bacillus thuringiensis* subsp. kurstaki, and spruce budworm (Lepidoptera: Tortricidae), on larval survival, growth and development. *Journal of Economic Entomology* 99: 2038–2047.
Bejer, B. 1988. The nun moth in European spruce forests. pp. 211–231. In: *Dynamics of Forest Insect Populations*. Springer, US.
Bentz, B.J., Régnière, J., Fettig, C.J., Hansen, E.M., Hayes, J.L. and others. 2010. Climate change and bark beetles of the western United States and Canada: direct and indirect effects. *BioScience* 60: 602–613.
Bergeron, Y., Leduc, A., Morin, H. and Joyal, C. 1995. Balsam fir mortality following the last spruce budworm outbreak in northwestern Quebec. *Canadian Journal of Forest Research* 25: 1375–1384.
Berryman, A.A. 1987. Equilibrium or non-equilibrium: Is that the question? *Bulletin Ecological Society of America* 68: 500–502.
Bjørnstad, O.N., Ims, R.A. and Lambin, X. 1999. Spatial population dynamics: analyzing patterns and processes of population synchrony. *Trends in Ecology and Evolution* 14: 427–432.
Bjørnstad, O.N., Peltonen, M., Liebold, A.M. and Baltensweiler, W. 2002. Larch bud moth population dynamics and forest defoliation in the European Alps. *Science* 298: 1020–1023.
Blais, J.R. 1983. Trends in the frequency, extent, and severity of spruce budworm outbreaks in eastern Canada. *Canadian Journal of Forest Research* 13: 539.
Boone, C.K., Aukema, B.H., Bohlmann, J., Carroll, A.L. and Raffa, K.F. 2011. Efficacy of tree defense physiology varies with bark beetle population density: a basis for positive feedback in eruptive species. *Canadian Journal of Forest Research* 41: 1174–1188.

Borden, J.H. 1974. Aggregation pheromones in the Scolytidae. In: *Pheromones*. M.C. Birch. Elsevier, North-Holland, Amsterdam, pp161–189.

Bouchard, M. and Auger, I. 2014. Influence of environmental factors and spatio-temporal covariates during the initial development of a spruce budworm outbreak. *Landscape Ecology* 29: 111–126.

Bouchard, M., Kneeshaw, D. and Bergeron, Y. 2006. Forest dynamics after successive spruce budworm outbreaks in mixedwood forests. *Ecology* 87: 2319–2329.

Buffo, E., Battisti, A., Stastny, M. and Larsson, S. 2007. Temperature as a predictor of survival of the pine processionary moth in the Italian Alps. *Agricultural and Forest Entomology* 9: 65–72.

Candau, J. and Fleming, R. 2011. Forecasting the response of spruce budworm defoliation to climate change in Ontario. *Canadian Journal of Forest Research* 41: 1948–1960.

Cooke, B.J., Nealis, V.G. and Régnière, J. 2007. Insect defoliators as periodic disturbances in northern forest ecosystems. In: *Plant Disturbance Ecology: The Process and the Response*. Edited by E.A. Johnson and K. Miyanishi. Elsevier Academic Press, Burlington, Mass., USA, pp. 487–525.

Cooke, B.J., MacQuarrie, C.J. and Lorenzetti, F. 2012 The dynamics of forest tent caterpillar outbreaks across east-central Canada. *Ecography* 35: 422–435.

Coulson, R.N. and Klepzig, K.D. 2011. Southern Pine Beetle II. *Gen. Tech. Rep. SRS-140*. U.S. Department of Agriculture Forest Service, Southern Research Station, Asheville, NC, US: 512 pp.

Dale, V.H., Joyce, L.A., McNulty, S., Neilson, R.P., Ayres, M.P. and others. 2001. Climate change and forest disturbances. *BioScience* 51: 723–734.

Damon, A. 2000. A review of the biology and control of the coffee berry borer, *Hypothenemus hampei* (Coleoptera: Scolytidae). *Bulletin of Entomological Research* 90: 453–465.

Dyer, L.A., Carson, W.P. and Leigh, E.G. Jr. 2012. Insect outbreaks in tropical forests: patterns, mechanisms and consequences. In: Barbosa, P., Letourneau, D.K. and Agrawal, A.A. *Insect Outbreaks Revisited*. Wiley Blackwell, Hoboken NJ, US, pp. 219–245.

Elton, C.S. 1958. *The Ecology of Invasions by Animals and Plants*. Methuen, London.

Esper, J., Buntgen, U., Frank, D.C., Nievergelt, D. and Liebhold, A. 2007. 1200 years of regular outbreaks in alpine insects. *Proc. R. Soc. B* 274: 671–679.

Eveleigh, E.S., McCann, K.S., McCarthy, P.C., Pollock, S.J., Lucarotti, C.J. and others. 2007. Fluctuations in density of an outbreak species drive diversity cascades in food webs. *Proceedings of the National Academy of Sciences* 104: 16976–16981.

FAO (Food and Agriculture Organization). 2014. Disturbance affecting forest health and vitality (1 000 ha) by FRA categories, year, country. *Forest Health, CountrySTAT, Food and Agriculture Organization of the United Nations*. Rome, Italy. Available on-line at http://countrystat.org/home.aspx?c=FOR&tr=5 [viewed 26 March 2014].

Fleming, R.A., Candau, J.N. and McAlpine, R.S. 2002. Landscape-scale analysis of interactions between insect defoliation and forest fire in central Canada. *Climatic Change* 55: 251–272.

Franceschi, V.R., Krokene, P., Christiansen, E. and Krekling, T. 2005. Anatomical and chemical defenses of conifer bark against bark beetles and other pests. *New Phytologist* 167: 353–376.

Gandhi, K.J.K. and Herms, D.A. 2010. Direct and indirect effects of alien insect herbivores on ecological processes and interactions in forests of eastern North America. *Biological Invasions* 12: 389–405.

Gilbert, M., Fielding, N., Evans, H.F. and Grégoire, J.C. 2003. Spatial pattern of invading *Dendroctonus micans* (Coleoptera: Scolytidae) populations in the United Kingdom. *Canadian Journal of Forest Research* 33: 712–725.

Gray, D.R. 2008. The relationship between climate and outbreak characteristics of the spruce budworm in eastern Canada. *Climatic Change* 87: 361–383.

Haynes, K.J., Allstadt, A.J. and Klimetzek, D. 2014. Forest defoliator outbreaks under climate change: effects on the frequency and severity of outbreaks of five pine insect pests. *Global Change Biology* 20: 2004–2018.

Heavilin, J. and Powell, J. 2008. A novel method of fitting spatio-temporal models to data with applications to the dynamics of mountain pine beetles. *Natural Resource Modeling* 21: 489–524.

Hébert, C., Berthiaume, R., Dupont, A. and Auger, M. 2001. Population collapses in a forecasted outbreak of *Lambdina fiscellaria* (Lepidoptera: Geometridae) caused by spring egg parasitism by *Telenomus* spp. (Hymenoptera: Scelionidae). *Environmental Entomology* 30: 37–43.

Hennigar, C.R., MacLean, D.A., Quiring, D.T. and Kershaw, J.A. Jr. 2008. Differences in spruce budworm defoliation among balsam fir and white, red, and black spruce. *Forest Science* 54: 158–166.

Hennigar, C.R., Wilson, J.S., MacLean, D.A. and Wagner, R.G. 2011. Applying a spruce budworm decision support system to Maine: projecting spruce-fir volume impacts under alternative management and outbreak scenarios. *Journal of Forestry* 109: 332–342.

Holling, C.S. 1973. Resilience and stability of ecological systems. *Annual Review of Ecology and Systematics* 1–23.

Holling, C.S. 1992. Cross-scale morphology, geometry, and dynamics of ecosystems. *Ecological Monographs* 62: 447–502.

Holling, C.S. 2001. Understanding the complexity of economic, ecological, and social systems. *Ecosystems* 4: 390–405.

Ims, R.A., Henden, J.A. and Killengreen, S.T. 2008. Collapsing population cycles. *Trends in Ecology and Evolution* 23: 79–86.

Iqbal, J., MacLean, D.A. and Kershaw, J.A. 2011. Impacts of hemlock looper defoliation on growth and survival of balsam fir, black spruce and white birch in Newfoundland, Canada. *Forest Ecology and Management* 261: 1106–1114.

Iqbal, J., Hennigar, C.R. and MacLean, D.A. 2012. Modeling insecticide protection versus forest management approaches to reducing balsam fir sawfly and hemlock looper damage. *Forest Ecology and Management* 265: 150–160.

Jactel, H. and Brockerhoff, E.G. 2007. Tree diversity reduces herbivory by forest insects. *Ecology Letters* 10: 835–848.

Janzen, D.H. 1970. Herbivores and the number of tree species in tropical forests. *The American Naturalist* 104: 510–528.

Jennings, D.T. and Crawford, H.S. 1985. Predators of the spruce budworm. *U.S. Dept. Agric., Agric. Handb.* No. 644.

Jepsen, J.U., Kapari, L., Hagen, S.B., Schott, T., Vindstad, O.P.L. and others. 2011. Rapid northwards expansion of a forest insect pest attributed to spring phenology matching with sub-Arctic birch. *Global Change Biology* 17: 2071–2083.

Jonasova, M. and Prach, K. 2004. Central-European mountain spruce (*Picea abies* (L.) Karst.) forests: regeneration of tree species after a bark beetle outbreak. *Ecological Engineering* 23: 15–27.

Jonsson, A., Appelberg, G., Harding, S. and Bärring, L. 2009. Spatio-temporal impact of climate change on the activity and voltinism of the spruce bark beetle, *Ips typographus*. *Global Change Biology* 15: 486–499.

Kaitala, V. and Ranta, E. 1998. Traveling wave dynamics and self-organisation in spatio-temporally structured populations. *Ecology Letters* 1: 186–192.

Kneeshaw, D.D. and Bergeron, Y. 1998. Canopy gap characteristics and tree replacement in the southeastern boreal forest. *Ecology* 79: 783–794

Kneeshaw, D., Bergeron, Y. and Kuuluvainen, T. 2011. Forest ecosystem dynamics across the circumboreal forest. In: A.C. Millington, M.A. Blumler, G. MacDonald, U. Shickhoff. (Eds.) *Handbook of Biogeography*. Sage, Washington, Chapter 14.

Koricheva, J., Vehvilainen, H., Riihimaki, J., Ruohomaki, K., Kaitaniemi, P. and Ranta, H. 2006. Diversification of tree stands as a means to manage pests and diseases in boreal forests: myth or reality? *Canadian Journal of Forest Research* 36: 324–336.

Kumbasli, M., Bauce, É., Rochefort, S. and Crépin, M. 2011. Effects of tree age and stand thinning related variations in balsam fir secondary compounds on spruce budworm *Choristoneura fumiferana* development, growth and food utilization. *Agricultural and Forest Entomology* 13: 131–141.

Larsson, S. 1989. Stressful times for the plant stress: insect performance hypothesis. *Oikos* 277–283.

Légaré, S., Bergeron, Y. and Paré, D. 2005. Effect of aspen (*Populus tremuloides*) as a companion species on the growth of black spruce (*Picea mariana*) in the southwestern boreal forest of Quebec. *Forest Ecology and Management* 201: 211–222.

Liebhold, A.M., Brockerhoff, E.G., Garrett, L.J., Parke, J.L. and Britton, K.O. 2012. Live plant imports: the major pathway for forest insect and pathogen invasions of the US. *Frontiers in Ecology and the Environment* 10: 135–143.

MacLean, D.A. 1980. Vulnerability of fir spruce stands during uncontrolled spruce budworm outbreaks: a review and discussion. *The Forestry Chronicle* 56: 213–221.

MacLean, D.A. and Ostaff, D.P. 1989. Patterns of balsam fir mortality caused by an uncontrolled spruce budworm outbreak. *Canadian Journal of Forest Research* 19: 1087–1095.

MacLean, D.A. and Ebert, P. 1999. The impact of hemlock looper (*Lambdina fiscellaria fiscellaria* (Guen.)) on balsam fir and spruce in New Brunswick, Canada. *Forest Ecology and Management* 120: 77–87.

MacLean, D.A., Erdle, T.A., MacKinnon, W.E., Porter, K.B., Beaton, K.P. and others. 2001. The spruce budworm decision support system: forest protection planning to sustain long-term wood supplies. *Canadian Journal of Forest Research* 31: 1742–1757.

Mallik, A.U. 2003. Conifer regeneration problems in boreal and temperate forests with ericaceous understory: role of disturbance, seedbed limitation, and keystone species change. *Critical Reviews in Plant Sciences* 22: 341–366.

Mattson, W.J. and Addy, N.D. 1975. Phytophagous insects as regulators of forest primary production. *Science* 190: 515–522.

Mattson, W.J. and Haack, R.A. 1987. The role of drought in outbreaks of plant-eating insects. *Bioscience* 37: 110–118.

Mattson, W.J., Herms, D.A., Witter, J.A. and Allen, D.C. 1991. Woody plant grazing systems: North American outbreak folivores and their host plants. In: Baranchikov, Y.N., Mattson, W.J., Haine, F.P. and Payne, T.L. Forest insect guilds: patterns of interactions with host trees. *U.S. Department of Agriculture, Forest Service, General Technical Report NE-153*, pp. 53–85.

Miller, A. and Rusnock, P. 1993 The Rise and fall of the silvicultural hypothesis in spruce budworm (*Choristoneura fumiferana*) management in Eastern Canada. *Forest Ecology and Management* 61: 171–189.

Moulinier, J., Lorenzetti, F. and Bergeron, Y. 2013. Effects of a forest tent caterpillar outbreak on the dynamics of mixedwood boreal forests of eastern Canada. *Ecoscience* 20: 182–193.

Muzika, R.M. and Liebhold, A.M. 2000. A critique of silvicultural approaches to managing defoliating insects in North America. *Agricultural and Forest Entomology* 2: 97–105.

Myers, J.H. and Cory, J.S. 2013. Population cycles in forest Lepidoptera revisited. *Annual Review of Ecology, Evolution, and Systematics* 44: 565–592.

Nair, K.S.S. 2012. *Tropical Forest Insect Pests: Ecology, Impact and Management*, Cambridge University Press, Cambridge, UK.

Novotny, V., Basset, Y., Miller, S.E., Weiblen, G.D., Bremer, B. and others. 2002. Low host specificity of herbivorous insects in a tropical forest. *Nature* 416: 841–844.

Palo, A. and Thomas, R. 1984. Distribution of birch (*Betula* spp.), willow (*Salix* spp.), and poplar (*Populus* spp.) secondary metabolites and their potential role as chemical defense against herbivores. *Journal of Chemical Ecology* 10: 499–520.

Paré, D. and Bergeron, Y. 1996. Effect of colonizing tree species on soil nutrient availability in a clay soil of the boreal mixedwood. *Canadian Journal of Forest Research* 26: 1022–1031.

Parry, D., Spence, J.R. and Volney, W.J.A. 1998. Budbreak phenology and natural enemies mediate survival of first-instar forest tent caterpillar (Lepidoptera: Lasiocampidae). *Environmental Entomology* 27: 1368–1374.

Peters, D.P., Pielke, R.A., Bestelmeyer, B.T., Allen, C.D., Munson-McGee, S. and Havstad, K.M. 2004. Cross-scale interactions, nonlinearities, and forecasting catastrophic events. *Proceedings of the National Academy of Sciences of the United States of America* 101: 15130–15135.

Powell, J.A. and Bentz, B.J. 2009. Connecting phenological predictions with population growth rates for mountain pine beetle, an outbreak insect. *Landscape Ecology* 24: 657–672.

Raffa, K.F., Phillips, T.W. and Salom, S.M. 1993. Strategies and mechanisms of host colonization by bark beetles. In: T.D. Schowalter and G.M. Filip (Eds.), *Beetle-Pathogen Interactions in Conifer Forests* (pp. 103–128). Academic Press, Ltd, San Diego, US.

Raffa, K., Aukema, B., Bentz, B., Carroll, A., Hicke, J. and others. 2008. Cross-scale drivers of natural disturbances prone to anthropogenic amplification: the dynamics of bark beetle eruptions. *Bioscience* 58: 501–517.

Ranta, E., Kaitala, V., Lindström, J. and Helle, E. 1997. The Moran effect and synchrony in population dynamics. *Oikos* 78: 136–142.

Reeve, J.D. 1997. Predation and bark beetle dynamics. *Oecologia* 112: 48–54.

Régnière, J. and Lysyk, T.J. 1995. Population dynamics of the spruce budworm, Choristoneura fumiferana. In J.A. Armstrong and W.G.H. Ive (Eds), *Forest Insect Pests in Canada*. (pp.95–105). Canadian Forest Service, Ottawa, Ontario, Canada.

Régnière, J. and Nealis, V.G. 2007. Ecological mechanisms of population change during outbreaks of the spruce budworm. *Ecological Entomology* 32: 461–477.

Régnière, J., St-Amant, R. and Duval, P. 2012. Predicting insect distributions under climate change from physiological responses: spruce budworm as an example. *Biological Invasions* 14: 1571–1586.

Reyes, G.P., Kneeshaw, D., DeGrandpre, L. and Leduc, A. 2010. Changes in woody vegetation abundance and diversity after natural disturbances causing different levels of mortality. *Journal of Vegetation Science* 21: 406–417.

Robert, L.E. 2014. Influence à l'échelle du paysage des legs associés à l'aménagement forestier sur les épidémies d'insectes. Thesis UQAM, Montreal, Canada.

Robert, L., Kneeshaw, D. and Sturtevant, B.R. 2012. Effects of forest management legacies on spruce budworm (*Choristoneura fumiferana*) outbreaks. *Canadian Journal of Forest Research* 42: 463–475.

Roland, J. 2005. Are the "seeds" of spatial variation in cyclic dynamics apparent in spatially-replicated short time-series? An example from the forest tent caterpillar. *Annales Zoologici Fennici* 42: 397–407.

Roland, J. and Kaupp, W.J. 1995. Reduced transmission of forest tent caterpillar (Lepidoptera: Lasiocampidae) nuclear polyhedrosis virus at the forest edge. *Environmental Entomology* 24: 1175–1178.

Roland, J. and Taylor, P.D. 1997. Insect parasitoid species respond to forest structure at different spatial scales. *Nature* 386: 710–713.

Royama, T. 1984. Population dynamics of the spruce budworm *Choristoneura fumiferana*. *Ecological Monographs* 54: 429–462.

Royama, T. 2005. Moran effect on nonlinear population processes. *Ecological Monographs* 75: 277–293.

Safranyik, L. and Carroll, A.L. 2006. The biology and epidemiology of the mountain pine beetle in lodgepole pine forests. In: *The Mountain Pine Beetle: A Synthesis of Biology, Management and Impacts on Lodgepole Pine*. L. Safranyik and W.R. Wilson. Natural Resources Canada, Canadian Forest Service, Pacific Forestry Centre, Victoria, BC, Canada, pp. 3–66.

Safranyik, L., Barclay, H., Thomson, A.J, and Riel, W.G. 1999. A population dynamics model for the mountain pine beetle, *Dendroctonus ponderosae* Hopk. (Coleoptera: Scolytidae). *Can. For. Serv., Pac. For. Cen. Inf. Rep.* BC-X-386. 35 pp.

Schelhaas, M.J., Nabuurs, G.J. and Schuck, A. 2003. Natural disturbances in the European forests in the 19th and 20th centuries. *Global Change Biology* 9: 1620–1633.

Sexton, J.P., McIntyre, P.J., Angert, A.L. and Rice, K.J. 2009. Evolution and ecology of species range limits. *Annual Review of Ecology, Evolution, and Systematics* 40: 415–436.

Seybold, S.J. and Tittiger, C. 2003. Biochemistry and molecular biology of de novo isoprenoid pheromone production in the scolytidae. *Annual Reviews Entomology* 48: 425–453.

Seybold, S.J., Huber, D.P.W., Lee, J.C., Graves, A.D. and Bohlmann, J. 2006. Pine monoterpenes and pine bark beetles: a marriage of convenience for defense and chemical communication. *Phytochemistry Reviews* 5: 143–178.

Shorohova, E., Kneeshaw, D.D., Kuuluvainen, T. and Gauthier, S. 2012. A comparison of old-growth forest dynamics across the circum-boreal forest zone. *Silva Fennica* 45: 785–806.

Spence, C.E. and MacLean, D.A. 2012. Regeneration and stand development following a spruce budworm outbreak, spruce budworm-inspired harvest, and salvage harvest. *Canadian Journal of Forest Research* 42: 1759–1770.

Stireman, J.O., Dyer, L.A., Janzen, D.H., Singer, M.S., Lill, J.T. and others. 2005. Climatic unpredictability and parasitism of caterpillars: implications of global warming. *Proceedings of the National Academy of Sciences of the United States of America* 102: 17384–17387.

Sturtevant, B.R., Miranda, B.R., Shinneman, D.J., Gustafson, E.J. and Wolter, P.T. 2012. Comparing modern and presettlement forest dynamics of a subboreal wilderness – Does spruce budworm enhance fire risk? *Ecological Applications* 22: 1278–1296.

Su, Q., MacLean, D.A. and Needham, T.D. 1996. The influence of hardwood content on balsam fir defoliation by spruce budworm. *Canadian Journal of Forest Research* 26: 1620–1628.

Svoboda, M., Fraver, S., Janda, P., Bače, R. and Zenáhlíková, J. 2010. Natural development and regeneration of a Central European montane spruce forest. *Forest Ecology and Management* 260: 707–714.

Swetnam, T.W. and Lynch, A.M. 1993. Multicentury, regional-scale patterns of western spruce budworm outbreaks. *Ecological Monographs* 63: 399–424.

Taylor, A.R. and Chen, H.Y.H. 2011. Multiple successional pathways of boreal forest stands in central Canada. *Ecography* 34: 208–219.

Taylor, S.L. and MacLean, D.A. 2007. Deadwood dynamics in declining balsam fir and spruce stands in New Brunswick, Canada. *Canadian Journal of Forest Research* 37: 750–762.

Tenow, O., Nilssen, A.C., Bylund, H., Pettersson, R., Battisti, A. and others. 2013. Geometrid outbreak waves travel across Europe. *Journal of Animal Ecology* 82: 84–95.

Townsend, P.A., Eshleman, K.N. and Welcker, C. 2004. Remote sensing of gypsy moth defoliation to assess variations in stream nitrogen concentrations. *Ecological Applications* 14: 504–516.

Wallner, W.E. 1987. Factors affecting insect population dynamics: differences between outbreak and non-outbreak species. *Annual Review of Entomology* 32: 317–340.

Walton, A. 2012. Update of the infestation projection based on the Provincial Aerial Overview Surveys of Forest Health conducted from 1999 through 2011 and the BCMPB model (year 9). *B.C. Forest Service Report*. http://www.for.gov.bc.ca/ftp/hre/external/!publish/web/bcmpb/year9/BCMPB. v9.BeetleProjection.Update.pdf

Wermelinger, B. 2004. Ecology and management of the spruce bark beetle *Ips typographus*—a review of recent research. *Forest Ecology and Management* 202: 67–82.

Werner, R.A., Holsten, E.H., Matsuoka, S.M. and Burnside, R.E. 2006. Spruce beetles and forest ecosystems in south-central Alaska: a review of 30 years of research. *Forest Ecology and Management* 227: 195–206.

Wesołowski, T. and Rowiński, P. 2006. Tree defoliation by winter moth *Operophtera brumata* L. during an outbreak affected by structure of forest landscape. *Forest Ecology and Management* 221: 299–305.

Williams, D.W. and Liebhold, A.M. 2000. Spatial synchrony of spruce budworm outbreaks in Eastern North America. *Ecology* 81: 2753–2766.

Wolda, H. 1987. Altitude, habitat and tropical insect diversity. *Biological Journal of the Linnean Society* 30: 313–323.

Wood, D.L. 1982. The role of pheromones, kairomones and allomones in the host selection and colonization of bark beetles. *Annual Review of Entomology* 27: 411–446.

Work, T.T., McCullough, D.G., Cavey, J.F. and Komsa, R. 2005. Arrival rate of nonindigenous insect species into the United States through foreign trade. *Biological Invasions* 7: 323–332.

Zhang, X., Lei, Y., Ma, Z., Kneeshaw, D.D. and Peng, C. 2014. Insect-induced tree mortality of boreal forests in eastern Canada under a changing climate. *Ecology and Evolution* 4: 2384–2394.

8

THE ROLE OF FIRE IN FOREST ECOSYSTEMS

David F. Greene and Sean T. Michaletz

Introduction

Natural disturbance not only kills plants and animals but also makes resources available for subsequent cohorts. Consequently, it plays a decisive role in births and deaths of individuals and is the major arbiter of demography. In this chapter, we review recent advances in our understanding of the processes linking climate, fire behavior, and fire effects on plants and animals. More specifically, we examine the ecological patterns that emerge from a coupling of fire behavior, physiology, and demographic processes. For example, fire effects on vegetation can be understood by considering the mechanisms governing fire behavior, heat transfer into plants, heat injury of plant tissues, and the physiology linking tissue injuries to whole-plant growth, mortality, and reproduction (Michaletz and Johnson 2007; Michaletz et al. 2012).

Fire is a combustion process in which rates of gaseous fuel production via pyrolysis (endothermic) are controlled by positive feedback of heat from the combustion process itself (exothermic; Quintiere 2006). We can distinguish between two types of combustion in forest fires: flaming combustion and smoldering combustion (Figure 8.1). These two types of combustion are distinguished by the rate of gaseous fuel supply; flaming combustion occurs when pyrolysis rates are high enough to support combustion in the air adjacent to the fuel elements, and smoldering combustion occurs when pyrolysis rates are low and/or limited by build up of ash and char so that combustion only occurs on the surface of fuel elements. The differences between these two types of combustion can be illustrated by comparing a candle (flaming combustion) and incense (smoldering combustion). Although the behaviors of flaming and smoldering combustion are quite different, both have important ecological effects in fire-prone areas.

Characterizing fire

Four conditions are necessary for forest fire initiation and spread (reviewed in Macias Fauria et al. 2011). First, there must be sufficient fuel to allow propagation of the burning front. Forest fuels are generally classified into three strata: (1) ground fuels comprising below-ground organic soil (duff) and plant roots near the soil surface; (2) surface fuels consisting of litter and small plants on the forest floor; and (3) crown fuels located within the forest canopy. In all strata, the

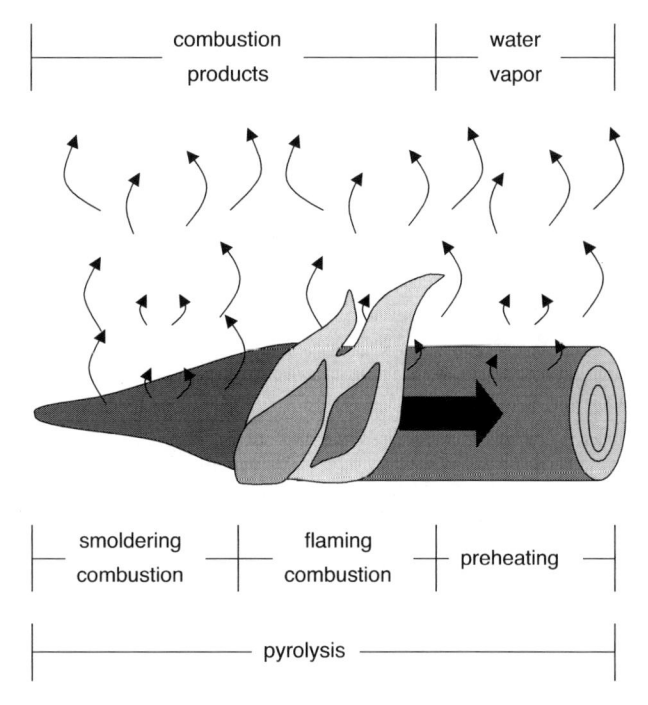

Figure 8.1 Diagram illustrating the phases and products of combustion for a forest fuel element. Fire spread progress from left to right with passage of the flaming combustion front followed by smoldering combustion

flaming fire front consumes only fine fuels less than approximately 6 mm in diameter. Second, there must be a period of at least 10–21 days with low relative humidity and little to no precipitation. This can dry the fuels sufficiently to sustain combustion because fine fuels have a large surface area to volume ratio and thus respond rapidly to changes in moisture conditions (Nelson 2001). In boreal and temperate areas, the drying period often results from persistent positive mid-tropospheric height anomalies that prevent precipitation by blocking zonal flow and encouraging meridional flow of warm, dry air. Third, there must be an ignition source, which is invariably lightning or human activity. Fourth, strong winds are necessary for creating high rates of spread and, consequently, large fire sizes (Figure 8.2). This is especially apparent in some regions adjacent to mountain ranges such as the California chaparral or the *Eucalyptus* forests of southeast Australia, where hot, dry downslope winds (e.g. katabatic winds and föhn winds) enhance fuel drying and fire spread (Keeley and Fotheringham 2001; Sharples *et al.* 2010).

Given these four conditions, it is clear why fire is unimportant in certain forested biomes. In temperate deciduous forests, high relative humidity and frequent convective storms preclude fuel drying. In other biomes, it is less clear why fire is unimportant. Tropical deciduous forests occur when there is a long period of drought during the low-sun season when the relative humidity is also low. Nonetheless, fires are extremely rare in this biome, possibly as a result of the high water content of live fuels on the constituent succulent tree species (Hayden and Greene 2008). Finally, there are some forested biomes, e.g. many tropical evergreen forests, where fire is presently quite common but mainly because of human activities associated with land clearance (Krawchuk *et al.* 2009). Here we focus on forests where lightning ignitions play

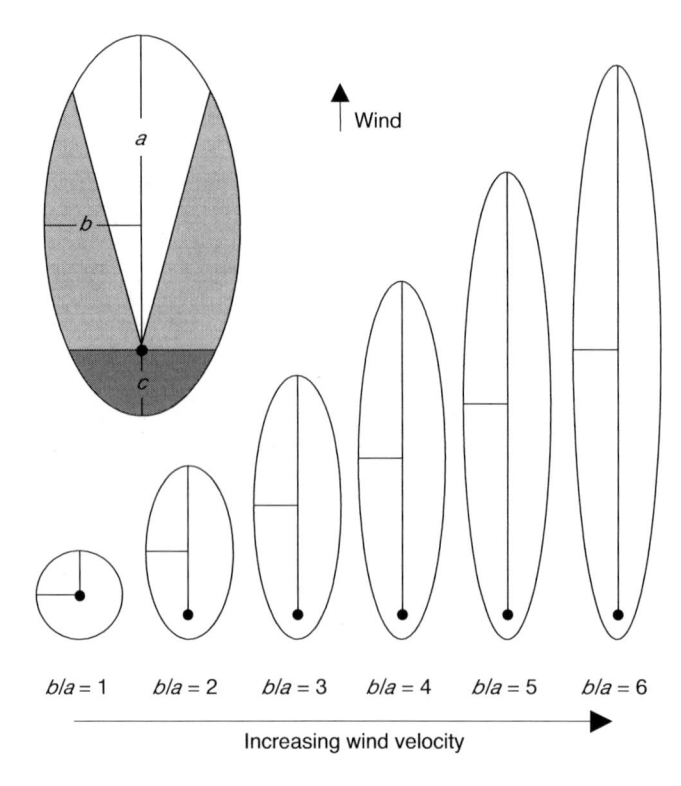

Figure 8.2 Fire shape is the outcome of fire spread rates in all directions from the ignition point (black points). Fire shapes can be idealized as ellipses characterized by the length-to-breadth ratio b/a, where a and b are the rates of spread for head and flank fires, respectively. Wind generally enhances spread rates in the direction of the wind (a) and suppresses spread rates perpendicular to (b) and against the wind (c), resulting in fire shapes with larger length-to-breadth ratios. Dark grey represents the area burned by a back fire, light grey the area burned by flank fire, and white the area burned by head fire. After Van Wagner (1969) and Finney (2001)

a large role in the total area burned. Transitional vegetation types such as savannas are beyond the scope of this chapter.

We distinguish between three classes of forest fires based on the fuels involved (ground, surface, or crown), the type of combustion (flaming or smoldering), and the fire intensity (the rate of heat production per unit length of the fireline). Ground fires involve smoldering combustion of duff and litter and generally occur after passage of the flaming fire front. Ground fires can persist for long periods—sometimes up to months or years. Surface fires involve flaming combustion of litter and live fine fuels of herbs and shrubs located on the forest floor. Surface fires generally have relatively low frontal intensities, as higher intensities lead to ignition of crown fuels (Cruz *et al.* 2006). Crown fires involve flaming combustion of fine fuels located in the forest canopy. Crown fires can be further classified as passive, active, or independent based on the degree to which ignition of unburned fuel relies on heat fluxes from the surface fire below (Michaletz and Johnson 2007).

As with so many other disturbance types, fire is a low frequency, high magnitude event. Although small fires may be more frequent, it is the rare, large fires (ranging from a few hundred to a few hundred thousand ha) that constitute most of the area burned in most fire-prone forest biomes. In Canada, for instance, about 3 percent of the fires accounted for 97 percent of the total area burned between 1959 and 1999 (Stocks *et al.* 2002). Since large fires burn most of the area, they have the most influence on forest structure and dynamics and are consequently of primary interest from an ecological perspective.

Burned areas are invariably elongated, reflecting the importance of wind speed and direction on fire shape (Figure 8.2). Fires also have characteristically large perimeter-to-area ratios (Anderson 1983). Further, small, unburned areas (residual stands) are common within large fires (Kolden *et al.* 2012). Including the edges of these residual stands, Greene (unpublished) has shown for large fires in Saskatchewan that the median distance from any edge to a randomly chosen point in the fire is only about 150 meters. While this still represents a serious constraint for seed dispersal, the situation is not as bleak as one might initially imagine for say a 50,000 ha fire.

Fire effects on plants and animals

The heat produced by combustion can cause physical injuries to plant roots, stems, or crowns that can affect post-fire growth, mortality, and/or reproduction (Figure 8.3; reviewed in Michaletz and Johnson 2007). This occurs as a result of heat transfer from the fire to key tissues such as meristems, vascular tissues, and embryos. Three heat transfer mechanisms link the heat produced by combustion to heat fluxes on plant surfaces: conduction, radiation, and convection. Root surfaces can be heated via conduction through soil, while stem and crown surfaces can be heated via radiation and convection from the flame and buoyant plume. Heat fluxes on plant surfaces can then drive heat conduction from the surface toward the interior of the plant. Injuries can have immediate effects on reproduction and mortality via necrosis of key plant parts like seeds or meristems, or can have delayed effects resulting from an interaction of injuries such as leaf and partial cambium necrosis (Michaletz and Johnson 2006, 2007; Michaletz *et al.* 2012).

Stem injury is governed by rates of heat conduction through the stem to tissues such as the vascular cambium, phloem, and xylem (Michaletz and Johnson 2007; Michaletz *et al.* 2012). Rates of conduction vary with the depth of the tissue in question and the thermal conductivity

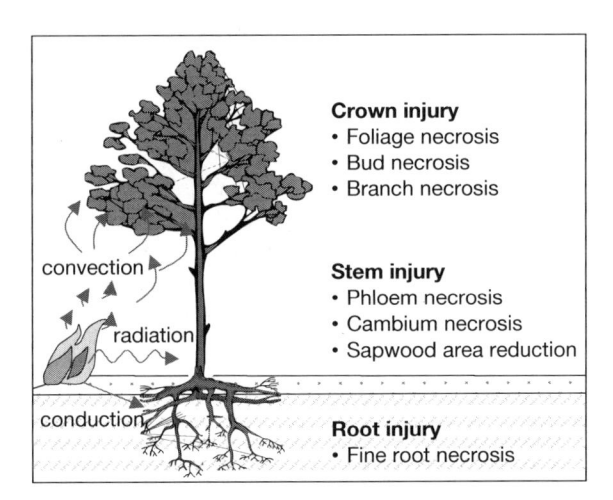

Figure 8.3 Mechanisms of heat transfer from a fire to plant root, stem, and crown surfaces. Surface heating drives heat conduction into the plant, resulting in injuries to several key tissues and organs

of the stem. Stem moisture content is a key variable influencing rates of stem heating, because water is a latent heat sink that limits stem temperature increases to a maximum of 100°C. For woody plants, bark provides additional resistance to conduction and helps insulate the underlying phloem, cambium, and xylem tissues. As larger stems have thicker bark, they are more insulated than smaller stems. However, stem size also influences air-flow patterns, with larger diameter stems producing a turbulent leeward wake that can increase flame residence times and leeward convection rates. This often causes uneven heating around the stem circumference, leading to increased injury on the leeward side (e.g. fire scars resulting from phloem and vascular cambium necrosis). In cases where phloem and cambium necrosis occur around the entire stem circumference (girdling), carbon translocation to roots is prevented and root growth must rely upon stored carbon reserves; when these reserves are depleted, fine root production will cease and the entire plant will die as a result of water stress (Balfour and Midgley 2006; Michaletz et al. 2012). Stem heating can also reduce the hydraulic conductivity of sapwood via enhanced air seed cavitation and conduit wall deformation of xylem (Michaletz et al. 2012), which results in more rapid mortality than girdling alone. Indeed, experiments with an African *Acacia* showed that stem heating caused rapid plant mortality that was associated with reduced sapwood area (Balfour and Midgley 2006).

Injuries to plant crown components (branches, buds, foliage, and seed-housing structures) occur via convective and radiative fluxes from flame and plume. In low intensity surface fires, convection is the dominant heat transfer process and height of crown injury varies with the 2/3 power of fireline intensity (Van Wagner 1973; Michaletz and Johnson 2006). In high-intensity crown fires, canopy fuels are consumed and injuries can be present at all heights. Small crown components such as leaves and buds are "thermally thin", meaning their rates of heating are controlled primarily by surface resistances to radiation and convection and not internal resistance to conduction (Michaletz and Johnson 2006); consequently, thermally thin objects generally do not have internal temperature gradients. The crown component width, surface area, shape, orientation, and degree of shielding by foliage control surface heat fluxes, and the mass, water content, and specific heat capacity determine how much energy is required to cause a temperature increase. For example, species with relatively large buds such as ponderosa pine and longleaf pine are far less susceptible to bud necrosis than are species with relatively small buds such as sugar maple and American beech. For larger crown components like branches and seed-housing structures, internal temperature gradients do exist and conduction within the component becomes important. In these cases, size plays an important role in helping to insulate key tissues against heat necrosis, as conduction rates decrease with distance. Thus, bark can help insulate key tissues like cambium and xylem. Likewise, many cones and fruits can insulate embryo tissues against heat necrosis in fires. This has been widely recognized in serotinous species that store seeds in aerial seed banks, but it can also be important for non-serotinous species provided the fire occurs during a temporal window when seeds are germinable but not yet abscised. For instance, it has been estimated that about 10 percent of *Picea glauca* seeds within closed cones can survive high-intensity crown fires when the fire event occurs between approximately 600 and 1,200 degree days (Michaletz et al. 2012). Finally, extreme vapor pressure deficits in fire plumes might also cause cavitation of xylem conduits in plant crowns that could lead to branch and leaf death (Kavanagh et al. 2010).

Despite our growing knowledge of how plant tissues are injured during fires, we still have a limited understanding of how they interact to control whole-plant function (Michaletz and Johnson 2007; Michaletz et al. 2012). It is generally accepted that stem girdling or death of all vegetative buds will cause mortality, but it is less clear how multiple injuries to multiple tissues interact to affect post-fire growth and mortality. This is of interest as some studies have used air photos of recent burns to argue that the proportion of tree mortality is extraordinarily variable within fires. However, interactions of multiple injuries can often lead to delayed mortality meaning that many plants suffering partial crown scorch will die within a few years.

Necrosis of root meristems may occur as a result of heat conduction through soil (Michaletz and Johnson 2007). Most root meristems are small and do not have internal temperature gradients, so the primary resistance to heating is provided by the overlying soil. Consequently, the key variables controlling rates of root heating are meristem depth, soil structure, and soil composition (water, organic, and mineral content; Hillel 1998). Root heating decreases with depth and soil water content and increases with heating duration, soil surface heat flux, and soil thermal conductivity. Meristem necrosis most commonly occurs as a result of smoldering combustion, as the period of flaming combustion is generally too short to cause lethal temperatures at appreciable depths (Hartford and Frandsen 1992). For a given soil type, the susceptibility of different sizes or species of plants can be predicted by comparing their vertical root distributions; plants with buds distributed to greater depths will be less susceptible to root meristem necrosis (Michaletz and Johnson 2007). Generally, at least some perennating tissues of most individual or clonal shrubs and perennial herbs survive a forest fire except in the upper few centimeters of organic material (which invariably smolders) and in those small patches (lateral scale of less than a meter or two) where the smoldering continues down to or very near the mineral soil. It is presumed that the hyphae of fungi likewise survive in the same parts of the organic layer and upper mineral soil. Mosses tend to be killed outright by smoldering although it has been noted that species with characteristically long setae (e.g. *Sphagnum*) can at times survive a fire (Greene *et al.* 2006). Most moss species in coniferous forests ("feathermosses") tend to be clustered under tree crowns—precisely where smoldering tends to consume organic material down to mineral soil.

We have discussed a number of plant traits that control injury of key tissues in forest fires. These traits are fire adaptive if they confer a fitness advantage to the plant, although traits that are fire adaptive in one fire regime are not necessarily fire adaptive in another (Keeley *et al.* 2011). For example, in fire regimes typified by frequent, low-intensity surface fires, adaptive traits include thick bark, elevated crowns, and large buds. These traits are adaptive in these fire regimes because they permit survival and reproduction for this suite of fire behavior characteristics. On the other hand, in fire regimes typified by high-intensity, stand-replacing fires, traits such as thin bark, low crowns, and small buds are adaptive because the biomass not allocated to structure can instead be invested in other areas that promote fitness such as seeds or root carbon reserves (resprouting). Finally, we stress that although fire adaptive traits confer a fitness advantage to the plant, they did not necessarily arise in response to fire as a selective agent (i.e. they are not necessarily adaptations; Keeley *et al.* 2011).

Large animals generally avoid fire. Most vertebrates—even toads—have the instinctive response to move away from smoke, and thus direct mortality from the fire is quite low. At the Yellowstone fires in 1988, half a million ha were burned and yet the mortality rate as measured immediately after the burns was about 0.5 percent for large mammal species such as elk, deer, and bison (Romme *et al.* 2011). Undoubtedly, the old and sick were well-represented among these casualties.

It is more difficult for small animals to avoid fire, because their movement rates can be smaller than fire spread rates. This does not necessarily mean that all small animals are killed by fire, however, and some fish, amphibians, and soil invertebrates can be relatively unaffected by the direct effects of flames or smoke because they are insulated by their surrounding soil or water media. In most other cases, however, post-fire mortality rates can be high and population densities can be strongly reduced. Ground-nesting bird populations can be dramatically reduced by fire, and one well-known example is *Tympanuchus cupido cupido* (heath hen), which was driven to the brink of extinction by a wildfire. Small mammals such as deer, mice and voles are not killed by heat but rather by asphyxiation, and thus generally experience high mortality rates from direct fire effects. Nonetheless, small mammal mortality rates are generally not measured directly, but instead estimated indirectly via comparisons of population densities in burned and unburned areas. However, this approach provides a rather poor estimate of mortality because: (1) surveys do not typically occur immediately after the fire, so recolonization is already under way; and (2) the distances from the burn sampling plots to the nearest unburned edges are not defined, which confounds density estimates. In consequence, we lack direct measurments of small mammal mortality rates as well as estimates on the pace of post-fire recolonization.

Recolonization of fires

The recolonization of plants can occur via resprouting or recruitment from seeds. Resprouting is rare among conifers but quite common among angiosperms. For angiosperm trees, the typical source of new shoots is from dormant buds around the root collar near the soil surface. These shoots are often initially numerous, but are reduced through self-thinning so that only one or two remain at maturity. A very minor number of angiosperm trees (but many shrubs and herbs) can also produce suckers from dormant buds along shallow roots or stolons. This latter form of asexual reproduction is far more useful than basal sprouts because it permits the clone to extend laterally. Some *Populus* (aspen) clones are among the largest organisms on Earth.

Likewise, much of the hyphal network of fungal clones survives in the duff and coarse woody debris. Many species, e.g. *Morchella* (morels) and *Geopyxis* (pixie cups) produce copious numbers of mushrooms immediately following wildfire, but almost exclusively where the organic soil (duff) has been substantially reduced by smoldering combustion. It is unclear why fungal fruiting bodies are mainly produced where hyphae have been pruned.

Sexual spread via seeds occurs even with the species relying primarily on asexual recruitment to persist *in situ*. However, given the tremendous dispersal constraints operating in large burns, dispersal by seed is far less reliable than asexual reproduction. Consider the case of *Picea glauca* (white spruce) in Figure 8.4. For an extensive array of conspecific seed sources (an area source) with a full display of leaves, seed densities 200 meters from the burn edge are reduced by more than 90 percent from those at the burn edge (Greene and Johnson 1996). Expected seed dispersal for an area source without leaves (resulting from crown scorch during the fire) is also shown in Figure 8.4. In this case, dispersal capacity is slightly improved because most seeds traveling well beyond the edge of the area source either began near the edge or spent most of the initial flight *above* the intact forest (i.e. they did not experience a long period in the low-speed environment of the forest).

Species with an aerial seedbank (i.e. delayed seed abscission leading to persistent seed crops) tend to produce a more or less equal number of seeds each year. By contrast, species without aerial seedbanks tend to instead exhibit masting, which results in tremendous inter-annual variation in crop size. It has been demonstrated from examination of tree ages that a mast year occurring one or two years after a large fire will subsequently produce the largest post-fire cohort (Peters *et al.* 2005).

Most woody species in fire-prone landscapes have wind-dispersed seeds and thus do not rely on animal dispersal. While the larger animals that can carry seeds beyond wind dispersal distances are found in recent fires because of the abundance of browse, they deposit seeds in infrequent clumps as they defecate or regurgitate. By contrast, while the wind cannot routinely

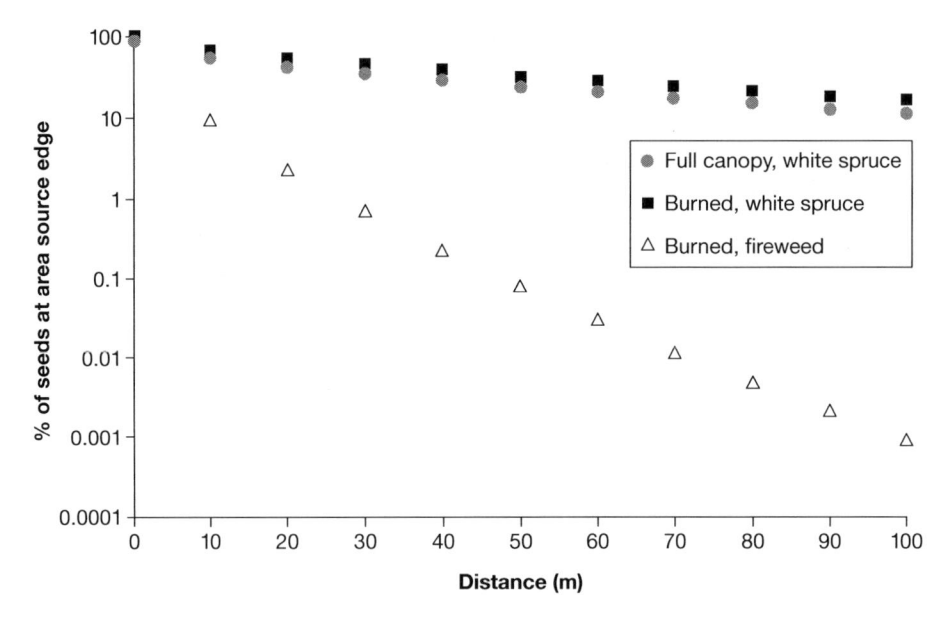

Figure 8.4 Seed dispersal by wind from the edge of an area source into a recent fire. "Full canopy" indicates that the leaf area index is that of an intact forest, while "burned" means that the area source has no leaves (e.g. dispersal of an aerial seedbank from burned trees). The expectations are based on the model of Greene and Johnson (1996). The herbaceous fireweed is poorly dispersed relative to the trees simply because its height is so much lower

disperse seeds more than a few hundred meters, it can insure that seeds are more evenly distributed across an area. Given the large size of burns, a more equitable spread of seeds is far more valuable than a scattering of seed-filled feces.

Germination success depends greatly on the seedbeds created by fire as the seeds, and therefore germinants, tend to be small. The wind-dispersed woody plants of fire-prone landscapes necessarily have small seeds because the rate of fall (and thus the distance traveled) is allometrically constrained by the ratio of seed mass to the area (including lift-producing wings or drag-producing fibers). For example, the fire-prone boreal forest has no species with seeds larger than about 10 mg. In turn, small seeds produce small germinants; given the isometric scaling of seed length, volume, and mass, germinant length generally scales as the 1/3 power of seed mass.

There are two main reasons why the transition from deposited seed to established seedling is so improbable. First, a few rainless days cause drying of the organic soil layer, which can lead to desiccation and death of the germinant unless its radicle reaches the underlying mineral soil or humus layers that can remain wet via capillary flow from below. Second, while the turgor pressure of a hydrating germinant is more than sufficient to push up dried leaves, deep litter layers can prevent photosynthetic tissues from reaching sufficient light which can lead to germinant death. For these reasons, establishment for small seeds requires a substrate with little to no organic layer (Figure 8.5).

Fires can create optimal seedbeds for small seeds via smoldering combustion of the duff. After the flames have burned a thin layer of the duff, smoldering combustion will subsequently reduce organic material down to the humus or, in some cases, the mineral soil (Miyanishi and Johnson 2002). These better seedbeds occur at the scale of a meter or two in a matrix of barely combusted organics, and germinants are effectively only found on these better microsites.

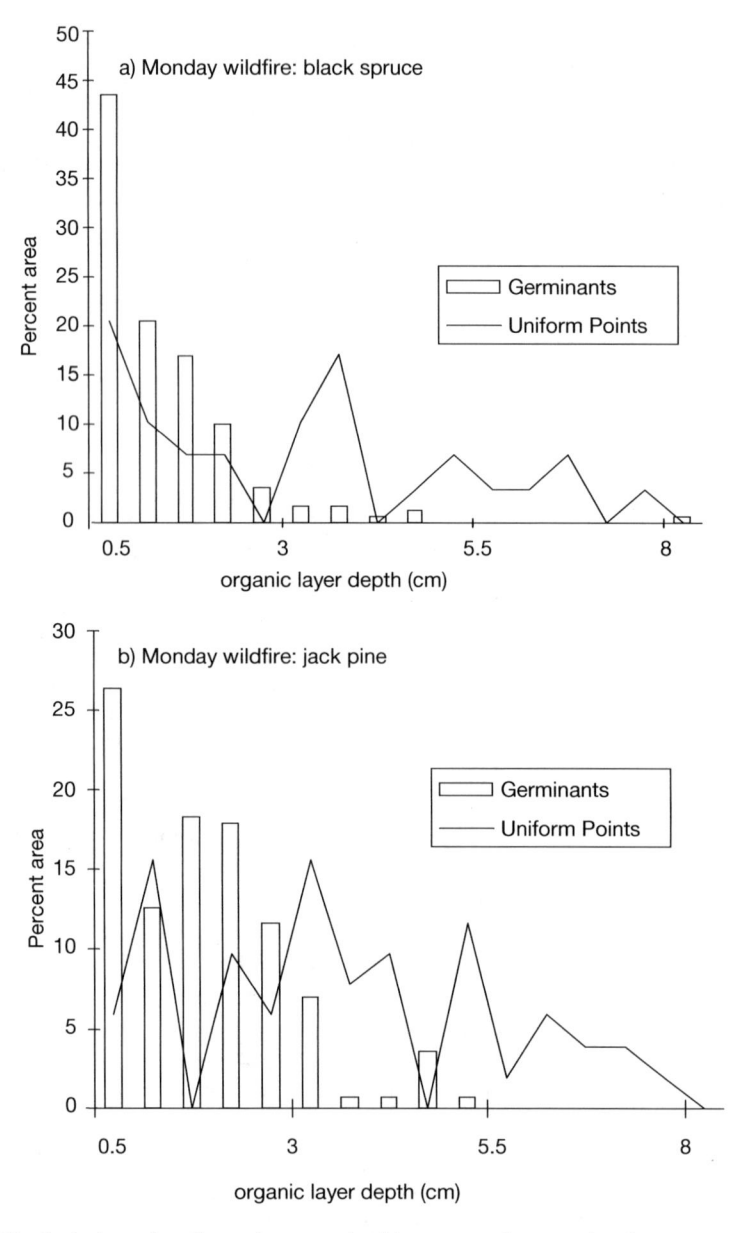

Figure 8.5 The limitation of small germinants to the thinnest post-fire organic substrates: an example from the Monday fire in the boreal forest of Saskatchewan. Germination success declines inversely with organic depth, and neither black spruce (*Picea mariana*) nor jack pine (*Pinus banksiana*) have many germinants at depths >3 cm. "Uniform points" refers to the actual percentage of the various organic layer depth classes; thus, the survivorship from the seed to the germinant stage is far higher on thin organic layers

While varying tremendously within or among forest fires, the average proportion of optimal seedbeds (defined as <3 cm depth following fire) is around 40 percent (Greene *et al.* 2007); by contrast it is only around 1 percent coverage in intact forest. On average, burned forests have a depth about half of the intact forest prior to the fire (Figure 8.5).

Moss and fungal spores are even smaller than the smallest trees seeds—only a few hundred microns in diameter—and consequently are even more dependent on substrate than are small seeds. The first arriving mosses, such as *Ceratodon purpureus*, tend to be tolerant of desiccation in open areas. Since spores are produced in such large numbers and spread so widely by the wind, one can use the bright green setae of, for example, new *Polytrichum* mosses as "phytometers" to measure seedbed quality for the seeds of conifers and angiosperms. Invariably, these mosses are found in the first few post-fire years not merely on mineral soil or thin humus, but on the more limited subsample of microsites that stay moist through a short drought. In a sense, then, the immediate post-fire substrate becomes a mosaic of mushrooms, mosses, and small-seeded, sexually recruited species on the well-combusted surfaces, with asexually responding plants on the adjacent, much less combusted, substrates.

These optimal seedbeds do not last long. As the herbaceous vegetation accrues, it deposits litter (much more rapidly than can the slower-growing woody plants) that begins to present a mechanical constraint to small seed germination. It has been argued, based on the few available studies, that this window of opportunity for small-seeded species is about four years (Charron and Greene 2002). Of course, the width of this window depends on herbaceous plant density and growth rates.

The foregoing argument refers to biomes such as the circumboreal where pre-fire organic layers are thick. In areas where ground fires are the predominant fire type, organic layers are typically much thinner, and thus smoldering combustion is not as crucial a process. In such forests, often lacking a closed canopy, the main inhibitor of germination is lack of water due to rainfall rather than evaporation from a porous, thick post-fire duff layer.

Aside from poor substrates, the other major cause of loss at the recruitment stage is granivory. Indeed, in an undisturbed forest, granivores typically take the majority of the seeds before they can germinate (Greene and Johnson 1998). However, in a large fire it is highly likely that initially the abundance of small mammals is low in the interior of the burn (Charron and Greene 2002). This is due to small mammal populations decimated by smoke inhalation, and subsequent re-invasion requires two to three years. By contrast, seed-eating birds have no serious initial dispersal constraint but typically occur in such low numbers that they are not a major source of pre-germination seed loss.

There are a limited number of plant strategies for coping with recurrent, large fires. The first, as already mentioned, is to possess a capacity for asexual reproduction, generally via apical meristems on stems or roots (e.g. axillary or adventitious buds) that are insulated against heating by bark or soil. The second is aerial seedbanks. A compelling trait associated with aerial seedbanks is that these are not masting species, and thus at the time of fire a population is assured of a large number of available seeds with no dramatic dispersal constraint. This requires that the seeds reside within a structure that can sufficiently insulate, provide sufficient resistance to heat transfer that some fraction of the seeds survives. It has been argued that in mast years non-serotinous species can produce such dense clusters of seed-bearing organs that the outer structures diminish the heat transfer sufficiently to permit the survival of seeds within the interior of the cluster. For example, seeds within small *Kunzea* (Myrtaceae) capsules of Australia can survive longer periods of heating when the capsules are clustered than when they are isolated (Pounden *et al.* in press). Persistence of serotinous species on a landscape is dependent on a fire return interval that is greater than the time required to produce the amount of seeds sufficient for subsequent self-replacement by the population. This may be the main arbiter for any species range boundary abutting the parkland transition to grassland invariably characterized by short return times (Brown and Johnstone 2012).

A third strategy for post-fire colonization is to diffuse seeds from surviving wind-dispersed plants at the fire edge or within residual stands, but this should really be viewed as a lack of a more refined adaptation. However, asexual reproduction and dispersal from aerial seedbanks are far more effective mechanisms for repopulating a large fire than is solely dispersing from an edge. Despite the well-invaginated, elongate shape of big fires, nonetheless only about 5 percent of wind-dispersed seeds of trees at the edge will travel more than 200 m from a surviving edge (Figure 8.4). The percentage decreases with distance from the burn edge. That is, asexual reproduction and dispersal from aerial seedbanks are far more effective mechanisms for repopulating a large fire than is solely dispersing from an edge.

Many herbaceous and shrub species mix these strategies. The modestly shade-intolerant fireweed (*Chamerion angustifolia*) persists in ever-dwindling numbers as a forested stand ages. Immediately after a fire, these few individuals then perennate from surviving organs within the forest floor and produce a first-summer seed crop of small seeds with a fibrous drag-producing appendage that is well-dispersed by wind. Despite the very low rate of fall (around 0.08 m s^{-1}), the dispersal distance from an area source of fireweed is poor compared to trees because the release height and wind speed is much lower. Given that most parts of the fire will have a few surviving individuals there will be a considerable number of new fireweed plants two years after a fire. These individuals disperse a far greater density of seeds than was seen the fall immediately following the fire. Three years after a fire brings the climax of the violet-flowered display on the burned landscape as the great majority of sites that can support fireweed have been colonized. Meanwhile, rapidly-growing woody plants have begun to shade out the herbaceous fireweed, and thus the populations begin their slow diminishment during the interval until the next fire.

Another connection to masting is exhibited by several Australian shrub species that resprout after fire and then produce a very large seed crop from carbohydrates stored in the roots. Experiments have shown that the mast crop produced after burning in these "fire-stimulated" species is indeed triggered by the burning of the plant (Lamont and Downes 2011).

Many animal species are reliant on large fires. Like fireweed, they readily colonize recent fires, rapidly increase their populations, and then begin a slow decline in abundance until the next large fire (Saint-Germain and Greene 2009). Such species include many saprophagous beetle species such as *Bupestris* or *Monochamus* that can sense wood smoke even from a great distance and move upwind toward the source. In turn, bird species specialized on saprophagous insects, most famously woodpeckers, increase their populations suddenly in the aftermath of a fire.

Before we leave the topic of recolonization by plants and animals, it is important to mention the very recent trend toward rapid salvage. As shown by a number of authors, having a second disturbance following immediately after a fire is a stress that most fire-adapted species cannot tolerate. Certainly, most plant species arriving primarily by seed are adversely affected (e.g. Leverkus *et al.* 2014 in Mediterranean forests), as are saprophagous species and their predators (Saint-Germain and Greene 2009).

References

Anderson, H. E. 1983. Predicting wind-driven wild land fire size and shape. USDA Forest Service Research Paper INT-305.

Balfour D. A., and J. J. Midgley. 2006. Fire induced stem death in an African acacia is not caused by canopy scorching. *Austral Ecology* 31: 892–896.

Brown, C. D., and J. F. Johnstone. 2012. Once burned, twice shy: Repeat fires reduce seed availability and alter substrate constraints on *Picea mariana* regeneration. *Forest Ecology and Management* 266: 34–41.

Charron, I., and D. F. Greene. 2002. Post-fire seedbeds and tree establishment in the southern mixedwood boreal forest. *Canadian Journal of Forest Research* 32: 1607–1615.

Cruz, M. G., B. W. Butler, M. E. Alexander, J. M. Forthofer, and R. H. Wakimoto. 2006. Predicting the ignition of crown fuels above a spreading surface fire. Part I: Model idealization. *International Journal of Wildland Fire* 15: 47–60.

Finney, M. A. 2001. Design of regular landscape fuel treatment patterns for modifying fire growth and behavior. *Forest Science* 47: 219–228.

Greene, D. F., and E. A. Johnson. 1996. Wind dispersal of seeds from a forest into a clearing. *Ecology* 77: 595–609.

Greene, D. F., and E. A. Johnson. 1998. Seed mass and juvenile survivorship of trees in clearings and shelterwoods. *Canadian Journal of Forest Research* 28: 1307–1316.

Greene, D. F., S. Gauthier, J. Noel, M. Rousseau, and Y. Bergeron. 2006. A field experiment to determine the effect of post-fire salvage on seedbeds and tree regeneration. *Frontiers in Ecology and the Environment* 4: 69–74.

Greene, D. F., S. E. Macdonald, S. Haeussler, S. Domenicano, J. Noël, and others. 2007. The reduction of organic layer depth by wildfire in the North American boreal forest and its effect on tree recruitment by seed. *Canadian Journal of Forest Research* 37: 1012–1023.

Hartford, R. A., and W. H. Frandsen. 1992. When it's hot, it's hot etc. or maybe it's not! (Surface flaming may not portend extensive soil heating). *International Journal of Wildland Fire* 2: 139–144.

Hayden, B., and D. F. Greene. 2008. 39 pages. The ecology of tropical dry forests. In: International Commission on Tropical Biology and Natural Resources, [Eds. K. Del Claro, P. S. Oliveira, V. Rico-Gray, A. A. Almeida Barbosa, A. Bonet, and others], *Encyclopedia of Life Support Systems (EOLSS)*, Developed under the Auspices of the UNESCO, EOLSS, Publishers, Oxford, UK, [http://www.eolss.net].

Hillel, D. 1998. *Environmental Soil Physics*. Academic Press, New York.

Kavanagh, K. L., M. B. Dickinson, and A. S. Bova. 2010. A way forward for fire-cased tree mortality prediction: Modeling a physiological consequence of fire. *Fire Ecology* 6: 80–94.

Keeley, J. E., and C. J. Fotheringham. 2001. Historic fire regime in southern California shrublands. *Conservation Biology* 15: 1536–1548.

Keeley, J. E., J. G. Pausas, P. W. Rundel, W. J. Bond, and R. A. Bradstock. 2011. Fire as an evolutionary pressure shaping plant traits. *Trends in Plant Science* 16: 406–411.

Kolden, C. A., J. A. Lutz, C. H. Key, J. T. Kane, and J. W. van Wagtendonk. 2012. Mapped versus actual burned area within wildfire perimeters: Characterizing the unburned. *Forest Ecology and Management* 286: 38–47.

Krawchuk, M. A., M. A. Moritz, M.-A. Parisien, J. Van Dorn, and K. Hayhoe. 2009. Global pyrogeography: The current and future distribution of wildfire. *PLoS ONE* 4: e5102.

Lamont, B. B. and K. S. Downes. 2011. Fire-stimulated flowering among resprouters and geophytes in Australia and South Africa. *Plant Ecology* 212: 2111–2125.

Leverkus, A. B., J. Lorite, F. B. Navarro, E. P. Sanchez-Canete, and J. Castro. 2014. Post-fire salvage logging alters species composition and reduces cover, richness, and diversity in Mediterranean plant communities. *Journal of Environmental Management* 133: 323–331.

Macias Fauria, M., S. T. Michaletz, and E. A. Johnson. 2011. Predicting climate change effects on wildfires requires linking processes across scales. *WIRES Climate Change* 2: 99–112.

Michaletz, S. T., and E. A. Johnson. 2006. A heat transfer model of crown scorch in forest fires. *Canadian Journal of Forest Research* 36(11): 2839–2851.

Michaletz, S. T., and E. A. Johnson. 2007. How forest fires kill trees: A review of the fundamental biophysical processes. *Scandinavian Journal of Forest Research* 22: 500–515.

Michaletz, S. T., E. A. Johnson, and M. T. Tyree. 2012. Moving beyond the cambium necrosis hypothesis of post-fire tree mortality: Cavitation and deformation of xylem in forest fires. *New Phytologist* 194: 254–263.

Miyanishi, K., and E. A. Johnson. 2002. Process and patterns of duff consumption in the mixedwood boreal forest. *Canadian Journal of Forest Research* 32: 1285–1295.

Nelson, R. M. Jr. 2001. Water relations of forest fuels. In: Johnson, E. A. and K. Miyanishi, eds. *Forest Fires: Behavior and Ecological Effects*. Academic Press, New York, NY, USA.

Peters, V. S., S. E. MacDonald, and M. R. T. Dale. 2005. The interaction between masting and fire is key to white spruce regeneration. *Ecology* 86: 1744–1750.

Pounden, E., D. F. Greene, and S. Michaletz. In press. Non-serotinous woody plants behave as aerial seed bank species when a fire late in the seed maturation period coincides with a mast seeding year. *Ecology and Evolution*.

Quintiere, J. G. 2006. *Fundamentals of Fire Phenomena*. John Wiley and Sons, New York.

Romme, W. H., M. S. Boyce, R. Gresswell, E. H. Merrill, G. W. Minshall, and others. 2011. Twenty years after the 1988 Yellowstone fires: Lessons about disturbance and ecosystems. *Ecosystems* 14: 1196–1215.

Saint-Germain, M., and D. F. Greene. 2009. Salvage logging in the boreal and cordilleran forests of Canada: Integrating industrial and ecological concerns in management plans. *Forestry Chronicle* 85: 120–134.

Sharples, J. J., G. A. Mills, R. H. D. McRae, and R. O. Weber. 2010. Foehn-like winds and elevated fire danger conditions in southeastern Australia. *Journal of Applied Meteorology and Climatology* 49: 1067–1095.

Stocks, B. J., J. A. Mason, J. B. Todd, E. M. Bosch, B. M. Wotton, and others. 2002. Large forest fires in Canada, 1959–1997. *Journal of Geophysical Research: Atmospheres* 108: 8149.

Van Wagner, C. E. 1969. A simple fire growth model. *Forestry Chronicles* 45: 103–104.

Van Wagner, C. E. 1973. Height of crown scorch in forest fires. *Canadian Journal of Forest Research* 3: 373–378.

9

ECOLOGICAL EFFECTS OF STRONG WINDS ON FORESTS

Stephen M. Turton and Mohammed Alamgir

Introduction

Natural and anthropogenic disturbances shape forest ecosystems by controlling their structure, species composition and functional processes (Dale *et al.*, 2001). Forest ecosystem dynamics are largely dependent on natural disturbances (like strong winds), that reshape ecosystem structure and composition, modulate ecosystem functioning, and reset and accelerate succession (Franklin *et al.*, 2002; Turton and Stork, 2008; Turner, 2010; Thom *et al.*, 2013). Strong winds – typically those above gale force or 61 km h^{-1} – are among the most important exogenic disturbance agents affecting forest ecosystems across the world, at a range of scales (Proctor *et al.*, 2001; Zhao *et al.*, 2006; Lugo, 2008; Turton, 2008; Wang and Xu, 2009; Yoshida *et al.*, 2011; Turton, 2012).

Tropical cyclones – also known as hurricanes and typhoons – affect wet and dry tropical forest regions adjacent to eight tropical ocean basins around the world: 1) northwest Pacific; 2) north Indian; 3) southwest Indian; 4) southeast Indian; 5) southwest Pacific; 6) northeast Pacific; 7) north Atlantic/Caribbean; and 8) south Atlantic (Turton, 2013). Forests between about 5–7 degrees north and south of the equator do not experience tropical cyclones due to the weak Coriolis effect near the equator. Nonetheless, there are many anecdotal reports of severe damage to forests over several square kilometres outside the typhoon belt in Southeast Asia (e.g. Whitmore and Burslem, 1998), and there is evidence that these events are common enough to have an influence on the structure of many of the forests in the region (e.g. Proctor *et al.*, 2001; Baker *et al.*, 2005).

Tropical cyclones also affect the forests of subtropical and temperate regions of both hemispheres as they often transition into intense extra-tropical cyclones as they move into higher latitudes (Everham III and Brokaw, 1996). Winter windstorms produced by intense mid-latitude depressions (extra-tropical lows) affect the temperate deciduous forests of western and central Europe, northeast Asia, eastern and northwest parts of North America and many temperate forest areas in the Southern Hemisphere, while tornadoes regularly affect the temperate forests of North America, and occasionally those in Europe and Asia (Fischer *et al.*, 2013). Furthermore, tornadoes often develop within the eye-wall and feeder bands of intense tropical cyclones and may produce severe local-scale effects on forests along their path (Turton, 2008).

Strong winds may affect forests at spatial scales ranging from a few hundred square metres (e.g. crown damage from a single severe thunderstorm) to thousands of square kilometres (e.g. landscape-scale damage from a large synoptic-scale intense low-pressure weather system). Other events may produce severe winds and forest damage at more intermediate spatial scales (e.g. a tornado). Winds may affect forests at temporal scales ranging from a few minutes (e.g. a single microburst from a severe thunderstorm) to several days (e.g. a slow-moving severe tropical cyclone or intense extra-tropical cyclone). Other events may produce strong winds that impact on forests at scales of a few hours (e.g. a slow-moving super-cell thunderstorm). Wind damage to forests may be due to both horizontal and vertical wind gusts, the latter often associated with downdrafts from severe thunderstorms.

There is little doubt that strong winds play an important role in shaping and moderating ecological processes in forests – from the tropics to the high latitudes. In this chapter we consider the ecological effects of strong winds on the main forest biomes of the world – tropical, temperate and boreal – at a range of scales. In particular we consider effects of strong winds on forest structure, forest function, forest succession and biodiversity. We conclude with an evaluation of the likely effects of projected global climate change on extreme wind events and the possible consequences for forest ecosystems.

Scales of forests disturbances

Wind is a universal phenomenon and 'average' winds rarely damage forests. However, there are some forests that experience 'chronic' wind stress from persistent winds of lower speeds that have a dramatic effect on forest structure. The Nanjenshan forest plot in southern Taiwan provides a good example of a lowland forest heavily influenced by prevailing monsoon winds (Chao et al., 2010). Many montane forests in the tropics also exhibit forest structures indicative of 'chronic' wind stress – they are typically lower in stature with canopies 'sheared' in the direction of the prevailing winds (Tanner and Bellingham, 2006).

Strong winds – associated with a number of weather systems – may affect forests across a range of spatial scales. In temperate and boreal forests, the disturbances regimes are relatively large in North American forests, small in European forests and intermediate to large in East Asian forests (Fischer et al., 2013). Severe wind events are responsible for more than 50 per cent of forest damage from catastrophic events by volume in Europe (Gardiner et al., 2011). A severe tropical cyclone (hurricane/typhoon) may affect thousands of square kilometres of forest area and can alter ecosystems significantly. By comparison, tornadoes usually affect small forest areas but cause the most severe localized damage in affected areas (Fischer et al., 2013).

Impacts of strong winds on forest ecosystems

Strong winds have both highly visible and cryptic impacts on forest ecosystems (Everham III and Brokaw, 1996; Lugo, 2008; Turton, 2008; Turton, 2012). Winds affect forests at levels ranging from the smallest functional unit to the largest structural unit. In Figure 9.1, we highlight the effects on various components of a forest ecosystem that has been subjected to a severe wind event. The figure also represents the changing dimensions of ecological effects from the short to long term that typically occur in forests in the aftermath (or recovery phase) of a strong wind event. The immediate ecological effects are closely related to changes in forest structure, whereas the long-term effects usually affect ecosystem functioning, forest succession and biodiversity.

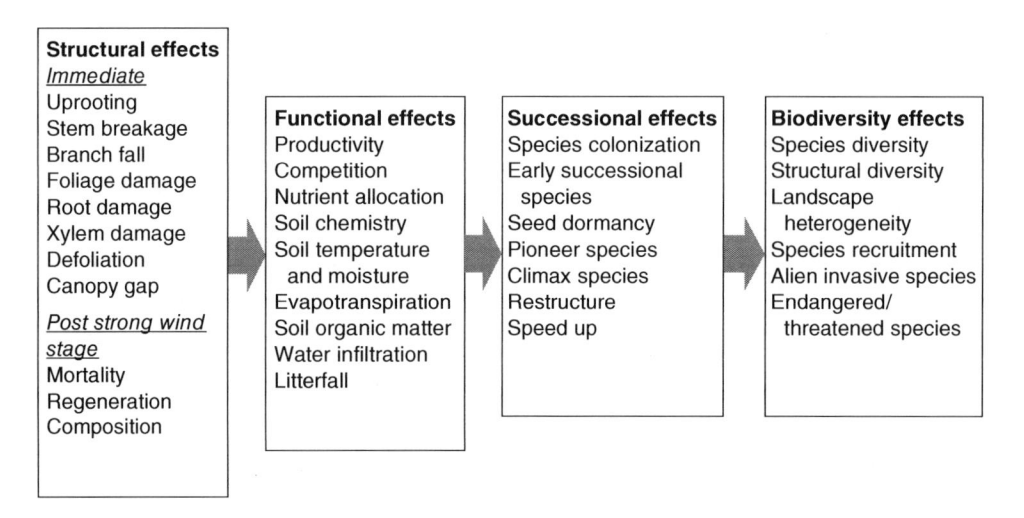

Structural effects	Functional effects	Successional effects	Biodiversity effects
Immediate Uprooting Stem breakage Branch fall Foliage damage Root damage Xylem damage Defoliation Canopy gap *Post strong wind stage* Mortality Regeneration Composition	Productivity Competition Nutrient allocation Soil chemistry Soil temperature and moisture Evapotranspiration Soil organic matter Water infiltration Litterfall	Species colonization Early successional species Seed dormancy Pioneer species Climax species Restructure Speed up	Species diversity Structural diversity Landscape heterogeneity Species recruitment Alien invasive species Endangered/ threatened species

Figure 9.1 Effects of strong winds on forests ecosystems (adapted from Seidl and Blennow, 2012)

Forest ecosystem structure

The main structural elements of forest ecosystems are trees. Strong winds substantially affect trees in any forest biome due to their prominence and density compared with other terrestrial biomes. The scale and intensity of effects are distributed from individual tree- to stand- to landscape- levels depending on wind intensity and the spatial extent of the impacting weather system. Individual tree-level effects include: 1) breakage of branches, foliage and main stem; 2) defoliation; 3) shedding of bark; 4) root damage; and 5) uprooting of the entire tree. Stand-level effects include: 1) creation of gaps in the canopy by uprooting a large tree or breakage of several canopy trees; 2) mortality of seedlings and saplings due to sudden exposure to unfavorable environmental factors such as more sunlight, higher temperatures and lower humidity; and 3) sometimes landslides. At the landscape-level, strong winds may create landscape heterogeneity (Foster and Orwig, 2006), e.g. when an intense tropical cyclone damages some patches of a primary forest/old growth secondary forest, there is always a possibility of changing the structure and composition of those patches during forest recovery.

Table 9.1 summarizes tree and forest attributes to be considered in relation to strong wind events. Trees along forest edges are more susceptible to snapping whereas interior forest trees – away from edges – are more susceptible to uprooting (Yoshida *et al.*, 2011). The regular exposure to wind makes edge trees more wind resistant and such trees are usually snapped before experiencing root damage. In a hurricane-effected area, defoliation was higher in canopy trees than those in the understory but after the hurricane higher re-foliation was found in the understory than canopy (Wunderle Jr. *et al.*, 1992; Brokaw *et al.*, 2004). Generally, damage is greater for the larger canopy trees compared with understory trees (Lugo, 2008). Apart from wind velocity, size is also a determining factor in tree damage due to strong winds (Webb, 1988; Zhao *et al.*, 2006). Generally, larger canopy trees (Webb, 1988; Lugo, 2008) and intermediate trees (Dyer and Baird, 1997) are more susceptible to damage than smaller trees (Webb, 1988). Alternatively, more established root systems might make larger trees more resistant to wind damage and less vulnerable to uprooting; consequently there is no disproportionate damage in larger trees (Dyer and Baird, 1997). However, more stem rot and root rot may result in large old (senescing) trees being more vulnerable to wind damage

Table 9.1 Tree and forest stand attributes and likely effects of strong winds

Attributes	Likely effects
Higher wood density trees	More uprooting, less snapping, more resistance
Lower wood density trees	Less uprooting, more snapping, less resistance
Slow growing trees	More resistance
Fast growing trees	Less resistance
Trees on forest edge	More snapping, less uprooting
Interior trees	More uprooting, less snapping
Larger trees (dbh* and height)	Less resistance**
Smaller trees (dbh and height)	More resistance
Trees with shallow root system	Less resistance
Trees with deep root system	More resistance
Canopy trees	More damage
Understory trees	Less damage
Denser forest stand	More tolerant*

* Diameter at breast height (1.3 m)
** Opposite arguments are also available in literature.

(Yoshida *et al.*, 2011). In the mangrove forests of Florida, larger diameter trees suffered from greater mortality and more snapping after Hurricane Andrew than smaller diameter trees (Baldwin *et al.*, 1995). In tropical forests, trees with higher wood density are more susceptible to uprooting than snapping, whereas trees with lower wood density are more susceptible to snapping and post-disturbance mortality due to pathogens and insect attack (Lugo, 2008).

The structural effects of strong winds on forest ecosystems are direct, immediate and highly visible. In the short term, visible damage occurs to the structural elements of the forest, but in the long term, structural damage substantially affects the species composition of the forest and its vertical structure, stand dynamics, landscape heterogeneity and ecosystem functioning (Figure 9.1). Severe structural damage may turn a carbon 'sink' forest into a carbon 'source' forest for a period of time, only to return to a carbon sink as regrowth ensues. Savanna forests are more exposed to strong winds due to their sparse distribution of trees; however, trees of savanna forests are more wind resistant due to their lower leaf area indices and lower stature. Dense forest stands are more tolerant to strong winds (Yoshida *et al.*, 2011), so tropical primary/old growth secondary forest are generally less susceptible to strong winds than more open forests. In general, temperate forests are more susceptible to strong winds than tropical forests (Webb and Scanga, 2001).

Fragmented forests (often within an agricultural matrix) are more vulnerable to strong winds than areas of contiguous forests (Laurance and Curran, 2008). This is due to their abrupt artificial margins and open and often exposed surrounding landscapes that enhance wind speeds near edges of forest remnants. However, forest structure, measured up to six months after Tropical Cyclone Larry in northeast Australia, did not differ between small (<40 ha) forest remnants and larger intact forest areas; the severity of effects in both was largely explained by distance from the path of the cyclone across the landscape (Catterall *et al.*, 2008). Hence, both local and regional contexts must be taken into account when evaluating the vulnerability of fragmented forest landscapes to strong wind events.

Forest ecosystem functioning and nutrient cycling

The effect of strong winds on forest ecosystem functioning is a critical element because disruption to any ecosystem functioning processes may have profound long-term effects on forests. After strong wind events residual trees in forests often suffer from root and xylem damage that ultimately reduces the overall productivity of affected forests in the following three to five years (Zhao *et al.*, 2006); however, it may take longer depending on the intensity of damage. In the long term the recovering forest might be more productive. This might be due to more light availability (Pascarella, 1997), reduced biotic competition, and increased supply of nutrients and rainwater in the residual trees. For example, Zhao *et al.* (2006) found that after 12 years of hurricane damage in the bottomland hardwood forests of South Carolina in the USA that stem density – both in the tree and sapling layers – exceeded pre-hurricane levels.

Uprooting of trees due to strong winds mixes the upper soil layer that is advantageous for site quality and plant growth, but uprooting creates pits and mounds in the local topography that may increase pest infestation at the site (Dyer and Baird, 1997). Partially damaged trees tend to become more susceptible to insect and fungal attack (Lugo, 2008).

If strong winds destroy the habitat of micro- and macro-fauna in a forest, they typically suffer for food and shelter – resulting in the death of invisible micro-fauna that are very important for nutrient cycling, while visible macro-fauna are very important for maintenance of food webs in any forest ecosystem. This phenomenon is very common in mangrove forests in Southeast Asia and northern Australia, and temperate forests of Europe, North America and Japan. Strong winds may also have an impact on soil temperature, moisture and soil compaction. Increased soil compaction after strong wind events reduces water infiltration capacity, increases surface run-off and erosion of sediment into catchments in the subsequent years after the event. Furthermore, river flows often increase in wind-disturbed forest catchments due to decreased evapotranspiration from the denuded forest cover (Foster and Orwig, 2006).

Strong wind disturbance produces high canopy litterfall in forests that eventually decomposes and adds nutrients to the soil, and ultimately has a strong influence on forest primary productivity. For example, in the subtropical humid forests of Puerto Rico – where hurricanes are somewhat infrequent – the sudden increase of nutrients in the soil from wind-induced litterfall significantly altered the nutrient patterns and forest responses in the understory (Lodge *et al.*, 1991). However, the impact may be very different in forest regions where tropical cyclones are more frequent, such as parts of Southeast and East Asia (Lin *et al.*, 2003). Tropical forest ecosystems are very efficient at decomposing litterfall and making the micronutrients available for trees within six months of a strong wind event (Ostertag *et al.*, 2003). In forests where decomposition rates are very slow, like temperate and boreal forests, litterfall – due to strong winds – may prevent germination on the forest floor by creating a 'space' between seeds and mineral soils. Such effects are not restricted to temperate and boreal forests, e.g. Shiels *et al.* (2010) reported the death of seedlings from the deposition of a layer of canopy-derived debris on the ground after a hurricane in Puerto Rico. Hence, strong winds have profound effects on ecosystem functioning and nutrient cycling but this varies among forest biomes. Most of the effects are negatively correlated with ecosystem health and forest stability. However, a few of them may be beneficial.

Forest ecosystem succession

Forest ecosystem dynamics substantially depend on natural disturbances like strong winds that play an important role in forest succession (Thom *et al.*, 2013). They may redirect forest succession (Lugo, 2008) and/or speed up forest succession – depending on the context (Zhao et al., 2006). Upper canopy layer trees are affected more in forests during tropical cyclones than intermediate canopy layer trees. So, tropical cyclones tend to accelerate growth of intermediate-sized trees, which are usually later successional species in forests (Webb and Scanga, 2001). Tropical cyclones also create opportunities for regeneration of early successional species on the forest floor, by creating canopy gaps and allowing more sunlight to reach the understory (Batista and Platt, 2003). They also promote growth of shade tolerant species within the seedling and sapling layer (Zhao et al., 2006). Larger trees in forests (usually shade intolerant) generally suffer more from tropical cyclone damage and post-disturbance mortality. Therefore, strong winds may facilitate colonization of shade intolerant species in some forests (Zhao et al., 2006).

Strong winds sometimes break the seed dormancy of successional species – which may have been inactive in the soil over many years. This process helps regeneration of dormant species at the site as well as restructuring forest succession. For example, increased regeneration of a dormant seed bank of *Lysiloma latisiliquum* – a late secondary canopy tree species – occurred in south Florida due to the strong diurnal soil temperature fluctuations associated with conditions after Hurricane Andrew (Pascarella, 1997).

Forest biodiversity

Strong winds may have profound impacts on forest biodiversity at a range of scales. Immediately after disturbance species diversity usually increases due to increased nutrients, sunlight and space availability – that together enhance regeneration and recruitment of new species (Lugo, 2008). Zhao et al. (2006) demonstrated that hurricane disturbances restructure species composition and enrich species diversity via this process. Fischer et al. (2013) reported that strong wind disturbance may kill large trees but may create necessary conditions for new tree cohorts – thereby increasing the species diversity in disturbed forests. Strong wind disturbances increase the abundance of climbers, vines and herbaceous plants (Lugo, 2008). These plants may suppress the growth of young trees and recruits – which may be a limiting factor for increasing species diversity after wind disturbances. Hence, following disturbance, abundance of late successional (climax) species may increase in forests (Weaver, 1989). The reason being that in most forests early successional (pioneer) species are generally light demanding, large and reach either the canopy or above the canopy as emergent trees – so they are impacted more than climax species by strong wind events.

Strong winds create structural diversity in forests by promoting growth from suppressed understory saplings and seedlings (Zhao et al., 2006). Foster and Terborgh (1998) discovered in the Amazon forest that selective killing of large trees due to wind storms contributed to structural heterogeneity. Similar affects were also reported from temperate forests due to severe wind damage from tornadoes (Glitzenstein and Harcombe, 1988).

Over longer time periods severe wind disturbances may create landscape diversity (heterogeneity) in forests (Dyer and Baird, 1997). This diversity may arise from landslides or from differential recovery rates of different forest patches after severe meteorological or other disturbances, e.g. fire, dieback due to disease. Forest fragmentation provides an opportunity for recruitment of new species into forest ecosystems, including non-native invasive species. Bellingham et al. (2005) found that strong wind disturbances accelerated the invasion of non-native plant species in tropical forests in Jamaica. The invasive plant species are the main cause of elimination of native biodiversity in some forest ecosystems. For example, after Hurricane Andrew in Florida, the frequency and cover of non-native species in forests was increased compared to native species (Horvitz et al., 1998). Tropical cyclones (hurricanes/typhoons), tornadoes and even windstorms in European forests may uproot large trees or break crowns of large trees. These large trees act as a seed bank for natural regeneration in forests and also provide important microhabitat for fauna and specialized flora, e.g. epiphytes and vines. So if there is no intermediate tree species in the seed bank, the forest suffers due to a lack of sufficient regeneration (Yoshida et al., 2011) – that ultimately affects the whole forested landscape. Animal diversity may also suffer following wind disturbance across a fragmented forest landscape due to a loss of food resources for vulnerable species (Turton, 2012).

Natural disturbances – like strong winds – are very important factors in maintaining high biodiversity in forests (Zhao et al., 2006). The influence of strong winds on biodiversity is mainly dependent on the scale of disturbance – both intensity and frequency. A rigorous debate exists about the scale of disturbance required to maintain the highest level of biodiversity in forests. Here we describe (Figure 9.2) one of the very early and most popular theories – the 'intermediate disturbance theory' developed by Connell (1978). This theory argues that ecosystems that experience small- and large-scale disturbances typically have lower levels of biodiversity compared with those that have intermediate levels of disturbance – that typically maintain the highest biodiversity. If there is no or low disturbance, there is more competition and ultimately low biodiversity. If there is a high level of disturbance there is more mortality and the forest has less time for recruitment and recovery, and consequently fewer species may survive (Connell 1978). Post-disturbance mortality also has an influence on biodiversity after a very large-scale disturbance. For example, strong winds may be catastrophic for animal diversity, particularly if the affected forests are habitat for endangered and threatened species because they may also die out during the post-disturbance stage. For example, Tropical Cyclone Larry in 2006 directly killed 35 per cent of the local endangered Southern Cassowary population at Mission Beach in northeast Australia. However, this flightless frugivorous bird suffered even higher mortality after the cyclone due to starvation and road kills in the highly fragmented forest landscape around Mission Beach (Turton 2012).

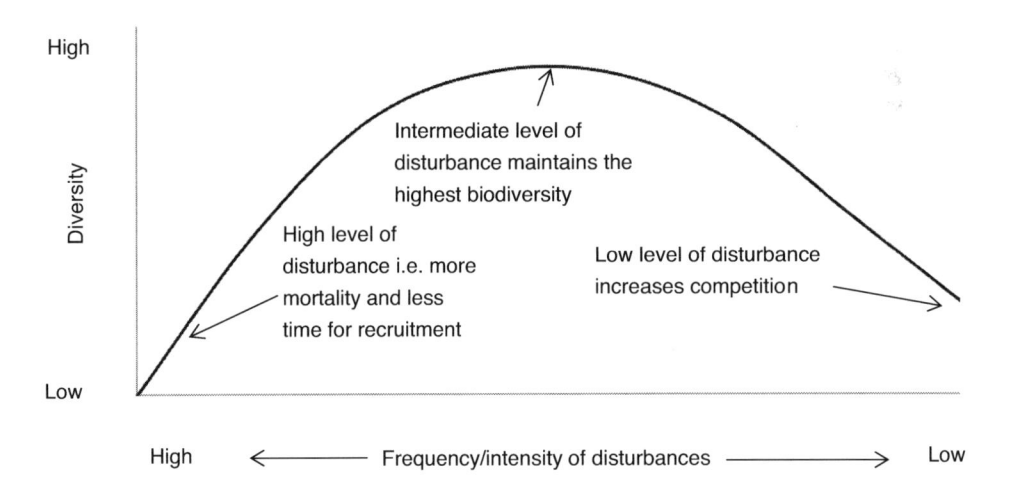

Figure 9.2 The intermediate disturbance theory (after Connell, 1978)

Forest recovery

Compared with other ecosystems (e.g. grasslands and wetlands), forest ecosystems usually require a long time for recovery after a strong wind event due to their relatively complex, slow growing and long-life characteristics. For example, it may take several decades after severe tropical cyclones for an affected forest to return to its original state (Metcalfe et al., 2008). Forest recovery after strong wind events depends on the scale and intensity of the disturbance regimes, composition and structure of the forest, site quality, forest resistance and resilience to strong winds and the adaptive capacity of the residual trees to the changing environment, notably the microclimate.

More resilient forest has more capacity to recover in the short term after disturbances, while more complex forests are more resistant to strong wind than less complex forests (Lugo et al., 1983; Zhao et al., 2006). Higher biodiversity is an indicator of a complex forest. In tropical forests recovery after cyclones is relatively rapid due to higher production rates associated with humid tropical environments (Metcalfe et al., 2008). However, Tanner and Bellingham (2006) reported for forests in Jamaica that higher elevation forests – with lower species diversity – were more resistant to strong winds than higher diversity, lower elevation forests. Following wind disturbance temperate and boreal forests recover more slowly than tropical forests due to their lower rates of primary productivity. Forests containing trees with sprouting ability may recover more rapidly than forests with trees lacking this function and are more dependent on other means of regeneration, e.g. seeding and sapling regrowth. In the early stage of recovery invasive species are abundant in disturbed forest areas. These invasive species limit the opportunity for regeneration and recruitment of many native species. They may delay forest recovery and change the species composition – particularly in large open areas and within riparian vegetation along river edges. However, in contiguous forest areas weed invasions tend to be ephemeral following severe wind disturbance. For example, Turton (2012) reported for the wet tropics of northeast Australia that most of the invasive weeds died out 12–24 months after severe forest damage from a tropical cyclone – as they were unable to persist following understory regrowth and infilling of the main forest canopy.

Notable strong wind events with forest disturbance regimes

Strong wind disturbances regularly affect forest biomes across the world. In European forest ecosystems strong wind is the most important disturbance factor in terms of the volume of timber damage (Schelhaas et al., 2003). In the forests of the eastern USA hurricanes are also one of the most important disturbance factors. For example, in terms of total area affected, hurricanes are the most important disturbance factor in the temperate New England forests in the USA (Foster and Orwig, 2006). Tropical cyclones are also one the most important disturbance factors for tropical forests in northeast Australia (Turton, 2008; Turton, 2012). Box 9.1 provides examples of significant wind disturbance events for three contrasting forest areas in the world.

> **Box 9.1 Examples of wind disturbance events for three contrasting forest biomes**
>
> It was found that in European forests the average structural damage due to strong winds was 1.8 million cubic metres from 1950 to 2000. For example, the storm Gudrun, from 8–9 January 2005, affected the spruce forests of Norway and Sweden, and was the worst storm on record for Sweden. It caused an estimated economic damage of 2.4 billion Euro, damaged over 75 million cubic metres of wood and resulted in significant growth reduction of residual trees (Schelhaas et al., 2003; Nilsson et al., 2004; Seidl and Blennow, 2012).
>
> A severe hurricane in 1938 affected the forest areas of central Massachusetts in the USA and was the most destructive storm in the last 175 years. It damaged 3 billion board feet of timber, increased river flow and caused major ecosystem changes and damage to 80 per cent of canopy trees (Patric, 1974; Cooper-Ellis et al., 1999; Foster and Orwig, 2006).
>
> Severe Tropical Cyclone Larry affected forests of northeast Australia in March 2006. It caused extensive damage to human communities, primary industries, remnant forests and the intact forests of the Wet Tropics of Queensland World Heritage Area (Turton, 2008; Turton, 2012).

Factors that determine effects of strong winds on forest ecosystems

Forest damage from strong winds largely depends on wind speed – or its velocity – as the damaging forces of wind are directly proportional to the fourth power of the velocity of the wind. Apart from the physical dimensions of wind, degree of damage also depends on a number of endogenic factors within forests. So a strong wind event with equivalent force affects different forest types at varying degrees depending on a range of endogenic factors:

- Forest structure and species composition
- Topographic conditions
- Soil characteristics.

Mangrove forests and coastal forests are more vulnerable to strong winds due to their proximity to the ocean where wind speeds are higher than adjacent land areas. Mangroves are possibly also vulnerable because they have shallow roots to avoid anoxic soils. Wang and Xu (2009) found that in Lower Pearl River Valley (USA) coastal wetland forests were more susceptible to strong wind damage than inland mixed forests and evergreen forests. The damage within a forest stand from strong winds also varies with tree size (Webb, 1989). Forests with trees of reduced diameter at breast height (dbh) and tree height are more resistant to strong winds, so generally incur less damage after strong wind events. For example, dry forest trees are more resistant to strong winds than wet forest trees because trees of dry forests usually have reduced dbh and height, whereas trees of wet forests have greater dbh and tree height (Van Bloem et al., 2005). Different tree species have different degrees of resistance to strong winds. Trees with higher wood densities are more wind resistant than those with lower wood densities. Higher wood density provides trees with more internal strength in the face of forces from strong winds. For example, in forests of the USA evergreen pines are more resistant to wind damage than many other tree species (Wang and Xu, 2009). Slow growing trees are more resistant to wind damage than fast growing trees. Slow growing trees may become wind resistant due to more wind exposure and disturbance during their longer life spans.

Topographic conditions have an influence on the intensity and scale of damage on forests due to strong winds. The primary topographic factors are: slope, aspect and elevation. Forests on windward slopes may experience more damage than forests on leeward slopes by virtue of their exposure to winds (Foster and Boose, 1992). However, because wind speeds usually peak at ridges, leeward slopes sometimes experience severe downslope turbulence due to gravity waves (Finnigan and Brunet, 1995; Turton, 2008). Interestingly, even though winds increase with elevation above sea level, forests at lower elevations generally receive more damage from strong winds than forests at higher elevations, mainly because upland forests are often lower in stature with aerodynamically smooth canopies (Reilly et al., 2002).

Soil characteristics have a profound impact on root development of trees (Nicoll et al., 2006) – which provide the anchoring strength against the effects of strong winds. Trees with shallow root systems are less resistant to strong winds than trees with deep root systems. Trees grown in poor drainage and seasonal waterlogging conditions usually develop shallow root systems (Mayer, 1989; Ray and Nicoll, 1998) resulting in their low resistance to strong winds; consequently, they often suffer from greater damage after a strong wind event.

Likely effects of climate change on winds and forests

Damage from strong winds may be increasing in some forest biomes because of increased intensity of winds due to global climate change (IPCC, 2013). Damage in European forests – due to strong winds – has increased by about 2.6 per cent yr^{-1} throughout the second half of the twentieth century (Schelhaas et al., 2003) with expectations that this trend will continue this century. In Australia, the mean total forest damage from wind was $3.125 \times 10^6\,m^3\,yr^{-1}$ over the period 2002–2010, demonstrating that wind is an effective change agent in forests (Thom et al., 2013). Damage to tropical forests (especially mangrove forests) of Southeast Asia is also increasing due to more intense tropical cyclones in that region (Alamgir and Turton, 2013). It is predicted that intensity of strong winds associated with tropical cyclones will increase significantly over the twenty-first century under projected climate change although there is more uncertainty of cyclone frequency in the future, except that is likely to vary significantly among tropical ocean basins (Knutson et al., 2010; Turton, 2013).

Given that climate change is projected to increase the risk of more extreme wind events in the future across all forest biomes (IPCC, 2013) it might be prudent to consider the following questions:

- Will there be a pole-ward shift in the range of tropical cyclones, bringing them into contact with forest ecosystems that currently do not experience severe cyclones?
- Will forest biodiversity decline in forested landscapes as a consequence of more extreme wind events as suggested by the 'intermediate disturbance theory'?
- Will forest structure change over time towards a generally lower stature?
- Will a shift in the severity spectrum for severe wind events across all forest biomes tend to favour species more resistant to strong winds, including less desirable non-native species?
- How will contiguous forest areas fare compared with forest remnants contained in agricultural or peri-urban matrices?
- How might projected increases in wind intensity act synergistically with other ongoing changes in forest ecosystems, such as those driving habitat loss and fragmentation, and associated desiccation and increased fire risk?

Synergies between climate change, more extreme weather events, forest habitat loss and fragmentation mean we must accept that our forest ecosystems may not recover to their original state in the aftermath of severe wind events. Instead, we should prepare ourselves for witnessing trajectories of responses that may culminate – over time – in forest ecosystems that are structurally and floristically very different to their pre-disturbance states.

Conclusions

- Strong winds – at a range of scales – are among the most important exogenic disturbance agents affecting forest ecosystems across the world. While extreme wind events are important to all forest biomes, we also need to consider the influences of 'chronic' high wind regimes on some forests.
- Strong winds may affect forests at spatial scales ranging from a few hundred square metres to thousands of square kilometres, while other events may produce severe winds and damage to forests at more intermediate spatial scales. Winds may affect forests at temporal scales ranging from a few minutes to several days, while other events may produce strong winds that impact on forests at scales of a few hours.
- Strong winds play an important role in shaping and moderating ecological processes in forests from the tropics to the high latitudes; notably they affect forest structure, forest function, forest succession and biodiversity at a range of scales.
- Climate change is affecting all the forest biomes on Earth – mainly due to an increase in average wind speeds and extreme wind events. Synergies between climate change, more extreme weather events, forest habitat loss and fragmentation mean we must accept that our forest ecosystems may not recover to their original state in the aftermath of severe wind events.

References

Alamgir, M. and Turton, S. M. (2013) 'Climate change and organic carbon storage in Bangladesh', in N. Tutja and S. S. Gill (eds) *Climate Change and Plant Abiotic Stress Tolerance*, Wiley-Blackwell, Germany

Baker, P. J., Bunyavejchewin, S., Oliver, C. D. and Ashton, P. S. (2005) 'Disturbance history and historical stand dynamics of a seasonal tropical forest in western Thailand', *Ecol. Monographs*, vol. 75, pp. 317–343

Baldwin, A. H., Platt, W. J., Gathen, K. L., Lessmann, J. M. and Rauch, T. J. (1995) 'Hurricane damage and regeneration in fringe mangrove forests of southeast Florida, USA', *J. Coastal Res.*, vol. 21, pp. 169–183

Batista, W. B. and Platt, W. J. (2003) 'Tree population responses to hurricane disturbance: syndromes in southeastern USA old growth forest', *J. Ecol.*, vol. 91, pp. 197–212

Bellingham, P. J., Tanner, E. V. J. and Healy, J. R. (2005) 'Hurricane disturbance accelerates invasion by the alien tree *Pittosporum undulatum* in Jamaican montane rain forests', *J. Veg. Sci.*, vol. 16, pp. 675–684

Brokaw, N., Fraver, S., Grear, J. S., Thompson, J. and Zimmerman, J. K. and others (2004) 'Disturbance and canopy structure in two tropical forests' in E. Losos and E. G. Leigh (eds) *Tropical Forest Diversity and Dynamism: Results from a Long-Term Tropical Forest Network*, University of Chicago Press, Chicago

Catterall, C. P., McKenna, S., Kanowski, J. and Piper, S. D. (2008) 'Do cyclones and forest fragmentation have synergistic effects? A before-after study of rainforest vegetation on the Atherton Tableland, Australia', *Austral Ecol.*, vol. 33, pp. 471–484

Chao, W.-C., Song, G.-Z. M., Chao, K.-J., Liao, C.-C, Fan, S.-W. and others (2010) 'Lowland rainforests in southern Taiwan and Lanyu, at the northern border of Paleotropics and under the influence of monsoon wind', *Plant Ecol.*, vol. 210, pp. 1–17

Connell, J. H. (1978) 'Diversity in tropical rain forests and coral reef', *Science*, vol. 199, no 4335, pp. 1302–1310

Cooper-Ellis, S. D., Foster, R., Carlton, G. and Lezberg, A. (1999) 'Response of forest ecosystems to catastrophic wind: evaluating vegetation recovery on an experimental hurricane', *Ecology*, vol. 80, pp. 2683–2696

Dale, V. H., Joyce, L. A., McNulty, S., Neilson, R. P., Ayres, M. P. and others (2001) 'Climate change and forest disturbances', *Bioscience*, vol. 51, pp. 723–734

Dyer, J. M. and Baird, P. R. (1997) 'Wind disturbance in remnant forest stands along the prairie-forest ecotone, Minnesota, USA', *Plant Ecology*, vol. 29, pp. 121–134

Everham III, E. M. and Brokaw, N. V. L. (1996) 'Forest damage and recovery from catastrophic wind', *Bot. Rev.*, vol. 62, no 2, pp. 113–185

Finnigan, J. J. and Brunet, Y. (1995) 'Turbulent airflow in forests on flat and hilly terrain', in M. P. Coutts and J. Grace (eds) *Wind and Trees*, Cambridge University Press, Cambridge, UK

Fischer, A., Marshall, P. and Camp, A. (2013) 'Disturbances in deciduous temperate forest ecosystems of the northern hemisphere: their effects on both recent and future forest development', *Biodivers. Conserv.*, vol. 22, pp. 1863–1893

Foster, D. R. and Boose, E. R. (1992) 'Patterns of forest damage resulting from catastrophic wind in Central New England, USA', *Journal of Ecology*, vol. 80, pp. 79–98

Foster, D. R. and Orwig, D. A. (2006) 'Preemptive and salvage harvesting of New England forests: when doing nothing is a viable alternative', *Conservation Biology*, vol. 20, pp. 959–970

Foster, M. S. and Terborgh, J. (1998) 'Impact of a rare storm event on an Amazonian forest', *Biotropica*, vol. 30, pp. 470–474

Franklin, J. F., Spies, T. A., Pelt, R., Van Carey, A. B., Thornburgh, D. A. and others (2002) 'Disturbances and structural development of natural forest ecosystems with silvicultural implications, using Douglas-fir forests as an example', *Forest Ecology and Management*, vol. 155, pp. 399–423

Gardiner, B., Blennow, K., Carnus, J-M., Fleischer, P., Ingemarson, F. and others (2011) *Destructive Storms in European Forests: Past and Forthcoming Impacts*, European Forest Institute, Final Report to European Commission

Glitzenstein, J. S. and Harcombe, P. (1988) 'Effects of December 1983 tornado on forest vegetation of the Big Thicket, South-east Texas, USA', *For. Ecol. Manage.*, vol. 25, pp. 269–290

Horvitz, C. C., Pascarella, J. B., McMann, S., Freedman, A. and Hofstetter, R. (1998) 'Functional roles of invasive non-indigenous plants hurricane-affected subtropical hardwood forests', *Ecol. Appl.*, vol. 8, pp. 947–974

IPCC. (2013) 'Summary for Policymakers: Climate Change 2013: The Physical Science Basis', *Contribution of Working Group I to the Fifth Assessment Report of the Intergovernmental Panel on Climate Change* (eds: T. F. Stocker, D. Qin, G-K. Plattner, M. Tignor, S. K. Allen, and others). Cambridge University Press, Cambridge, UK and New York, NY, USA

Knutson, T. R., McBride, J. L., Chan, J., Emanuel, K., Holland, G. and others (2010) 'Tropical cyclones and climate change', *Nat. Geosci.*, vol. 3, pp. 157–163

Laurance, W. F. and Curran, T. J. (2008) 'Impacts of wind disturbance on fragmented tropical forests: an international review', *Austral Ecol.*, vol. 33, pp. 399–408

Lin, K., Hamburg, S. P., Tang, S., Hsia, Y. and Lin, T. (2003) 'Typhoon effects on litterfall in a subtropical forest', *Can. J. For. Res.*, vol. 33, pp. 2184–2192

Lodge, D. J., Scatena, F. N., Asbury, C. E. and Sánchez, M. J. (1991) 'Fine litterfall and related nutrient inputs resulting from hurricane Hugo in subtropical wet and lower montane rain forests of Puerto Rico' *Biotropica*, vol. 23, no. 4a, pp. 336–342

Lugo, A. E. (2008) 'Visible and invisible effects of hurricanes on forest ecosystems: an international review', *Austral Ecol.*, vol. 33, pp. 368–398

Lugo, A. E., Applefield, M., Pool, D. J. and McDonald, R. B. (1983) 'The impact of Hurricane David on the forests of Dominica', *Can. J. For. Res.*, vol. 13, pp. 201–211

Mayer, H. (1989) 'Windthrow', *Philosophical Transactions of the Royal Society of London Series B, Biological Sciences*, vol. 324, pp. 267–281

Metcalfe, D. J., Bradford, M. G. and Ford, A. J. (2008) 'Cyclone damage to tropical rain forests: species- and community-level impacts', *Austral Ecol.*, vol. 33, pp. 432–441

Nicoll, B. C., Gardiner, B. A., Rayner, B. and Peace, A. J. (2006) 'Anchorage of coniferous trees in relation to species, soil type, and rooting depth', *Canadian Journal of Forest Research*, vol. 36, pp. 1871–1883

Nilsson, C., Stjernquist, I., Bärring, L., Schlyter, P., Jonsson, A. M. and Samuelsson, H. (2004) 'Recorded storm damage in Swedish forests 1901–2000', *Forest Ecology and Management*, vol. 199, no 1, pp. 165–173

Ostertag, R., Scatena, F. N. and Silver, W. L. (2003) 'Forest floor decomposition following hurricane litter inputs in several Puerto Rican forests', *Ecosystems,* vol. 6, pp. 261–273

Pascarella, J. B. (1997) 'Hurricane disturbance and the regeneration of *Lysiloma latisiliquum* (Fabaceae): a tropical tree in south Florida', *For. Ecol. Manage,* vol. 92, pp. 97–106

Patric, J. H. (1974) 'River flow increases in central New England after the hurricane of 1938', *Journal of Forestry,* vol. 72, pp. 21–25

Proctor, J., Brearley, F. Q., Dunlop, H., Proctor, K., Supramono and Taylor, D. (2001) 'Local wind damage in Barito Ulu, Central Kalimantan: a rare but essential event in a lowland dipterocarp forest?', *Journal of Tropical Ecol.,* vol. 17, pp. 473–475

Ray, D. and Nicoll, B. C. (1998) 'The effect of soil water-table depth on root-plate development and stability of Sitka spruce', *Forestry,* vol. 71, pp. 169–182

Reilly, J., Mayer, M. and Harnisch, J. (2002) 'The Kyoto Protocol and non-CO_2 greenhouse gases and carbon sinks', *Environmental Modeling and Assessment,* vol. 7, pp. 217–229

Schelhaas, M-J., Nabuurs, G-J. and Schuck, A. (2003) 'Natural disturbances in the European forests in the 19th and 20th centuries' *Global Change Biology,* vol. 9, no. 11, pp. 1620–1633

Seidl, R. and Blennow, K. (2012) 'Pervasive growth reduction in Norway spruce forests following wind disturbance', *PLoS ONE,* vol. 7, no. 3, p. e33301

Shiels, A. B., Zimmerman, J. K., García-Montiel, D. C., Jonckheere, I., Holm, J. and others (2010) Plant responses to simulated hurricane impacts in a subtropical wet forest, Puerto Rico, *Journal of Ecol,* vol. 98, pp. 659–673

Tanner, E. V. J. and Bellingham, P. J. (2006) 'Less diverse forest is more resistant to hurricane disturbance: evidence from montane rain forests in Jamaica' *J. Ecol.,* Vol. 94, pp. 1003–1010

Thom, D., Seidl, R., Steyrer, G., Krehan, H. and Formayer, H. (2013) 'Slow and fast drivers of the natural disturbance regime in Central European forest ecosystems', *Forest Ecology and Management* vol. 307, pp. 293–302

Turner, M. G. (2010) 'Disturbance and landscape dynamics in a changing world', *Ecology,* vol. 91, pp. 2833–2849

Turton, S. M. (2008) 'Landscape-scale impacts of Cyclone Larry on the forests of northeast Australia including comparisons with previous cyclones impacting the region between 1858 and 2006', *Austral Ecol,* vol. 33, pp. 409–416

Turton, S. M. (2012) 'Securing landscape resilience to tropical cyclones in Australia's wet Tropics under a changing climate: lessons from Cyclones Larry (and Yasi)', *Georgr Res,* vol. 50, pp. 15–30

Turton, S. M. (2013) 'Tropical cyclones and forests dynamics under a changing climate: what are the long-term implications for tropical forest canopies in the cyclone belt?', in M. Lowman, S. Devy and T. Ganesh (eds) *Treetops at Risk: Challenges of Global Canopy Ecology and Conservation,* Springer, New York

Turton, S. M. and Stork, N. E. (2008) 'Impacts of tropical cyclones on forests in the Wet Tropics of Australia', in N. E. Stork and S. M. Turton (eds) *Living in a Dynamic Tropical Forest Landscape,* Blackwell Publishing, Oxford, UK

Van Bloem, S. J., Murphy, P. G., Lugo, A. E., Ostertag, R., Costa, M. R. and others (2005) 'The influence of hurricane winds on Caribbean dry forest structure and nutrient pools', *Biotropica,* vol. 37, pp. 571–583

Wang, F. and Xu, Y. J. (2009) 'Hurricane Katrina-induced forest damage in relation to ecological factors at landscape scale', *Environmental Monitoring and Assessment,* vol. 156, pp. 491–507

Weaver, P. L. (1989) 'Forest changes after hurricanes in Puerto Rico's Luquillo Mountains'. *Interciencia,* vol. 14, pp. 181–192

Webb, S. L. (1988) 'Windstorm damage and microsite colonization in two Minnesota forests. *Canadian J. of Forest Res.,* vol. 18, pp. 1186–1195

Webb, S. L. (1989) 'Contrasting windstorm consequences in two forests, Itasca State Park, Minnesota', *Ecology,* vol. 70, pp. 1167–1180

Webb, S. L. and Scanga, S. E. (2001) 'Windstorm disturbance without patch dynamics: 12 years of change in Minnesota'. *Forest Ecology,* vol. 82, pp. 893–897

Whitmore, T. C. and Burslem, D. F. R. P. (1998) 'Major disturbances in tropical rain forests', in D. M. Newbery, H. H. T. Prins and N. Brown (eds), *Dynamics of Tropical Communities,* Blackwell Science, Oxford, UK

Wunderle Jr., J. M., Lodge, D. J. and Waide, R. B. (1992) 'Short-term effects of Hurricane Gilbert on terrestrial bird populations on Jamaica', *The Auk.,* vol. 109, no. 1, pp. 148–66

Yoshida, T., Noguchi, M., Uemura, S., Yanaba, S., Miya, H. and Hiura, T. (2011) 'Tree mortality in a natural mixed forest affected by stand fragmentation and by a strong typhoon in northern Japan', *Journal of Forest Research*, vol. 16, pp. 215–222

Zhao, D., Allen, B. and Sharitz, R. R. (2006) 'Twelve year response of old-growth southeastern bottomland hardwood forests to disturbance from Hurricane Hugo', *Canadian Journal of Forest Research*, vol. 36, pp. 3136–3147

10
FOREST SUCCESSION AND GAP DYNAMICS

Rebecca A. Montgomery and Lee E. Frelich

Overview

Succession is a foundational concept in forest ecology. While succession has been defined variously, the most common definition is the sequence of communities or species that successively replace each other through time after a disturbance. The term is also used to describe the process of change itself. Many include change in forest and community structure and system-level properties such as diversity and productivity in their definition. Succession can be divided into two major types: primary succession, change that occurs on a previously unvegetated site, and secondary succession, change that occurs after disturbance to an existing ecosystem. In secondary succession, biological legacies left by past vegetation remain and influence succession. This chapter focuses on secondary succession, as it is the most common and widespread in forested ecosystems. We use a broad definition of succession discussing compositional, structural and system-level change. We also discuss management strategies for forested ecosystems that are based on principles of ecological succession and challenges associated with global change.

A brief history of the concept

Henry Chandler Cowles pioneered the study of ecological succession with work on the patterns and processes that transform sand dunes along the shores of Lake Michigan, USA into hardwood forests (Cowles 1899). Frederick Clements created a general classification of causes that is still applied today (Clements 1916). His scheme involved: (1) *nudation*, the removal of vegetation by disturbance; (2) *migration*, the arrival of organisms at the open site; (3) *ecesis*, the establishment of organisms at the open site; (4) *competition*,[1] the interaction of organisms at the site; (5) *reaction*, the alteration of the site by the organisms. The result of these five was *stabilization* in the form of the climax. For Clements, the climax was the major unit of vegetation and the permanent and final stage of succession (Clements 1916, 1936). The nature of the climax was intimately intertwined with the climate. He considered the climax an expression of climate and a superorganism with its own particular trajectory of growth and development. The growth and

---

1 Clements called it competition but other interactions are generally recognized today.

development or succession terminated in the climax and more specifically in the mono-climax, which could maintain itself indefinitely if not disturbed. Although the conception of succession as a process where species successively replace each other through time has endured, the concept of a mono-climax, the primary role of climate as mechanism and the climax as a super-organism proved too restrictive for a number of early ecologists (Cooper 1913; Gleason 1927). One of the prominent critics was Henry A. Gleason who asserted that succession is an 'extraordinarily mobile phenomenon' that may not be repeatable or predictable (Gleason 1927). He recognized multiple causes that could lead succession down multiple pathways. In other words, Gleason recognized succession as a complex process without a fixed endpoint. Moreover, he argued that any predictions of successional trajectories must recognize that multiple forces operate and that some may act in opposing directions. Gleason espoused an individualistic view of communities that was later championed by Whittaker (cf. Whittaker 1957) in which species respond to their abiotic and biotic environment in an individualistic way and the patterns in succession reflect the ebb and flow of populations of species through time.

Compositional and structural change

The process of succession involves changes in both the species composition and the physical structure of a forest through time. Forest development describes change in forest structure through time. In general, all forested ecosystems go through a predictable sequence of changes in structure as they develop following disturbances that kill or level the forest canopy (Figure 10.1). These changes include two even-aged stages—stand initiation or establishment phase and the stem exclusion or thinning phase; a third stage involving transition to an uneven-age canopy or understory reinitiation, and a final multi-aged stage, sometimes referred to as old-growth or steady-state (Oliver 1981; Frelich 2002; but see Franklin et al. 2002 for critique). These changes may or may not be associated with compositional change. It is important to note that structural change and compositional change can occur together or separately.

The stand initiation or establishment phase is defined as the period of time immediately after disturbance when new stems are establishing and filling available growing space. This is a time of very high density of small stems. These stems invade and grow until all growing space is occupied and one or more resources becomes limiting. At this point, new stems do not establish. The stem exclusion phase is one of intense competition among established individuals during which one or several species become dominant. Shade-intolerant species that do not become dominant die while those tolerant of shade can persist in the understory. Over time,

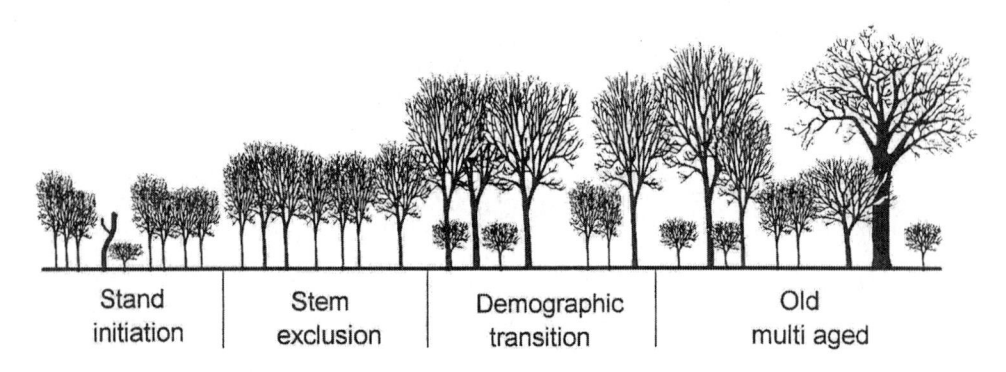

Figure 10.1 Stages of stand development (from Frelich 2002)

as trees grow larger and use more resources there is growing space for fewer stems and those less vigorous or slower growing die while more vigorous stems expand their crowns into the space left behind. Eventually, neighboring trees cannot fill the space opened by death of canopy trees and existing suppressed or newly established stems begin to grow in gaps made by canopy tree death. This is the transition or understory reinitiation phase. The characteristics of the multi-aged stage of development vary within and among forest types, and include rare steady-state stands with a constant rate of canopy turnover in the form of small canopy gaps, as well as more common stands where a variety of partial disturbances creates a constantly changing mix of gap sizes, age class structures and mixtures of shade-tolerant and mid-tolerant tree species.

At the landscape scale, the frequency of major disturbances that kill or level the forest canopy regulates the proportion of stands among the stages of succession and development, a concept proposed by Watt (1947). Boreal forests with high-intensity fires at intervals less than tree lifespan may be composed mostly of stands in initiation and stem exclusion stages with rare multi-aged stands, whereas in regions where stand-leveling disturbance is rare (e.g. mesic cold-temperate forests or tropical rainforest), the distribution of stands among stages of development is the opposite, with a high proportion of multi-aged or old-growth stands.

Stand-level structural changes during forest development and succession include changing diameter distributions, stem density, maximum tree height and stand basal area that follow general patterns across global forested ecosystems (Guariguata and Ostertag 2001; Frelich 2002; Powers *et al.* 2009; Whitfeld *et al.* 2014). During forest succession, stands increase in overall height and basal area, eventually reaching an asymptotic maximum height and basal area. The time to reach asymptotic height and basal area differs with forest types, climate, edaphic factors and disturbance history. Bormann and Likens (1979) proposed a similar model for biomass accumulation post-disturbance. Density changes appear more variable but generally show a more peaked pattern, rising and then declining, though the degree of decline varies among forests. Generally, the range of diameters present in a stand increases and becomes more heterogeneous as stands move from even- to multi-aged composition and stands become more diverse in vertical and horizontal structure.

An important aspect of post-disturbance structure is the existence of legacy structures (Franklin *et al.* 2002). Legacy structures include living organisms that survive the disturbance such as mature trees and dead materials such as standing dead snags or downed wood. These legacy structures are important to post-disturbance successional processes providing seed sources of new colonists, habitat for animals, substrate for seed germination, bud banks for rapid recolonization through resprouting and much more. The abundance and nature of biological legacies varies with disturbance type and severity. For example, a windstorm may selectively kill species that are not wind firm leaving other species undamaged and releasing suppressed individuals in the understory (Rich *et al.* 2007).

Compositional changes during succession in many ways have defined the concept. Often but not always the early successional community is dominated by fast-growing, relatively short-lived species. As a relatively even-aged early successional canopy self thins, the species that reinitiate the understory are often but not always more shade-tolerant species than in the canopy. These species may have been present from the point of disturbance or they may establish as resource conditions change. These two models were originally described as initial floristics and relay floristics, respectively (Egler 1954). Relay floristics describes a process by which species at any given stage either facilitate the establishment of the next or inhibit colonization leading to new individuals and species establishing and dominating sequentially. In contrast, initial floristics describes a process by which most species establish early after disturbance but due to life history characteristics, species dominate at different points along the successional sequence.

The propensity for species with particular life history characteristics and functional traits to be associated with different successional stages has led to the widespread use of the terms early, mid and late successional to describe species. In forests, there is often the dichotomy between shade-intolerant pioneers and shade-tolerant late successional species (Swaine and Whitmore 1988). Early successional tree species or shade-intolerant pioneers tend to produce many, small and wind dispersed seeds that can lie dormant for many years. Although, in the tropics, birds dominate dispersal of Asian pioneers (Macaranga, etc.) while fruit bats seem to be as or more important in the neotropics (e.g. Cecropia). They have fast growth rates; suffer considerable herbivory; are short-lived; and have high rates of photosynthesis, nutrient-rich rapidly decomposing leaves and low wood density. In contrast, shade-tolerant late successional species have larger seeds that either lack or show short periods of dormancy. They are slow growing; possess defenses against structural damage and herbivory; are long-lived and tall; and have relatively low rates of photosynthesis, lower leaf nutrient levels, slower decomposition rates and high wood density. Of course, this is a simplified view of the diversity of forest plant strategies. In reality, trees and other forest plants fall along a continuum of strategies that involves a mix of colonizing strategies, growth rates, demographic strategies and positions in the vertical and horizontal structure of the forest. A recent exploration of tropical forest secondary succession used growth rates, canopy height and colonization groups to classify species and follow abundance of these through succession. They found that tree stature and growth rates captured considerable functional variation that was related to secondary successional dynamics (Chazdon et al. 2010).

Mechanisms

A general theory of ecological succession has failed to emerge despite many efforts. Such theory rests on laying out causes or mechanisms of succession, which are manifold. In attempts to develop theory, Connell and Slatyer (1977) laid out successional pathways that invoke the mechanisms of facilitation, tolerance and inhibition. The Clementsian view of succession asserted the primacy of facilitation by early colonists who modify the environment in such a way that later colonists can establish. Evidence in forests for this mechanism comes from succession on old-fields in the temperate zone and abandoned pastures in the tropics. Tolerance has been interpreted as both an active and passive mechanism (Pickett et al. 1987) that involves either ability to endure low resources and slowly grow in stature, eventually shading the early dominants (active) or life cycle complementarity (passive) where two species with different growth rates establish simultaneously and the slow grower eventually succeeds the fast grower. In contrast, inhibition invokes structural or competitive dominance by early colonists that prevents other species from establishing or becoming dominant. Evidence for inhibition in forests is poor. Moreover, inhibition is also intertwined with tolerance: late successional species accumulate because they can tolerate low resources and then just wait for a disturbance to release them from suppression. This is demonstrated in forests that feature small-scale disturbances such as gaps (Uhl et al. 1988; Abe et al. 1995).

For some, Connell and Slatyer's 'models' were too rigid and reductionist and a general hierarchical framework of causes promised a more broadly applicable approach to creating a predictive framework and a general theory of ecological succession (Finegan 1984; Pickett et al. 1987).

The general hierarchical framework proposed by Pickett et al. (1987) remains one of the best syntheses of the causes and mechanisms of succession. The framework starts with the question 'What causes succession?' and answers with the following three responses: open sites are available, species are differentially available at the open sites, and species differ in the capacity to deal with one site or another (Pickett *et al.* 1987). The second question or level of hierarchy asks, 'What interactions, processes or conditions contribute to the general causes of succession?' These include dispersal, resource availability, ecophysiology, competition, herbivory, and disease. Finally, a third question asks, 'What site-specific factors or behaviors determine outcome of interactions?' This question invokes interactions of the organism and site-specific features. We illustrate the framework with the following example. Open sites are made available by a crown-destroying fire. Post-disturbance stands are initiated from several sources: mobile seeds that come from outside the stand or from reproductive individuals that survived the fire (legacy trees); seeds in the seed bank; and individuals that survived and resprout. The nature of the regenerating forest in part depends on the relative importance of these different sources of colonists (differential species availability), which may depend on disturbance severity (a site-specific factor) and on the interaction of species with characteristics of the site (differential species performance as a result of post-disturbance resource availability).

Gap dynamics

Gaps are openings in the forest created by small-scale disturbance generated by mortality of one or multiple trees due to density independent mechanisms such as windthrow or insect damage (Brokaw 1982). Gaps alter resource availability and microclimate, increasing light, soil moisture and nutrients and the extremes of temperature and humidity, especially near the forest floor. Gap dynamics have received the most attention in forests where they dominate late successional processes, such as mesic temperate and tropical forests (Runkle 1981, 1982; Denslow 1987). However, there is growing evidence that under certain disturbance regimes gaps play a role in boreal forest dynamics in North America and Eurasia (Kneeshaw and Bergeron 1998; Shorohova *et al.* 2009).

Gaps play several roles in forest dynamics depending in part on gap size and other conditions at the time of gap formation. Gaps provide sites for the colonization and establishment of pioneer species that require high light for seed germination and growth. Usually, the establishment of pioneers requires large gaps created by more than a single tree. These regenerating pioneers (*sensu* Chazdon *et al.* 2010) are maintained within the forest ecosystem by the existence of regeneration sites in large tree fall gaps. In this role, gaps create patches of early successional forest embedded in late successional forests (Watt 1947). Gaps also release suppressed shade-tolerant trees that have formed the advanced regeneration layer. There is good evidence that small-scale disturbances associated with gaps in late successional forests are sufficient to maintain the landscape species composition in a relatively steady state (Runkle 1981).

Gaps are considered one of the mechanisms that maintain diversity in forested ecosystems (Denslow 1987; Brokaw and Busing 2000); however, how gaps promote coexistence remains poorly characterized (Gravel *et al.* 2010). An early articulation of the mechanism was the gap-partitioning hypothesis, which posits that species colonize gaps of different size and different areas of gaps depending on their resource requirements (the regeneration niche; Grubb 1977). In a large gap, pioneers that require high light colonize and dominate the center, mid-tolerant species dominate parts of the gap closer to the edges and shade-tolerant species dominate the edges (Denslow 1987; Schnitzer and Carson 2000). A similar colonization and dominance scheme applies to gaps that range in size from small, single-tree falls to very large gaps (Runkle 1982). This hypothesis applies not only to the tree component of forests but also the forest understory layer of shrubs and herbs. However, gap colonization or capture can also be a function of stochastic processes. For example, gap capture may be dependent on the species that happen to be in the advance regeneration layer at time of canopy disturbance (Brokaw and Busing 2000). There is evidence for a significant role of stochastic processes in the tree replacement process in systems with high numbers of shade-tolerant trees (Hubbell *et al.* 1999; Gravel *et al.* 2010).

Boreal forests, whose disturbance regimes are dominated by fire, have generally not been considered ecosystems where gaps play a role. However, recent work in North America and Eurasia suggest that small disturbances can influence forest dynamics between larger high-intensity fires. Such canopy gap processes are important in certain regions of the boreal biome where intervals between fires are greater than 200 years, such as eastern Canada and parts of Scandinavia (Kneeshaw and Bergeron 1998; Shorohova *et al.* 2009).

Greater appreciation of heterogeneity and complexity in forest ecosystems coupled with changing management goals had led to reconsideration of traditional silvicultural systems that often assume and prescribe homogeneity with the goal of timber production (Coates and Burton 1997). In particular, applying lessons from ecological research on gap dynamics to achieve ecosystem management goals through silviculture has become more common in systems involving partial harvest of the ecosystem (Coates and Burton 1997). Ecosystem management refers to an approach that delivers goods and services while sustaining forest function, diversity and structure in perpetuity. Increasingly, it has been adopted in forestry to achieve multiple goals that include not only timber production, but also biodiversity conservation, wildlife habitat, and sustaining ecosystem functions services (e.g. nutrient cycling, productivity, water quality). In this framework, if the goal of a management intervention is to increase the life history diversity of the forest or to increase understory biodiversity while still producing timber, then a range of harvested patch sizes might be used to simulate a range of gap sizes and concomitant effects on resource availability and tree regeneration.

Diversity and forest succession

Although patterns of change in composition through succession have been widely described and are a hallmark of the process, system-scale measures such as diversity have yielded conflicting patterns. Three major hypotheses regarding change in diversity through succession have emerged, all with some support. The classic treatment of patterns of species diversity through succession is that of low species diversity in early succession and increasing diversity through time due to increasing structural complexity or higher rates of immigration than extinction in successional stands. For example, in forests of subtropical China, species richness increases through successional time and changes are largely due to continuous immigration of new species that enrich the sites (Bruelheide *et al.* 2011). Alternately, diversity could decrease if initial floristic composition dominates, with species being eliminated through time as succession proceeds (Egler 1954). Species diversity could also decline through succession as forest density declines due to the positive relationship between density and richness. A third model suggests that diversity patterns tend to shift with forest development stages. Diversity is initially relatively high as species resprout or recolonize available sites after disturbance. Diversity of trees drops as one or several species attains canopy dominance, creating a low resource environment in the understory that can also lead to declines in ground layer diversity. Diversity increases as the forest becomes more vertical and horizontally stratified, eventually reaching a mid-successional peak. However, given the general importance of small disturbances in later successional forests, gap formation may increase and maintain diversity (Runkle 1982)—with no mid-successional peak. However, if small-scale disturbances are not prevalent, diversity can decline if one or several species are able to dominate and exclude others. This pattern of diversity through succession was key to the formulation of the intermediate disturbance hypothesis, first described in relation to tropical rainforests and coral reefs (Connell 1978). It posits that most communities are kept in a non-equilibrium state such that most patches are in highly diverse mid-successional stages. Frequent disturbance or large disturbance moves communities towards dominance by a few early successional species, while rare or small and low severity disturbance allows dominance by those species that can competitively exclude other members (Connell 1978; Figure 10.2). One possibility for these different models may be due to the difficulty in actually sampling a full successional sequence and hence any one study only sees part of the story (Howard and Lee 2003). It is also possible that no single model describes changes in species diversity because the causes of succession may differ among forests in different locations, or in the same forest over time.

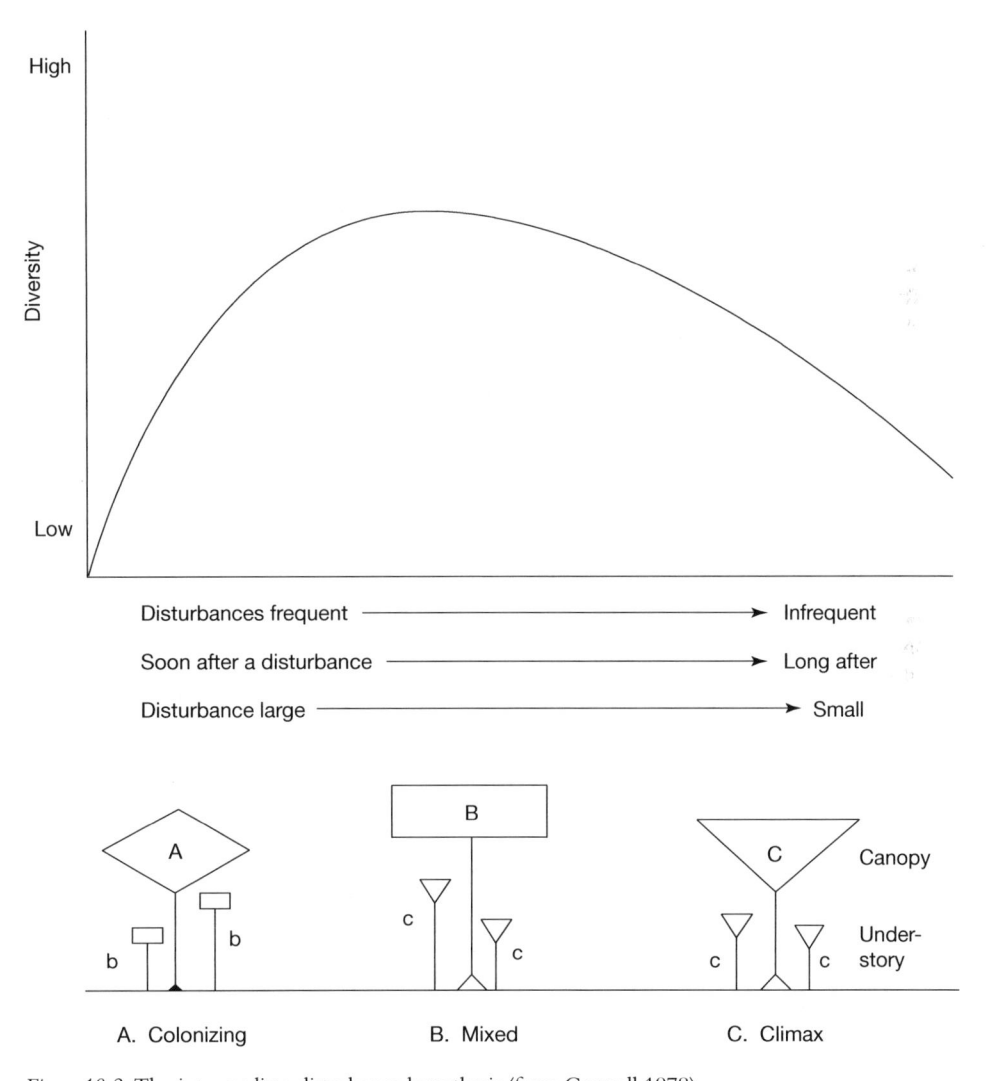

Figure 10.2 The intermediate disturbance hypothesis (from Connell 1978)

Stochasticity, complexity and multiple pathways

The preceding sections have presented general patterns of change associated with succession. Often, those patterns were accompanied by caveats such as 'but not always'. In essence, for every pattern presented in this chapter there is undoubtedly an exception illustrating the complex nature of succession and the myriad pathways of change that can occur due to variation in the dominant causes in each place-based successional sequence. Myriad pathways also result, in part, from the action of stochastic processes.

Recruitment limitation is an example of a stochastic process; it implies that species fail to recruit into sites most favorable for their growth (Hubbell *et al.* 1999). Thus, after major or small-scale disturbances, new colonists often are a function of chance, related to the species producing seeds at the time of disturbance or the species already present, rather than species that are most competitive on the site. There is ample evidence that recruitment limitation plays an important role in forest and gap dynamics (cf. Brokaw and Busing 2000).

To illustrate complexity and multiple pathways we use two examples. In the first, we explore how complexity usually associated with later successional stages can occur much earlier in sites where early and late successional species establish together after disturbance and in which establishment is sparse (Donato *et al.* 2012). The 'precocious succession model' describes a system where structural complexity develops early in a natural successional sequence rather than only during late successional states (Figure 10.3). In this pathway, after major disturbance the tree establishment phase is protracted due to large patch sizes, long distances to seed sources, unfavorable environment conditions for seed establishment, and early competition from other vegetation (e.g. shrubs). An open community is established early and the low density and/or competition from shrubs stratifies tree establishment and can create an early interspecific competitive exclusion phase or forgo canopy closure altogether (Donato *et al.* 2012). In such young stands, structural aspects that imbue old-growth stands with complexity are present early including: clumped, widely-spaced trees; vertical heterogeneity in the canopy and among tree crowns; coexistence of under-, mid- and overstory; and facilitation of shade-tolerant species (Donato *et al.* 2012).

Our second example involves interactions among species traits, herbivory and multiple disturbances that lead to divergent successional pathways that maintain white pine dominance versus transition stands to maple dominance in forests in the Lakes States, USA (Tester *et al.* 1997; Frelich 2002). In the pine forests of the Lakes States, fire and wind are dominant disturbance agents. In mature white pine stands, young cohorts of pine must establish in the understory and grow large enough to survive the next surface fire. They can fail to do so for many reasons including deer browse, drought, poor soils or taller maple or spruce saplings in the understory. If they fail to grow large enough before the occurrence of surface fires then they, along with maple and spruce saplings, will be killed by surface fire. The longer it takes to grow large enough, the more likely a surface fire will occur that kills them. Moreover, longer times for saplings to grow large enough also increase the likelihood that the adult white pine will be killed by a windstorm, leaving no seed sources for white pine to recolonize. In this compound disturbance regime, white pine stands transition to hardwoods. In contrast, if the white pine saplings grow large enough, they survive surface fire while maples are killed. Moreover, if canopy individuals are killed by windstorm, subcanopy individuals of white pine are released to become the future canopy and seed source.

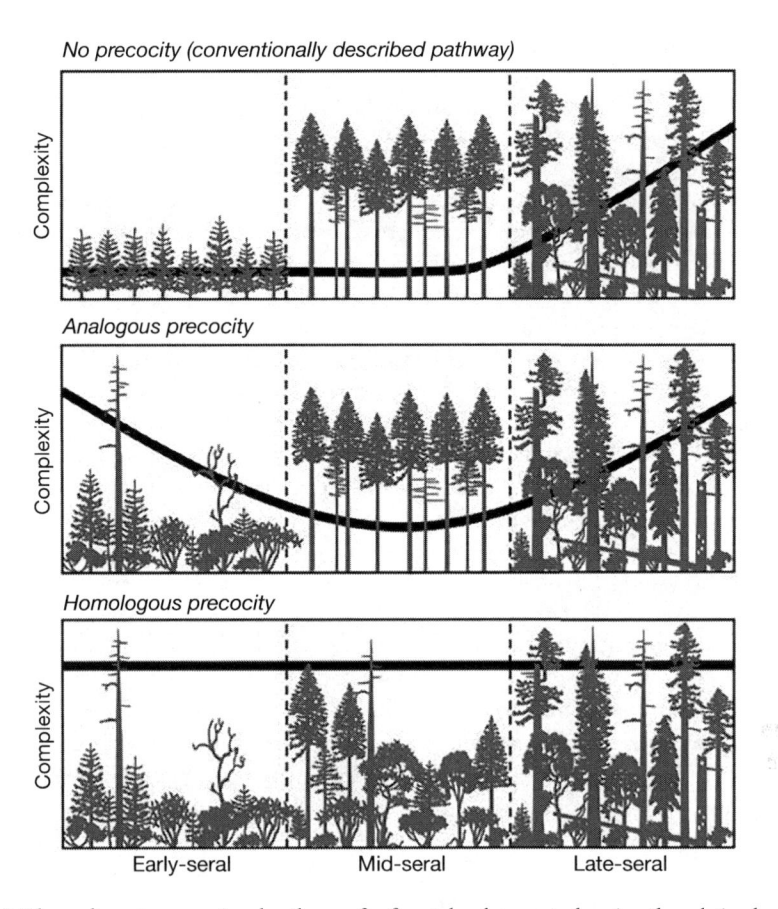

Figure 10.3 Three alternate successional pathways for forest development, showing the relative levels of structural complexity exhibited in each seral stage. In the conventional successional model, both early- and mid-seral conditions are dominated by a relatively even-aged tree cohort, and structural complexity does not arise until the latest stage of development. In the case of analogous precocity, early-successional stands exhibit structural complexity in some ways similar to that in old stands, but canopy closure results in reduced complexity during mid-succession. In the case of homologous precocity, the lack of a tree canopy-closure phase results in a continuity of complexity throughout forest development (from Donato *et al.* 2012)

Succession and global change

Global change, such as elevated CO_2, warming temperatures, altered disturbance and browsing regimes and global movement of species has and will continue to impact ecological systems, often in ways detrimental to the goods and services they provide. To understand the consequences of global change requires prediction of forest dynamics into the future. This must be attempted despite uncertainty over exact environmental conditions or the nature of biotic interactions and disturbance regimes. Predicting future dynamics is one of the hallmarks of research on succession and thus succession provides a foundation for understanding impacts of global change on forested ecosystems.

An excellent example is provided by recent work in boreal forests of central Alaska, USA (Johnstone *et al.* 2010). Many boreal North American ecosystems are dominated by high-severity fire regimes and can show repeatable successional cycles that are characterized by re-establishment of dominant pre-fire vegetation. The majority of the current landscape is occupied by black spruce whose serotinous seeds help ensure its post-fire dominance. In sites with deciduous dominance, successional shifts to conifers occur at time intervals considerably longer than average disturbance cycles. Thus, at present patches of black spruce and patches of deciduous species tend to be self-replacing. What might happen to successional trajectories and landscape patterns of spruce and deciduous forests in the future? Research after a severe fire year in central Alaska, USA suggests that high fire severity and frequent fires, predicted by climate change, will favor successional pathways to deciduous species especially on moderate to well-drained sites. They predict that climate change may shift the balance at the landscape level from predominantly spruce forests towards deciduous dominated forests. Such a shift would have regional consequences for local climate, forest flammability and human subsistence activities (Johnstone et al. 2010).

Another pattern of change that will undoubtedly alter forest succession is the increased prevalence of drought induced tree mortality and dieback (Allen et al. 2010). An increase in temperature during the growing season also increases evaporation; this is expected to increase drought stress for forests in relatively dry climates adjacent to deserts and grasslands, however, even in wet climates, increased evaporation in a warmer climate can outpace precipitation or projected increases in precipitation, causing increased drought-related mortality in places like the Amazon rainforest (Phillips et al. 2009). Despite this, water savings associated with growing at elevated CO_2 could reduce drought stress due to climate warming. Secondary effects of drought and heat stress will also raise tree mortality rates—these include increased likelihood of fire, insect herbivores and pathogens (Ayres and Lombardero 2000). Across North America, recent decline episodes in trembling aspen (*Populus tremuloides*) in certain parts of its range have been linked to drought (Worrall et al. 2013). Severe drought has the potential to catastrophically affect trees through both hydraulic failure and carbon starvation (Sevanto et al. 2014) and species differ in drought tolerance. Such differences in species performance may lead to shifting dominance patterns associated with climate change induced droughts (Cavin et al. 2013).

Arrested succession has been noted in some systems where ground layer and understory vegetation retards establishment and growth of trees. Royo and Carson (2006) reviewed studies of species that form dense and persistent understory layers. They found that such layers were often the result of anthropogenic factors. Specifically, recalcitrant understory layers appear to result when there is a high level of canopy disturbance, when there are high levels of browsing by large vertebrate herbivores, and when fire regimes are altered. These anthropogenic factors appear unlikely to change in the future and suggest that arrested succession may become more common in the future.

Lastly, climate change and the global movement of species may alter successional pathways by introducing novel biotic interactions into forested ecosystems. Climate change could alter the relative competitive ability of species at range limits through alleviation of cold-limitation of species at the cold-edge of the range. To detect range shifts most studies have focused on the regeneration layer, looking at composition, relative abundances and performance of the juveniles with the implicit assumption that these individuals are expected to become members of the canopy in the future. Such studies have shown evidence for enhanced recruitment of currently cold-limited species at range edges and declines in recruitment of species at their warm range edge (e.g. Fisichelli et al. 2014). Shifts in recruitment suggest a high potential for turnover in species composition in forests in regions where species with different ranges overlap (e.g. ecotones).

The global movement of species has resulted in an increase in invasive plants and animals. These species can alter successional pathways by eliminating tree species or altering the environment so that species from one or more successional stages are no longer successful. Hemlock woolly adelgid (*Adelges tsugae*), an insect pest from Asia, has eliminated eastern hemlock (*Tsuga canadensis*), a dominant late successional tree species, from forests in parts of the eastern USA (Ford et al. 2012). European earthworms invading the Great Lakes Region of the USA have changed the seedbed conditions so that *Acer saccharum*, a late successional dominant over large areas, cannot compete successfully at the seed germination and seedling growth stages of its life cycle (Hale et al. 2006). It remains to be seen what new successional pathways will develop in response to changes wrought by these invasive species.

Summary

Succession represents a foundational concept in forest ecology and management with a long history of scholarship and application. It guides the design of silviculture prescriptions and forest management plans. As forests face growing challenges associated with global change, studies of forest succession also guide our understanding of the potential trajectory of forests into the future. Despite common threads of cause and consequence in forest succession across the globe, it is clear that complexity is the rule rather than the exception. Moreover, as an inherently place- and time-based subject, there will always be more to learn.

References

Abe, S., Masaki, T. and Nakashizuka, T. (1995) 'Factors influencing sapling composition in canopy gaps of a temperate deciduous forest'. *Vegetatio*, vol. 120, no. 1, pp. 21–31.

Allen, C.D., Macalady, A.K., Chenchouni, H., Bachelet, D., McDowell, N., and others (2010) 'A global overview of drought and heat-induced tree mortality reveals emerging climate change risks for forests.' *Forest Ecology and Management*, vol. 259, pp. 660–684.

Ayres, M.P. and Lombardero, M.J. (2000) 'Assessing the consequences of global change for forest disturbances for herbivores and pathogens'. *The Total Science of the Environment*, vol. 262, pp. 263–286.

Bormann, F.H. and Likens, G.E. (1979) *Pattern and Process in a Forested Ecosystem*. Springer-Verlag, New York, NY.

Brokaw, N.V.L. (1982) 'The definition of treefall gap and its effects on measures of forest dynamics'. *Biotropica*, vol. 14, pp. 158–160.

Brokaw, N.V.L. and Busing R.T. (2000) 'Niche versus chance and tree diversity in forest gaps'. *Trends in Ecology and Evolution*, vol. 15, no. 5, pp. 183–188.

Bruelheide, H., Bohnke, M., Both, S., Fang, T., Assmann, T., and others (2011) 'Community assembly during secondary forest succession in a Chinese subtropical forest'. *Ecological Monographs*, vol. 81, no. 1, pp. 25–41.

Cavin, L., Mountford, E.P., Peterken, G.F. and Jump, A.S. (2013) 'Extreme drought alters competitive dominance within and between tree species in a mixed forest stand' *Functional Ecology*, vol. 27, no. 6, pp. 1424–1435.

Chazdon, R.L., Finegan, B., Capers, R.S., Salgado-Negret, B., Casanoves, F. and others (2010) 'Composition and dynamics of functional groups of trees during tropical forest succession in Northeastern Costa Rica'. *Biotropica*, vol. 42, no. 1, pp. 31–40.

Clements, F.E. (1916) *Plant Succession: An Analysis of the Development of Vegetation*. Carnegie Institute of Washington, Publication 242.

Clements, F.E. (1936) 'Nature and structure of the climax'. *Journal of Ecology*, vol. 24, pp. 252–284.

Coates, K.D. and Burton, P.J. (1997) 'A gap-based approach for development of silvicultural systems to address ecosystem management objectives'. *Forest Ecology and Management*, vol. 99, pp. 337–354.

Connell, J. H. (1978) 'Diversity in tropical forests and coral reefs' *Science* 199: 1302–1310.

Connell, J.H. and Slatyer, R.O. (1977) 'Mechanisms of succession in natural communities and their role in community stability and organization'. *American Naturalist*, vol. 111, pp. 1119–1144.

Cooper, W.S. (1913) 'The climax forests of Isle Royale, Lake Superior and its development'. *Botanical Gazette*, 55: I 1–44, II 115–140, III 189–235.

Cowles, H.C. (1899) 'The ecological relations of the vegetation on the sand dunes of Lake Michigan'. *The Botanical Gazette* 27: 1–388.

Denslow, J.S. (1987) 'Tropical forest gaps and tree species diversity' *Annual Review of Ecology and Systematics*, vol. 18, pp. 431–451.

Donato, D.C., Campbell, J.L. and Franklin, J.F. (2012) 'Multiple successional pathways and precocity in forest development: can some forests be born complex' *Journal of Vegetation Science*, vol. 23, pp. 576–584.

Egler, F. E. (1954) 'Vegetation science concepts. I. Initial floristic composition, a factor in old field vegetational development.' *Vegetatio*, vol. 4, pp. 412–417.

Finegan, B. (1984) 'Forest succession'. *Nature*, vol. 312, pp. 109–114.

Fisichelli, N.A., Frelich, L.E. and Reich, P.B. (2014) 'Temperate tree expansion into adjacent boreal forest patches facilitated by warmer temperatures' *Ecography*, vol. 37, pp. 52–161.

Ford, C.R., Elliot, K.J., Clinton, B.D., Kloeppel, B.D. and Vose, J.M. (2012) 'Forest dynamics following eastern hemlock mortality in the southern Appalachians'. *Oikos*, vol. 121, pp. 523–536.

Franklin, J.F., Spies, T.A., Van Pelt, R., Carey, A.B., Thornburgh, D.A. and others (2002) 'Disturbances and structural development of natural forest ecosystems with sivicultural implications, using Douglas-fir forests as an example'. *Forest Ecology and Management*, vol. 155, pp. 399–423.

Frelich, L.E. (2002) *Forest Dynamics Disturbance Regimes: Studies from Temperate Evergreen–Deciduous Forests*, Cambridge University Press, Cambridge, UK.

Gleason, H.A. (1927) 'Further views on the succession concept'. *Ecology*, vol. 8, pp. 299–326.

Gravel, D., Canham, C.D., Beaudet, M. and Messier, C. (2010) 'Shade tolerance, canopy gaps and mechanisms of coexistence of forest trees' *Oikos*, vol. 119, pp. 475–484.

Grubb, P.J. (1977) 'The maintenance of species-richness in plant communities: the importance of the regeneration niche' *Biological Review*, vol. 52, pp. 107–145.

Guariguata, M. R. and Ostertag, R. (2001) 'Neotropical secondary forest succession: changes in structural and functional characteristics'. *Forest Ecology and Management*, vol. 148, pp. 185–206.

Hale, C.M., Frelich, L.E. and Reich, P.B. (2006) 'Changes in cold-temperate forest understory plant communities in response to invasion by European earthworms'. *Ecology*, vol. 87, pp. 1637–1649.

Howard, L.F. and Lee, T.D. (2003) 'Temporal patterns of vascular plant diversity in southeastern New Hampshire forests' *Forest Ecology and Management*, vol. 185, pp. 5–20.

Hubbell, S.P., Foster, R.B., O'Brien, S.T., Harms, K.E., Condit, R. and others (1999) 'Light-gap disturbances, recruitment limitation, and tree diversity in a neotropical forest'. *Science*, vol. 283, pp. 554–557.

Johnstone, J.F., Hollingsworth, T.N., Chapin, F.S. III and Mack, M.C. (2010) 'Changes in fire regime break the legacy lock on successional trajectories in Alaskan boreal forest'. *Global Change Biology*, vol. 16, no. 4, pp. 1281–1295.

Kneeshaw, D.D. and Bergeron Y. (1998) 'Canopy gap characteristics and tree replacement in the southeastern boreal forest'. *Ecology*, vol. 79, pp. 783–794.

Oliver, C.D. (1981) 'Forest development in North America following major disturbances'. *Forest Ecology and Management*, vol. 3, pp. 153–168.

Phillips, O.L., Aragão, L.E.O.C., Lewis, S.L., Fisher, J.B., Lloyd, J. and others (2009) 'Drought sensitivity of the Amazon rainforest'. *Science*, vol. 323, pp. 1344–1347.

Pickett, S.T.A., Collins, S.L. and Armesto, J.J. (1987) 'Models, mechanisms and pathways of succession'. *Botanical Review*, vol. 53, no. 3, pp. 335–371.

Powers, J.S., Becknell, J.M., Irving, J. and Perez-Aviles, D. (2009) 'Diversity and structure of regenerating tropical dry forests in Costa Rica: geographic patterns and environmental drivers'. *Forest Ecology and Management*, vol. 258, pp. 959–970.

Rich, R.L., Frelich, L.E. and Reich, P.B. (2007) 'Wind-throw mortality in the southern boreal forest: effects of species, diameter and stand age' *Journal of Ecology*, vol. 95, pp. 1261–1273.

Royo, A.A. and Carson, W.P. (2006) 'On the formation of dense understory layers in forests worldwide: consequences and implications for forest dynamics, biodiversity, and succession'. *Canadian Journal of Forest Research*, vol. 36, pp. 1345–1362.

Runkle, J.R. (1981) 'Gap regeneration in some old growth forests of the eastern United States'. *Ecology*, vol. 62, pp. 1041–1051.

Runkle, J.R. (1982) 'Patterns of disturbance in some old-growth mesic forests eastern North America'. *Ecology*, vol. 63, pp. 1533–1546.

Schnitzer, S.A. and Carson, W.P. (2000) 'Have we forgotten the forest because of the trees?' *Trends in Ecology and Evolution*, vol. 15, pp. 375–376.

Sevanto, S., McDowell, N.G., Dickman, L.T., Pangle, R. and Pockman, W.T. (2014) 'How do trees die? A test of the hydraulic failure and carbon starvation hypotheses'. *Plant, Cell and Environment*, vol. 37, pp. 153–161.

Shorohova, E., Kuuluvainen, T., Kangur, A. and Jõgiste, K. (2009) 'Natural stand dynamics, disturbance regimes and successional dynamics in the Eurasian boreal forests: a review with special reference to Russian studies'. *Annals of Forest Science*, vol. 66, no. 201, pp. 1–20.

Swaine, M.D. and Whitmore, T.C. (1988) 'On the definition of ecological species groups in tropical rain forests'. *Vegetatio*, vol. 75, pp. 81–86.

Tester, J., Starfield, A. and Frelich, L.E. (1997) 'Modeling for ecosystem management in Minnesota pine forests'. *Biological Conservation*, vol. 80, pp. 313–324.

Uhl, C., Clark, K. and Maquirno, P. (1988) 'Vegetation dynamics in Amazonian treefall gaps' *Ecology*, vol. 69, pp.751–763.

Watt, A.S. (1947) 'Pattern and process in the plant community'. *Journal of Ecology*, vol. 35, pp. 1–22.

Whitfeld, T.J.S., Lasky, J.R., Damas, K., Sosanika, G., Molem, K. and Montgomery, R.A. (2014) 'Species richness, forest structure, and functional diversity during succession in the New Guinea Lowlands'. *Biotropica*, 46: 538–548. doi: 10.1111/btp.12136

Whittaker, R.G. (1957) 'Recent evolution of ecological concepts in relation to the eastern forests of North America'. *American Journal of Botany*, vol. 44, pp. 197–206.

Worrall, J.J., Rehfeldt, G.E., Hamann, A., Hogg, E.H., Marchetti, S.B. and others (2013) 'Recent declines of *Populus tremuloides* in North America linked to climate'. *Forest Ecology and Management*, vol. 299, pp. 35–31.

11
TREE GENETIC DIVERSITY AND GENE FLOW IN FOREST ECOSYSTEMS

Francine Tremblay

Introduction

Genetic diversity represents the potential to adapt and evolve in response to a changing environment for all living organisms. Forest tree species are recognized for their high level of genetic diversity. This source of variation is expressed in a large tree-to-tree variation called the phenotype which is a combination of a genotype and its growing environment. In other words, "it is the tree that you see". This source of variation has been exploited by foresters to implement tree genetic improvement programs all over the world. The selected material is widely used in forest plantations as it helps increase forest productivity, timber supply and the provision of several ecological services to the community (Bauhus *et al.* 2010).

Trees are not an easy subject for geneticists. Their biological attributes, long juvenile phase, life-span and predominant outcrossing mating system all imposed constraints on experimental genetic studies. However, it is important to understand the genetics of trees in order to get insight into the evolution, conservation and sustainability of forest diversity around the world. In this chapter, our aim is to synthesize the biological and ecological determinants that impact the genetic diversity of forest tree's species in boreal and tropical ecosystems. We review tree main life-history traits, reproduction, dispersal potential and demography with respect to genetic dynamics of tree populations. The relative importance of adaptive and neutral levels of genetic diversity in tree populations will also be examined. The relationship between these traits and forest tree evolutionary response to climate changes will be discussed as well.

We focused our attention on boreal and tropical ecosystems because they represent two contrasting cases to illustrate the potential interaction of biological constraints, environmental conditions and human influence on the genetic diversity of trees. The boreal zone is characterized by the presence of a small number of tree species (generally less than six) in any one stand, with large, monotype stands being quite common. In moist tropical forests, the number of tree species is usually high but their population density is low (Finkeldey and Hattemer 2006). In tropical Asia, for example, between 150 and 250 tree species can be found in one hectare with about 26 to 36 percent of all tree species being represented by one individual tree. At a global scale, the boreal forest, at least in North America, represents a zone of minimal human impact in terms of urban and agricultural developments while the pressures for land use are usually very high in the tropics. The current conversion of tropical forest to agricultural systems for instance, is recognized as one of the major causes of global biodiversity loss (Pereira et al. 2010).

Basic concepts of population genetics

Genetic variation in an individual tree is measured by the presence of distinct alleles at a given locus, located on a pair (diploid organism) of homologous chromosomes. Therefore, if you consider a group of individuals belonging to the same species, the differences between genomes (sets of genes) within this group is designated as intraspecific genetic diversity. The term intraspecific diversity reflects the fact that the individuals from one species are not all genetically identical. The level of genetic diversity can be reflected in the variation in phenotypic traits in trees growing in similar ecological conditions. However it may be more subtle when variation in non-coding DNA (or genes) is unexpressed in the phenotype.

Population genetics

The aim of population genetics is to understand how genetic variability is transmitted from one generation to the next and to examine how and why this variability evolves over generations. The level, structure and partitioning of genetic diversity within and between populations can be characterized using different tools (e.g. neutral markers) and indices of diversity. The Hardy-Weinberg model is commonly applied in population genetic studies. It is based on the principle that genotype frequencies can be predicted from gene/allele frequencies which remained constant over generations in random mating populations. It is easy to use, but the underlying assumptions (no mutation, selection or migration) for its use are rarely met simultaneously.

What are the sources of genetic variation and population differentiation?

Evolution is the change of the genetic (allelic or genotypic) structure of a population at one or several gene loci. The evolutionary forces and their impacts for shaping the genetic variation patterns within and among populations are briefly summarized below (see also Table 11.1).

Genetic drift-definition and consequences

Genetic drift is defined as random changes in the frequency of alleles in a gene pool. The frequency of alleles gradually deviates from their expected frequency under random mating (deviation from the Hardy-Weinberg equilibrium). Drift is expected to occur in finite populations that are subjected to drastic reduction of their original sizes or in newly founded populations that are derived from a small number of individuals. When genetic drift continues for many generations it leads eventually to the random fixation of one allele and the loss of the other. The expected time of losing or fixing an allele is dependent on the initial population size. Genetic drift is accentuated by inbreeding (crossing between related individuals), eventually causing an increase in the frequency of homozygotes and a reduction in the genetic variation of the population. Genetic drift is also influenced by correlated mating that occurs when only a part of the reproductive population produces a greater proportion of the new generations.

Current evidence shows that large effective population sizes combine with life-history traits such as long-distance pollination and seed dispersal (migration) prevents most tree species from genetic isolation and its consequences (drift, inbreeding). However, there are some exceptions like red pine (*Pinus rubra*). Despite its wide geographic range and its life-history traits, red pine is considered to be one of the most genetically depauperate conifer species in North America (Mosseler *et al.* 1992). This pattern may have been caused by a severe genetic bottleneck (drastic reduction in population size) experienced during glacial episodes of the Holocene.

Mutation

Mutation is the source of new alleles and produced changes that may be heritable. The rate of spontaneous mutation is generally low, in the order of 10^{-4} to 10^{-6} mutations per gene per generation. In conifers, the nucleotide substitution rate is even lower and estimated to be 0.68×10^{-9} synonymous substitutions per site per year. Longer generation time organisms are associated with lower rates of mutation (calibrated by year) and molecular evolution. The *rate of mitosis* hypothesis established a link between height and rate of genome copying in plants. According to Lanfear *et al.* (2013) "Genetic changes that occur during cell division in plant shoots could potentially get passed on to future generations." Over the long term, the rate of cell division and genome copying slows down in taller plants. The consequences are that somatic mutations accumulate faster in rapid-growing but shorter lived plant species that have faster cell cycles than in relatively slow growing but long-lived species like trees.

Mutations are therefore very important on an evolutionary timescale but are unlikely to be important at an ecological timeframe (e.g. climate change). If a mutation has no effect on the function of the modified gene it is considered as selectively neutral. The rate of fixation of non-neutral mutations generally occurs at low frequencies because many non-synonymous changes are deleterious and are removed from the population. Therefore their consequences are almost negligible at our timescale.

Somatic mutations can influence a tree's capacity to adjust to a changing environment over its lifetime. In the mosaic tree, *Eucalyptus melliodora,* small changes in DNA within a single tree (in the order of ten nucleotide differences between two branches) lead to the differential up-regulation of a secondary metabolism from one branch to another and differential susceptibility to herbivory (Padovan *et al.* 2013). Whether these mutations are heritable and transmitted to the next generation remains an open question.

Migration

Migration is the passage of genes from one population to another through pollen, seeds, propagules or plant parts and the source of gene flow between populations. At the intra-population level, migration increases genetic diversity and the effective population size by bringing new alleles. However, at the same time, it limits genetic divergence between populations and opposes local adaptation by preventing the fixation of new, more adaptive, alleles. A small level of migration can have significant impact on intra-population diversity and prevent marked divergence between populations, even those subjected to genetic drift. Migration also gives birth to new populations when a seed produces a viable individual in a site where the species was not present before. This phenomenon is more frequently observed at the front or the rear edge of a species range. One side effect of migration is that it is difficult to clearly establish the exact genetic size of a population and delineate its boundary.

Selection

Each tree in a population does not necessarily contribute equally to the next generation. The fitness of an individual depends on its viability, its probability of reaching the age of reproduction and its fertility. Natural selection acts on the phenotypes in interaction with its environment by changing the average value of a population for characteristics that are subject to selection. When the environments are very similar from one population to another within a species, natural selection will prevent population divergence because the *filter* for adaptation is identical across all populations.

Selection acts on phenotype traits that are governed by monogenic (one gene) or polygenic (two or more genes) factors. The resistance of sugar pine (*Pinus lambertiana*) and western white pine (*Pinus monticola*) to the introduced disease white pine blister rust (*Cronartium ribicola*) is a good example of a trait determined by a single gene (Kinloch 1992). White pine blister rust is an agent of mortality for all five-needled white pines in the Pacific Northwest. The resistance is controlled by a dominant allele for a single gene called major gene resistance (MGR). Genotypes that are either heterozygous or homozygous for this allele (i.e. have either one or two copies of the resistance allele) are largely resistant to this disease.

Table 11.1 Impacts of evolutionary forces on population genetic diversity

	Selective forces	
	Drift/Mutation/Migration Random	Selection Directional
Differentiation among populations	+/+/-	+
Variation within populations	-/+/+	-
Heterozygosity	-/+/+	+/-

Redrawn from Musch *et al.* (2004).

Population genetic diversity

More than 90 percent of the total genetic diversity of tree species is found within populations with low levels of genetic differentiation between populations that are commonly observed (Hamrick 2004). Mean differentiation among populations (G_{ST} based on allozymes) is estimated to 8.9 percent for trees compared to 35.5 percent and 25.6 percent for annuals and herbaceous perennials respectively (Hamrick 2004). The high level of gene flow occurring between populations is the most common explanation for the high level of within population diversity (Petit and Hampe 2006). Pollen is often wind-dispersed in boreal and temperate trees species which are predominantly outcrossing. In their review, Kremer *et al.* (2012) reported that wind-dispersed viable pollen can travel up to 600 km. However, pollination distances that lead to successful mating (as determined by genetic parentage analysis) are shorter, ranging between 3 to 100 km (Table 11.2). In the tropics, pollen is predominantly animal dispersed and dispersal can be effective, reaching a scale of tens of kilometers. Mean distances of seed dispersal mediated by animals are in the range of 100 m to 1 km and those for pollen movement are in the range of 100 m to 14 km (Table 11.2).

The life-cycle of trees also plays a role in the partitioning of genetic diversity among and within populations, with their long juvenile phase playing an especially important role (Austerlitz *et al.* 2000). The presence of individuals from several age or size classes with overlapping generations is characteristic of tree stands. Following a colonization event and

Table 11.2 Examples of observed long-distance pollen and seed dispersal in trees (> 3 km for pollen and > 1 km for seeds). The table is arranged first by propagule type (pollen or seed), then by vector type (wind, insects) and dispersal type (potential, viable or effective).

Species	Propagule	Vector	Type[1]	Location	Dispersal system Method	Maximum Range	Dispersal distance Proportion ≥ threshold[2]	Reference
Pinus banksiana and Picea glauca	Pollen	Wind	Potential	Canada	Aerobiologic analysis	3,000 km		Campbell et al. (1999)
Pinus sylvestris	Pollen	Wind	Potential	Northern Europe	Aerobiologic analysis and phenological analysis	600 km		Varis et al. (2009)
Populus trichocarpa	Pollen	Wind	Effective	Western North America	Genetic parentage analysis		5% > ~ 5–10 km	Slavov et al. (2009)
Ficus spp.	Pollen	Wind	Potential	Central America	Genetic parental reconstruction	14 km (isolated mother trees)		Nason et al. (1998)
Ficus sycomorus	Pollen	Insects	Effective	Namibia	Genetic parentage analysis	165 km	40–80% ‡ 4 km (in small fragments)	Ahmed et al. (2009)
Swietenia humilis	Pollen	Insects	Effective	Central America	Genetic paternity analysis			White et al. (2002)
Xylopia hypolampra and seven other species	Seeds	Birds	Potential	Cameroon	Empirically based simulations of vector movements and seed passage time	6.9 km		Holbrook and Smith (2000)
Tamarindus indica	Seed	Elephant	Potential	Myanmar (Burma)	Empirically based simulations of vector movements and seed passage time	5.4 km	50% > 1.2 km	Campos-Arceiz et al. (2008)
Duroia duckei and two other species	Seed	Fish	Potential	Peru	Empirically based simulations of vector movements and seed passage time	5.5 km	5% > 1.7 km	Anderson et al. (2011)
Aucoumea klaineana	Seed Pollen	Wind	Potential	Gabon Central Africa	Parentage and sex assignment analyses	118 ± 62 m		Born et al. (2011b)
Swietenia humilis	Pollen	Bees, moths, thrips	Potential	Costa Rica	Between forest fragments	up to 4.5 km		White et al. (2002)
Ficus sp.	Pollen	Fig wasps	Potential	Panama		5.8–14.2 km		Nason et al. (1998)

[1] Dispersal types are respectively: 1) potential or the distance dispersed by a propagule at any condition; 2) effective, the pollen that gave rise to seeds, or seeds that established, yielding seedlings, saplings or young/adult plants.

[2] The proportion (percentage) of propagules dispersed to equal or greater distances than the specified threshold. The threshold distances were defined by the authors of each study, often arbitrarily or according to features of the study landscape and/or populations (adapted from Kremer et al. 2012).

before the first cohort of trees reaches reproductive age (during the first decades), there is a constant arrival of new migrants (seed flow) and no in-stand reproduction. Therefore, in a newly founded population, there is always a non-negligible part of the space that is already occupied by juveniles from seeds that arrived years earlier. In a recent study, Lesser et al. (2013) showed that alleles accumulated rapidly in ponderosa pine (*Pinus ponderosa*) populations following initial colonization and that contemporary levels of genetic diversity are formed early in tree population development. At population sizes of approximately 100 individuals, allele accumulation saturated. High levels of gene flow in the early stages of population growth results in a rapid accumulation of alleles and creates relatively homogenous genetic patterns among populations.

Boreal forest

There is a huge number of studies that have documented the partitioning of genetic diversity within and between populations of boreal species (Table 11.3). Most of them were conducted with economically important conifers that are widespread with typical F_{ST} values < 0.1 (mean value of ~ 0.050 for boreal species with a continuous range). In black spruce (*Picea mariana*) and white spruce (*Picea glauca*), no more than 1 percent of the genetic variation occurs among populations (Gamache et al. 2003; Jaramillo-Correa et al. 2001). Even a clonal species like trembling aspen (*Populus tremuloides*) maintains a high level of within population variability at continental and local scale and a low level of differentiation among populations (mean F_{ST} for microsatellite loci at continental scale = 0.086 [Callahan et al. 2013] and local scale = 0.032 [Wyman et al. 2003]). High levels of genetic diversity are also observed in fragmented northern marginal populations of coniferous species (Gamache et al. 2003; Xu et al. 2012).

Tropical forest

According to the Baker-Fedorov hypothesis (BFH) genetic drift would have promoted speciation over relatively limited spatial scales leading to the typical pattern of community structure in tropical forests. This hypothesis was based on the belief that tropical trees are highly inbred due to restricted gene flow. Several decades of research have invalidated this hypothesis, at least for canopy trees. In fact, many tropical tree species are self-incompatible with high rates of outcrossing and long-distance pollen dispersal (Bawa 1990). Typical values of F_{ST} range from 0.034 to 0.17 for populations separated by hundreds of meters to kilometers (Hamrick 1994; Dick et al. 2008) and are in the same range as those reported for boreal tree species (Table 11.3). Species with abiotic means of seed dispersal (e.g. barochorous, anemochorous) show on average a much higher differentiation among populations (G_{ST} = 0.138) than zoochorous species (G_{ST} = 0.050) (Finkeldey and Hattemer 2006). A small level of genetic differentiation is indicative of extended gene flow between tropical tree populations. The impact of low tree density on gene exchange is apparently compensated by efficient pollinators or animal dispersers with large home ranges that are able to cross significant distances between individuals.

Spatially explicit analysis of genetic variation for tree species occurring at low density in tropical forests may be more informative than the traditional approach that divides the overall genetic variation within and among populations. It is often difficult to determine which trees in a species-rich tropical forest belong to the same population. The non-random spatial patterns of gene dispersal imposed by animal pollinators and seed dispersers may provoke highly structured genetic diversity within populations of animal dispersed species (Jordano and Godoy 2002). The work of Sezen et al. (2005) revealed that the genetic composition of seedlings

Table 11.3 Examples of genetic diversity (microsatellite markers[1]) studies in populations of tree species

Species	A	A_R	H_O	H_E	F_{IS}	F_{ST}/G_{ST}	Reference
Protium subserratum	4.8 (2.8–6.0)	—	0.43 (0.05–0.66)	0.53 (0.36–0.67)	0.071 (0.01–0.16)	(0.022–0.46)	Misiewicz and Fine (2014)
Vriesea gigantea	2.83 (1.18–3.49)	—	0.430 (0.039–0.614)	0.579 (0.065–0.727)	0.276 (0.053–0.489)	0.211	Palma-Silva et al. (2009)
Shorea xanthophylla	8.07 (5.0–12.3)	—	0.616 (0.405–0.795)	0.662 (0.461–0.83)	0.069 (−0.135–0.317)	—	Kettle et al. (2010)
Parashorea tomentella	10.67 (4.92–17.52)	—	0.572 (0.342–0.770)	0.608 (0.345–0.869)	0.059 (−0.008–0.151)	—	Kettle et al. (2010)
Dipterocarpus grandiflorus	13.67 (9.63–16.66)	—	0.634 (0.550–0.734)	0.680 (0.550–0.763)	0.067 (−0.039–0.266)	—	Kettle et al. (2010)
Aucoumea klaineana	—	—	0.455 (0.349–0.539)	0.503 (0.442–0.599)	0.383 (−0.050–0.647)	(0.039–0.162)	Born et al. (2011b)
Shorea javanica	—	—	0.423	0.477	0.383 (−0.050–0.647)	0.064	Rachmat et al. (2012)
Swietenia macrophylla	9.5 (7.6–10.7)	—	0.750 (0.68–0.81)	0.781 (0.754–0.812)	0.038 (−0.004–0.10)	0.097	Lemes et al. (2003)
Carapa guianensis	—	—	0.66	0.64	(−0.02–0.14)	(0.001–0.087)	Dayanandan et al. (1999)
Acer saccharum	8.2 (6.6–9.0)	7.0 (5.8–7.6)	0.597 (0.496–0.716)	0.693 (0.637–0.715)	0.138 (−0.051–0.302)	0.016 (0.009–0.041)	Graignic (2014)
Populus tremuloides	8.83 (7.58–10.08)	—	0.465 (0.45–0.48)	0.67 (0.61–0.73)	0.30 (0.21–0.39)	—	Namroud et al. (2005)
Populus tremuloides	—	5.99 (3.34–6.83)	—	0.758 (0.613–0.801)	0.019 (−0.12–0.19)	0.086	Callahan et al. (2013)
Populus tremuloides	7.44 (6.25–8.2)	—	0.556 (0.478–0.704)	0.725 (0.691–0.767)	0.201 (−0.054–0.325)	0.032	Wyman et al. (2003)
Pinus strobus	9.43 (9.23–9.62)	—	0.522 (0.505–0.538)	0.607 (0.599–0.615)	—	—	Rajora et al. (2000)★
Pinus strobus	—	6.7	0.47	0.48	0.01	—	Marquardt and Epperson (2004)★
Thuja occidentalis	9.58 (7.83–11.17)	9.21 (7.66–10.68)	0.590 (0.505–0.640)	0.600 (0.519–0.662)	0.019 (−0.025–0.050)	—	Pandey and Rajora (2012a)
Thuja occidentalis	7.3 (5.67–9.33)	6.8 (5.16–8.51)	0.601 (0.492–0.662)	0.611 (0.490–0.678)	0.013 (−0.063–0.105)	0.078	Pandey and Rajora (2012b)
Thuja occidentalis	7.8 (5.0–10.0)	5.9 (4.6–6.9)	0.734 (0.463–0.883)	0.773 (0.712–0.840)	0.145	0.065	Xu et al. (2012)

[1] Microsatellites or single sequence repeats (SSRs) are short DNA fragments of usually only two or three base pairs in length which are repeated several times in a particular location of the DNA.

★ Study used logging or non-natural forest, we only reported populations from old-growth and natural forests. A, mean number of alleles per locus; A_R, mean allelic richness; H_O, mean observed heterozygosity; H_E, mean expected heterozygosity; F_{IS}, inbreeding coefficient; F_{ST}, mean pairwise F_{ST}, G_{ST}, mean pairwise G_{ST}. Range values are given in parentheses (adapted from Graignic [2014])

regenerating in tropical forests is influenced both by the behavior of dispersal agents and by spatial patterns of genetic variation in the parental pool.

Phylogeographic pattern of variation

The cool and dry glacial period characteristics have shaped the genetic structure of tree species on all continents. These episodes have severely constricted forest ranges in North America, Europe and in the tropics. Analysis of the geographic pattern of genetic diversity, also called phylogeography, is studied using different types of markers. Cytoplasmic DNA is often used to trace lineages that have spread across continents following glacial retreat. In conifers, mtDNA (mitochondrial DNA) is maternally inherited and cpDNA (chloroplastic DNA) is paternally inherited. There are two exceptions, the *Taxodiaceae* and *Cupressaceae* where mtDNA is strictly inherited from the paternal parent. In angiosperm, cpDNA and mtDNA are both inherited through the mother. In contrast with ncDNA (nuclear DNA), cytoplasmic DNAs are highly conservative.

Boreal trees species

The boreal region was covered with ice during the late-Quaternary and previous glaciations. At that time, the organisms were largely displaced south of their current limits. Many species migrated northward when the climate warmed and when the glaciers receded. The study of the transcontinental boreal distribution of white spruce has revealed the presence of three white spruce (*Picea glauca*) glacial refugia; Beringian, Mississippian and east Appalachia (de Lafontaine *et al.* 2010). Analysis of the distribution of mitotypes (mtDNA) in black spruce (*Picea mariana*) showed the presence of four distinct groups: two restricted to the western and eastern part of Canada, one large group distributed along most of the species range, and a fourth group observed exclusively in the north-east (Jaramillo-Correa *et al.* 2001). A range-wide study of balsam poplar (*Populus balsamifera*) populations in North America has identified three geographically separated demes found in the northern, central and eastern portions of the species' range (Keller *et al.* 2010). The results suggest a recent population growth and massive expansion from the center to the north and east.

Tropical tree species

To date, most phylogeographic studies have been conducted on temperate and boreal tree species and despite their importance, studies on tropical tree species are comparatively scarce. The impact of Pleistocene climate remained poorly understood partly because of the very sparse palynological records. The climate was dry and cool in the tropics during glacial episodes and close to the actual during interglacial episodes. Climatic oscillations have certainly led to contraction/expansion of the forest as well as transitions between different types of forest communities. Both genetic and species diversities were shaped by the biogeographic processes triggered by the climatic oscillations of the Pleistocene. The analysis of nuclear DNA and cpDNA haplotypes in Central Africa revealed the presence of Pleistocene forest refuges for the Bush Mango, *Irvingia gabonensis*, and a north–south phylogeographic disjunction in *Greenwayodendron suaveolens* (Lowe *et al.* 2010). In Central Africa, concerning *Aucoumea klaineana* Pierre (*Burseraceae*), an African pioneer tropical rainforest tree species, Born *et al.* (2011a), showed the presence of four genetic units that resulted from the expansion of subdivided source populations, a pattern consistent with the hypothesis of population fragmentation during the LGM.

A phylogeographic pattern of variation has also been documented in South America. In their analysis of natural populations of mahogany (*Swietenia macrophylla* King [*Meliaceae*]) ranging from Mexico to Panama, in Central America, and across 2,100 km of the southern arc of the Amazon basin, Lemes *et al.* (2010) found a strong break between Central and South America mahogany populations. In addition, they reported the presence of high levels of population differentiation in the Amazon basin and, in contrast, relatively low differentiation across Central America. Comparative phylogeographical analysis has shown a congruent signal of population size decline in Panamanian rainforest trees (Turchetto-Zolet *et al.* 2013).

Geographic variation in quantitative traits in forest trees

Neutral markers are poor indicators of the level of variation in adaptive traits affecting tree fitness. Multiple-sites provenance trials have been used to examine the plastic responses of tree populations to environmental conditions. In these trials, several provenances (in the broader sense, i.e. the source of seed from a defined geographic area) were tested in common garden experiments. Provenance trials were established in various stands with different temperatures and water availability. The amount of differentiation for quantitative traits (growth, phenology, survival) often showed a continuous (clinal) variation along environmental gradients (altitudinal, latitudinal, longitudinal) (Morgenstern 1996). This clinal variation is generally assumed to result from natural selection. The mean Q_{ST} estimate or the proportion of total quantitative genetic variation due to differences among populations for large populations in northern areas is ten times higher than the average F_{ST} (mean Q_{ST} = 0.463; mean F_{ST} = 0.044). The Q_{ST} estimates for growth in diameter and height range from very low (0) to close to 1, while many estimates for traits relating to bud set timing and cold hardiness, which are climate-related and more critical for survival and growth, are over 0.5 (Alberto *et al.* 2013).

Boreal tree species

The most commonly reported pattern of variation for boreal tree species is high genetic diversity with clinal variation in quantitative traits (Figure 11.1a). For example, bud flush is triggered by the accumulation of cold sums followed by heat sums above a threshold temperature sum. These genetically determined critical temperature sums and thresholds may vary among species and, to a lesser extent, among populations of the same species. Bud set is largely controlled by photoperiod, and modulated by temperatures and drought. In a warming climate, spring phenology can likely respond and advance without much genetic change, as long as the chilling requirement has been met. In the fall, a change in bud set date is more likely to require a genetic change in photoperiodic responses.

This clinal variation in phenological (timing of bud set, bud flush), growth (height) and physiological characteristics can occur over short or long distances. For example, in Douglas fir (*Pseudotsuga menziesii*) variation is correlated with elevation of the seed source (Rehfeldt 1983). Genetic variation is organized into numerous local specialist populations adapted to a relatively narrow subset of environments. Other species, like western white pine (*Pinus monticola*) and western red cedar (*Thuja plicata*), are adaptive generalists with populations adapted to a broad range of environments through phenotypic plasticity. A given genotype can produce a variety of phenotypes depending on the environment either through multiple biochemical pathways associated with high genetic diversity (i.e. heterozygosity) or through natural selection, favoring alleles that tolerate broad environmental differences (Morgenstern 1996).

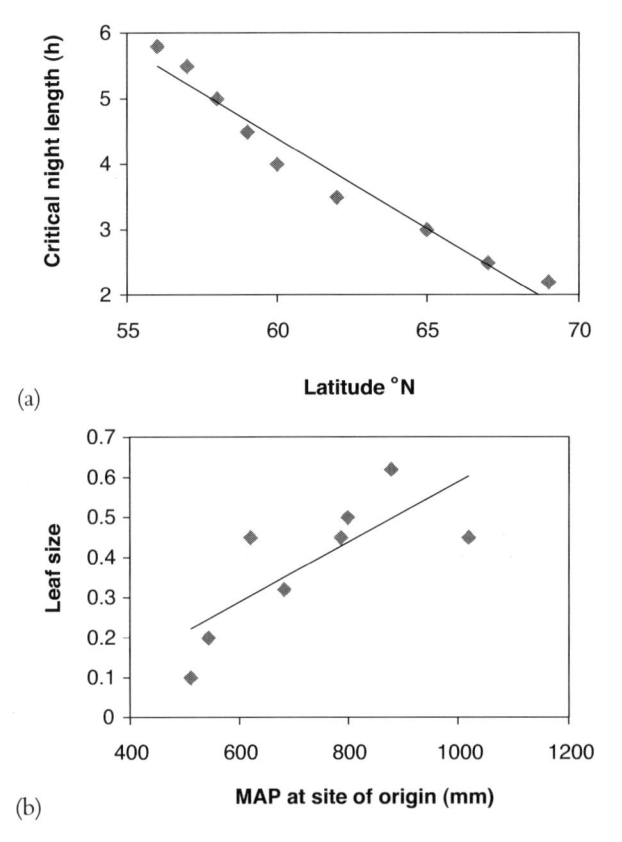

Figure 11.1 Clinal variation in quantitative traits; (a) relationship between critical night length (h) and timing of growth cessation in *Picea abies* populations from different latitudes; (b) relationships between trait plasticity (relative trait range; RTR) of nine *Eucalyptus tricarpa* provenances and the mean annual precipitation (MAP) at their site of origin, plasticity of leaf size showed a positive linear correlation with MAP. The figures are redrawn from data from (a) Savolainen *et al.* (2007) and (b) McLean *et al.* (2014)

Tropical tree species

Clinal variation in the tropics is often associated with the response to drought (Figure 11.1b). In the West African Sahel, there is a transitional zone between the relatively humid savannah woodlands to the south and the Sahara Desert to the north. In the first provenance/progeny test of *Prosopis africana*, growth and survival are related to rainfall gradients (Weber *et al.* 2008). Provenances from drier parts of the region had better growth and survival than provenances from more humid parts of the region (when tested in a relatively dry site in Niger). Local adaptation to drought has been reported for *Eucalyptus tricarpa*, a species occurring across a climatic gradient in south-eastern Australia, a region of increasing aridity (McLean *et al.* 2014). Provenances from drier locations are more plastic for several functional traits and exhibit differential growth rates in the common gardens. Specialization across edaphically heterogeneous environments may also promote population-level divergence even when plant populations are not geographically isolated. Misiewicz and Fine (2014) found significant morphological variation correlated to soil type in the Mesoamerican mahogany *Swietenia macrophylla* King (Big-leaf mahogany). They reported higher levels of genetic differentiation and lower migration rates between adjacent populations found on different soil types than between geographically distant populations on the same soil type.

Impacts of forest harvesting

Boreal forest landscapes change with succession and disturbance. In North America it is characterized by a low number of tree species over a wide range and includes species such as trembling aspen, jack pine and black spruce. These species regenerate following natural (fire, insect, windthrow) or anthropogenic (logging) disturbances. Clear-cutting is the most common harvesting system in the boreal forests but other types such as partial-cut, selection-cut or shelterwood can also be applied. Small population differentiation was observed between managed and wild stands in lodgepole pine (*Pinus contorta*) (MacDonald et al. 2001). Harvest-origin stands have significantly lower average population genetic diversity than unmanaged (fire-origin) stands. The removal of 75 percent of trees (seed tree cut) had a negative impact on genetic diversity in eastern white pine (*Pinus strobus*) (Buchert et al. 1997). However, in most studies, clear-cutting followed by natural or artificial regeneration had no negative genetic effect on several coniferous species. Fageria and Rajora (2013) found similar genetic structure in white spruce between un-harvested control or pre-harvest old-growth and post-harvest natural regeneration. The populations present in the landscape contribute (through gene flow) to the preservation of genetic diversity in harvested stands.

The conservation of genetic diversity in trees in a tropical ecosystem is especially challenging because harvesting in this part of the world often goes along with deforestation, habitat destruction and population fragmentation. Partial cutting (selective cut – with minimum diameter cutting limits) is the most common harvesting practice in the tropical forest. Logging (or deforestation) may affect population density but most importantly, the abundance and behavior of pollinator and seed disperser communities (Dick 2010). Some reports found significant differences between outcrossing rates between undisturbed and logged plots and a significant increase in self-fertilization rates in fragmented forest patches. For example, in Costa Rica, reduced outcrossing rates were documented in disturbed habitats that contained low population densities of reproductive *Symphonia globulifera* (Aldrich and Hamrick 1998) and *Pachira quinata* (Fuchs et al. 2003). Cloutier et al. (2007) found no genetic impact from selective logging on gene diversity, inbreeding, pollen dispersal and spatial genetic structure in *Carapa guianensis* populations in Brazil. However, genetic diversity was reduced while outcrossing and gene flow was maintained after selective logging of *Hymenaea courbaril* from the same forest area in Brazil (Carneiro et al. 2011). Results from model simulations supported the idea that genetic information at species level is essential to predict the long-term impacts of selective logging on Brazilian Amazon tree species and derive sustainable management scenarios for tropical forests (Vinson et al. 2014).

Impact of climate change

We cannot predict the fate of natural populations of trees in the context of global change. In the Northern Hemisphere, most models predict a shift north in the current range of many tree species. Will trees be able to cope with the anticipated rapid onset of change from a genetic point of view (for review see Alberto et al. 2013)? Trees exhibiting a high degree of phenotypical plasticity are better able to persist and survive when exposed to climatic variation. However, current evidence suggests that tree mortality is rising in many populations and forest health is declining (Clark 2007; Peng et al. 2011). For instance, direct climate impacts of heat and drought have been identified as the primary causes of forest dieback at the southern edge of the boreal forests (Allen *et al*. 2010). In addition, common boreal tree species have shown increased mortality over the last two decades (Peng *et al*. 2011). In the tropics, it is difficult to disentangle the specific effect of ongoing global change from the historical context (human activities) in study sites (Clark 2007).

The long-generation times of trees and their slow rate of mutation may impose large constraints on their capacity to adapt. On the other hand, forest trees possess a high level of genetic diversity within populations that, theoretically, provide the basic materials for adaptation to new climate conditions. Trees also often have very large effective population sizes, which makes selection more effective (relative to drift). Long-distance gene flow and heterogeneous selection contribute to maintaining genetic variability within populations. Therefore, gene flow increases the genetic variance of a population available for selection (Hamrick and Nason 2000) and may provide pre-adapted genotypes that facilitate adaptation (Kremer *et al*. 2012). Numerous observations have shown that genes can move, through seeds and pollen, over spatial scales larger than the habitat shifts predicted by climate change models within one generation. Recently, Kremer and Le Corre (2012) proposed that adaptation may occur without a detectable change in allele frequency. Their simulations showed that allelic associations were rapidly modified under moderate to strong divergent selection. The response elapsed during less than 10–20 generations. They suspect that the build-up of allelic associations is enhanced by the very large genetic diversity residing within populations which is sustained by a high gene flow.

Epigenetic effects can also play a role in tree adaptation to new environmental conditions. Significant epigenetic effects on the phenotypes have been reported in Norway spruce (*Picea abies*) (Yakovlev *et al*. 2014). These effects are, on average, about 20 percent as large as the true additive genetic variation for quantitative traits. This could have significant implications for the interpretation of provenance trial data, explaining some of the phenotypic variation among populations that are commonly interpreted as genetic variation. The work of Raj *et al*. (2011) has shown that genetically identical poplar clones have slightly different stomatal behavior under drought conditions in relation to their growing maternal environment. The study, conducted with a genetically depauperated species, *Pinus pinea*, revealed a high level of cytosine methylation in the genome as well as a high level of variation of methylation between trees (Saez-Laguna *et al*. 2014). This suggests the potential role played by cytosine methylation in the regulation of gene expression and variation in phenotypic traits in absence of detectable genetic variation that improves stone pine fitness under different environmental conditions. Some of these epigenetic effects seem to provide the immediate ability to adapt to new climatic conditions as well as provide sufficient time for an evolutionary response. It is not clear, however, whether or not these effects will be passed on to progeny.

Different forest management strategies have been proposed to mitigate the potential negative impact of climate change on forest ecosystems. Resistance and resilience strategies may be applied to highly valued resources to aim at maintaining or restoring current ecosystems if possible (Millar *et al*. 2007). Another approach is to facilitate transitions to new ecosystems that are adapted to a change in climatic conditions and which may involve the human-assisted migration of species. However, the most commonly-cited approach consists of a better match of genotypes of commercial species for future climates in a reforestation program. In other words, assisted migration (moving seed source to a new climate) is restricted to the movement of populations within a species' range. For these purposes, data from the provenance trials will be used and combined with statistical models (response function) to predict a species' response to climate change. The underlying hypothesis for these models is using geographic variables as a surrogate for elusive climates that govern adaptation on a local scale. Seed zone maps are produced with the assumption that environmental variation and genetic variation in adaptive traits are largely correlated. However, the degree to which patterns of genetic variation are reflected by environmental differences varies among species. These species-specific seed zones illustrate the extent to which species are adaptive generalists or specialists by nature. However, with only 15–20 percent of the among-population variation (or 2–3 percent of the total variation) possibly linked to the climate, it brings the following question: will assisted migration prescriptions be necessary to address climate change? Additionally, the intentional movement of species does not come without criticism. The debate is largely focused on the assessment of risks and benefits. Uncertainty about future climate conditions and possible impairment of ecological function and structure are major constraints to its implementation (Aitken and Whitlock 2013).

Forest management should also consider the practice of *composite provenancing* to buffer the great uncertainty of climate change. These "composite provenances" would be predominantly composed of locally sourced material, that would also incorporate proximate and ecogeographically matched sources. In addition, a smaller proportion of material (somewhere between 10 percent and 30 percent depending on the inferred gene flow dynamics of the species) should include material from much further afield. This approach mitigates the relative risks of assuming populations can adapt *versus* the risks of missing different gene pools in populations.

Future perspectives

The advent of fast and cost-effective next-generation-sequencing (NGS) technology has opened opportunities to apply new approaches to the study of tree genomes. More sequencing data is becoming available that can be used to understand the extent of standing variation for quantitative traits in trees. Poplar (*Populus trichocarpa*) was the first tree for which the genome was sequenced (Tuskan *et al.* 2006), establishing this species as the model research system for long-lived woody perennials. The genome size of poplar is small relative to conifers. Genome sizes in conifers are among the largest (typically 20–30 gigabases of pairs (Gb) among all organisms) making genome-wide analyses particularly challenging for gymnosperms. Despite this, the draft genome of Norway spruce (*Picea abies*) was recently published (Nystedt *et al.* 2013) and full genome sequences are also available for woody perennial species such as *Eucalyptus* (Myburg *et al.* 2014), *Papaya* (Ming *et al.* 2008) and *Theobroma cacao* (Argout *et al.* 2011). The steady increase in the number of sequences available supports the development of powerful genomic tools. There is a huge body of research literature on this subject which is too large to cover here (for reviews see González-Martínez *et al.* 2006; Groover and Jansson 2014). For example, expressed sequence tag collections (ESTs) have been developed that are used for selecting candidate genes or detecting single nucleotide polymorphism (SNPs) that are potentially up for selection. Recently, a set of 17 SNPs was associated with serotiny in natural populations of maritime pine (*Pinus pinaster*) growing in a fire-prone environment (Budde *et al.* 2014). This study is a good illustration of the interest in identifying genes that are specifically associated with ecologically relevant traits under strong selection. Holliday *et al.* (2010) identified SNPs with phenotype associations for bud set and cold hardiness in the widely distributed conifer Sitka spruce (*Picea sitchensis*). Many SNPs with phenotypic associations were also correlated with at least one climate variable. This represents a significant step forward in the characterization of the genomic basis of adaptation to local climate in conifers. The widespread use of population genomic approaches in the future would therefore help to increase our understanding of traits that could be used to model capacity for adaptation responses as well as provide tools for molecular breeding strategies in trees.

References

Ahmed, S., Compton, S.G., Butlin, R.K., Gilmartin, P.M. (2009) Wind-borne insects mediate directional pollen transfer between desert fig trees 160 kilometers apart. *Proceedings of the National Acadamy of Sciences of the USA* 106, 20342–20347

Aitken, S.N., Whitlock, M.C. (2013) Assisted gene flow to facilitate local adaptation to climate change. *Annual Review of Ecology Evolution and Systematics* 44: 367–388

Alberto, F.J., Aitkens, S.N., Alía, R., González-Martínez, S.C., Hänninen, H. and others (2013) Potential for evolutionary responses to climate change – evidence from tree populations. *Global Change Biology*, doi: 10.1111/gcb.12181

Aldrich, P.R., Hamrick, J.L. (1998) Reproductive dominance of pasture trees in a fragmented tropical forest mosaic. *Science* 281: 103–105

Allen, C.D., Macalady, A.K., Chenchouni, H., Bachelet, D., McDowell, N. and others (2010) A global overview of drought and heat-induced tree mortality reveals emerging climate change risks for forests. *Forest Ecology and Management* 259: 660–684

Anderson, J.T., Nuttle, T., Saldanã Rojas, J.S., Pendergast, T.H., Flecker, A.S. (2011). Extremely long-distance seed dispersal by an overfished Amazonian frugivore. *Proceedings of the Royal Society B: Biological Sciences* 278: 3329–3335

Argout, X., Salse, J., Aury, J.M., Guiltinan, M.J., Droc, G. and others (2011) The genome of *Theobroma cacao*. *Nature Genetics* 43(2): 101–108. doi: 10.1038/ng.736

Austerlitz, F., Mariette, S., Machon, N., Gouyon, P.H., Godelle, B. (2000) Effects of colonization processes on genetic diversity: differences between annual plants and tree species. *Genetics* 154: 1309–1321

Bauhus, J., van der Meer, P., Kanninen, M. (2010) *Ecosystem Goods and Services from Plantation Forests*. The Earthscan Forest Library, ISBN: 978-1-84971-168-5

Bawa, K.S. (1990) Plant-pollinator interactions in tropical rainforests. *Annual Review of Ecology, Evolution and Systematics* 21: 399–422

Born, C., Alvarez, N., McKey, D., Ossari, S., Wickings, E.J. and others (2011a) Insights into the biogeographical history of the Lower Guinea Forest Domain: evidence for the role of refugia in the intraspecific differentiation of *Aucoumea klaineana*. *Molecular Ecology* 20: 131–142

Born, C., Kjellberg, F., Chevallier, M.H., Vignes, H., Dikangadissi, J.-T. and others (2011b) Colonization processes and the maintenance of genetic diversity: insights from a pioneer rainforest tree, *Aucoumea klaineana*. *Proceedings of the Royal Society B: Biological Sciences* 275: 2171–2179

Buchert, G.P., Rajora, O.P., Hood, J.V., Dancik, B.P. (1997) Effects of harvesting on genetic diversity in old-growth eastern White Pine in Ontario, Canada. *Conservation Biology* 11: 747–758

Budde, K.D., Heuertz, M., Hernandez-Serrano, A., Pausas, J.G., Vendramin, G.G. and others (2014) In situ genetic association for serotiny, a fire-related trait, in Mediterranean maritime Pine (*Pinus pinaster*). *New Phytologist* 201: 230–241 doi: 10.1111/nph.12483

Callahan, C.M., Rowe, C.A., Ryel, R.J., Shaw, J.D., Madritch, M.D., Mock, K.E. (2013) Continental-scale assessment of genetic diversity and population structure in quaking aspen (Populus tremuloides). *Journal of Biogeography* 40: 1780–1791

Campbell, I.D., McDonald, K., Flannigan, M.D., Kringayark, J. (1999) Long distance transport of pollen into the Arctic. *Nature* 399: 29–30

Campos-Arceiz, A., Larringa, A.R., Weerasinghe, U.R., Takatsuki, S., Pastorini, J. and others (2008) Behavior rather than diet mediates seasonal differences in seed dispersal by Asian elephants. *Ecology* 89: 2684–2691

Carneiro, F.S., Lacerda, A.E.B., Lemes, M.R., Gribel, R., Kanashiro, M. and others (2011) Effects of selective logging on the mating system and pollen dispersal of *Hymenaea courbaril* L. (*Leguminosae*) in the Eastern Brazilian Amazon as revealed by microsatellite analysis. *Forest Ecology and Management* 262: 1758–1765

Clark, D. (2007) Detecting tropical forests' responses to global climatic and atmospheric change: current challenges and a way forward. *Biotropica* 39(1): 4–10

Cloutier, D., Kanashiro, M., Ciampi, A.Y., Schoen, D.J. (2007) Impact of selective logging on inbreeding and gene dispersal in an Amazonian tree population of *Carapa guianensis* Aubl. *Molecular Ecology* 16: 797–809

Dayanandan, S., Dole, J., Bawa, K. and Kessel, R. (1999) Population structure delineated with microsatellite markers in fragmented populations of a tropical tree, *Carapa Guianensis* (Meliaceae). *Molecular Ecology* 8: 1585–1592

de Lafontaine, G., Turgeon, J., Payette, S. (2010) Phylogeography of White Spruce (*Picea glauca*) in eastern North America reveals contrasting ecological trajectories. *Journal of Biogeography* 37: 741–751

Dick, C.W. (2010) Phylogeography and population structure of tropical trees. *Tropical Plant Biology* 3: 1–3 DOI 10.1007/s12042-009-9039-0

Dick, C.W., Hardy, O.J., Jones, F.A., Petit, R.J. (2008) Spatial scales of pollen and seed-mediated gene flow in tropical rain forest trees. *Tropical Plant Biology* 1: 20–33

Fageria, M.S., Rajora, O.M. (2013) Effects of harvesting of increasing intensities on genetic diversity and population structure of White Spruce. *Evolutionary Applications* 6(5): 778–794

Finkeldey, R., Hattemer, H.H. (2006) *Tropical Forest Genetics*. Springer Science and Business Media, 328 p.

Fuchs, E.J., Lobo, J.A., Quesada, M. (2003) Effects of forest fragmentation and flowering phenology on the reproductive success and mating patterns of the tropical dry forest tree *Pachira quinata*. *Conservation Biology* 17: 149–157

Gamache, I., Jaramillo-Correa, J.P., Payette, S., Bousquet, J. (2003) Diverging patterns of mitochondrial and nuclear DNA diversity in subarctic black Spruce: imprint of a founder effect associated with postglacial colonization. *Molecular Ecology* 12: 891–901

González-Martínez, S.C., Krutovsky, K.V., Neale, D.B. (2006) Forest-tree population genomics and adaptive evolution. *New Phytologist* 1.70: 227–238

Graignic, N. (2014) Impact de la fragmentation sur la capacité reproductrice et la diversité génétique de l'érable à sucre (*Acer saccharum* Marshall) au Québec. PhD thesis, Institut de recherche sur les forêts, UQAT, Rouyn-Noranda, Québec

Groover, A., Jansson, S. (2014) *Comparative and Evolutionary Genomics of Forest Trees Challenges and Opportunities for the World's Forests in the 21st Century Forestry Sciences* Volume 81, Springer Science and Business Media, pp. 597–614

Hamrick, J.L. (1994) Genetic diversity and conservation in tropical forests. In: Drysdale, R.M., John, S.E.T., Yopa, A.C.O. (Eds.), *Proceedings of the International Symposium on Genetic Conservation and Production of Tropical Forest Tree Seed*, pp. 1–9. ASEAN-Canada Forest Tree Seed Centre Project, Muak-Lek, Saraburi, Thailand, pp. 1–9

Hamrick, J.L. (2004) Response of forest trees to global environmental changes. *Forest Ecology and Management* 197: 323–335.

Hamrick, J.L., Nason, D. (2000) Gene flow in forest trees. In: *Forest Genetics and Conservation: Principles and Practice*. (eds Young, A., Boshier, D., Boyle, T.) pp. 81–90. CSIRO Publishing, Collingwood, Australia

Holbrook, K.M., Smith, T.B. (2000) Seed dispersal and movement patterns in two species of *Ceratogymna* hornbills in a West African tropical lowland forest. *Oecologia* 125: 249–257

Holliday, J.A., Ritland, K., Aitken, S.N. (2010) Widespread, ecologically relevant genetic markers developed from association mapping of climate-related traits in Sitka Spruce (*Picea sitchensis*). *New Phytologist* 188(2): 501–514

Jaramillo-Correa, J.P., Beaulieu, J., Bousquet, J. (2001) Contrasting evolutionary forces driving population structure at expressed sequence tag polymorphisms, allozymes and quantitative traits in White Spruce. *Molecular Ecology* 10: 2729–2740

Jordano, P., Godoy, J.A. (2002) Frugivore-generated seed shadows: a landscape view of demographic and genetic effects. In D.J. Levey, W.R. Silva, and M. Galetti, (eds) *Seed Dispersal and Frugivory: Ecology, Evolution and Conservation*. CAB International, Wallingford, UK, pp. 305–321

Keller, S.R., Olson, M.S., Silim, S., Schroeder, W.R., Tiffin, P. (2010) Genomic diversity, population structure, and migration following range expansion in the Balsam Poplar, *Populus balsamifera*. *Molecular Ecology*, 19(6): 1212-26. doi: 10.1111/j.1365-294X.2010.04546.x

Kettle, C.J., Hollingsworth, P.M., Burslem, D.F.R.P., Maycock, C.R., Khoo, E., Ghazoul, J. (2010) Determinants of fine-scale spatial genetic structure in three co-occurring rainforest canopy trees in Borneo. *Perspectives in Plant Ecology, Evolution and Systematics*, doi:10.1016/j.ppees.2010.11.002

Kinloch, B.B. Jr (1992) Distribution and frequency of a gene for resistance to white pine blister rust in natural populations of sugar pine. *Canadian Journal of Botany* 70: 1319–1323

Kremer, A., Le Corre, V. (2012) Decoupling of differentiation between traits and their underlying genes in response to divergent selection. *Heredity* 108: 375–385

Kremer, A., Ronce, O., Robledo-Arnuncio, J.J., Guillaume, F., Bohrer, G. and others (2012) Long-distance gene flow and adaptation of forest trees to rapid climate change. *Ecology Letters* 15: 378–392

Lanfear, R., Ho, S.M.Y., Davies, T.J., Moles, A.T., Aarssen, L. and others (2013) Taller plants have lower rates of molecular evolution. *Nature Communications* 4, (1879) doi:10.1038/ncomms2836

Lemes, M.R., Girbel, R., Proctor, J., Gratapaglia, D. (2003) Population genetic structure of mahogany (*Swietenia macrophylla* King, Meliaceae) across the Brazilian Amazon, based on variation at microsatellite loci: implications for conservation. *Molecular Ecology* 12, 2875–2883

Lemes, M.R., Dick, C.W., Navarro, C., Lowe, A.J., Cavers, S., Gribel, R. (2010) Chloroplast DNA microsatellites reveal contrasting phylogeographic structure in mahogany (*Swietenia macrophylla* King, Meliaceae) from Amazonia and Central America. *Tropical Plant Biology* 3: 40–49

Lesser, M.R., Parchman, T.L., Jacksons, S.T. (2013) Development of genetic diversity, differentiation and structure over 500 years in four ponderosa pine populations. *Molecular Ecology* 22: 2640–2652 doi: 10.1111

Lowe, A.J., Harris, D., Dormontt, E., Dawson, I.K. (2010) Testing putative African tropical forest refugia using chloroplast and nuclear DNA phylogeography. *Tropical Plant Biology* 3(1): 50–58

MacDonald, S.E., Thomas, B.R., Cherniawsky, D.M., Pudy, B.G. (2001) Managing genetic resources of lodgepole pine in west-central. Alberta: patterns of isozyme variation in natural populations and effects of forest management. *Forest Ecology and Management* 152: 45–58

McLean, E.H., Prober, S.M., Stock, W.D., Steane, D.A., Potts, B.M. and others (2014) Plasticity of functional traits varies clinally along a rainfall gradient in *Eucalyptus tricarpa*. *Plant, Cell and Environment*, doi: 10.1111/pce.12251

Marquardt, P.E., Epperson, B.K. (2004) Spatial and population genetic structure of microsatellites in white pine. *Molecular Ecology* 13: 3305–3315

Millar, C.I., Stephenson, N.L., Stephens, S.L. (2007) Climate change and forests of the future: Managing in the face of uncertainty. *Ecological Applications* 17: 2145–2151

Ming, R., Hou, S., Feng, Y. (2008) The draft genome of the transgenic tropical fruit tree papaya (*Carica papaya* Linnaeus). *Nature* 24; 452(7190): 991–6. doi: 10.1038/nature06856

Misiewicz, T.M. and Fine, P.V.A. (2014) Evidence for ecological divergence across a mosaic of soil types in an Amazonian tropical tree: *Protium subserratum* (Burseraceae) *Molecular Ecology* 23: 2543–2558

Morgenstern, E.K. (1996) *Geographic Variation in Forest Trees: Genetic Basis and Application of Knowledge in Silviculture*. University of British Columbia Press, Vancouver, BC, Canada, 209 pp.

Mosseler, A., Egger, K.N., Hughes, G.A. (1992) Low levels of genetic diversity in Red Pine confirmed by random amplified polymorphic DNA markers. *Canadian Journal of Forest Research* 22: 1332–1337

Musch, B., Valadon, A., Oddou-Mutatorio, S. (2004) A propos de génétique des populations. *Rendez-vous techniques de l'ONF*, hors-série n° 1, Diversité génétique des arbres forestiers, 6–15

Myburg, A., Grattapaglia, D., Tuskan, G.A., Hellsten, U., Hayes, R.D. and others (2014) The genome of *Eucalyptus grandis*. *Nature* 510, 356–362 doi:10.1038/nature13308

Namroud, M.-C., Park, A., Tremblay, F., Bergeron, Y. (2005) Clonal and spatial genetic structures of aspen (*Populus tremuloides* Michx.). *Molecular Ecology* 14: 2969–2980

Nason, J.D., Herre, E.A., Hamrick, J.L. (1998) The breeding structure of a tropical keystone plant species. *Nature* 391: 685–687

Nystedt, B., Street, N., Wetterbom, A., Zuccolo, A., Lin, Y.-C. and others (2013) The Norway Spruce genome sequence and conifer genome evolution. *Nature* 497(7451): 579–584 doi.10.1038/nature12211

Padovan, A., Kesze, A., Foley, W.J., Külheim, C. (2013) Differences in gene expression within a striking phenotypic mosaic *Eucalyptus* tree that varies in susceptibility to herbivory. *BMC Plant Biology* 13: 57

Palma-Silva, C., Lexer, C., Paggi, G.M., Barbara, T., Bered, F., Bodanese-Zanettini, M.H. (2009) Range-wide patterns of nuclear and chloroplast DNA diversity in *Vriesea gigantea (Bromeliaceae)*, a neotropical forest species. *Heredity* 103: 503–512

Pandey, M. and Rajora, O. (2012a) Higher fine-scale genetic structure in peripheral than in core populations of a long-lived and mixed-mating conifer – eastern white cedar (*Thuja occidentalis* L.). *BMC Evolutionary Biology* 12: 48

Pandey, M. and Rajora, O.P. (2012b) Genetic diversity and differentiation of core vs. peripheral populations of eastern white cedar, *Thuja occidentalis* (*Cupressaceae*). *American Journal of Botany* 99: 690–699

Peng, C., Ma, Z., Lei, X., Zhu, H., Chen, W. and others (2011) A drought-induced pervasive increase in tree mortality across Canada's boreal forests. *Nature Climate Change* 1: 467–471

Pereira, H.M., Leadley, P.W., Proenca, V., Alkemade, R., Scharlemann, J.P. and others (2010) Scenarios for global biodiversity in the 21st century. *Science* 330(6010), 1496–1501

Petit, R.J., Hampe, A. (2006) Some evolutionary consequences of being a tree. *Annual Review of Ecology, Evolution and Systematics* 37: 187–214

Rachmat, H.H., Kamiya, K., Harada, K. (2012) Genetic diversity, population structure and conservation implication of the endemic Sumatran lowland dipterocarp tree species (*Shorea javanica*). *International Journal of Biodiversity and Conservation* 4(14): 573–583

Raj, S., Bräutigama, K., Hamanishi, E.T., Wilkins, O., Thomas, B.R. and others (2011) Clone history shapes *Populus* drought responses. *Proceedings of the National Academy of Sciences of the USA* 108 (30): 12521–12526

Rajora, O.P., Rahman, M.H., Buchert, G.P., Dancik, B.P. (2000) Microsatellite DNA analysis of genetic effects of harvesting in old-growth eastern white pine (*Pinus strobus*) in Ontario, Canada. *Molecular Ecology* 9, 339–348

Rehfeldt, G.E. (1983) Ecological adaptations in Douglas-Fir (*Pseudotsuga menziesii* var. *Glauca*) populations. III. Central Idaho. *Canadian Journal of Forest Research* 13: 626–632

Saez-Laguna, E., Guevara, M.A., Dıaz, L.M., Sanchez-Gomez, D., Collada, C. and others (2014) Epigenetic variability in the genetically uniform forest tree species *Pinus pinea* L. *PLoS ONE* 9(8): e103145. doi:10.1371/journal.pone.0103145

Savolainen, O., Pyhäjärvi, T., Knürr, T. (2007) Gene flow and local adaptation in trees. *Annual Review of Ecology, Evolution and Systematics* 38: 595–619

Sezen, U.U., Chazdon, R., Holsinger, K.E. (2005) Genetic consequences of tropical second-growth forest regeneration. *Science* 307: 891

Slavov, G.T., Leornardi, S., Burczyck, J., Adams, W.T., Strauss, S.H., Di Fazio, S.P. (2009) Extensive pollen flow in two ecologically contrasting populations of *Populus trichocarpa*. *Molecular Ecology* 18: 357–373

Turchetto-Zolet, A., Pinheiro, F., Salgueiro, F., Palma-Silva, C. (2013) Phylogeographical patterns shed light on evolutionary process in South America. *Molecular Ecology* 22: 1193–1213. doi: 10.1111/mec.12164

Tuskan, G.A., DiFazio, S., Jansson, S., Bohlmann, J., Grigoriev, I. and others (2006) The genome of Black Cottonwood, *Populus trichocarpa* (Torr. and Gray). *Science* 313: 1596–1604

Varis, S., Pakkanen, A., Galofre, A., Pulkkinen, P. (2009) The extent of south north pollen transfer in Finnish Scots pine. *Silva Fennica* 43: 717–726

Vinson, C.C., Kanashiro, M., Sebbenn, A.M., Williams, T.C.R., Harris, S.A. and others (2014) Long-term impacts of selective logging on two Amazonian tree species with contrasting ecological and reproductive characteristics: inferences from Eco-gene model simulations. *Heredity*, doi: 10.1038/hdy.2013.146

Weber, J.C., Larwanou, M., Abasse, T.G., Kalinganire, A. (2008) Growth and survival of *Prosopis africana* provenances tested in Niger and related to rainfall gradients in the West African Sahel. *Forest Ecology and Management* 256 (4): 585–592. doi: 10.1016/j.foreco.2008.05.004

White, G.M., Boshier, D.H., Powell, W. (2002) Increased pollen flow counteracts fragmentation in a tropical dry forest: an example from *Swietenia humilis* Zuccarini. *Proceedings of the National Academy of Sciences of the USA* 99: 2038–2042

Wyman, J., Bruneau, A., Tremblay, F. (2003) Microsatellite analysis of genetic diversity in four populations of *Populus tremuloïdes* in Quebec. *Canadian Journal of Botany* 81(4): 350–367

Xu, H., Tremblay, F., Bergeron, Y., Paul, V., Chen, C. (2012) Genetic consequences of fragmentation in the "arbor vitae", the Eastern White Cedar (*Thuja occidentalis* L.), towards the northern limit of its distribution range. *Ecology and Evolution*. 2(10): 2506–2520. DOI: 10.1002/ece3.371

Yakovlev, I.A., Lee, Y., Rotter, B., Jorunn, E.O., Skrøppa, T. and others (2014) Temperature-dependent differential transcriptomes during formation of an epigenetic memory in Norway spruce embryogenesis. *Tree Genetics and Genomes* 10 (2): 355–366

Glossary

Adaptive divergence: the differentiation among the mean phenotypes of populations subject to different selective pressures.

Adaptive trait: a phenotypic trait that enhances the fitness of an individual in a particular environment. Examples for trees are the time of flushing and bud set, and physiological traits determining water use efficiency.

Allele: one of the different forms of a gene that can exist at a single locus.

Allelic richness (A_R): the total number of alleles in a population.

Bud set: formation of a terminal bud at the end of the vegetative period, the timing of which is heritable and determines cold tolerance in boreal and temperate trees.

Epigenetic: all inheritable changes in gene expression that are produced without any changes in the DNA sequence. Differences in gene expression affect the functioning of traits coded by these genes and produce different phenotypes.

Expected heterozygosity (H_E): the most frequently used measure of genetic diversity within populations. It is based on the assumptions that alleles are randomly combined to genotypes.

Fixation index (F_{ST} or G_{ST}): the proportion of the variance among subpopulations relative to the total variance.

Hardy-Weinberg equilibrium: a principle stating that both allele and genotype frequencies in a randomly mating population remain constant – and remain in this equilibrium across generations – unless a disturbing influence is introduced.

Heterosis: the higher fitness of progeny obtained through crosses between populations rather than within the same population.

Inbreeding coefficient (F_{IS}): the mean reduction in heterozygosity of an individual due to non-random mating within a subpopulation, i.e. a measure of the extent of genetic inbreeding within populations that range from −1.0 (all individuals heterozygous) to +1.0 (no observed heterozygotes).

Inbreeding depression: reduced fitness of inbred individuals.

Migration load: the contribution of immigrant genes to genetic load.

Observed heterozygosity (H_O): the proportion of all heterozygotes from all individuals (homozygotes and heterozygotes) investigated in a population.

Parentage analysis: the probabilistic determination of the parents of an individual, frequently using genetic markers.

Phenotypic cline: a continuous change of a phenotypic trait along an environmental and/or geographical gradient.

Phenotypic plasticity: the capacity of a genotype to produce distinct phenotypes.

Provenance: the original geographic source of a population or group of individuals (used also to refer to such a population or group).

Provenance test: a common garden experiment, in one or more locations, where the genetic variation of different provenances is evaluated (*see* provenance).

12
CHANGING FOREST DYNAMICS
Plot-based evidence

Simon Willcock and Nikée E. Groot

Introduction

Since the Middle Ages, when intensive forest use led to wood shortages in some regions, decision-makers have had to monitor forests to ensure an adequate supply of resources for future use. In these early days of forestry, visual estimations of forest characteristics dominated, until the 1830s when Scandinavian foresters developed more rigorous monitoring techniques whereby stems were measured in systematic strips of forest, contributing to the very first national forest inventories (Kangas and Maltamo 2006). However, it was soon recognised the systematic use of strips was not the most efficient method to monitor forest characteristics, so this was replaced by measuring stems within small, square or circular plots at set intervals. Up to this point, established plots were temporary and it was not until the mid-twentieth century that continuous forest inventory systems were introduced (Stott 1947). This newly developed method relied on permanent fixed plots in which all trees were numbered and measured annually, allowing components of forest growth to be estimated for the first time, e.g. tracking growth and death of individual stems over time. The continuous forest inventory system was refined from annual measurements to 5–10 year intervals, often being used in combination with temporary plots, much like forest inventory techniques used across the world today. Forest monitoring has continually been adapted to reflect current issues as the role of forests in society has evolved. In the 1980s attention turned from managing timber stock to maintaining forest health and conserving biodiversity, with methods altering to reflect this. More recently, the United Nations Framework Convention on Climate Change has attempted to reduce the impacts of human-induced climate change by reducing greenhouse gas emissions; many forest monitoring methods now include this component, enabling managed forests to have a positive effect on the atmospheric CO_2 balance (see Chapter 37).

The past century has seen a wide variety of global change, with humans having an increasing influence over most ecosystems through direct effects (e.g. logging) or the perturbation of biogeochemical cycles (see Chapters 23–25). Whilst the impacts from direct human disturbance are somewhat familiar, shifting forest structure before ecological succession continues towards the climax community (see Chapters 7–10), the impacts of indirect effects are much less certain. Permanent forest plots are a critical tool, providing inferences of how the long-term shifts of biogeochemical cycles affect forest dynamics. However, utilising historical plot datasets may cause substantial problems for interpretation, since 1) permanent forest plots were not specifically established to address such research questions, and thus may lack the appropriate spatial distribution and/or temporal resolution; 2) research objectives and methodologies have both varied over time. Phillips *et al.* (2002) evaluated the potential biases of using long-term forest plots to monitor changes in biomass, emphasising the importance of standardised protocols, but concluding any systematic biases introduced by historical data were small and calculable, each bias leaving a unique 'footprint' in the data (see Table 12.1).

In the remainder of the chapter, we evaluate the plot-based evidence for changing forest dynamics, focussing on changes within pristine 'old-growth' forests. First, we discuss aboveground changes, many of which have been observed using long-term plots. Second, we focus on belowground shifts, primarily using short-term experimental plots. Finally, we summarise the evidence by which the drivers of the observed changes in forest dynamics can be identified and suggest which possible changes in forest dynamics may be expected in the future.

Table 12.1 Possible sources of bias when using plot-based evidence to provide estimates of forest basal area change (adapted from Phillips *et al.* 2002)

Source of error/bias	Specific error/bias description	Footprint
Negative bias		
Site selection bias	**Majestic forest bias** – Biased selection of mature phase, gap-free sites in the landscape	• Decline in number of big trees with increasing time • Mortality and recruitment increase with increasing time • Basal area correlates negatively with plot size
Site selection bias	**Progressive fragmentation and edge effects** – Biased selection of accessible sites vulnerable to fragmentation and edge effects	• Mortality increases over time • Mortality and negative changes in basal area correlate with fragment size and/or distance to edge
The process of the research itself	**Methodological impact on vegetation** – For example, researchers compacting soil, tagging trees, climbing trees and collecting vouchers, drawing attention of others to plot, etc.	• Growth negatively correlates with time • Mortality positively correlates with time • Climbed or sampled trees have depressed growth and elevated mortality • Infection rates of climbed/sampled trees positively correlate with time
Field measurement errors	**Incomplete recensusing** – New recruits may be missed, and some surviving trees may be missed and assumed dead ('ghost mortality')	• Apparent sudden 'recruitment' of large trees

Table 12.1 continued

Source of error/bias	Specific error/bias description	Footprint
Field measurement errors	**Improved measurements of buttressed trees** – Methods for measuring buttressed trees typically improve with time, with researchers more likely to measure around buttresses in initial censuses	• Apparent sudden loss of basal area of some individual large trees
Post-measurement data checking	**Reducing extreme increments** – Exceptional increments (e.g. 0.75 mm/yr) eliminated a priori or reduced in case measurement is in error	• Affects only the latest census interval (since most trees discovered to have been rounded-down incorrectly previously may be recorrected at subsequent censuses)
Publication bias	**Selective reporting of plots** – Catastrophic disturbances (cyclones, fire, flood, etc.) during census period increases interest in reporting results	• Negative changes in basal area explicitly linked by authors to catastrophic events
Positive bias		
Site selection bias	**Immature-forest bias** – Biased selection of successional forest	• Stem density declines as basal area increases (self-thinning)
The process of the research itself	**Methodological impact on vegetation** – Increasing swelling around nail used to place tag on tree	• Effect increases with time • No evidence of increase in recruitment • No evidence of researchers moving point of measurement
Field measurement errors	**Buttress creep** – Bole irregularities move up with time, becoming more likely to affect point of measurement with increasing time	• Effect increases with time • Effect especially marked in trees with large diameter • No evidence of increase in recruitment • No evidence of researchers moving point of measurement
Field measurement errors	**Basal area inflation** – Disproportionately rapid radial increment of buttresses	• Effect especially marked in trees with large diameter • Effect increases with increasing time • Some trees with implausibly large diameters
Post-measurement data checking	**Rounding-up negative increments** – In evaluating changes in diameter, false negatives are rounded up to zero, but false positives are kept because they cannot be distinguished from trees that have genuine increases in diameter	• Effect size small and diminishes with increasing length of interval • No negative increments in researchers' tree-by-tree datasets • Effect mostly on smallest size classes (since small understory light-limited trees most likely to show little or no real growth)

Changing forest dynamics

Aboveground dynamics

Long-term studies of permanent plots suggest aboveground biomass of old-growth forests might have increased globally during recent decades. Alongside this potential trend of an increasing global forest carbon sink, increased dynamics (recruitment and mortality), increased disturbances, and changes in forest composition have been observed. These changes are not consistent across all forest biomes (boreal, temperate and tropical), so we will discuss evidence for changing dynamics in each of the biomes below.

In the boreal zone, forest dynamics are disturbance-driven, with the impact of fires, windthrows and insect outbreaks ranging from local to complete stand replacement (see Chapters 7–9). This disturbance-recovery regime creates forests with a mosaic of cohorts with a wide variety of time intervals since last disturbance (from freshly disturbed to hundreds of years), which can include even-aged patches of up to stand-level scale, potentially making them more vulnerable to future large-scale disturbances. An important component of the aboveground biomass in the boreal forest is the relatively large dead biomass pool, which decomposes more slowly than in other biomes, due to prevailing cold, moist conditions. Recently, large-scale disturbances have increased both in frequency and severity in boreal forests, decreasing their primary productivity sink and increasing the amount of dead aboveground carbon (though less so in the case of fire-induced mortality).

Across the boreal zone, forest composition and ranges have been changing. Alaskan and Russian forests have advanced on tundra vegetation, moving northwards. In Canadian forests, different successional species are establishing after recent fire disturbances, similarly extending their range northwards. Compositional changes have also been observed following a combination of insect outbreak and moose grazing in Scandinavia (Soja *et al.* 2007). Severe disturbances are increasingly observed across the boreal zone. In Canadian boreal forests, the area burnt annually has been increasing, with five of the eight most severe fire years being observed in the last 17 years of an 84-year monitoring period, despite increased and improved fire suppression (Soja *et al.* 2007). A 56-year inventory of Alaskan forest fires exhibits a similar increase in the frequency of severe fire years, with seven of the eleven most severe fire years occurring after 1988 (Soja *et al.* 2007). An increase in severe fire events has also been observed in Siberia, with seven extreme fire years (years during which half the area is burnt; over twice as much as the area burnt in 'normal' fire years) occurring between 1998 and 2006 (Soja *et al.* 2007). Additionally, Alaskan forests are becoming increasingly plagued by insect outbreaks, with the spruce beetle killing 90 per cent of spruce trees, affecting an area of approximately 1 Mha between 1992 and 2000 (Soja *et al.* 2007). Forests in Fennoscandia are likely to experience an increase in frequency and severity of insect outbreaks, as rising temperatures favour egg survival. In western Canada, even in forests unaffected by fires and insect outbreaks, drought stress has resulted in increased tree mortality and decreased productivity rates (Ma *et al.* 2012). Despite experiencing increased losses through disturbance, boreal forests exhibit an overall net increase in growth rates, thus providing a net sink for atmospheric CO_2 (Pan *et al.* 2013).

Overall, during recent decades, temperate forests have increased their productivity and live aboveground biomass stocks. Since large-scale disturbances are not as inherent in the temperate forest biome, insect- or fire-induced losses following drought or heat stress have mainly been observed in drier areas, along with decreased growth rates and live aboveground stocks, with no significant change in recruitment rates (Pan *et al.* 2013); the Mediterranean and the western United States illustrate this. Losses caused by post-drought disturbances can be very severe, e.g. in the piñon-juniper forests of the south-western United States the former species experienced losses of up to 97 per cent during the period 2000–2007 (McDowell et al. 2008). Consecutive extreme drought years, combined with prolonged heat waves, have also affected forests across less arid European regions in the past decade, resulting in increased tree mortality rates and compositional changes (Bréda et al. 2006; Weber et al. 2007). Furthermore, a 30-year study of New Zealand's mountain beech forests failed to detect any increase in net carbon accumulation; overall, the system functioned as a carbon source, being predominantly driven by large-scale disturbances by insects, earthquakes and extreme weather events, which mainly affected larger trees (Coomes et al. 2012). Despite vast areas of secondary or (former) production forests, plot-based evidence of changing dynamics in undisturbed old-growth temperate forests is scarce, especially for Asia.

Evidence from a pan-tropical permanent plot dataset indicates tropical forests have been increasing their aboveground live biomass since at least the late 1970s. However, the pan-tropical signal is dominated by the vast majority of plots being located in the Neotropics. The pan-Amazonian RAINFOR plots show that increased gains in recruitment and growth, have outpaced increased mortality losses (which have increased dead biomass stocks), both in terms of basal area and number of stems, even when losses following several recent El Niño-induced droughts are included (Phillips et al. 2008). Additionally, liana biomass and dynamics are also increasing in Amazonian forests (Chave et al. 2008). However, different changes in tropical forest dynamics have been observed in Central America (Panama, Costa Rica) and South-East Asia (Malaysia), where tree species exhibited a decreased growth rate over the last two decades, though these regions have experienced a greater increase in temperature, relative to most of the Amazon Basin, along with an increase in the frequency and severity of droughts, thus not necessarily refuting the Amazonian trends (Pan et al. 2013). A smaller-scale study of 18 Central Amazonian plots (1984–1999) found no significant change in live aboveground biomass, despite confirming the pan-Amazonian Basin trend of increased dynamics. Furthermore, compositional changes were registered in these plots, generally favouring faster-growing non-pioneer canopy and emergent species, and larger individuals over mainly slower-growing, smaller sub-canopy species (Laurance et al. 2004). A compositional shift was also observed in Ghana, where a two-decade-long drought increased the abundance of more drought-tolerant canopy species (Fauset et al. 2012). Ghanaian forests are a carbon sink, through increased live aboveground biomass and basal area, suggesting that tropical forests might be more resilient to droughts than previously thought (Fauset et al. 2012). However, Australian forests experienced such severe disturbances (soil pathogens, erosion, cyclones, droughts) during a 40-year study, that it is impossible to detect any potential increase in productivity, though the frequency and severity of inherent cyclone disturbances have been increasing, with six severe cyclones being registered in the forests during 1971–2012 (Murphy et al. 2013).

Belowground dynamics

Given the timber-focussed history of forest plots and the difficulties associated with direct measurements of soil and root characteristics, it is perhaps not surprising that data describing soil and root dynamics are scarcer than aboveground dynamics data. Thus, ongoing changes in belowground dynamics typically go unobserved, although exceptions do exist (e.g. Zhou et al. 2006). However, through applying experimental treatments to tree inventory plots, changes in the belowground dynamics' response to specific drivers are investigated. In the remainder of this section we discuss evidence from experimental plots investigating changes in ambient CO_2 concentrations, temperature, precipitation, nutrient availability and disturbance.

Under the free-air CO_2 enrichment (FACE) experiments, CO_2 concentrations have been elevated to 535 ppm in coniferous and deciduous temperate forest plots across the globe. In response to this, an increased rate of root loss has been observed in coniferous forests, however the response of relative turnover has been inconsistent (Norby and Jackson 2000). Deciduous trees show a greater response to CO_2 enrichment than their coniferous counterparts, showing net increases in fine root density from 60 per cent to 140 per cent in elevated CO_2 (Norby et al. 1999). However, these impacts may be limited to soils where nitrogen is freely available. Experimental plots have also suggested elevated CO_2 results in greater ectomycorrhizal fungi concentrations. Field studies indicate that increased carbon availability results in increased levels of mycorrhization. However, this response is unlikely to be observed in boreal forests, where tree root tip colonisation by ectomycorrhizal fungi approaches 100 per cent (Ekblad et al. 2013). Shifts in ectomycorrhizal communities have also been recorded, with mycorrhizal types possessing thick mantles and rhizomorphs being favoured at higher CO_2 concentrations. Increases in soil CO_2 result in increased rates of soil acidification and mineral weathering, as well as significant increases in stable soil carbon. Soil carbon increased at all sites between 1 and 54 g C m^{-2} y^{-1} in the upper mineral soil layers, but the effect was only significant in two of four FACE experiments (both of which are situated in deciduous forests), perhaps due to the relatively high analytical error associated with the methodologies involved (Hoosbeek 2010).

Increased atmospheric CO_2 concentrations are associated with shifts in temperature and precipitation regimes. Most ecosystems increase soil respirations by 20 per cent and net nitrogen mineralisation by 46 per cent in response to warming between 0.3 and 6.0°C, with forested ecosystems showing a more pronounced response (Rustad et al. 2001). However, these effects may be short-lived, since roots acclimatise quickly, enabling rapid adaptation to new temperatures, although the degree of acclimation varies among species, and may be further hampered by seasonal and diurnal temperature fluctuations (Norby and Jackson 2000). It is likely that a significant increase in soil temperature in boreal forests will result in a net carbon emission, the thaw allowing for the decomposition of a large proportion of labile carbon (9–41 per cent) (Lal 2005). Precipitation changes would also be expected to affect belowground forest dynamics, either through microbial stress, substrate limitation, or through increased root allocation as a response to drought (Brando et al. 2008). These plausible responses differ in sign and may help explain the lack of consistent and significant treatment effects of the throughfall displacement/reduction experiments. Experimental plots in both temperate and tropical regions have demonstrated that forests have an extraordinary resilience in maintaining a relatively constant active root system, with little or no significant change in root respiration and biomass, despite imposed variations in soil moisture (Norby and Jackson 2000; Brando et al. 2008).

Globally, terrestrial ecosystems are receiving unprecedented amounts of nitrogen compounds, primarily from gaseous deposition via soil pools of nitrate and ammonium. Fertilisation experiments have been conducted to understand how this increased nitrogen deposition may be affecting belowground forest dynamics. However, different methodological approaches have resulted in conflicting conclusions; as such the impact of increased nitrogen availability is uncertain. However, there are some suggestions that fine root biomass shows a general decrease in response to increased nitrogen availability, whereas turnover increases, which could lead to impaired root function if increased nitrogen depositions occur for a prolonged period (Norby and Jackson 2000). Similarly, the effect of increased nitrogen availability on ectomycorrhizal fungi is also inconsistent, although most responses involve a shift in community composition. The inconsistent results may stem from complex interactions between nitrogen and carbon cycles, with the direction of changes in belowground forest dynamics being determined by both nitrogen and carbon availability. Further research into these interactions is urgently needed.

Successional theory indicates that disturbance events are associated with losses in biomass, followed by a gradual recovery until the system has returned to its stable state. Indeed, some results seem to support this, with soil carbon accumulating at approximately 33.8 g C m^{-2} y^{-1} (Post and Kwon 2000). However, this effect may be limited to the upper soil layers. In re-establishing forest on previously farmed land in South Carolina, USA, there were no significant increases in soil carbon deeper than 7.5 cm (Richter et al. 1999). Fire is a major source of disturbance and may have a long-term impact on belowground dynamics, particularly in boreal regions where climate change may lead to a significant change in levels of burning. The impact of fire depends on fire intensity, frequency and duration; furthermore the effects of the fire on belowground carbon stocks are not always negative (Lal 2005). Forested systems may not always recover from such disturbances, and even small shifts in ecological drivers may result in large changes in dynamics and/or biome structure if a tipping point were reached.

Possible drivers and future changes in dynamics

Besides increased human pressures, various environmental changes have been observed over the last few decades, or in some cases for over a century, which can influence forest dynamics and composition. These variations, most of which are increases, include changes in atmospheric CO_2 concentrations, temperature, solar radiation, precipitation patterns (including El Niño–Southern Oscillation droughts), nitrogen availability and acid depositions. Each of these changes can have a different impact on forest dynamics across and within biomes and can interact with other changing environmental factors, leading to uncertainties as to how forests might be affected presently and in the future, and how changed forest dynamics can feedback to the changing environmental conditions.

Generally speaking, enhanced productivity is expected to result from enhanced CO_2 concentrations in global forests, which are not water-, light- or nutrient-limited, though eventually the fertilisation effect will level off, as other resources become more limiting. In regions where productivity is constrained by low temperatures, increasing temperatures can improve carbon gains, provided they are not coupled with a decrease in available moisture, since respiration increases at higher temperatures. However, if temperatures rise beyond the species-specific optimum, proteins critical to cellular cycles may denature, leading to adverse effects. Temperatures can also enhance decomposition of dead organic stocks, increasing availability of nutrients, which could increase productivity. Air pollution caused solar irradiance to drop from 1960 to 1980, but it has been increasing in a similar fashion since then (Wright 2005). At low solar irradiance levels, sub-canopy trees in closed canopy forests will experience decreased productivity. Similarly, reduced rainfall can either limit productivity, by extending the dry season length or introducing drought stress, or increase it, through a reduction in cloud cover, thus increasing solar radiation. Nitrogen deposition can enhance productivity in forests that are N-limited, though N-deposition usually coincides with deposition of other pollutants (ozone, sulphates, etc.), which may cause stress when deposited in great quantities (Wright 2005).

In the boreal zone, the increase in temperature has been most pronounced, which, combined with increased atmospheric CO_2 concentrations, is believed to be the main driver of recent net increases in boreal biomass stocks. Temperature increases are also believed to be the main driving force of the increased frequency and severity of fire and drought stress, the latter of which can predispose trees to insect damage, as well as the decrease in the above- and belowground dead organic matter pools, though this has been offset by increased inputs into these pools from the disturbance-driven losses (Pan et al. 2013).

Changes in boreal forests dynamics are expected to feedback positively to the environmental drivers, enhancing future changes. Continued increase in severity and frequency of fire, biotic agents and wind-throw disturbances, decomposition of substantial dead organic matter pools due to increased temperatures, and decreased forest cover (since range-shifts to the north are expected to be smaller than the movement of the southernmost boundaries) are all expected to result in net carbon losses, thus further increasing atmospheric CO_2 concentrations and global temperatures. It should be noted that regrowth vegetation after fire tends to be a lot less prone to burning, thus dampening any potential runaway positive feedback loops. In addition, disturbance-cleared snow-covered surfaces in the boreal zone can have a cooling effect, due to the increased surface albedo. Following a tipping point, some boreal forests may show a compositional shift to a grassland biome under increased heat-induced stress or to a steppe biome under increased water stress. Such stand-replacing disturbances may facilitate adaptation of boreal vegetation to changed environmental conditions, or could possibly result in the treeline actually moving south (Chapin *et al.* 2004).

Temperate forests seem to have benefited mostly from fertilisation through nitrogen deposition, though increased CO_2 levels have also contributed to increased productivity. Increased temperatures were mainly found to have a negative effect, through creating drought and heat stress, thus increasing susceptibility to fire and insect outbreaks (Pan *et al.* 2013). The frequency and severity of droughts are expected to keep increasing in temperate forests, with a greater effect on evergreen species, since their permanent foliage intercepts more rainwater than deciduous species under equal rainfall conditions. Since coniferous species tend to have less structural resistance to cavitation, those with lower wood densities are expected to decrease in prevalence. Thus, compositional shifts to stands with more drought-tolerant dense-wooded deciduous species are expected (Bréda *et al.* 2006).

Increases in tropical forest carbon dynamics and sinks could be driven by increased resource availability (e.g. enhanced CO_2 concentrations), although, due to increased temperatures, coupled with decreased total precipitation, this trend is not observed in Central American and South-East Asian plots (Pan *et al.* 2013). The increased presence and size of lianas in the Central American plots can also have a negative impact on tree dynamics (Wright 2005; Feeley *et al.* 2007). There has been a lot of discussion over whether observed increases in forest biomass and dynamics are driven by CO_2 or any other kind of fertilisation (irradiance, nitrogen, changed rainfall), rather than it merely being an artefact caused by the plots being in locations which are recovering from past disturbances, or through unsound methodological approaches (Phillips *et al.* 2002; Lewis *et al.* 2004; Lewis *et al.* 2009). As previously discussed, any issues arising from the methodology can be detected and addressed. If all of the plots in this pan-tropical network and in other unrelated plots were indeed all recovering from past disturbances, it would have to have been a disturbance of such a magnitude, that signs of it would be readily discernible in the historical record (Lewis *et al.* 2009). Furthermore, the current dynamics of increasing gains (growth and recruitment) outpacing increasing mortality would not be expected, nor would the increase in basal area manifest itself mainly in fast-growing species, which tend to profit more from CO_2 fertilisation (Laurance *et al.* 2004). Since the relative importance of carbon gains and losses has not changed during the monitoring period, the observed trend is unlikely to be driven by any environmental conditions changing during the field measurements, like increased solar radiation. Equally, the trend occurs uniformly, suggesting a changing global variable, rather than something localised like N- or P-fertilisation from fires or industry (Lewis *et al.* 2009).

Until recently, the Amazon forest was believed to be at risk from large-scale drought-induced dieback, whereas presently there are predictions of the current increasing trend continuing until other resources become limiting, though droughts and fires can still be expected to cause large-scale losses (Pan et al. 2013). Continued increases in temperature are bound to change forests from a sink to a source eventually, through increasing both soil and tree respiration, combined with eventual levelling off of the CO_2 fertilisation effect, though it is uncertain when this will happen. Future climatic conditions are believed to result in rainfall becoming more seasonal in south-eastern Amazonia, which could favour further increases in liana biomass and change the forest composition to favour more light-demanding, faster-growing species, or, combined with increasing temperatures, could even lead the forests to shift to a savannah system (Phillips et al. 2002). Furthermore, continued temperature rises may result in an enhanced decomposition of coarse woody debris, turning it into a source of further environmental change (Iwashita et al. 2013).

Conclusions

Old-growth forests across the globe (from boreal, temperate and tropical biomes) have been documented as undergoing substantial changes in their ecological dynamics. Here, we provided evidence for this from experimental and long-term monitoring plots. However, plot-based evidence should not be considered alone. Satellite data, micrometeorological measurements, direct measurements of atmospheric CO_2 and atmospheric transport models all support the assertion that global forest dynamics have, and continue to be, altered (Lewis 2006). At present, the net effect of changes in mortality, growth and recruitment result in global forests acting as a carbon sink (Pan et al. 2013), providing a stabilising effect on global environmental conditions and reducing the impacts of climate change (see Chapter 37). However, this negative feedback may be under threat in the future. Boreal forests are particularly likely to provide a positive feedback to climate change in the future, with some studies suggesting that North American boreal forests have already become a carbon source (Pan et al. 2011), though all forest biomes hold this potential (e.g. if global temperatures or disturbance rates rise high enough). Multiple environmental drivers could be altering forest dynamics, though current evidence suggests that boreal forest dynamics are driven by disturbance (often as a result of increased temperatures), temperate forest dynamics are linked to nutrient deposition and tropical forest dynamics are driven by a fertilisation effect. Further understanding of the drivers is needed in order to determine what each biome's future is likely to be, and whether anything can still be done to prevent it.

Acknowledgments

SW undertook this work funded with support from the United Kingdom's Ecosystem Services for Poverty Alleviation programme (ESPA; www.espa.ac.uk). ESPA receives its funding from the Department for International Development (DFID), the Economic and Social Research Council (ESRC) and the Natural Environment Research Council (NERC). NEG is funded by a Microsoft Research Cambridge PhD Scholarship at the Ecology and Global Change cluster of the University of Leeds's School of Geography.

References

Brando, P.M., Nepstad, D.C., Davidson, E.A., Trumbore, S.E., Ray, D. and Camargo, P., 2008. Drought effects on litterfall, wood production and belowground carbon cycling in an Amazon forest: results of a throughfall reduction experiment. *Philosophical Transactions of the Royal Society of London. Series B, Biological sciences*, 363(1498), pp. 1839–1848. Available at: http://rstb.royalsocietypublishing.org/content/363/1498/1839.short [Accessed 27 January 2014].

Bréda, N., Huc, R., Granier, A. and Dreyer, E., 2006. Temperate forest trees and stands under severe drought: a review of ecophysiological responses, adaptation processes and long-term consequences. *Annals of Forest Science*, 63(6), pp. 625–644. Available at: http://dx.doi.org/10.1051/forest:2006042 [Accessed 22 January 2014].

Chapin, F.S., Callaghan, T.V., Bergeron, Y., Fukuda, M., Johnstone, J.F. and others, 2004. Global change and the boreal forest: thresholds, shifting states or gradual change? *AMBIO: A Journal of the Human Environment*, 33(6), pp. 361–365. Available at: http://dx.doi.org/10.1579/0044-7447-33.6.361 [Accessed 27 January 2014].

Chave, J., Condit, R., Muller-Landau, H.C., Thomas, S.C., Ashton, P.S. and others, 2008. Assessing evidence for a pervasive alteration in tropical tree communities. *PLoS Biol*, 6(3), p. e45. Available at: http://dx.doi.org/10.1371/journal.pbio.0060045 [Accessed 8 August 2013].

Coomes, D.A., Holdaway, R.J., Kobe, R.K., Lines, E.R. and Allen, R.B., 2012. A general integrative framework for modelling woody biomass production and carbon sequestration rates in forests. *Journal of Ecology*, 100(1), pp. 42–64. Available at: http://doi.wiley.com/10.1111/j.1365-2745.2011.01920.x [Accessed 8 August 2013].

Ekblad, A., Wallander, H., Godbold, D.L., Cruz, C., Johnson, D. and others, 2013. The production and turnover of extramatrical mycelium of ectomycorrhizal fungi in forest soils: role in carbon cycling. *Plant and Soil*, 366(1-2), pp. 1–27. Available at: http://link.springer.com/10.1007/s11104-013-1630-3 [Accessed 20 January 2014].

Fauset, S., Baker, T.R., Lewis, S.L., Feldpausch, T.R., Affum-Baffoe, K. and others, 2012. Drought-induced shifts in the floristic and functional composition of tropical forests in Ghana. *Ecology letters*, 15(10), pp. 1120–1129. Available at: http://www.ncbi.nlm.nih.gov/pubmed/22812661 [Accessed 23 January 2014].

Feeley, K.J., Wright, S.J., Supardi, M.N.N., Kassim, A.R. and Davies, S.J., 2007. Decelerating growth in tropical forest trees. *Ecology Letters*, 10(6), pp. 461–469. Available at: <Go to ISI>://000246364500003.

Hoosbeek, M.R., 2010. Soil carbon stabilization under increased atmospheric CO_2. In Proceedings of the COST Action FP0803 *Conference on Belowground Carbon Turnover in European Forests – State of the Art*, Birmensdorf, Switzerland, 26–28 January 2010. Birmensdorf, Switzerland. pp. 35–38.

Iwashita, D.K., Litton, C.M. and Giardina, C.P., 2013. Coarse woody debris carbon storage across a mean annual temperature gradient in tropical montane wet forest. *Forest Ecology and Management*, 291, pp. 336–343. Available at: http://www.sciencedirect.com/science/article/pii/S0378112712007190 [Accessed 29 January 2014].

Kangas, A. and Maltamo, M., 2006. *Forest Inventory: Methodology and Applications*, Springer. Available at: http://books.google.co.uk/books?id=zF7DOgm6MbEC.

Lal, R., 2005. Forest soils and carbon sequestration. *Forest Ecology and Management*, 220, pp. 242–258.

Laurance, W.F., Oliveira, A.A., Laurance, S.G., Condit, R., Nascimento, H.E.M. and others, 2004. Pervasive alteration of tree communities in undisturbed Amazonian forests. *Nature*, 428(6979), pp. 171–175. Available at: <Go to ISI>://000220103600046.

Lewis, S.L., 2006. Tropical forests and the changing earth system. *Philosophical Transactions of the Royal Society B-Biological Sciences*, 361(1465), pp. 195–210. Available at: <Go to ISI>://000234928400013.

Lewis, S.L., Malhi, Y. and Phillips, O.L., 2004. Fingerprinting the impacts of global change on tropical forests. *Philosophical Transactions of the Royal Society of London Series B-Biological Sciences*, 359(1443), pp. 437–462. Available at: <Go to ISI>://000220545100010.

Lewis, S.L., Lloyd, J., Sitch, S., Mitchard, E.T.A. and Laurance, W.F., 2009. Changing ecology of tropical forests: evidence and drivers. *Annual Review of Ecology, Evolution, and Systematics*, 40(1), pp. 529–549. Available at: http://www.annualreviews.org/doi/abs/10.1146/annurev.ecolsys.39.110707.173345.

Ma, Z., Peng, C., Zhu, Q., Chen, H., Yu, G. and others, 2012. Regional drought-induced reduction in the biomass carbon sink of Canada's boreal forests. *Proceedings of the National Academy of Sciences of the United States of America*, 109(7), pp. 2423–2427. Available at: http://www.pnas.org/content/109/7/2423.short [Accessed 29 January 2014].

McDowell, N., Pockman, W.T., Allen, C.D., Breshears, D.D., Cobb, N. and others, 2008. Mechanisms of plant survival and mortality during drought: why do some plants survive while others succumb to drought? *The New Phytologist*, 178(4), pp. 719–739. Available at: http://www.ncbi.nlm.nih.gov/pubmed/18422905 [Accessed 28 January 2014].

Murphy, H.T., Bradford, M.G., Dalongeville, A., Ford, A.J. and Metcalfe, D.J., 2013. No evidence for long-term increases in biomass and stem density in the tropical rain forests of Australia, F. Gilliam, ed.

Journal of Ecology, 101(6), pp. 1589–1597. Available at: http://doi.wiley.com/10.1111/1365-2745.12163 [Accessed 29 January 2014].

Norby, R.J. and Jackson, R.B., 2000. Root dynamics and global change: seeking an ecosystem perspective. *New Phytologist*, 147(1), pp. 3–12. Available at: http://doi.wiley.com/10.1046/j.1469-8137.2000.00676.x [Accessed 27 January 2014].

Norby, R.J., Wullschleger, S.D., Gunderson, C.A., Johnson, D.W. and Ceulemans, R., 1999. Tree responses to rising CO_2 in field experiments: implications for the future forest. *Plant, Cell and Environment*, 22(6), pp. 683–714. Available at: http://doi.wiley.com/10.1046/j.1365-3040.1999.00391.x [Accessed 27 January 2014].

Pan, Y., Birdsey, R.A., Fang, J., Houghton, R., Kauppi, P.E. and others, 2011. A large and persistent carbon sink in the world's forests. *Science*, 333(6045), pp. 988–993. Available at: http://www.sciencemag.org/content/333/6045/988.abstract.

Pan, Y., Birdsey, R.A., Phillips, O.L. and Jackson, R.B., 2013. The structure, distribution, and biomass of the world's forests. *Annual Review of Ecology, Evolution, and Systematics*, 44(1), pp. 593–622. Available at: http://www.annualreviews.org/doi/abs/10.1146/annurev-ecolsys-110512-135914 [Accessed 24 January 2014].

Phillips, O.L., Malhi, Y., Vinceti, B., Baker, T., Lewis, S.L. and others, 2002. Changes in growth of tropical forests: evaluating potential biases. *Ecological Applications*, 12(2), pp. 576–587. Available at: <Go to ISI>://000174457800022.

Phillips, O.L., Lewis, S.L., Baker, T.R., Chao, K.-J. and Higuchi, N., 2008. The changing Amazon forest. *Philosophical Transactions of the Royal Society B: Biological Sciences*, 363(1498), pp. 1819–1827. Available at: http://rstb.royalsocietypublishing.org/content/363/1498/1819.abstract.

Post, W.M. and Kwon, K.C., 2000. Soil carbon sequestration and land-use change: processes and potential. *Global Change Biology*, 6(3), pp. 317–327. Available at: http://dx.doi.org/10.1046/j.1365-2486.2000.00308.x.

Richter, D.D., Markewitz, D., Trumbore, S.E. and Wells, C.G., 1999. Rapid accumulation and turnover of soil carbon in a re-establishing forest, 400(6739), pp. 56–58. Available at: http://dx.doi.org/10.1038/21867 [Accessed 27 January 2014].

Rustad, L., Campbell, J., Marion, G., Norby, R., Mitchell, M. and others, 2001. A meta-analysis of the response of soil respiration, net nitrogen mineralization, and aboveground plant growth to experimental ecosystem warming. *Oecologia*, 126(4), pp. 543–562. Available at: http://link.springer.com/10.1007/s004420000544 [Accessed 22 January 2014].

Soja, A.J., Tchebakovab, N.M., Frenchc, N.H.F., Flannigand, M.D., Shugarte, H.H. and others, 2007. Climate-induced boreal forest change: Predictions versus current observations. *Global and Planetary Change*, 56(3), pp. 274–296. Available at: http://www.sciencedirect.com/science/article/pii/S0921818106001883 [Accessed 27 January 2014].

Stott, C.B., 1947. Permanent growth and mortality plots in half the time. *Journal of Forestry*, 45(9), pp. 669–673(5). Available at: http://www.ingentaconnect.com/content/saf/jof/1947/00000045/00000009/art00010 [Accessed 29 January 2014].

Weber, P., Bugmann, H. and Rigling, A., 2007. Radial growth responses to drought of *Pinus sylvestris* and *Quercus pubescens* in an inner-Alpine dry valley. *Journal of Vegetation Science*, 18(6), pp. 777–792. Available at: http://doi.wiley.com/10.1111/j.1654-1103.2007.tb02594.x [Accessed 31 January 2014].

Wright, S.J., 2005. Tropical forests in a changing environment. *Trends in Ecology and Evolution*, 20(10), pp. 553–560. Available at: <Go to ISI>://000232606900007.

Zhou, G., Liu, S., Li, Z., Zhang, D., Tang, X. and others, 2006. Old-growth forests can accumulate carbon in soils. *Science* (New York, N.Y.), 314(5804), p. 1417. Available at: http://www.sciencemag.org/content/314/5804/1417.abstract [Accessed 23 January 2014].

PART III

Forest flora and fauna

13
LIANAS IN FOREST ECOSYSTEMS

Stefan A. Schnitzer

Introduction

Lianas are found in forests worldwide where they can have a wide range of effects on community and ecosystem dynamics and functioning. Lianas are woody climbing plants that remain rooted in the ground throughout their lifetime (as opposed to hemi-epiphytes and many epiphytes) and take advantage of the architecture of other plants, typically trees, to ascend to the top of the forest canopy (Schnitzer and Bongers 2002). The characteristics of lianas are readily identifiable; they have specialized organs for climbing (e.g., tendrils, hooks, twining, adhesive roots), thin stems relative to the trees that they climb, and often a large number of leaves that convert sunlight into carbohydrates (Schnitzer and Bongers 2002). Lianas are a polyphyletic group with nearly 40 percent of the dicot plant families having at least one climbing species (including herbaceous climbers), and more than 75 percent of all dicot orders containing at least one climbing species (Gianoli 2015). Lianas are particularly abundant in forests, especially in the lowland tropics (Figure 13.1), where they can commonly comprise more than one-third of the woody species (lianas, shrubs, and trees; e.g., Schnitzer et al. 2012, Thomas et al. 2015).

Lianas can have both positive and negative effects in forest ecosystems. Lianas contribute substantially to tropical forest diversity, often adding hundreds of species to a given forest community (e.g., Schnitzer et al. 2012, Gianoli 2015, Thomas et al. 2015). They provide copious amounts of food in the form of nectar, pollen, fruits, leaves, or sap for many animal species (Arroyo-Rodriguez et al. 2015, Yanoviak 2015). The nutrition that lianas provide may be especially important during seasonal droughts, when some animal species, such as primates, switch their diets to rely on lianas as a food resource (Arroyo-Rodriguez et al. 2015). Lianas also serve as bridges that connect tree canopies, thus providing a complex structural environment that may help maintain animal diversity and provide shelter and transportation routes to many animal species (e.g., Yanoviak 2015).

In contrast, lianas reduce tree growth, regeneration, and fecundity, thus altering ecosystem functioning and potentially tree community composition. Lianas recruit rapidly and in high densities following disturbance (DeWalt et al. 2000, Schnitzer et al. 2000), where they can blanket the existing vegetation and reduce the recruitment and diversity of regenerating tree species (Schnitzer and Carson 2010). Lianas compete intensely with trees and thus significantly reduce tree growth and survival (Ingwell et al. 2010, Toledo-Aceves 2015). Intense competition

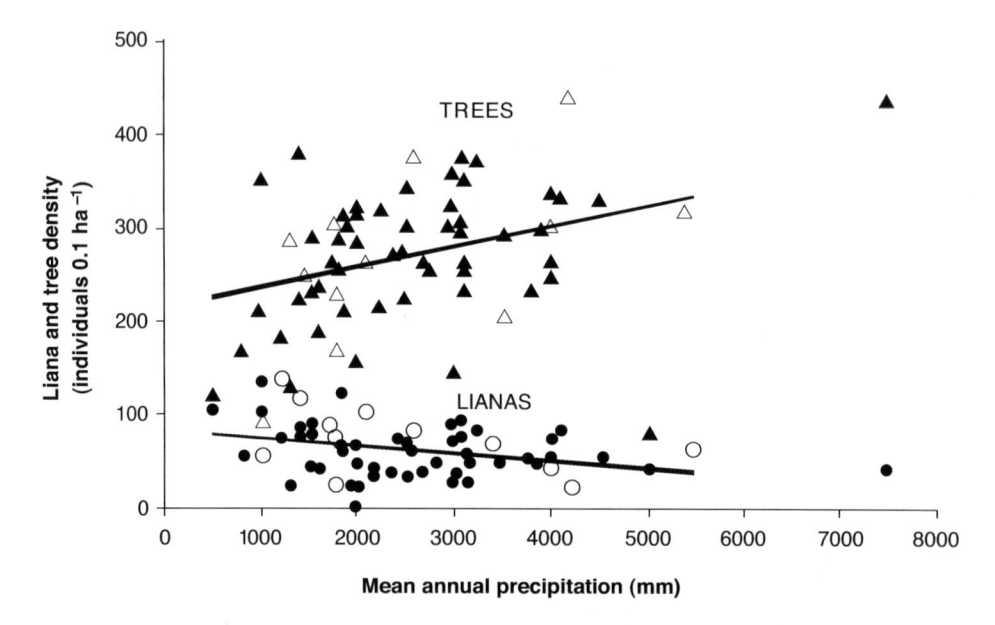

Figure 13.1 The change in liana and tree density with mean annual precipitation in lowland tropical forests. Data were collected by A.H. Gentry in ten small (2 × 50 m) transects (0.1 ha total) in each forest, and the minimum diameter cutoff was ≥ 2.5 cm diameter. Sites are from Africa (n=8), Asia (n=4), Central America (n=9), and South America (n=45). Triangles represent trees and circles represent lianas. Closed symbols are neotropical sites (Central and South America) and open symbols are old-world sites (Africa and Asia). Figure is based on Schnitzer (2005)

from lianas reduces the amount of carbon that trees sequester and store, thereby lowering the carbon storage capacity of forests (Durán and Gianoli 2013, van der Heijden *et al.* 2013, Schnitzer *et al.* 2014). Lianas may also deplete soil nutrients and moisture, further affecting tree communities and ecosystem functioning. By suppressing tree reproduction and fecundity (e.g., Kainer *et al.* 2014), lianas reduce per-capita tree seed production and amplify the effects of dispersal limitation. Moreover, lianas are now increasing in abundance and biomass in many forests, particularly in the neotropics, an increase that will likely have substantial consequences for forest diversity, composition, and ecosystem-level functioning (Schnitzer and Bongers 2011, Schnitzer 2015).

While lianas are now recognized as having many profound ecological effects in forests, until recently the study on the ecology of lianas has lagged far behind that of trees. However, over the past two decades, the study of liana ecology has grown exponentially, revealing many important contributions of lianas to forest ecology (Schnitzer *et al.* 2015). In this chapter, I review the state of knowledge about the ecology of lianas and their contribution to forest ecosystems. I divide the chapter into three main sections: 1) The global distribution of lianas and the scale-dependent factors that control their distribution. Determining the factors that control the distribution of organisms is one of the fundamental goals of ecology, and determining where lianas are most abundant (and why) is key to understanding in what forest types lianas are likely to have the greatest effect. 2) The positive and negative effects of lianas on forest ecosystems. In this section, I review the literature of the competitive effects of lianas on tree performance and diversity, as well as the mechanisms whereby lianas alter tree community composition and forest carbon accumulation. I also review some of the positive contributions of lianas to forest plant and animal diversity, factors which are often overlooked. 3) In the third major section, I review the recently documented pattern of increasing liana abundance in neotropical and subtropical forests – a phenomenon that has substantial potential effects on tropical forest diversity and functioning. Although lianas can be an important component in the ecology of temperate forests (Ladwig and Meiners 2015), lianas are far more abundant and diverse in the tropics, where they likely have a much greater effect on the forest ecosystem. Furthermore, the ecology of tropical lianas has been studied far more extensively than that of their temperate counterparts, and thus I draw most of the examples in the chapter from studies on the ecology of lianas in neotropical forest ecosystems.

The global distribution of lianas

The abundance and diversity of lianas vary considerably within and among forests, and the factors that determine liana distribution appear to depend largely on spatial scale (Schnitzer 2005). That is, the factors that control liana abundance and diversity differ among local, regional, altitudinal, and latitudinal scales. However, a single unified mechanism may explain how liana distribution is regulated at all scales, which I explain in this section.

The distribution of local (within-forest) tropical liana density and diversity

There is now compelling evidence to support the hypothesis that disturbance controls the local (within-forest) scale distribution of liana species and maintains liana diversity in tropical forests. Lianas are known to be highly responsive to such disturbances as treefall gaps and forest edges in both temperate and tropical forests (e.g., Schnitzer *et al.* 2000, Laurance *et al.* 2001, Londré and Schnitzer 2006). For example, in the 50-ha forest dynamics plot on Barro Colorado Island, Panama (BCI), the local distribution of more than 50 percent of the liana species was positively related to disturbance – primarily treefall gaps resulting from tree mortality (Dalling *et al.* 2012, Schnitzer *et al.* 2012, Ledo and Schnitzer 2014). Disturbance combined with positive density dependence (a measure of liana spatial clumping) explained the local distribution of 75 percent of the liana species on the BCI 50-ha plot (Ledo and Schnitzer 2014).

The rapid response of lianas to disturbance appears to be driven primarily by the ability of lianas to colonize via clonal reproduction following disturbance. More than 50 percent of the liana species produced significantly more clonal stems in disturbed areas than in intact, undisturbed forest (Ledo and Schnitzer 2014). Furthermore, gaps maintain liana diversity beyond the effect of high liana density, and liana diversity was higher in treefall gaps even when controlling for stem density (Schnitzer and Carson 2001, Ledo and Schnitzer 2014). Ledo and Schnitzer (2014) contrasted the role of disturbance with two other putative diversity maintenance mechanisms (negative density dependence and niche partitioning) and found that disturbance was the major driver of liana diversity and distribution, whereas trees responded in the exact opposite way. Consequently, there is now compelling evidence that disturbance controls local liana distribution and maintains liana diversity in tropical forests.

The distribution of pan-tropical liana density and diversity

Among lowland tropical forests, lianas increase in density with decreasing mean annual precipitation, peaking in seasonally dry areas, where mean annual rainfall is relatively low and the number of dry months is high (Schnitzer 2005, DeWalt *et al.* 2010). This pattern for liana density is contrary to that of trees, which increase with increasing precipitation in tropical forests (Figure 13.1; Schnitzer 2005). The unique pattern of decreasing liana density with increasing rainfall was first demonstrated with datasets collected in forests around the world by A.H. Gentry, in which he surveyed lianas (≥ 2.5 cm diameter) in ten small (2 × 50 m) transects (0.1 ha total) in each forest (Schnitzer 2005). The sites included: Africa (n=8), Asia (n=4), Central America (n=9), and South America (n=45). The Gentry data, even considering the very low level of within-forest sampling (and thus high variation among forests), revealed a clear decrease in liana density with increasing mean annual rainfall (Figure 13.1).

DeWalt et al. (2015) used the recently assembled Global Liana Database to confirm the pattern of decreasing liana density with increasing mean annual precipitation using 29 relatively well-sampled forest plots around the world: Africa (n=3), Asia (n=9), Mexico (n=2), Central America (n=4), and South America (n=11). The authors confirmed that liana abundance was negatively correlated with mean annual precipitation and also found that liana abundance was positively correlated with increasing dry season length. Consequently, liana abundance peaks in highly seasonal, tropical forests, but liana abundance decreases with increasing precipitation: thus resulting in a pattern of relatively few lianas in wet, aseasonal, ever-wet tropical forests.

In contrast to what is known about the density of lianas, the variation in the diversity of lianas with precipitation is not yet resolved, and there may be several patterns depending on how diversity is calculated. For example, liana diversity when calculated relative to tree diversity may decrease with increasing precipitation. In Ghana, lianas varied from 30 percent of the vascular plant species in the wetter forests (2,000 mm annual rainfall) to 43 percent of the vascular plant species in the drier forests (1,000 mm annual rainfall; Swaine and Grace 2007). The absolute number of liana species (i.e., not a percentage of vascular plant species), however, may vary in a hump-shaped pattern with precipitation. For example, using the Global Liana Database, liana diversity was lower in the very dry and very wet forests compared to the moderately wet, seasonal forests (DeWalt et al. 2015). Consequently, the contribution of lianas to vascular plant species diversity may increase with decreasing mean annual precipitation, but liana diversity independent of trees may exhibit more of a hump-shape distribution with respect to mean annual precipitation. To more accurately determine the pattern of liana diversity with rainfall and seasonality, additional datasets that are of sufficient size (and number of individuals sampled) are necessary to provide a reliable estimate of liana and tree species richness.

The latitudinal distribution of lianas

Lianas are primarily a tropical phenomenon, and their abundance and diversity are sharply lower outside of the tropics than within the tropics (Schnitzer 2005). For example, although there are some exceptional temperate forests that contain many liana individuals (Ladwig and Meiners 2015), more commonly, liana diversity, density, and basal area in the northern hemisphere are substantially lower than in the tropics (Figure 13.2; Schnitzer 2005). Lianas typically constitute around 10 percent or less of the vascular plant flora in temperate forests, which is a substantially lower percentage than the distribution of lianas in the tropics (Ladwig and Meiners 2015). Liana abundance and diversity can be highly variable in temperate forests – even those that seem to share similar rainfall and seasonality (Londré and Schnitzer 2006). However, lianas are typically more abundant along forest edges and in heavily disturbed forests, and they are less prevalent in intact, undisturbed areas (Londré and Schnitzer 2006). Nonetheless, when lianas are in high abundance, they likely have a large effect on temperate forest communities and functioning, and thus determining the processes that control temperate liana distributions remains an important question that has only recently been explored (Ladwig and Meiners 2015).

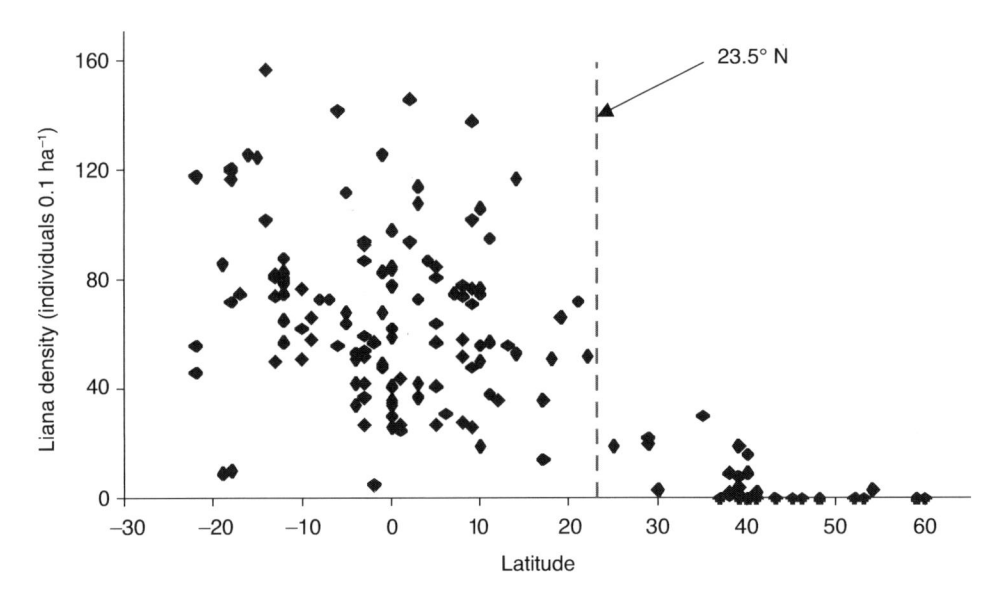

Figure 13.2 The change in liana density with increasing latitude from the southern edge of the tropics to the northern hemisphere. Data were collected by A.H. Gentry in ten small (2 × 50 m) transects (0.1 ha total) in each forest, and the minimum diameter cutoff was ≥ 2.5 cm diameter. The dashed line denotes the northern edge of the tropics. The data demonstrate a sharp decrease in liana density outside of the tropics. Figure is based on Schnitzer (2005)

A unified mechanistic explanation for the distribution of lianas at multiple scales

The peak in tropical liana abundance with high seasonality and rainfall in the tropics, and the decrease in liana density and diversity with increasing latitude (and altitude), are patterns that may both be explained by the unique ecological, anatomical, and physiological traits of lianas (Schnitzer 2005). Lianas tend to have thin stems with a relatively large mass of leaves that are typically deployed at the very top of the forest canopy. The ratio of leaf area to conductive stem area tends to be higher for lianas than trees (Wyka *et al.* 2013). For thin-stemmed lianas to be able to supply adequate amounts of water to their leaves, they need to have extraordinarily large vessel elements modified for highly efficient water movement (e.g., Wyka *et al.* 2013). Theoretically, however, plants with such large and efficient vascular systems would generally avoid dry areas, because large vessels make them vulnerable to cavitation and embolism. When the plant water column is exposed to such high negative internal pressures (derived from the plant losing much more water at the leaf surface than it can pull up from the soil via its roots) the water column breaks, which causes a permanent loss of water transport capacity. The challenge of dry conditions and maintaining a healthy internal water balance is compounded by the fact that lianas deploy most of their leaves at the top of the forest canopy, which can be a particularly stressful place to grow, especially during the dry season, because of the high temperatures and low relative humidity, resulting in high vapor pressure deficits (VPD). High VPD causes plants to lose water rapidly because of the large gradient of water inside the leaf to that of the surrounding air. Thus, lianas are faced with a significant challenge of maintaining their internal water balance throughout the year, particularly during seasonal droughts.

It thus seems paradoxical for lianas to persist outside of ever-wet, aseasonal tropical forests that do not experience an extended dry season when lianas have the potential to suffer from seasonal drought more than trees. One potential strategy for lianas to avoid water stress and suffering potentially catastrophic embolism is to lose their leaves during the dry season. This strategy would allow them to persist throughout the most stressful months. However, few liana species in moist seasonal forests lose their leaves, and lianas in extremely dry forests tend to retain their leaves far longer into the dry season than do co-occurring trees (see Schnitzer 2005 and references therein). In fact, most liana species appear to grow more than trees in seasonal tropical forests (Schnitzer 2005, Cai et al. 2009, Zhu and Cao 2009).

The apparent paradox of high liana abundance in seasonal forests may be resolved if lianas have deep and efficient root and vascular systems, which allow them to acquire more soil water and nutrients and suffer less water stress during the dry season compared to many tree species. There is some (albeit limited) evidence that lianas have particularly deep root systems (e.g., Restom and Nepstad 2004), which they use to uptake water from deep in the soil profile (Andrade et al. 2005, Chen et al. 2015). These deep root systems may allow lianas to suffer less water stress during the dry season, when the amount of direct radiation (due to lower cloud cover) is high. If so, lianas may grow more than competing trees during the dry season, which may confer a competitive advantage for lianas (Schnitzer 2005, Cai et al. 2009, Zhu and Cao 2009, Chen et al. 2015). In contrast, in aseasonal, wet forests, where water is rarely limiting and radiation is relatively low year-round, lianas gain no such advantage (Schnitzer 2005). Over decades, this dry season growth advantage may allow lianas to increase in abundance relative to trees in seasonal forests compared to non-seasonal wet forests, thus explaining their pan-tropical distribution.

At the local scale, high liana abundance and diversity in treefall gaps may also be explained by their ability to grow in hot, dry, and sunny environments with high VPD. When a tree falls in a tropical forest, lianas recruit into the resulting treefall gap as seedlings and as advanced regeneration (established plants in the forest understory). However, many lianas are also pulled into the gap along with the falling tree, where they survive the treefall and proliferate rapidly (Putz 1984, Schnitzer et al. 2000, Schnitzer and Carson 2001, 2010, Ledo and Schnitzer 2014). The established adult canopy lianas that are pulled into the gaps are already adapted for the hot, dry conditions found at the top of the canopy and have fully established root systems. These transplanted lianas can proliferate immediately in the newly established gap, which provides conditions that are similar to the forest canopy. In contrast, trees that were present in the dark understory prior to gap creation must replace their leaves with new leaves that can tolerate high light. These trees may also have to allocate additional resources to roots, which had not been necessary for survival in the shaded understory. Therefore, the ability of lianas to grow in the arid environment of the forest canopy, survive treefalls, and produce copious numbers of clonal stems in treefall gaps may give them an advantage in gaps that is unavailable to trees.

The explanation for the precipitous decrease in liana abundance (and species richness) with increasing latitude (and altitude) is also based on liana anatomy – particularly the well-developed liana vascular system (Schnitzer 2005). Large vessel elements and relatively thin stems that lack insulation make lianas particularly vulnerable to prolonged freezing conditions. The vessels of tropical lianas exposed to freezing temperatures could rupture when water in the vessels turns to ice, or they could suffer freeze-induced embolism, wherein air bubbles form in the vessels when ice is converted back to water when temperatures rise above freezing (Sperry et al. 1987, Jiménez-Castillo and Lusk 2013). Both scenarios could render the vascular system inoperable and ultimately kill the plant (Schnitzer 2005). Likewise, the decrease in lianas with altitude in the tropics may also be driven by the lack of resistance to cold temperatures (Jiménez-Castillo and Lusk 2013), combined with copious rainfall, and the relatively low light, cloudy/misty environment – all of which may be unfavorable to lianas. Thus, the striking decrease in liana abundance and diversity outside of the tropics (and with increasing altitude within the tropics) appears to be driven by tradeoffs of fast growth, slender stems, and a highly efficient vascular system, all of which are beneficial in warm climates but detrimental in cold ones.

Effects of lianas in forest ecosystems

Competition between lianas and trees

Over the past two decades, more than 20 experimental and observational studies in numerous tropical forests have demonstrated that lianas compete intensely with trees (reviewed by Toledo-Aceves 2015). These studies have shown unequivocally that lianas reduce tree performance using a variety of metrics, such as recruitment, growth, fecundity, and survival (e.g., Schnitzer *et al.* 2000, 2005, Ingwell *et al.* 2010, Schnitzer and Carson 2010, Kainer *et al.* 2014). Lianas compete aboveground by deploying their large canopy of leaves over that of their host tree, thus reducing light. Lianas also compete for belowground resources, primarily for water and nutrients. Competition for water during the dry season may be particularly strong between lianas and trees, and belowground competition has been reported in a variety of studies (Schnitzer *et al.* 2005, Tobin *et al.* 2012, Toledo-Aceves 2015, Alvarez-Cansino *et al.* 2015). Lianas also cause mechanical stress by adding considerable weight to the tree crown, thus possibly forcing trees to increase stem diameter at the expense of height (Schnitzer *et al.* 2005).

Examples of the effects of lianas on trees include work by Grauel and Putz (2004) who found that, five years after removing lianas in a seasonally inundated monodominant forest in the Darien region of Panama, tree diameter growth had doubled compared to control plots. In a forest in Côte d'Ivoire in West Africa, tree saplings grew more than five times faster in plots when lianas were removed compared to saplings competing with lianas (Schnitzer *et al.* 2005). Two studies in central Panama demonstrated that short-term sap velocity and longer-term growth of canopy trees increased significantly following liana cutting (Tobin *et al.* 2012, Alvarez-Cansino *et al.* 2015). Both of these studies also reported that the competitive effect from lianas on canopy trees was stronger during the dry season, suggesting that lianas compete more intensely during seasonal drought – presumably for soil moisture.

Competition from lianas may also limit tree species diversity. Using an eight-year liana-removal experiment in Panama, Schnitzer and Carson (2010) found that shade-tolerant tree species diversity increased over 60 percent faster in treefall gaps without lianas than in control gaps with lianas. Gaps had once been thought to benefit shade-tolerant trees by providing a heterogeneous and resource-rich environment; however, gaps also promote the density and diversity of lianas (e.g., Putz 1984, Schnitzer and Carson 2001, Ledo and Schnitzer 2014), which compete intensely with trees in gaps. Lianas appear to limit shade-tolerant tree diversity by suppressing their colonization, survival, density, and ultimately their diversity. Schnitzer and Carson (2010) suggested that by limiting tree colonization, "biotic interference" from lianas reduced tree species number in gaps to the point that it prevented tree community level niche partitioning.

Lianas and tree community composition

Lianas can alter tree community composition by competing more intensely with some tree species than others (Schnitzer and Bongers 2002). A number of studies have now demonstrated that lianas substantially reduce the performance of shade-tolerant trees, but that pioneer trees appear to largely escape competition from lianas (e.g., van der Heijden *et al.* 2015, Schnitzer and Carson 2010). Shade-tolerant trees tend to grow slowly with many branches that are designed to maximize light interception. This architecture, however, makes them particularly vulnerable to lianas, which rapidly climb and blanket shade-tolerant trees, particularly in treefall gaps (Schnitzer *et al.* 2000). Within the shade-tolerant tree guild, lianas tend to have a negative effect on most species, and there is little evidence for lianas having a much larger effect on some shade-tolerant species than on others (e.g., Ingwell *et al.* 2010, Alvarez-Cansino *et al.* 2015). Thus, there has been much speculation that lianas may alter tree community dynamics by competing more intensely with some species than others; to date, however, empirical support for this claim has been restricted largely to the pioneer–shade-tolerant tree dichotomy (e.g., Ingwell *et al.* 2010, Schnitzer and Carson 2010).

Lianas and forest carbon dynamics

By competing with trees, which typically store more than 90 percent of aboveground forest carbon, lianas can dramatically reduce rates of forest carbon uptake (Durán and Gianoli 2013, van der Heijden *et al.* 2013). Forests with high liana densities tend to have far lower biomass than forest areas with low liana densities, suggesting that lianas reduce whole-forest carbon accumulation and storage (e.g., Durán and Gianoli 2013). For example, in a liana-removal study in central Panama, Schnitzer *et al.* (2014) reported that lianas reduced tree biomass accumulation in gaps by nearly 300 percent over an eight-year period (a 35 percent annual reduction in biomass accumulation). Since lianas themselves, however, accumulate and store very little carbon, lianas were able to compensate for only 24 percent of the biomass accumulation that they displaced in trees (Schnitzer *et al.* 2014). Scaling up to the whole-forest level, the annual loss in biomass due to lianas in gaps can be as much as 18 percent of total forest biomass (Schnitzer *et al.* 2014). In an observational study in the Peruvian Amazon, lianas appeared to reduce annual tree biomass increase by around 10 percent; however, annual liana biomass increase compensated for only around 29 percent of the biomass displaced in trees (van der Heijden and Phillips 2009).

The finding that competition from lianas can have a net negative effect on tropical forest carbon uptake is surprising. Most theories assume that competition among plants is a zero-sum game in terms of carbon uptake. That is, one plant's carbon loss is another's carbon gain. This assumption, however, breaks down with competition among contrasting growth forms that inherently store different amounts of carbon, and competition from lianas will result in a net carbon loss for tropical forests (Schnitzer *et al.* 2014). Considering that tropical forests store around one-third of all aboveground terrestrial forest carbon, with the vast majority of this carbon stored in trees, the effects of lianas on tree carbon accumulation and storage affects the global carbon budget. Furthermore, the ongoing increase in liana abundance throughout the neotropics (see below) will amplify the effects of lianas on global carbon dynamics.

Positive contributions of lianas in forest ecosystems

Lianas also have positive effects in forest ecosystems, where they substantially increase plant diversity (e.g., Schnitzer *et al.* 2012, Gianoli 2015) and provide important resources for many forest animals (Schnitzer *et al.* 2015). Many animals consume liana leaves, flowers, and fruits, and the ability of lianas to produce during the dry season may provide a critical resource for many animal species. Indeed, many primate species switch their diet to include far more lianas during the dry season (reviewed by Arroyo-Rodriguez et al. 2015). Lianas weave through the forest canopy, connecting the crowns of many trees, and these connections provide critical inter-crown pathways for many arboreal animals (e.g., ants, sloths, rodents, monkeys; Arroyo-Rodriguez et al. 2015, Yanoviak 2015). The unique growth strategy of lianas, which allows them to fill spaces between trees, adds considerable structure to tropical forests, which benefits a wide variety of animal species, from ants, to birds, to small mammals. Lianas can also bind the forest canopy together, thus potentially reducing treefalls when liana connectivity is high (Garrido-Pérez et al. 2008). In the understory, lianas can form dense tangles of looping stems (e.g., Schnitzer et al. 2000), which provide structure, forage, and an escape from predators for many bird and small mammal species (e.g., Lambert and Halsey 2015). Lianas also provide a wide variety of important products that are utilized by humans, including food (passion fruits, melons, gourds, legumes), stimulants (e.g., guarana from *Paullinia cupana*), medicines (e.g., curare from *Strychnos toxifera* and *Chondrodendron tomentosum*), and the raw ingredients for the production of wine and cognac from grapes (*Vitis spp.*) and beer from hops (*Humulus lupulus*). Thus, although lianas are usually considered to be detrimental to forest ecosystems, they also have a variety of positive effects on forest biodiversity and wildlife, including humans.

Increasing liana abundance in neotropical forests: patterns, causes, and consequences

Lianas' density, biomass, and productivity appear to be increasing in neotropical forests and there are now 12 published studies documenting this increase (Schnitzer and Bongers 2011, Schnitzer 2015). The pattern of increasing liana abundance has been reported in a number of neotropical forests, including the Bolivian Amazon, Brazilian Amazon (two separate studies), Costa Rica (two separate studies), French Guiana, Panama (four separate studies), and in the subtropical southeastern USA (Schnitzer 2015). Increasing liana abundance seems to be consistent across tropical forest types, and liana increases have been reported in mature wet, moist, and dry forests (Schnitzer 2015).

For example, in various forests in northwest South America, Amazonia, and Central America, Phillips et al. (2002) found that both small and large lianas had increased significantly in stem density and basal area, both in absolute values and relative to trees. The authors also found that the most rapid increase occurred in the last decade of the study, thus suggesting that liana increases will likely continue into the future. In central Amazonia, liana stem density (stems ≥ 2 cm diameter) increased an average of 1 percent per year from 1999 until 2012 (Laurance et al. 2014). In this same forest, liana seedling density increased 500 percent from 1993 to 1999, while tree and herbaceous plant seedling recruitment decreased over this same period (Benítez-Malvido and Martínez-Ramos 2003). In French Guiana, the density of large (≥ 10 cm diameter) lianas increased 1.8 percent, while large tree abundance decreased 4.6 percent from 1992 until 2002 (Chave et al. 2008).

On Barro Colorado Island, Panama, liana productivity (measured as leaf litter production) increased by 57 percent from 1986 until 2002 (Wright et al. 2004), liana flower production increased more than 125 percent faster than that of trees (Wright and Calderon 2006), the percentage of trees with liana infestation increased from 32 percent in 1967–1968 to 47 percent in 1979, and then to nearly 75 percent in 2007 (Ingwell et al. 2010). Over a 30-year period ending in 2007, liana density increased 75 percent for stems ≥ 1 cm diameter and 140 percent for stems > 5 cm diameter (Schnitzer et al. 2012). In Costa Rica, liana increases have been reported in both dry and wet forests (San Emilio, Guanacaste Provence and La Selva Biological Station, respectively). For example, in a wet forest at La Selva Biological Station in Costa Rica,

Yorke *et al.* (2013) found a 2.9 percent annual increase in mean liana basal area and a 2.0 percent annual increase in mean liana density (stems ≥ 1 cm diameter) in six old-growth plots over an eight-year period from 1999 to 2007. To date, empirical evidence for increasing liana abundance is restricted to neotropical forests, and there are currently too few long-term datasets from tropical Africa or Asia to adequately test the pattern of increasing liana abundance in these regions.

The main hypotheses proposed to explain increasing liana abundance in neotropical forests include increases in forest disturbance (including both natural disturbance and changes in land use and fragmentation), greater intensity of seasonal drought and lower annual rainfall, elevated atmospheric CO_2, and elevated nutrient deposition. To date, these putative mechanisms have not been rigorously tested and linked to ongoing global changes. However, there is support for a mechanistic response of lianas to increasing forest disturbance and increasing seasonal drought intensity. Lianas respond strongly to disturbance, which may provide an important regeneration niche for liana establishment (Schnitzer *et al.* 2000, 2012, Dalling *et al.* 2012, Ledo and Schnitzer 2014). Lianas also grow more rapidly than trees during seasonal drought, thus allowing lianas to increase proportionally more than trees if the duration and intensity of seasonal droughts increase. Elevated atmospheric CO_2 and nutrient deposition may also contribute to liana increases by enabling lianas to grow more efficiently, both during seasonal drought and in the hot, dry conditions that follow disturbance; however, these latter processes remain largely untested.

Summary

Lianas are now recognized as a key component of forests worldwide, particularly in highly seasonal tropical areas, where lianas peak in abundance. Lianas compete intensely with trees, decreasing tree recruitment, growth, fecundity, and diversity. Lianas also significantly reduce tree biomass accumulation. Lianas themselves, however, have thin stems and low wood volume, and thus they cannot compensate for the biomass that they displace in trees. The effects of lianas on forest community and ecosystem dynamics are becoming more pronounced in neotropical forests, where liana abundance is increasing relative to trees. However, not all of the effects of lianas are negative. Lianas contribute significantly to vascular plant diversity, particularly in tropical forests. Lianas are also an important source of food and shelter for animals. By crossing from one canopy to another, lianas provide aerial pathways for many animals to traverse the forest canopy without having to descend to the forest floor. In summary, lianas have both positive and negative effects in forests worldwide, and regardless of whether you love them, hate them, or have never thought about them, lianas are an important part of the dynamic ecology of the forest ecosystems.

Acknowledgments

This work was made possible by financial support from US National Science Foundation grants DEB-0613666, NSF-DEB 0845071, and NSF-DEB 1019436. I thank Richard Corlett for inviting me to write this chapter and K. Barry, A. Ercoli, and S. Turton for helpful comments on the manuscript.

References

Alvarez-Cansino, L., S.A. Schnitzer, J. Reid and J.S. Powers (2015). Liana competition with tropical trees varies with seasonal rainfall and soil moisture, but not tree species identity. *Ecology* 96: 39–45.

Andrade, J.L., F.C. Meinzer, G. Goldstein and S.A.Schnitzer (2005). Water uptake and transport in lianas and co-occuring trees of a seasonally dry tropical forest. *Trees: Structure and Function* 19: 282–289.

Arroyo-Rodriguez, V., N. Asensio, J.C. Dunn, J. Cristóbal-Azkarate and A. Gonzalez-Zamora (2015). The use of lianas by primates: more than a food resource. Pages 407–426. In: *The Ecology of Lianas* (eds S.A. Schnitzer, F. Bongers, R.J. Burnham, F.E. Putz). Wiley-Blackwell, Oxford, UK.

Benítez-Malvido J. and M. Martínez-Ramos (2003). Impact of forest fragmentation on understory plant species richness in Amazonia. *Conservation Biology* 17: 389–400.

Cai Z.-Q., S.A. Schnitzer and F. Bongers (2009). Seasonal differences in leaf-level physiology give lianas a competitive advantage over trees in a tropical seasonal forest. *Oecologia* 161: 25–33.

Chave J., J. Olivier, F. Bongers, P. Châtelet, P.M. Forget, and others (2008). Aboveground biomass and productivity in a rain forest of eastern South America. *Journal of Tropical Ecology* 24: 355–366.

Chen, Y.-J., Z.-X. Fan, S.A. Schnitzer, J.-L. Zhang, K.-F. Cao (2015) Water-use advantage of lianas over trees in seasonal tropical forests. *New Phytologist* 205: 128–136.

Dalling, J.W., S.A. Schnitzer, C. Baldeck, K.E. Harms, R. John, and others (2012). Resource-based habitat associations in a neotropical liana community. *Journal of Ecology* 100: 1174–1182.

DeWalt, S.J., S.A. Schnitzer and J.S. Denslow. (2000). Density and diversity of lianas along a chronosequence in a central Panamanian tropical forest. *Journal of Tropical Ecology* 16: 1–19.

DeWalt S.J., S.A. Schnitzer, J. Chave, F. Bongers, R.J. Burnham and others (2010). Annual rainfall and seasonality predict pan-tropical patterns of liana density and basal area. *Biotropica* 42: 309–317.

DeWalt, S.J., S.A. Schnitzer, L.F. Alves, F. Bongers, R.J. Burnham and others (2015). Biogeographical patterns of liana abundance and diversity. Pages 131–146. In: *The Ecology of Lianas* (eds S.A. Schnitzer, F. Bongers, R.J. Burnham, F.E. Putz). Wiley-Blackwell, Oxford, UK.

Durán, S.M. and E. Gianoli (2013). Carbon stocks in tropical forests decrease with liana density. *Biology Letters* 9: 20130301.

Garrido-Pérez, E.I., J.M. Dupuy, R. Durán-García, G. Gerold, S.A. Schnitzer, M. Ucan-May (2008). Structural effects of lianas and Hurricane Wilma on trees in Yucatan peninsula, Mexico. *Journal of Tropical Ecology* 24: 559–562.

Gianoli, E. (2015). Evolutionary implications of the climbing habit in plants. Pages 239–250. In: *The Ecology of Lianas* (eds S.A. Schnitzer, F. Bongers, R.J. Burnham, F.E. Putz). Wiley-Blackwell, Oxford, UK.

Grauel, W.T. and F.E. Putz (2004). Effects of lianas on growth and regeneration of *Prioria copaifera* in Darien, Panama. *Forest Ecology and Management* 190: 99–108.

Ingwell, L.L., S.J. Wright, K.K. Becklund, S.P. Hubbell, S.A. Schnitzer. (2010). The impact of lianas on 10 years of tree growth and mortality on Barro Colorado Island, Panama. *Journal of Ecology* 98: 879–887.

Jiménez-Castillo, M. and C.H. Lusk (2013). Comparative vascular anatomy and function of woody plants in a temperate rainforest: lianas suffer higher levels of freeze-thaw embolism than associated trees. *Functional Ecology* 27: 403–412.

Kainer, K.A., L.H.O. Wadt and C.L. Staudhammer (2014). Testing a silvicultural recommendation: Brazil nut responses 10 years after liana cutting. *Journal of Applied Ecology*, doi: 10.1111/1365-2664.12231.

Ladwig, L. and S. Meiners (2015). The role of lianas in temperate tree communities. Pages 188–202. In: *The Ecology of Lianas* (eds S.A. Schnitzer, F. Bongers, R.J. Burnham, F.E. Putz). Wiley-Blackwell, Oxford, UK.

Lambert, T.D. and M.K. Halsey (2015). Relationship between lianas and arboreal mammals: examining the Emmons–Gentry hypothesis. Pages 398–406. In: *Ecology of Lianas* (eds S.A. Schnitzer, F. Bongers, R.J. Burnham, F.E. Putz). Wiley-Blackwell Publishing, Oxford, UK.

Laurance, W.F., D. Pérez-Salicrup, P. Delamônica, P.M. Fearnside, S. D'Angelo and others. (2001). Rain forest fragmentation and the structure of Amazonian liana communities. *Ecology* 82: 105–116.

Laurance, W.F., A.S. Andrade, A. Magrach, J.L.C. Camargo, J.J. Valsko and others (2014). Long-term changes in liana abundance and forest dynamics in undisturbed Amazonian forests. *Ecology* 95: 1604–1611. http://dx.doi.org/10.1890/13-1571.1

Ledo, A. and S.A. Schnitzer (2014). Disturbance, not negative density dependence or habitat specialization maintains liana diversity in a tropical forest. *Ecology* 95: 2169–2178.

Londré, R.A. and S.A. Schnitzer (2006). The distribution of lianas and their change in abundance in temperate forests over the past 45 years. *Ecology*, 87, 2973–2978.

Phillips, O.L., R. Vasquez Martinez, L. Arroyo, T.R. Baker, T. Killeen, and others (2002). Increasing dominance of large lianas in Amazonian forests. *Nature* 418: 770–774.

Putz, F.E. (1984). The natural history of lianas on Barro Colorado Island, Panama. *Ecology* 65: 1713–1724.

Restom, T.G. and D.C. Nepstad (2004). Contribution of vines to the evapotranspiration of a secondary forest in eastern Amazonia. *Plant and Soil* 236: 155–163.

Schnitzer, S.A. (2005). A mechanistic explanation for global patterns of liana abundance and distribution. *The American Naturalist* 166: 262–276.

Schnitzer, S.A. (2015). Increasing lianas abundance in neotropical forests: causes and consequences. Pages 451–464. In: *The Ecology of Lianas* (eds S.A. Schnitzer, F. Bongers, R.J. Burnham, F.E. Putz). Wiley-Blackwell, Oxford, UK.

Schnitzer, S.A. and W.P. Carson (2001). Treefall gaps and the maintenance of species diversity in a tropical forest. *Ecology* 82: 913–919.

Schnitzer, S.A. and F. Bongers (2002). The ecology of lianas and their role in forests. *Trends in Ecology and Evolution* 17: 223–230.

Schnitzer, S.A. and W.P. Carson (2010). Lianas suppress tree regeneration and diversity in treefall gaps. *Ecology Letters* 13: 849–857.

Schnitzer, S.A. and F. Bongers (2011). Increasing liana abundance and biomass in tropical forests: emerging patterns and putative mechanisms. *Ecology Letters* 14: 397–406.

Schnitzer, S.A., J.W. Dalling, and W.P. Carson (2000). The impact of lianas on tree regeneration in tropical forest canopy gaps: evidence for an alternative pathway of gap-phase regeneration. *Journal of Ecology* 88: 655–666.

Schnitzer, S.A., M. Kuzee, and F. Bongers (2005). Disentangling above- and below-ground competition between lianas and trees in a tropical forest. *Journal of Ecology* 93: 1115–1125.

Schnitzer, S.A., S.A. Mangan, J.W. Dalling, C.A. Baldeck, S.P. Hubbell, and others. (2012). Liana abundance, diversity, and distribution on Barro Colorado Island, Panama. *PLoS ONE* 7: e52114.

Schnitzer, S.A., G.M.F. van der Heijden, J. Mascaro, and W.P. Carson (2014). Lianas reduce biomass accumulation in a tropical forest. *Ecology* 95: 3008–3017.

Schnitzer, S.A., F.E. Putz, F. Bongers and K. Kroening (2015). The past, present, and potential future of liana ecology. Pages 3–10. In: *The Ecology of Lianas* (eds S.A. Schnitzer, F. Bongers, R.J. Burnham, F.E. Putz). Wiley-Blackwell, Oxford, UK.

Sperry J.S., N.M. Holbrook, M.H. Zimmerman and M.T. Tyree (1987). Spring filling of xylem vessels in wild grapevine. *Plant Physiology*, 83, 414–417.

Swaine, M.D. and J. Grace (2007). Lianas may be favoured by low rainfall: evidence from Ghana. *Plant Ecology* 192: 271–276.

Thomas, D., R.J. Burnham, G .Chuyong, D. Kenfack, M. Nsangy Sainge (2015). Liana abundance and diversity in Cameroon's Korup National Park. Pages 13–22. In: *The Ecology of Lianas* (eds S.A. Schnitzer, F. Bongers, R.J. Burnham, F.E. Putz). Wiley-Blackwell, Oxford, UK.

Tobin, M.F., A.J. Wright, S.A. Mangan and S.A. Schnitzer. (2012). Lianas have a greater competitive effect than trees of similar biomass on tropical canopy trees. *Ecosphere* 3, Article 20: 1–11. http://dx.doi.org/10.1890/ES11-00322.1

Toledo-Aceves, T. (2015). Above and belowground competition between lianas and trees. Pages 149–163. In: *The Ecology of Lianas* (eds S.A. Schnitzer, F. Bongers, R.J. Burnham, F.E. Putz). Wiley-Blackwell, Oxford, UK.

van der Heijden, G.M.F. and O.L. Phillips (2009). Liana infestation impacts tree growth in a lowland tropical moist forest. *Biogeosciences* 6: 2217–2226.

van der Heijden, G.M.F. O.L. Phillips and S.A. Schnitzer (2015). Effects of lianas on forest level biomass. Pages: 164–174. In: *Ecology of Lianas*, Schnitzer, S.A., F. Bongers, R.J. Burnham, F.E. Putz, editors. Wiley-Blackwell Publishing, Oxford.

van der Heijden, G.M.F., S.A. Schnitzer, J.S. Powers and O.L. Phillips (2013). Liana impacts on carbon cycling, storage and sequestration in tropical forests. *Biotropica* 45: 682–692.

Wright, S.J. and O. Calderon (2006) Seasonal, El Nino and longer term changes in flower and seed production in a moist tropical forest. *Ecology Letters* 9: 35–44.

Wright S.J., O. Calderón, A. Hernandéz and S. Paton (2004). Are lianas increasing in importance in tropical forests? A 17-year record from Panamá. *Ecology* 85: 484–489.

Wyka, T.P., J. Oleksyn, P. Karolewski and S.A. Schnitzer. (2013). Phenotypic correlates of the lianescent growth form – a review. *Annals of Botany* 112: 1667–1681.

Yanoviak, S. (2015). Effects of lianas on canopy arthropod community structure. Pages 345–361. In: *The Ecology of Lianas* (eds S.A. Schnitzer, F. Bongers, R.J. Burnham, F.E. Putz). Wiley-Blackwell, Oxford, UK.

Yorke, S.R., S.A. Schnitzer, J. Mascaro, S. Letcher and W.P. Carson (2013). Increasing liana abundance and biomass in a tropical forest: the contribution of long distance clonal colonization. *Biotropica* 45: 317–324.

Zhu S.-D. and K.-F. Cao (2009). Hydraulic properties and photosynthetic rates in co-occuring lianas and trees in a seasonal tropical rainforest in southwestern China. *Plant Ecology* 204: 295–304.

14
VASCULAR EPIPHYTES IN FOREST ECOSYSTEMS

David H. Benzing

Between 8 and 10 percent of the vascular plants are epiphytic, which means that they spend at least part of their lives anchored on the aerial scaffolding provided by woody hosts (phorophytes). About one in ten is either a xylem-tapping or a holo-parasitic (taps xylem and phloem) type mistletoe, the balance being entirely self-nourished, i.e., autotrophic or free-living. Members of this second category differ most fundamentally by fidelity to arboreal (aerial) versus terrestrial (soil-based) substrates (the obligate versus facultative types) and whether or not the life cycle is spent entirely above the forest floor (the holo- versus hemi-epiphytes respectively). Primary hemi-epiphytes germinate in the canopy and produce root systems that eventually access the ground. The secondary types do the opposite (Figure 14.1). For many species the distinction between liana/vine versus hemi-epiphyte varies depending on local growing conditions. Modifying words such as twig, bark, and ant nest distinguish the holo-epiphytes that mostly always occur on narrowly defined kinds of substrates.

Epiphytism occurs across the vascular plant kingdom, but not uniformly (Benzing 2012). Whereas few gymnosperms regularly root on bark or in the soil-like humus that often accumulates on the surfaces of trunks and large branches, fully one-third of the ferns and a somewhat more modest fraction of the lycophytes do. Among the flowering plants, the monocots greatly outnumber the eudicots in aerial habitats, largely because Orchidaceae contributes over half of the total species with Bromeliaceae and Araceae ranking second and third. Apocynaceae, Ericaceae, Gesneriaceae, Melastomataceae, and Rubiaceae field up to 500 eudicot-type epiphytes apiece. Piperaceae assures that the basal angiosperms contribute species as well. The branch parasites constitute a more tightly related assemblage, the single angiosperm order Santalales accounting for all but a few of the mistletoes.

Comparisons of the higher taxa that contain large contingents of arboreal species versus those with many fewer canopy-dwellers suggest why this lifestyle occurs unevenly. Characteristics that appear to have predisposed certain terrestrial lineages to generate epiphytes include cheap construction (little or no secondary body), substantial drought tolerance, small seeds, and pollen exchanged by fauna rather than wind (Benzing 2012). The exceptions include Ericaceae and Melastomataceae, the memberships of both of which tend to be robustly woody yet include hundreds of epiphytes. It is also worth noting that epiphytism has failed to emerge as a prominent life strategy in every candidate clade (e.g., Asteraceae, Crassulaceae).

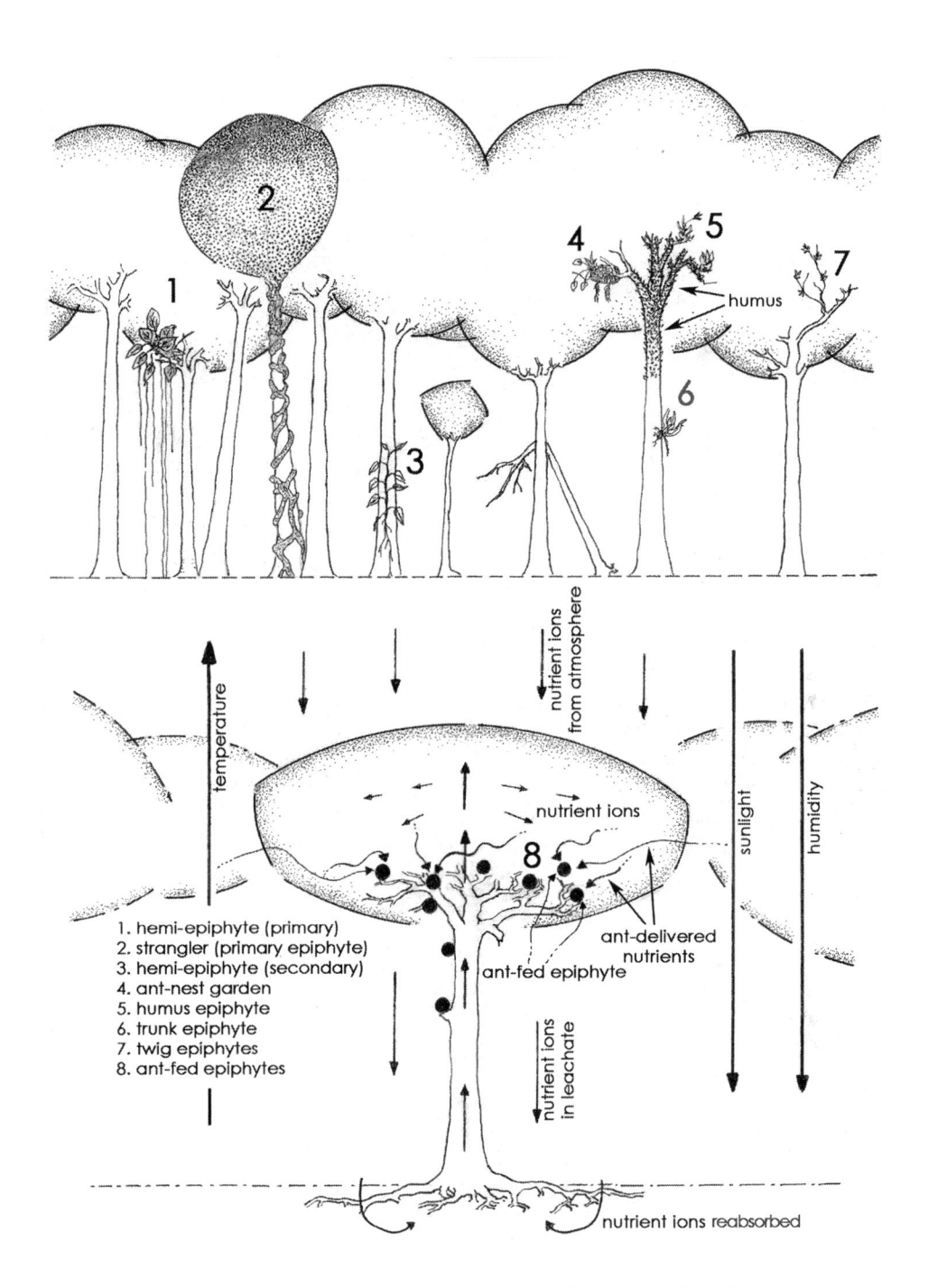

Figure 14.1 A mature forest viewed in profile depicting eight different kinds of epiphytes and important environmental gradients, specifically humidity, sunlight, and temperature. Also indicated are the paths followed by nutrient ions as they move between soil and canopy and within a canopy that includes ant-house epiphytes

1. hemi-epiphyte (primary)
2. strangler (primary epiphyte)
3. hemi-epiphyte (secondary)
4. ant-nest garden
5. humus epiphyte
6. trunk epiphyte
7. twig epiphytes
8. ant-fed epiphytes

Although the epiphytes are not noteworthy sources of folk or modern pharmaceuticals, or providers of human food or fiber, many have become horticultural favorites. Modest size and easily-met requirements for cultivation account for the growing popularity of numerous arboreal bromeliads, ferns and orchids and their domesticated derivatives, a number of which already are regular inventory items in large retail outlets across the developed world. Their propagation for export by the tens of millions annually generates substantial returns in countries as far flung as Colombia, the Netherlands, and Thailand. Many additional, no less appealing epiphytes in families such as Ericaceae and Gesneriaceae have yet to enter the global marketplace.

Geographic occurrence

Epiphytism in both its free-living and parasitic versions is best developed at low latitudes (Benzing 1990, 2012). Deepest frost hardiness prevails among the dwarfed mistletoes (*Arceuthobium*), a few of which range far enough northward to attack boreal conifers (Watson 2001). Most of the extra-tropical autotrophs inhabit rainforests, particularly those located in the Pacific Northwest of the United States and adjacent Canada, in central Chile and on New Zealand's South Island (Zotz 2005). None of their arboreal floras, however, equals in diversity those native to many similarly moist, but warmer woodlands. Pteridophytes, ferns, but fewer lycophytes, tend to dominate and any angiosperms present are usually ecological outliers within predominantly terrestrial, cool growing families (e.g, Liliaceae *sensu lato*, Ranunculaceae).

Drought powerfully sorts the tropical epiphytes assuring that the highest densities of individuals, species, and families occur in ever-wet forests (Figure 14.2; Gentry and Dodson 1987). Heat, as a driver of evaporation, magnifies the impact of scarce precipitation and especially so in upland regions where much aerial flora are confined to life zones that measure in as few as tens of meters in depth (Nadkarni and Solano 2002). Conversely, only the most

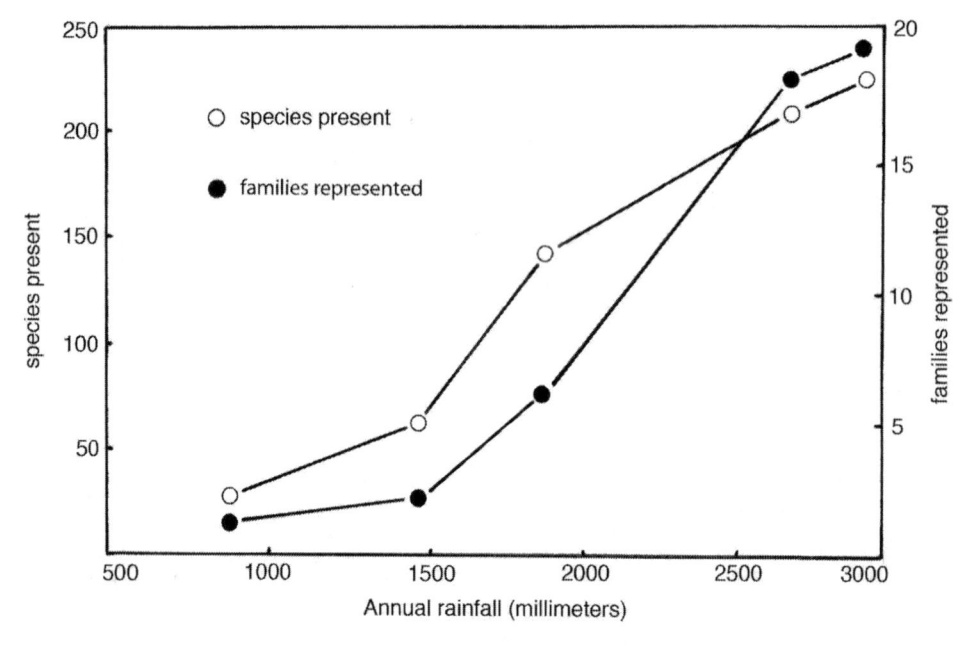

Figure 14.2 The relationship between annual rainfall and the species-richness of epiphytes at five locations in tropical America according to Gentry and Dodson (1987)

xerophytic members of Araceae, Bromeliaceae and Orchidaceae, plus some similarly stress-tolerant species in a few additional angiosperm families and a scattering of ferns possess sufficient drought-hardiness to inhabit sites characterized by hot, prolonged dry seasons. The high cost in water of producing xylem also explains why so few of the woody holo-epiphytes inhabit other than humid forests.

Epiphyte diversity peaks in north Andean South America and the Atlantic Rainforest and throughout similarly accommodating mid-elevation regions of Central America, New Guinea, and the southern flanks of the eastern Himalayas (Gentry and Dodson 1987). Biogeography bears the stamp of historical coincidence as well: a large majority of the flowering plant families that field substantial numbers of epiphytes happen to have undergone their most exuberant expansions above ground in the New World (Benzing 2012). Most notable of the exceptions are Apocynaceae, Rubiaceae and Melastomataceae. Bromeliaceae and Cactaceae, two families that include about 2000 and 130 epiphytes respectively, are wholly tropical American except for one Old World species each. Orchidaceae, in which epiphytism has evolved repeatedly, is well represented in canopies across the low latitudes. Much the same applies for the pteridophytes.

The aerial habitat

The substrates that sustain the free-living epiphytes differ by physical structure, water-holding capacity, fertility and much more. Naked bark constitutes the most challenging of the many options, whereas more complex alternatives such as the composite "carton" material manufactured by certain arboreal ants to construct runways and nests provides more hospitable growing conditions (Davidson 1988; Figure 14.1). At another level, epiphytes densely infest the crowns of some woody species while those of others remain little or never colonized (Callaway *et al.* 2002; Bartels and Chen 2012). Host specificity is more pronounced among the mistletoes compared to the free-living epiphytes, but few members of either group have but one kind of phorophyte (Watson 2001; Benzing 2012).

Physical gradients greatly increase the environmental complexity of a forest canopy habitat, four of which far exceed the others for influence on epiphytes (Figure 14.1). Sunlight and temperature diminish downward while humidity does the opposite. Nutrients follow multiple, more circuitous paths because they arrive from two sources: the atmosphere and soil, in the second instance by way of xylem conduits in tree trunks followed by release as leafy crowns leach ions and shed spent organs (Wania *et al.* 2002). Animals, particularly ants, and the debris-impounding epiphytes that include the tank-forming bromeliads and ferns further complicate deliveries (Figures 14.1, 14.3). These four variables, with lesser influences from the rest, variously intersect and interact to create for epiphytes deeply three-dimensional mosaics comprised of widely disparate microhabitats.

Epiphytes in turn modify woodland ecosystems and most strikingly where cool, wet climates encourage algae, lichens, and bryophytes to overgrow most bark surfaces, and as epiphylls, even those of long-lived foliage. The blurred boundaries between aerial and terrestrial realms that result expand the former at the expense of the latter and often enough to permit normally soil-restricted fauna such as earthworms to invade parts of the canopy that drier conditions would render too hostile (Paoletti *et al.* 1991; Frank and Lounibos 2009). It is not uncommon in ever-wet, tropical forests to encounter many meters above ground diverse communities that include what normally are soil-dwellers. However, aridity sufficient to leave most bark surfaces naked need not deny the same biota equal opportunity where tank-forming bromeliads create veritable hanging wetlands complete with additional fauna, some of which occur nowhere else as described below.

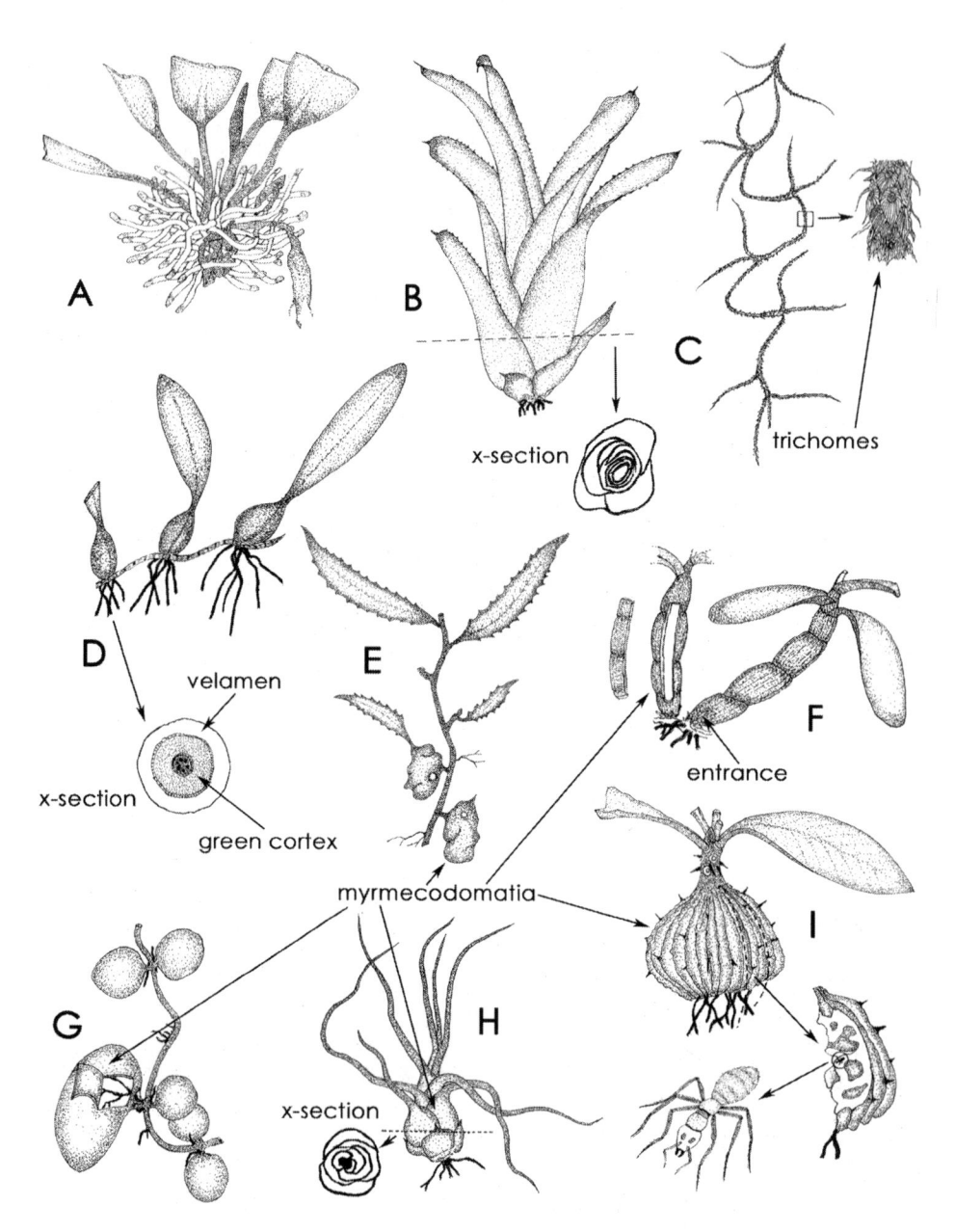

Figure 14.3 Epiphytes that illustrate morphological specializations that permit or enhance access to moisture and nutrients located in sources other than terrestrial soil. A. bird's nest or trash basket *Anthurium* sp. (Araceae); B. shoot of tank-forming bromeliad with lower portion shown in x-section; C. *Tillandsia usneoides* (Bromeliaceae) illustrating absorbing trichomes and rootless condition; D. *Bulbophyllum* sp. (Orchidaceae) with velamentous root in x-section; E. *Solenopteris bifrons* (Polypodiaceae) with myrmecodomatia; F. *Myrmecophyllum* sp. (Orchidaceae) hollow pseudobulb cut away with entrance shown in adjacent pseudobulb; G. *Dischidia simplex* (Apocynaceae) showing normal leaves and foliar myrmecodomatium invaded by adventitious roots; H. *Tillandsia bulbosa* (Bromeliaceae) intact and in x-section; I. *Myrmecodia* sp. (Rubiaceae) with myrmecodomatium sectioned. The ant pictured in I provides a rough approximation of scale

Requirements and tolerances for variables other than those related to humidity, temperature, and substrates further determine how communities of epiphytes are organized in space. Species restricted to the lower reaches of dense evergreen forests typically photo-saturate in moderate shade, in addition to withstanding drought less successfully than those that anchor on more exposed surfaces (Benzing 1990, 2012). Many of the hemi-epiphytic scramblers, particularly members of Araceae, being vine-like, can adjust shoot form and function to match conditions, particularly photon flux, that can shift markedly across short distances. Dangling roots produced by these same plants and more robust types characteristic of the woody "stranglers" (e.g., hemi-epiphytic *Ficus*) help tie together canopy layers and the canopy as a whole to the ground (Figure 14.1).

In essence, the more chemically and physically heterogeneous an aerial habitat the more elaborate its community of epiphytes can be, humidity usually being the most decisive determinant. Even a rainforest, despite receiving super abundant precipitation, subjects to desert-like conditions residents perched on its thinnest twigs while others that root in ant nests, mats of bryophytes and lichens (the "humus" epiphytes), or debris-filled knotholes draw on more plentiful supplies (Figure 14.1). However, the notion that the degree to which key resources can be partitioned determines how many species a living space can host may not always apply in the same way in aerial as opposed to some other kinds of habitats. Quite often, closely related sympatric species share substrates perhaps owing primarily to non-overlapping aspects of reproduction such as flowering phenology and pollination biology (e.g., many bromeliads and orchids).

Adaptations for aerial life

The adaptations that underlie epiphytism range from conspicuous to cryptic depending on prevailing climate. Many a canopy-dweller encountered in an ever-wet forest displays little to nothing that explains its aerial habit, and others of its kind may be growing just as successfully on the ground. Much the same pattern prevails in certain hyper-arid woodlands, particularly those situated on rocky terrain. Here, also, humidity apparently fosters growing conditions that favor facultative over obligate performance. This being the case, it seems reasonable to anticipate that species native to moderately moist rather than wetter or drier forest will provide the deepest insights into the basis of obligate epiphytism. These are the species that fail to thrive after being dislodged from a canopy, or they die outright within a few weeks to several months (Pett-Ridge and Silver 2002).

Observations of the type just described prompt two hypotheses. First, it is unlikely that life spent suspended above ground free of contact with soil requires attributes that evolved exclusively to support autotrophic epiphytism. Second, a majority of the adaptations that enable this lifestyle probably emerged among ancestors native to similarly challenging terrestrial habitats—challenging owing to extremes of humidity lethal to plants lacking corresponding adaptations. Why else would the incidence of facultative versus obligate epiphytism peak where growing conditions within and below the canopy converge, whether due to excessively wet or dry conditions? A key question seems to be how much the environment on a forest floor must differ from that above to oblige obligate epiphytism.

Adaptations that relax the threat posed to arboreal flora by drought fall into two categories: those that promote the capture and uptake of scarce precipitation and a second suite, not entirely distinct from the first that favors its sparing economic use. A number of attributes improve access, the most thoroughly studied being the velamen-equipped roots of the orchids, leafy cisterns that are most impressively illustrated by the tank bromeliads, and dense coverings of hydrophilic, absorbing foliar hairs that again are best exemplified by certain Bromeliaceae, but this time by its so-called "atmospheric" members (Benzing and Ott 1981; Benzing et al. 1983; Figure 14.3).

Aspects of anatomy and physiology that increase water use efficiency include CAM-type photosynthesis, which, although usually supported by less water storage capacity than that serving many of the desert-dwelling succulents, occurs widely among the vascular epiphytes (Winter et al. 1983). Seasonal deciduousness, a drought-avoiding rather than drought-enduring arrangement, is far less common among arboreal flora, being restricted to a handful of canopy-dwelling Bromeliaceae and Zingiberaceae, considerably more species of orchids, and dozens of ferns. Still other pteridophytes are poikilohydric (e.g., the desiccation-tolerant polypodiums). Arguably, just about all of the strategies employed by vascular plants to counter pronounced aridity do so for at least a few of the epiphytes (Heitz and Briones 1998; Benzing 2012).

Nutrients are plentiful in canopy habitats, but unevenly available to epiphytes. Those present in phytomass constitute by far the largest of the reservoirs, where, except for the mistletoes, they cannot be accessed directly. Access for the free-living species often involves abiotic and biotic intermediaries, which explains much unusual epiphyte structure and function and several novel associations with fauna (Figure 14.3). Supplies more immediately obtainable by the non-parasites include those sequestered in humus-rich suspended media, sources that can be as fertile as many terrestrial soils, but tend to be shallower and more subject to rapid drying (Ingram and Nadkarni 1993; Cardelús et al. 2009; Figure 14.1). Many of the epiphytes that grow with the least contact with their substrates employ the high-affinity roots and foliar hairs mentioned above (Figure 14.3). Demand reduced by slow growth accompanies their impressive capacities to harvest essential ions from dilute sources.

Mycorrhizas appear to play relatively minor roles in all but the wettest, humus-rich settings, and even here little is known about what transpires in the aerial rhizosphere (Benzing 2012). What, for example, are the microbes routinely present within the velamen of the roots of orchids doing that might benefit their hosts? Nitrogen fixers may be important suppliers for certain types of epiphytes, the most likely recipients being those that promote intimate contact with microbes by impounding moist debris in cistern-like shoots or masses of tangled roots (Richardson 1999; Figure 14.3). Deeper inquiry into the dynamics of the microcosms that develop within the phytotelms (water-filled plant cavities) of bromeliads and other impounding epiphytes and the canopy microbiome in general should help explain how an aerial flora can accumulate enough nutrients to add multiple metric tons of phytomass per hectare to a forest canopy (Hofstede et al. 1993).

Epiphytes probably far exceed terrestrial flora for direct reliance on animals for nutrition. A host of plant-maintained invertebrates assist the tank-formers to extract supplies from what otherwise would be more recalcitrant debris (Richardson 1999). A second, less numerous assemblage of species features hollow leaves and stems (myrmecodomatia) attractive as nest sites to arboreal ants (Dejean et al. 1995; Figure 14.3). Essential elements absorbed from excrement and the remains of prey arrive from afar just as if a hosting plant's root system explored the same extent of canopy through which its benefactors forage (Figure 14.1). Mutualisms of this description occur across the tropics and among the memberships of arboreal Apocynaceae, Bromeliaceae, Melastomataceae, Orchidaceae, Rubiaceae, plus several of the major fern clades, compelling evidence of their high value for both the participating ants and epiphytes (Figure 14.3). A second ant-mediated arrangement described below and known as the ant nest-garden mutualism has broader implications within ecosystems.

Certain features displayed by some of the most specialized of the epiphytes reflect their adaptation to what borders on being a single rather than a biphasic living space (atmosphere versus soil). It seems that the more complete its exposure to air the more likely a subject and where its vegetative functions operate deviates from the root–shoot body plan that emerged more than 400 million years ago (Figure 14.3). Abbreviated morphology such as that demonstrated by certain bromeliads and the "shootless" orchids definitely promotes resource use economy, but could something more important be going on as well (Benzing and Ott 1981)? The least substrate-dependent of the bromeliads (e.g., Spanish moss; Figure 14.4) produce no roots beyond those needed to anchor the seedling; holdfast being accomplished by lax, elongated shoots or tendril-like foliage. More mundane morphologies beneficial for life suspended above ground, such as pendent inflorescences and cascading shoots, abound.

Little about how the epiphytes reproduce differentiates their kind from the memberships of many of the other ecologically defined groups of plants (Benzing 2012). Wind pollination remains virtually unreported, but fauna representing every major category of pollen vectors serve at least a few of the arboreal species. Flowers and inflorescences adapted to attract birds predominate in epiphytic Bromeliaceae, Ericaceae, and Gesneriaceae, but whether more so compared to co-occurring terrestrials in the same families remains to be seen. Nor have claims been tested that self-compatibility is over represented, or that the attending insects are mostly "trapline" fliers. Animals of many descriptions disperse the seeds of the epiphytes, but not

Figure 14.4 Dense growth of the bromeliad Spanish moss (*Tillandsia usneoides*) in the crown of debilitated *Citrus* in central Florida

to the exclusion of wind carriage. Minute size assists aerial delivery for the orchid "microsperm" and the spores of the pteridophytes. Hairy comas, and less commonly membranous appendages, increase buoyancy for the heavier diaspores shed from capsules (e.g., some canopy-dwelling members of *Dischidia* and *Hoya* of Apocynaceae).

Influences on other biota

Although vascular epiphytes dominate numerically the inventories of vascular species at many tropical locations, their substantive contribution to biodiversity dwarfs this mere statistic. Resources provided include food, shelter, and breeding and hiding space for an immense array of consumers ranging from microbes to mammals (e.g., Nadkarni and Matelson 1989; Richardson 1999; Diesel and Schuh 1993). The consequences of certain other interactions are not so easily assessed. One recent analysis revealed the remarkable fact that shrubby *Amborella trichopoda*, the most basal of the extant angiosperms, possesses a mitochondrial genome multiplied many fold by horizontal transfers from its epiphytic algae, leafy liverworts, and flowering plants (Rice *et al.* 2013). Another study demonstrated that three years following the removal of its mistletoes an Australian forest had lost more than 25 percent of its species of woodland-dependent birds and approximately 20 percent of all the assayed biota (Watson and Herring 2012).

Much is known about a variety of intimate relationships that involve arboreal flora and fauna, some of which border on being absent among terrestrial plants. Space being short, the following treatment deals only with pairings that combine epiphytes and ants, epiphytes and phorophytes and tank-equipped bromeliads with the diverse users of their leafy phytotelms (Figure 14.3). Mutualisms with ants come first, this arrangement having a particularly powerful influence on the inter-workings of humid lowland forests across the tropics. Diet and behavior assure pivotal roles for these arthropods as does their sheer abundance. Ants alone can exceed in mass that of all of the other canopy-dwelling invertebrates combined (Wilson 1987).

Epiphytes that engage in mutualisms with ants belong either to the ant-house or ant nest-garden categories (Figures 14.1, 14.3). In the first instance, the myrmecodomatium (ant house) is either a modified leaf or stem, of which the most alluring to gravid queens recruit the most faithful of the plant-associating ant species (Figure 14.3). Cruder, often less spacious quarters, for example, those afforded to partners by bromeliads and orchids, attract more sporadic users, most of which also favor plant cavities of other descriptions and maintain small, simply organized colonies. Members of *Hydnophytum* and *Myrmecodia* (Rubiaceae) top the list for elegant accommodations by offering their associates lodging in swollen hypocotyles (seedling stems), some of whose labyrinthine chambers are lined with stubby roots (Figure 14.3).

All of the ant-house epiphytes tested have proven capable of absorbing nutrients across the inner surfaces of their myrmecodomatia (Benzing 2012). Moreover, fixed nitrogen putatively deposited by mutualists in Costa Rica was recovered from inside the hollow bulbs of myrmecophytic *Tillandsia caput-medusae* (Bromeliaceae) specimens. None of these discoveries, however, revealed how much of an epiphyte's nutrient budget is satisfied by ants in nature. A second benefit accrues where occupants (e.g., *Odontomachus*) deter herbivores. But service of this second kind pales compared to that provided by the most aggressive of the nest-gardening ants. It is reasonable to assume from the number of times that the ant-house arrangement has evolved that cavities suitable for ant use are often scarce and that substantial benefits accrue to plants able to exploit this shortfall (Figure 14.3).

Separate sets of ants and epiphytes distinguish the nest-garden from the ant-house mutualism, the insects and plants being the more interdependent in the first of these two arrangements. The former also involves more elaborate ant behavior but at the same time yields more benefits.

Plants endowed with extensive root systems mechanically reinforce cartons, and acting as transpiration driven sump pumps, keep them dry enough to spare sensitive brood exposure to lethal humidity (Yu 1994). Nest-garden epiphytes further reward their keepers by providing pasturage for ant-tended aphids and scale insects. How the botanical partners fare in this multidirectional exchange depends in part on whether any larger herbivores deterred by pugnacious guards would have caused more damage than that inflicted by herds of captive sap-sucking insects.

Some of the fiercest of the New Worlds' arboreal ants (e.g., *Azteca*, *Camponotus*) maintain gardens populated by carton-requiring aroids, bromeliads, cacti, ferns, and orchids among others (Davidson 1988). Less is known about comparable associations in the Old World tropics, but they probably occur as widely, if not as conspicuously, in similarly wet, warm woodlands (Kaufman and Maschwitz 2006). Ant behavior through much of Amazonas suggests why the more elaborate of its nest gardens are so luxuriant. Even a slight nudge will elicit a roiling mass of hundreds to thousands of stinging, biting workers, many sufficiently agitated to leap into space to attack tormentors. Smaller ants, frequently members of *Crematogaster* and *Solenopsis* that often share the same quarters (parabiosis), are seldom seen during these encounters.

Diets heavily supplemented by sugar-rich phloem sap obtained from tended insects capable of piercing the tough bundle sheaths that enclose plant vascular tissue help, if not allow, the nest-gardening ants to dominate canopy habitats through much of Amazonas (Wilson 1987). Most of the required calories come from adjacent woody flora, but the essential services received by ants from their smaller botanical nest mates elevate the latter to high importance. Perhaps the most fascinating aspect of the ant nest-garden phenomenon is the manner of its perpetuation. How is it that epiphytes that require such a narrowly defined, widely scattered substrate, and the ants that produce it manage to link up often enough to sustain a mutually obligatory relationship?

The ant nest-garden epiphytes were first thought to colonize developing cartons by way of berries offered to frugivorous birds. No so, or at least a second kind of carrier also contributes after seeds have passed through avian guts. About a dozen ant-garden species from almost as many families provision their seeds with methyl-6-methyl-salicylate to help disperse offspring to far-flung substrates. Those produced by the tropical American ant nest-garden endemic *Peperomia macrostachys* (Piperaceae) contain a suite of compounds that together attract workers of *Camponotus femoratus* (Youngsteadt *et al.* 2008). These same volatiles elicited negative responses from some other kinds of ants. Tropical American *Codonanthe crassifolia* (Gesneriaceae) further solidifies ant engagement by equipping its seeds with edible appendages (arils). Interestingly, some of what appear to be ant nest-garden users from South Asia eastward to Australia produce dry, wind-carried seeds (e.g., *Aeschynanthus*: Gesneriaceae, *Dischidia*: Apocynaceae).

Epiphytes display another broadly convergent trait: capacity to create soil substitutes that in addition to providing moisture and nutrients for their creators elevate their value as symbiotic partners. Numerous tank-forming bromeliads and some additional angiosperms and ferns intercept litter in leafy phytotelms attractive enough to lure the invertebrates and microbes necessary to process it (Paoletti *et al.* 1991; Richardson 1999; Frank and Lounibos 2009). Dense tangles of aerial roots serve the same purpose for the so-called trash basket aroids and orchids and a number of ferns (Figure 14.3). The bromeliads in particular create veritable perched wetlands that under favorable circumstances suspend thousands of liters of water per hectare of canopy (Hofstede *et al.* 1993). Bromeliads such as *Aechmea bracteata* harbor especially diverse communities of moisture-loving invertebrates plus ant and termite colonies in dryer portions of the same vase-shaped shoots (Dejean *et al.* 1995).

Little evidence suggests that the impounding epiphytes have co-evolved with their animal associates, the shapes of bromeliad shoots, for example, appearing to be more attuned to microclimate, specifically exposure and humidity, and perhaps to patterns of litter fall (Sugden and Robins 1979). Conversely, many of the most dedicated of the tank users, the majority being insects, salamanders, and frogs, exhibit behaviors, colors, and shapes, and reproduce in ways that indicate long associations with botanical partners (e.g., Diesel and Schuh 1993). A number of frogs breed exclusively in water-filled leaf bases and some arthropods spend entire lives there. Conditions vary from one type of impoundment to another and within the same shoot as it ages. Water-holding capacity is probably the most critical variable.

The relationship between epiphyte and phorophyte is more intimate for the mistletoes than for their autotrophic counterparts (Watson 2001). Being a parasite, the former possesses clear means to debilitate, but what about the free-livers whose roots never penetrate a host's vasculature? Why is their presence on orchard and ornamental trees so often considered deleterious, and why do expensive efforts to remove them prove worthwhile? More to the point, how might the condition of a phorophyte whose crown is densely populated by nutritionally self-sufficient epiphytes be adversely impacted by that circumstance when the offending colonists supposedly require hosts only for mechanical support?

A dense colony of free-living epiphytes casts extensive shade and can weigh down to the breaking point twigs and entire branches. A short walk through any epiphyte-rich woodland will demonstrate how frequently heavily burdened axes end up on the ground. Impacts may be especially severe at sites such as cloud forests where high humidity promotes vigorous over growth by vascular and nonvascular flora alike and sunlight is much diminished before reaching the canopy. It also stands to reason that even a host growing in a much drier, sunnier location must be affected by an usually prolific epiphyte such as Spanish moss (*Tillandsia usneoides*; Figure 14.4). Less clear is any harm that a colony of epiphytes might impose by encouraging the kinds of pathogens that attack woody plants. On the other hand, a heavily infested canopy might host more beneficial microbes than otherwise.

Chemical warfare is yet another plausible avenue for intervention, in this case not by a host to ward off harmful invaders, but the other way around. Reports from South America describe how certain sun-loving members of the genus *Tillandsia* (Bromeliaceae) appear to chemically defoliate the canopies of their supports, coincidentally or perhaps adaptively, improving microclimates for themselves (Benzing 2012). No toxins or modes of delivery were identified. Testing for yet another possibility will require assessing epiphyte nutrition more comprehensively than has been done so far. Recall that rather than acting as parasites the autotrophs acquire essential ions from phorophytes following their exit from intact foliage and bark, and in litter, and by delivery by mutualists. In effect, the free-living epiphytes by various means deny their supports vital substances that otherwise could be recycled after reaching the ground. They essentially "pirate" essential elements, and by doing so can further stress shrubs and trees already debilitated by disease or by growth on impoverished soils.

Broader roles in ecosystems

Epiphytes do more than provide sustenance and shelter for fauna, affect the welfare of botanical hosts and create above ground opportunities for soil-adapted invertebrates. Being powerful humidifiers and coolers of air as well has prompted one observer to describe them as the "air conditioners" of lowland tropical forests (Stuntz *et al.* 2002). The tank-forming bromeliads are particularly well suited for this role, although bryophytes and lichens probably suspend even more moisture aloft at exceptionally mossy sites. In addition to moderating climate, arboreal flora can heighten a canopy's capacity to intercept occult precipitation with positive local and off-site consequences (Benzing 1997). Many a cloud forest would be less able to irrigate a more arid downstream ecosystem were it epiphyte-free.

Much as epiphytes can boost the capacity of a canopy to capture cloud water, the same applies for airborne solutes (Nadkarni 1981; Clark *et al.* 1998). Moreover, their involvement exceeds simple accumulation (Chia-chun *et al.* 2002). Epiphytes also trap falling litter, which along with self-generated humus encourages the building of suspended, soil-like substrates useful for themselves and other flora, including phorophytes able to produce canopy roots (Nadkarni 1981; Ingram and Nadkarni 1993; Cardelús *et al.* 2009). Epiphytes further influence within-system distributions of nutrients by straddling the path followed by ions as they flux between canopy and forest floor (Figure 14.1). Mention has already been made of the nonparasitic epiphytes' pirate-like way of removing from circulation for a time nutrients that otherwise would be available sooner to the plants that normally recycle them.

Most of the plant material present in a forest ecosystem belongs to its terrestrial flora with the epiphytes accounting for only a few percent of the on-site total. However, influences on system performance may not always follow suit. For instance, a well-developed community of epiphytes might exceed the dominant phototrophs for caloric output. Or their effect could be negative. Trees adapted for hyper- to moderately humid sites typically possess more drought-vulnerable, productive, and shorter-lived foliage than that of the majority of their epiphytes (Benzing 2012). This disparity reflects the fact that being soil-rooted ensures the former type of vegetation access to more plentiful and continuous supplies of moisture and nutrients than available above ground except possibly to epiphytes served by tanks or other similarly high quality aerial sources. As a result, aggregate photosynthesis might diminish to the extent that epiphyte foliage displaces that borne by the local terrestrials with corresponding shifts in the efficiencies with which nitrogen and other scarce resources are used.

Likely responses to global change

Temperatures are rising faster at high latitudes than closer to the Equator, which suggests that altered humidity attributable to climate change more than warming per se threatens the epiphytes (Bloir *et al.* 2013). More precisely shifts in annual rainfall—especially its seasonality—along with mounting thermal stress, will reshape the nature and geographic distributions of the biomes that host most of the vascular epiphytes. This observation is a near certainty given the fact that while plants overall tolerate climatic extremes of every kind, the individual species does not owing to tradeoffs mandated by adaptation to the more narrowly defined growing conditions that characterize specific spatial niches. The vascular epiphytes, being specialized for demanding circumstances already, probably rank among the most vulnerable to climate change of the larger of the ecologically defined categories of flora.

Hypersensitivity to multiple agencies explains why the tropical epiphytes, compared to their terrestrial companions, are poised to over-respond to climate change whether acting directly or through third parties (Figure 14.2; Table 14.1; Gentry and Dodson 1987). Imagine the fate of a drought-deciduous, bark-dwelling orchid that relies on photoperiod to cue its most water-expensive activity, which happens to be C_3-type photosynthesis mediated by short-lived, transpiration-prone foliage. How would a population of such individuals fare were the leaves of its members to continue to flush and abscise on about the same calendar dates while those marking the beginnings and endings of dry and wet seasons change? Plants routinely adjust to naturally evolving climates, but keeping pace with the runaway change currently under way probably exceeds the migratory capacities of most species (Colwell *et al.* 2008; Diffenbaugh and Fields 2013).

Shifting climates acting through other organisms will create problems for the epiphytes to the extent that they disrupt vital relationships. For example, numerous pollen and nectar seekers only fly during a few weeks to months each year, which means that they and the plants dependent upon their visits must coordinate in multiple ways to exchange indispensable materials and services. Should climate change, for instance, alter either party's phenology or geographic distribution too much, one or both participants will pay a price determined by who depends on whom for which service or product and to what degree. Numerous epiphytes also manipulate bats and birds to disperse their seeds, but at what risk is not known.

Epiphytes owe their physiological vulnerability to climate change to performance tradeoffs required to survive where supplies of moisture tend to be either scarce and unpredictable or continuous and hyper-abundant. Seasonal drought would need to be no more than fleeting to eliminate species native to what had been pluvial woodland (Figure 14.2; Table 14.1). Similarly, mounting humidity would assure the eventual demise of their more stress-tolerant counterparts. Both predictions accord with current distributions: most of the atmospheric *Tillandsias*, for example, tend to grow in dry forests, or where precipitation is more plentiful, only on substrates well exposed to sun and moving air. Conversely, the tank-formers equipped to conduct C_3 rather than CAM-type photosynthesis never inhabit regions that experience extended dry seasons.

Cloud forest epiphytes operate under constraints that may further heighten vulnerability to global change. In fact these species may be some of the most climate-sensitive of all because shade already limits photosynthesis and bathing mists constitute their major source of moisture (Nadkarni and Solano 2002). Rising Earth surface temperatures destined to drive cloud banks up mountainsides almost certainly will prove disruptive. Photo-injury inflicted on fundamentally shade-adapted foliage by the accompanying more intense (less filtered) sunlight alone might be sufficient to devastate populations. Add heightened rates of transpiration driven by warmer, less humid air to much-diminished opportunity for recharge and lethal consequences become even more likely.

Predictions about how mounting supplies of CO_2 and reactive nitrogen will impact the epiphytes must rest on less information than available for climate change. Judging from data obtained from flora of other descriptions, the canopy-dwellers should vary in how they respond to the fertilizing effects of both substances. Possible effects include altered growth rate, changed vulnerabilities to disease and herbivores, and diminished to enhanced water use efficiency, whereas plant-based characteristics poised to influence which outcomes materialize include the type of photosynthetic pathway present, tolerance for certain physical stresses, and growth responsiveness to nutrient enrichment. Context will play a role as well, for example, in the way

Table 14.1 Number of species of vascular epiphytes present in three types of forest in Ecuador compared to those with other growth habits (after Gentry and Dodson 1987)

Growth form	Wet	Moist	Dry
Epiphytes	35	8	2
Trees	31	38	28
Shrubs	11	10	6
Climbers	10	25	34
Herbs	14	11	29

that arid compared to more humid conditions will shape the consequences for water balance of modified guard cell behavior.

Many hundreds of cultivated epiphytes, mostly members of Bromeliaceae, Cactaceae, and Orchidaceae, long have had extensive opportunity to escape into the wild from managed landscapes, but so far few have done so. Why woodland canopies remain largely free of naturalized exotics, a few hemi-epiphytic figs being the most notable of the exceptions (e.g., Asian *Ficus micrantha* in Florida), remains unaddressed. Perhaps the generally stringent requirements for epiphytism and the ephemeral and patchy nature of the substrates that its practitioners use not only preclude invasions of aerial habitats by alien species, they constrain competition among endemics.

Not every epiphyte grows slowly, however. For example, *Tillandsia recurvata*, although certainly sluggish by precedents set by many weedy terrestrials, can colonize aerial substrates at extraordinary rates (Figure 14.5). Is it reasonable to describe a species such as this bromeliad invasive? Does it ever competitively exclude other epiphytes, and even if it does not should such behavior be requisite for invasive status? If the answers to both questions are negative then what label should apply where the foliage of an epiphyte has replaced most of that of its host (Figure 14.4)? Putting considerations of labels and comparative vigor aside, one fact remains indisputable: among the numerous threats to native woodlands posed by global change, invasions by alien epiphytes ranks close to, if not, dead last.

Land use change unquestionably poses the greatest of the immediate threats to canopy-based flora. Surveys following forest clearing indicate that the crowns of trees left standing in pastures and along roadsides and where conditions such as deep ravines deny human use of any kind often continue to host epiphytes (Heitz-Seifert *et al.* 1996). Recruitment sometimes follows, but usually only by the most stress-tolerant members of arboreal communities that, prior to clearing, included species that require the more forgiving conditions that only intact evergreen forest can provide. Diversity also plummets where plantations poorly suited to host epiphytes (e.g., *Eucalyptus* and *Pinus*, but not *Coffea* or *Theobroma*) replace natural woody ecosystems.

In essence, the epiphytes in addition to being exceptionally diverse within the bounds of their peculiar lifestyle, differ as well by vulnerability to change imposed by human activity in a rapidly developing world. Species endemic to the most physically demanding and/or threatened habitats face the bleakest futures. Of ultimate concern of course is the question of how many of their kind are prepared to accommodate the conditions that prevail in the altered woodlands that increasingly represent what remains of the refuges for plants that require aerial versus terrestrial anchorages (Barthlott *et al.* 2001; Woods and DeWalt 2012).

It would be futile at this point to attempt an estimation of the planet-wide extinction debt (sum of moribund yet still extant species) for the vascular epiphytes (Jackson and Sax 2009), but it must be substantial. Too much remains unknown about the responses of populations and communities to the many kinds and degrees of disturbance across the full range of forest types and climates. What, for instance, are the consequences of selective logging and other extractive practices? How much time is required for a disturbed, mature community of epiphytes to collapse, return to its former condition, or stabilize somewhere in-between (Benavides *et al.* 2006)? And what about ecological and other services? To what extent can a forest be used to cultivate for trade high-value epiphytes without compromising its indigenous aerial flora? Given the importance of the myriad roles played by epiphytic vegetation across the tropics and the commercial importance of much of its membership, questions like these warrant serious consideration whether prompted by desire to conserve or wisely manage native forests.

Figure 14.5 *Tillandsia recurvata* on a small crape myrtle tree (*Lagerstroemia*) in central Florida

References

Bartels, S. F. and H. Y. H. Chen. 2012. Mechanisms regulating epiphytic plant diversity. *Reviews in Plant Sciences* 31: 391–400.

Barthlott, W., V. Schmidt-Neuerbury, J. Niedler, and S. Engwald. 2001. Diversity and abundance of vascular epiphytes: a comparison of secondary vegetation and primary montane rain forest in the Venezuelan Andes. *Plant Ecology* 152: 145–156.

Benavides, A., J. H. D. Wolf, and J. F. Duivenvooden. 2006. Recovery and succession of epiphytes in upper Amazonian fallows. *Journal of Tropical Ecology* 152: 145–156.

Benzing, D. H. 1990. *Vascular Epiphytes*. Cambridge University Press. New York.

Benzing, D. H. 1997. Vulnerabilities of tropical forests to climate change: the significance of resident epiphytes. *Climate Change* 39: 519–540.

Benzing, D. H. 2012. *Air Plants: Epiphytes and Aerial Gardens*. Comstock Publishing Associates. Cornell University Press. Ithica, N. Y. and London.

Benzing, D. H. and D. W. Ott. 1981. Vegetative reduction in epiphytic Bromeliaceae and Orchidaceae: its origin and significance. *Biotropica* 13: 131–140.

Benzing, D. H., W. E. Friedman, G. Petersen, and A. Renfrow. 1983. Shootlessness, velamentous roots and the pre-eminence of Orchidaceae in the epiphytic biotope. *American Journal of Botany* 70: 121–133.

Bloir, J. L., P. L. Zarnetske, C. Fitzpatric, and S. Finnegan. 2013. Climate change: past, present, and future of biotic interactions. *Science* 341: 499–504.

Callaway, R. M., K. O. Reinhart, G. W. Moore, D. J. Moore, and S. C. Pennings. 2002. Epiphyte host preference and host traits: mechanism for species-specific interactions. *Oecologia* 132: 221–230.

Cardelús, C. L., M. C. Mack, C. L. Wood, J. DeMarco, and K. K. Treseder. 2009. The influence of tree species on canopy nutrient status in a tropical lowland wet forest in Costa Rica. *Plant and Soil* 318: 47–61.

Chia-chun, H., F. W. Hong, and C. M. Kuo. 2002. Epiphyte biomass and nutrient capital of a moist subtropical forest in Northeast Taiwan. *Journal of Tropical Ecology* 18: 659–670.

Clark, K. L., N. M. Nadkarni, D. Schaeffer, and H. L. Gholz. 1998. Atmospheric deposition and net retention of ions by the canopy of a tropical rainforest, Monteverde, Costa Rica. *Journal of Tropical Ecology* 14: 127–145.

Colwell, R. K., G. Brehm, C. L. Cardelús, A. C. Gilman, and J. T. Longino. 2008. Global warming, elevation range shifts and lowland biotic attrition in the wet tropics. *Science* 322: 258.

Davidson, D. W. 1988. Ecological studies of Neotropical ant gardens. *Ecology* 69: 1138–1152.

Dejean, A., I. Olmstead, and R. R. Schnelling. 1995. Tree-epiphyte-ant relationships in low inundated forest of Sian Ka'an Bioreserve, Quintana Roo, Mexico. *Biotropica* 27: 57–70.

Diesel, R., and M. Schuh. 1993. Maternal care in the bromeliad crab *Metopaulias depressus* (Decopoda) maintaining oxygen, pH, and calcium levels optimal for larvae. *Behavioral Ecology and Sociobiology* 32: 1–15.

Diffenbaugh, H., and C. B. Fields. 2013. Changes in critical terrestrial climate conditions. *Science* 342: 486–492.

Frank, J. H. and I. P. Lounibos. 2009. Insects and allies associated with bromeliads: a review. *Terrestrial Arthropod Reviews* 1: 125–153.

Gentry, A. H. and C. H. Dodson. 1987. Diversity and biogeography of Neotropical vascular epiphytes. *Annals of the Missouri Botanical Gardens* 69: 577–593.

Heitz, P. and O. Briones. 1998. Correlation between water relations and within canopy distributions of epiphytic ferns in a Mexican cloud forest. *Oecologia* 114: 305–316.

Heitz-Seifert, U., P. Heitz, and S. Guevara. 1996. Epiphyte vegetation and diversity on remnant trees after forest clearance in southern Vera Cruz, Mexico. *Biological Conservation* 75: 103–111.

Hofstede, R. G. M., J. D. H. Wolf, and D. H. Benzing. 1993. Epiphytic biomass and nutrient status of a Colombian upper montane rainforest. *Selbyana* 13: 37–45.

Ingram, S. W. and N. M. Nadkarni. 1993. The composition and distribution of epiphytic organic matter in a Neotropical cloud forest, Costa Rica. *Biotropica* 25: 370–383.

Jackson, S. T. and D. F. Sax. 2009. Balancing biodiversity in a changing environment: extinction debt and immigration credit and species turnover. *Trends in Ecology and Evolution* 25: 153–160.

Kaufman, E. and U. Maschwitz. 2006. Ant gardens in tropical Asian forests. *Naturwissenschafften* 93: 216–227.

Nadkarni, N. M. 1981. Canopy roots: convergent evolution in rainforest nutrient cycles. *Science* 214: 1023–1024.

Nadkarni, N. M. and T. J. Matelson. 1989. Bird use of epiphytic resources in Neotropical montane cloud forest, Monteverde, Costa Rica. *Condor* 91: 891–907.

Nadkarni, N. M. and R. Solano. 2002. Potential effects of climate change on canopy communities in a tropical forest: an experimental approach. *Oecologia* 131: 580–586.

Paoletti, M., R. Taylor, B. Stinner, D. Stinner, and D. H. Benzing. 1991. Diversity of soil fauna in the canopy and forest floor of a Venezuelan cloud forest. *Journal of Tropical Ecology* 7: 373–383.

Pett-Ridge, J. and W. L. Silver. 2002. Survival, growth, and ecosystem dynamics of displaced bromeliads in a montane cloud forest. *Biotropica* 34: 211–224.

Rice, D. W., A. J. Alverson, A. O. Richardson, G. T. Young, V. M. Sanchez-Puerta, and others. 2013. Horizontal transfer of entire genomes via mitochondrial fusion in the angiosperm *Amborella*. *Science* 342: 1468–1473.

Richardson, B. A. 1999. The bromeliad microcosm and the assessment of faunal diversity in a Neotropical forest. *Biotropica* 31: 321–336.

Stuntz, S., U. Simon, and G. Zotz. 2002. Rain forest air-conditioning: the moderating influences of epiphytes on the microclimate in tropical tree crowns. *International Journal of Biometeorology* 46: 53–59.

Sugden, M. A. and R. J. Robins. 1979. Aspects of the ecology of vascular epiphytes in Colombian cloud forest. 1. The distribution of epiphytic flora. *Biotropica* 11: 173–188.

Wania, R., P. Heitz, and W. Wanek. 2002. Natural abundance of N^{15} depends on position within the forest canopy: sources, signals, and isotopic fractionation. *Plant Cell and Environment* 25: 581–589.

Watson, D. M. 2001. Mistletoe—a keystone resource in forests and woodlands world wide. *Annual Review of Ecology and Systematics* 32: 219–249.

Watson, D. M., and M. Herring. 2012. Mistletoe as a keystone resource: an experimental test. *Proceedings of the Royal Society B* 279: 3853–3860.

Wilson, E. O. 1987. The arboreal ant fauna of Peruvian Amazonian forests: a first assessment. *Biotropica* 19: 245–282.

Winter, K., B. J. Wallace, G. C. Stocker, and Z. Roksandic. 1983. Crassuleceqan Acid Metabolism in Australian epiphytes and some related species. *Oecologia* 57: 129–141.

Woods, C. L. and S. J. DeWalt. 2012. The conservation value of secondary forests for vascular epiphytes. *Biotropica* 45: 119–127.

Youngsteadt, E., S. Nojima, C. Häberlein, S. Schult, and C. Schal. 2008. Seed odor mediates an obligate ant-plant mutualism in Amazonian rain forests. *Proceedings of the National Academy of Sciences USA* 105: 4571–4575.

Yu, W. 1994. The structural role of epiphytes in ant gardens *Biotropica* 26: 217–221.

Zotz, G. 2005. Vascular epiphytes of the temperate zone—a review. *Plant Ecology* 176: 173–183.

15
INSECTS IN FOREST ECOSYSTEMS

Andrea Battisti

Introduction

The greatest part of the terrestrial planet has been covered by forest biomes for a long time and over a large variety of climatic conditions. This has created enormous possibilities for the animal community to diversify and to adapt to local conditions at different spatial scales. Arthropods, and in particular insects, play a major role in the community because of their great potential to adapt and to build up trophic networks linking below-ground and above-ground communities. Latitude and elevation are the main factors affecting the distribution and abundance of animal communities in terrestrial habitats, being related by the specific lapse rate of temperature (0.6°C every 100 m of elevation and 1° of latitude) (Hodkinson, 2005). The distribution and abundance of insects are mainly driven by temperature gradients associated with latitude and elevation, although topography may play a decisive role at local scale (Figure 15.1). As a consequence, communities associated with mountain forest ecosystems are generally more fragmented than those associated with forests of low elevation, even if growing across a wide latitudinal span.

The human impact on the structure of forests and the animal communities associated with them has increased progressively and dramatically from the last glaciation to the current, ongoing climate change. In the temperate zone, transformation of forest into agricultural land and intensive silviculture have changed the landscape dramatically, resulting in fragmentation of the distribution range which is ultimately responsible for the present structure of insect communities. In addition, the plantation of tree species outside their native range, either intentional or unintentional, has promoted the spread and the abundance of particular species which found very favourable conditions in the new stands, becoming pests with great economic impact. The human-induced changes in forest ecosystems have also affected the communities of arthropods associated with soil and responsible for the turnover of nutrients, with different types of feedback on tree growth and ecosystem persistence. A typical example being the case of forest ants, 'ecosystem engineers' whose activity is directly affected by human activity and climate, becoming less abundant in simplified ecosystems and in colder habitats (Hölldobler and Wilson, 1990; Underwood and Fisher, 2006).

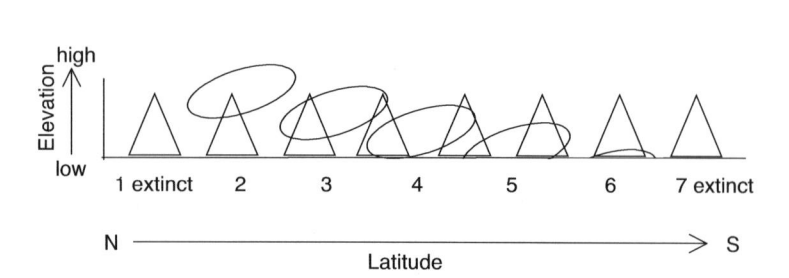

Figure 15.1 Effects of latitude and elevation on the distribution of organisms in the northern hemisphere (modified from Gorodkov, 1985). The horizontal axis represents the latitude and the vertical axis the elevation. The triangles along the horizontal line identify mountains occurring at different latitudes. The oblique lenses represent the potential range of the organism as determined by climate (both inclination and width of the lens may vary depending on the reaction norms of the individual species). The potential range may shift from south to north, and from low to high elevation, and vice versa, depending on the warming and cooling of the climate. Gorodkov exemplified different types of distribution which can be observed in relation to the position of the lens along the gradient, i.e. 2: montane local (endemic); 3: montane wide; 4: disjunct plain-montane; 5: continuous plain-montane; 6: plain local (endemic). Types 1 and 7 tend to extinction. Types 4 and 5 are the most common and are associated with wider geographic distribution

Biodiversity

There is little doubt that insects represent the bulk of biodiversity of forest ecosystems, occupying with a huge number of species all the available ecological niches associated with trees (Figure 15.2). It is noteworthy that estimates on the total number of species inhabiting the planet are made by censuses in especially rich tropical forests (Basset *et al.*, 2012), where many taxa are still undescribed. The functional role of biodiversity and the ecosystem services associated with it are acquiring more importance, together with the consciousness that human activities and climate change are threatening the survival of many important species. Fragmentation and ecotonization of the habitats are thought to be the main reasons for reduction of biodiversity, especially for species characterized by narrow niches and reduced mobility.

The intrinsic value of biodiversity is easy to understand, however its study requires large efforts and the availability of expertise which is becoming rare. The census of the species and their abundance, the knowledge of the relationships with the other organisms are essential points for monitoring the status of forest ecosystems in space and time. The phylum of arthropods is by far the most diverse with more than 1.2 million species described, mainly from terrestrial ecosystems, with an estimated number of species between 2.9 and 12.7 million (Hamilton *et al.*, 2013). The success of this group relies on small size (mm to a few cm), high access to oxygen, fast reproduction both sexual and asexual, high adaptive potential to microniches and to specialization in relation to feeding habit; its origin goes back in time to the beginning of diversification of plants in the primary era. Insects are the most species-rich group among arthropods, followed by mites, and both are very important for the functioning of the forest ecosystems, and in general for the survival of many species of plants and animals which are totally dependent on them, including man (Wilson, 1992).

1	2	3	4	5	6	7

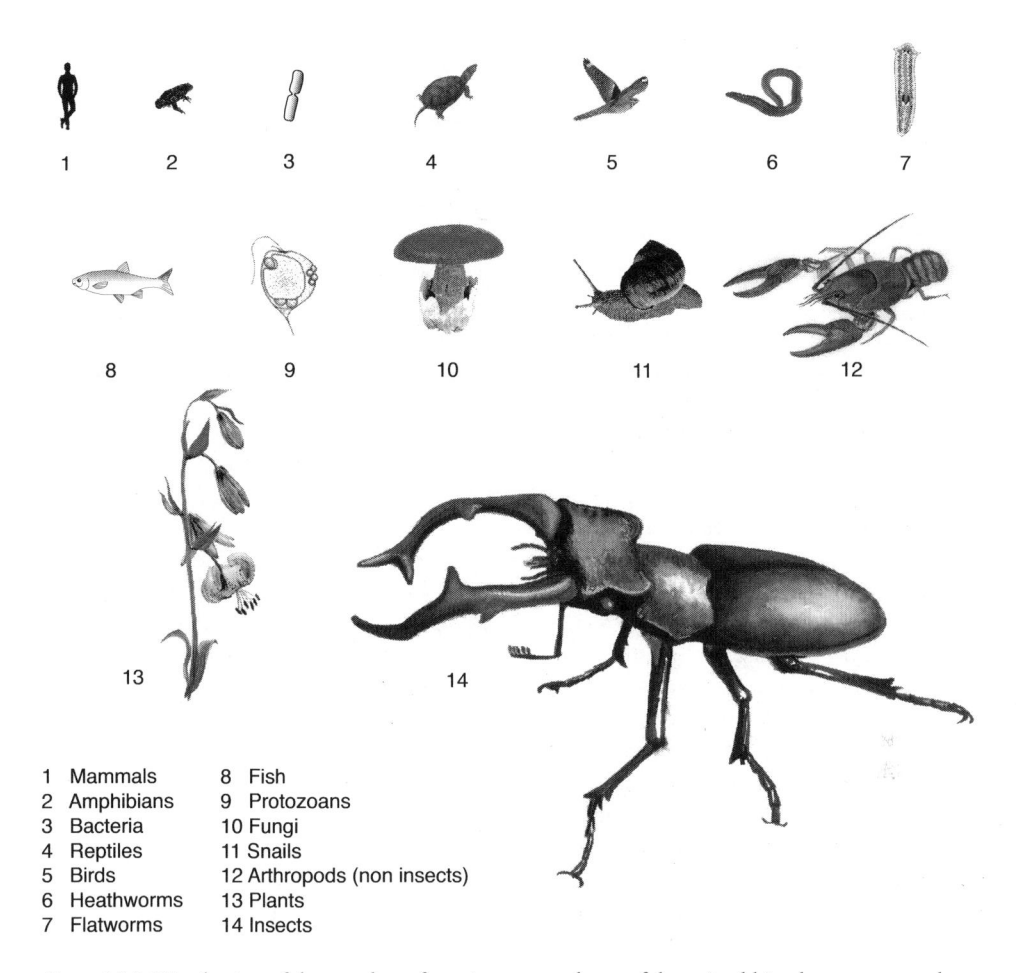

8	9	10	11	12	

13 14

1 Mammals	8 Fish
2 Amphibians	9 Protozoans
3 Bacteria	10 Fungi
4 Reptiles	11 Snails
5 Birds	12 Arthropods (non insects)
6 Heathworms	13 Plants
7 Flatworms	14 Insects

Figure 15.2 Distribution of the number of species among classes of the animal kingdom represented through the relative size of a component of each group (redrawn from Speight *et al.*, 2008)

Insects and forests

The easiest and more functional way to classify herbivorous insects associated with forest trees is their grouping by feeding guilds (Table 15.1) and then by the level of specialization. In general, monophagous species are associated exclusively with one genus of host plants, oligophagous with one family, and polyphagous with more than one family. A similar type of specialization and classification can be used for the higher trophic level, i.e. that of predators and parasitoids of herbivorous insects, although the specialization gets weaker at higher levels of the trophic network.

The specialization and niche adaptation of forest insects is generally explained by the avoidance of competition among organisms interested in exploiting the same part of the plant. This can be seen in space as well as in time, for example with the shifting of the distribution range or the phenological window of activity, respectively. It may concern also the relationships between insect herbivores and their natural enemies, which often act as major regulating factors (regulation from above or top-down). In general, however, the first hurdle to being an insect herbivore in a forest is related to the scarce nutrient content of plant tissues, or to the occurrence

Table 15.1 Major feeding guilds of main insect orders

Feeding guilds	Orthoptera	Heteroptera	Homoptera	Lepidoptera	Diptera	Coleoptera	Hymenoptera
Sap suckers		X	X				
Gall makers			X		X		X
Leaf feeders	X			X	X	X	X
Shoot feeders				X		X	
Wood borers				X		X	X
Cone and seed		X	X	X	X	X	X
Parasitoids					X		X
Predators	X	X			X	X	X

of compounds which may deter insect feeding (bottom-up regulation). Lignin and cellulose are the most abundant components of plant tissues and are not exploited by the large majority of forest insects, with a few exceptions observed when complex symbiotic interactions with micro-organisms occur. Green tissues such as leaves and shoots are generally more palatable but they occur for a short period of time, and often become unpalatable because of hardening or replenishing with secondary metabolites such as resins and phenols. Liquids occurring inside the cells or in the vascular systems are generally rich in nutrients, although these are difficult to access and highly specialized mouthparts are required. Other plant tissues produced for the sake of reproduction, such as flowers, seed, and fruits, are also exploited by herbivorous insects, which have to face a highly irregular occurrence of this resource in space and in time, often becoming very limiting for their success.

As plants are unpromising food for insects, because they largely comprise indigestible structural compounds, insects have evolved many different strategies to feed on plants including associations with mutualistic symbionts, which can be important mediators of direct and indirect interactions between herbivorous insects and their host plants (Schoonhoven et al., 1998). Answers as to how micro-organisms can tilt the balance in favour of the insect are provided by recent advances in our mechanistic understanding of the signalling pathways and biochemical responses involved in plant defence and insect counterattack, and how these responses can be modulated by both pathogenic or non-pathogenic micro-organisms. Finally, how insect mutualists interact with plants is potentially of major applied significance. The invasion of new pests has often been facilitated by their mutualists and some novel interactions have resulted in new and more virulent insect pests. Manipulating symbionts is an open avenue to improve pest control (Frago et al., 2012).

In spite of these limitations, insects may sometimes become so abundant that their host plants can be completely deprived of their functional tissues such as the leaves and may die, or the insects may directly attack the vascular system and cause a quick dieback. In general, however, the consumption of plant tissues by insects is limited and not clearly perceived by a non-expert until a threshold is reached which is between 5 and 10 per cent of the biomass, depending on the tree species. It is thought that this amount of herbivory is functional to the nutrient cycling in the ecosystem and in this way can be considered as beneficial for tree growth (Mattson and Addy, 1975). Consumption of roots certainly exists but it is unfortunately little documented, with very few species ranked as pests (Wainhouse, 2005). Arthropods, however, are extremely abundant in the soil and play an essential role in the decomposition of soil organic matter and dead wood. They may contribute, together with the earthworms (Annelida), to the breaking down of the litter and its incorporation in soil, where the process will be completed by mites (Acarina) and springtails (Collembola), in cooperation with a large number of soil micro-organisms (Speight *et al.*, 2008). In the well-studied experimental basin of Hubbard Brook (New Hampshire USA) it has been shown that recurrent interannual insect defoliation causes a variance in the distribution of biomass across trophic levels, implying a variation in plant nutritional quality. Synchronous fluctuations of a main consumer guild (such as foliage feeding Lepidoptera) involved interannual variation in soil mineral nutrient availability (Stange *et al.*, 2011), which in turn is able to modify the leaf chemistry and thus the susceptibility to herbivores (Frost and Hunter, 2004 and 2008). The variation in the amount and quality of litter is likely to affect the structure and functionality of the soil fauna, although this is an aspect of forest ecology which has not been thoroughly considered and deserves attention in the future. In general, there is a need to characterize better the links between above-ground and below-ground communities of insects and associated organisms, which are mediated by the host plant responses (Wardle *et al.*, 2004).

Ecosystem services are also affected by forest insect pests (Boyd *et al.*, 2013) in different ways. A reduction in tree health affects the value placed on the forest resources by the society. This may happen in a number of ways and can have very different and unexpected impacts. For example, iconic tree species such as chestnuts (*Castanea* spp.) and elms (*Ulmus* spp.) can be swept away by combinations of endemic and exotic pests and pathogens, or an outbreak of urticating insects may affect humans even far away from defoliated areas (Battisti *et al.*, 2011). The non-monetary nature of cultural services means it is difficult to track their value, and this can lead to unexpected and apparently rapid shifts in the public value placed on them. Social amplification is a term used to describe dramatic switches in public concern, often caused by positive feedback through new and old media. Such events can greatly increase the pressure on policy-makers to respond to threats to cultural services (Boyd *et al.*, 2013).

Population dynamics

Outbreaks of forest insects are common in all types of forest and may cause dramatic consequences for conservation of the habitat and for the human economy (Barbosa *et al.*, 2012). Insect abundance can be generally explained by both endogenous and exogenous factors. The endogenous factors generally indicated are those related to the biology of the species (e.g. fecundity, voltinism, sex ratio, dispersal), the host plant (e.g. susceptibility, tolerance, resistance), the competitors and facilitators (e.g. other herbivores of the same guild, associated micro-organisms), and natural enemies (e.g. parasitoids and predators), while the only exogenous but very important factor is the weather. Of course, interactions between endogenous factors are numerous and sometimes very complex, in addition all of them are affected by the weather. The idea that the weather is a major driver of the upsurge of population density over large and different areas was developed by Moran (1953) and is still in need of study (Liebhold *et al.*, 2004).

Endogenous factors may generate predictable responses if they are density-dependent, and for this reason are of great value to managers for taking decisions about possible actions to counteract the population growth and to limit the impacts on the ecosystems. Density-dependence may occur over a different time (e.g. months or years) in relation to the nature of the regulation mechanism involved. There are models which try to combine general density-dependent regulation with variations of the weather, especially in relation to climate-change predictions (Barbosa *et al.*, 2012).

Outbreaks of forest insects have been studied for several species and various explanations have been attempted. In some cases it has been possible to link the increase of the population density with a change in some environmental variables which were considered to positively affect the growth, or to negatively affect the action of regulation factors. In other cases a periodicity of the outbreak event has been observed and explained with complex mechanisms, with a period variable between a few years up to decades. Density-dependence has been shown in a few cases to be responsible for such a periodicity, although the precise mechanisms are often unknown. Cyclic population behaviour may be manifested as synchronized or non-synchronized fluctuations between disjunct populations. A third behaviour that has recently received much attention is travelling waves in populations of periodic species. Travelling waves means that the population peak densities move in one main direction across space. In the winter moth (*Operophtera brumata*) a continental-scale wave has been shown in Europe, where outbreaks occur every nine to ten years over a distance of nearly 4,000 km (Tenow et al., 2013). Similarly, the larch budmoth (*Zeiraphera diniana*) population waves travel across the European Alps with a wave speed of 220–250 km per year and a cycle length of eight to nine years (Bjørnstad et al., 2002). Although periodic travelling waves in ecology have become increasingly recognized in several fields of study, their origin is unknown. Theory concerns 'reaction–diffusion' models, and in the insect case trophic interactions (e.g. predator-prey or herbivore–plant) and dispersal (spatial movement of reactants or individuals) are involved (Tenow et al., 2013). In the larch budmoth model, an interaction with parasitoids predicts directional waves if dispersal is unidirectional or habitat productivity varies across the landscape (Bjørnstad et al., 2002). For the same system, a tri-trophic model demonstrated that landscape gradients in favourable habitats alone could induce waves from epicentres (Johnson, Bjørnstad and Liebhold, 2004). Both are applicable to the flying *Z. diniana* but not to the flightless *O. brumata*. In the latter case, the connectivity of stands across Europe and the resistance of host plants irrespective of species, interplaying with defoliator enemies, should be key interactants in creating both the nine- to ten-year cyclicity and the travelling waves (Tenow et al., 2013). It should be noted that these regular cycles have been observed in relatively 'simple' treeline forest ecosystems, such as European larch forests in the Alps (Figure 15.3) and mountain birch forests in Fennoscandia, whereas they are less regular in other types of forest ecosystems.

Sometimes the outbreak may become destructive for the ecosystem and start a succession phase which may last for several decades. In these cases the insects generally kill the trees over large areas, and often become so numerous that they overcome the resistance mechanisms and spread over large territories. The largest event of this type ever recorded is the outbreak of the mountain pine beetle *Dendroctonus ponderosae* in north-western America, which has affected more than 15 million ha of coniferous forest (Kurz et al., 2008) and is expanding further.

Forest insects may also participate in outbreaks of plant pathogens, especially when they act as vectors. The association can be purely neutral or may benefit one component of the system, or both (mutualistic symbiosis). Numerous examples are associated with diebacks of tree species over large areas, as in the case of elms (*Ulmus* spp.) and cypress (*Cupressus* spp.). The relationships seem to be very important, especially for pathogens outside of their native range, as in the recent case of pine wood nematode, *Bursaphelenchus xylophilus*.

Natural enemies are very important for the regulation of populations at both low and high density, and they are often taken into consideration for pest management programmes. Insect parasitoids are extremely common and diverse; they differ from true parasites because they always cause the host to die, being in this way more close to predators. As they need the host for the development of their progeny, they have developed sophisticated mechanisms of host location and finding, which makes them rather specific and reliable for biological control actions. Because of the particular life history, they are characterized by a delay in the regulation, which is largely predictable in population models. Parasitoids can be specialized on every developmental stage of the host, adjusting their size accordingly. Most species are included in

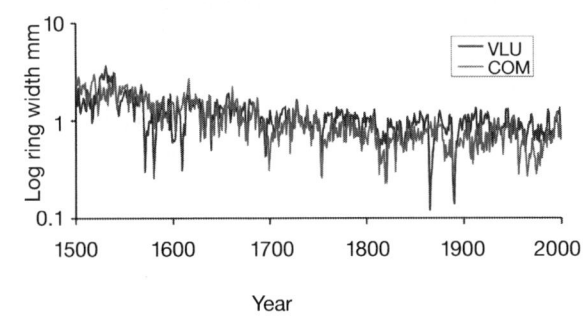

Figure 15.3 Section of a larch trunk from the Italian Central Alps with the indication of the growth rings periodically reduced, as shown by the graph below representing two different sites. Reductions correspond to outbreaks of the larch budmoth *Zeiraphera griseana* (data kindly provided by Marco Carrer, University of Padova)

only two insect orders, i.e. Diptera and Hymenoptera, where there are families which are entirely specialized in this behaviour. Conversely, insect predators are less specialized and occur in very many taxonomic groups. Some common predators like ants may behave as such only occasionally, so that they are less reliable than parasitoids for biological control. Other predators of insect herbivores are vertebrates, especially birds. These animals seem to play an important role to keep populations at low density, while they are unable to take numbers down in cases of outbreak. Finally, insect pathogens such as viruses, bacteria, fungi, and nematodes, may affect the population dynamics and can also be used for direct control at high host density. Viruses and bacteria need to be ingested with the food and are generally specific, especially for viruses. They may cause spectacular collapses of the host populations when densities are high. Entomopathogenic fungi are common agents of mortality in the soil or under the bark, and more generally when the insects are protected under high moisture conditions. The mode of action is based on the penetration of insect cuticle and the mechanism is generic. Entomopathogenic nematodes are associated with the water film in the soil or in the wood, where they actively search for their hosts. These are killed by bacteria carried by the nematodes which ultimately thrive as saprophagous. They are quite generic in their action so that they can be multiplied on laboratory hosts and then released against target insects in the field.

Insects in a changing environment

In addition to the intrinsic vulnerability of the trees (see Chapters 7 and 16 of this volume), the forest insect communities can be strongly affected by several environmental factors, some of which are linked to forest management. Accumulating evidence of climate change over the past decades has fuelled debate and research regarding its ecological consequences. Meteorological data confirm general increases in ambient temperatures and changes in precipitation patterns, which have been connected to changes in species' ranges (Parmesan et al., 1999; Battisti et al., 2005), advanced phenology (Walther et al., 2002), and phenological asynchrony among interacting species (Visser and Both, 2006). Much less is known about consequences for population dynamics, but it seems likely from first principles that there would be effects for some species on reproductive success, life histories, mean abundance, and variation in abundance.

Insect herbivores, being generally poikilothermic, are highly sensitive to changes in their thermal environment (Uvarov, 1931). Furthermore, responses of insect herbivores are likely based on their life cycle in combination with the life cycle of their host plant (Bale et al., 2002). Development rate generally increases with increasing temperature to some maximum, above which there are frequently sharp increases in mortality associated with the decreasing development rate (Uvarov, 1931). The development rate of insects in mid- to high latitude systems should increase with climate warming as ambient temperatures are well below that which permits maximum rates of feeding and growth (Hodkinson, 2005).

An increased development rate could lead to increased voltinism in facultative multivoltine species, as predicted for several bark beetle species. Recent research has found increased voltinism in macrolepidopteran species thought to be strictly uni- or bivoltine, as a consequence of climate warming (Altermatt, 2010). Winter mortality is likely to decrease under increasing temperatures (Ayres and Lombardero, 2000), although decreased snow cover (and therefore decreased insulation of overwintering sites) can reverse the pattern. Warmer winters may permit some non-diapausing species to continue feeding and developing during months that were previously too cold. For example, larvae of the pine processionary moth (*Thaumetopoea pityocampa*) have a higher probability of survival if winter temperatures do not often fall below specific feeding thresholds (Battisti et al., 2005).

A first case of indirect effects goes through the host plant, with the lengthening of the growing season (especially in spring) for plants at mid- to high latitudes. As many insect species evolved to match their feeding activity with certain developmental stages in the host plant, it is likely that insect herbivores are quite efficient at matching the start of their feeding to the time of budburst (Visser and Both, 2006). How well this synchrony is maintained under climate warming will depend on how well the physiological controls on insect development match those of their host plants (Ayres, 1993). The primary and secondary metabolites of woody plants can be highly responsive to environmental factors (Herms and Mattson, 1992). Climate change will produce situations where trees are poorly matched to the new conditions, leading to an increase in frequency of stressful conditions in plants. Stressed plants frequently undergo biochemical changes (Koricheva et al., 1998) which may affect survival and reproduction of some forest insects. That plant stress can trigger insect outbreaks is a long-standing hypothesis in forest entomology, although there are variable consequences for insects because of partly nonlinear responses of plant defences to water and nutrient availability (Herms and Mattson, 1992) and to differential effects of environmental conditions on constitutive versus inducible plant defences (Lombardero et al., 2000). Finally, the increase in atmospheric carbon dioxide (CO_2) promotes photosynthesis and leads to an increase of the C/N ratio which makes the plant tissues less attractive to pests. In addition, increasing CO_2 suppresses the production of jasmonates and ethylene and increases the production of salicylic acid, and these differential responses of plant hormones affect specific secondary chemical pathways. In addition to changes in secondary chemistry, elevated CO_2 decreases rates of water loss from leaves, increases feeding rates, and alters nutritional content (Zavala et al., 2013). The completion of the cycle of some insects may therefore take longer than under a milder climate, and the time during which the pests remain exposed to natural enemies is longer.

A second example of indirect effects involves their natural enemies, which can exert powerful forces on the population dynamics of herbivorous insects. Assessing their response to climate warming, and understanding the consequences for density-dependent feedback systems, is crucial to understanding possible shifts in outbreak frequencies in the future. The distribution and abundance of natural enemies, like that of their prey, can be influenced by direct climatic effects, but temperature sensitivity might increase with trophic level (Berggren et al., 2009), in which case natural enemies would be more affected than the herbivores on which they feed. For arthropod parasitoids and predators, phenological synchrony can affect host/prey availability and host/prey size at the time of attack. Specialist enemies should be under strong selection to track phenological changes in their prey, which might make them less likely to become temporally uncoupled from their prey (Klapwijk et al., 2012). Increases or decreases in the phenological match between parasitoids and hosts could change the population dynamics of enemies through effects on prey attack rates and conversion of attacks into fecundity. Insect pathogens, e.g., fungi, bacteria and viruses, play an important role in the dynamics of many insect populations. Temperature can be important in both infection rate and defence responses within the host, with different thermal optima for host and pathogen. Fungal pathogens tend to be favoured by increases in humidity, especially when temperatures are high, but not too high. Outside of their host, resting stages of nuclear polyhedral viruses are vulnerable to UV-radiation, so increases in UV-radiation could influence viral controls on some insect populations. As with specialist predators and parasitoids, the most pronounced consequences of climatic effects on herbivore–pathogen interactions may be in the decline phase of outbreaks.

In conclusion, the direct effects of climatic change on the herbivorous insect will be generally positive as a result of increased winter survival, faster development rates, and sometimes increased number of generations per year. Some insect herbivores will benefit from an increased frequency of stressful events for plants, but others will not, depending on how tree defences are affected by the changing climate. Insects that exploit young, nutritionally favourable foliage may be sensitive to climatically driven changes at the start of the growing season and to temperatures that influence the relative rates of leaf maturation and insect development after budburst. The nature of physiological controls on plant versus insect phenology will influence the extent to which plant and insect phenology remains coupled under climate change. Increased activity in arthropod natural enemies could lead to increased strength of top-down control on the herbivorous insect. The magnitude of the effect, however, will depend on the attack strategy of predators and host usage in parasitoids. Pathogens are likely to increase their infection rate under increased humidity and temperature. An important step to consider is how climatic effects on insect individuals can be scaled up to effects on species range and population dynamics. Predictions regarding the dynamics of populations under climate change require integration of the direction and strength of direct and indirect effects.

The geographic range of herbivorous insects can change over years and decades based on changes in climate, among other things (Gaston, 2003). Although adequate data are only available for some species in some regions, there are quite clear signals of recent pole-ward range expansion in numerous insect species (Parmesan et al., 1999), including outbreaking forest insects, such as the pine processionary moth and bark beetles in North America (Battisti and Larsson, in press). Whether the new outbreaks occur within or beyond the current range, candidate explanations are the same: release from adverse climatic conditions, natural enemies, competition with other herbivores, and host plant resistance.

The overall consequences of climate change on the vulnerability of forests to attack by insect pests are still under investigation and are difficult to predict. There is still considerable uncertainty, as it is difficult to connect the individual to the population responses, which is a crucial step for predicting the extent of possible attacks. According to the most reliable theoretical models, climate change may lead to increases in both frequency and magnitude of insect outbreaks, depending on the characteristics of each species and the type of forest, while it is evident that climate change is leading to considerable changes in the distribution of all organisms, including trees.

Climate change also has consequences that seem to be going in the opposite direction. The increase in atmospheric carbon dioxide promotes photosynthesis and leads to an increase of the C/N ratio which makes the plant tissues less attractive to pests. The completion of the cycle of some insects may therefore take longer than under a milder climate, and the time during which the pests remain exposed to natural enemies is longer. Natural enemies have also been predicted to benefit from climate change more than herbivorous insects, leading to better control. In addition, it is therefore appropriate to consider thoroughly the actions relating to the choice or selection of tree species in the forests of temperate zones in the near future. For the same reasons it is necessary to intensify the attention on plant health and increase the knowledge about both native and invasive pest species.

References

Altermatt, F. (2010) 'Climatic warming increases voltinism in European butterflies and moths', *Proceedings of the Royal Society B – Biological Sciences*, vol 277, pp. 1281–1287

Ayres, M. P. (1993) 'Plant defense, herbivory and climate change', in P. Kareiva, J. G. Kingsolver and R. B. Huey (eds) *Biotic Interactions and Global Change*, Sinauer, Sunderland MA, US

Ayres, M. P. and Lombardero, M. J. (2000) 'Assessing the consequences of global change for forest disturbance from herbivores and pathogens', *Science of the Total Environment*, vol 262, pp. 263–286

Bale, J. S., Masters, G. J., Hodkinson, I. D., Awmack, C., Bezemer, T. M. and others (2002) 'Herbivory in global climate change research: direct effects of rising temperature on insect herbivores', *Global Change Biology*, vol 8, pp. 1–16

Barbosa, P., Letourneau, D. K. and Agrawal, A. A. (2012) *Insect Outbreaks Revisited*, Academic Press, New York

Basset, Y., Cizek, L., Cuenoud, P. (2012) 'Arthropod diversity in a tropical forest', *Science*, vol 338, pp. 1481–1484

Battisti, A. and Larsson, S. (in press) 'Climate change and pest insect distribution range', in C. Bjorkman and P. Niemela (eds) *Climate Change and Insect Pests*, CABI International

Battisti, A., Stastny, M., Netherer, S., Robinet, C., Schopf, A. and others (2005) 'Expansion of geographic range in the pine processionary moth caused by increased winter temperatures', *Ecological Applications*, vol 15, pp. 2084–2096

Battisti, A., Holm, G., Fagrell, B. and Larsson, S. (2011) 'Urticating hairs in arthropods – their nature and medical significance', *Annual Review of Entomology*, vol 56, pp. 203–220

Berggren, Å., Björkman, C., Bylund, H. and Ayres, M. P. (2009) 'The distribution and abundance of animal populations in a climate of uncertainty', *Oikos*, vol 118, pp. 1121–1126

Bjørnstad, O. N., Peltonen, M., Liebhold, A. M. and Baltensweiler, W. (2002) 'Waves of larch budmoth outbreaks in the European Alps', *Science*, vol 298, pp. 1020–1023

Boyd, I. L., Freer-Smith, P. H., Gilligan, C. A. and Godfray, H. C. J. (2013) 'The consequence of tree pests and diseases for ecosystem services', *Science*, vol 342, p. 823

Frago, E., Dicke, M. and Godfray, H. C. J. (2012) 'Insect symbionts as hidden players in insect–plant interactions', *TREE*, vol 27, pp. 705–711

Frost, C. J. and Hunter, M. D. (2004) 'Insect canopy herbivory and frass deposition affect soil nutrient dynamics and export in oak mesocosms', *Ecology*, vol 85, pp. 3335–3347

Frost, C. J. and Hunter, M. D. (2008) 'Insect herbivores and their frass affect *Quercus rubra* leaf quality and initial stages of subsequent litter decomposition', *Oikos*, vol 117, p. 1322

Gaston, K. J. (2003) *The Structure and Dynamics of Geographic Ranges*, Oxford University Press, New York

Gorodkov, K. B. (1985) 'Three-dimensional climatic model of potential range and some of its characteristics. I.', *Entomologicheskoye Obozrenyie* (translated: Entomological Review, Scripta Technica 1986), vol 64, pp. 295–310

Hamilton, A. J., Novotny, V., Waters, E. K., Basset, Y., Benke, K. K. and others (2013) 'Estimating global arthropod species richness: refining probabilistic models using probability bounds analysis', *Oecologia*, vol 171, pp. 357–365

Herms, D. A. and Mattson, W. J. (1992) 'The dilemma of plants: to grow or defend', *Quarterly Review of Biology*, vol 67, pp. 283–335

Hodkinson, I. D. (2005) 'Terrestrial insects along elevation gradients: species and community responses to altitude', *Biological Reviews of the Cambridge Philosophical Society*, vol 80, pp. 489–513

Hölldobler, B. and Wilson, E. O. (1990) *The Ants*, Harvard University Press, Cambridge MA, US

Johnson, D. M., Bjørnstad, O .N. and Liebhold, A. M. (2004) 'Landscape geometry and travelling waves in the larch budmoth', *Ecology Letters*, vol 7, pp. 967–974

Klapwijk, M. J., Ayres, M. P., Battisti, A. and Larsson, S. (2012) 'Assessing the impact of climate change on outbreak potential', in P. Barbosa, D. K. Letourneau and A. A. Agrawal (eds) *Insect Outbreaks Revisited*, Academic Press, New York

Koricheva, J., Larsson, S. and Haukioja, E. (1998) 'Insect performance on experimentally stressed woody plants: a meta-analysis', *Annual Review of Entomology*, vol 43, pp. 195–216

Kurz, W. A., Dymond, C. C., Stinson, G., Rampley, G. J., Neilson, E. T. and others (2008) 'Mountain pine beetle and forest carbon feedback to climate change', *Nature*, vol 452, pp. 987–990

Liebhold, A., Koenig, W. D. and Bjørnstad, O. A. (2004) 'Spatial synchrony in population dynamics', *Annual Review Ecology Evolution Systematics*, vol 35, pp. 467–490

Lombardero, M. J., Ayres, M. P., Lorio, P. L. Jr. and Ruel, J. J. (2000) 'Environmental effects on constitutive and inducible resin defences of *Pinus taeda*', *Ecology Letters*, vol 3, pp. 329–339

Mattson, W. J. and Addy, N. D. (1975) 'Phytophagous insects as regulators of forest primary production', *Science*, vol 190, pp. 515–522

Moran, P. A. P. (1953) 'The statistical analysis of the Canadian lynx cycle', *Australian Journal of Zoology*, vol 1, pp. 163–173

Parmesan, C., Ryrholm, N., Stefanescu, C., Hill, J. K., Thomas, C. D. and others (1999) 'Poleward shifts in geographical ranges of butterfly species associated with regional warming', *Nature*, vol 399, pp. 579–583

Schoonhoven, L.M., Jermy, T. and van Loon, J. J. A. (1998) *Insect-plant biology*, Chapman & Hall, London

Speight, M., Hunter, M. and Watt, A. (2008) *Ecology of Insects: Concepts and Applications*, Second edition, Wiley-Blackwell, Oxford, UK

Stange, E. E., Ayres, M. P. and Bess, J. A. (2011) 'Concordant population dynamics of Lepidoptera herbivores in a forest ecosystem', *Ecography*, vol 34, pp. 772–779

Tenow, O., Nilssen, A., Bylund, H., Pettersson, R., Battisti, A. and others (2013) 'Geometrid outbreak waves travel across Europe', *Journal of Animal Ecology*, vol 82, pp. 84–95

Underwood, E. C. and Fisher, B. L. (2006) 'The role of ants in conservation monitoring: if, when, and how', *Biological Conservation*, vol 132, pp. 166–182

Uvarov, B. P. (1931) 'Insects and climate', *Transactions of the Royal Entomological Society of London*, vol 79, pp. 1–232

Visser, M. E. and Both, C. (2006) 'Shifts in phenology due to global climate change: the need for a yardstick', *Proceedings of the Royal Society of London Series B – Biological Sciences*, vol 272, pp. 2561–2569

Wainhouse, D. (2005) *Ecological Methods in Forest Pest Management*, Oxford University Press, Oxford, UK

Walther, G. R., Post, E., Convey, P., Menzel, A., Parmesan, C. and others (2002) 'Ecological responses to recent climate change', *Nature*, vol 416, pp. 389–395

Wardle, D. A., Bardgett, R. D., Klironomos, J. N., Setälä, H., van der Putten, W. H. and Wall, D. H. (2004) 'Ecological linkages between aboveground and belowground biota', *Science*, vol 304, pp. 1629–1633

Wilson, E.O. (1992) *The Diversity of Life*, Harvard University Press, Harvard, US

Zavala, J. A., Nabity, P. D. and DeLucia, E. H. (2013) 'An emerging understanding of mechanisms governing insect herbivory under elevated CO_2', *Annual Review of Entomology*, vol 58, pp. 79–97

16
PATHOGENS AND PESTS IN NORTH AMERICAN FOREST ECOSYSTEMS

Louis Bernier and Sandy M. Smith

Introduction

Trees are usually thought of as the organisms that define forests because of their overriding biomass. Trees, however, account for only a small part of the biodiversity of forests. Microbes and invertebrates are much larger contributors to biodiversity. Collectively, they also perform a variety of functions that are critical to the continuous functioning of forest ecosystems. Bacteria, fungi and invertebrates are key players in the development and fertility of forest soils. Living trees are hosts to and form a range of associations with a wide variety of endophytic microbes. Sexual reproduction of trees is greatly aided by insect pollinators. Once trees die, a succession of wood boring invertebrates, bacteria, and decay fungi allows for the recycling of the lignocellulosic complex, and in so doing, forms the widely recognized complexity of biodiversity associated with the world's forests.

Living trees are also impacted by a variety of disease-causing microbial pathogens and by invertebrates (often referred to as pests) that feed on organs or tissues, either aerial or below ground. Depending on the combination of host, pathogen/pest and environmental conditions, the outcome may vary from no observable effect to decreased growth to tree death. The effect of pathogens and pests may be observable at different scales, from individuals to species, from communities to forest stands. Losses caused by pathogens and pests range from the inconspicuous to the spectacular, from short-term catastrophic tree mortality to long-term successional changes in forest ecosystems. Although this may look counterintuitive to forest managers, pathogens and pests are important drivers of forest composition, succession and evolution through disturbance. They contribute to stand structure, availability of deadwood, and creation of suitable habitat for plants and wildlife. Consequently, the older utilitarian concept of healthy forests characterized by low tree mortality and an absence of forest diseases and pests has gradually given way to an ecosystem-based concept, one that recognizes that disease and pests are natural components of healthy forests, and in fact define them (Kolb *et al.*, 1994; Castello *et al*, 1995; Ostry and Laflamme, 2009). Today, the diversity of these biological communities, often formally termed biodiversity, is quantified and converted to indices that measure forest health.

Unfortunately, pathogens and invertebrate pests may also jeopardize the resilience of forest ecosystems. This phenomenon is becoming increasingly apparent with the intensification of management and anthropogenic activities that disturb natural ecosystems. Forestry practices, such as silviculture, management and harvesting, all have a direct effect on the structure and function of forest ecosystems. Indirect anthropogenic effects are also a cause for concern, particularly in the case of invasive pathogens, pests and weeds. When these diverse communities are disrupted and replaced with simpler, more uniform complexes, the forest ecosystem is destabilized and its resilience compromised. Already, a combination of direct and indirect factors has pushed some forest ecosystems over the tipping point, after keystone tree species were destroyed or cascading effects were set in motion (Kenis et al., 2009).

In this chapter, we discuss how pathogens and pests shape the structure and functions of forest ecosystems. We address this issue both in the 'traditional' context in which natural forests are interacting with native pathogens and insect pests with which they have co-evolved, as well as in the current context of intensive forest management and plantation-based forestry, biological invasions, and climate change.

Biological features of pathogens and pests

Prior to discussing the effect of pathogens and pests, it is important to recall some of the salient features of these organisms. This is also a good opportunity to introduce terminology used in the forest health disciplines to readers who may not be familiar with this field.

Pathogens are biotic agents that cause disease. They include viroids, viruses, bacteria, phytoplasmas, fungi, nematodes and parasitic plants. With the exception of nematodes, invertebrates are not considered pathogens. Disease refers to any malfunctioning of host cells or tissues that result from continuous irritation by a pathogen or environmental factor and lead to the development of symptoms (Agrios, 2005). Fungi figure prominently among the most successful pathogens of plants and are responsible for epidemics that have decimated natural populations of several tree species during the last centuries. Dutch elm disease, chestnut blight, beech bark disease, white pine blister rust, and maple decline are examples of diseases that are well documented.

Deleterious herbivores, often referred to more generically as 'pests', are the other major group of biotic agents that can have negative impacts on forest ecosystems. A pest refers to any animal that interferes with the survival and successful development or reproduction of trees. Invertebrate pests include insects, acari (mites), molluscs (slugs/snails), as well as exotic annelids (earthworms). Well-recognized vertebrate pests, especially in plantations, include rodents (mice, rabbits, squirrels, porcupines), ungulates (deer, moose) and infrequently, depending on the context, birds (sapsuckers, cormorants). Most commonly, pests are restricted to specific stages of tree growth, with some attacking only young seedlings, others only early successional stages or mature closed canopies, while some are found only on older, declining trees (Coulson and Witter, 1984). Invertebrate pests are usually specialized and classified, according to the part of the tree they feed on or the type of localized damage they inflict, as leaf feeders, tip and shoot feeders, xylem and phloem feeders, root feeders or seed and cone feeders.

Pests are typically mobile and locate their hosts primarily through chemotaxis (i.e. smell, taste, etc.), initially orienting to the optimal habitat (habitat location), and then refining their search to susceptible targets within that area (host location). Most invertebrates rely on specific chemical cues (kairomones) emitted from their host plants and can recognize healthy or stressed trees from considerable distances, in some cases kilometers. Some species use other cues such as sound for drought-responsive scolytid beetles (Mattson and Haack, 1987) and recently burned areas for fire-responsive species such as cerambycids (long-horned beetles) (McCullough, 1998).

Unlike invertebrates, pathogens that attack aerial parts of plants are unable to actually 'locate' their host. Instead, they are dispersed passively by wind, rain splash, running water, or by biological vectors. Many pathogenic fungi produce very large quantities of spores (through asexual reproduction, sexual reproduction, or both) that are released in the air when environmental conditions are favourable, e.g. under high moisture conditions. The vast majority of spores land on host plants that are not suitable for infection and colonization. When spores land on a suitable host species, two outcomes are possible: either the pathogen is able to establish itself in the host plant (compatible interaction) or it fails to do so (incompatible interaction). Success or failure is determined by a combination of genetically-controlled internal plant and microbe factors and is mediated by external, environmentally-controlled factors.

Pathogens rely heavily on free water movement and air currents for dispersal, and usually produce sporulating structures directly on the surface of infected plants. Pathogens that are dispersed by biological vectors, however, may produce their sporulating structures inside host tissues. For example, ophiostomatoid fungi produce their sporulating structures in wood galleries made by bark beetles. Masses of fungal spores found in these structures are often coated with mucilage, thus promoting their adherence to insects that exit the galleries on their way to colonizing new host trees. Other fungal species emit volatiles that attract insects, and sometimes vertebrates. Some interactions between fungi and vectors are highly specific. For example, female southern pine beetles (*Dendroctonus frontalis*) carry the mutualistic fungi *Ceratocystiopsis ranaculosus* and *Entomocorticium* sp. in specialized structures known as mycangia. These fungi are inoculated by the insect into the inner bark and phloem of host trees, and their development in galleries has been associated with increased fitness of developing larvae of *D. frontalis*. Mycangial associates of *D. frontalis* also includes actinomycetes that selectively inhibit the sapstaining fungus *Ophiostoma minus*, which is transported phoretically by the insect and has an antagonistic effect on the development of beetle larvae (Scott et al. 2008). Another group of wood colonizing insects, ambrosia beetles, also have highly specific interactions with fungi and are, in fact, considered to actively grow and maintain fungus gardens (Biedermann et al. 2013).

Pathogens use various means to penetrate and invade host plant tissues. Specialized structures called appressoria enable some fungal pathogens to physically penetrate host tissue. Furthermore, several pathogens secrete lytic enzymes that weaken host tissues or components thereof, and toxins that allow them to kill host tissue ahead of their progression. Additional weapons in the arsenal of pathogens include molecules that may block the production of defense molecules by plants, or detoxify such molecules if they are produced. Some pathogens also secrete homologs of plant growth hormones, which allow them to alter the physiology of their host (Agrios, 2005).

Fungal pathogens are subdivided into three broad classes based on how they interact with their host. Necrotrophs kill tissues as they grow through them and complete their life cycle (including reproduction) on dead tissue. Several species of canker-causing agents fall in this category. Biotrophs need host tissue to remain alive for completion of their life cycle. They are often host-specific and include aggressive pathogens that cause stem rusts and leaf rusts. While the host must remain alive for the pathogen to complete its life cycle, diseases caused by biotrophs are nevertheless highly destructive. The third class of pathogens, termed hemibiotrophs, exhibit an intermediate lifestyle characterized by an initial growth phase in living host tissue, followed by a reproductive phase on dead tissue.

Typically, pests are localized to either the dead or living parts of the tree; the former would include saproxylic species such as most wood boring long-horned or buprestid beetles and carpenter ants, while the latter would include the majority of invertebrate species that attack the phloem, photosynthetic or meristematic tissues of trees (i.e. leaf-feeding caterpillars, leafminers, shoot and phloem feeders, gallformers, seed and root feeders) as well as vertebrate animals such as ungulates, rodents and sapsuckers (Coulson and Witter, 1984). Pests can kill trees quickly in a year or less by breaking the vascular transport of the tree (phloem feeders), or more slowly through repeated removal of photosynthetic parts that limit overall tree growth and nutrient storage (e.g. leaf feeders causing defoliation).

Pests play an important role in the death spiral of trees and can act as the primary or secondary agents in this process. Agents that create physical openings for pathogens to invade living trees (e.g. beech bark scale, elm and other scolytid bark beetles) act indirectly while those who play an obligatory role in the pathogen's life cycle include leaf viruses and phytoplasmas introduced into the tree through the sucking mouthparts of insects such as aphids and scales (Weintraub and Beanland, 2006).

Pests usually display a preference for feeding on a specific part of the tree, with invertebrates often being solely dependent on this localization for their complete life cycle. Defoliators are the most common guild, and include all pests that feed on living leaf or needle tissues. They can be external feeders such as caterpillars (lepidopteran, coleopteran and hymenopteran immatures), adult beetles, vertebrates such as ungulates and rodents, and usually consume the photosynthetic material of the tree. Many invertebrates such as leafminers and gallformers also feed internally on these leaf tissues and are recognized by their characteristic feeding patterns (blotch leafminer, serpentine leafminer, etc.); all have chewing mouthparts and mechanically destroy plant tissue, as do vertebrate pests.

Tip and shoot feeding pests rarely cause whole tree mortality. Their impact changes the form and growth of the tree, and may cause serious timber defects in plantation forests. Under natural growing conditions, species such as balsam fir adelgid and white pine weevil cause weakened trees to be out-competed by their neighbours, thus altering stand succession. Root-feeding pests spend the majority of their time underground and can lead to overall tree decline (in some cases tree mortality). Their impact in natural forest dynamics is unknown but is thought to be similar to that in managed plantations, albeit at a lower frequency (Coulson and Witter, 1984).

Trunk-feeding pests that attack phloem or vascular tissues of the tree are usually considered the most damaging of all feeding guilds, leading quickly to whole tree mortality if the trunk or main branches become completely girdled by feeding. Aggressive *Dendroctonus* bark beetle species such as the mountain pine beetle (*D. ponderosae*) and southern pine beetle (*D. frontalis*) are two well-known pests that have caused major stand- and landscape-level mortality. Recent research shows that they, along with widespread defoliators such as the spruce and jack pine budworms, are linked to stand break up and increased fire probability under natural conditions (McCullough, 1998). In all cases, these pests are considered nature's 'silviculturalists', as they remove slow-growing, declining trees and stands to allow for vigorous understory succession.

Pests that feed on the reproductive parts of trees (flowers, seeds and cones) can almost completely deplete a tree's annual reproductive potential, with often over 90 percent of a crop lost (Coulson and Witter, 1984). In most cases, this is of minor ecological consequence as trees either overproduce reproductive structures or vary their annual output through masting (e.g. family Fagaceae). In so doing, they ensure that pests are unable to track this unpredictable, yet highly valuable food resource every year.

Ecological effect of pathogens and pests on forest ecosystems

Natural pathogen and invertebrate–host interactions that have co-evolved tend to be at equilibrium (Dinoor and Eshed, 1984). However, this does not imply that the equilibrium persists at all times or over the entire host range. For example, pests that damage or kill individual trees contribute to gap formations in the stand dynamics of temperate forests, while those impacting large forested areas are often precursors of fire and stand replacement. Natural forest stands can be strongly impacted by insect defoliators, such as spruce budworm (*Choristoneura fumiferana*) and jack pine budworm (*C. pinus pinus*) in coniferous forests and forest tent caterpillar (*Malacosoma disstria*) and gypsy moth (*Lymantria dispar*) in hardwood stands. In most cases, trees do not die immediately from defoliation but instead become weakened and susceptible to attack by other agents (Coulson and Witter, 1984). Whole tree or stand mortality occurs if this is repeated over time. Although these epidemics may be spectacular, they do not compromise forest ecosystems as they are often the disturbance factor needed to transition a mature forest canopy into a new cycle of succession. Studies show an increasing number of native pest species that interact with forest systems on such large scales, causing extensive, periodic landscape-level stand mortality and subsequent forest renewal. For example, dendroecological evidence suggests that the larch sawfly (*Pristiphora erichsonii*), an endemic defoliator causing extensive tree mortality, has been a major driver in larch dynamics in eastern North America (Jardon *et al.*, 1994).

Epidemics caused by introduced pathogens and/or invertebrates may alter significantly the composition of forest ecosystems, impact ecosystem services, and eventually lead to ecosystem meltdown. Chestnut blight, caused by the fungus *Cryphonectria parasitica*, almost eradicated American chestnut (*Castanea dentata*) within the few decades after the pathogen was accidentally introduced from Asia to the United States. The impact of the disease has been especially visible in northeastern USA, where chestnuts accounted for more than 20 percent of the canopy in some areas. The widespread death of these trees allowed other deciduous tree species, such as pignut hickory (*Carya glabra*), red maple (*Acer rubrum*), and sugar maple (*A. saccharum*), to colonize those areas (McCormick and Platt, 1980). A similar shift in forest composition has occurred following the past century of invasion (from Europe) by gypsy moth in oak-dominated forests across eastern North America, with a significant loss of the oak complement (Davidson *et al.*, 1999); concomitant with this change has been responses at multiple trophic levels in native faunal communities. Such widespread forest restructuring is also being repeated with the loss of ash following the introduction from Asia of the emerald ash borer (*Agrilus planipennis*) into North America.

Diseases, pests, and fire are often investigated separately by pathologists, entomologists and fire specialists, respectively. Many important forest and tree health issues, however, result from associations between at least two disturbance agents. For example, the pandemics of Dutch elm disease (DED) that have devastated native elm populations in Europe and North America resulted from the acquisition of the exotic fungal pathogens *Ophiostoma ulmi* and *O. novo-ulmi* by elm bark beetles, specifically the native elm bark beetle (*Hylurgopinus rufipes*) and the smaller European elm bark beetle (*Scolytus multistriatus*). In fact, in the absence of the DED fungi, elm bark beetles can hardly be considered as pests, since they cause very little, if any, damage to elm trees. DED illustrates how otherwise benign invertebrates become pests once they serve as vectors for aggressive pathogens.

Diseases of trees may result from successions of attacks by different disturbance agents. This is the case for beech bark disease which occurs when the bark of trees that have been attacked and damaged by the beech scale insect, *Cryptococcus fagisuga*, is colonized by spores of the fungi *N. ditissima* or *N. coccinea*. While beech have developed a degree of tolerance to the fungus, especially while young and actively growing, the disease has intensified with the arrival of beech bark scale, which feeds by puncturing the bark with sucking mouthparts, allowing the pathogen to enter the vascular system of the tree (Houston, 1994). This has led to widespread beech mortality where the two disturbance agents co-occur.

There are also several cases where insect attacks follow initial attack by pathogens. For example, trees that have been weakened by root pathogens, such as *Armillaria mellea* and *Heterobasidion annosum*, are highly attractive to secondary insects. Stress compounds emitted by weakened trees are the main cues used by all saproxylic beetles to locate potential hosts for colonization. The larger pine shoot beetle, *Tomicus piniperda*, is one recent North American example of the impact of sequential stress factors, where the combination of poor site conditions, root pathogens and beetle attack on both the shoots and trunks of trees lead to rapid stand mortality (Paine *et al.*, 1997). Some authors have hypothesized that pathogens that rely on arthropods for dissemination may manipulate their host in order to make it more attractive to vectors. An interesting case was presented by McLeod *et al.* (2005) who suggested the DED pathogens modify the biochemistry of their host in order to make it more attractive to elm bark beetles through the release of volatile terpenes acting as aggregation pheromones.

A final example of complex interaction is provided by forest declines which have been proposed to result from interactions among successive predisposing, inciting and contributing factors (reviewed in Manion, 2003). Predisposing factors are usually abiotic and long-term acting. They alter the ability of trees to withstand or respond to pathogens or pests. Inciting factors can be abiotic or biotic and are characterized by high intensity and short duration. They further weaken the trees and usually result in dieback. These trees may not survive the effect of contributing factors such as canker fungi, decay fungi or wood boring insects. Central to the concepts proposed by Manion (2003) is the notion that some forest declines are part of natural cycling of populations and are not expected to result in the death of ecosystems. Thus, forest declines might, in fact, act as a stabilizing selection agent, whereby competitive dominant trees are selectively killed, whereas stress-tolerant dominant individuals survive and gain the opportunity to contribute to the gene pool.

Effect of changing environments on pathogens and pests

Traditional forest management

So far, we have focused on the impact of pathogens and pests on forest ecosystems. We will now address how shifts in the environment, either natural or anthropogenic in nature, affect biotic disturbance agents.

The indirect role of abiotic factors in predisposing trees to attack by pathogens and invertebrate pests has already been noted. Environmental factors can also directly affect populations of pathogens and pests. For instance, populations of fungal leaf pathogens fluctuate widely according to climatic conditions. Under conditions of higher moisture, large populations may arise swiftly through rapid and repeated cycles of asexual reproduction. In addition, the impact of fires in reducing inoculum of parasitic dwarf mistletoes (*Arceuthobium* spp.) in coniferous forests of interior western North America is well documented (Parker *et al.* 2006). Likewise, high-intensity wildfires may lower populations of pathogens and arthropods that inhabit forest soils, litter and coarse woody debris (Parker *et al.* 2006).

Silvicultural practices may increase tree mortality incited by both pathogens and pests. In many localities in the northern hemisphere, hardwood or mixed stands have been converted into conifer plantations. After a few years of growth, high levels of mortality are sometimes observed in these plantations. Pronos and Patton (1978) reported that mortality caused by the root pathogen *Armillaria mellea* in young red pine plantations in Wisconsin was more noticeable in plantations established on sites that had previously supported oak (*Quercus rubra* and *Q. alba*). *Armillaria mellea* was present on the roots of living oaks, on which it grew as an epiphyte, without being able to invade the cambial area and cause diseases. However, oaks that were herbicide-killed when the pine plantations were established were heavily colonized by the fungus and became reservoirs for further attacks on red pine saplings. Due to its ability to produce large and long-lived underground rhizomorphs, as well as its capacity to grow saprophytically on dead trees and coarse woody debris, *A. mellea* can persist for decades on suitable sites. In particular, the presence of large stumps provides the fungus with a food base that it may exploit for decades.

Another example of silvicultural prescriptions that may lead to unforeseen changes in tree growth, form and structure has been reported when young conifer stands were thinned very early in stand establishment, thereby promoting the survival of the white pine weevil (*Pissodes strobi*). This major shoot and tip pest kills the tree leader and reduces tree height growth by one to three years. Open, sunny microhabitat conditions favour overwintering survival of adult weevils in the ground and promote larval feeding and development in the leaders. When tree densities are reduced and stands opened up early, weevil populations thrive and attack the remaining tree leaders, and do not decline again until the stand canopy closes (Paine et al., 1997).

Inadequate logging practices may also promote the spread of pathogens within forest stands. Annosus root and butt rot of coniferous species is caused by the basidiomycete fungus *Heterobasidion annosum* which is an efficient colonizer of fresh wounds (Hodges, 1969). Although the disease can be expected in unmanaged forest stands, its occurrence in managed stands can increase dramatically if thinning operations are carried out during periods when high concentrations of *H. annosum* spores are present in the air. Freshly cut stumps will rapidly be colonized by the fungus, which will later move to adjacent, living trees through root contacts. Mycelium of *H. annosum* will then infect and kill the roots of newly infected trees. As a result, several new infection centers around stumps may be created and expand radially over ensuing years. Similarly, high rates of tree damage, or failure to remove infested material during logging, may result in outbreaks of pest species that can build up in high enough populations to kill healthy living trees in adjacent areas.

In many parts of the world, industrial forestry and the need to protect human dwellings have resulted in aggressive suppression of wildfires. This has led to significant alterations of the landscape, including changes in tree species composition, distribution patterns, and increases in stand density. These changes have, in turn, altered the balance between fire, pathogens and invertebrate pests. According to Teale and Castello (2011), the rapid expansion of fusiform rust of pines in the southeastern United States in the twentieth century is, in part, due to fire suppression. The latter favoured the expansion of slash pine (*Pinus elliottii*) and loblolly pine (*P. taeda*), both more sensitive to fire and fusiform rust than longleaf pine (*P. palustris*). Fire suppression also stimulated the regeneration of oak, which is the alternate host required by the fusiform rust fungus (*Cronartium quercuum* f. sp. *fusiforme*) to complete its life cycle, thus allowing pathogen populations to expand.

Fire suppression has also been associated with a suite of diseases and pests in forests of western North America (Parker et al., 2006). Root pathogens and bark beetles appear to have been the main beneficiaries from this situation. The massive outbreak of mountain pine beetle (*Dendroctonus ponderosae*) during the past decade is another example of secondary pest outbreaks thought to be due, at least partially, to fire suppression. Although many factors are involved, including warmer winters and drier summers, one of the main drivers is thought to be overmature, declining pine that would have been removed by natural wildfire. Large contiguous stands on the landscape under fire suppression become susceptible to lightning strikes and fuel ignition (Parker et al., 2006). This is also true in eastern North America where mature conifer stands maintained under fire suppression lead to increased susceptibility to spruce budworm (Blais, 1983).

Modern forestry also relies increasingly on plantations. The deployment of monospecific, even-aged plantations poses phytosanitary risks, as it may promote outbreaks that would not otherwise be observed in natural stands characterized by a broader diversity and heterogeneous age structure. For example, the epidemic of *Gremmeniella* canker that started in several US states and Canadian provinces in the 1970s was in good part fueled by the establishment of red pine (*Pinus resinosa*) plantations. The epidemic was worsened by the fact that several tree nurseries that supplied red pine seedlings were unknowingly exporting contaminated, yet asymptomatic material. This led to a rapid expansion of the disease zone, before proper quarantine measures were put in place to restrict movement of red pine seedlings (Warren *et al.*, 2011). Similarly, the recent outbreak of emerald ash borer (*Agrilus planipennis*) has been fueled in part by the overabundance of susceptible ash cultivars in eastern North America (Aukema *et al.*, 2010). All urban areas in this region have been planting their parks, streets and new suburbs with nursery stock of green ash (*Fraxinus pennsylvanica*) grafted from clonal material that is highly susceptible to this recently introduced pest. This has allowed emerald ash borer populations to establish rapidly and expand spectacularly throughout the region.

Monospecific plantations are not necessarily composed of genetically uniform material. In fact, it is a recommended practice to deploy a variety a genotypes, to minimize the risk that an entire plantation is destroyed by a given pathogen or pest. This strategy has been used for decades in Europe for managing poplar leaf rust caused by *Melampsora larici-populina*. Epidemiological studies of European populations of *M. larici-populina* have shown that this pathogen was evolving rapidly in response to the deployment of new varieties of poplars that carried specific resistance genes that had been selected for by breeders (Xhaard *et al.*, 2011). Thus, over the years, increasingly complex races of the pathogen that had acquired the genes necessary to overcome poplar resistance genes were recovered. This observation confirms observations that had been made earlier in agricultural systems, where strong selection pressure for crop resistance is a major driver of pathogen and pest rapid evolution (Stukenbrock and McDonald, 2008).

The introduction of exotic tree species may have unexpected consequences on pathogens and pests, as shown by the rise of *Septoria* canker in North America. Poplar breeding programs initiated in the mid-1900s in the USA and Canada relied extensively on interspecific crosses among native and exotic poplars. Although breeders had selected material with high resistance against the known diseases of aspens and poplars, entire plantations were decimated by a new canker disease. Investigations revealed that it was caused by a native fungus (*Mycosphaerella populorum*) responsible for a relatively benign leaf spot disease on native poplars (Feau *et al.*, 2010). Although resistance to *Septoria* canker has been successfully integrated into poplar breeding programs, this episode is a potent reminder of the genetic and behavioral plasticity of pathogens and pests. In the case of pests, resistance breeding has been pursued for effective management, however, in most cases, this tactic is still in its infancy. Recent research using molecular tools to identify hybridized lodgepole and jack pine under attack by mountain pine beetle suggests that tree resistance may have some promise (Cullingham *et al.*, 2011); however, tree lifespans and the specialized attack patterns of pests are likely to limit the success of this strategy in the long run.

Invasive species

Early cases of introduced pathogens and pests led to legislation and quarantine measures aimed at preventing the intercontinental spread of these agents. Unfortunately, the efficiency of these measures has been in good part diminished by the explosion in international trade and movements of people during the last decades. As a result, new invasive pathogens and pests are being encountered annually in various countries (Aukema et al., 2010; Boyd et al., 2013). The introduction of novel pathogens and pests into new geographic areas threatens the integrity of global forest ecosystems. In many cases, we have seen epidemic outbreaks of exotic pathogens and pests on native tree species, to the detriment and sometimes extinction of the latter. In others, both deliberate and accidental introductions have led to major concerns over potential invasion meltdown of native ecosystems, those already threatened by shrinkage, intensified management and urbanization.

Some of our most significant introduced pathogens and pests have resulted in large-scale and costly management programs, including chestnut blight, DED, gypsy moth, many species of conifer sawflies, and now hemlock woolly adelgid, emerald ash borer and the Asian long-horned beetle. Hundreds of exotic forest species are currently regulated in Canada and the USA based on their potential to cause economic loss. Pimentel et al. (2005) reported about 50,000 foreign species in the USA causing losses of up to $120 billion per year, with 42 percent of endangered or threatened species at risk primarily due to these invaders. Lovett et al. (2006) provide an overview assessing the potential ecological impact of both groups in North American forests based on mode of action, host specificity, virulence, importance of host, uniqueness of host and phytosociology of host. Without question, this is the single most important factor affecting the future ecological integrity of North American forests.

Although the impact of exotic invasive pests on the landscape can be dramatic, affected tree species may recover if they exhibit genetic diversity for traits such as tolerance and/or resistance to invaders. Since resistant individuals occur at very low frequency, tree species that include large populations stand a better chance of surviving than those in low numbers. Butternut (*Juglans cinerea*) is one example of tree species put at risk of extinction due to a single pathogen, *Ophiognomonia clavigignenti-juglandacearum*, believed to have been introduced into North America during the twentieth century (Woeste et al., 2009). Although the geographic range of butternut is large, encompassing most of temperate eastern North America, it is not an abundant species. The fate of butternut is further impacted by loss of habitat due to anthropic activities and 'genetic invasion' due to hybridization with exotic congeners (*J. ailantifolia* and *J. regia*). Interestingly, since these interspecific hybrids usually are highly resistant to butternut canker, controlled interspecific crosses followed by introgression of resistance through backcrosses have been proposed as a strategy against the disease. A similar path had been chosen to save American chestnut from possible extinction caused by the fungus *C. parasitica* (Anagnostakis, 2012). In the case of butternut, however, the observation of disease resistance within nature has led researchers to advocate resistant germplasm from single-tree intraspecific selection and breeding, rather than developing interspecific hybrids (Michler et al., 2006).

Dramatic ecosystem changes may occur as a result of pathogen or pest introductions. For example, whitebark pine (*Pinus albicaulis*), considered a keystone species in subalpine ecosystems of western North America, is under threat across its range as a result of white pine blister rust (caused by the introduced pathogen *Cronartium ribicola*), attacks by mountain pine beetle, fire exclusion and climate change. Cascade effects from the potential loss of whitebark pine have been predicted to include provision of high-energy food for wildlife, nurse trees for other species in open terrain, and retention of snowpack (Smith et al., 2013). Broad-scale ecosystem concerns have also been expressed with respect to other invasive insect species, such as the hemlock woolly adelgid. Although at different spatial and temporal scales, Kizlinski et al. (2002) showed profound changes in stand structure, composition, and ecosystem function in forests dominated by hemlock following invasion by the adelgid.

Similarly, the slow advance of native and introduced earthworm populations has shown significant ecological impacts on forests in northeastern North America. The non-selective feeding, characteristic mucus channels, and soil compaction activities of these newly-arrived annelids have been linked to shifts in soil communities, away from native mycorrhizae in favour of bacterial and microbial communities (Hendrix, 2006). As species co-evolved with native forest flora are disrupted and fungi lost from the forest floor, small mammal communities shift from fungal-based voles to omnivorous deer mice, which have been linked to higher tick populations and the spread of human lyme disease in forest stands of eastern North America. These cascading effects are complex and the true impact of such ecological changes on long-term forest vegetation and succession in native forests remains poorly known.

A single invasive species may alter the landscape over a large area without necessarily inducing ecosystem meltdown, as in the case of chestnut blight. However, some areas have suffered regular invasions by different invasive species, which may eventually negatively affect the resilience of forests. This is clearly the case in many urban forests in North America that were first hit by chestnut blight, and then DED in the twentieth century. In many of these cities and towns, street chestnuts were replaced with elm and then with ash (*Fraxinus* spp.) or Norway maple (*Acer platanoides*). Ash is now being decimated by the emerald ash borer, an invasive beetle in North America, while maple is the prime target of the Asian long-horned beetle (*Anoplophora glabripennis*), a new species that attacks over ten species of tree and may never be eradicated from North America (Dodds and Orwig, 2011).

Climate change

Several decades ago, plant pathologists began to speculate that climate change might influence the fate of tree–pathogen/pest interactions at the landscape level (Coakley *et al.*, 1999). Evidence is now accumulating that this is happening in various regions of the world (Sturrock, 2012). For instance, shifts to warmer and wetter climate have been associated with outbreaks of *Diplodia* shoot blight of pines in France (Fabre *et al.*, 2011) and *Dothistroma* needle blight in western North America (Welsh *et al.*, 2014). This result is not surprising, given that such environmental conditions are known to promote the expansion of plant pathogen populations.

Similar insights have been gleaned by those making predictions about the future impact of climate change on invasive pests in forest systems. Most models point to increasing tree mortality under rising temperatures, however, the complexity of the models and high degree of uncertainty means that no specific predictions can be made (Dukes *et al.*, 2009). General consensus is that the increasing variability and extremes in weather patterns will favour some pest species and not others. The mobility and short generation times of most pests relative to their hosts means that even though they can disperse and adapt quickly in the short term, shifts in temperature and moisture will have significant effects on survival and reproduction in the long run, especially along range edges where intensive selection pressure occurs. Ultimately, the success of highly specific invertebrate pests will be heavily dependent on the ability of their host itself to survive the effects of rapidly changing, extreme weather patterns.

Conclusion

Forest insects and pathogens are the most pervasive and important agents of disturbance in North American forests, affecting an area almost 50 times larger than fire and with an economic impact nearly five times as great (Aukema *et al.*, 2010). Disturbances caused by pathogens and pests are a constant challenge to forest managers but are nevertheless a vital driver of forest composition, succession and evolution. In light of these complex ecological processes and natural disturbances, our response to pest and pathogen challenges must be cautious. We need to develop a strong understanding of these processes, our human footprint on them, and the large temporal and spatial scales on which they operate, especially in the remaining intact forests. We must avoid focusing on specific isolated pathogen and pest problems, and instead, respond strategically with integrative thinking that supports the development of broad forest resilience.

Many factors are converging that will constrain our ability to retain healthy, resilient forests. Increasing intensification of forest practices for timber production and land use changes that result in deforestation for agricultural and urban development will dramatically restrict most tree species. This, combined with the rapidly expanding introduction of invasive alien species to most of the world's forests, will exacerbate the effect of pathogens and pests. On top of this is the large unknown of climate change, a new factor that will have major implications for pathogen–insect interactions in our forests. We must be cautious in believing we can predict, let alone manage, the outcome of disturbances under these complex scenarios.

If we are to successfully retain ecological balance in our forests, our response must become more considered than simply removing every individual (the naturally resistant along with the susceptible) when faced with a pathogen or pest problem. The concept of a healthy forest requires thinking about gene pools and species interactions at the population level in order to achieve successful adaptation under a changing environment. This means working with forest systems to avoid ecological tipping points and supporting natural biodiversity that enables a return to natural states. Moving forward, we need to aim for ecosystem resilience in our forests, where forest health is defined not by the lack of specific pests or pathogens, but by the complex ecological interactions and functioning of diverse forest organisms and communities.

References

Agrios, G. N. (2005) *Plant Pathology*, 5th Edition. Elsevier Academic Press, Amsterdam.

Anagnostakis, S. L. (2012) Chestnut breeding in the United States for disease and insect resistance. *Plant Disease*, vol 96, no 10, pp. 1392–1403.

Aukema, J. E., McCullough, D. G., Von Holle, B., Liebhold, A. M., Britton, K., and Frankel, S. J. (2010) Historical accumulation of nonindigenous forest pests in the continental United States. *Bioscience*, vol 60, no 11, pp. 886–897.

Biedermann, P. H. W., Klepzig, K. D., Taborsky, M. and Six, D. L. (2013) Abundance and dynamics of filamentous fungi in the complex ambrosia gardens of the primitively eusocial beetle *Xyleborinus saxesenii* Ratzeburg (Coleoptera: Curculionidae, Scolytinae). *FEMS Microbiology Ecology*, vol 83, no 3, pp. 711–723.

Blais, J. R. (1983) Trends in the frequency, extent, and severity of spruce budworm outbreaks in eastern Canada. *Canadian Journal of Forest Research*, vol 13, no 4, pp. 539–547.

Boyd, I. L, Freer-Smith, P. H., Gilligan, C. A. and Godfray, H. C. J. (2013) The consequence of tree pests and diseases for ecosystem services. *Science*, vol 342, no 6160, pp. 823+.

Castello, J. D., Leopold, D. J. and Smallidge, P. J. (1995) Patterns and processes in forest ecosystems. *BioScience*, vol 45, no 1, pp. 16–24.

Coakley, S. M., Scherm, H. and Chakraborty, S. (1999) Climate change and plant disease management. *Annual Review of Phytopathology*, vol 37, pp. 399–426.

Coulson, R. N. and Witter, J. A. (1984) *Forest Entomology: Ecology and Management*. Wiley-Interscience Publications, New York.

Cullingham, C. I., Cooke, J. E., Dang, S., Davis, C. S., Cooke, B. J. and Coltman, D. W. (2011) Mountain pine beetle host-range expansion threatens the boreal forest. *Molecular Ecology*, vol 20, no 10, pp. 2157–2171.

Davidson, C. B., Gottschalk, K. W. and Johnson, J. E. (1999) Tree mortality following defoliation by European gypsy moth (*Lymantria dispar* L.) in the United States: A review. *Forest Science*, vol 45, no 1, pp. 74–84.

Dinoor, A. and Eshed, N. (1984) The role and importance of pathogens in natural plant communities. *Annual Review of Phytopathology*, vol 22, pp. 443–466.

Dodds, K. J. and Orwig, D. A. (2011) An invasive urban forest pest invades natural environments – Asian longhorned beetle in northeastern US hardwood forests. *Canadian Journal of Forest Research*, vol 41, no 9, pp. 1729–1742.

Dukes, J. S., Pontius, J., Orwig, D., Garnas, J. R., Rodgers, V. L. and others (2009) Responses of insect pests, pathogens, and invasive plant species to climate change in the forests of northeastern North America: What can we predict? *Canadian Journal of Forest Research*, vol 39, no 2, pp. 231–248.

Fabre, B., Piou, D., Desprez-Loustau, M. L. and Marçais, B. (2011) Can the emergence of pine *Diplodia* shoot blight in France be explained by changes in pathogen pressure linked to climate change? *Global Change Biology*, vol 17, no 10, pp. 3218–3227.

Feau, N., Mottet, M.-J., Périnet, P., Hamelin, R. C. and Bernier, L. (2010) Recent advances related to poplar leaf spot and canker caused by *Septoria musiva*. *Canadian Journal of Plant Pathology*, vol 32, no 2, pp. 122–134.

Hendrix, P. F. (2006) Biological invasions belowground – earthworms as invasive species. *Biological Invasions*, vol 8, no 6, pp. 1201–1204.

Hodges, C. S. (1969) Modes of infection and spread of *Fomes annosus*. *Annual Review of Phytopathology*, vol 7, pp. 247–266.

Houston, D. R. (1994) Major new tree disease epidemics: Beech bark disease. *Annual Review of Phytopathology*, vol 32, pp. 75–87.

Jardon, Y., Filion, L. and Cloutier, C. (1994) Tree-ring evidence for endemicity of the larch sawfly in North America. *Canadian Journal of Forest Research*, vol 24, no 4, pp. 742–747.

Kenis, M., Auger-Rozenberg, M-A., Roques, A., Timms, L., Pere, C. and others (2009) Ecological effects of invasive alien insects. *Biological Invasions*, vol 11, no 1, pp. 21–45.

Kizlinski, M. L., Orwig, D. A., Cobb, R. C. and Foster, D. R. (2002) Direct and indirect ecosystem consequences of an invasive pest on forests dominated by eastern hemlock. *Journal of Biogeography*, vol 29, nos 10–11, pp. 1489–1503.

Kolb, T. E., Wagner, M. R. and Covington, W. W. (1994) Utilitarian and ecosystem perspectives: concepts of forest health. *Journal of Forestry*, vol 92, no 7, pp. 10–15.

Lovett, G. M., Caham, C. D., Arthur, M. A., Weathers, K. C. and Fitzhugh, R. D. (2006) Forest ecosystem responses to exotic pests and pathogens in eastern North America. *Bioscience*, vol 56, no 5, pp. 395–405.

Manion, P. D. (2003) Evolution of concepts in forest pathology. *Phytopathology*, vol 93, no 8, pp. 1052–1055.

Mattson, W. J. and Haack, R. A. (1987) The role of drought in outbreaks of plant-eating insects. *Bioscience*, vol 37, no 2, pp. 110–118.

McCormick, J. F. and Platt, R. B. (1980) Recovery of an Appalachian forest following the chestnut blight or Catherine Keever – you were right! *American Midland Naturalist*, vol 104, no 2, pp. 264–273.

McCullough, D. G. (1998) Fire and insects in northern and boreal forest ecosystems of North America. *Annual Review of Entomology*, vol 43, pp.107–127.

McLeod, G., Gries, R., von Reuß, S. H., Rahe, J. E., McIntosh, R. and others (2005) The pathogen causing Dutch elm disease makes host trees attract insect vectors. *Proceedings of the Royal Society B-Biological Sciences*, vol 272, no 1580, pp. 2499–2503.

Michler, C. H., Pijut, P. M., Jacobs, D. F., Meilan, R., Woeste, K. E. and Ostry, M. E. (2006) Improving disease resistance of butternut (*Juglans cinerea*), a threatened fine hardwood: a case for single-tree selection through genetic improvement and deployment. *Tree Physiology*, vol 26, no 1, pp. 121–128.

Ostry, M. E. and Laflamme, G. (2009) Fungi and diseases – natural components of healthy forests. *Botany*, vol 87, no 1, pp. 22–25.

Paine, T. D., Raffa, K. F. and Harrington, T. C. (1997) Interactions among scolytid bark beetles, their associated fungi, and live host conifers. *Annual Review of Entomology*, vol 42, pp. 179–206.

Parker, T. J., Clancy, K. M. and Mathiesen, R. L. (2006) Interactions among fire, insects and pathogens in coniferous forests of the interior western United States and Canada. *Agricultural and Forest Entomology*, vol 8, no 3, pp. 167–189.

Pimentel, D., Zuniga, R. and Morrison, D. (2005) Update on the environmental and economic costs associated with alien-invasive species in the United States. *Ecological Economics*, vol 52, no 3, pp. 273–288.

Pronos, J. and Patton, R. F. (1978) Penetration and colonization of oak roots by *Armillaria mellea* in Wisconsin. *European Journal of Forest Pathology*, vol 8, no 4, pp. 259–267.

Scott, J. J., Oh, D. C., Yuceer, M. C., Klepzig, K. D., Clardy, J. and Currie, C. R. (2008) Bacterial protection of beetle-fungus mutualism. *Science*, vol 322, no 5898, p. 63.

Smith, C. M., Shepherd, B., Gillies, C. and Stuart-Smith, J. (2013) Changes in blister rust infection and mortality in whitebark pine over time. *Canadian Journal of Forest Research*, vol 43, no 1, pp. 90–96.

Stukenbrock, E. H. and McDonald, B. A. (2008) The origins of plant pathogens in agro-ecosystems. *Annual Review of Phytopathology*, vol 46, pp. 75–100.

Sturrock, R. N. (2012) Climate change and forest diseases: using today's knowledge to address future challenges. *Forest Systems*, vol 21, no 2, pp. 329–336.

Teale, S. A. and Castello, J. D. (2011) Regulators and terminators: the importance of biotic factors to a healthy forest, in J. D. Castello and S. A. Teale (eds) *Forest Health: An Integrated Perspective*, Cambridge University Press, Cambridge, UK.

Warren, G. R., Harrison, K. J. and Laflamme, G. (2011) New and updated information on Scleroderris canker in the Atlantic Provinces. *The Forestry Chronicle*, vol 87, no 3, pp. 382–390.

Weintraub, P. G. and Beanland, L. (2006) Insect vectors of phytoplasmas. *Annual Review of Entomology*, vol 51, pp. 91–111.

Welsh, C., Lewis, K. J. and Woods, A. J. (2014) Regional outbreak dynamics of *Dothistroma* needle blight linked to weather patterns in British Columbia, Canada. *Canadian Journal of Forest Research*, vol 44, no 3, pp. 212–219.

Woeste, K., Farlee, L., Ostry, M., McKenna, J. and Weeks, S. (2009) A forest manager's guide to butternut. *Northern Journal of Applied Forestry*, vol 26, no 1, pp. 9–14.

Xhaard, C., Fabre, B., Andrieux, A., Gladieux, P., Barrès, B. and others (2011) The genetic structure of the plant pathogenic fungus *Melampsora larici-populina* on its wild host is extensively impacted by host domestication. *Molecular Ecology*, vol 20, no 13, pp. 2739–2755.

17
BRYOPHYTES IN FOREST ECOSYSTEMS

Nicole J. Fenton, Kristoffer Hylander and Emma J. Pharo

Introduction

Bryophytes are a polyphyletic group that have existed for millions of years, with over 17,000 species of bryophytes found globally (Vanderpoorten and Goffinet 2009). This includes species in three divisions that differ from one another morphologically and reproductively. The recent suggestion is that there are 5,000 species of liverworts, the most ancient group of bryophytes, 12,000 species of mosses, the most diverse and abundant group which also includes the peat mosses, and 300 species of hornworts, the most recent branch (Vanderpoorten and Goffinet 2009). Despite their separation into different taxonomic groups, together they differ from vascular plants in a few fundamental ways that make a significant difference to their distribution and diversity. Specifically, bryophytes have:

- no reinforced vascular tissue, which means they are rarely taller than a few centimetres;
- like all plants, bryophytes alternate between haploid and diploid generations. In contrast to vascular plants, the haploid gametophyte stage is dominant;
- spores as the means of sexual reproduction, most of which are about 10–50 µm and are easily borne by the wind;
- various means of vegetative reproduction including specialized structures and plant fragments;
- a poikilohydric habit, which means that many bryophytes can fully dehydrate and suspend function in times of drought.

In this chapter, we look at patterns of bryophyte diversity at different spatial and temporal scales, variation in bryophyte function across major forest biomes and the influence of human disturbance on forest bryophyte diversity.

Species diversity

Global and continental scales

Bryophyte diversity does not mirror the pattern of a decrease in local and regional diversity from the tropics towards the poles, as is well established for many groups (Möls *et al.* 2013). In fact, many temperate and tropical areas of the same size have similar numbers of moss species (Möls et al. 2013). For pleurocarpous mosses (certain families with a branching growth form, often growing hanging or in carpets) it has been showed that species turnover among similar sized tropical areas is higher than among the same sized temperate areas, which leads to higher total diversity in the tropics (Hedenäs 2007). Less is known of global liverwort diversity patterns, but it is worth noting that certain families have radiated more in tropical areas and others more in temperate areas. For example, the families Plagiochilaceae and Lejeuneaceae are very species rich in the tropics, while Lophoziaceae and Scapaniaceae are diverse in temperate and montane areas (Vanderpoorten and Goffinet 2009).

Bryophytes are found in most forests around the world. In general, bryophytes seem to have the highest biomass and diversity in relatively cool and moist forests, such as tropical montane cloud forests and temperate rain forests (e.g. Frahm and Gradstein 1991). However, boreal forests can also be species rich, especially in areas with high humidity (Dynesius et al. 2009, Fenton and Bergeron 2008). Compared to montane cloud forests, tropical lowland forests have been assumed to be species poor (Frahm and Gradstein 1991), but it has been recently suggested that their bryophyte diversity is confined to the previously un-inventoried canopies (Sporn et al. 2010). However, bryophytes are not as abundant or as large on tree bases and lower tree trunks in lowland rain forests compared to cloud forests (Frahm and Gradstein 1991). Bryophytes do not seem to play an important role in tropical deciduous forests, where they can be very scarce, both in terms of cover and species richness (K. H., personal observation). Climatic gradients across continents clearly influence the composition and richness of forest bryophytes (Möls et al. 2013). Along the continental to oceanic gradient it is apparent that the number of liverworts increases with higher oceanity (Vanderpoorten and Goffinet 2009).

Interestingly, a number of species in forests in such diverse systems as moist Afromontane forests in south-west Ethiopia, temperate forests embedded in Mediterranean type vegetation in South Africa, and boreal forests along streams in Sweden, all show rather high levels of species richness or alpha diversity at the scale of 0.02 ha, even if the tropical example have the highest numbers (Figure 17.1). The relative order of species richness among different forest ecosystems at the scale of 0.02 ha is also mirrored in the order when pooling the cumulative

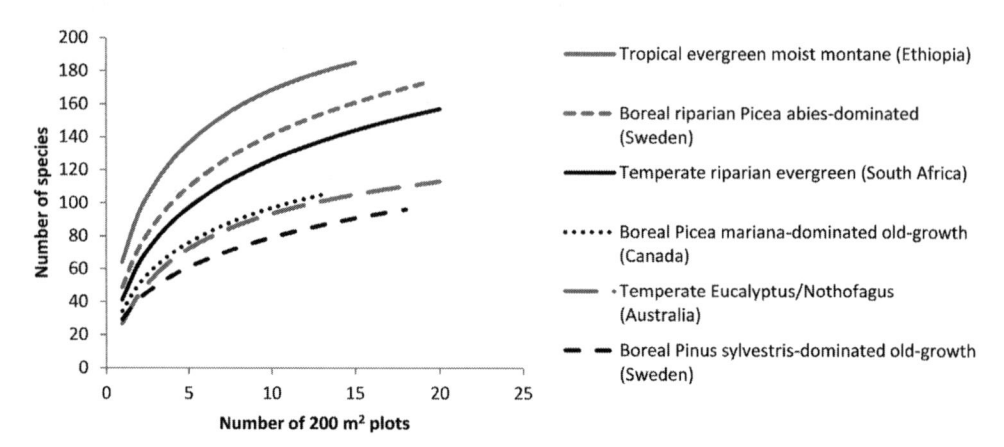

Figure 17.1 Species accumulation curves for bryophytes in scattered 0.02 ha plots (in most cases 10 × 20 m) in different biomes and forest environments across the world. Data unpublished or reanalysed from: Dynesius (2001); Hylander and Hedderson (2007); Hylander and Dynesius (2006); Hylander and Nemomissa (2009); Pharo et al. (2013)

number of species over 10–20 sites (compare the curves in Figure 17.1). Interestingly, the diversity is quite similar across three continents if the curves for the datasets from Australia, non-stream boreal Sweden and boreal Canada are compared.

Landscape scale

There are many gradients in environmental variables that regulate forest types as well as bryophyte communities across landscapes. Typically the topography of the landscape determines the variation in solar radiation and can create contrasting moisture conditions. These gradients will directly influence the distribution of bryophytes across landscapes, with typically higher species richness in landscape positions with a cool and moist microclimate, even if some species are confined to sunny and dry conditions (Dynesius et al. 2009, Tng et al. 2009). Topography also affects bryophyte communities indirectly through its regulation of soil conditions and the dominant vegetation. For example, gradients in pH and nutrient availability, and the dichotomy between podzols and brown soils are particularly important in boreal and boreonemoral settings. Similarly, gradients in precipitation and mist (often along with altitude and temperature) have a marked effect on the structure of tropical bryophyte communities, with generally higher cover and diversity associated with more mist (Frahm and Gradstein 1991). The unusual occurrence of mists in one lowland rain forest area in South America is also the reason why this area has higher diversity than average lowland rain forests (Gradstein 2006).

Differences in environmental conditions and forest types across landscapes also determine disturbance regimes, which in turn affect bryophyte community composition and diversity. Forest regenerated after different types of disturbances, e.g. fires and insect disturbances, can have different community composition depending on the amount of dead wood or light regime (Schmalholz et al. 2011). Another illustration of the importance of disturbances for bryophytes at the landscape level is the change in composition and diversity among sites differing in time since stand-replacing wildfire in eastern Canada, which determines the abundance and richness of peat mosses (*Sphagnum* spp.) and liverworts on the forest floor (Fenton and Bergeron 2008). In Fennoscandinavia, wet forests can be species rich and/or species poor depending on hydrology, pH and forest type (cf. Hylander and Dynesius 2006).

Stand scale

The importance of the variability in type and quality of microhabitats within the forest as a controlling factor for bryophyte diversity at the stand scale is difficult to understate. In dry forest types, bryophytes tend to be confined to certain microhabitats, e.g. seepage areas along streams unless they are dry habitat specialist epiphytes. A relative scarcity of forest floor bryophytes can also be typical in certain temperate deciduous forests, but here the reason is that it is difficult for bryophytes to establish and survive the leaf litter that accumulates on the forest floor. Similarly, in lowland and montane tropical moist to wet forests bryophytes are generally rather scarce on the forest floor. In certain other forest types, bryophytes cover the ground (Hylander et al. 2005).

Overall, different bryophyte species utilize different sets of microhabitats such as dead wood, shaded rock surfaces, mineral soil, and seepage areas in boreal forests (Hylander et al. 2005), while in moist temperate and tropical forests epiphytic species generally dominate, even if many other microhabitats are also utilized. Even leaves have their own bryophyte flora in moist tropical forests that could be species rich (epiphylls; Zartman et al. 2012). Consequently the presence of microhabitats of various kinds generally increases bryophyte species richness.

However, bryophyte richness can also depend on the size of the regional species pool for certain substrate types. For example, there seem to be more species that can potentially colonize moderate to high pH riparian forests than low pH riparian forests in Fennoscandia (Hylander and Dynesius 2006).

On an even smaller scale, gradients in pH, surface orientation and moisture holding capacity can correlate with a turnover in species composition within pieces of dead wood, rock surfaces or tree bark. Leachate from deciduous trees with higher pH leaves could influence underlying bryophyte vegetation on, for example, boulders in a profound way (Weibull and Rydin 2005). Similarly, the bark characteristics (e.g. pH, bark structure) of tree species can result in quite different levels of bryophyte diversity (Löbel et al. 2006). Less is known from tropical forest systems about turnover of bryophyte communities among tree species, even if it is obvious that there are differences along vertical gradients and bark structure gradients, with more drought tolerant species on smooth bark and higher up in the canopy.

Temporal scale

Since many bryophytes can expand clonally and die off in some parts while growing in other parts, bryophyte colonies could, theoretically, be very old. Therefore bryophyte population dynamics in forests are tightly bound to the dynamics of their substrates and display patch-tracking meta-population dynamics (Snäll et al. 2005). One could thus assume that bryophytes on dead wood are more dynamic than epiphytes, which are still more dynamic than those on the forest floor or on boulders. Epiphyllous communities are probably the most dynamic, with reported generations of liverworts within a year in lowland Amazonian forests (Zartman et al. 2012). Another rather temporary substrate in forests is newly exposed mineral soil, for example in tree uprooting. This microhabitat can house a quite distinct bryophyte community, in which at least some components are quite short lived (Jonsson and Esseen 1990). Several of those species are also adapted to brighter conditions and can dominate early successional stages after large-scale disturbances. The dispersal ability of forest bryophytes is largely unknown, but over longer time periods most species seem to be able to colonize substrates across the landscape (over 10 km; Fenton and Bergeron 2008). However, it has also been proposed that there are many species which are rather dispersal limited over shorter timescales (Snäll et al. 2005, Löbel et al. 2006).

Variation in bryophyte function across forest biomes

Bryophytes influence a variety of ecosystem functions in different biomes, particularly in many forest ecosystems where bryophytes accumulate significant biomass (e.g. over 20 kg/m^2 in paludified boreal forests, Lecomte et al. 2006, see also Vanderpoorten and Goffinet 2009). For example, superficially different systems such as boreal forest floors (Turetsky et al. 2012), epiphytic bryophyte mats on tree branches in humid temperate forests (Pypker et al. 2006), and epiphytic and epiphyllous bryophyte mats in humid tropical forests, can all be seen as analogous (Zartman et al. 2012). These different systems provide a variety of different ecosystem functions including water retention, nutrient cycling, and habitat for other organisms.

Water retention

All plants are dependent on water for the turgor of their tissues, and for photosynthesis (Glime 2007). Bryophytes, similar to lichens (Chapter 18), are poikilohydric plants that must absorb and retain the water available to them in the above ground environment, either within a shoot or within a colony. Therefore, when a significant mass of bryophytes are present within an ecosystem, they have a significant impact on the water balance of this system (Turetsky et al. 2012, Pypker et al. 2006; Figure 17.2).

The water retained within bryophyte colonies has a multitude of effects on ecosystem function, which vary with the position of the bryophyte colony. In forests where bryophyte colonies cover the mineral soil, the presence and the composition of the bryophytes significantly changes water availability in the mineral soil (Price et al. 1997). Deep bryophyte colonies can intercept rainfall, limiting percolation to the mineral soil underneath. In addition, species composition will influence the retention capacity of the bryophytes, as species vary significantly in their ability to retain water. An extreme example of this is seen during paludification, where *Sphagnum* spp. mosses replace *Pleurozium schreberi* and other feather mosses in some boreal forests. As a consequence of this replacement, the mineral soil is eventually isolated from the surface of the organic layer (generated by the mosses) by a perpetually flooded anoxic zone (Lecomte et al. 2006).

The role of bryophyte colonies growing over coarse woody debris, branches, or leaves, while functionally similar, results in different consequences. In these environments, the water retained by bryophytes facilitates the development of other plants. On coarse, woody debris and branches bryophytes promote vascular plant establishment and growth (Clark et al. 2005). For example, epiphytic bryophytes in cloud forests can harvest water droplets (Clark et al. 2005). Vascular plants are then able to access this water source, which with their waxy cuticle they normally would not be able to do (Clark et al. 2005, Pypker et al. 2006). Similarly, epiphytes in the upper canopy transfer water (and nutrients) to epiphytes and epiphylls lower in the canopy (Clark et al. 2005).

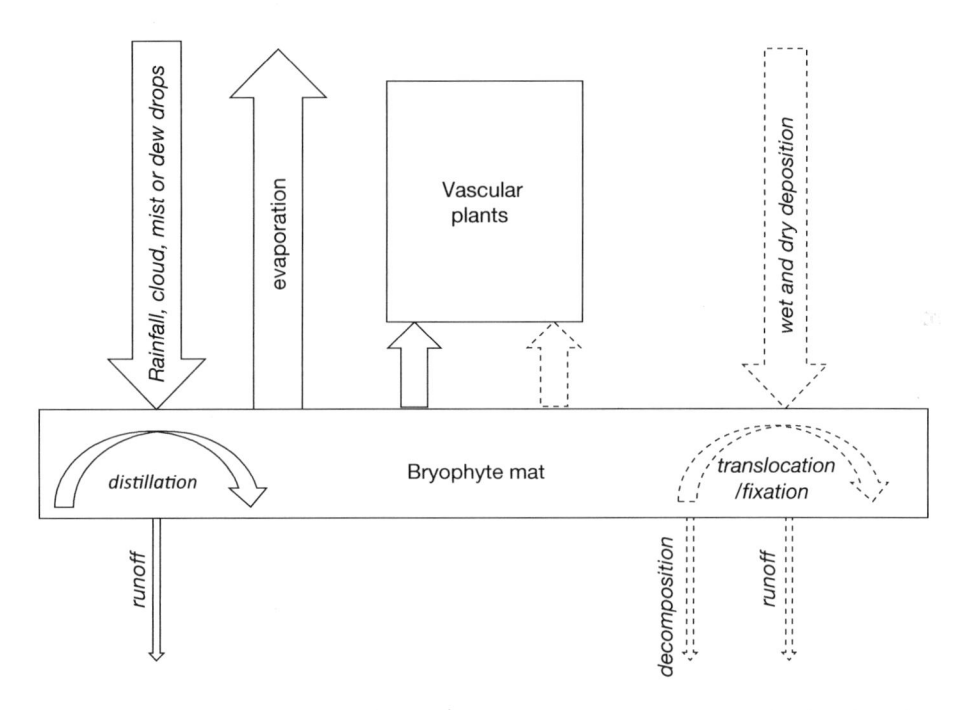

Figure 17.2 Summary of effects of bryophyte mats in various positions (over mineral soil, on branches or on leaves) on water (solid lines) and nutrient (dashed lines) fluxes

Nutrient cycling

Bryophytes influence nutrient cycles in a variety of ways, but most can be mediated through one of three processes: retention, decomposition, and nitrogen fixation (Figure 17.2). One of the simplest ways in which bryophytes influence nutrient cycles is simply by retention of nutrients. Bryophytes obtain a significant portion of their mineral requirements from either dry (dust) or wet (rain and leachate) deposition (Glime 2007). Nutrient availability is typically less concentrated in these sources than in the mineral soil, and consequently bryophytes are well adapted to retain the nutrients that are available to them (Glime 2007). Once adsorbed or absorbed, the nutrients can remain within their tissues for significant periods, as several studies have indicated that bryophytes, like lichens, are able to recycle nutrients within their tissues (i.e. from old towards young; Glime 2007). For example, in one study where marked nitrogen was applied to a bryophyte mat, nitrogen was moved from the senescent segments to the growing segments, but none was released to the environment (Eckstein and Karlsson 1999, in Glime 2007). Similarly, epiphytic bryophytes have been shown to retain 50 per cent of atmospheric deposition in tropical montane forests (Clark *et al.* 2005).

The decomposition rate of bryophytes, both on the forest floor and in the canopy (Lang *et al.* 2009) is notoriously low. Many studies have compared the decomposition rates of bryophytes to vascular plants and lichens (e.g. Lang *et al.* 2009), and all come to the conclusion that bryophytes decompose much more slowly than these other groups. The cause for this slower rate is less clear, however. Bryophytes lack lignin, the structural component that slows decomposition in woody plants, however, their structural carbohydrates (pentosans) and aromatic compounds do seem to resist decomposition (Lang *et al.* 2009). In addition, bryophytes have a variety of secondary metabolites that seem to inhibit microbial growth, fungal growth (Opelt *et al.* 2007) and grazing by invertebrates (Glime 2007), three pathways that lead to vascular plant decomposition. Consequently, bryophytes seem to be primarily decomposed by a subset of tolerant fungi (Davey *et al.* 2012). Little is known about bryophyte–fungi interactions; however both mycorrhizal and saprophytic fungi have been identified on and within green and senescent bryophytes (Davey *et al.* 2012).

In addition to their retention capacity and low decomposition rate, bryophytes also host a variety of bacteria species, including cyanobacteria (DeLuca *et al.* 2002). Cyanobacteria, some of whom have the capacity to fix nitrogen via heterocysts (specialized cells), have been found in symbiosis (Bay *et al.* 2013) with a wide variety of bryophyte species, including forest floor species (e.g. *Pleurozium schreberi*), *Sphagnum* spp., epiphylls and epiphytes (Opelt *et al.* 2007, Lindo and Whiteley 2011). Long considered inconsequential in the nitrogen budget of ecosystems, a variety of studies have now indicated that the moss–cyanobacterial unit contributes significantly to the overall budget of forests (DeLuca *et al.* 2002). This is particularly true in nitrogen poor areas such as boreal forests, as nitrogen fixation rate appears to increase with a decrease in available nitrogen, either naturally (via succession) or anthropogenically (Lindo *et al.* 2013).

When these three mechanisms are taken together (retention, slow decomposition and nitrogen fixation), bryophytes clearly can have a significant impact on the ecosystem nutrient cycles, particularly in nutrient poor environments. The consequences for ecosystem functions are numerous. Bryophyte colonies (terrestrial, aerial, or epiphyllous) retain nutrients that otherwise would have been lost to the ecosystem via leaching (Clark *et al.* 2005, Lindo and Whiteley 2011, Turetsky *et al.* 2012). These colonies are nutrient storehouses, with different nutrients either adsorbed to their surfaces, or converted into complex compounds and stored within bryophyte cells.

The fate of nutrients captured by bryophytes is less clear. Several studies have demonstrated that the nutrients retained by the bryophytes or fixed by cyanobacterial epiphylls are subsequently used by vascular plants. This occurs either through absorption via mycorrhizal hyphae, which are particularly abundant around senescent mosses, through leaching as cellular contents of the bryophytes are lost during wet/dry cycles, or via a more direct transfer as was found for nitrogen fixed by cyanobacteria living in epiphyllous liverworts (Glime 2007). However, most studies examining the residence time of nutrients within bryophytes find that nutrients are retained for significant periods, years in nearly all cases (Turetsky et al. 2012). In these situations, the nutrients contained within the bryophyte would be functionally unavailable, and some have argued that bryophyte contribute to decrease nutrient availability in many ecosystems. However, while the rate of nutrient release from bryophyte mats is quite likely slow, and although the relationship with mycorrhizal fungi remains unresolved, these nutrients do remain in the system and are eventually released. Without bryophyte mats, these nutrients may well have been definitively lost from the system via leaching, and the amount of nitrogen would have been significantly less. This is particularly true of aerial bryophyte mats that retain nutrient rich layers on branches, effectively generating a soil within the canopy environment (Lindo and Whiteley 2011).

Habitat

While bryophytes may seem diminutive, when observed at the appropriate scale, they become a forest unto themselves and thus represent a three-dimensional habitat that is inhabited and used by a variety of organisms. Glime (2007) reviews the variety of different organisms that live in bryophyte mats; these include cyanobacteria, protists, fungi and mesofauna. In general these groups are very poorly studied. However, the few studies that have looked at the different inhabitants of bryophytes indicate significant diversity at different spatial scales, and some levels of specialization for different bryophyte species. Fungi from a wide variety of groups are associated with all types of bryophytes (Döbbeler and Hertel 2013), and Davey et al. (2012) found that different species of fungi were associated with different bryophyte species.

Forest bryophytes and habitat change

Most of the studies of human use and our effect on forests examine the impacts of fire, timber harvesting and fragmentation. We focus here on aspects of forest management that are particular to bryophytes. For many issues that have been important in the forest management literature over the last 20–30 years, such as the question of remnant size and edge effects, bryophytes follow a similar pattern to other species, with results depending on the interaction between the particular bryophyte community, their environment and the characteristics of the disturbance (Fischer and Lindenmayer 2007).

Bryophyte response to human disturbance of forests needs to be understood in the context of the earlier discussion about the importance of suitable microhabitats. The importance of microhabitat cannot be overstated in seeking to understand bryophyte diversity in both natural and human-altered forests. While microhabitat and macrohabitat are often closely correlated, the history of events within any one particular area of forest can lead to a rich variety of substrates for bryophytes to inhabit or a relatively poor environment. Because bryophytes are small, suitable microhabitat may be available in a landscape that, on first impression, would appear to be highly unsuitable, such as exotic pine plantations with thick needle litter (Pharo and Lindenmayer 2009).

Forest fragmentation

Fragmentation is well understood as a significant landscape change for many taxa. Many bryophytes are widely distributed and therefore, in theory at least, have the potential to disperse between areas of suitable habitat. Bryophytes generally have small, light propagules, which suggest that successful dispersal in a fragmented landscape might be possible for many species. There is evidence to suggest that this is indeed the case, with some populations of bryophytes able to spread their propagules into fragmented patches in the landscape (e.g. Hylander 2009). However, in at least some forest landscapes, dispersal appears to be more of a controlling factor than in open biomes (Löbel et al. 2006).

Bryophytes follow a pattern that is common to many other types of organisms in fragmented landscapes. For example, large remnants tend to be superior habitat compared with small remnants (Oishi and Morimoto 2013). As for other taxa, the results are variable, depending on the life history strategy of the species in question. The changes in humidity, light and temperature associated with forest edges is generally negatively associated with bryophyte richness, but this is variable, with some species being much less affected or positively changed. Species that you might expect to be resilient to fragmentation are not necessarily able to adapt to the disturbance. For example, tropical epiphylls might be expected to be resilient to changes since they are adapted to high turnover environments and regularly produce spores (Zartman et al. 2012). However, even among these bryophytes adapted to the ephemeral habitat of canopy leaves, researchers have found evidence for dispersal limitation (Zartman et al. 2012).

Timber, pulp and biofuel production in forests

Bryophytes' response to timber harvesting is largely determined by the severity of the disturbance. Clear felling of forests is obviously a severe disruption and both cover and diversity declines in the first years after clear felling (Hylander et al. 2005). Boreal forests that have had between 30 and 70 years to recover from clear felling have less diverse forest bryophyte communities compared with adjacent old stands, not least because of the lower amount of dead wood per area (cf. Dynesius et al. 2009). However, for both clear fell and selective harvesting, retention of microhabitats to 'lifeboat' some species through the disturbance can soften the effect, especially for species growing on concave surfaces (Hylander et al. 2005, Perhans et al. 2009).

The relative contribution of different factors, such as microclimate, dispersal ability, and substrate availability, in forest recovery is often not particularly clear. It would be fair to say that given the importance of coarse, woody debris for boosting bryophyte diversity in some forest ecosystems, repeated extraction of forest products needs careful management in order to prevent or minimize ongoing erosion of that resource. Epiphytic and epiphyllous species are obviously strongly affected by timber harvesting. On the other hand some of them might be adapted to certain successional stages, which means that a total focus on old-growth stages might put some species at a disadvantage (cf. the species composition in aspen stands regenerated after fires versus the dead wood rich stands developed after spruce budworm outbreaks; Schmalholz et al. 2011).

The conversion of more natural forests to plantations of exotic trees is an interesting and important issue. Plantations are often found to be hostile to bryophytes, but the severity of the impact can be softened through the retention of microhabitats. The replacement of native trees with exotic 'crop' trees tends to leave little in the way of native biodiversity, although biological legacies such as logs and green tree retention can soften the effect (Pharo and Lindenmayer

2009). However, in south-west Ethiopia, plantations of *Cupressus lusitanica* did contain quite high bryophyte diversity (Hylander and Nemomissa 2009).

Climate change

Our current understanding of bryophyte physiology with respect to climate change is informed by Tuba *et al.* (2011) who detailed the implications of our changing climate for bryophytes in particular environments. While they do not include a section on forests, there are discussions of bryoflora responses in Britain and Hungary, and Europe more generally, which are relevant to understanding forest bryophyte response to climate change. The main conclusions are that many bryophytes are adapted to conditions of intermittent water availability by being desiccation tolerant and avoiding drought through the development of specialist storage organs (Tuba *et al.* 2011). While bryophytes may have some inherent advantages compared with other taxa, Tuba *et al.* (2011) believe that desiccation will become a problem under changing climates. Some recent work suggests that this is already the case for more vulnerable species. For example, Ruete *et al.* (2012) concluded that populations of the moss *Buxbaumia viridis*, would decline by more than 65 per cent under even low-emission scenarios. This work does not take into account species interactions and other higher order effects, making it difficult to accurately predict what the consequences of changing climates might be.

Conclusion

Bryophyte ecology is similar to invertebrate ecology in that a thorough inventory of a particular area of forest tends to be highly labour-intensive and requires specialist knowledge that takes years to acquire. For this reason, there is not a large amount of literature on the effects of large-scale disturbances, such as timber harvesting, on bryophyte diversity. Canopy species are not well understood, nor are some of the most species-rich environments in tropical countries. However, we do know enough to conclude that some of the same general rules of thumb apply to bryophytes as for other organisms:

- There is a large variation among bryophyte response to disturbance, depending on their life history strategy.
- Some bryophyte species are resilient to disturbances, such as timber harvesting, within limits on the magnitude of the disturbance and intensity of the change, while others seems very sensitive to even slight changes in microclimate or substrate quality.
- Ongoing landscape-scale fragmentation and homogenization of stand ages and substrate availabilities tend to result in the decrease of bryophyte diversity.
- Climate change is an additional challenge over and above ongoing landscape change.

Given that bryophytes are difficult to identify, an important approach for informing managers is to emphasize that a variation in microhabitats generally will favour bryophyte diversity including certain species that are sensitive to alteration of habitats. Important microhabitats for bryophytes in forests could be: unaltered springs and small streams, shaded rocky outcrops, dead wood and a variation in tree species with different bark structures and chemistry. A complementary approach based on bryophyte life forms (colony sizes and shapes) could be possible to use in certain forest ecosystems. A manager without expert knowledge could, in tropical areas, identify stands with high cover of epiphyllous species or plenty of large pendulous or fan-like bryophytes growing on trees and shrubs. In temperate areas it could be more useful to look out for a variety of life forms growing over woody debris and rocky outcrops, sometimes epiphytic indicating a favourable microclimate, and therefore probable high bryophyte diversity.

References

Bay, G., Nahar, N., Oubre, M., Whitehouse, M. J., Wardle, D. A. and others (2013) 'Boreal feather mosses secrete chemical signals to gain nitrogen', *New Phytologist*, vol 200, pp. 54–60

Clark, K. L., Nadkarni, N. M. and Gholz, H. L. (2005) 'Retention of inorganic nitrogen by epiphytic bryophytes in a tropical montane forest', *Biotropica*, vol 37, pp. 328–336

Davey, M. L., Heegaard, E., Halvorsen, R., Ohlson, M. and Kauserud, H. (2012) 'Seasonal trends in the biomass and structure of bryophyte-associated fungal communities explored by 454 pyrosequencing', *New Phytologist*, vol 195, pp. 844–856

DeLuca, T., Zackrisson, O., Nilsson, M. and Sellstedt, A. (2002) 'Quantifying nitrogen-fixation in feather moss carpets of boreal forests', *Nature*, vol 419, pp. 917–920

Döbbeler, P. and Hertel, H. (2013) 'Bryophilous ascomycetes everywhere: Distribution maps of selected species on liverworts, mosses and Polytrichaceae', *Herzogia*, vol 26, pp. 361–404

Dynesius, M. (2001) Spatial and evolutionary aspects of species diversity, species traits, and human impact with examples from boreal riparian and forest plant communities. PhD-thesis. Umeå University, Sweden

Dynesius, M., Hylander, K. and Nilsson, C. (2009) 'High resilience of bryophyte assemblages in streamside compared to upland forests', *Ecology*, vol. 90, pp. 1042–1054

Eckstein, R. and Karlsson, P. (1999) Recycling of nitrogen among segments of Hylocomnium splendens as compared with Polytrichum commune: Implications for clonal integration in an ectohydric bryophyte, *Oikos*, vol 86, pp. 87–96

Fenton, N. J. and Bergeron, Y. (2008) 'Does time or habitat make old-growth forests species rich? Bryophyte richness in boreal *Picea mariana* forests', *Biological Conservation*, vol 141, pp. 1389–1399

Fischer, J. and Lindenmayer, D. B. (2007) 'Landscape modification and habitat fragmentation: A synthesis', *Global Ecology and Biogeography*, vol 16, pp. 265–280

Frahm J-P. and Gradstein, S. R. (1991) 'An altitudinal zonation of tropical rain forests using bryophytes' *Journal of Biogeography*, vol 18, pp. 669–678

Glime, Janice M. (2007) *Bryophyte Ecology*, Ebook sponsored by Michigan Technological University and the International Association of Bryologists. Accessed on 3 December 2013

Gradstein, S.R. (2006) 'The lowland cloud forest of French Guiana—a liverwort hotspot', *Cryptogamie Bryologie*, vol 27, pp. 141–152

Hedenäs, L. (2007) 'Global diversity patterns among pleurocarpous mosses', *Bryologist*, vol 110, pp. 319–331

Hylander, K. (2009) 'No increase in colonization rate of boreal bryophytes close to propagule sources', *Ecology*, vol 90, pp. 160–169

Hylander, K. and Dynesius, M. (2006) 'Causes of the large variation in bryophyte species richness and composition among boreal streamside forests', *Journal of Vegetation Science*, vol 17, pp. 333–346

Hylander, K. and Hedderson, T. A. (2007) 'Does the width of isolated ravine forests influence moss and liverwort diversity and composition? A study of temperate forests in South Africa', *Biodiversity and Conservation*, vol 16, pp. 1441–1458

Hylander, K., and Nemomissa, S. (2009) 'Complementary roles of home gardens and exotic tree plantations as alternative habitats for plants of the Ethiopian montane rainforest', *Conservation Biology*, vol 23, pp. 400–409

Hylander, K., Dynesius, M., Jonsson, B. G. and Nilsson, C. (2005) 'Substrate form determines the fate of bryophytes in riparian buffer strips', *Ecological Applications*, vol 15, pp. 674–688

Jonsson, B. G. and Esseen, P. (1990) 'Treefall disturbance maintains high bryophyte diversity in a boreal spruce forest', *Journal of Ecology*, vol 78, pp. 924–936

Lang, S. I., Cornelissen, J. H. C., Klahn, T., van Logtestijn, R. S. P., Broekman, R. and others (2009) 'An experimental comparison of chemical traits and litter decomposition rates in a diverse range of subarctic bryophyte, lichen and vascular plant species', *Journal of Ecology*, vol 97, pp. 886–900

Lecomte, N., Simard, M., Fenton, N. and Bergeron, Y. (2006) 'Fire severity and long-term biomass dynamics in coniferous boreal forests of eastern Canada', *Ecosystems*, vol 9, pp. 1215–1230

Lindo, Z. and Whiteley, J. (2011) 'Old trees contribute bio-available nitrogen through canopy bryophytes', *Plant and Soil*, vol 342, pp. 141–148

Lindo, Z., Nilsson, M. C. and Gundale, M. J. (2013) 'Bryophyte-cyanobacteria associations as regulators of the northern latitude carbon balance in response to global change', *Global Change Biology*, vol 19, pp. 2022–2035

Löbel, S., Snäll, T. and Rydin, H. (2006) 'Species richness patterns and metapopulation processes – Evidence from epiphyte communities in boreo-nemoral forests', *Ecography*, vol 29, pp. 169–182

Möls, T., Vellak, K. and Ingerpuu, A. N. (2013) 'Global gradients in moss and vascular plant diversity', *Biodiversity and Conservation*, vol 22, pp. 1537–1551

Oishi, Y. and Morimoto, Y. (2013) 'Identifying indicator species for bryophyte conservation in fragmented forests', *Landscape and Ecological Engineering*, vol. XX, pp. 1–8

Opelt, K., Berg, C., Schonmann, S., Eberl, L. and Berg, G. (2007) 'High specificity but contrasting biodiversity of Sphagnum-associated bacterial and plant communities in bog ecosystems independent of the geographical region', *Isme Journal*, vol 1 (6), pp. 502–516.

Perhans, K., Appelgren, L., Jonsson, F., Nordin, U., Söderström, B. and Gustafsson, L. (2009) 'Retention patches as potential refugia for bryophytes and lichens in forest landscapes', *Biological Conservation*, vol 142, pp. 1125–1133

Pharo, E. J. and Lindenmayer, D. B. (2009), 'Biological legacies soften pine plantation effects for bryophytes', *Biodiversity and Conservation*, vol 18, pp. 1751–1764

Pharo, E. J., Meagher, D. A. and Lindenmayer, D. B. (2013) 'Bryophyte persistence following major fire in eucalypt forest of southern Australia', *Forest Ecology and Management*, vol 296, pp. 24–32

Price, A., Dunham, K., Carleton, T. and Band, L. (1997) 'Variability of water fluxes through black spruce (*Picea mariana*) canopy and feather moss (*Pleurozium schreberi*) carpet in the boreal forest of Northern Manitoba', *Journal of Hydrology*, vol 196, pp. 310–323

Pypker, T. G., Unsworth, M. H. and Bond, B. J. (2006) 'The role of epiphytes in rainfall interception by forests in the Pacific Northwest. II. Field measurements at the branch and canopy scale', *Canadian Journal of Forest Research*, vol 36, pp. 819–832

Ruete, A., Yang, W., Bärring, L., Stenseth, N. C. and Snäll, T. (2012) 'Disentangling effects of uncertainties on population projections: Climate change impact on an epixylic bryophyte', *Proceedings of the Royal Society B: Biological Sciences*, vol 279, pp. 3098–3105

Schmalholz, M., Hylander, K. and Frego, K. (2011) 'Bryophyte species richness and composition in young forests regenerated after clear-cut logging versus after wildfire and spruce budworm outbreak', *Biodiversity and Conservation*, vol 20, pp. 2575–2596

Snäll, T., Ehrlén, J. and Rydin, H. (2005) 'Colonization-extinction dynamics of an epiphyte metapopulation in a dynamic landscape', *Ecology*, vol 86, pp. 106–115

Sporn, S. G., Bos, M. M., Kessler, M. and Gradstein, S. R. (2010) 'Vertical distribution of epiphytic bryophytes in an Indonesian rainforest', *Biodiversity and Conservation*, vol 19, pp. 745–760

Tng, D. Y. P., Dalton, P. J. and Jordan, G. J. (2009) 'Does moisture affect the partitioning of bryophytes between terrestrial and epiphytic substrates within cool temperate rain forests?', *Bryologist*, vol 112, pp. 506–519

Tuba, Z., Slack, N. G. and Stark, L. R. (2011) *Bryophyte Ecology and Climate Change*, Cambridge University Press, Cambridge, UK

Turetsky, M. R., Bond-Lamberty, B., Euskirchen, E., Talbot, J., Frolking, S. and others (2012) 'The resilience and functional role of moss in boreal and arctic ecosystems', *New Phytologist*, vol 196, pp. 49–67

Vanderpoorten, A. and Goffinet, B. (2009) *Introduction to Bryophytes*, Cambridge University Press, Cambridge, UK

Weibull, H. and Rydin, H. (2005) 'Bryophyte species richness on boulders: Relationship to area, habitat diversity and canopy tree species', *Biological Conservation*, vol 122, 71–79

Zartman, C. E., Nascimento, H. E., Cangani, K. G., Alvarenga, L. D. and Snäll, T. (2012) 'Fine-scale changes in connectivity affect the metapopulation dynamics of a bryophyte confined to ephemeral patches', *Journal of Ecology*, vol 100, pp. 980–986

18
LICHENS IN FOREST ECOSYSTEMS

Per-Anders Esseen and Darwyn Coxson

Introduction

Lichens are unique organisms that consist of a partnership (symbiosis) between a heterotrophic fungus and an autotrophic green alga (chlorolichens) or cyanobacterium (cyanolichens). The fungus (mycobiont, mostly an ascomycete) dominates and its hyphae envelop the algae or cyanobacteria (photobiont). The photobiont provides carbohydrates by photosynthesis while the mycobiont provides both protection from high light and desiccation and facilitates water storage. This close association has proved to be extremely successful and lichens have evolved a large diversity of life forms and adaptations that are widely distributed in forests throughout the world.

Lichens exhibit extreme variability in colour, growth form and size, ranging from less than 1 cm in many crustose species to over 1 m in length in some epiphytic species (e.g. *Ramalina menziesii*, *Usnea longissma*). The vegetative body (thallus) is usually classified into three main growth forms (Figure 18.1): crustose (forming a crust), foliose (flattened) and fruticose (erect or pendent, often richly branched), but intermediate types are frequently found.

Unlike plants, lichens lack root systems and therefore lack active mechanisms for regulation of uptake and loss of water (poikilohydry). Nonetheless, many lichens show morphological traits that facilitate regulation of hydration status. These include rhizines on the lower surface of *Peltigera* that wick moisture from the forest floor surface, development of cushions by reindeer lichens that facilitate retention of water, or the filmy thallus of understory rainforest species such as *Sticta filix* that promotes drying after rainfall events. These traits are linked to the main water sources in lichens (rain, humid air, dew) and help to explain lichen distribution and function (Gauslaa 2014).

The photobiont plays a critical role in lichen growth strategies. Cyanolichens must be wetted by liquid water for activation of photosynthesis while chlorolichens can be activated by humid air. This has major implications for the distribution of these lichens in relation to climate. Epiphytic cyanolichens, for example, dominate in oceanic areas with frequent rainfall, but in more continental areas are most abundant in sites with a more humid microclimate (e.g. sprayzones near waterfalls).

The functional role of lichens in forest ecosystems depends on the interaction of the traits in each species with forest structure and dynamics. In this chapter we give a broad overview of

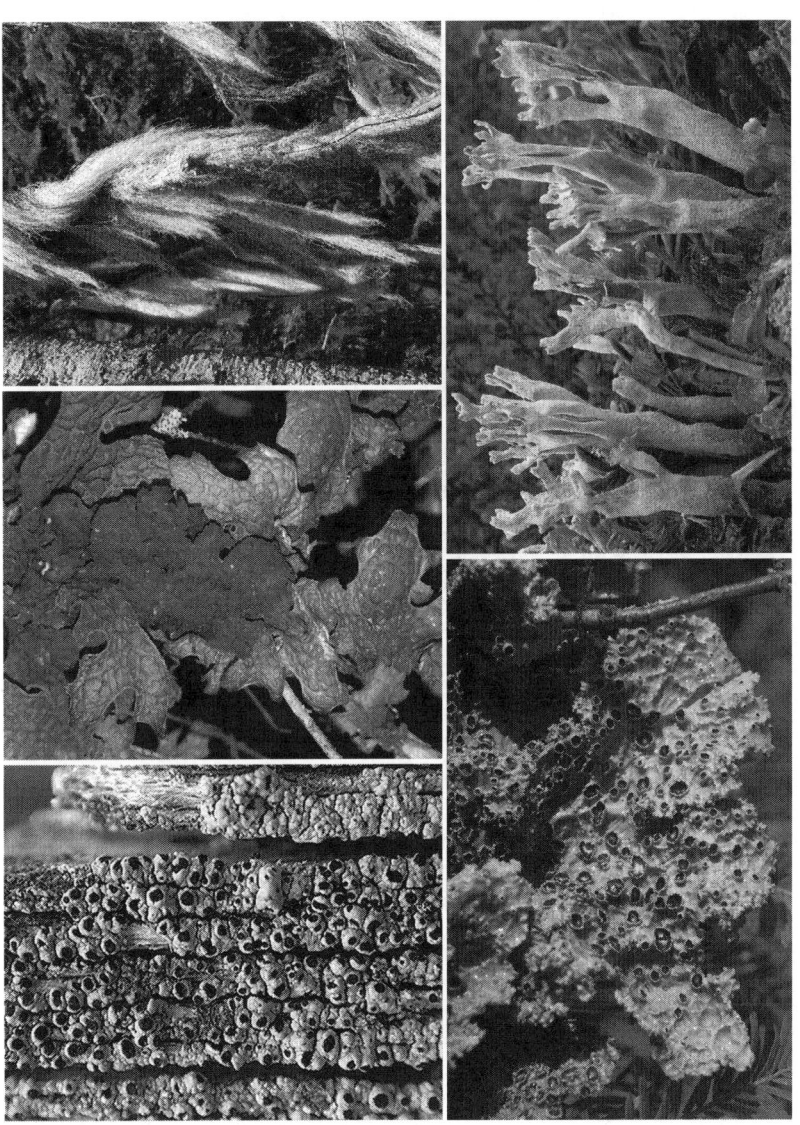

Figure 18.1 The crustose lichen *Cyphelium tigillare* on wood in boreal forest, Norway (top left). The foliose lichens *Lobaria pulmonaria* and *L. scrobiculata* (dark grey, in centre) epiphytic on *Picea glauca* in sub-boreal forest in northern British Columbia, Canada (top middle). The fruticose, pendulous lichens *Alectoria sarmentosa* (light colored) and *Bryoria* spp. (dark colored) on *Abies lasiocarpa* in subalpine forest, British Columbia, Canada (top right). The foliose lichen *Pseudocyphellaria coronata* on *Dacrycarpus dacrydioides* in south temperate New Zealand rainforest (bottom left). The fruticose lichen *Cladonia sulphurina* growing on boreal forest floor, Norway (bottom right). Photos (top left) and (bottom right) by Einar Timdal ©, (top middle) and (bottom left) by Darwyn Coxson ©, and (top right) by Per-Anders Essen ©

lichens in forest ecosystems throughout the world, their roles and interactions with other components in forests, and exemplify their role in selected forest ecosystems. For general information on lichens see Brodo *et al.* (2001) or Nash (2008). For more detailed reviews on forest lichens we recommend Sillett and Antoine (2004) or Ellis (2012).

Distribution of forest lichens

Lichens represent a significant component of biodiversity in forest ecosystems. Although some 20,000 lichen species have presently been described, the total number may be as high as 28,000 species, with an estimated 14,000 species found in the tropics (Lücking *et al.* 2009). By comparison, 5,000 species may be found in the boreal zone (G. Thor, pers. comm.). Microlichens (crustose) dominate by species number, and constitute about two-thirds of all lichens, while macrolichens (foliose + fruticose) make up the bulk of lichen biomass in most forests. Lichens can be found in a range of microhabitats within forests but epiphytes dominate by species numbers. Lücking *et al.* (2009) estimated that over 60 per cent of global lichen species richness was represented by species on bark or on leaves (foliicolous lichens).

Boreal forests

Boreal forests support abundant forest floor and canopy lichens but also host species-rich communities on rocks and woody debris (Esseen *et al.* 1997). They are less variable in terms of species composition than temperate and tropical forests, containing many widespread circumpolar macrolichens (Hauck 2011). Among factors contributing to the lower variability of boreal lichen communities is a similarity in ecological conditions, with cool and/or nutrient poor soils often limiting plant growth. Reoccurring fires also play a major role in maintaining the dominance of coniferous forests. Boreal lichens vary in response to fire and can be classified as:

- Fire-dependent (e.g. *Carbonicola* spp., *Hertelidea botryosa* on fire-scarred wood)
- Fire-related, associated with post-fire successions (e.g. *Collema* spp.)
- Fire-sensitive, preferring fire-refugia (e.g. *Usnea longissma*).

Terricolous lichens can dominate the forest floor in dry boreal forests. Lichen woodlands are especially well developed on coarse or well-drained soils, often growing under pine (*Pinus*), but also spruce (*Picea*), larch (*Larix*) and birch (*Betula*). The forest floor in lichen woodlands is dominated by chlorolichens, including reindeer lichens (mat-forming species in *Cladonia* subgenus *Cladina*) and cup-bearing *Cladonia* lichens, with *Cetraria* and *Stereocaulon* abundant in some regions.

Epiphytic lichen communities in the boreal forest reflect the dominance of conifers (*Abies, Larix, Picea, Pinus* spp.) and low tree species diversity. Crustose lichens dominate by species numbers, many of which are microhabitat specialists. A large number of crustose lichens are associated with bark and decaying wood surfaces (Figure 18.1 top left). They are often host-specific, particularly calicioid species ('pin lichens', crustose species that often have thin stalked fruiting bodies), many of whom are old-growth indicators (Thor 1998).

A striking feature of old coniferous forests throughout the boreal zone is the abundant growth of fruticose pendulous (alectorioid) lichens in the genera *Alectoria*, *Bryoria* (Figure 18.1 top right) and *Usnea*. These lichens can be thin and hairlike, which is an adaptation to rapid uptake of water from humid air. Alectorioid lichens show a typical vertical zonation, with dark

Bryoria spp. in the upper canopy and light, *Alectoria sarmentosa* in the lower canopy. Several factors are involved in this zonation. For example, *Bryoria* is sensitive to prolonged lower canopy hydration (Coxson and Coyle 2003) while dark fungal pigments (melanins) act as a sun-screen (Färber *et al.* 2014) and allow *Bryoria* to grow in well-lit upper canopies. The vertical zonation of canopy lichens is further described below in the section 'Interactions with climate'.

In younger boreal forests foliose chlorolichens, in contrast, constitute most of the epiphyte biomass on boreal conifers, including species of *Hypogymnia*, *Platismatia* and *Tuckermannopsis*. Rich cyanolichen communities can develop in boreal rainforests such as in western Norway, the northern half of temperate rainforests in western North America, wet forests of eastern Canada, and in inland southern Siberia (DellaSala 2011). Boreal rainforests often contain *Lobarion* community lichens, a distinct community of lichen epiphytes common to cool, temperate rainforests, including conspicuous foliose cyanolichens such as *Lobaria* (Figure 18.1 top middle), *Nephroma* and *Pseudocyphellaria* (Ellis and Coppins 2007, Rolstad *et al.* 2001).

Temperate forests

Lichen communities in temperate forests are more variable in species richness than in boreal forests due to higher tree species diversity, warmer climate and greater variation in precipitation and elevation. Major temperate forest ecosystems can be found along the Pacific Coast and inland north-western regions of North America, eastern North America, Chile and Argentina, Europe, China, Korea and Japan, and Australasia. Temperate forests have experienced significant deforestation from conversion to agricultural and/or urban landscapes (Nascimbene *et al.* 2013).

Temperate European forests in areas with cool and wet climates were historically known for rich canopy lichen communities, including members of the *Lobarion* community. *Lobarion* lichens are currently much more restricted in distribution due to the cumulative impacts of air pollution and habitat loss (Ellis and Coppins 2007). Many old-growth associated lichens in European temperate forests are red-listed (Nascimbene *et al.* 2013). Some rare/red-listed species occurring in gaps or other open forest environments are now found mainly on old trees in agricultural landscapes, parks and religious sites.

As in the boreal forest, lichen diversity in deciduous forests is strongly influenced by continuity of forest cover. Selva (2003) was able to distinguish between 'young old-growth stands' and 'ancient old-growth forests' in the Acadian Forest ecoregion of the Canadian Maritimes using calicioid lichen indicator species. The rich lichen communities in deciduous forests of the southern Appalachian Mountains in eastern North America are also thought to reflect their presumed role as refugia during past glacial episodes (Brodo *et al.* 2001).

Temperate rainforests in western North America, reaching from northern California to south-eastern Alaska, host diverse lichen communities (DellaSala 2011). Many of these lichens, including representatives of the *Lobarion* community, are associated primarily with old-growth forests, where site continuity and protected microclimates result in the accumulation of species that are otherwise rare in regional landscapes (Sillett and Antoine 2004).

High lichen diversity can also be found in the temperate rainforests of New Zealand, Australia and Chile, where foliose *Lobarion* community genera such as *Pseudocyphellaria* (Figure 18.1 bottom left) show high species diversity, reflecting their presumed origin in ancestral Gondwanaland forests (Nash 2008, p. 315). Temperate forests in China and Japan are also very species rich. Lichen floras in temperate forests in Asia are relatively unknown and in need of further study.

Tropical forests

Tropical forests support high lichen species richness with thousands of undescribed species (Lücking *et al.* 2009). The rapid conversion of species-rich primary forests to species-poor secondary forests or other land uses (Wolseley and Aguirre-Hudson 1997) makes the documentation of tropical lichen communities an urgent priority. Largely unexplored communities include mangrove forests, dry savanna forests and subtropical forests in Asia (Li *et al.* 2013).

The greatest lichen diversity and biomass within tropical forests occurs in canopy environments. About three-quarters of tropical lichens grow on bark or on leaves; this proportion is higher than in temperate and boreal forests (Lücking *et al.* 2009). Tropical forests stand out with a higher proportion of foliicolous lichens than in other forests (> 1,000 species), representing about 90 per cent of the global diversity in this group.

In general, lichens are a minor component of the forest floor flora in tropical lowland forests as the dense canopy restricts light penetration. The lichen flora in lowland forests is dominated by crustose epiphytic chlorolichens, with diverse communities in habitats such as mangrove forests. Little is known about the ecology of tropical crustose lichens, but annual growth rates may be as high as 5 mm – among the highest reported for crustose lichens (Zotz 1999). Foliicolous must complete their life-cycle in a short time period as leaves are ephemeral. The scarcity of macrolichens in lowlands is probably due to most species not being well adapted to the high temperatures, constant high humidity and low light levels in these forests (Zotz 1999).

Diversity and biomass of macrolichens increase with altitude in tropical forests (Wolf 1993). Lichen communities are best developed in montane forests, particularly in cloud forests, where broad-lobed cyanolichens, e.g. *Lobaria*, *Pseudocyphellaria* and *Sticta*, form a conspicuous epiphyte component.

Function in forest ecosystems

Lichens interact with forests in numerous ways and support many important ecosystem functions that we are now only beginning to understand. They are involved in biogeochemical cycles, provide food and habitat for many species of animals and interact with forest microclimates (Nash 2008, p. 274) as discussed below.

Biogeochemical cycling

Lichens and bryophytes (see Chapter 17) represent only a small proportion of the biomass in most forested ecosystems; nonetheless they can have a disproportionate impact on ecosystem function and nutrient cycling. Lichens, however, by virtue of their active uptake of nutrients from atmospheric sources and potentially high rates of turnover can sequester and release nutrients that might otherwise be unavailable.

Lichens play an especially important role in nitrogen cycling. Lichens with cyanobacterial symbionts are capable of fixing nitrogen from atmospheric sources. This ability to fix atmospheric nitrogen enriches the lichen thallus; fixed nitrogen is then available as further leaching and decomposition occur. Nitrogen fixation in lichens can have significant ecosystem impacts, especially where cyanolichens have a high standing biomass. In subarctic and boreal ecosystems terricolous lichens such as *Stereocaulon* are a major source of newly fixed nitrogen (Crittenden 2000), while in wet temperate rainforests *Lobarion* lichens provide up to 16 kg of N per hectare annually (Nash 2008, p. 232, Sillett and Antoine 2004). The amount of nitrogen fixed by lichens is small compared to the amount of nitrogen held in wood and other organic components, but this nitrogen is largely unavailable for growth and metabolism, being tightly bound in complex organic molecules. A more appropriate comparison is with other new nitrogen sources, such as inputs from precipitation, which are typically less than 5 kg per ha per year in pristine environments, or weathering from minerals.

Lichens are often regarded as slow growing and long-lived. However, growth and turnover of forest lichens can be significant. Standing biomass of reindeer lichens can range from 1,000–2,000 kg ha^{-1} in dry continental sites (Coxson and Marsh 2001) to over 9,000 kg ha^{-1} in wet oceanic sites (Auclair and Rencz 1982). Epiphytic lichens can likewise achieve substantial biomass accumulation. Stevenson and Coxson (2003) documented 1,200 kg ha^{-1} standing biomass of alectorioid lichens in old-growth subalpine coniferous forests, with annual turnover near 10 per cent. The decomposition of lichen litterfall in forests releases nutrients to the soil, however, lichen decomposition is still poorly understood (Cornelissen *et al.* 2007).

Mat-forming lichens in boreal forests provide an interesting example of nutrient cycling within individual lichen mats. Mats of *Cladonia stellaris* and *Stereocaulon paschale* often develop a thick layer of necrotic material at the base of the mat and it has been suggested that key elements are recycled internally to support new growth at the apices (Crittenden 2000). Although throughflow leachates from *Cladonia* mats have long been postulated to have allelopathic properties, recent studies suggest that phenolic compounds in *Cladonia* may function as substrates for soil microorganisms and have little influence on seedling germination (Favero-Longo and Piervittori 2010).

Interactions with animals

Lichens constitute forage, habitat, shelter and nest material for invertebrates and vertebrates in forests but animals, in turn, can be important as dispersal-vectors for lichens. Herbivory is low in most lichens, probably because of their low nutrient content, structural characteristics and defence compounds produced by the mycobiont (Solhaug and Gauslaa 2012). One example of such an antiherbivore compound is vulpinic acid, a bright yellow pigment that occurs in the wolf lichen (*Letharia vulpina*).

Among invertebrates, insects, mites and molluscs are the main lichen feeders (Seaward 1977, Nash 2008, p. 288). Cyanolichens with their higher nutrient content are preferred food for gastropods (reviewed by Solhaug and Gauslaa 2012). One study showed that gastropod grazing on three species of *Lobaria* increased near the ground on tree trunks, suggesting that gastropods may play a role in shaping the vertical distribution of lichens.

Chlorolichens in canopies are also important for invertebrates. Branches in old boreal *Picea abies* forests support much higher lichen biomass than in younger forests (Figure 18.2a). Branches with abundant lichens supported both greater diversity and number of invertebrates (Figure 18.2b) by providing more complex habitat structure than branches with few lichens. Large invertebrates (e.g. flies and spiders) are in turn important as forage for overwintering birds such as tits and the Siberian Jay (*Perisoreus infaustus*). Epiphytic lichens thus play a key role in the function of boreal forest canopies and changes in their biomass may have cascading effects on higher trophic levels (Pettersson *et al.* 1995).

Among vertebrates, some tree frogs, reptiles and birds are well camouflaged against a background of lichens but these animals do not eat lichens (Seaward 1977). A large number of forest birds use lichens for nest-building. In contrast, both ground and canopy lichens constitute an important part of the diet of several mammals. Reindeer and caribou (subspecies of *Rangifer tarandus*) represent the best examples of a mammal–lichen interaction (see below), but lichens are also preferred food by some species of deer, voles and others such as the northern flying

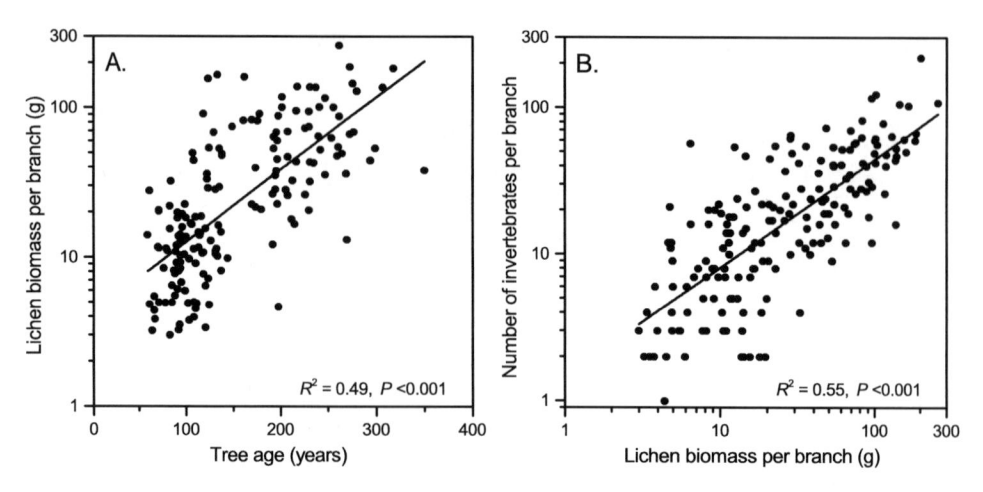

Figure 18.2 Relationship between biomass of epiphytic macrolichens per branch and tree age (a), and between number of invertebrates per branch and lichen biomass (b) in lower canopy of *Picea abies* forests in northern Sweden. Drawn from data in Esseen *et al.* (1996) and Pettersson *et al.* (1995).

squirrel (*Glaucomys sabrinus*). One interesting example is the bank vole (*Clethrionomys glareolus*), which prefers *Bryoria* spp. during winter.

Forage lichens for reindeer and caribou

Subarctic and boreal forests are the main habitats of reindeer and caribou during winter, and lichens are their dominant forage source, often constituting more than 50 per cent of the diet (Heggberget *et al.* 2002). Reindeer and caribou can locate lichens under snow pack by scent, and then use their front hooves to excavate craters. The most important terricolous forage lichens are 'reindeer lichens', particularly *Cladonia arbuscula*, *C. rangiferina*, and *C. stellaris* (Figure 18.3a), but *Cetraria*, *Stereocaulon* spp. and other species are also eaten. Forage lichens are generally rich in easily digestible carbohydrates, but poor in protein and minerals and the winter diet must be complemented with mushrooms, low shrubs and graminoids to be nutritionally balanced.

The selection of suitable lichen pastures takes place on several spatial scales: regional, stand and patches within stands suitable for cratering. Lichen availability is not just a function of the distribution of forests suitable for growth of forage lichens, it also depends on snow conditions, mainly depth and hardness (Heggberget *et al.* 2002).

High reindeer and caribou densities during winter may lead to overgrazing delaying the recovery of lichen mats. Although *Cladonia* lichens can regenerate well following fragmentation, the recovery of overgrazed lichen woodlands (Figure 18.3b) may take several years or even decades (Crittenden 2000).

When forest floor lichens are unavailable, reindeer and caribou often feed on epiphytic lichens. Highly digestible pendulous *Bryoria* spp. are preferred, but other species, such as *Alectoria* and some foliose lichens are also consumed. *Bryoria* spp. are an important winter forage for both woodland and mountain caribou ecotypes in western North America. The management of forests for caribou habitat in North America has become a critical issue for forest managers, especially for mountain caribou populations, which are in decline (Stevenson *et al.* 2001).

(a)

(b)

Figure 18.3 The lichen *Cladonia stellaris*, an important forage lichen for reindeer and caribou (a). Contrast between heavily grazed (left) and ungrazed forest floor (b), enclosure from 1949 with reindeer lichens in dry, boreal *Pinus sylvestris* forest in Sweden. Photos taken 2013 by P.-A. Esseen

Interactions with climate

Terricolous lichen mats such as *Cladonia arbuscula*, *C. stellaris* and *Stereocaulon paschale* play a major role in mediating above- and below-ground energy exchange in mid- to late-successional boreal and subarctic forests. Reflectance of shortwave radiation increases as terricolous lichen mats dominate the forest floor in late post-fire successional stands (Kershaw 1985).

Epiphytic lichens respond strongly to vertical and horizontal gradients in temperature, moisture and light within forest canopies. McCune (1993) noted that as forests age, lichen communities associated with xeric canopy conditions tended to move upwards in the canopy. McCune (1993) proposed the similar gradient hypothesis, namely that: 'epiphyte species are ordered similarly on three distinct spatial and temporal gradients':

1   Vertical differences in species composition within a given stand;
2   Species compositional differences among stands differing in moisture regime but of the same age;
3   Changes in species composition through time in a given stand.

Evidence for this hypothesis can be seen in comparisons between wet coastal and drier interior temperate forests (Figure 18.4) where foliose cyanolichens, which require mesic canopy conditions, occupy only the lower part of the dry interior forests (Benson and Coxson 2002), but are found in mid- to upper canopy in wet coastal forests, where bryophytes instead dominate the wet lower canopy (McCune 1993). The interactions between climatic and successional gradients for old forest lichens are reviewed by Gauslaa et al. (2007).

Interactions with light and humidity can also play a large part in the response of epiphytic lichens to forest edges. The drier conditions experienced by lichens near forest edges and the greater potential for wind scouring, can have significant negative impacts on epiphytic lichens (Esseen and Renhorn 1998). These impacts can be highly detrimental in fragmented landscapes, where edge influence reduces the proportion of forest area with interior microclimate.

Dynamics of lichen communities

The development of lichen communities is tightly linked to forest dynamics, where many lichens can be characterized as 'patch-tracking' organisms that follow tree population dynamics. Changes in biomass and species richness (see 'Distribution of forest lichens' above) are driven by the functional traits (morphology, reproduction, growth, defence compounds) in local populations as they respond to microclimate, chemical and physical environments (Ellis 2012), as well as the larger-scale impact of forest disturbance regimes (type, severity, size and interval between disturbances). For example, growth form may change during succession with crustose lichens being pioneers on bark followed by foliose chlorolichens and later by bryophytes and cyanolichens.

Epiphytic lichens

Epiphyte communities are controlled by factors that operate at tree, stand and landscape level (Ellis 2012), but mechanisms are complex as factors are often confounded.

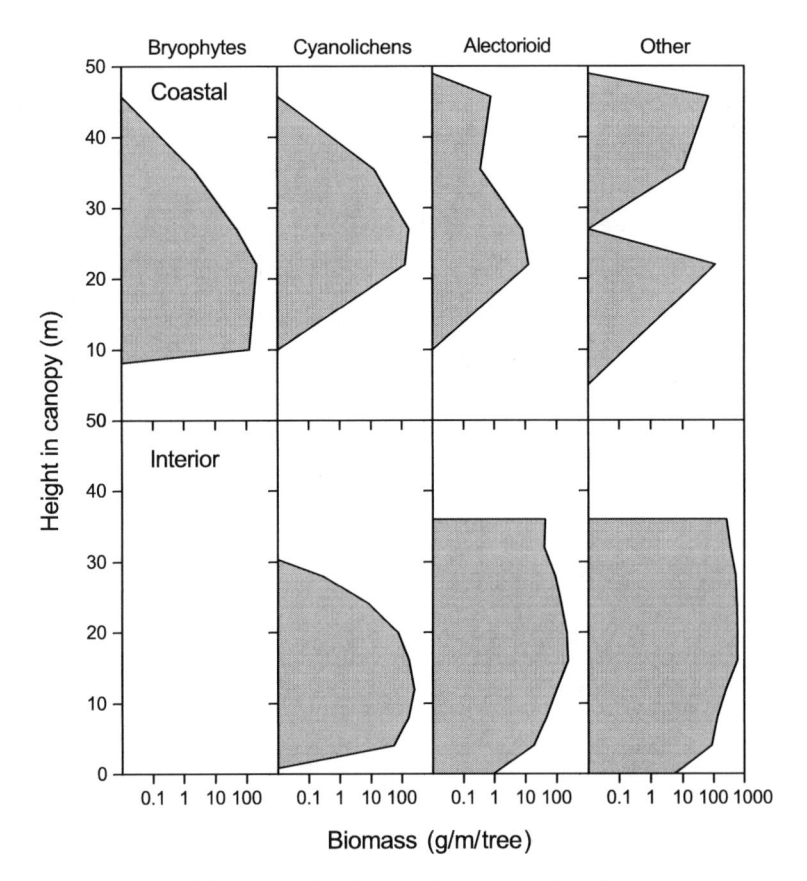

Figure 18.4 A comparison of the vertical distribution of canopy epiphyte functional groups in old-growth (> 400 years old) coastal (redrawn from McCune 1993) and interior (redrawn from Benson and Coxson 2002) wet temperate rainforests in western North America. No bryophyte data available for interior forests

Key factors for each level include:

- Tree level: tree species, bark chemistry (bark pH, mineral nutrients), bark structure (texture, stability, water-holding capacity), tree age, diameter, height and crown structure. The variability at this level is fundamental for species richness during succession as most lichens show some degree of microhabitat specificity. The time scale for epiphyte succession can vary greatly, but generally it is a slow process taking decades to centuries (Figure 18.2a).
- Stand level: tree species composition, horizontal (gaps, clumps of trees) and vertical canopy structure (foliage height profile), age distribution, stand size and shape. Forest structure changes during succession and modifies canopy microclimates, particularly openness and humidity. This, in turn affects lichen biomass accumulation over time (Sillett and Antoine 2004).
- Landscape level: position in relation to elevation, topography and water bodies, landscape composition and spatial configuration. Landscape heterogeneity affects microclimate, availability of lichen propagules and edge effects (Dettki and Esseen 1998).

Lichens on dead wood

The dynamics of lichen communities on dead wood varies with decay rate, competition with bryophytes, fungi and vascular plants as well as with microclimate (Spribille *et al.* 2008, Ellis 2012). The rapid turnover of woody debris in most tropical forests often hinders development of lichen communities. In contrast, dead wood may persist for decades or centuries in temperate and boreal forests and can host diverse communities on snags, stumps and logs. On treefalls, epiphytes on bark remains for some time and are then followed by wood-inhabiting and finally forest floor species.

Terricolous lichens

The successional pattern of terricolous lichens after fire is similar in nutrient poor lichen woodlands across the boreal zone with five stages (Kershaw 1985, p. 9, Seaward 1977, p. 165):

1   Bare soil stage
2   Crustose lichen stage (e.g. *Trapeliopis granulosa*)
3   Cup-lichen stage (e.g. *Cladonia cornuta, C. sulphurina;* Figure 18.1 bottom right)
4   First reindeer lichen stage (*C. arbuscula, C. rangiferina, C. uncialis*)
5   A second reindeer lichen stage often develops where *C. stellaris* dominate about 80–120 years after fire in northern lichen woodlands.

In practice, these stages overlap and the sequence is often interrupted by reindeer/caribou grazing or reset by fire. Fire maintains lichen woodlands within landscapes by creating a mosaic of early- to mid-successional conditions that favour lichen growth; however, as stands age and canopy closure develops terricolous lichens can be replaced by feather-moss mats (Coxson and Marsh 2001). Jonsson Čabrajić *et al.* (2010) found that growth of *Cladonia stellaris* dropped sharply when canopy openness was less than 40 per cent.

Indicators of forest ecosystem health

Epiphytic lichens have long been used as indicators of environmental quality. Their interception of nutrients from atmospheric sources, including entrapment of particles and active uptake of ions, can result in the accumulation of environmental contaminants (Nash 2008). This, combined with their differential sensitivity to pollutants, makes epiphytic lichens an important group from which to select indicator species. Terricolous lichens are less often used for these purposes due to the covariate influences of soil contaminants. Standardized methods for lichen indicators have been developed for many regions (Nimis *et al.* 2002).

Lichens have also been used to develop Indices of Ecological Continuity (IEC), based on the strong association with old stands and the need for site continuity (Nimis *et al.* 2002), and in several other contexts: as 'signal' species for stands with red-listed species, indicating biodiversity hotspots (woodland key habitats) and specific environmental conditions (e.g. high air humidity). *Lobaria pulmonaria* is a widely applied lichen indicator and is used as a model species for lichen conservation (Scheidegger and Werth 2009).

Methods of using lichens for conservation surveys have been developed in several countries (Nimis *et al.* 2002). A comprehensive system is used in Sweden, with species selected to indicate different forest types and microhabitats (Thor 1998). *Alectoria sarmentosa* and *Usnea longissima* are two well-known old-growth indicators on conifers in Fennoscandia (Esseen *et al.* 1997). *Usnea longissima* has faced severe declines throughout Europe.

The application of lichen indicators is, however, not without problems. For example, the links between lichens and their habitats are sometimes not fully understood and few studies separate long-term forest continuity from tree age (Rolstad *et al.* 2001). Ellis (2012) discussed three scenarios that could explain why epiphyte indicators need long forest continuity:

1. They are specialized on certain microhabitats and/or microclimates;
2. They are dispersal-limited (Sillett *et al.* 2000) and need a continuous supply of suitable habitats;
3. A combination of both.

The problem is exemplified by *U. longissima*, where tree age is a poor predictor of abundance, as most dispersal is local.

Conservation biology of forest lichens

Deforestation, landscape fragmentation, forestry, habitat degradation, pollution and climate change have all had dramatic effects on lichens, with many species declining in abundance or regionally extirpated (e.g. Esseen *et al.* 1996, Thor 1998, Scheidegger and Werth 2009). The boreal felt lichen (*Erioderma pedicellatum*), only found in Alaska, eastern Canada, Kamchatka and Scandinavia, is a prime example of a critically endangered boreal lichen.

A major problem in managed forests is the widespread adoption of clear-cutting with short rotations (100 years or less), which hinders the development of rich lichen communities (Esseen *et al.* 1997, Nascimbene *et al.* 2013). Industrial forestry drives forests well outside of their natural range of variability, shifting age–class distributions towards fewer older forest stands and reduced structural heterogeneity. Given the major changes in lichen communities as forest stands age it is important to maintain the diversity of all successional stages across landscapes.

In conclusion, key priorities for conservation of forest lichens and the ecosystem functions they support in a global perspective are to:

- Identify and protect lichen biodiversity hotspots at all spatial scales
- Maintain tree species diversity and its link to natural disturbance regimes
- Preserve old trees in all kinds of ecosystems.

Priorities in managed forests are to:

- Develop novel forest management methods (e.g. partial cutting, variable thinning, long rotations) as alternatives to clear-cutting
- Preserve lichen-rich forests that can function as sources of lichen propagules
- Retain and restore structural elements in managed forests such as native tree species, dead wood as well as old and large trees.

Acknowledgements

This study was supported by Formas, Sweden. We thank Göran Thor, the late Susan Stevenson and two anonymous reviewers for comments and Einar Timdal for permission to use his photos.

References

Auclair, A. N. D. and Rencz, A. N. (1982) 'Concentration, mass, and distribution of nutrients in a subarctic *Picea mariana-Cladonia alpestris* ecosystem', *Canadian Journal of Forest Research* vol 12, pp. 947–968

Benson, S. and Coxson, D. S. (2002) 'Lichen colonization and gap structure in wet-temperate rainforests of northern interior British Columbia', *Bryologist* vol 105, pp. 673–692

Brodo, I. M., Sharnoff, S. D. and Sharnoff, S. (2001) *Lichens of North America*, Yale University Press, New Haven, CT, US

Cornelissen, J. H. C., Lang, S. I., Soudzilovskaia, N. A. and During, H. J. (2007) 'Comparative cryptogam ecology: a review of bryophyte and lichen traits that drive biogeochemistry', *Annals of Botany* vol 9, pp. 987–1001

Coxson, D. S. and Marsh, J. (2001) 'Lichen chronosequences (postfire and postharvest) in lodgepole pine (*Pinus contorta*) forests of northern interior British Columbia', *Canadian Journal of Botany* vol 79, pp. 1449–1664

Coxson, D. S. and Coyle, M. (2003) 'Niche partitioning and photosynthetic response of alectorioid lichens from subalpine spruce-fir forest in north-central British Columbia, Canada: the role of canopy microclimate gradients', *Lichenologist* vol 35, pp. 157–175

Crittenden, P. D. (2000) 'Aspects of the ecology of mat-forming lichens', *Rangifer* vol 20, pp. 127–139

DellaSala, D. A. (ed.) (2011) *Temperate and Boreal Rainforests of the World: Ecology and Conservation*, Island Press, Washington, DC, US

Dettki, H. and Esseen, P.-A. (1998) 'Epiphytic macrolichens in managed and natural forest landscapes: a comparison at two spatial scales', *Ecography* vol 21, pp. 613–624

Ellis, C. J. (2012) 'Lichen epiphyte diversity: a species, community and trait-based review', *Perspectives in Plant Ecology, Evolution and Systematics* vol 14, pp. 131–152

Ellis, C. J. and Coppins, B. J. (2007) 'Changing climate and historic-woodland structure interact to control species diversity of the "*Lobarion*" epiphyte community in Scotland', *Journal of Vegetation Science* vol 18, pp. 725–734

Esseen, P.-A. and Renhorn, K.-E. (1998) 'Edge effects on an epiphytic lichen in fragmented forests', *Conservation Biology* vol 12, pp. 1307–1317

Esseen, P.-A., Renhorn, K.-E. and Pettersson, R. B. (1996) 'Epiphytic lichen biomass in managed and old-growth boreal forests: effect of branch quality', *Ecological Applications* vol 6, pp. 228–238

Esseen, P.-A., Ehnström, B., Ericson, L. and Sjöberg, K. (1997) 'Boreal forests', *Ecological Bulletins* vol 46, pp. 16–47

Färber, L., Solhaug, K. A., Esseen, P.-A., Bilger, W. and Gauslaa, Y. (2014) 'Sunscreening fungal pigments influence the vertical gradient of pendulous lichens in boreal forest canopies', *Ecology* vol 95, pp. 1464–1471

Favero-Longo, S. E. and Piervittori, R. (2010) 'Lichen-plant interactions', *Journal of Plant Interactions* vol 5, pp. 163–177

Gauslaa, Y. (2014) 'Rain, dew, and humid air as drivers of morphology, function and spatial distribution in epiphytic lichens', *Lichenologist* vol 46, pp. 1–16

Gauslaa, Y., Palmqvist, K., Solhaug, K. A., Holien, H., Hilmo, O. and others (2007) 'Growth of epiphytic old forest lichens across climatic and successional gradients', *Canadian Journal of Botany* vol 37, pp. 1832–1845

Hauck, M. (2011) 'Site factors controlling epiphytic lichen abundance in northern coniferous forests', *Flora* vol 206, pp. 81–90

Heggberget, T. M., Gaare, E. and Ball, J. P. (2002) 'Reindeer (*Rangifer tarandus*) and climate change: importance of winter forage', *Rangifer* vol 22, pp. 13–31

Jonsson Čabrajić, A. V., Moen, J. and Palmqvist, K. (2010) 'Predicting growth of mat-forming lichens on a landscape scale—comparing models with different complexities', *Ecography* vol 33, pp. 949–960

Kershaw, K. A. (1985) *Physiological Ecology of Lichens*, Cambridge University Press, Cambridge, UK

Li, S., Liu, W. Y., and Li, D.-W. (2013) 'Epiphytic lichens in subtropical forest ecosystems in southwest China: species diversity and implications for conservation', *Biological Conservation* vol 159, pp. 88–95

Lücking, R., Rivas Plata, E., Chaves, J. L., Umaña, L. and Sipman, H. J. M. (2009) 'How many tropical lichens are there... really?', in A. Thell, M. R. D. Seaward and T. Feuerer (eds), *Diversity of Lichenology – Jubilee Volume*, Bibliotheca Lichenologica vol 100, pp. 399–418, J. Cramer, Berlin, Germany

McCune, B. (1993) 'Gradients in epiphyte biomass in three *Pseudotsuga-Tsuga* forests of different ages in western Oregon and Washington', *Bryologist* vol. 96, pp. 405–411

Nascimbene, J., Thor, G. and Nimis, P. L. (2013) 'Effects of forest management on epiphytic lichens in temperate deciduous forests of Europe—a review', *Forest Ecology and Management* vol 298, pp. 27–38

Nash, T. H. L. III. (ed.) (2008) *Lichen Biology* (2nd edn), Cambridge University Press, Cambridge, UK

Nimis, P. L., Scheidegger, C. and Wolseley, P. A. (eds) (2002) *Monitoring with Lichens – Monitoring Lichens*, NATO Science Series vol 7, Kluwer Academic Publishers, Dordrecht, the Netherlands

Pettersson, R. B., Ball, J. P., Renhorn, K.-E., Esseen, P.-A. and Sjöberg, K. (1995) 'Invertebrate communities in boreal forest canopies as influenced by forestry and lichens with implications for passerine birds', *Biological Conservation* vol 74, pp. 57–63

Rolstad, J., Gjerde, I., Storaunet, K. O. and Rolstad, E. (2001) 'Epiphytic lichens in Norwegian coastal spruce forest: historic logging and present forest structure', *Ecological Applications* vol 11, pp. 421–436

Scheidegger, C. and Werth, S. (2009) 'Conservation strategies for lichens: insights from population biology', *Fungal Biology Reviews* vol 23, pp. 55–66

Seaward, M. R. D. (ed.) (1977) *Lichen Ecology*, Academic Press, London, UK

Selva, S. B. (2003) 'Using calicioid lichens and fungi to assess ecological continuity in the Acadian Forest Ecoregion of the Canadian Maritimes', *Forestry Chronicle* vol 79, pp. 550–558

Sillett, S. C. and Antoine, M. E. (2004) 'Lichens and bryophytes in forest canopies', in M. D. Lowman and H. B. Rinker (eds), *Forest Canopies* (2nd edn), pp. 151–174, Elsevier Academic Press, Burlington, MA, US

Sillett, S. C., McCune, B., Peck, J. E., Rambo, T. R. and Rutchy, A. (2000) 'Dispersal limitations of epiphytic lichens result in species dependent on old-growth forests', *Ecological Applications* vol 10, pp. 789–799

Solhaug, K. A., and Gauslaa, Y. (2012) 'Secondary lichen compounds as protection against excess solar radiation and herbivores', in U. Lüttge, W. Beyschlag, B. Büdel and D. Francis (eds), *Progress in Botany* vol 73, pp. 283–304, Springer-Verlag, Berlin, Germany

Spribille, T., Thor, G., Bunnell, F. L., Goward, T. and Björk, C. R. (2008) 'Lichens on dead wood: species-substrate relationships in the epiphytic lichen floras of the Pacific Northwest and Fennoscandia', *Ecography* vol 31, pp. 741–750

Stevenson, S. K. and Coxson, D. S. (2003) 'Litterfall, growth and turnover of arboreal lichens after partial cutting in an Engelmann spruce-subalpine fir forest in north-central British Columbia', *Canadian Journal of Forest Research* vol 33, pp. 2306–2320

Stevenson, S. K., Armleder, H. M., Jull, M. J., King, D. G., McLellan, B. N. and Coxson, D. S. (2001) *Mountain Caribou in Managed Forests: Recommendations for Managers* (2nd edn), Ministry of Environment, Lands and Parks, Wildlife Report no R-26, Victoria, BC, Canada

Thor, G. (1998) 'Red-listed lichens in Sweden: habitats, threats, protection, and indicator value in boreal coniferous forests', *Biodiversity and Conservation* vol 7, pp. 59–72

Wolf, J. H. D. (1993) 'Diversity patterns and biomass of epiphytic bryophytes and lichens along altitudinal gradients in the northern Andes', *Annals of the Missouri Botanical Garden* vol 80, pp. 928–960

Wolseley, P. A. and Aguirre-Hudson, B. (1997) 'The ecology and distribution of lichens in tropical deciduous and evergreen forests of northern Thailand', *Journal of Biogeography* vol 24, pp. 327–343

Zotz, G. (1999) 'Altitudinal changes in diversity and abundance of non-vascular epiphytes in the tropics—an ecophysiological explanation', *Selbyana* vol 20, pp. 256–260

19
MAMMALS IN FOREST ECOSYSTEMS

Richard T. Corlett and Alice C. Hughes

Introduction

The more than 5,400 mammal species are divided between 28 orders (Figure 19.1), almost all of which include species that occur in forests. Mammals occupy every substantial area of forest worldwide and play multiple ecological roles (Table 19.1). The diversity of most large orders peaks in the tropics (Rolland *et al.*, 2014), and more than 120 species of mammals have been recorded from the richest tropical rainforest sites (e.g. Happold, 1996; Voss and Emmons, 1996; Figure 19.2). Diversity declines slowly with latitude but rapidly with isolation on oceanic islands, reflecting the poor cross-water dispersal of most taxa. Diversity has also declined since the Pleistocene origin and spread of modern humans. The majority of terrestrial ecosystems outside Africa have lost large mammals since the middle Pleistocene (Corlett, 2013) and declines in vulnerable species have accelerated in recent decades (Di Marco *et al.*, 2014).

Bats are the most widespread mammals and the only native terrestrial species on Hawaii, New Zealand and many other Pacific islands. Rodents now equal this distribution, but no native species occurs on the more isolated islands. Artiodactyls (even-toed ungulates) and carnivores occur in most forests, and primates in most warm regions, but they did not reach New Guinea or Australia, or most small islands, until introduced by humans. Proboscids (elephants and their relatives) were almost as widespread until the late Pleistocene, but suffered massively during the megafaunal extinctions mentioned above and are now confined to the tropical and subtropical forests of Asia and Africa. The true insectivores (Eulipotyphla) are also widespread in forests, but did not reach New Guinea, Australia, or Madagascar, and are recent arrivals in South America, where they are restricted to the northern Andes. The other orders are all geographically more or less restricted, although many are regionally important (Figure 19.2).

The literature on forest mammals is strongly biased taxonomically, while we have tried to provide a comprehensive coverage of all forest mammals, including those small, nocturnal and/or uncharismatic taxa that have attracted only a few specialists. An inevitable corollary of this is that we have had space for only a superficial summary of the vast literature on, for example, forest primates or cats. We have focused on diversity and function (i.e. ecological roles) in the order-level accounts below, which are structured phylogenetically, and then cover other topics at a more general level.

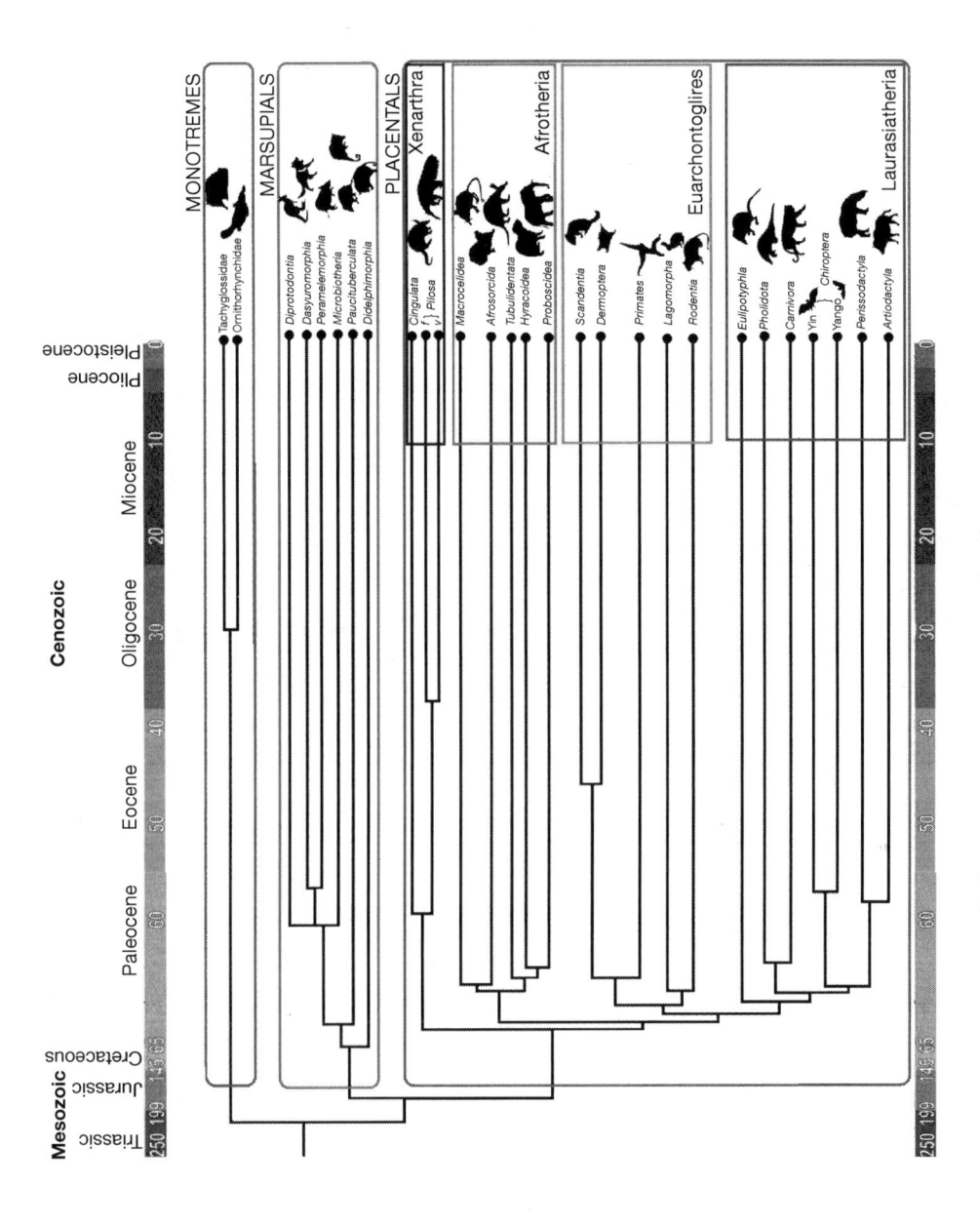

Figure 19.1 A phylogeny of the extant mammalian orders based on morphological and molecular data, re-drawn from O'Leary *et al.* (2013) for placentals and Springer *et al.* (2009) for marsupials. Note that the suggested post-Cretaceous age of the initial radiation of placental mammals is controversial, although consistent with a literal interpretation of the current fossil record

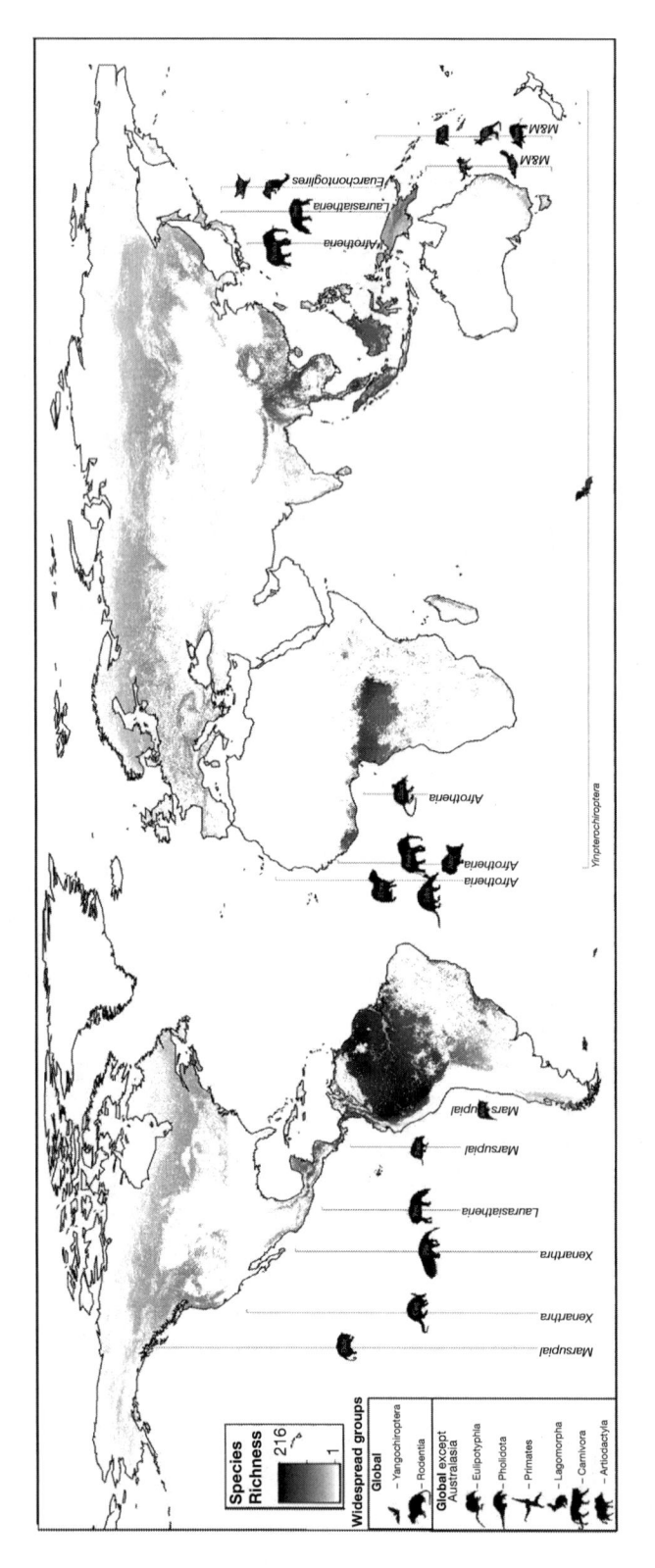

Figure 19.2 The global distribution of the mammalian orders in forests and the pattern of mammalian species diversity. The icons for the orders are the same as used in Figure 19.1. Forest cover for 2012 was extracted from the MODIS dataset (MCD12Q1, with forest layers as defined by the IGBP system). The mammal data with a spatial resolution of 10 × 10 km is from Jenkins et al. (2013), who assembled it from IUCN range maps. Forested areas were used to extract mammal richness to display the number of species of mammal within each 10 km grid for forested areas

Table 19.1 Major ecological roles played by the mammalian orders in forests. The number of crosses indicates the probable significance of the order in this role: X – present; XX – at least locally significant; XXX – at least locally dominant. Minor roles are omitted. See the text for details. 'Others' refers to other specialized diets such as sap or blood. Orders which feed on fruits may play an important role in seed dispersal while those that feed on nectar may be involved in pollination.

Mammal Order	Ecosystem engineer	Invertebrates	Social insects	Small vertebrates	Large vertebrates	Leaves	Browse	Bark/ cambium	Fruits	Seeds	Nectar/pollen	Others
Monotremata		XX	XXX									
Didelphimorphia		XX		X					XX		X	X
Paucituberculata		XX		X								
Microbiotheria		X							X		X	
Dasyuromorphia		XX		XX	XX							
Peramelemorphia		XX		X								
Diprotodontia						XXX	XXX		XX		X	X
Afrosoricida		XXX		X								
Macroscelidea		XX										
Tubulidentata			XX									
Hyracoidea						XX						
Proboscidea	XXX					XXX	XXX	XX	XX			
Cingulata		XXX										
Pilosa			XXX			XXX						
Scandentia		XX		XX					XX			
Dermoptera						XX						
Primates		XX				XXX	X	X	XXX	XX		X
Rodentia	XXX	XX				X		XX	XX	XXX	X	
Lagomorpha						XX	XXX					
Eulipotyphla		XXX		X								
Chiroptera		XXX	X	X					XXX		XX	
Pholidota			XXX									
Carnivora		XX	X	XX	XXX				XX		X	X
Perissodactyla	X					XX	XXX		XX			
Artiodactyla	X					XXX	XXX	X	XX			

Mammalian orders in forests

Monotremes

Order Monotremata

A single order and five species remain in this subclass, restricted to Australia and New Guinea. The short-beaked echidna (*Tachyglossus aculeatus*) is found all over Australia and southern New Guinea and is a specialist on ants and termites, while the three species of long-beaked echidnas (*Zaglossus*) occur only in New Guinea and have broader diets of earthworms and other subterranean invertebrates. The platypus (*Ornithorhynchus anatinus*) occurs in forest streams in Australia, where it feeds on small aquatic vertebrates and invertebrates.

Marsupials

Order Diprotodontia

This order, with around 120 species, includes the most important forest herbivores of Australia and New Guinea. Up to 20 species can coexist in the same landscape, although with specialization by habitat (e.g. Vernes *et al.*, 2006). The family Macropodidae consists of the terrestrial kangaroos and wallabies in addition to arboreal tree-kangaroos. These are browsers and grazers with a ruminant-like digestive system, playing similar ecological roles to the artiodactyls on the ground and colobine primates and sloths in the canopy. The family Phalangeridae includes the largely arboreal brushtail possums and cuscuses, which feed mostly on leaves and fruits. Other families include the Pseudocheiridae (ringtail possums), which are specialized leaf eaters, and the omnivorous Petauridae, which includes the striped-possums and several species of gliders. The order also includes the koala, a specialized folivore, and three species of omnivorous, burrowing wombats. The impacts of native herbivores at natural densities are difficult to assess, but the potential roles of diprotodontid herbivory in Australian forests are illustrated by the effects of overabundant swamp wallabies (*Wallabia bicolor*) on the regeneration of palatable species following exotic predator control (Dexter *et al.*, 2013). Similarly, brushtail possums (*Trichosurus vulpecula*) introduced to New Zealand, where they lack predators, reduce the canopy biomass of preferred tree species and increase mortality (Gormley *et al.*, 2012).

Order Dasyuromorphia

Dasyuromorphia includes three recent families, the Dasyuridae, with 72 species, the Mymercobiidae, with one, the termite-specialist numbat (*Myrmecobius fasciatus*), and the Thylacinidae, whose sole survivor, the dog-sized thylacine, became extinct in 1936. The biggest surviving dasyurid, the 6–8 kg Tasmanian devil, is also extinct on the Australian mainland. In addition to devils, six species of quolls (*Dasyurus* spp.; 0.3–7 kg) feed mainly on other vertebrates, and are capable of killing prey several times their own size (Jones and Barmuta, 2000; Glen *et al.*, 2010). Devils are also bone-crushing scavengers. Up to four large dasyurid species coexisted in some Australian forests before the introduction of the placental dingo, around 4,000 years ago, and the feral cat and European fox in the last 200 years. The other members of this order are smaller and feed mostly on invertebrates, but even species < 50 g kill and eat small vertebrates.

Order Microbiotheria

The only living member of this order, *Dromiciops gliroides*, is a small arboreal marsupial found in the temperate forests of South America (Fontúrbel *et al.*, 2012). It is omnivorous, but consumes large amounts of fruit and appears to be a key dispersal agent (Amico *et al.*, 2009).

Order Paucituberculata

This order consists of seven species of poorly known shrew-opossums, confined to cold and humid environments in the Andes (Ojala-Barbour *et al.*, 2013). They feed on invertebrates and small vertebrates, and spend much of their time in burrows.

Order Didelphimorphia

This order includes most of the surviving American marsupials (> 100 species), known as opossums. They are confined to South and Central America, with only the largest species, the Virginia opossum, extending north into the United States and southern Canada. Amazonian rainforests can support at least 12 coexisting species (Voss and Emmons, 1996). Most are more or less arboreal and more or less omnivorous, but some focus more on fruit, leaves, exudates, or animal prey, and some spend more time on the ground.

Xenarthra

Order Cingulata

This order was formerly much more diverse than the present 21 armadillo species and included a number of extinct families, such as the glyptodonts and pampatheres, which formed part of the late Pleistocene megafauna of the Neotropics (Corlett, 2013). Diets vary but the forest armadillos feed predominantly on invertebrates. At least four species can coexist in Amazonian rainforests (Voss and Emmons, 1996),

Order Pilosa

The two suborders, sloths (Folivora; six species) and anteaters (Vermilingua; four species), are mostly confined to tropical forests in South and Central America. In the late Pleistocene, the sloths included terrestrial species the size of modern elephants. Both suborders have specialist adaptations to their diets, with sloths feeding on leaves, and anteaters on ants and termites. All extant sloths are arboreal but the anteaters include the entirely arboreal *Cyclopes*, two semi-arboreal species of *Tamandua* and the terrestrial *Myrmecophaga*.

Afrotheria

Order Macroscelidea

The 17 species of elephant shrews (sengis) are confined to Africa and mostly occupy non-forest habitats, but a few species occur in forests (e.g. Rovero *et al.*, 2013), where they feed on the ground, mostly on invertebrates.

Order Afrosoricida

The tenrecs (Tenrecidae) are diverse and abundant in Madagascan forests, where they co-dominate the small-mammal community with rodents (e.g. Dammhahn *et al.*, 2013). Local species richness varies from 10–15 in humid montane forests to 2–7 in dry forests (Stanley *et al.*, 2011). They range in size from 5 g to 2.5 kg and include climbing, ground-dwelling, swimming and burrowing species. Little is known about their diets or other aspects of their ecology, although all species appear to feed predominantly on invertebrates and the larger species take some small vertebrates. The three species of otter shrews are African tenrecs that feed on aquatic invertebrates and vertebrates. Southern African forests also support several species of burrowing, insectivorous, golden moles (Chrysochloridae).

Order Tubulidentata

The only surviving species in this order is the aardvark, which is better adapted to more open habitats but also occurs in forests across much of Africa. It is a specialist feeder on ant and termite nests and shows evolutionary convergence with other myrmecophagous taxa (Delsuc *et al.*, 2014).

Order Hyracoidea

The two species of tree hyraxes are small (< 5 kg) arboreal folivores in African forests (Milner and Harris, 1999).

Order Proboscidea

Although there are only three species extant today (two African and one Asian), the proboscideans were much more diverse and widespread in the middle to late Pleistocene (Corlett, 2013). Among these recently extinct taxa, at least the many Asian *Stegodon* species and the North American mastodon (*Mammut americanum*) inhabited forests, as probably did at least some gomphotheres of Central and South America. The African and Asian forest elephants selectively consume a variety of plant materials in large amounts and are important—sometimes apparently unique—dispersal agents for trees with very large, large-seeded, fruits (Campos-Arceiz and Blake, 2011). They also share with beavers the ability to re-shape their habitat as ecosystem engineers (Haynes, 2012).

Euarchontoglires

Order Scandentia

This order consists of 20 species of tree shrews found in Asian tropical and subtropical forests, with up to five species coexisting at one site (Emmons, 2000). They range from largely terrestrial to fully arboreal and their diets include various combinations of invertebrates, fruits and small vertebrates. Some species may be significant dispersal agents for small seeds in soft fruits, such as figs (Shanahan and Compton, 2000).

Order Dermoptera

This order consists of two species of colugos, one in the southern Philippines and the other occupying much of Southeast Asia. They are poorly studied, nocturnal, gliding folivores.

Order Primates

The primates are the third largest mammalian order, with 480 species, all but a few associated with forests. Although now largely confined to the tropics, gibbons, colobines and macaques were widespread in subtropical Asia into historical times and macaques still extend north into temperate deciduous forests in several places. Maximum diversities are in wet tropical forests, where up to a dozen co-occur in the richest sites (Corlett and Primack, 2011). Most primate communities include specialist leaf eaters, seed-predators, frugivores and insectivores, and many include specialists on plant exudates. Where present, primates make up a large proportion of the canopy vertebrate biomass and the frugivorous species are probably third in importance to birds and bats as seed dispersal agents.

Order Lagomorpha

Although most species in this order are found in non-forest habitats, there are a few forest specialists and others that occupy open forests with a grassy understory. Snowshoe hares (*Lepus americanus*) occupy boreal and montane forests across North America, where they are active year-round (Wirsing and Murray, 2002). In summer they eat herbs and new woody growth, but in winter they depend on a few species of woody browse, and can leave well-defined browse-lines. These hares in turn are an important food supply for several predators, including lynx, owls and diurnal birds of prey. Other *Lepus* species play similar roles in northern Eurasia (Lyly *et al.*, 2014). A single species of rabbit (*Sylvilagus brasiliensis*) is widespread in South American rainforests and there are three localized species in tropical East Asian forests, but little is known about any of these tropical forest rabbits.

Order Rodentia

Most non-flying mammals are rodents (> 2,000 species), but their local diversities are generally lower in forests than those of bats, with a maximum of 20–30 species in some tropical forests (Happold, 1996; Voss and Emmons, 1996). Rodents have chisel-shaped, continually growing, gnawing incisors, coupled with large and complex jaw muscles, which gives them access to mechanically protected plant foods, such as seeds and cambium, although many also eat softer plant foods and invertebrates. There are also specialist folivores and insectivores, particularly on islands, and even one species on Sulawesi that lacks molars and specializes on soft-bodied earthworms (Esselstyn *et al.*, 2012). Rodent communities utilize the full volume of forests, with gliding squirrels in Asia and North America and the unrelated gliding anomalures (Anomaluridae) in Central Africa, arboreal members in several families, and a variety of ground-dwelling species, some of which burrow.

In addition to consuming diverse plant materials and invertebrates, forest rodents impact plant reproduction through their consumption of large seeds. Most of this is predation, which may occur before or after seed dispersal, but some terrestrial rodents cope with the fluctuating supply of seeds by storing some of the surplus in scattered caches of one or a few seeds buried in the soil—scatter-hoarding. This can be a very effective means of dispersal, even if most cached seeds are eventually retrieved and eaten, and at least one highly successful plant family, the Fagaceae (oaks, chestnuts etc.) depends on it, as do many taxa in other families (Vander Wall, 2010). Both Old and New World porcupines feed on the cambium of preferred tree species and can cause significant damage to plantations. Despite their small sizes, several species of voles can also cause significant damage to seedlings of preferred tree species in boreal forests, with damage most obvious during the peak phase of their population cycles (e.g. Lyly et al., 2014). The two species of beaver, found in North America and western Eurasia respectively, are unique in their roles as ecosystem engineers, cutting trees for use as food and in the construction of dams and lodges (Nummi and Kuuluvainen, 2013). Rodents dominate the diets of numerous medium-sized carnivores, including mammals, birds and snakes, and form an essential link in forest food chains.

Laurasiatheria

Order Eulipotyphla

This order includes the shrews (Soricidae), the hedgehogs, moonrats and gymnures (Erinaceidae), the moles, desmans and shrew-moles (Talpidae), and the solenodons (Solenodontidae). The order is diverse and abundant in forests across Eurasia and the Americas, south to the northern Andes. They range in size from around 2 g to more than 1 kg, although the few larger taxa have restricted distributions. The highest shrew diversities are in African tropical forests (eight to eighteen species: Dudu et al., 2005; Stanley et al., 2011), where they are the only members of the order present and eat overlapping combinations of arthropods, earthworms and snails. Diversities are lower in the temperate zone (two to nine species), but with a clearer ecological separation on the basis of body size (Churchfield et al., 1999).

The Erinaceidae are mostly larger and much less diverse than the shrews, and are absent from the Americas and from most forested areas in Africa. Hedgehogs occur in forests across northern Eurasia, while the moonrats and gymnures are confined to moist forests in Southeast Asia (He et al., 2012). Diets, where known, are similar to the larger species of shrew, but they consume more vertebrates and, in some species, plant material. Eurasian and North American forests also often support one to two species of subterranean moles (Talpidae). The two species of *Solenodon* are relatively large (c. 1 kg), venomous, burrowing animals, found only on the Caribbean islands of Cuba and Hispaniola. They have similar diets to the other large members of the order.

Order Pholidota

The eight species of pangolins are confined to tropical and subtropical Asian and African forests. Pangolins are terrestrial or semi-arboreal specialists feeding on ants and termites.

Order Carnivora

Excluding the semiaquatic pinnipeds, the Carnivora includes 245 species currently placed in 13 families, with the African palm civet, Asian linsangs, Malagasy carnivores, skunks and red panda now each having their own families (Eizirik et al., 2010). All families but the hyenas have representatives in forests. Local diversity of Carnivora communities is highest in the tropics, with 15–25 coexisting species in lowland Southeast Asia, < 15 in the Amazon and Central America, and < 10 in Africa (Corlett, 2011). Carnivora reached Madagascar once, but did not reach New Guinea and Australia until the recent introduction of dogs to both and cats and foxes to Australia. The largest species in each region, such as tigers in Asia, can kill all but the largest forest herbivore. The loss of large carnivores can have major consequences for the rest of the food web, increasing browsing pressure on plants and releasing populations of medium-sized carnivores, which are the major predators of small vertebrates (Ripple et al., 2014). Most Carnivora, except the hyper-carnivorous cats, also eat some fruit, and the most frugivorous species, including bears, many mustelids, civets (in Asia and Africa) and procyonids (in the Americas), are important dispersal agents for large seeds in large fruits (Corlett and Primack, 2011).

Order Chiroptera

Bats make up a quarter to a fifth of known mammal species with 1,306 species currently described; 120 since 2005 (Tsang et al., in press). Within forests, bats are frequently the most species-rich mammal group, and this is especially marked in the tropics. Peak diversities are attained in tropical forests in South America and Southeast Asia, which can support > 60 species in one area. Diversity declines with distance from the equator (Buckley et al., 2010), but bats are found from 70°N to 54°S.

Around a third of bat species, largely in tropical forests, feed on fruit and/or nectar, with separate clades involved in the New World (Phyllostomidae) and Old World (Pteropodidae) (Corlett and Primack, 2011). Most other species are insectivores, but forest-dwelling bats have a wide variety of diets, including blood, fish and small vertebrates, as well as specialists on particular invertebrate groups. Bats play a number of important roles within forest ecosystems, including the regulation of insect populations. They pollinate many plant species and for some are the major or only pollinator. Forest bats also have a major role in seed dispersal. On oceanic islands, large pteropodid fruit bats can be essential mutualists for many plant species (Scanlon et al., 2014).

Order Perissodactyla

This order includes five species of tapir, four in South and Central America, and one in Southeast Asia, and two forest rhinoceroses in Southeast Asia, plus additional non-forest rhinoceroses and horses. The forest species are large browsers and consumers of fallen fruits. The forest rhinos are large enough to push over saplings and small trees, and at natural densities (which no longer occur anywhere) may have influenced forest structure in a similar way to elephants. All species disperse seeds from fallen fruits (Corlett, 1998; Campos-Arceiz et al., 2012).

Order Artiodactyla

The even-toed ungulates include 247 species in ten families, seven of which are commonly found in forests: the non-ruminant pigs (Suidae) and peccaries (Tayassuidae), and the ruminant chevrotains (Tragulidae), okapi (Giraffidae), deer (Cervidae), musk deer (Moschidae) and cattle (Bovidae). They did not reach New Guinea, Australia, or Madagascar until introduced by humans. They were also absent from remote oceanic islands, although at least pigs are now present on most and *Bubalus* reached Sulawesi and several Philippine islands. Artiodactyls are most diverse in the moist tropical forests of Central and West Africa, where up to seven species of duikers (small forest antelopes; Bovidae) coexist with several larger taxa (Corlett and Primack, 2011), followed by tropical Asia, with fewer species in the forests of South and Central America

(< 4) and outside the tropics. Being confined to the forest floor limits their ecological roles to browsing plants within reach, trampling plants underfoot, grazing grass in forest gaps and edges, and consuming fruits and other plant material that fall from the canopy (Corlett and Primack, 2011). Browsing impacts can be dramatic when populations are high, with long-term effects on the structure and species composition of the forest. Some species also eat tree bark and the pigs excavate roots and consume some small animals. Many species disperse small seeds from the fruits they eat in their feces and deer disperse large, hard seeds by regurgitation (Prasad et al., 2006). Artiodactyls are the major prey of the largest forest carnivores and their feces support a diversity of specialist beetles and other invertebrates.

Mammals vs. other animal groups: how important are they in forests?

Most ecological roles played by forest mammals are shared with other taxa, including birds and snakes as predators of vertebrates, numerous small vertebrate and invertebrate species as predators of invertebrates, herbivorous insects as consumers of most plant materials, and birds as consumers of fleshy fruits. Within these broadly defined roles, however, there is usually a part that is exclusive to mammals. At most mainland forest sites with intact faunas, the apex predator is a mammal (e.g. the tiger in Asia), although birds or snakes may be apex predators on islands (Corlett, 2011). Mammals also appear to dominate predation on large colonies of social insects, which are well defended against invertebrate predators, and bats are the major predators on nocturnal flying insects. Total folivory by mammals is probably a small fraction of that by insects, but browsing of woody shoots by large mammalian herbivores has an impact on plant growth and form that no other animal equals. Most forests have fruit types that are taken exclusively or mainly by mammals, and even where fruits are shared with birds, mammals may provide complementary seed dispersal services. In the tropics, there are also many plants with flowers specialized for bat pollination (Corlett and Primack, 2011).

Management issues

Overexploitation

The recent literature on tropical forest mammals is dominated by the impacts of hunting: mainly for bushmeat, but also elephants for ivory and, particularly in Asia, a wide range of species for traditional medicine. Even subsistence hunting by low density populations using traditional methods depletes populations of preferred species, but these impacts are potentially reversible if the hunters move to a new area. Over the last 20 years in particular, however, hunting has drastically reduced the diversities and densities of large and medium-size mammals in all but the most remote and/or best-protected areas, resulting in local and regional extirpations and threatening global extinctions (e.g. Laurance et al., 2012; Abernethy et al., 2013; Harrison et al., 2013). The causes are similar across the tropics and much of the subtropics: increasing commercial trade fueling hunting beyond subsistence needs; increased accessibility along new roads and from new settlements; improved hunting technology, including access to guns in many areas; and increased income and demand, particularly in urban areas. Hunting is initially selective for large-bodied species, but overhunting fuels a demand for progressively smaller taxa, until small-bodied rodents dominate the catch. Intermediate levels of hunting selectively remove the species responsible for modification of forest structure and the dispersal of large seeds, with consequences for the tree community that can be rapid and dramatic, including increased clustering of saplings and a relative decline in the recruitment rates for animal-dispersed species (e.g. Harrison et al., 2013). A scattering of success stories across the tropics show that enforcement, combined with education and community engagement, can largely exclude hunters from protected areas, but these lessons are not being applied more widely and the situation in most areas is bleak.

Herbivore and mesopredator release

The impact of selective human persecution of large carnivores is very different from the generalized overexploitation described in the previous section. In at least some ecosystems, large carnivores limit the populations of both large herbivores, by predation, and medium-sized carnivores, by predation and competition (Ripple et al., 2014). Many more complex direct and indirect impacts on lower trophic levels have been documented in particular systems. One of the most widespread and conspicuous impacts of carnivore declines has been the massive expansion of deer populations in both North America and Eurasia, with consequent increases in browsing pressure and impacts on both forest structure and the populations of numerous plants and smaller animals. Mesopredator release is also widespread in the same ecosystems, although the impacts are typically less visible (Prugh et al., 2009). Overall, the elimination of large carnivores appears to be one of the most significant human impacts in some temperate and boreal forests (Ripple et al., 2014).

Alien species

Many mammals, from shrews to elephants, have been deliberately or accidentally introduced outside their natural ranges, but most of the literature on impacts refers to ungulates (predominantly pigs, deer and goats), small carnivores (largely cats, foxes and mongooses), or rodents (particularly rats, mice and squirrels). Alien rats are established on almost all islands and many islands also have alien ungulates and small carnivores (e.g. Corlett, 2010), but introductions are also becoming increasingly widespread in mainland forests. Introductions started millennia ago (e.g. at least 7,000 years ago on Flores; van den Bergh, 2009) and may be hard to recognize without archaeological data. Initial impacts can be dramatic and devastating, including the extinction of prey species (especially flightless bird species) and competitors, but in the longer term the alien species becomes part of a new equilibrium and its subsequent elimination may have complex and unpredictable effects (e.g. the Australian dingo; Ripple et al., 2014). The eradication of invasive rodents and goats from small oceanic islands is becoming routine (Howald et al., 2007) and it is likely that an equal effort applied to many other taxa would be successful.

Restoration and reintroduction

If local extirpation of mammal populations is the problem, then reintroduction is a potential solution. Even when surplus stock is available from wild or captive populations, however, the success rate of reintroductions is low and the process can be prolonged and expensive (Reading et al., 2013). In general, translocations from wild populations have a higher rate of success, since captivity is associated with an almost inevitable loss of genetic diversity as well as the acquisition of maladaptive behavioral traits. Pre-release behavioral enrichment may be essential for captive-born animals, at least in primates and carnivores, where many behaviors needed for survival in the wild (locomotion, predator avoidance, foraging, social skills etc.) must be learned (Reading et al., 2013).

As noted in the introduction, many large mammal species have become extinct since the middle Pleistocene, so reintroduction is no longer an option. In such cases, it may be worth considering the introduction of a close relative or even an unrelated taxon with a similar ecological role, but such 'taxon substitutions' are highly controversial and should not be undertaken lightly. An additional term, 'rewilding', is increasingly used in the literature to refer to large-scale restoration of fully functioning ecosystems, with the focus often on reintroducing apex predators. The *de facto* rewilding of parts of Europe over the last 50 years shows what can be achieved in a densely populated region with economic incentives and public support (Deinet et al., 2013).

Conclusions

The Cenozoic is the Age of Mammals and it is hard to imagine forests without them. The calorific demands of an endothermic lifestyle make mammals peculiarly vulnerable to death by starvation, although torpor allows some species to wait for better times. High food demand as a result of an endothermic lifestyle is behind the major impacts of mammals in forests, making them quantitatively dominant as consumers of most items that they eat. The same demands make introduced mammals particularly destructive. As mammals ourselves, we are biased towards our closest wild relatives, although this has done little to stem the tide of anthropogenic extinctions from the late Pleistocene onwards (Turvey and Fritz, 2011; Corlett, 2013). It is not just species that are threatened, but their evolutionary history since whole clades are under threat (Isaac et al., 2007). While conservationists are often accused of focusing on charismatic mammals in preference to other deserving taxa, the future for many species still looks grim.

References

Abernethy, K. A., Coad, L., Taylor, G., Lee, M. E. and Maisels, F. (2013) 'Extent and ecological consequences of hunting in Central African rainforests in the twenty-first century', *Philosophical Transactions of the Royal Society B: Biological Sciences,* vol 368, no. 1625.

Amico, G. C., Rodríguez-Cabal, M. A. and Aizen, M. A. (2009) 'The potential key seed-dispersing role of the arboreal marsupial *Dromiciops gliroides*', *Acta Oecologica*, vol 35, pp. 8–13.

Buckley, L. B., Davies, T. J., Ackerly, D. D., Kraft, N. J. B., Harrison, S. P. and others (2010) 'Phylogeny, niche conservatism and the latitudinal diversity gradient in mammals', *Proceedings of the Royal Society B*, vol 277, pp. 2131–2138.

Campos-Arceiz, A. and Blake, S. (2011) 'Megagardeners of the forest – the role of elephants in seed dispersal', *Acta Oecologica,* vol 37, pp. 542–553.

Campos-Arceiz, A., Traeholt, C., Jaffar, R., Santamaria, L. and Corlett, R. T. (2012) 'Asian tapirs are no elephants when it comes to seed dispersal', *Biotropica*, vol 44, pp. 220–227.

Churchfield, S., Nesterenko, V. A. and Shvarts, E. A. (1999) 'Food niche overlap and ecological separation amongst six species of coexisting forest shrews in the Russian Far East', *Journal of Zoology*, vol 248, pp. 349–359.

Corlett, R. T. (1998) 'Frugivory and seed dispersal by vertebrates in the Oriental (Indomalayan) Region', *Biological Reviews of the Cambridge Philosophical Society,* vol 73, pp. 413–448.

Corlett, R. T. (2010) 'Invasive aliens on tropical East Asian islands', *Biodiversity and Conservation*, vol 19, pp. 411–423.

Corlett, R. T. (2011) 'Vertebrate carnivores and predation in the Oriental (Indomalayan) Region', *Raffles Bulletin of Zoology*, vol 59, pp. 325–360.

Corlett, R. T. (2013) 'The shifted baseline: prehistoric defaunation in the tropics and its consequences for biodiversity conservation', *Biological Conservation,* vol 163, pp. 13–21.

Corlett, R. T. and Primack, R. B. (2011) *Tropical Rain Forests: An Ecological and Biogeographical Comparison, 2nd Edition.* Wiley-Blackwell, Oxford, U.K.

Dammhahn, M., Soarimalala, V. and Goodman, S. M. (2013) 'Trophic niche differentiation and microhabitat utilization in a species-rich montane forest small mammal community of eastern Madagascar', *Biotropica*, vol 45, pp. 111–118.

Deinet, S., Ieronymidou, C., McRae, L., Burfield, I. J., Foppen, R. P. and others (2013) *Wildlife Comeback in Europe: The Recovery of Selected Mammal and Bird Species*. Zoological Society of London, London, U.K.

Delsuc, F., Metcalf, J. L., Wegener Parfrey, L., Song, S. J., González, A. and Knight, R. (2014) 'Convergence of gut microbiomes in myrmecophagous mammals', *Molecular Ecology*, vol 23, pp. 1301–1317.

Dexter, N., Hudson, M., James, S., MacGregor, C. and Lindenmayer, D. B. (2013) 'Unintended consequences of invasive predator control in an Australian forest: overabundant wallabies and vegetation change', *PLoS One*, vol 8, e69087.

Di Marco, M., Boitani, L., Mallon, D., Hoffmann, M., Iacucci, A. and others (2014) 'A retrospective evaluation of the global decline of carnivores and ungulates', *Conservation Biology*, vol 28(4), pp. 1109–1118.

Dudu, A., Churchfield, S. and Hutterer, R. (2005) 'Community structure and food niche relationships of coexisting rain-forest shrews in the Masako Forest, northeastern Congo'. In: *Advances in the Biology of Shrews II* (eds. J. F. Merritt, S. Churchfield, R. Hutterer and B. I. Sheftel). International Society of Shrew Biologists, New York, pp. 229–239.

Eizirik, E., Murphy, W. J., Koepfli, K.-P., Johnson, W. E., Dragoo, J. W. and others (2010) 'Pattern and timing of diversification of the mammalian order Carnivora inferred from multiple nuclear gene sequences', *Molecular Phylogenetics and Evolution*, 56, pp. 49–63.

Emmons, L. H. (2000) *Tupai: A Field Study of Bornean Treeshrews*. University of California Press, Berkeley, California, U.S.

Esselstyn, J. A., Achmadi, A. S. and Rowe, K. C. (2012) 'Evolutionary novelty in a rat with no molars', *Biology Letters*, vol 8, pp. 990–993.

Fontúrbel, F. E., Franco, M., Rodríguez-Cabal, M. A., Daniela Rivarola, M. and Amico, G. C. (2012) 'Ecological consistency across space: a synthesis of the ecological aspects of *Dromiciops gliroides* in Argentina and Chile', *Naturwissenschaften*, vol 99, pp. 873–881.

Glen, A. S., Wayne, A., Maxwell, M. and Cruz, J. (2010) 'Comparative diets of the chuditch, a threatened marsupial carnivore, in the northern and southern jarrah forests, Western Australia', *Journal of Zoology*, vol 282, pp. 276–283.

Gormley, A. M., Holland, E. P., Pech, R. P., Thomson, C. and Reddiex, B. (2012) 'Impacts of an invasive herbivore on indigenous forests', *Journal of Applied Ecology*, vol 49, pp. 1296–1305.

Happold, D. C. D. (1996) 'Mammals of the Guinea–Congo rain forest', *Proceedings of the Royal Society of Edinburgh Section B: Biological Sciences*, vol 104, pp. 243–284.

Harrison, R. D., Tan, S., Plotkin, J. B., Slik, F., Detto, M. and others (2013) 'Consequences of defaunation for a tropical tree community', *Ecology Letters*, vol 16, pp. 687–694.

Haynes, G. (2012) 'Elephants (and extinct relatives) as earth-movers and ecosystem engineers', *Geomorphology*, vols 157–158, pp. 99–107.

He, K., Chen, J., Gould, G. C., Yamaguchi, N., Ai, H. and others (2012) 'An estimation of Erinaceidae phylogeny: a combined analysis approach', *PLoS One* vol 7, e39304.

Howald, G., Donlan, C. J., Galván, J. P., Russell, J. C., Parkes, J. and others (2007) 'Invasive rodent eradication on islands', *Conservation Biology*, vol 21, pp. 1258–1268.

Isaac, N. J. B., Turvey, S. T., Collen, B., Waterman, C. and Baillie, J. E. M. (2007) 'Mammals on the EDGE: conservation priorities based on threat and phylogeny', *PloS One*, vol 2.

Jenkins, C. N., Pimm, S. L. and Joppa, L. N. (2013) 'Global patterns of terrestrial vertebrate diversity and conservation', *Proceedings of the National Academy of Sciences of the USA*, vol 110, E2602-E2610.

Jones, M. E. and Barmuta, L. A. (2000) 'Niche differentiation among sympatric Australian dasyurid carnivores', *Journal of Mammalogy*, vol 81, pp. 434–447.

Laurance, W. F., Carolina Useche, D., Rendeiro, J., Kalka, M., Bradshaw, C. J. A. and others (2012) 'Averting biodiversity collapse in tropical forest protected areas', *Nature*, vol 489, pp. 290–294.

Lyly, M., Klemola, T., Koivisto, E., Huitu, O., Oksanen, L. and Korpimäki, E. (2014) 'Varying impacts of cervid, hare and vole browsing on growth and survival of boreal tree seedlings', *Oecologia*, vol 174, pp. 271–281.

Milner, J. M. and Harris, S. (1999) 'Habitat use and ranging behaviour of tree arboreus, in the Virunga Volcanoes, Rwanda', *African Journal of Ecology*, vol 37, pp. 281–294.

Nummi, P. and Kuuluvainen, T. (2013) 'Forest disturbance by an ecosystem engineer: beaver in boreal forest landscapes', *Boreal Environment Research*, vol 18, pp. 13–24.

Ojala-Barbour, R., Miguel Pinto, C., Brito, J. M., Albuja, L. V., Lee, T. E., Jr. and Patterson, B. D. (2013) 'A new species of shrew-opossum (Paucituberculata: Caenolestidae) with a phylogeny of extant caenolestids', *Journal of Mammalogy*, vol 94, pp. 967–982.

O'Leary, M. A., Bloch, J. I., Flynn, J. J., Gaudin, T. J., Giallombardo, A. and others (2013) 'The placental mammal ancestor and the post-K-Pg radiation of placentals', *Science*, vol 339, pp. 662–667.

Prasad, S., Krishnaswamy, J., Chellam, R. and Goyal, S. P. (2006) 'Ruminant-mediated seed dispersal of an economically valuable tree in Indian dry forests', *Biotropica*, vol 38, pp. 679–682.

Prugh, L. R., Stoner, C. J., Epps, C. W., Bean, W. T., Ripple, W. J. and others (2009) 'The rise of the mesopredator', *BioScience*, vol 59, pp. 779–791.

Reading, R. P., Miller, B. and Shepherdson, D. (2013) 'The value of enrichment to reintroduction success', *Zoo Biology*, vol 32, pp. 332–341.

Ripple, W. J., Estes, J. A., Beschta, R. L., Wilmers, C. C., Ritchie, E. G. and others (2014) 'Status and ecological effects of the world's largest carnivores', *Science,* vol 343, 1241484 [DOI:10.1126/science.1241484].

Rolland, J., Condamine, F. L., Jiguet, F. and Morlon, H. (2014) 'Faster speciation and reduced extinction in the tropics contribute to the mammalian latitudinal diversity gradient', *PLoS Biology*, vol 12, e1001775.

Rovero, F., Collett, L., Ricci, S., Martin, E. and Spitale, D. (2013) 'Distribution, occupancy, and habitat associations of the gray-faced sengi (*Rhynchocyon udzungwensis*) as revealed by camera traps', *Journal of Mammalogy*, vol 94, pp. 792–800.

Scanlon, A. T., Petit, S., Tuiwawa, M. and Naikatini, A. (2014) 'High similarity between a bat-serviced plant assemblage and that used by humans', *Biological Conservation*, vol 174, pp. 111–119.

Shanahan, M. and Compton, S. G. (2000) 'Fig-eating by Bornean tree shrews (*Tupaia* spp.): evidence for a role as seed dispersers', *Biotropica*, vol 32, 759–764.

Springer, M. S., Krajewski, C. W. and Meredith, R. W. (2009) 'Marsupials (Metatheria)'. In: *The Timetree of Life* (eds. S. B. Hedges and S. Kumar). Oxford University Press, Oxford, U.K. pp. 466–470.

Stanley, W. T., Goodman, S. M. and Hutterer, R. (2011) 'Small mammal Inventories in the East and West Usambara Mountains, Tanzania. 2. Families Soricidae (Shrews) and Macroscelididae (Elephant Shrews)', *Fieldiana Life and Earth Sciences*, vol 4, pp. 18–33.

Tsang, S. M., Cirranello, A. L., Bates, P. J. J. and Simmons, N.B. (In press) 'The roles of taxonomy and systematics in bat conservation', In: *Bats in the Anthropocene: Conservation of bats in a changing world* (eds. T. Kingston and C.C. Voigt). Springer, New York.

Turvey, S. T. and Fritz, S. A. (2011) 'The ghosts of mammals past: biological and geographical patterns of global mammalian extinction across the Holocene', *Philosophical Transactions of the Royal Society B: Biological Sciences,* vol 366, pp. 2564–2576.

van den Bergh, G. D., Meijer, H. J., Due Awe, R., Morwood, M. J., Szabó, K. and others (2009) 'The Liang Bua faunal remains: a 95 k.yr. sequence from Flores, East Indonesia', *Journal of Human Evolution*, vol 57, pp. 527–537.

Vander Wall, S. B. (2010) 'How plants manipulate the scatter-hoarding behaviour of seed-dispersing animals', *Philosophical Transactions of the Royal Society B: Biological Sciences*, vol 365, pp. 989–997.

Vernes, K., Green, S., Howes, A. and Dunn, L. (2006) 'Species richness and habitat associations of non-flying mammals in Gibraltar Range National Park', *Proceedings of the Linnean Society of New South Wales*, vol 127, pp. 93–105.

Voss, R. S. and Emmons, L. H. (1996) 'Mammalian diversity in Neotropical lowland rainforests: A preliminary assessment', *Bulletin of the American Museum of Natural History*, vol 230, pp. 3–115.

Wirsing, A. J. and Murray, D. L. (2002) 'Patterns in consumption of woody plants by snowshoe hares in the northwestern United States', *EcoScience,* vol 9, pp. 440–449.

20
BIRDS IN FOREST ECOSYSTEMS

Jeffrey A. Stratford and Çağan H. Şekercioğlu

Avian diversity in forests

Birds have been associated with forests as long as there have been birds (Sereno and Chenggang 1992). Since their origin, birds have diversified to occupy a remarkable array of habitats and foraging strategies, unparalleled by any other terrestrial vertebrate (Naish 2014). Over the eons, birds have formed intimate relationships between their habitats, their prey, and formed tight symbiotic relationships, such as flower-pollinator symbiosis. Because the majority of birds are conspicuous and relatively easy to study, they are among the best studied animals in forested ecosystems (Şekercioğlu 2006b; Şekercioğlu et al. in press).

Despite their relative ease of being detected, new birds are still being discovered, particularly in the rainforests of the Neotropics and Southeast Asia (Jenkins et al. 2013; Lohman et al. 2010). Currently, there are between approximately 10,300 (www.birdlife.org/datazone/info/taxonomy; www.birds.cornell.edu/clementschecklist) to 10,546 (http://www.worldbirdnames.org/) accepted extant species of birds. The exact number of avian species is unknown since we still debate species concepts, new species are still being discovered, and the loss of species through extinctions is happening in real time (Newton 2003; Sodhi et al. 2011). The importance of forests to birds cannot be overstated: forests are home to about 75 percent of avian species and comprise the primary habitat of the majority of bird species (Şekercioğlu et al. 2004). The highest diversity of birds (> 5,000 species) occurs in lowland tropical and subtropical forests near the Equator in the Americas and Africa and 25°N in Southeast Asia and declines towards the poles (Birdlife International 2014; Newton 2003).

Lowland tropical forests have the greatest number of species and, among tropical forest sites, the Neotropics have the greatest number of species. At least 30 bird families are endemic to the Neotropics (not including Pluvianellidae or Thinocoridae). The understory of Neotropical forests is dominated by approximately 1,100 suboscine (suborder Tyranni) species that include the endemic antbirds (Formicariidae), treecreepers (Dendrocolaptidae), ovenbirds (Furnariidae) and the incredibly colorful manakins (Pipridae) and contingas (Cotingidae). Non-passerines that are endemic to the Neotropics include the tinamous (Tinamidae), motmots (Momotidae), toucans (Ramphastidae) and others. Oscines (suborder Passeri), such as the tanagers (Thraupini) dominate the Neotropical forest canopy. Only a few suboscines are found in the canopy or in bright treefall gaps and only 52 suboscine species are found outside the Neotropics (Corlett and

Primack 2011). In the Afrotropical forests, the dominant groups include the cuckoos (Cuculidae), shrikes (Laniidae) and others. More than 20 families are endemic to the Afrotropics, most of which are found in forests. Tropical forests of Southeast Asia have at least ten endemic families including the ioras (Aegithinidae), leafbirds (Chloropseidae) and fairy bluebirds (Irenidae). Of the roughly 30 families endemic to Australia, New Guinea, and surrounding tropical islands, approximately ten are found only in tropical and subtropical forests. Some of these endemic families include the large and unusual cassowaries (Casuariidae), beautiful birds-of-paradise (Paradisaeidae), terrestrial logrunners (Orthonychidae) and others. There are fewer endemic species in the Australian, Afrotropical and Southeast Asian forests, but many shared families between these biogeographic areas. Madagascar, despite its relatively small size and proximity to Africa, has five endemic families.

The highest species richness is in lowland rainforests and decreases with increasing elevation (Able and Noon 1976; Rahbek 1997; Terborgh 1977), latitude, and decreasing productivity such as in seasonal tropical and boreal forests (Rahbek and Graves 2001). The Palaearctic Region is the largest biogeographic region and includes temperate forest, boreal forest and tundra but has just over 900 species, many of which are shared with the Indomalaysian and Afrotropical Regions (Newton 2003).

Forests affect birds

Forests provide shelter and sustenance

Forests provide the essential resources necessary for the completion of life cycles, including food for adults and nestlings and nesting sites. Birds occur on various trophic levels in forests; from primary consumers to vertebrate predators, as well as omnivores and scavengers. Because birds are endotherms, their caloric requirements are higher than equivalently-sized ectotherms, and hence their demands for food are higher and are likely to be more sensitive to changes in forest resources.

As primary consumers, birds get nutrients from nectar, fruits, seeds and vegetative tissues (roots, shoots and leaves). Birds that consume the vegetative parts of plants may supplement their diet with other sources of protein such as insects (Karasov 1990; López-Calleja and Bozinovic 1999). A strictly folivorous diet is rare among birds; only 3 percent of avian species are strictly herbivorous (Şekercioğlu *et al.* 2004) and these tend to be large (> 1 kg) non-forest birds (López-Calleja and Bozinovic 1999) with most species in Asia (Kissling *et al.* 2012). Though forests have an abundance of leaves, secondary plant compounds and indigestible fibers (e.g., cellulose) make a diet of mature leaves an unusable food source for most birds (López-Calleja and Bozinovic 1999).

Granivory (seed-eating) and frugivory (fruit-eating) are much more common among herbivorous birds. Granivorous birds get most of their calories from the starches in seeds and there are just over 1,000 avian species that are primarily granivores (Şekercioğlu *et al.* 2004). Though not numerically the predominant guild, granivores can make up the greatest proportion of avian biomass in the Amazon (Terborgh *et al.* 1990).

In subtropical and deciduous forests, acorns and beech nuts are an important source of lipids and starches for a number of gallinaceous birds (fowl), corvids, woodpeckers and titmice. Acorns for example, make up a significant portion of the diet in turkeys in all seasons (Steffen *et al.* 2002). Nuts are also a key food source during the winter months for birds that cache seeds. In coniferous forests of the Northern Hemisphere, the seed crop influences several species of birds that forage on the seeds of firs and spruces (Petty *et al.* 1995).

Nearly 600 species of birds are nectivorous (Şekercioğlu *et al.* 2004) and are concentrated in the tropics (Brown and Hopkins 1995). In the Americas, hummingbirds (Trochilidae) and flowerpiercers (genus *Diglossa*) are the primary nectivores. Throughout Africa and Southeast Asia, sunbirds (Nectariniidae) are the primary nectivores, and honeyeaters (Meliphagidae) are the main nectivores in Australia. All of these groups include species that forage in forested habitats. Many other species may also consume nectar including orioles, *Phrygilus* finches, bulbuls and white-eyes. Flowers are often brightly colored to attract pollinators and offer nectar as a reward.

Nectar is an energy-rich food source although the quantities are small to promote movements between flowers (McCallum *et al.* 2013). The composition of nectar is variable, but typically the flowers of avian pollinated plants contain sucrose and amino acids (Baker *et al.* 1998). Other sugars are often found in nectar and provide energy for pollinators, which are often among the most energy-demanding taxa in forests. Hummingbirds, for instance, because of their small size and hovering behavior, have the highest mass-specific metabolic rates for any animal and require energy-rich foods.

Frugivory, in some form, is relatively common in forest birds and consists of birds consuming a fleshy pulp associated with a seed or seeds (Howe and Smallwood 1982). Nearly one in seven bird species are frugivores (Şekercioğlu *et al.* 2004) and 23 families of birds include fruit in at least half of their diet. For many frugivores, the proportion of fruit in the diet varies seasonally as fruit abundance changes (Jordano 2000). Another 16 families are mostly frugivores (Jordano 2000; Kissling *et al.* 2012; Wenny *et al.* in press), however only a few species are exclusively frugivorous (Izhaki and Safriel 1989; Jordano 2000; Wenny *et al.* in press). Fruits are such a key resource for birds that the diversity of fruiting plants may play a role in determining avian diversity (Kissling *et al.* 2007). In tropical and subtropical forests, figs (*Ficus* spp.) are particularly important and eaten by over 1,200 species of birds in 92 families, including birds that are typically carnivorous.

Fruit availability increases with proximity to the Equator and with increasing moisture. Consequently, tropical rainforests tend to have the highest biomass of fruit available and lower seasonal variation in abundance compared to temperate forests (Jordano 2000). On local scales, fruit availability is variable in both space and time, with a trend for fruits to be spatially and temporally aggregated (Jordano 2000). For example, fruit availability in temperate and tropical forests is greater in gaps, such as treefalls (Blake and Hoppes 1986; Levey 1988; Willson *et al.* 1982).

As a food source, fruits are highly variable in quality as a consequence of several traits including nutrients, secondary plant metabolites, total size, relative seed size and water content (Jordano 2000). These traits, however, are partially constrained by phylogeny and show considerable overlap within families and genera (Jordano 1995). Generally, fruit content falls into three categories based on sugar, fiber and lipid content with fruits tending to be either lipid or sugar rich. Fruits also tend to be low in nitrogen and proteins, which probably explains why there are few birds that are exclusively frugivorous (Snow 1981). In the temperate zone, fruits provide a source of lipids and other resources that allow birds to put on fats required for successful migration (Bairlein 2003; Stiles 1980). Though fruits might be a high source of energy, the lack of protein and the presence of secondary plant metabolites, such as tannins, may reduce the nutrient value of fruits (Cipollini and Levey 1997). In the tropics, birds may move on smaller scales to track fruit availability (Blake and Loiselle 1991; Blake and Loiselle 1992). There are also tropical birds that are nomadic and search large areas for adequate fruit crops for reproduction (Stouffer and Bierregaard 1993).

Insectivory is a commonly used term to describe a diet based on insects but this is probably too narrow a term for many birds that should be called invertivores, since their diet includes other invertebrates, such as spiders and gastropods (Poulin *et al.* 1994). Insectivores are typically divided into aerial insectivores and terrestrial/arboreal insectivores. This division is based on foraging strategy with the former, such as swifts and swallows, catching insects while staying in flight for long periods of time. There are far fewer aerial insectivores (228 species) than terrestrial/arboreal insectivores (4,900 species), which is the largest guild of birds (Kissling *et al.* 2012). Insects and other invertebrates provide proteins and nitrogen in bird diets, which are particularly important for growing birds. Insectivores have diverse strategies for finding insects. Terrestrial insectivores, for instance, forage in the leaf litter but there are various methods of foraging in litter including leaf tossing, searching the surface or searching under living leaves while walking across the surface of leaf litter (Stratford and Stouffer 2013). A number of tropical species are obligate ant-followers and only forage for invertebrates escaping from army ant swarms (Willis and Oniki 1978).

In temperate forests, lepidopteran larvae are important food sources that are fed to nestlings. Their importance is underscored by the fact that the timing of reproduction in many birds is correlated with the highest abundance of lepidopteran larvae in forests (Martin 1987). North American *Coccyzus* cuckoos may track gypsy moth caterpillar (*Lymantria dispar*) outbreaks (Barber *et al.* 2008). Other birds that forage for invertebrates on the ground, the terrestrial insectivores, are influenced by insects in the leaf litter, that are, in turn, influenced by microclimate (Johnston and Holberton 2009).

Forests are also home to many carnivorous birds, including hawks, falcons and owls. There are also several carnivorous birds of other orders found in forests including wood rails, ground cuckoos and hornbills. Carnivores, like insectivores, eat animals, but carnivory typically refers to species that eat vertebrates. There are some 300 species of carnivorous birds (Kissling *et al.* 2012). Other prey items found in forests include small mammals, lizards, snakes and amphibians. Thirty-six species of avian scavengers (Şekercioğlu *et al.* 2004) exclusively consume dead animals and other organic material. New World vultures (Cathartidae), in particular, forage frequently in forests and play an integral role in nutrient cycling therein (Houston 1985; Şekercioğlu 2006a).

Most food items, by themselves, do not provide a complete diet so that frugivores, nectivores or insectivores will often supplement their diet with alternative food items. These are often taken opportunistically, such as abundant fruits during fall migration or termite emergences. Though many birds take multiple types of food items, there are only 500 omnivores proper (sensu Kissling *et al.* 2012) and even fewer species in forests (Kissling *et al.* 2012). Part of the reason for the lack of omnivores is the need for specialized anatomy or physiology to capture or procure food items and digest them (Gill 2007). Omnivores pay the cost of increased handling time or decreased digestive efficiency but have the benefit of higher encounter rates of potential food items. Omnivores may also be less susceptible to the effects of forest fragmentation (Blake 1983; Henle *et al.* 2004; Willson *et al.* 1994).

Forests provide nesting sites

Forests also provide nesting sites for birds. Nests can be found in all forest strata: from the ground, in the shrubs, and in the treetops. Forest structure, such as canopy openness, influences reproductive success of forest birds. For many forest species, nesting success generally increases with increasing canopy closure (Bakermans *et al.* 2012).

Cavity nesting birds use hollowed out parts of trees or other tall plants as nesting sites. Primary cavity-nesters are species that excavate the cavities and typically include woodpeckers. Secondary cavity-nesters use the cavities created by primary cavity-nesters or use naturally-occurring cavities and include a taxonomically diverse assemblage. Secondary cavity-nesters include ducks (e.g., mergansers), falcons (e.g., kestrels), parrots, several owls, swifts, swallows, trogons, several flycatchers, tits, wrens, bluebirds and many others. The availability of dead trees and branches (snags) in forests can limit populations of both primary and secondary cavity nesting birds (Kroll et al. 2012; Newton 1994).

Forests provide migratory stopover habitat and wintering sites

Forests are important habitats for migrating birds in all the major flyways (Kirby et al. 2008) where birds forage on insects and fruit and replenish fat reserves used to cross seas or non-forested habitat (Moore and Kerlinger 1987). In southern North America, riparian forests are important stopover habitats (Buler et al. 2007; Hutto 1998) and forests are selected even when less representative than other habitats in the landscape, such as areas in the Midwest, US (Grundel and Pavlovic 2007). Many migrant birds then join tropical forest species for several months (Wunderle and Waide 1993), primarily in Neotropical and Indomalaysian forests, and become part of tropical food webs (Bauer and Hoye 2014). Habitat quality on the tropical wintering grounds affects reproductive success in the temperate breeding grounds and influences populations of many long-distance migrants (Norris et al. 2004).

Forests provide thermal refugia

Forests can provide microclimates that are refugia to physiologically challenging temperatures. For example, small passerines will move into trees and forested habitats during the winter to be in an environment that is sheltered from winds. The implication is that birds do not expend as much energy maintaining body temperature (Wolf and Walsberg 1996; Wolf et al. 1996). Temperate birds will move towards the ground to minimize exposure to wind (Dolby and Grubb 1999). On the other extreme, forests can be cool refugia places when temperatures are high enough to cause thermal stress (Seavy 2006). During the nesting season, canopy cover may affect the temperature of chicks during development and influence reproductive success, at least for cavity nesting birds (Dawson et al. 2005).

Forest structure affects avian communities

The scale of measuring diversity has a key importance in elucidating drivers of diversity. Across the globe, evolutionary history, plate tectonics and other historical factors influence patterns of diversity. Rainfall and temperature influence biomes and create plant types such as the different forests, grasslands and deserts. Within these different biomes, increased structural complexity of vegetation is associated with increased avian species richness (MacArthur et al. 1966; MacArthur and MacArthur 1961; MacArthur et al. 1962; Orians and Wittenberger 1991). One measure of forest structure is foliage height diversity and is defined by the variation in the layers of a forest. Increasing foliage height diversity is associated with increasing avian diversity, particularly insectivores (MacArthur and MacArthur 1961; MacArthur et al. 1962). Increasing foliage height diversity is associated with increasing foraging sites and increased niches available to exploit (MacArthur et al. 1966). The diversity of the tropics might be increased by increasing specialization of forest birds. For example, there are a number of birds that specialize on dead leaves that gather on branches (Rosenberg 1997) or foraging with raiding army ants (Johnson et al. 2013; Willis and Oniki 1978).

Birds affect forests

Just as forests provide several resources for birds, birds influence forests. This interaction is both direct (e.g., seed dispersal) and indirect (e.g., eating phytophagous insects). Birds are now widely recognized as having several positive interactions with forests, providing several ecological services (Şekercioğlu 2006a, b; Şekercioğlu et al. in press; Wenny et al. 2011).

Dispersal agents

Seed dispersal is a key aspect of plant ecology. Seeds that disperse away from the parent plant are less likely to suffer mortality from herbivores and pathogens (Janzen et al. 1976). Moreover, seeds landing farther away from the parent plant will be less likely to compete and reproduce with relatives. Because of their abundance, breadth of taxonomic interactions and mobility, birds are important as seed dispersers in forested ecosystems (Jordano 2000), particularly the wet tropics where nearly 90 percent of tree species produce fruits (Howe and Smallwood 1982). Birds disperse seeds for a taxonomically diverse range of plants including angiosperms and gymnosperms (Jordano 2000; Stephens and Fry 2005).

Birds disperse seeds by eating fruits (endozoochory) then vomiting or defecating seeds or by having seeds stick to them (epizoochory). At least during migration, endozoochory far outnumbers epizoochory (Costa et al. 2014). Even though epizoochory by birds might be relatively rare, these long-distance events might be incredibly important from the perspective of the plants (and, incidentally, small invertebrates and parasites [Darwin 1859]). The same is true for endozoochorous seeds, which may need to be consumed to germinate (Jordano 2000). Many seeds are deposited where germination is favored, such as fields (Carlo et al. 2013). Moreover, endozoochory by birds can remove fungi and bacteria that can damage seeds and add a small amount of nutrients in feces (Fricke et al. 2013).

There are however, few examples of close associations between birds and fruits. That is, many species of birds eat many species of fruiting plants but there are few cases of specialized relationships between birds and their fruits (Blüthgen et al. 2007; Herrera 2002). A general relationship has the advantages of resiliency. Birds would be able to use alternative fruits should the abundance of any plant decrease. Likewise, fruiting plants do not rely on any particular species for dispersal.

The lack of specialization between frugivores and plants also means that invasive plants can find a dispersal mechanism in novel areas (Gosper and Vivian-Smith 2006; Merow et al. 2011). Birds may, in some situations, even prefer fruits of invasive species to native species and play a key role in their expansion (Greenberg and Walter 2010).

Much less is known about avian dispersal of nuts compared to avian dispersal of fruit but recent work on jay–acorn dynamics has shown the importance of European, Blue and Scrub Jays in oak dispersal, particularly long-distance dispersal (Gómez 2003). Though birds that consume nuts are primarily seed predators, they are also key dispersal agents when seeds are not recovered after caching them in soil (Steele et al. 2010). Jays remove seeds from individual oaks by the thousands and cache sites can be several kilometers away and often in sites where seedling establishment is high, such as fields (Steele et al. 2002). Thus jays are incredibly important in forest succession. Their propensity for long-distance dispersal of acorns may explain the rapid northern advancement of oaks after glaciers retreated in North America and Europe (Johnson and Webb III 1989).

Insectivory

Birds that consume phytophagous insects can potentially reduce the effects of leaf damage and increase plant growth. These tritrophic interactions have been shown to exist experimentally in temperate (Barber and Marquis 2009; Marquis and Whelan 1994; Mols and Visser 2002) and tropical forests (Van Bael *et al.* 2008). Birds can reduce damage to commercially important trees including cocoa (Van Bael *et al.* 2007) and coffee plantations (Johnson *et al.* 2010), preventing up to US$310/ha/yr in damage to coffee crops (Johnson *et al.* 2010).

Threats

Directly or indirectly, humans have been causing avian extinctions in the wake of our species settling the globe. Human-caused extinctions are best documented starting in the sixteenth century, though forensic genetics can demonstrate human-induced extinctions before this time (Allentoft *et al.* 2014; Pimm *et al.* 2006). Most extinctions have occurred on islands (Szabo *et al.* 2012) though future extinctions are predicted to occur in biodiversity hotspots where they are often associated with high rates of deforestation (Orme *et al.* 2005). Deforestation rates are increasing in many tropical forest biodiversity hotspots as a result of increasing production and global trade of agricultural crops such as oil palm, soy beans and corn, as human food (Phalan *et al.* 2011), livestock feed (Anonymous 2014), biofuel (Danielsen *et al.* 2009), and industrial use (Fitzherbert, *et al.* 2008).

Deforestation, forest fragmentation and degradation

Deforestation can occur by removing a single area of habitat and leaving the remaining habitat intact. Forest fragmentation results from a combination of forest removal and the creation of forest patches amid a non-forest matrix. Strictly speaking, deforestation can occur without the creation of forest fragments. More typically though, deforestation is accompanied by forest fragmentation and both have negative consequences for avian populations. Forest degradation can occur as a result of selective logging, which results in bird species losses at medium to high intensities (Burivalova, *et al.* 2014). Moreover, logging and deforestation opens areas to hunting and development. Consequently, habitat loss, high land-use intensity, forest degradation and fragmentation are major drivers of avian species endangerment (Birdlife International 2014; Fahrig 2003; Kerr and Deguise 2004; Newbold *et al.* 2014; Sodhi *et al.* 2011).

Typically, the number of forest interior species is reduced in forest fragments (Banks-Leite *et al.* 2012; Leck *et al.* 1988; Whitcomb *et al.* 1981) consistent with species–area relationships (MacArthur and Wilson 1967). Globally, the most fragmentation-sensitive groups are tropical insectivores and large frugivores (Bregman *et al.* 2014). Less sensitive are small-bodied canopy frugivores and pollinators (Stouffer and Bierregaard Jr 1995). Though the species richness of these groups is not strongly affected by forest fragmentation – other aspects, such as behavior, are affected by fragmentation (Hadley and Betts 2009). Temperate species appear less sensitive to forest fragmentation and the declines are less predictable (Bregman *et al.* 2014). Causes of sensitivity are likely to vary geographically (Stratford and Robinson 2005b). In North America, for example, forest fragmentation increases exposure to brood parasites (Robinson *et al.* 1995) and nest predation (Rodewald 2002). In Neotropical forests, birds appear to be more sensitive to light conditions (Patten and Smith-Patten 2012), changes in vegetation structure (Stratford and Stouffer 2013) and many are unwilling or unable to cross open habitats that separate forest from forest fragments (Ibarra-Macias *et al.* 2011; Moore *et al.* 2008; Powell *et al.* 2013).

The loss of species as a result of forest fragmentation is likely to affect ecosystem services that birds provide (Saunders *et al.* 1991; Şekercioğlu 2006a, b). Fruit removal rates tend to be lower in forest fragments (Rey and Alcántara 2014), which implies reduced seed dispersal. Other processes may be less affected or balanced by other effects. For example, population genetics of *Heliconia* plants appear to not be strongly affected by fragmentation because their avian pollinators and frugivores are able to move long distances between fragments and continuous forests (Suarez-Montes *et al.* 2011). Remnant native trees are important for sustaining some tropical forest bird species and their ecological services in agricultural landscapes (Şekercioğlu *et al.* 2007; Douglas *et al.* 2014). Restoration efforts in these landscapes can result in the return of some forest bird species (Catterall *et al.* 2012).

Over-exploitation

Forest birds, especially large, edible species like guans, curassows, pigeons and pheasants, or those that are victims of the pet trade, such as parrots and finches, are greatly over-exploited and many of them are threatened with extinction as a result (Birdlife International 2014; Sodhi *et al.* 2011). In some forests, especially in western Africa and Southeast Asia, this has resulted in the elimination of large birds and mammals, leading to the phenomenon of the "empty forest" (Redford 1992) where the forest is devoid of many large vertebrates and their ecosystem services (Kurten 2013; Şekercioğlu 2010). Globally, over 400 bird species are threatened with extinction due to over-exploitation (Birdlife International 2014). These comprise a third of all threatened bird species, the majority of which are found in forests (IUCN 2014). The cage bird trade is increasingly understood to be far more widespread for Southeast Asian bird species than previously thought and is pervasive even in many protected areas in the region (Kai *et al.* 2014; pers. observ.; Bert Harris, pers. comm.).

Invasive species and emergent infectious diseases

Invasive trees may compete with native trees and alter food resources and nesting sites. Invasive plants are associated with lower species richness (Ortega *et al.* 2014) and higher nest predation rates (Borgmann and Rodewald 2004; Borgmann and Rodewald 2005). Invasive species can affect birds not only directly but indirectly by affecting forest structure. Numerous invasive pests have altered the North American landscape. For example, chestnut blight, Dutch elm disease and the wooly adelgid have affected the amount of American chestnut, American elm and eastern hemlock, respectively, and gypsy moths are currently reducing the proportion of oaks. The loss of American chestnut may have contributed to the extinction of the Passenger Pigeon (*Ectopistes migratorius*) (Bucher 1992).

In general, invasive bird species have failed to colonize most tropical and subtropical forests, but there are some major exceptions on islands, including Hong Kong, Singapore and many oceanic islands, such as Hawaii and New Zealand, where invasive birds have fundamentally altered bird communities (Sol *et al.* 2015). Other invasive vertebrates have profoundly affected avian populations. The extent to which cats have affected avian populations has only recently been appreciated (Loss *et al.* 2013). The effect of invasive vertebrates on islands is particularly damaging (Blackburn *et al.* 2004; Towns *et al.* 2006). The brown tree snake (*Boiga irregularis*), for instance, has resulted in the extinction and population declines of at least 25 native species on Guam, where it was introduced (Wiles *et al.* 2003).

Emergent infectious diseases may also have negative impacts on forest birds, such as West Nile virus in North America (Kilpatrick *et al.* 2007; LaDeau *et al.* 2007), and mycoplasma conjunctivitis (Fischer *et al.* 1997).

Urbanization

Urbanization is the conversion of forested habitats to human-created impervious surfaces. Urbanization is a particularly pernicious form of land-use change since urban areas are rarely converted back to a natural state. Another feature of urbanization is the frequent increase in pollution (Pickett *et al.* 2001). Associated with increasing urbanization is a loss of native species, particularly insectivores (Chace and Walsh 2006). However, granivores and omnivores increase in abundance and species richness, though these increases do not offset the loss of other species (Marzluff 2001; Zhou *et al.* 2012). In eastern North America, long-distance migrants are highly sensitive to urbanization and most species are not found in landscapes that are more than 20 percent urbanized (Stratford and Robinson 2005a). The effects of urbanization can be mitigated by preserving urban forests, which can be important stopover sites for migrants (Kohut *et al.* 2009).

Climate change

As with all ecosystems and taxa, climate change is rapidly approaching habitat loss in the magnitude of the threat it poses for forest birds (Harris *et al.* 2011; Şekercioğlu *et al.* 2012; Wormworth and Şekercioğlu 2011). Climate change particularly threatens bird species endemic to tropical montane forests, as most of these species are endemic to narrow elevational ranges and have specific microhabitat and microclimate requirements. Species without access to higher elevations, coastal forest birds and restricted-range species are also highly vulnerable (Şekercioğlu, *et al.* 2012). Some forest birds are especially susceptible to increased rainfall seasonality and to extreme weather events, such as heat waves, cold spells and tropical cyclones. Protected forest areas will be more important than ever, but they need to be designed with climate change in mind. Networks of protected forests need to incorporate extensive topographical diversity, cover wide elevational ranges, have high connectivity and integrate human-dominated landscapes into conservation schemes (Şekercioğlu *et al.* 2012).

The future of birds in forests

The threats listed above to forest birds are not mutually exclusive. Forest birds can simultaneously face increased habitat loss, invasion from introduced species, and novel diseases. Managing forest birds, therefore, requires detailed knowledge of interacting threats as well as understanding of long-term trends in avian abundance. Two long-standing tools that have given ornithologists great insight are citizen science programs, such as the Breeding Bird Survey (Link and Sauer 1998) and the Christmas Bird Count (Butcher *et al.* 1990). Other programs such as Monitoring Avian Productivity and Survivorship (MAPS: www.birdpop.org/maps.htm) are providing greater detail on age structure and reproductive success in different habitats. Involving the public in forest bird research and conservation should not be optional. The advent of eBird (Sullivan *et al.* 2009) has demonstrated the power of citizen science in generating data on bird distributions and movements. As is the case in conservation and research in general, increased public involvement increases the sense of public ownership, augments public participation in conservation policy and funding, and improves the chances of success for bird conservation programs (Şekercioğlu 2012). Comprising the majority of the world's bird species threatened with extinction, forest birds are in urgent need of much greater levels of public and political support if we are to have any chance of reducing the ever growing numbers of threatened and extinct bird species.

References

Able, K. P. and B. R. Noon (1976) Avian community structure along elevational gradients in the northeastern United States, *Oecologia,* vol 26, no 3, pp. 275–294.

Allentoft, M. E., R. Heller, C. L. Oskam, E. D. Lorenzen, M. L. Hale and others (2014) Extinct New Zealand megafauna were not in decline before human colonization, *Proceedings of the National Academy of Sciences,* vol 111, no 13, pp. 4922–4927.

Anonymous (2014) Empire of the pig. URL: http://www.economist.com/news/christmas-specials/21636507-chinas-insatiable-appetite-pork-symbol-countrys-rise-it-also *The Economist.* Haywards Heath, U.K.

Bairlein, F. (2003) The study of bird migrations – some future perspectives: Capsule routes and destinations have been unveiled but modern techniques offer the chance to explore much more, *Bird Study,* vol 50, no 3, pp. 243–253.

Baker, H. G., I. Baker and S. A. Hodges (1998) Sugar composition of nectars and fruits consumed by birds and bats in the tropics and subtropics 1, *Biotropica,* vol 30, no 4, pp. 559–586.

Bakermans, M. H., A. D. Rodewald and A. C. Vitz (2012) Influence of forest structure on density and nest success of mature forest birds in managed landscapes, *The Journal of Wildlife Management,* vol 76, no 6, pp. 1225–1234.

Banks-Leite, C., R. M. Ewers and J. P. Metzger (2012) Unraveling the drivers of community dissimilarity and species extinction in fragmented landscapes, *Ecology,* vol 93, no 12, pp. 2560–2569.

Barber, N. A. and R. J. Marquis (2009) Spatial variation in top-down direct and indirect effects on white oak (*Quercus alba* L.), *The American Midland Naturalist,* vol 162, no 1, pp. 169–179.

Barber, N. A., R. J. Marquis and W. P. Tori (2008) Invasive prey impacts the abundance and distribution of native predators, *Ecology,* vol 89, no 10, pp. 2678–2683.

Bauer, S. and B. J. Hoye (2014) Migratory animals couple biodiversity and ecosystem functioning worldwide, *Science,* vol 344, no 6179

BirdLife International (2014) Datazone. URL: http://www.birdlife.org/datazone Cambridge, U.K.

Blackburn, T. M., P. Cassey, R. P. Duncan, K. L. Evans and K. J. Gaston (2004) Avian extinction and mammalian introductions on oceanic islands, *Science,* vol 305, no 5692, pp. 1955–1958.

Blake, J. G. (1983) Trophic structure of bird communities in forest patches in east-central Illinois, *Wilson Bulletin,* vol 95, pp. 416–430.

Blake, J. G. and W. G. Hoppes (1986) Influence of resource abundance on use of tree-fall gaps by birds in an isolated woodlot, *The Auk,* vol 103, pp. 328–340.

Blake, J. G. and B. A. Loiselle (1991) Variation in resource abundance affects capture rates of birds in three lowland habitats in Costa Rica, *Auk,* vol 108, pp. 114–130.

Blake, J. G. and B. A. Loiselle (1992) Fruits in the diets of Neotropical migrant birds in Costa Rica, *Biotropica,* vol 24, pp. 200–210.

Blüthgen, N., F. Menzel, T. Hovestadt, B. Fiala and N. Blüthgen (2007) Specialization, constraints, and conflicting interests in mutualistic networks, *Current Biology,* vol 17, no 4, pp. 341–346.

Borgmann, K. L. and A. D. Rodewald (2004) Nest predation in an urbanizing landscape: The role of exotic shrubs, *Ecological Applications,* vol 14, no 6, pp. 1757–1765.

Borgmann, K. L. and A. D. Rodewald (2005) Forest restoration in urbanizing landscapes: Interactions between land uses and exotic shrubs, *Restoration Ecology,* vol 13, no 2, pp. 334–340.

Bregman, T. P., C. H. Şekercioğlu and J. A. Tobias (2014) Global patterns and predictors of bird species responses to forest fragmentation: Implications for ecosystem function and conservation, *Biological Conservation,* vol 169, pp. 372–383.

Brown, E. D. and M. Hopkins (1995) A test of pollinator specificity and morphological convergence between nectarivorous birds and rainforest tree flowers in New Guinea, *Oecologia,* vol 103, no 1, pp. 89–100.

Bucher, E. (1992) The causes of extinction of the Passenger Pigeon. In D. Power (Ed.). *Current Ornithology,* Springer, US.

Buler, J. J., F. R. Moore and S. Woltmann (2007) A multi-scale examination of stopover habitat use by birds, *Ecology*, vol 88, no 7, pp. 1789–1802.

Burivalova, Z., C.H. Şekercioğlu and L.P. Koh, (2014) Thresholds of logging intensity to maintain tropical forest biodiversity. *Current Biology*, vol 24, 1893–1898.

Butcher, G. S., M. R. Fuller, L. S. McAllister and P. H. Geissler (1990) An evaluation of the Christmas Bird Count for monitoring population trends of selected species, *Wildlife Society Bulletin*, vol 19, pp. 129–134.

Carlo, T. A., D. Garcia, D. Martinez, J. M. Gleditsch and J. M. Morales (2013) Where do seeds go when they go far? Distance and directionality of avian seed dispersal in heterogeneous landscapes, *Ecology*, vol 94, no 2, pp. 301–307.

Catterall, C. P., A. N. D. Freeman, J. Kanowski and K. Freebody (2012) Can active restoration of tropical rainforest rescue biodiversity? A case with bird community indicators, *Biological Conservation*, vol 146, no 1, pp. 53–61.

Chace, J. F. and J. J. Walsh (2006) Urban effects on native avifauna: A review, *Landscape and Urban Planning*, vol 74, no 1, pp. 46–69.

Cipollini, M. L. and D. J. Levey (1997) Secondary metabolites of fleshy vertebrate-dispersed fruits: Adaptive hypotheses and implications for seed dispersal, *The American Naturalist*, vol 150, no 3, pp. 346–372.

Corlett, R. T. and R. B. Primack (2011) *Tropical Rain Forests: An Ecological and Biogeographical Comparison*, John Wiley and Sons, Oxford, UK.

Costa, J. M., J. A. Ramos, L. P. da Silva, S. Timoteo, P. M. Araújo and others (2014) Endozoochory largely outweighs epizoochory in migrating passerines, *Journal of Avian Biology*, vol 45, no 1, pp. 59–64.

Danielsen, F., H. Beukema, N. D. Burgess, F. Parish, C. A. Bruehl and others (2009). Biofuel plantations on forested lands: Double jeopardy for biodiversity and climate. *Conservation Biology*, vol 23, pp. 348–358.

Darwin, C. (1859) *Origin of Species*, John Wiley, New York, US.

Dawson, R. D., C. C. Lawrie and E. L. O'Brien (2005) The importance of microclimate variation in determining size, growth and survival of avian offspring: experimental evidence from a cavity nesting passerine, *Oecologia*, vol 144, no 3, pp. 499–507.

Dolby, A. S. and T. C. J. Grubb (1999) Effects of winter weather on horizontal and vertical use of isolated forest fragments by bark-foraging birds, *Condor*, vol 101, pp. 408–412.

Douglas, D. J. T., D. Nalwanga, R. Katebaka, P. W. Atkinson, D. E. Pomeroy and others (2014) The importance of native trees for forest bird conservation in tropical farmland, *Animal Conservation*, vol 17, no 3, pp. 256–264.

Fahrig, L. (2003) Effects of habitat fragmentation on biodiversity, *Annual Review of Ecology and Systematics*, vol 34, pp. 487–515.

Fischer, J. R., D. E. Stallknecht, P. Luttrell, A. A. Dhondt and K. A. Converse (1997) Mycoplasmal conjunctivitis in wild songbirds: The spread of a new contagious disease in a mobile host population, *Emerging Infectious Diseases*, vol 3, no 1, p. 69.

Fitzherbert, E. B., M. J. Struebig, A. Morel, F. Danielsen, C. A. Brühl and others (2008) How will oil palm expansion affect biodiversity? *Trends in Ecology and Evolution*, vol 23, pp. 538–545.

Fricke, E. C., M. J. Simon, K. M. Reagan, D. J. Levey, J. A. Riffell and others (2013) When condition trumps location: Seed consumption by fruit-eating birds removes pathogens and predator attractants, *Ecology Letters*, vol 16, no 8, pp. 1031–1036.

Gill, F. B. (2007) *Ornithology*, Freeman, New York, US.

Gómez, J. M. (2003) Spatial patterns in long-distance dispersal of *Quercus ilex* acorns by jays in a heterogeneous landscape, *Ecography*, vol 26, no 5, pp. 573–584.

Gosper, C. R. and G. Vivian-Smith (2006) Selecting replacements for invasive plants to support frugivores in highly modified sites: A case study focusing on *Lantana camara*, *Ecological Management and Restoration*, vol 7, no 3, pp. 197–203.

Greenberg, C. H. and S. T. Walter (2010) Fleshy fruit removal and nutritional composition of winter-fruiting plants: A comparison of non-native invasive and native species, *Natural Areas Journal*, vol 30, no 3, pp. 312–321.

Grundel, R. and N. B. Pavlovic (2007) Distinctiveness, use, and value of midwestern oak savannas and woodlands as avian habitats, *Auk*, vol 124, no 3, pp. 969–985.

Hadley, A. S. and M. G. Betts (2009) Tropical deforestation alters hummingbird movement patterns, *Biology Letters*, vol 5, pp. 207–210.

Harris, J. B. C., C. H. Şekercioğlu, N. S. Sodhi, D. A. Fordham, D. C. Paton and B. W. Brook (2011) The tropical frontier in avian climate impact research, *Ibis,* vol 153, no 4, pp. 877–882.

Henle, K., K. Davies, M. Kleyer, C. Margules and J. Settele (2004) Predictors of species sensitivity to fragmentation, *Biodiversity and Conservation,* vol 13, no 1, pp. 207–251.

Herrera, C. M. (2002) Seed dispersal by vertebrates. In C. M. Herrera and O. Pellmyr (Eds.). *Plant-Animal Interactions: An Evolutionary Approach* Blackwell Science, Oxford, UK.

Houston, D. C. (1985) Evolutionary ecology of Afrotropical and Neotropical vultures in forests. In M. Foster (Eds.), *Neotropical Ornithology,* American Ornithologists' Union Monographs, Washington, US.

Howe, H. F. and J. Smallwood (1982) Ecology of seed dispersal, *Annual Review of Ecology and Systematics,* vol 13, pp. 201–228.

Hutto, R. L. (1998) On the importance of stopover sites to migrating birds, *Auk,* vol 115, pp. 823–825.

Ibarra-Macias, A., W. D. Robinson and M. S. Gaines (2011) Experimental evaluation of bird movements in a fragmented Neotropical landscape, *Biological Conservation,* vol 144, no 2, pp. 703–712.

IUCN (2014) *The IUCN Red List of Threatened Species,* International Union for Conservation of Nature (now World Conservation Union) URL: http://www.iucnredlist.org

Izhaki, I. and U. N. Safriel (1989) Why are there so few exclusively frugivorous birds? Experiments on fruit digestibility, *Oikos,* vol 54, pp. 23–32.

Janzen, D. H., G. Miller, J. Hackforth-Jones, C. Pond, K. Hooper and D. P. Janos (1976) Two Costa Rican bat-generated seed shadows of *Andira inermis* (Leguminosae), *Ecology,* vol 57, pp. 1068–1075.

Jenkins, C. N., S. L. Pimm and L. N. Joppa (2013) Global patterns of terrestrial vertebrate diversity and conservation, *Proceedings of the National Academy of Sciences,* vol 110, no 28, pp. E2602–E2610.

Johnson, E. I., P. C. Stouffer and C. F. Vargas (2013) Diversity, biomass, and trophic structure of a central Amazonian rainforest bird community, *Revista Brasileira de Ornitologia* 19, 1–16.

Johnson, M. D., J. L. Kellermann and A. M. Stercho (2010) Pest reduction services by birds in shade and sun coffee in Jamaica, *Animal Conservation,* vol 13, no 2, pp. 140–147.

Johnson, W. C. and T. Webb III (1989) The role of blue jays (*Cyanocitta cristata* L.) in the postglacial dispersal of fagaceous trees in eastern North America, *Journal of Biogeography,* vol 16, pp. 561–571.

Johnston, J. C. and R. L. Holberton (2009) Forest management and temporal effects on food abundance for a ground-foraging bird (*Catharus guttatus*), *Forest Ecology and Management,* vol 258, no 7, pp. 1516–1527.

Jordano, P. (1995) Angiosperm fleshy fruits and seed dispersers: A comparative analysis of adaptation and constraints in plant-animal interactions, *American Naturalist,* vol 145, pp. 163–191.

Jordano, P. (2000) Fruits and frugivory. In M. Fenner (Ed.). *Seeds: The Ecology of Regeneration in Plant Communities,* pp. 125–166, CAB International Publishing, London.

Kai, Z., T. S. Woan, L. Jie, E. Goodale, K. Kitajima and others (2014) Shifting baselines on a tropical forest frontier: Extirpations drive declines in local ecological knowledge, *PLoS ONE,* vol 9, no 1, p. e86598.

Karasov, W. H. (1990) Digestion in birds: Chemical and physiological determinants and ecological implications, *Studies in Avian Biology,* vol 13, no 39, pp. 1–4.

Kerr, J. T. and I. Deguise (2004) Habitat loss and the limits to endangered species recovery, *Ecology Letters,* vol 7, no 12, pp. 1163–1169.

Kilpatrick, A. M., S. L. LaDeau and P. P. Marra (2007) Ecology of West Nile virus transmission and its impact on birds in the Western Hemisphere, *The Auk,* vol 124, no 4, pp. 1121–1136.

Kirby, J. S., A. J. Stattersfield, S. H. M. Butchart, M. I. Evans, R. F. A. Grimmett and others (2008) Key conservation issues for migratory land- and waterbird species on the world's major flyways, *Bird Conservation International,* vol 18, pp. S49–S73.

Kissling, W. D., C. Rahbek and K. Böhning-Gaese (2007) Food plant diversity as broad-scale determinant of avian frugivore richness, *Proceedings of the Royal Society B: Biological Sciences,* vol 274, no 1611, pp. 799–808.

Kissling, W. D., C. H. Şekercioğlu and W. Jetz (2012) Bird dietary guild richness across latitudes, environments and biogeographic regions, *Global Ecology and Biogeography,* vol 21, no 3, pp. 328–340.

Kohut, S., G. Hess and C. Moorman (2009) Avian use of suburban greenways as stopover habitat, *Urban Ecosystems,* vol 12, no 4, pp. 487–502.

Kroll, A. J., S. D. Duke, M. E. Hane, J. R. Johnson, M. Rochelle and others (2012) Landscape composition influences avian colonization of experimentally created snags, *Biological Conservation,* vol 152, pp. 145–151.

Kurten, E. L. (2013) Cascading effects of contemporaneous defaunation on tropical forest communities, *Biological Conservation,* vol 163, pp. 22–32.

LaDeau, S. L., A. M. Kilpatrick and P. P. Marra (2007) West Nile virus emergence and large-scale declines of North American bird populations, *Nature,* vol 447, no 7145, pp. 710–713.

Leck, C. F., B. G. Muray and J. Swinebroad (1988) Long-term changes in the breeding bird populations of a New Jersey forest, *Biological Conservation,* vol 46, pp. 145–157.

Levey, D. J. (1988) Tropical wet forest treefall gaps and distributions of understory plants and birds, *Ecology,* vol 69, pp. 1076–1089.

Link, W. A. and J. R. Sauer (1998) Estimating population change from count data: Application to the North American Breeding Bird Survey, *Ecological Applications,* vol 8, no 2, pp. 258–268.

Lohman, D. J., K. K. Ingram, D. M. Prawiradilaga, K. Winker, F. H. Sheldon and others (2010) Cryptic genetic diversity in "widespread" Southeast Asian bird species suggests that Philippine avian endemism is gravely underestimated, *Biological Conservation,* vol 143, no 8, pp. 1885–1890.

López-Calleja, M. V. and F. Bozinovic (1999) Feeding behavior and assimilation efficiency of the Rufous-tailed Plantcutter: A small avian herbivore, *Condor,* vol 101, pp. 705–710.

Loss, S. R., T. Will and P. P. Marra (2013) The impact of free-ranging domestic cats on wildlife of the United States, *Nature communications,* vol 4, p. 1396.

MacArthur, R. H. and J. W. MacArthur (1961) On bird species diversity, *Ecology,* vol 42, pp. 594–598.

MacArthur, R. H. and E. O. Wilson (1967) *The Theory of Island Biogeography,* Princeton University, New Jersey, US.

MacArthur, R. H., J. W. MacArthur and J. Preer (1962) On bird species diversity. II. Prediction of bird census from habitat measurements, *American Naturalist,* vol 96, pp. 167–174.

MacArthur, R., H. Recher and M. Cody (1966) On the relation between habitat selection and species diversity, *American Naturalist,* vol 100, pp. 320–332.

Marquis, R. J. and C. J. Whelan (1994) Insectivorous birds increase growth of white oak through consumption of leaf-chewing insects, *Ecology,* vol 75, pp. 2007–2014.

Martin, T. E. (1987) Food as a limit on breeding birds: A life-history perspective, *Annual Review of Ecology and Systematics,* vol 18, pp. 453–487.

Marzluff, J. M. (2001) Worldwide urbanization and its effects on birds. In J. Marzluff, R. Bowman and R. Donnelly (Eds.), *Avian Ecology and Conservation in an Urbanizing World,* Kluwer Academic Publishers, Norwell, US.

McCallum, K., F. McDougall and R. Seymour (2013) A review of the energetics of pollination biology, *Journal of Comparative Physiology B,* vol 183, no 7, pp. 867–876.

Merow, C., N. LaFleur, J. A. Silander Jr, A. M. Wilson and M. Rubega (2011) Developing dynamic mechanistic species distribution models: Predicting bird-mediated spread of invasive plants across northeastern North America, *The American Naturalist,* vol 178, no 1, pp. 30–43.

Mols, C. M. M. and M. E. Visser (2002) Great tits can reduce caterpillar damage in apple orchards, *Journal of Applied Ecology,* vol 39, pp. 888–899.

Moore, F. and P. Kerlinger (1987) Stopover and fat deposition by North American wood-warblers (Parulinae) following spring migration over the Gulf of Mexico, *Oecologia,* vol 74, no 1, pp. 47–54.

Moore, R. P., W. D. Robinson, I. J. Lovette and T. R. Robinson (2008) Experimental evidence for extreme dispersal limitation in tropical forest birds, *Ecol Lett,* vol 11, no 9, pp. 960–968.

Naish, D. (2014) The fossil record of bird behaviour, *Journal of Zoology,* vol 292, no 4, pp. 268–280.

Newbold, T., J. Scharlemann, S. Butchart, C. H. Şekercioğlu, L. Joppa and others (2014) Functional traits, land-use change and the structure of present and future bird communities in tropical forests, *Global Ecology and Biogeography,* vol 23, pp. 1073–1084.

Newton, I. (1994) The role of nest sites in limiting the numbers of hole-nesting birds: A review, *Biological Conservation,* vol 70, no 3, pp. 265–276.

Newton, I. (2003) *Speciation and Biogeography of Birds,* Academic Press.

Norris, D. R., P. P. Marra, T. K. Kyser, T. W. Sherry and L. M. Ratcliffe (2004) Tropical winter habitat limits reproductive success on the temperate breeding grounds in a migratory bird, *Proceedings of the Royal Society of London. Series B: Biological Sciences,* vol 271, no 1534, pp. 59–64.

Orians, G. H. and J. F. Wittenberger (1991) Spatial and temporal scales in habitat selection, *American Naturalist,* vol 137, pp. 29–49.

Orme, C. D. L., R. G. Davies, M. Burgess, F. Eigenbrod, N. Pickup and others (2005) Global hotspots of species richness are not congruent with endemism or threat, *Nature,* vol 436, no 7053, pp. 1016–1019.

Ortega, Y. K., A. Benson and E. Greene (2014) Invasive plant erodes local song diversity in a migratory passerine, *Ecology,* vol 95, no 2, pp. 458–465.

Patten, M. A. and B. D. Smith-Patten (2012) Testing the microclimate hypothesis: Light environment and population trends of Neotropical birds, *Biological Conservation*, vol 155, pp. 85–93.

Petty, S., I. Patterson, D. Anderson, B. Little and M. Davison (1995) Numbers, breeding performance, and diet of the sparrowhawk *Accipiter nisus* and merlin *Falco columbarius* in relation to cone crops and seed-eating finches, *Forest Ecology and Management*, vol 79, no 1, pp. 133–146.

Phalan, B., M. Onial, A. Balmford, R. E. Green. (2011) Reconciling food production and biodiversity conservation: Land sharing and land sparing compared, *Science*, vol 333, pp. 1289–1291.

Pickett, S. T. A., M. L. Cadenasso, J. M. Grove, C. H. Nilon, R. V. Pouyat and others (2001) Urban ecological systems: Linking terrestrial ecological, physical, and socioeconomic components of metropolitan areas, *Annual Review of Ecology and Systematics*, vol 32, pp. 127–157.

Pimm, S., P. Raven, A. Peterson, Ç. H. Şekercioğlu and P. R. Ehrlich (2006) Human impacts on the rates of recent, present, and future bird extinctions. *Proceedings of the National Academy of Sciences (PNAS)* vol 103, pp. 10941–10946.

Poulin, B., G. Lefebvre and R. McNeil (1994) Diets of landbirds from northeastern Venezuela, *Condor*, vol 96, pp. 354–367.

Powell, L. L., P. C. Stouffer and E. I. Johnson (2013) Recovery of understory bird movement across the interface of primary and secondary Amazon rainforest, *Auk*, vol 130, no 3, pp. 459–468.

Rahbek, C. (1997) The relationship among area, elevation, and regional species richness in Neotropical birds, *American Naturalist*, vol 149, pp. 875–902.

Rahbek, C. and G. R. Graves (2001) Multiscale assessment of patterns of avian species richness, *Proceedings of the National Academy of Sciences*, vol 98, pp. 4534–4539.

Redford, K. H. (1992) The empty forest, *BioScience*, vol 42, pp. 412–422.

Rey, P. J. and J. M. Alcántara (2014) Effects of habitat alteration on the effectiveness of plant-avian seed dispersal mutualisms: Consequences for plant regeneration, *Perspectives in Plant Ecology, Evolution and Systematics*, vol 16, no 1, pp. 21–31.

Robinson, S. K., F. R. Thompson, III, T. M. Donovon, D. R. Whitehead and J. Faaborg (1995) Regional forest fragmentation and the nesting success of migratory birds, *Science*, vol 267, pp. 1987–1990.

Rodewald, A. D. (2002) Nest predation in forested regions: Landscape and edge effects, *Journal of Wildlife Management*, vol 66, no 3, pp. 634–640.

Rosenberg, K. V. (1997) Ecology of dead-leaf foraging specialists and their contribution to Amazonian bird diversity, *Ornithological Monographs*, vol 48, pp. 673–700.

Saunders, D. A., R. J. Hobbs and C. R. Margules (1991) Biological consequences of ecosystem fragmentation: A review, *Conservation Biology*, vol 5, pp. 18–29.

Seavy, N. E. (2006) Physiological correlates of habitat association in East African sunbirds (Nectariniidae), *Journal of Zoology*, vol 270, no 2, pp. 290–297.

Şekercioğlu, Ç. H. (2006a) Ecological significance of bird populations. In J. del Hoyo, A. Elliot and D. A. Christie (Eds.), *Handbook of the Birds of the World*, Lynx Press and Birdlife International, Barcelona, Spain.

Şekercioğlu, Ç. H. (2006b) Increasing awareness of avian ecological function, *Trends in Ecology and Evolution*, vol 21, no 8, pp. 464–471.

Şekercioğlu, Ç. H. (2010) Ecosystem functions and services. In N. S. Sodhi and P. R. Ehrlich (Eds.), *Conservation Biology for All*. Oxford University Press, Oxford, UK.

Şekercioğlu, Ç. H. (2012) Promoting community-based bird monitoring in the tropics: Conservation, research, environmental education, capacity-building, and local incomes. *Biological Conservation*, vol 151, pp. 69–73.

Şekercioğlu, Ç. H., G. C. Daily and P. R. Ehrlich (2004) Ecosystem consequences of bird declines, *Proc Natl Acad Sci USA*, vol 101, no 52, pp. 18042–18047.

Şekercioğlu, Ç. H., S. R. Loarie, F. O. Brenes, P. R. Ehrlich and G. C. Daily (2007) Persistence of forest birds in the Costa Rican agricultural countryside, *Conservation Biology*, vol 21, no 2, pp. 482–494.

Şekercioğlu, Ç. H., R. B. Primack and J. Wormworth (2012) The effects of climate change on tropical birds, *Biological Conservation*, vol 148, no 1, pp. 1–18.

Şekercioğlu, Ç. H., D. Wenny and C. J. Whelan, (Eds.). In press. *Why Birds Matter*. University of Chicago Press. Chicago, US.

Sereno, P. C. and R. Chenggang (1992) Early evolution of avian flight and perching: New evidence from the lower cretaceous of China, *Science*, vol 255, no 5046, pp. 845–848.

Snow, D. W. (1981) Tropical frugivorous birds and their food plants: A world survey, *Biotropica*, vol 13, pp. 1–14.

Sodhi, N.S., Ç. H. Şekercioğlu, S. Robinson, J. Barlow (2011) *Conservation of Tropical Birds*. Wiley-Blackwell, Oxford, UK.

Sol, D., T. M. Blackburn, P. Casey, R. Duncan and J. Clavell (2015) The ecology and impact of non-indigenous birds. In J. del Hoya, A. Elliot and D. A. Christie (Eds.). *Handbook of the Birds of the World* Lynx and Birdlife International, Barcelona, Spain.

Steele, M. A., and P. D. Smallwood (2002) Acorn dispersal by birds and mammals. In W. McShea and W. Healy (Eds.) *Oak Forest Ecosystems: Ecology and Management for Wildlife,* pp. 182–195, Johns Hopkins University Press, Baltimore, US.

Steele, M. A., N. Lichti and R. K. Swihart (2010) Avian-mediated seed-dispersal: An overview and synthesis with an emphasis on temperate forests of Central and Eastern U.S. In S. K. Majumbar, T. L. Master, M. C. Brittingham, R. M. Ross, R. S. Mulvihill and J. E. Huffman (Eds.), *Avian Ecology and Conservation: A Pennsylvania Focus with National Implications,* Pennsylvania Academy of Sciences, US.

Steffen, D. E., N. W. Lafon and G. W. Norman (2002) Turkeys, acorns, and oaks. In W. McShea and W. Healy (Eds.), *Oak Forest Ecosystems: Ecology and Management for Wildlife,* Johns Hopkins University Press, Baltimore, US.

Stephens, S. and D. Fry (2005) Spatial distribution of regeneration patches in an old-growth Pinus Jeffrey-mixed conifer forest in northwestern Mexico, *Journal of Vegetation Science,* vol 16, no 6, pp. 693–702.

Stiles, E. W. (1980) Patterns of fruit presentation and seed dispersal in bird-disseminated woody plants in the eastern deciduous forest, *American Naturalist,* vol 116, pp. 670–688.

Stouffer, P. C. and R. O. Bierregaard Jr (1995) Use of Amazonian forest fragments by understory insectivorous birds, *Ecology,* vol 76, pp. 2429–2445.

Stouffer, P. C. and R. O. J. Bierregaard (1993) Spatial and temporal abundance patterns of Ruddy Quail-Doves (*Geotrygon montana*) near Manaus, Brazil, *Condor,* vol 95, pp. 896–903.

Stratford, J. A. and W. D. Robinson (2005a) Distribution of Neotropical migratory bird species across an urbanizing landscape, *Urban Ecosystems,* vol 8, pp. 59–77.

Stratford, J. A. and W. D. Robinson (2005b) Gulliver travels to the fragmented tropics: Geographic variation in mechanisms of avian extinction, *Frontiers in Ecology and the Environment,* vol 3, pp. 91–98.

Stratford, J. A. and P. C. Stouffer (2013) Microhabitat associations of terrestrial insectivorous birds in Amazonian rainforest and second-growth forests, *Journal of Field Ornithology,* vol 84, no 1, pp. 1–12.

Suarez-Montes, P., J. Fornoni and J. Nunez-Farfan (2011) Conservation genetics of the endemic Mexican *Heliconia uxpanapensis* in the Los Tuxtlas tropical rain forest, *Biotropica,* vol 43, no 1, pp. 114–121.

Sullivan, B. L., C. L. Wood, M. J. Iliff, R. E. Bonney, D. Fink and S. Kelling (2009) eBird: A citizen-based bird observation network in the biological sciences, *Biological Conservation,* vol 142, no 10, pp. 2282–2292.

Szabo, J. K., N. Khwaja, S. T. Garnett and S. H. Butchart (2012) Global patterns and drivers of avian extinctions at the species and subspecies level, *PloS One,* vol 7, no 10, p. e47080.

Terborgh, J. (1977) Bird species diversity on an Andean elevational gradient, *Ecology,* vol 58, pp. 1007–1019.

Terborgh, J., S. K. Robinson, T. A. I. I. I. Parker, C. A. Munn and N. Pierpont (1990) Structure and organization of an Amazonian forest bird community, *Ecological Monographs,* vol 60, pp. 213–238.

Towns, D., I. E. Atkinson and C. Daugherty (2006) Have the harmful effects of introduced rats on islands been exaggerated? *Biological Invasions,* vol 8, no 4, pp. 863–891.

Van Bael, S. A., P. Bichier and R. Greenberg (2007) Bird predation on insects reduces damage to the foliage of cocoa trees (*Theobroma cacao*) in western Panama, *Journal of Tropical Ecology,* vol 23, pp. 715–719.

Van Bael, S. A., S. M. Philpott, R. Greenberg, P. Bichier, N. A. Barber and others (2008) Birds as predators in tropical agroforestry systems, *Ecology,* vol 89, no 4, pp. 928–934.

Wenny, D. G., T. L. Devault, M. D. Johnson, D. Kelly, Ç. H. Şekercioğlu and others (2011) The need to quantify ecosystem services provided by birds, *The Auk,* vol 128, no 1, pp. 1–14.

Wenny, D. G., C. H. Şekercioğlu, N. Cordeiro, H. S. Rogers and D. Kelly (In Press) Seed dispersal by fruit-eating birds. In C. H. Şekercioğlu, D. G. Wenny and C. J. Whelan (Eds.), *Why Birds Matter,* University of Chicago Press, Chicago, US.

Whitcomb, R. F., C. S. Robbins, J. F. Lynch, B. L. Whitcomb, M. K. Klimkiewicz and D. Bystrak (1981) Effects of forest fragmentation on avifauna of the eastern deciduous forest. In R. L. Burgess and D. M. Sharpe (Eds.), *Forest Island Dynamics in Man-Dominated Landscapes,* Springer-Verlag, New York, US.

Wiles, G. J., J. Bart, R. E. Beck and C. F. Aguon (2003) Impacts of the brown tree snake: Patterns of decline and species persistence in Guam's avifauna, *Conservation Biology,* vol 17, no 5, pp. 1350–1360.

Willis, E. O. and Y. Oniki (1978) Birds and army ants, *Annual Review of Ecology and Systematics,* vol 9, pp. 243–263.

Willson, M. F., T. L. de Santo, C. Sabag and J. J. Armesto (1994) Avian communities of fragmented south-temperate rainforests in Chile, *Conservation Biology,* vol 8, pp. 508–520.

Willson, M. F., E. A. Porter and R. S. Condit (1982) Avian frugivores activity in relation to forest light gaps, *Caribbean Journal of Science,* vol 18, pp. 1–6.

Wolf, B. O. and G. E. Walsberg (1996) Thermal effects of radiation and wind on a small bird and implications for microsite selection, *Ecology,* vol 77, pp. 2228–2236.

Wolf, B. O., K. M. Wooden and G. E. Walsberg (1996) The use of thermal refugia by two small desert birds, *Condor,* vol 98, pp. 424–427.

Wormworth, J. and C. H. Şekercioğlu (2011) *Winged Sentinels: Birds and Climate Change,* Cambridge University Press. Cambridge, UK.

Wunderle, J. M., Jr. and R. B. Waide (1993) Distribution of overwintering nearctic migrants in the Bahamas and Greater Antilles, *Condor,* vol 95, pp. 904–933.

Zhou, D., T. Fung and L. M. Chu (2012) Avian community structure of urban parks in developed and new growth areas: A landscape-scale study in Southeast Asia, *Landscape and Urban Planning,* vol 108, nos 2–4, pp. 91–102.

21
GLOBAL PATTERNS OF BIODIVERSITY IN FORESTS

Christine B. Schmitt

Introduction

The global forest ecosystems are of immense importance due to their high biodiversity and the ecosystem services they provide. They constitute some of the last large, and hence unfragmented, ecosystems of the world, yet some forest ecosystems are considered severely threatened (Hansen *et al.*, 2013). For instance, although the vast majority of the global biodiversity hotspots harbour forest, only small areas of intact natural vegetation remain (Schmitt *et al.*, 2009; Sloan *et al.*, 2014). Knowledge of the biodiversity patterns in forests worldwide is crucial for strategic conservation planning on the global scale and for monitoring global conservation targets (Pereira *et al.*, 2013, Pimm *et al.*, 2014). Furthermore, there is strong international interest in the conservation of forest cover and forest biodiversity due to the important role of forests in global climate change mitigation (Thompson *et al.*, 2012; Schmitt, 2013).

It is well known that global species richness is highest in the tropics, and that the tropical rainforests represent some of the most diverse ecosystems on Earth (Mace *et al.*, 2005). The latitudinal diversity gradient is assumed to be positively correlated with area size, long-term environmental stability, kinetics and productivity, and be negatively correlated with climatic variability (Kreft and Jetz, 2007; Brown, 2014). Evolutionary processes such as rates of speciation and extinction, and range expansion also play a crucial role (Lomolino *et al.*, 2010; Brown, 2014). However, the latitudinal diversity gradient does not apply to certain taxa and ecosystems. For instance, the diversity of species of wasps, conifers and mosses is high at mid- or high latitudes (Geffert *et al.*, 2013; Brown, 2014), and plant diversity is exceptionally high in the southern African Cape region and the Mediterranean regions of Chile (Kreft and Jetz, 2007; Bannister *et al.*, 2011).

This chapter aims to look beyond the latitudinal diversity gradient and to identify further vital patterns of biodiversity in forests worldwide. Species are useful units for the assessment of diversity (Mace *et al.*, 2005); however, a comprehensive evaluation of global biodiversity patterns needs to consider more than just simple species numbers (e.g., Olson and Dinerstein, 2002; Jenkins *et al.*, 2013). Hence, this chapter presents an overview of the criteria, indicators and data commonly used to describe biodiversity, followed by a brief summary of the ecosystem classifications adopted widely in reporting global biodiversity patterns. Against this background, the major global patterns of biodiversity in forests are outlined. It should be noted that both differences in biodiversity patterns between managed and unmanaged forests and threat and conservation issues are beyond the scope of this chapter.

Biodiversity criteria and indicators

Biodiversity refers to the diversity of genes, species and ecosystems, and can be measured with a wide range of criteria and indicators on different taxonomic and spatial scales (Mace *et al.*, 2005). There are three major criteria for describing the biodiversity of a given area: variety, composition and distribution (Table 21.1). Gobal analyses often define variety in terms of the number of species, i.e., species richness, or the number of other taxonomic units (e.g., Table 21.2). Many studies highlight the number of species with particular ranges, e.g., restricted range species, also called endemics (Kier *et al.*, 2009; Pimm *et al.*, 2014). This information is important, especially at the global level, because it helps to distinguish areas with unique ecological features, such as the global centres of endemism. Information on genetic or functional variety is often difficult to obtain at the global level, and is therefore typically used in studies on smaller geographical scales.

Biodiversity in terms of composition is also usually assessed in studies with a smaller geographic scope because data on species communities and individual species populations are usually not available globally, or are too specific for large-scale analyses (Mace *et al.*, 2005). Likewise, diversity indices can be useful when comparing different sites within a region, but they are not suitable for global studies which cover different ecosystems and which are based on strongly heterogeneous data sets (Magurran, 2004). In contrast, maps of species richness and distribution maps for particular taxa or species groups (for birds see Figure 21.1) are common methods for illustrating global biodiversity patterns. Finally, the spatial distribution of ecological units, such as global forest types, is a useful indicator of biodiversity and ecological processes expected at a local level, especially where more comprehensive information on species numbers, identities and ranges is lacking (Schmitt, 2013; Tuanmu and Jetz, 2014).

Table 21.1 Biodiversity criteria and indicators (modified Mace *et al.* 2005)

Criterion	Indicator (examples)
Variety	= *the number of different types*
Genetic	• number of alleles and allele combinations (genotypes)
Taxonomic★	• number of species, taxa, families and major lineages (orders)
Range-related★	• number of restricted range (endemic) and migratory species
Functional	• number of life forms, guilds and ecological groups
Composition	= *quantity and quality*
Populations	• population structure and genetic diversity of individual species
Communities	• species lists, presence/absence data, abundance data
Diversity indices	• Shannon, Jaccard, evenness
Distribution	= *spatial pattern*
Genetic	• maps of gene flow, evolutionary history and phylogeny
Variety★	• maps of species richness
Species★	• maps of the distribution of individual species and species groups
Ecological units★	• maps of ecological zones, habitat types, biomes and ecoregions

★ frequently used in global-level analyses

Table 21.2 Number of species of mammals, birds, amphibians and conifers in the IUCN forest habitats. The four taxa have been completely assessed globally (IUCN, 2014b); total global species numbers in brackets. Percentage of species is defined as the proportion of all species in this group occurring in that particular habitat. Species can occur in more than one habitat. (IUCN Red List Version 2014.1, last updated 12 June 2014). Subtropical/Tropical abbreviated as (Sub-)tropical.

IUCN habitat	Mammals (5,514 species)		Birds (10,064 species)		Amphibians (6,410 species)		Conifers (605 species)	
	No of species	% of species	No of species	% of species	No of species	% of species	No of species	% of species
(Sub-)tropical moist Lowland forest	2,276	41%	5,105	51%	2,977	46%	114	19%
(Sub-)tropical moist Montane forest	1,558	28%	3,651	36%	2,878	45%	187	31%
(Sub-)tropical dry forest	1,016	18%	1,872	19%	430	7%	24	4%
(Sub-)tropical swamp forest	267	5%	694	7%	132	2%	15	2%
(Sub-)tropical mangrove forest*	173	3%	916	9%	15	< 1%	1	< 1%
Temperate forest	773	14%	880	9%	546	9%	320	53%
Boreal forest	104	2%	226	2%	24	< 1%	16	3%
Subarctic forest	38	1%	8	< 1%	7	< 1%	5	1%
Subantarctic forest	9	< 1%	6	< 1%	18	< 1%	1	< 1%
All forest habitats	3,854	70%	7,655	76%	5,354	84%	545	90%

* Subtropical/tropical mangrove forest vegetation above high tide level

Global species data

There are a number of organisations which make global data sets available for a wide range of individual species or species groups (Jetz et al., 2012). The most renowned global database is the Red List of Threatened Species by the International Union for Conservation of Nature (IUCN), which provides free, regular updates of global species data. The Red List Version 2014.1 contained assessments of just over 73,000 species. Information on ecology, population size and threats was available for about 60,000 species, and distribution maps based on known species occurrences and expert opinion were available for about 45,000 of these species (IUCN, 2014b). However, the distribution maps may overestimate a species range of occurrence and do not reflect fluctuations in species abundances across the range, especially at coarse resolutions (Jenkins et al., 2013). While previous analyses often relied on a 1-degree cell resolution (ca. 100 × 100 km, compare Figure 21.1), Jenkins et al. (2013) used a spatial grain of 10 × 10 km in their analysis of global vertebrate diversity.

Despite the expanding database, the species covered by the IUCN Red List represent only a small fraction of the estimated 1.9 million known species described globally. Yet, vast numbers of species have still not been scientifically documented (Pimm et al., 2014). For instance,

(a)

(b)

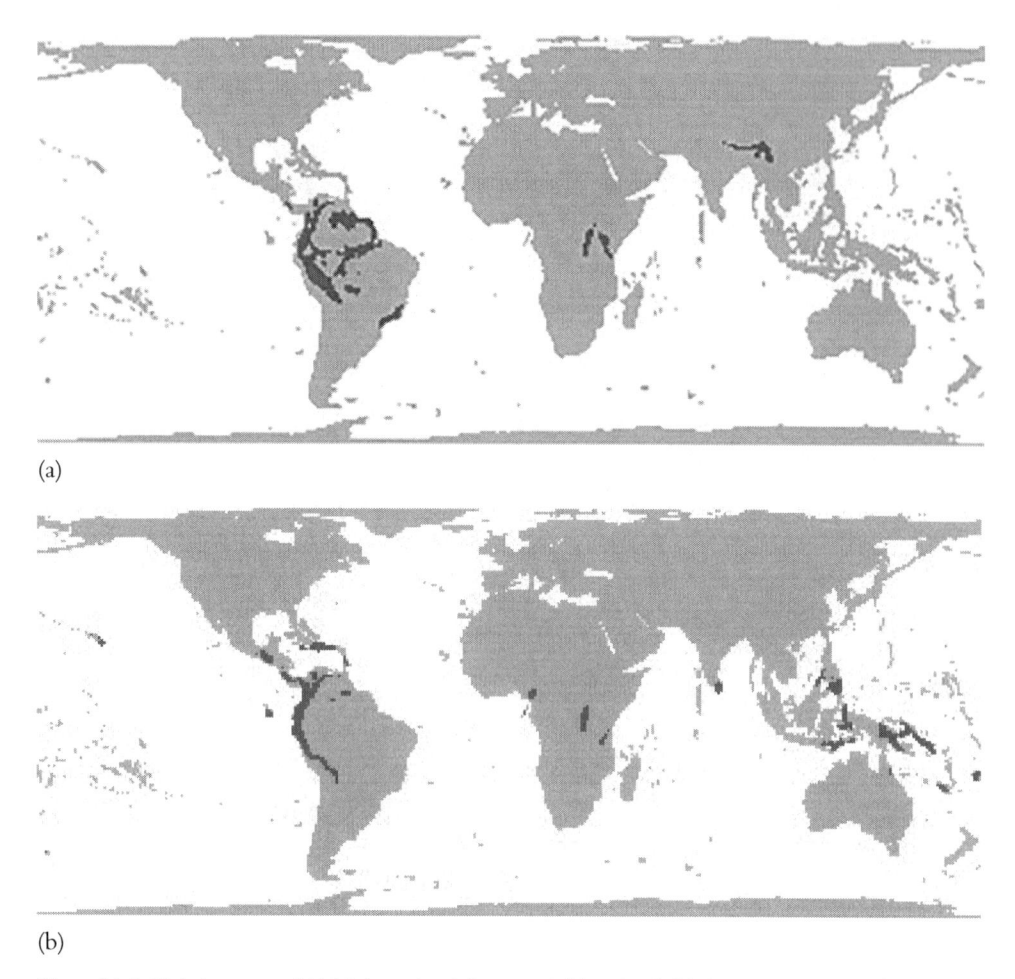

Figure 21.1 Global centres of (a) bird species richness and (b) endemic bird species. Centres (shown in black) are defined as the richest 2.5 per cent of grid cells for both measures of diversity (Orme *et al.*, 2005). Endemic species (n = 2,421) are those with ranges restricted to fewer than 30 grid cells (Behrmann projection, cell resolution = 96.486 km, approximately equivalent to a one degree cell resolution at the equator)

incomplete inventories and taxonomic revisions are major problems for insects and fungi (Pimm *et al.*, 2014). The best available IUCN data for terrestrial vertebrates are for amphibians, mammals and birds, and include virtually all species described globally for these taxa (IUCN, 2014b) with the exception of a large number of amphibian species not yet formally described (Jenkins *et al.*, 2013). For plants, only the conifer and cycad species have now been fully documented (IUCN, 2014b). Even for the well-documented taxa, the present distribution maps may not cover areas where species may be present, but have not yet been recorded, and data can be biased by large variations in sampling intensity between the different regions of the world (Jenkins *et al.*, 2013).

In addition to the IUCN Red List, there are an increasing number of biodiversity initiatives on the worldwide web which collate species data for particular taxa, ecosystems or geographic regions. For instance, the Global Biodiversity Information Facility has become a key clearing-house for worldwide species data from a wide range of sources. Yet, data quality control remains problematic (Jetz *et al.*, 2012).

Distribution of the global forest ecosystems

In order to evaluate global patterns of species diversity in forests, the global species data need to be combined with information on global forest distribution. Global forest cover maps based on remote-sensing technology aim to represent actual forest cover at a given point in time. The latest series of global forest cover maps show global forest cover change from 2000 to 2012 at a spatial resolution of 30 m (Hansen et al., 2013). However, they do not distinguish between natural forests and plantation forests, which makes an important difference in terms of expected forest diversity, and which could be achieved by incorporating information from global land cover products (Sloan et al., 2014; Tuanmu and Jetz, 2014).

In addition, there are global ecosystem templates based on predefined, spatially-explicit ecological units. While they can provide an indication of the geographical location of major forest ecosystems, they lack information on actual forest cover. For instance, the WWF developed the global terrestrial ecoregions framework to prepare a biodiversity map reflecting the distribution of Earth's natural species assemblages and communities prior to major land-use changes (Olson et al., 2001). It identifies 8 realms, 14 major biomes and 825 terrestrial ecoregions. The realms, such as the Neotropic and Afrotropic, represent a division of the landmasses of the world according to large-scale biogeographical differences in species distributions. Adopting the biogeographical approach, the framework combines the tropics and subtropics and distinguishes azonal biomes shaped by edaphic and hydrological factors, e.g., flooded grasslands and savannas, and mangroves. In contrast, the Global Ecological Zones, defined by the Food and Agriculture OrganiZation of the United Nations (FAO), are based primarily on climatic factors instead of species assemblages (FAO, 2012).

Figure 21.2 shows an example of an analysis of forest cover by major biomes (Schmitt et al., 2009). The study used the 2005 MODIS Vegetation Continuous Field products, which represent forest cover as continuous tree cover densities from 0 to 100 per cent at a spatial resolution of 500 m. Pixels with more than 10 per cent tree cover were classified as forest in order to comply with international forest definitions such as those adopted by the FAO and the United Nations Framework Convention on Climate Change. In a next step, the forest cover map was overlaid with the Global Land Cover 2000 product to exclude forests areas modified by humans, such as tree plantations and certain types of agro-ecosystems (Schmitt et al., 2009). Further overlay with the WWF ecosystems framework showed that the tropical and subtropical, hereafter referred to as '(sub-)tropical', moist broadleaf forests biome comprised the largest forest cover globally, followed by the boreal forests/taiga biome (Figure 21.2). Yet in 2005, the percentage area of forest cover in these biomes was only 60 per cent and 68 per cent of the full biome area, respectively, reflecting the massive conversion of forests to other land uses (compare Mace et al., 2005; Hansen et al., 2013). It should be noted that the grassland biomes also contained some forest cover at the selected 10 per cent tree cover threshold (Figure 21.2). This is particularly conspicuous in the (sub-)tropical grasslands, savannas and shrublands biome because some savanna areas, e.g., parts of the woody savannas in central and southern Africa, have a relatively closed tree cover, while maintaining a significant grass component (Murphy and Bowman, 2012; Schmitt, 2013).

Finally, the IUCN has been developing its own habitat classification scheme in an ongoing iterative process; it is not spatially explicit and requires expert knowledge in the assignment of species to habitats. The classification scheme differentiates 18 major terrestrial, marine and human modified habitats on the basis of biogeography and latitudinal zonation (IUCN, 2014a). The forest and woodland habitat includes nine forest types (Table 21.2). Due to the ad hoc assignment of species records to different habitats, the IUCN scheme considers actual habitat

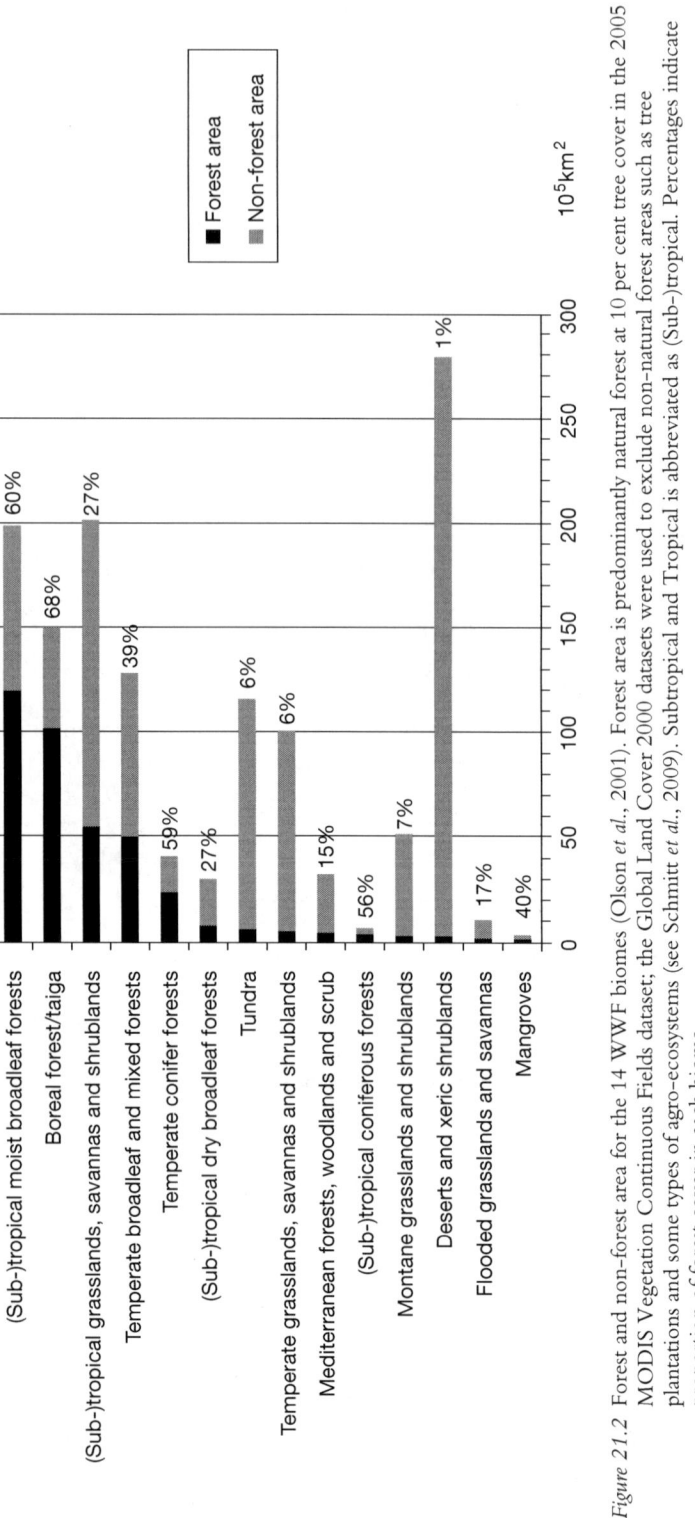

Figure 21.2 Forest and non-forest area for the 14 WWF biomes (Olson *et al.*, 2001). Forest area is predominantly natural forest at 10 per cent tree cover in the 2005 MODIS Vegetation Continuous Fields dataset; the Global Land Cover 2000 datasets were used to exclude non-natural forest areas such as tree plantations and some types of agro-ecosystems (see Schmitt *et al.*, 2009). Subtropical and Tropical is abbreviated as (Sub-)tropical. Percentages indicate proportion of forest cover in each biome

status, but lacks a global map of habitat distribution. While there is relatively good agreement in the definition of boreal and temperate forest types between the WWF and IUCN classification systems, the subdivisions of (sub-)tropical forests are not congruent (Figure 21.2; Table 21.2). For instance, the WWF emphasises the difference between broadleaf and coniferous forest biomes, whereas the IUCN highlights the difference between lowland and montane forest habitats. Furthermore, Mediterranean forests are included with the Mediterranean forests, woodlands and scrub biome in the WWF framework, but may be considered as temperate forest or (sub-)tropical dry forest habitat in the IUCN classification system (IUCN, 2014a).

Global patterns of biodiversity in forests

Global maps of species richness for vascular plants, birds, mammals and amphibians clearly indicate the well-known concentration of species numbers in the tropics (Barthlott *et al.*, 2005; Orme *et al.*, 2005; Jenkins *et al.*, 2013). Generally, the (sub-)tropical moist broadleaf forests biome has a much higher species richness and endemism than all other biomes, and many tropical rainforest taxa have a long independent evolutionary history, which is a measure of taxonomic uniqueness and irreplaceable genetic diversity (Mace *et al.*, 2005). It could be that the outstanding species diversity of this biome is related to its large area size (see Figure 21.2), but species richness and area are not statistically related for the 14 terrestrial biomes (Mace *et al.*, 2005).

The enormous diversity in areas of tropical rainforest is illustrated by the 473 tree species recorded in 1-ha plots in Amazonian Ecuador (diameter at breast height (dbh) ≥ 5 cm) (Valencia *et al.*, 1994). Generally, the number of tree species in 1-ha plots in the Upper Amazonian forests can vary between 149 and 307 species (dbh ≥ 10 cm) (Valencia *et al.*, 1994). Lowland Amazonia as a whole is estimated to harbour a total of 16,000 tree species (ter Steege *et al.*, 2013), which is in stark contrast to the 53 tall tree species known to occur in the Central European broadleaf and mixed forests (Ellenberg and Leuschner, 2010). Conifers are an exception because they have highest species richness in the temperate forest habitat (Table 21.2).

Tropical communities are usually characterised by a much higher proportion of rare species than communities at higher latitudes, a pattern explained to some extent by niche specialisation and dispersal mechanisms (Brown, 2014). For instance, ter Steege *et al.* (2013) estimated that 227 tree species, i.e., 1.4 per cent of all tree species, account for half of all trees in the Amazon. Similarly only four woody species, i.e. 1.5 per cent of all woody species, constitute more than 50 per cent of all woody individuals in Ethiopian moist montane forest. This is illustrated by the long tail of species in the ranked species-abundance distribution (Figure 21.3). For high latitude assemblages, a similar graph would typically be steep and approximately linear (Brown, 2014).

Similar to plants, the number of animal species is higher in the (sub-)tropical moist broadleaf forests biome than in all other biomes (Mace *et al.* 2005). The latest IUCN data for mammals, birds and amphibians illustrate the large number of species in the (sub-)tropical forest habitats (Table 21.2). For instance, 84 per cent of the global amphibian species occur in forest habitat, and amphibians are clearly most diverse in the (sub-)tropical moist lowland and montane forests. In contrast, mammals and birds are more diverse in moist lowland forest than in moist montane forest, and a considerable number of these species occur in dry forest (Table 21.2). In fact the (sub-)tropical dry broadleaf forests biome holds many endemic species (Mace *et al.*, 2005). Individual dry forest areas are highly variable in forest structure and floristic diversity, with up to 121 woody species in a 0.1-ha plot in Colombia (Gillespie *et al.*, 2000). Thus, considering the relatively small global area of (sub-)tropical dry forests which still remains (Figure 21.2), this forest habitat harbours an important portion of the world's biodiversity.

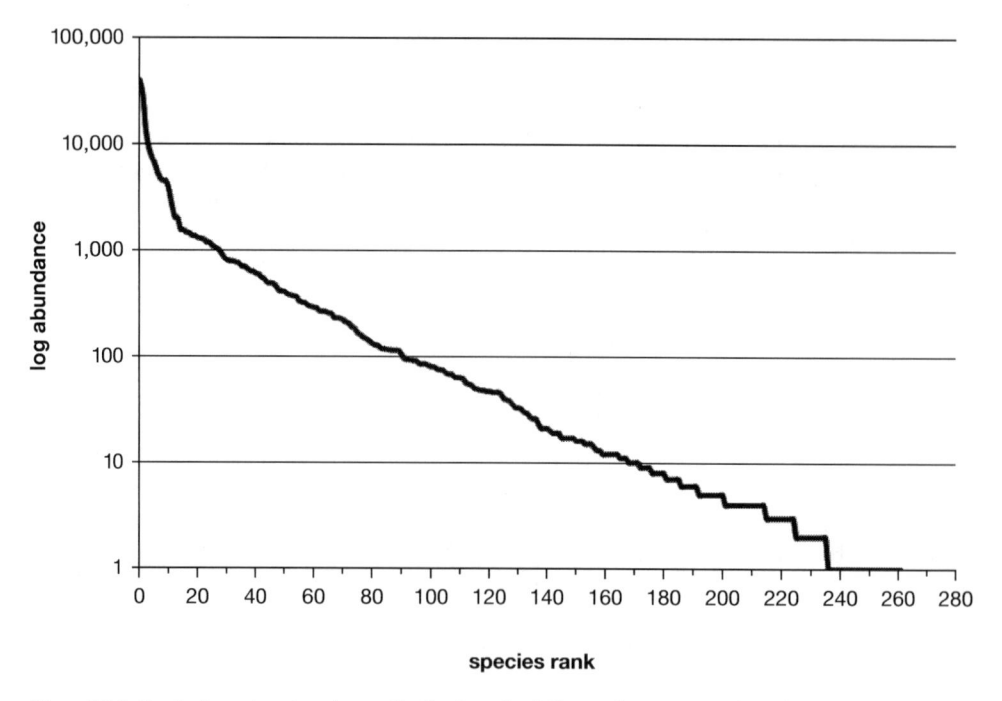

Figure 21.3 Ranked species-abundance distributions for 262 woody species in the Ethiopian moist montane forests (logarithmic scale). Abundance is defined as the number of individuals for trees, shrubs and woody lianas with height > 0.5 m in 180 study plots (20 m × 20 m) (see Schmitt *et al.*, 2013). Four species account for more than 50 per cent of all woody individuals

Although both species richness and endemism peak in the tropics, their global patterns are not congruent. As a general rule the global centres of species richness are mostly located on the continents, whereas the centres of endemism are more strongly correlated with large tropical islands and island archipelagos (Orme *et al.*, 2005; Kier *et al.*, 2009; Jenkins *et al.*, 2013). For birds, the centres of species richness are concentrated in the moist forest of the Amazon, the Brazilian Atlantic Forest and the continental (sub-)tropical mountain areas (Figure 21.1). The same areas are important centres of diversity for vascular plants and mammals, and furthermore, these two taxa are very species rich on some of the Southeast Asian islands (Barthlott *et al.*, 2005; Jenkins *et al.*, 2013). Although endemism is high on large tropical islands for birds and mammals, continental mountain areas such as Mount Cameroon and the Eastern Arc in Africa are also important (Figure 21.1 and Jenkins *et al.*, 2013). In contrast, amphibian species mostly have small ranges and thus there are no major centres of endemism globally (Jenkins *et al.*, 2013). The Andes show an outstanding congruence of species richness and endemism for plants and many vertebrate species, which is partly explained by the steep altitudinal and climatic gradients stimulating high species turnover (Orme *et al.*, 2005; Kier *et al.*, 2009).

Vast differences in the distribution ranges and endemism patterns of vertebrates and invertebrates are apparent. However, the species data for the latter are still extremely sparse (Mace *et al.*, 2005). A recent study has proven that there can also be large variations in global diversity patterns between vascular plants and bryophytes (Geffert *et al.*, 2013). The study was based on 11,388 moss species and showed that the tropical rainforests are relatively species-poor with high species numbers only in the tropical mountain ranges and cloud forest ecosystems such as in the northern Andes. In contrast, temperate broadleaf forests, boreal forests and tundra have relatively high moss species richness, e.g., in Central Europe, Scandinavia and Japan. It cannot be ruled out, however, that the absence of a latitudinal gradient for moss species richness is related to both a sampling bias towards temperate regions and to pending taxonomic revisions for many tropical taxa (Geffert et al., 2013).

Next to the global variations in species richness and endemism, there are large differences in species diversity between the biogeographic realms for each forest biome. For instance, the Neotropical rainforests have a much higher plant species richness (ca. 93,500 species) than the rainforests in the Asia-Pacific region (ca. 61,700 species) and the African tropics (ca. 20,000 species) (Corlett and Primack, 2011). Local species richness also differs with an average of ca. 180, 150 and 95 large tree species per hectare in the Neotropical, Asian and African rainforests, respectively (Corlett and Primack, 2011). While the Neotropical rainforests are renowned for their high species richness in birds, bats, amphibians and Bromeliad epiphytes, the Afrotropical rainforests have highest browser diversity, and the Southeast Asian forests exhibit the highest diversity in gliding animals and dipterocarp tree species (Corlett and Primack, 2011; Jenkins et al., 2013). Underlying reasons for these variations in species diversity include continental drift over the past 200 million years, the glacial cycles during the Pleistocene and current climate patterns (Lomolino et al., 2010; Corlett and Primack, 2011).

In the temperate regions, beech forests cover large areas of the Palearctic, Nearctic and Australasia but differ in the dominant genus, i.e., *Fagus* in the northern hemisphere and *Nothofagus* in the southern (Lomolino et al., 2010). Furthermore, with a total of around 124 tree species, the temperate broadleaf and mixed forests of eastern North America are more species rich than those of Central Europe (n = 53) (Ellenberg and Leuschner, 2010). In Europe, mountain ranges such as the Alps blocked the latitudinal shift of species in response to cooler temperatures, and led to major species extinction events, whereas in North America species shifts were facilitated by north–south running rivers and mountain ranges (Lomolino et al., 2010). Furthermore, warmer summer temperatures in eastern North America today support higher species richness than in Europe (Ellenberg and Leuschner, 2010).

In the boreal forests of Europe, Asia and North America, species typically have wider distribution ranges and species assemblages are relatively homogeneous across larger regions (Olson and Dinerstein, 2002). In the WWF ecoregions framework, this is reflected by the large size of the northern ecoregions, which were delineated with an emphasis on dynamics and processes, such as major variations in climate, fire disturbance regimes, and large vertebrate migrations (Olson et al., 2001). Conifers are the dominant species in the northern boreal forests where they cover vast areas with only five to ten species co-occurring in units of 10.000 km^2 (Mutke and Barthlott, 2005). In contrast, the global centres of conifer species richness are located in temperate and (sub-)tropical forest areas (Table 21.2), especially in Southeast Asia, western North America and Mexico (Mutke and Barthlott, 2005). Temperate conifer forests are recognised as an individual biome by the WWF due to their unique species assemblages (see Figure 21.2). In particular, the temperate rainforests along the coast of western North America, New Zealand and Chile are dominated by huge, slow-growing trees, have high epiphyte diversity and many endemic species (Olson and Dinerstein, 2002; Lomolino et al., 2010). Similarly, the Mediterranean forests, woodlands and shrubs cover only a small area globally (Figure 21.2), but are recognised for their distinct plant species diversity (Kier et al., 2009; Bannister et al., 2011).

There are notable variations in forest diversity along the altitudinal gradient. In temperate forests, broadleaf tree species are replaced by conifers at higher altitudes (Olson et al., 2001), and temperate coniferous forests can extend as far south as southern Mexico at high elevations (Lomolino et al., 2010). Individual (sub-)tropical moist montane forests are global centres of species richness and can harbour a substantial number of endemic species along altitudinal gradients (Barthlott et al., 2005; Jenkins et al., 2013). They often form part of biodiversity hotspots, e.g., in the Western Ghats and Sri Lanka, the tropical Andes and the mountains of southwest China (Sloan et al., 2014). The forests in the Eastern Arc Mountains of Tanzania and Kenya support at least 96 endemic vertebrate species and 800 species of endemic vascular plants (Burgess et al., 2007). This exceptional diversity cannot be explained by a single mechanism, and more recent speciation due to heterogeneous habitats, a minimal influence of Pleistocene climatic changes and long evolutionary histories in isolation seem to play a role (Burgess et al., 2007; Tolley et al., 2011). The Ethiopian moist montane forests (Figure 21.3) are floristically related to the Eastern Arc forests, but have fewer endemic species (Schmitt et al., 2013). Overall, the global (sub-)tropical moist montane forests support large proportions of the worldwide vertebrate diversity (Table 21.2) despite the fact that their area is much smaller than that of the (sub-)tropical moist lowland forests (compare Thompson et al., 2012).

Finally, soil type can strongly influence the species diversity in forests worldwide. For example, tree diversity in the Amazon is partly determined by soil type (ter Steege et al., 2013). Furthermore, the highly diverse (sub-)tropical moist broadleaf forests biome comprises areas of relatively low plant species richness such as the (sub-)tropical freshwater and peat swamp forests. An assessment of tree species diversity (dbh $\geq$ 10 cm) in 1-ha plots revealed that mean species richness in Ecuadorian swamp forest was 74 as opposed to 239 in the adjacent non-flooded forest (Pitman et al., 2014). Southeast Asian peat swamp forests and well-drained forests comprised between 30 and 122 species and between 100 and 290 species per hectare, respectively (Posa et al., 2011). Fewer tree species is presumably related to problems of dispersal, germination, establishment and growth caused by seasonal extremes in water levels (Corlett and Primack, 2011). Mangrove forests also contain relatively low plant species richness and are formed by a highly specialised group of trees and shrubs which thrive in saline soils subject to tidal flooding (Spalding et al., 2010). However, the (sub-)tropical mangrove and swamp forests constitute important habitats for global bird species richness (Table 21.2). In the temperate broadleaf and mixed forests, species diversity, especially of understorey and herbaceous plants, can be highly variable depending on local climate, soil type and fire frequency (Lomolino et al., 2010).

Conclusions

With the increasing wealth of global species data, high-resolution data on forest cover and biogeographical forest classifications at hand, the evaluation of global patterns of biodiversity in forests would seem a straightforward task. This review shows that we do know much about global patterns of species richness and endemism of particular groups of vertebrates and plants. There are also many studies which assess, describe and evaluate biodiversity in individual forest areas worldwide. Yet, the analysis of these datasets with different scope and resolution in combination still presents a major challenge today (compare Jetz et al., 2012).

There has been much emphasis internationally on the outstanding biodiversity of the tropical rainforests, which are included within the (sub-)tropical moist broadleaf forests biome in biogeographical terms. This review shows that not all tropical rainforests are equally rich in species, and that there are large variations in species diversity between rainforests in the Neotropics, the Afrotropics and the Asian-Pacific region. Furthermore, variation in species diversity is also evident along edaphic and altitudinal gradients within the (sub-)tropical moist broadleaf forests biome in each realm. In addition, global diversity patterns may differ between different taxa, and there are large data gaps still for many species groups, e.g., for invertebrates.

A focus on species numbers is bound to overlook globally unique forest ecosystems other than the (sub-)tropical moist forests. For instance locally, the (sub-)tropical dry broadleaf forests, Mediterranean forests, woodlands and scrub, and the temperate coniferous forests biomes can host a large number of species, including many endemics, yet they do not nearly approach the total species richness of the (sub-)tropical moist broadleaf forests biome. Besides, waterlogged forest areas such as the global mangrove forests can harbour highly specialised species assemblages. The boreal forest ecosystems are relatively poor in species, but functional in a harsh environment, and they support exceptional ecological processes such as large vertebrate migrations.

Most studies on global diversity patterns of particular species or species groups do not consider the spatial distribution of forest cover and forest ecosystems. Even if forest ecosystems are considered, the prevalence of different global classification systems complicates comparisons of biodiversity patterns, especially for the (sub-)tropical forests. Therefore the clear documentation of the source and spatial resolution of species data, forest cover data and forest classification systems is crucial when comparing studies on forest biodiversity. Finally, the evaluation of global forest biodiversity patterns requires a balanced approach that takes into account the different groups of species, species richness and endemism as well as distinctive species assemblages and adaptations. In addition, biogeographical ecosystem classifications can help to ensure that smaller, unique forest ecosystems are not overlooked in the global comparison.

References

Bannister, J. R., Vidal, O. J., Teneb, E. and Sandoval, V. (2011) 'Latitudinal patterns and regionalization of plant diversity along a 4270-km gradient in continental Chile', *Austral Ecology*, vol 37, no 4, pp. 500–509

Barthlott, W., Mutke, J., Rafiqpoor, D., Kier, G. and Kreft, H. (2005) 'Global centers of vascular plant diversity', *Nova Acta Leopoldina NF*, vol 92, no 342, pp. 61–83

Brown, J. H. (2014) 'Why are there so many species in the tropics?', *Journal of Biogeography*, vol 41, pp. 8–22

Burgess, N. D., Butynski, T. M., Cordeiro, N. J., Doggart, N. H.; Fjeldsa, J. and others (2007) 'The biological importance of the Eastern Arc Mountains of Tanzania and Kenya', *Biological Conservation*, vol 134, pp. 209–231

Corlett, R. T. and Primack, R. B. (2011) *Tropical Rain Forests: An Ecological and Biogeographical Comparison*, Wiley-Blackwell, Oxford, UK

Ellenberg, H. and Leuschner, C. (2010) *Vegetation Mitteleuropas mit den Alpen, 6th Edition*, Eugen Ulmer KG, Stuttgart, Germany

FAO (2012) *Global Ecological Zones for FAO Forest Reporting: 2010 Update*, Working Paper 179, Food and Agriculture Organization of the United Nations (FAO), Rome

Geffert, L. J., Frahm, J.-P., Barthlott, W. and Mutke, J. (2013) 'Global moss diversity: spatial and taxonomic patterns of species richness', *Journal of Bryology*, vol 35, no 1, pp. 1–11

Gillespie, T. W., Grijalva, A. and Farris, C. N. (2000) 'Diversity, composition, and structure of tropical dry forests in Central America', *Plant Ecology*, vol 147, pp. 37–47

Hansen, M. C., Potapov, P. V., Moore, R., Hancher, M., Turubanova, S. A. and others (2013) 'High-resolution global maps of 21st-century forest cover change', *Science*, vol 342, pp. 850–853

IUCN (2014a) 'The IUCN Red List of Threatened Species. Version 2014.1. Habitats Classification Scheme (Version 3.1)', www.iucnredlist.org/technical-documents/classification-schemes/habitats-classification-scheme-ver3, accessed 11 July 2014

IUCN (2014b) 'The IUCN Red List of Threatened Species. Version 2014.1. Overview', www.iucnredlist.org/about/overview, accessed 11 July 2014

Jenkins, C. N., Pimm, S. L. and Joppa, L. N. (2013) 'Global patterns of terrestrial vertebrate diversity and conservation', *PNAS*, doi: 10.1073/pnas.1302251110

Jetz, W., McPherson, J. M. and Guralnick, R. P. (2012) 'Integrating biodiversity distribution knowledge: toward a global map of life', *Trends in Ecology and Evolution*, vol 27, no 3, pp. 151–159

Kier, G., Kreft, H., Lee, T. M., Jetz, W., Ibisch, P. L. and others (2009) 'A global assessment of endemism and species richness across island and mainland regions', *PNAS*, vol 106, no 23, pp. 9322–9327

Kreft, H. and Jetz, W. (2007) 'Global patterns and determinants of vascular plant diversity', *PNAS*, vol 104, no 14, pp. 5925–5930

Lomolino, M. V., Riddle, B. R., Whittaker, R. J. and Brown, J. H. (2010) *Biogeography*, 4th Edition, Sinauer Associates, Inc., Sunderland, MA, US

Mace, G., Masundire, H., Baillie, J., Ricketts, T., Brooks, T. and others (2005) 'Biodiversity', in R. Hassan and N. Ash (Eds) *Ecosystems and Human Well-being: Current State and Trends, Volume 1*, Millennium Ecosystem Assessment, Island Press, Washington, DC

Magurran, A. E. (2004) *Measuring Biological Diversity*, Blackwell Publishing, Oxford, UK

Murphy, B. P. and Bowman, D. M. J. S. (2012) 'What controls the distribution of tropical forest and savanna?', *Ecology Letters*, vol 15, no 7, pp. 748–758

Mutke, J. and Barthlott, W. (2005) 'Patterns of vascular plant diversity at continental to global scales', *Biologiske Skrifter*, vol 55, pp. 521–531

Olson, D. M. and Dinerstein, E. (2002) 'The Global 200: Priority ecoregions for global conservation', *Annals of the Missouri Botanical Garden*, vol 89, no 2, pp. 199–224

Olson, D. M., Dinerstein, E., Wikramanayake, E. D., Burgess, N. D., Powell, G. V. N. and others (2001) 'Terrestrial ecoregions of the world: A new map of life on Earth', *BioScience*, vol 51, no 11, pp. 933–938

Orme, D. L., Davies, R. G., Burgess, M., Eigenbrod, F., Pickup, N. and others (2005) 'Global hotspots of species richness are not congruent with endemism or threat', *Nature*, vol 436, pp. 1016–1019

Pereira, H. M., Ferrier, S., Walters, M., Geller, G. N., Jongman, R. H. G. and others (2013) 'Essential biodiversity variables', *Science*, vol 339, pp. 277–278

Pimm, S. L., Jenkins, C. N., Abell, R., Brooks, T. M., Gittleman, J. L. and others (2014) 'The biodiversity of species and their rates of extinction, distribution, and protection', *Science*, vol 344, doi: 10.1126/science.1246752

Pitman, N. C. A., Andino, J. E. G., Aulestia, M., Cerón, C. E., Neill, D. A. and others (2014) 'Distribution and abundance of tree species in swamp forests of Amazonian Ecuador', *Ecography*, vol 37, pp. 1–14

Posa, M. R. C., Wijedasa, L. S. and Corlett, R. T. (2011) 'Biodiversity and conservation of tropical peat swamp forests', *BioScience*, vol 61, pp. 49–57

Schmitt, C. B. (2013) 'Global tropical forest types as support for the consideration of biodiversity under REDD+', *Carbon Management*, vol 4, no 5, pp. 501–517

Schmitt, C. B., Burgess, N. D., Coad, L., Belokurov, A., Besançon, C. and others (2009) 'Global analysis of the protection status of the world's forests', *Biological Conservation*, vol 142, pp. 2122–2130

Schmitt, C. B., Senbeta, F., Woldemariam, T., Rudner, M. and Denich, M. (2013) 'Importance of regional climates for plant species distribution patterns in moist Afromontane forest', *Journal of Vegetation Science*, vol 24, no 3, pp. 553–568

Sloan, S., Jenkins, C. N., Joppa, L. N., Gaveau, D. L. A. and Laurance, W. F. (2014) 'Remaining natural vegetation in the global biodiversity hotspots', *Biological Conservation*, vol 177, pp. 12–24

Spalding, M. D., Kainuma, M. and Collins, L. (2010) *World Mangrove Atlas* Earthscan, with International Society for Mangrove Ecosystems, FAO, The Nature Conservancy, UNEP World Conservation Monitoring Centre, United Nations Scientific and Cultural Organization, United Nations University, London

ter Steege, H., Pitman, N. C. A., Sabatier, D., Baraloto, C., Salomão, R.P. and others (2013) 'Hyperdominance in the Amazonian Tree Flora', *Science*, vol 342, doi: 10.1126/science.1243092

Thompson, I. D., Ferreira, J., Gardner, T., Guariguata, M., Koh, L. P. and others (2012) 'Forest biodiversity, carbon and other ecosystem services: relationships and impacts of deforestation and forest degradation', in J. A. Parrotta, C. Wildburger and S. Mansurian (eds) *Understanding Relationships between Biodiversity, Carbon, Forests and People: The Key to Achieving REDD+ Objectives. A Global Assessment Report. Prepared by the Global Forest Expert Panel on Biodiversity, Forest Management and REDD+*. IUFRO World Series Volume 31, Vienna

Tolley, K. A., Tilbury, C. R., Measey, G. J., Menegon, M., Branch, W. R. and Matthee, C. A. (2011) 'Ancient forest fragmentation or recent radiation? Testing refugial speciation models in chameleons within an African biodiversity hotspot', *Journal of Biogeography*, vol 38, pp. 1748–1760

Tuanmu, M.-N. and Jetz, W. (2014) 'A global 1-km consensus land-cover product for biodiversity and ecosystem modelling', *Global Ecology and Biogeography*, doi: 10.1111/geb.12182

Valencia, R., Balslev, H. and Paz y Miño, C. G. (1994) 'High tree alpha-diversity in Amazonian Ecuador', *Biodiversity and Conservation*, vol 3, pp. 21–28

PART IV

Energy and nutrients

22
MYCORRHIZAL SYMBIOSIS IN FOREST ECOSYSTEMS

Leho Tedersoo

Introduction

Mycorrhiza is a term for symbiotic interaction between plant roots and fungal mycelium in which the tissues of both partners are specifically differentiated for improved exchange of nutrients. In addition to plants and fungi, a wide array of prokaryotes and protists participate in the establishment and functioning of mycorrhizal symbiosis (Bueé et al., 2009). The outcome of symbiosis is not always beneficial to growth because both partners fight for their individual benefits (Jones and Smith, 2004) and there are other benefits unrelated to growth *per se*. Plants have evolved a mechanism to promote preferential association with the most beneficial mycorrhizal fungi at least to some extent (Kiers et al., 2011). In natural conditions, ca. 94 per cent of vascular plant species, most hepatics and bryophytes are typically colonized by mycorrhizal fungi. Notably, most members of the Sphagnaceae, Caryophyllaceae, Polygonaceae, Cyperaceae and Brassicaceae families as well as various parasitic, hydrophilous, carnivorous and cluster-rooted plants have secondarily reverted to the non-mycorrhizal habit (Brundrett, 2009). In fungi, the evolution of mycorrhizal strategy is as complex as it is in plants. In natural conditions, the association with root symbiotic fungi is obligatory in most of the mycorrhizal plants and in nearly all mycorrhizal fungi, depending on mycorrhiza type.

The benefits of mycorrhizal symbiosis are many-fold for both plants and fungi. Mycorrhizal fungi are more efficient than plant roots and their root hairs in capturing dissolved mineral nutrients due to their 10-fold smaller diameter compared with root hairs and greater plasticity in growth and branching. By exuding organic acids from hyphal tips, fungi are able to dissolve minerals such as biotite and apatite in search for phosphorus (Gadd, 2007). Ecto- and ericoid mycorrhizal fungi have limited enzymatic capacities to release micro- and macronutrients from organic matter and to take up simple organic compounds such as amino acids and oligopeptides (Courty et al., 2010). Nitrogen and phosphorus are typically among the most limiting elements in plants due to the chemical build-up of photosynthesis apparatus, proteins and DNA (Smith and Read, 2008). Besides nutrition, mycorrhizal fungi protect plants from hazardous chemicals (PAH, allelochemicals, etc.) and heavy metals by means of immobilizing these molecules with siderophores or organic acids in soil, on the surface of hyphae or inside vacuoles within the living mycelium (Joner and Leyval, 2003; Gadd, 2007). Mycorrhizal fungi may also protect plants from various pathogens and herbivores by alerting neighbouring plants via hyphal connections, inducing systemic resistance, producing antibiotics and allelochemicals in the soil environment and occupying the physical niche for pathogens (Whipps, 2004; Babikova et al., 2013). Deep-rooted trees and mycorrhizal fungi collectively retain critical water potentials in soil to ameliorate drought stress. At night when stomata are open, trees use transpiration flow to suck water from deep soil and spread part of it to lateral roots, mycorrhizas and eventually to fungal hyphae in soil. In the daytime when stomata are closed, water is re-absorbed by mycelium and used for biochemical processes in both partners (Querejeta et al., 2003). It is likely that other soil microbes retain some functional activity by tapping into this system in dry periods.

Fungi are heterotrophic organisms and their growth and reproduction are largely limited by the availability of carbon energy. In mycorrhizal associations, fungi receive photosynthetically fixed carbon via plant roots. In addition, roots of perennial plants and trees provide a relatively stable habitat compared with decomposing litter. Plants and fungi collectively defend the mycorrhizal root system from grazers and pathogens by use of chemical weapons. Deposition of salt crystals and development of sharp cystidia are also believed to deter herbi- and fungivorous soil biota (Taylor and Alexander, 2005).

Types of mycorrhiza

Based on the anatomy of symbiosis, four principal types of mycorrhiza are distinguished, viz. arbuscular mycorrhiza (AM), ectomycorrhiza (EcM), ericoid mycorrhiza (ErM) and orchid mycorrhiza (OM) (Figure 22.1). Taxonomically distinct groups of plants and fungi are associated

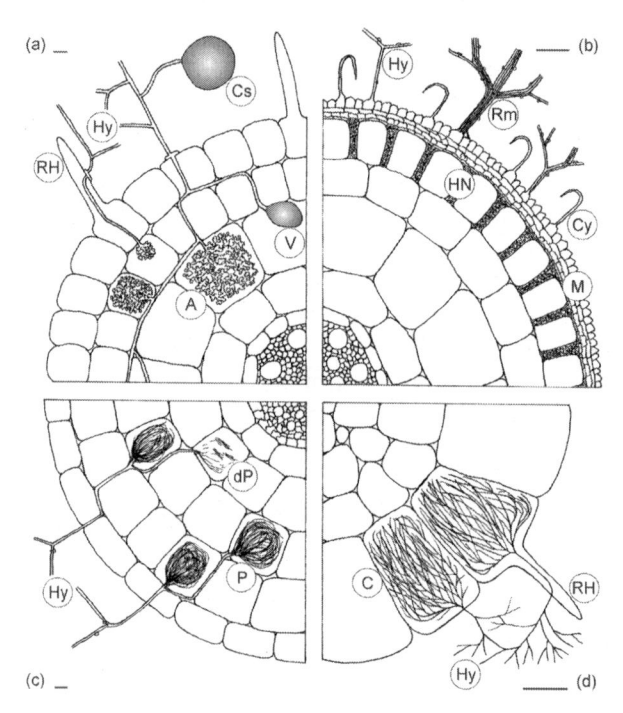

Figure 22.1 Structural differences among four major mycorrhiza types: (a) arbuscular mycorrhiza; (b) ectomycorrhiza; (c) orchid mycorrhiza; (d) ericoid mycorrhiza. Abbreviations: A, arbuscules; V, vesicles; RH, root hair; Hy, extraradical hyphae; Cs, chlamydospores; M, mantle; HN, Hartig net; Cy, cystida; Rm, rhizomorphs; P, pelotons; dP, digested pelotons; C, coils; Bar, 20 μm

with different mycorrhiza types. These types are widely distributed in various ecosystems and continents except Antarctica. The associated fungi differ in their access to organic nutrients (Figure 22.2).

In AM, fungal colonization of plant roots is mostly intracellular (Figure 22.1a). Fungi typically enter the root cells through root hairs and form complex bush-like or clew-like structures that fill the cells and maximize surface contact with the plant's plasma membrane. These so-called arbuscules constitute the interface for nutrient transfer between the symbionts. AM is thought to be the oldest type of mycorrhiza that facilitated land colonization by early plants in the Ordovician. Therefore, AM is present in most groups of terrestrial plants, but it can be secondarily lost or replaced by other mycorrhizal types (Brundrett, 2009). In fungi, AM associations have evolved only once in the ancestors of the fungal phylum Glomeromycota. These relatively primitive fungi exhibit multinucleate hyphae and lack sexual propagation, which is compensated by the ability of conspecific individuals to exchange nuclei. The Glomeromycota comprise ca. 500 species (Kõljalg *et al.*, 2013; Öpik *et al.*, 2013), a mere 0.03 per cent of the estimated 1.5 million fungal species (Hawksworth, 2001). Dispersal of

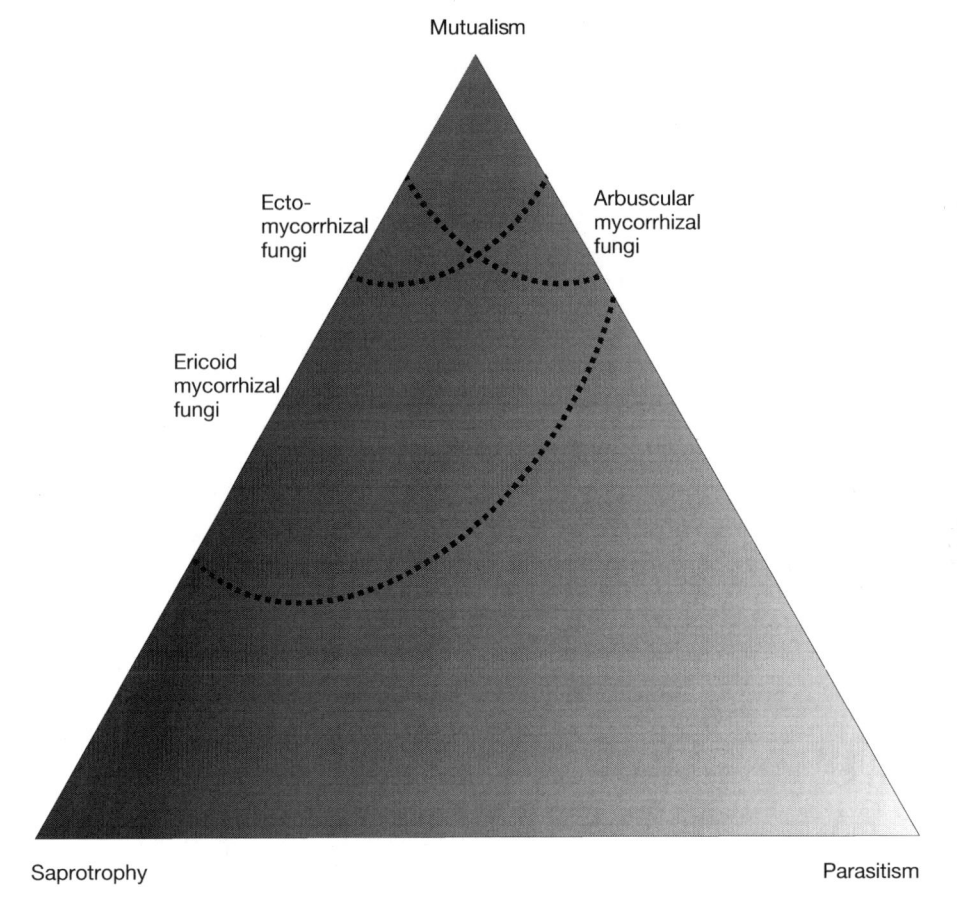

Figure 22.2 Relative placement of mycorrhizal fungi in the parasitism–mutualism–saprotrophy continuum (above dashed line). Orchid mycorrhizal fungi may belong to all trophic types, depending on plant species

Glomeromycota occurs via root fragments and particularly chlamydospores that develop on the extraradical mycelium. Compared with other types of mycorrhiza, hyphae of AM fungi are multinucleate, relatively broad (10–20 µm) and sparsely distributed in soil. Altogether 73 per cent of all plant species are obligately AM and 8 per cent are facultatively AM. AM is predominant in herbs and in tropical plants, including most tropical trees (Brundrett, 2009).

EcM is characterized by the presence of a fungal mantle, which covers root tips, and a Hartig net that constitutes plant root epidermal cells densely surrounded by fine, highly branched hyphae (Figure 22.1b). The Hartig net plays a key role in nutrient exchange between the partners. EcM has evolved >20 times repeatedly in AM or non-mycorrhizal (NM) plants in the Carboniferous (Pinaceae) to Tertiary (most other taxa). In fungi, EcM root associations have independently evolved in ca. 80 groups (lineages) in the same time frame (Ryberg and Matheny, 2012; Tedersoo and Smith, 2013). EcM fungi differ strongly in the amount of extraradical mycelium, presence and structure of rhizomorphs and capacity to produce degradation enzymes (Courty et al., 2010). EcM fungi include taxa that produce either stipitate (agaricoid), resupinate (crusts on litter), ramarioid (bush-like or club-like) or sequestrate (closed hypogeous) fruit-bodies as well as groups that reproduce only asexually. Ascomycota and Basidiomycota exhibit monokaryotic (haploid) or dikaryotic (two separate nuclei) mycelium and most taxa undergo complicated mating procedures. Tiny spores are effectively dispersed by wind or by animals that are attracted by a conspicuous smell (e.g. truffles). EcM occurs in ca. 2 per cent of all vascular plant species, but these plants include both ecologically and economically important timber species in the northern temperate (Pinaceae, Fagaceae, Salicaceae), tropical (Dipterocarpaceae, Caesalpinioideae of Fagaceae) and southern temperate (Myrtaceae, Nothofagaceae) ecosystems (Brundrett, 2009). EcM fungi comprise 20,000–25,000 species that account for ca. 1.5 per cent of the estimated fungal richness (Rinaldi et al., 2008). There are several subtypes of EcM such as arbutoid mycorrhiza (in Arbutoideae, Ericaceae), monotropoid mycorrhiza (Monotropoideae, Ericaceae) and ectendomycorrhiza (Pinaceae involving certain ascomycete mycobionts) that are anatomically distinct with differences in mantle thickness and/or presence of intracellular colonization, but the fungal partners include typical EcM symbionts (Brundrett, 2004). Both EcM and AM associations are obligatory to the fungal partners involved.

Orchid mycorrhizal symbiosis is characterized by dense intracellular colonization of root cells by fungal hyphae that form clew-like structures termed pelotons (Figure 22.1c). When aged, these pelotons are digested by plants in many orchid groups. Most orchids associate with saprotrophic fungi from the polyphyletic form genus *Rhizoctonia* that comprises Ceratobasidiaceae, Tulasnellaceae and Sebacinaceae families of the Basidiomycota. Certain terrestrial orchids have shifted to associate with typical EcM fungi that has led to partial or full loss of photosynthesis capacity in many unrelated orchid groups. These so-called myco-heterotrophs parasitize mycorrhizal networks. Orchids have minute seeds with limited reserves and therefore all orchid species depend on fungi in their nutrition, at least in the early stages of development (Dearnaley et al., 2012). Orchids comprise ca. 27,000 species that account for around 10 per cent of plant diversity. There are no estimates on the fungal richness involved in OM and ErM associations because of insufficient knowledge of the breadth and functionality of orchid–fungus interactions. Although orchids are widespread from subarctic to tropical ecosystems, they form perhaps the most important ecosystem component in montane rain forests as epiphytes. Collection of wild orchid flowers for decoration and rhizomes for food has had a substantial economic impact and conservational concern in many countries including China and Zambia. Besides orchids, partly and fully mycoheterotrophic plants include several minor taxonomic groups that associate with EcM fungi (Pyroleae and Monotropeae tribes of

Ericaceae) or AM fungi (mostly tropical groups, e.g. certain members of the Burmanniaceae and Triuridaceae families). Because mycoheterotrophic associations are certainly non-beneficial to fungi and it is difficult to reliably measure costs and benefits in associations involving green orchids, consideration of OM and mycoheterotrophic associations as mycorrhizal is questionable (Leake, 2004).

In ErM, fungal colonization occurs exclusively inside root cells, where fungi form clew-like structures termed as coils (Figure 22.1d). In some cases, fungi form a sparse and thin (one cell layer) mantle-like structure on the surface of roots. ErM has evolved once in Ericaceae and probably once in *Schizocodon* (Diapensiaceae) (Okuda *et al.*, 2011). There is a lot of uncertainty in the mycorrhiza forming ability and functionality of the wide array of fungal species detected from roots of ErM plants (Cairney and Meharg, 2003). The associated fungi mostly belong to Helotiales, Chaetothyriales and Sebacinales that also act as root endophytes (asymptomatic root-associated fungi) in non-Ericaceae plants and as peat and humus saprobes. Therefore, ErM fungi can be viewed as a subset of root endophytes 'domesticated' by the ancestors of the Ericaceae family that allowed formation of coils inside the root cells. Root endophytes in turn seem to form a subset of generalist saprotrophs that lack strong ligninolytic activities. Limited saprotrophic capacity probably enables the plant to control the mycorrhizal system and benefit from access to the pool of micro- and macronutrients in highly acidic and infertile soils. ErM accounts for 1.4 per cent of plant species due to the extensive speciation in the Ericaceae family (Brundrett, 2009).

Distribution of mycorrhizal plants and fungi

The distribution of mycorrhizal types has conspicuous patterns on a global scale and across local nutrient gradients (Allen *et al.*, 1995; Read *et al.*, 2004; Lambers *et al.*, 2008). Arctic grasslands and heathlands are dominated by ErM (*Empetrum* spp., *Cassiope* spp.), EcM (dwarf *Salix* and *Betula* spp., *Bistorta vivipara*, *Dryas octopetala*) and NM (*Carex* spp.) plants. Very similar communities occur above the treeline in the Alps, the Himalayas, the Rocky Mountains and many other mountain chains as relicts of historical glacial connections. Subarctic and subalpine forest tundra is dominated by small EcM trees and their bush-like forms (families Betulaceae and Pinaceae) with an understorey of ericoid mycorrhizal plants (*Vaccinium* spp.). The same EcM families and ErM genera also dominate in the vast belt of boreal forests. AM plants become increasingly abundant southward and in more productive habitats.

Temperate forests are dominated by both EcM (Fagaceae, Pinaceae, Betulaceae, Salicaceae, Tiliaceae, etc.) and AM (Oleaceae, Ulmaceae, Rosaceae, Aceraceae, Lauraceae, etc.) trees with almost exclusively AM understorey. AM vegetation tends to be more common on clayey soils, moist habitats and phosphorus-limited soils. The same plant families, along with many additional groups of tropical origin, are also common in subtropical forests, where AM usually gains dominance over EcM in lowland habitats.

In tropical lowland and montane rain forest, most of the trees and understorey are AM. Relative abundance of various AM plant families and particularly EcM families strongly differs among continents. Of EcM plants, Dipterocarpaceae are the most important plant family in India and SE Asia, where this group often forms monodominant stands in lowland and lower montane habitats in both moist tropical forests and dry tropical forests biomes. Dipterocarpaceae are also present in African rain forests and miombo woodlands, where they occur as subdominant trees (*Marquesia*) or understorey (*Monotes*). EcM groups related to Dipterocarpaceae are also present in Madagascar (the Sarcolaenaceae family) and South America (*Pakaraimea*, *Pseudomonotes*) but with highly localized distribution. African woodlands receiving <600 mm annual rainfall are solely comprised of AM vegetation. In semi-open miombo woodlands, EcM vegetation is clearly dominant and represented by Amhersteae tribe of the Fabaceae and *Uapaca* (Uapacaceae). While *Uapaca* is a widespread understorey tree in African rain forests, the distribution of Amhersteae is more localized. Both groups may form monodominant stands especially in the Congo basin. African montane rain forests are substantially dominated by AM trees with occasional *Uapaca* and *Berlinia* (Amhersteae) specimens. The South American EcM tree legume genus *Dicymbe* (Amhersteae) originates from Africa and it dominates certain areas in the Orinoco basin. Another EcM legume tree genus *Aldina* (Papilionoideae, Fabaceae) is mostly distributed in seasonally flooded areas on white sands, where it may gain monodominance. Throughout tropical lowland and lower montane rain forests of South America, EcM *Coccoloba* (Polygonaceae), *Neea* and *Guapira* (Nyctaginaceae) are widely distributed, but usually occur in low numbers as understorey trees. In coastal areas, *Coccoloba* species may form monodominant stands on sandy soils. Taken together, both in Africa and South America, dominance of EcM trees is related to nutrient-poor sandy soils. In tropical and subtropical Australia, vast semiarid areas are dominated by EcM Myrtaceae (*Eucalyptus*, *Melaleuca*), *Acacia* (Mimoisoideae, Fabaceae) and Casuarinaceae. These taxa are present in Australian remnant rain forest patches as a minor component. Due to its geological history and position on the migration routes, the island of New Guinea has floristic elements of both Australia and Southeast Asia. While the lowland rain forests are dominated by AM families with occasional EcM Dipterocarpaceae species, EcM eucalypts dominate savanna ecosystems in the rain shadow areas, whereas Fagaceae or Nothofagaceae sometimes form pure stands in the montane forest vegetation. Temperate forests of the Southern Hemisphere are dominated by Nothofagaceae and *Eucalyptus* in Australia including Tasmania. Nothofagaceae is the dominant tree family in montane forests of New Zealand and southern South America. However, the vast majority of tree genera including Podocarpaceae and Araucariaceae gymnosperms and the forest understorey is AM in southern temperate forests. ErM Ericaceae are also common in the understorey of southern temperate forests, tropical montane rain forests and above the treeline. In montane cloud forests, both Ericaceae and Orchidaceae are common as epiphytes and their mycorrhizal symbionts probably enable acquisition of nutrients from bark and accumulating debris.

In general, distribution of mycorrhizal fungi largely follows that of their host plants. Major families of the AM Glomeromycota are present in all forest ecosystems and there is surprisingly great overlap at the fungal species level (Öpik *et al.*, 2013). In EcM fungi, the major lineages are also shared among all forest ecosystems, but smaller phylogenetic groups are mostly restricted to northern temperate forests (Tedersoo and Smith, 2013). In contrast to other terrestrial organisms, species richness of EcM fungi peaks in northern temperate forests (Tedersoo *et al.*, 2012). Very little is known about the distribution of ErM and OM fungi at the global scale, but the key saprotrophic groups involved seem to be the same. Associations with EcM fungi and development of mycoheterotrophy in orchids is prominent in temperate and subtropical forest ecosystems but relatively rare in tropical, boreal and arctic habitats.

Phylogenetic constraints of plants, dispersal limitation and edaphic conditions strongly affect the global distribution patterns of mycorrhizal types. EcM plants and fungi usually form obligate associations, which renders their establishment after long-distance dispersal an unlikely event. Therefore, many remote islands are colonized solely by AM plants. Certain EcM trees, however, are also able to form AM associations that may facilitate colonization of distant sites and coping with a variety of habitat conditions (e.g. Salicaceae). At the local scale, soil texture and nutrients determine the relative competitive abilities of mycorrhizal types and their prevalence. Soil texture determines the water holding capacity and moisture regime of soil. In the most nutrient-poor habitats, ErM and NM plants dominate on organic and mineral soils, respectively. ErM fungi are able to obtain nutrients in the organic form, but it is difficult to understand the benefits of being NM. Probably fungi become nutrient-limited in very poor soils and compete with plants for these. NM cluster roots developed by Cyperaceae and Proteaceae may enable sufficient nutrient uptake (Lambers et al., 2008). In less severe nutrient limitation, EcM associations seem to be more beneficial compared with AM. EcM fungi perform better in acquisition of simple organic nutrients due to their enzymatic capacity and formation of rhizomorphs for efficient nutrient transportation. These benefits are, however, metabolically costly to plants and therefore AM plants tend to dominate in more productive and phosphorus-rich soils where competition for light becomes increasingly important (Read et al., 2004; Lambers et al., 2008). There are numerous exceptions to this generalization and distribution models of mycorrhizal types should include non-nutritional benefits.

It is notable that EcM trees tend to form monodominant stands (Torti et al., 2001). This is, however, another generalization, because there are many monodominant AM trees and sparsely distributed EcM tree species. Clumped distribution of individuals that are conspecific or belong to closely related species poses certain benefits and risks that, taken together, question the long-term perspective of the monodominant habit. Monodominant vegetation patches accumulate taxon-specific pathogens that diminish the establishment of seedlings (Liu et al., 2012). Furthermore, monodominance enhances inbreeding, because pollen and seed dispersal among patches is limited due to geographic distance. Monodominance has also certain benefits such as more efficient signalling to trigger mast fruiting at irregular time intervals to reduce the load of pathogens. Many monodominant EcM plants exhibit recalcitrant litter and, via positive feedback from mycorrhizal fungi, monodominant plants may secure internal cycling of nutrients (Newbery et al., 1997). The presence of abundant mycorrhizal inoculum may outweigh the negative feedback from soil pathogens (McGuire, 2007).

Functional role of mycorrhizal fungi in forest ecosystems

Mycorrhizal fungi exhibit a fundamental role in nutrient cycling in forest ecosystems. The boreal forest ecosystems are limited by nitrogen and sometimes additionally by phosphorus. To acquire these critical nutrients, one-quarter to two-thirds of the primary production is allocated below ground. Up to 20 per cent of the fixed carbon is supplied to EcM fungi (Hobbie, 2006) that account for up to 40 per cent of soil microbial biomass especially in low-nutrient soils (Högberg et al., 2010). These large amounts of carbon are required for the development and maintenance of extensive extraradical mycelium and fruit bodies, and production of extracellular exudates that are composed of degradation enzymes, organic acids and simple sugars. A wide variety of mycorrhizosphere protists and prokaryotes rely on these compounds for their nutrition and in return these organisms fix dinitrogen, synthesize vitamins, signal molecules, enzymes and allelochemicals that suppress pathogens (Bueé et al., 2009). These heterotrophic microbes, EcM fungi and roots are estimated to account for ca. 60 per cent, 25 per cent and 15 per cent of soil respiration, respectively (Heinemeyer et al., 2007). The relative respiration of EcM root tips is driven by their nitrogen content that in turn depends on fungal species (Trocha et al., 2010). Fungal species also differ markedly in their capacity to produce extracellular enzymes (Courty et al., 2010) that at least partly accounts for their differential ability to take up and provide nitrogen or phosphorus to plants (van der Heijden and Kuyper, 2003). In addition, species of EcM fungi partition soil nitrogen by source and depth in the soil profile (Hobbie et al., 2014). After death, recalcitrant cell walls of EcM and ErM fungal hyphae account for a major component of soil humus (Clemmensen et al., 2013). Unfortunately, the vast majority of studies on mycorrhizal roles in ecosystem functioning has been performed in boreal forests or arctic tundra involving EcM symbiosis. Models of nutrient cycling developed in these ecosystems may not hold in temperate and tropical forests in which edaphic processes differ markedly. Periodical rainy and dry seasons certainly intensify erosion and leaching in tropical forests. Termites rapidly translocate and reduce reserves of woody litter in tropical ecosystems. Nearly annual fires in seasonally dry ecosystems also reduce the amount of soil carbon.

Compared with EcM fungi, acquisition of phosphorus constitutes the major nutritional role of AM symbiosis. Grassland plants transport 2–20 per cent of fixed carbon to AM fungi that account for 20–30 per cent of soil microbial biomass (Leake et al., 2004), but information about AM trees and non-EcM forest plants in general is lacking.

Mycorrhizal fungi and plants are integrated by multiple links to the meta-networks of ecosystems (Figure 22.3; Simard et al., 2013). Besides providing exudates to mycorrhizosphere microbial consortia, both plants and fungi provide a rich source for direct consumers. Plants are grazed by a variety of herbivorous animals above and below ground and are degraded by pathogenic and saprotrophic fungi. Hyphae of mycorrhizal fungi are attacked by enzymes of mycoparasitic bacteria and other fungi. Fungal mycelium is actively grazed by soil mesofauna such as nematodes, mites and springtails. In addition to these organisms, fungal fruit bodies are consumed by larvae of gnats and flies, beetles, snails, rodents and large ungulates such as elk and wild boar. Truffles and truffle-like fungi with sequestrate fruit bodies attract mycophagous mammals by a specific odour of ripening sporocarps. Both the arthropods and mammals provide an active means of dispersal for the fungi as many spores remain intact or their germination is triggered after passing through the digestive tract (Claridge and Trappe, 2005). Moreover,

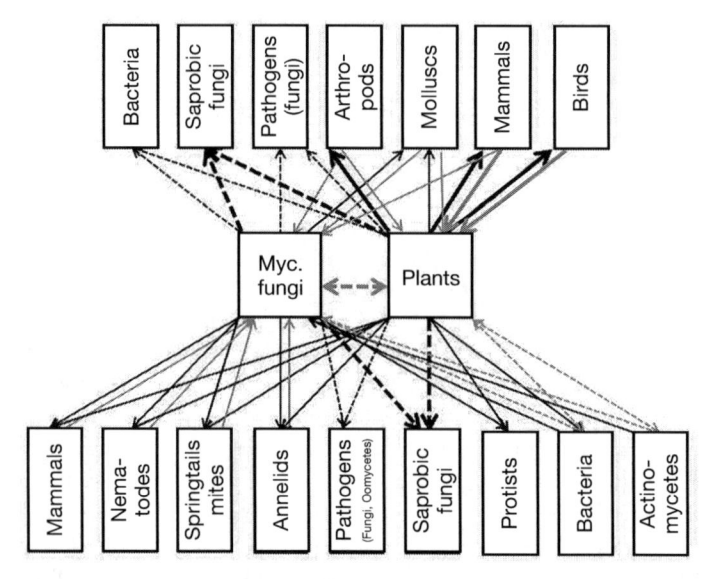

Figure 22.3 Simplified structure of meta-networks in forest ecosystems emphasizing the relationships of mycorrhizal fungi and plants with other major organism groups above ground (top) and below ground (bottom). Interactions among other organisms are excluded for clarity. Arrowheads indicate benefit from the relationship. Thick lines indicate ecologically most important relationships. Black solid lines, food web interactions; black dashed lines, decomposition; grey solid lines, dispersal; grey dashed lines with double arrows, mutualistic nutrient exchange

ornamented spores are easily adhered to the fur, appendices or slime of soil animals. Spores of Glomeromycota are similarly ingested by soil animals and can be transported over long distances. Mammals act as dispersal agents for plant seeds that remain undigested or that are hidden as a reserve. If the seeds germinate away from their natural substrate, the seedlings may readily benefit from the mycorrhizal inoculum present in scats. Animal-vectored dispersal is important, because wind dispersal can be relatively inefficient; the vast majority of spores are deposited within a few metres of the fruit-body (Li, 2005).

Forest disturbance

The effects of natural disturbance and forest management on mycorrhizal associations strongly depend on the type and severity of disturbance, mycorrhiza types, particular ecosystem and the interaction of these factors. Forests are naturally affected by various disturbance events such as wildfire, herbivore attack and pathogen outbreak. The magnitude of these effects depends on ecosystem resilience, i.e. adaptation of organisms to these disturbance events by self-protection and regeneration ability. Disturbance usually results in death or weakening of plants that limits photosynthesis and thereby shakes the nutritional balance with mycorrhizal fungi below ground. Typically, the loss of above ground biomass results in restructuring carbon flow to secure rapid regeneration above ground, which causes reduction of mycorrhizal biomass and activity (Štursová *et al.*, 2014).

Intensive fires are one of the most destructive types of natural disturbance that destroy most or all above ground biomass and the forest floor. The heat itself probably kills up to 90 per cent of the mycorrhizal biomass and the rest is likely to be lost within 1–2 years, depending on the re-sprouting ability of vegetation, a widespread adaptation in fire-prone ecosystems. Fires induce seedling establishment from the dormant seed bank that rapidly develops mycorrhizal associations with mycelium that has persisted in deeper soil or simultaneously germinated from spores. Similar to plants, many fungi are adapted to periodic fires by evolving thickened cell walls and germination triggered by heat or rise in soil pH (Peay *et al.*, 2009). Wildfires may result in nearly complete turnover of the resident fungal communities, depending on ecosystem adaptation, intensity and interval of fires. Low-intensity periodical fires often have a negligible effect on soil fungal communities in fire-adapted ecosystems (Cairney and Bastias, 2007).

Compared with fires, attacks of pathogens and herbivores have little effect on the soil's physical and chemical properties, except an altered quality and quantity of litter input. The reduction in photosynthesis results in a decrease in plant and fungal reproductive effort and loss in soil fungal biomass, disfavouring mycorrhizal species that produce large amounts of mycelium and are thus costly to maintain (Kuikka *et al.*, 2003). Harmful soil pathogens such as *Heterobasidion parviporum* and *Phytophthora cinnamomi* or bark beetles may cause the death of entire patches of trees. Because the root systems of adult trees extend >20 metres, soils under such patches probably retain some of the mycorrhizal biomass, but its relative proportion compared with saprotrophs and opportunistic pathogens is reduced (Štursová *et al.*, 2014).

Many exposed sites in forest ecosystems are affected by wind damage, which may be particularly severe in tropical coastal regions and next to clear-cut areas. The landscape of fallen trees, root mounds and resulting pits nearby provides suitable microsites for seedling establishment due to lower competition for light and soil resources. In boreal and temperate forests, fallen trunks and stumps degrade for decades, while termites rapidly destroy much of the downed wood in tropical ecosystems. Therefore, strongly decayed wood acts as an important seed bed in non-tropical forests long after the disturbance events. Besides the direct effects of reduced competition, seedlings regenerating on dead wood benefit from the lower abundance of soil-borne pathogens and greater water holding capacity of the fibrous woody substrate. Ectomycorrhizal fungi explore the relatively sound dead wood with their mycelium, providing recruitment networks for seedlings (Simard *et al.*, 2012). Compared with AM plants, EcM and ErM plants seem to gain relatively more benefits from their association with dead wood due to their limited capacity to take up organic nutrients. EcM plants establishing on elevated logs associate mostly with the dominant fungal species in soil, suggesting that dead wood provides a suitable source of regeneration for mycorrhizal symbiosis (Tedersoo *et al.*, 2009).

Waterlogging of soil may result from annual periodic rain events, snow melt, construction work, etc. Excess water reduces the oxygen content and renders the soil habitat periodically anaerobic. Mycorrhizal fungi seem to be relatively tolerant of waterlogging, because they can withstand periodic inundation. Many wetland plants have mechanisms of providing oxygen below ground to ameliorate anaerobiosis. Persistent waterlogging of mesic forests usually results in high mortality of vegetation that clearly has a detrimental effect on symbionts below ground.

Harvesting of trees has severe impacts on mycorrhizal fungi due to the disruption of a carbon source, exposition of forest floor to solar radiation and disturbance to soil. The extent of the disturbance depends on time and machinery used for harvesting. In clear-cuts, the vast majority of mycorrhizal organisms usually die within one year of harvesting. The remaining seedlings are unable to sustain the energy-demanding mycorrhizal networks. Thus, they retain depauperate mycota comprised of fungi that require low amounts of carbon (Durall *et al.*, 2005). After harvesting, soil water circulation is substantially altered due to soil exposure to the sun, highly reduced transpiration flow and disrupted deep water uptake. This usually results in standing water in humid sites and excessive soil desiccation in dry sites. An altered water regime may have severe consequences for natural and assisted forest regeneration. Deposition of harvesting residues, especially large-diameter dead wood on the forest floor, is useful both for maintaining soil moisture and continuation of soil development. However, piles of debris may hamper regeneration of shade-intolerant trees and seedling recruitment of small-seeded plants. Tree retention is very important for securing the seed source of selected tree species and for maintaining both the soil water balance and mycorrhizosphere organisms in the soil (Durall *et al.*, 2005). Selective harvesting is of substantial ecological importance in forest sites that are prone to desiccation and erosion. Tropical rain forests are especially intolerant of clear-cutting due to rapid loss of soil to erosion and reduction of humidity in tree canopies that is required for optimal functioning and retention of epiphyte and animal communities. Highly valued EcM tropical timber trees such as Dipterocarpaceae are particularly sensitive to heavy cutting and disturbance due to their patchy occurrence and rapid loss of vitality of EcM mycelium in soil. For AM trees, consequences of harvesting are usually less severe, because the understorey vegetation usually provides abundant mycorrhizal inoculum. The ErM Ericaceae also regenerate rapidly, because they obtain their suitable mycorrhizal fungi from soil and the roots of neighbouring plants. Many Ericaceae species are adapted to pioneer strategy.

Even negligible mechanistic disturbance such as branch removal and thinning affect mycorrhizal fungi. Fruit body production – i.e. the reproductive output in fungi – is dramatically reduced for several years. While the direct cause for this is unclear, both the decrease in above ground biomass and altered balance of soil moisture, nutrients and competition may play a role.

Climate change

Industrialization has led humankind to pollute the environment with litter and various chemicals. Carbon dioxide and methane are the main drivers of global climate warming; large amounts of these gases are annually released into the atmosphere. Altered air temperature affects circulation of air masses and precipitation patterns that cause shifting of biomes and result in extinction of specialist organisms that are unable to migrate at the pace of climate change. Changes in both temperature and precipitation patterns are predicted to result in further aridification of relatively dry southern temperate and subtropical areas that will become more susceptible to fire and lose the tree cover. Due to warming and humidification, boreal forests will greatly expand into northern arctic and tundra ecosystems (Beaumont et al., 2011). Past glacial cycles caused massive extinction of plants in Europe, because the Mediterranean Sea and the Alps represented efficient dispersal barriers. Similarly, African rain forest has a relatively depauperate flora compared with South America and Southeast Asia that is ascribed to a substantial loss of rain forest habitat during the glacial maxima (Linder, 2001). Changes in vegetation may or may not result in changes of dominant mycorrhizal types. For example, aridification has led to the contraction of EcM-dominated oak savannas in North Africa and miombo woodlands in north Central and south Central Africa in the Holocene and these processes are likely to continue along with climate change (Linder, 2001). In the Mediterranean biome, ErM Ericaceae and AM plants characteristic of semi-deserts will probably expand at the expense of EcM-dominated tree communities. The direct effect of prolonged dry periods on mycorrhizal fungi has remained unclear. There is conflicting evidence from moderate drought stress on mycorrhizal fungal communities in the short term (Richard et al., 2011; Hawkes et al., 2011). EcM trees experiencing dieback from extensive drought periods have depauperate mycorrhizal fungal communities (Swaty et al., 2004).

Although we can relatively well predict the future alterations in plant communities based on the expected climate change scenarios, there is relatively little information about potential shifts in soil and root fungal communities. EcM fungi respond more strongly to shifts in plant species than to climate conditions on a regional geographic scale (Bahram et al., 2012). The relative effects of climate and host plant taxa remain unknown in AM systems, but host-generalist AM fungi do exhibit differences in species distribution across biomes (Öpik et al., 2013). Pathogenic fungi and EcM fungi that are often specialized on a phylogenetically narrow host taxon, have a relatively greater risk of extinction compared with other fungal groups when the population size of suitable host shrinks due to human activities and climate change. In response to changing climate, mycorrhizal fungi and plants co-migrate, facilitating each other's range expansion (Wilkinson, 1998; Põlme et al., 2013). During natural succession, the arriving plant species also benefit from pre-established mycorrhizal networks that facilitate recruitment of their seedlings (Nara, 2006).

To physiologically cope with changing climate, mycorrhizal systems optimize their nutrient allocation, whereas plants select for the most tolerant or efficient fungal genotypes. In most ecosystems, global warming is expected to enhance primary production, which renders soil nutrients more limiting for growth. Experiments manipulating global change on a small scale have revealed accelerated carbon and nitrogen cycling below ground partly due to the activities of mycorrhizal fungi and greater allocation to mycorrhizal biomass but smaller effects on mycorrhizal community composition (Johnson et al., 2013). Changes in the dominant mycorrhizal types will strongly affect nutrient cycling (see above).

Mycorrhiza applications in forestry

Application of mycorrhizal fungi has received great attention both in agriculture and forestry. Inoculation with mycorrhizal fungi promotes growth and survival of seedlings, especially when planted in harsh site conditions or in non-native soil. In practice, the greatest growth benefits of inoculation occur with EcM trees, because their natural inoculum in soil is patchy and it degrades rapidly after severe disturbance such as clear-felling. Several economically important EcM tree families – Pinaceae, Amhersteae of Fabaceae and Dipterocarpaceae – exhibit obligate associations with their fungi and therefore plantations of these trees may completely fail without pre-inoculation. Inoculation with EcM fungi is usually performed by adding spore slurry or pieces of axenically grown mycelium to sown seeds or young containerized seedlings. Compressed beads comprised of dried spores are commercially available for inoculation programmes. However, only a few fungal taxa are commonly used for re-forestation. The selected strains may not be well suited to a particular host or environment. For example, the same fungal inoculum has been frequently used to produce EcM seedlings of pines, eucalypts and dipterocarps in southern Asia. These tree species do not co-occur naturally and neither do their fungi. On non-native hosts, mycorrhiza development is usually poor and the association may be only partly functional, resulting in no or limited benefits (Brundrett et al., 2005). Therefore, it is recommended to use multiple strains for inoculation and let both tree genotypes and environment select the best suitable fungi. Long-term monitoring programmes have revealed that the originally inoculated fungi may become replaced by indigenous species within a few years after plantation. Due to a high risk of contamination with moulds and bacteria, production of mycelium-based mycorrhizal inoculum is costly and time consuming. Most mycorrhizal fungi fail to grow in pure culture and others grow very slowly. Nurseries that consider fungal inoculation have to compromise between fertilization and above ground biomass on one hand and mycorrhization and normal root development on the other hand. Reduced fertilization results in a longer rotation period but healthier seedlings that are better prepared for re-planting and thus have greater potential for survival. Inoculation with forest soil results in good mycorrhiza development, but also co-introduces soil-borne pathogens (Brundrett et al., 1996).

Throughout the twentieth century, vast areas of tropical and subtropical land have been replanted with exotic EcM forestry trees, mostly fast-growing pine, eucalypt or wattle species for pulp and construction wood. As trees were usually imported as containerized seedlings, they retained their mycorrhizal fungi from the potting soil (Schwartz et al., 2006). In some regions of the world, these co-introduced mycorrhizal fungi are surprisingly diverse because of multiple introductions from various sources. In several regions such as New Zealand, South Africa and central South America, the introduced trees have become strongly invasive and pose a threat to the indigenous biota and ecosystem services (Pringle et al., 2009). Co-introduced mycorrhizal fungi probably play a key role in the invasiveness of EcM trees. Invasive EcM trees and their symbiotic fungi alter soil nutrient cycling by reduced rate of litter decomposition that also results in greater flammability. Deep-rooted eucalypts, oaks and their mycorrhizal fungi may cause desiccation of topsoil due to more efficient water uptake and greater transpiration rates compared with native vegetation. The altered habitats and dramatically increased fire frequency reduce the establishment of native vegetation and fauna that are not adapted to intensive fires. Species-poor plantations and invaded land also provide limited shelter and food for consumers (Richardson et al., 2000). The invasion of AM vegetation usually has lower impact on ecosystem services, because these plants link to the native AM fungal networks. Introduced AM and EcM plants have often escaped their specific pathogens and pests, which further stimulates their invasion. By using allelochemicals, certain NM herbs from the Brassicaceae family inhibit the development and function of local mycorrhizal networks that further reduce the competitive ability of native vegetation. Besides invasion of exotic plants, introduced fungal species may also broaden their host range and become established in local forest ecosystems. So far, this phenomenon has been documented in fly agarics (*Amanita* species) that form conspicuous fruit bodies and are easily determined (Pringle et al., 2009). *Amanita* species may, however, represent the tip of the iceberg, because most EcM fungal species and all AM fungal species produce no or inconspicuous fruit bodies and can thus only be tracked by use of molecular identification methods. Spread of farming practices and trade with agricultural and forestry products over several millennia has probably resulted in unintended introduction of non-native AM fungi to virtually all inhabitable places on Earth (Schwartz *et al.*, 2006). Tracing these human-assisted introductions and recognizing historical dispersal by natural means poses a nearly impossible task.

While most mycorrhiza inoculation applications intend to improve plant productivity, mushroom production of mycorrhizal fungi is becoming increasingly important. Edible mushrooms, in particular truffles, represent a high-revenue crop that pays off long-term investments into forestry plantations. Natural crops of highly prized edible forest mushrooms are declining because of increasing loads of pollution, non-optimal management, changes in land use and climate that all degrade natural productive habitats. Truffles are inoculated by use of spores and the seedlings are transplanted into sites that preferably lack pre-existing EcM vegetation. Native EcM fungi may have superior competitive abilities and therefore, tree plantations with European trees and truffles are commonly established in the Southern Hemisphere. For many highly prized edible forest mushrooms such as matsutake (*Tricholoma matsutake*), porcini (*Boletus edulis*) and the golden chanterelle (*Cantharellus cibarius*), cost-effective inoculation and seedling propagation methods are yet to be developed. Therefore, proper management of productive stands is essential to secure the harvest. In rural areas, sustainable harvesting and trade of wild mushrooms offer revenues to people in all parts of the world (Härkönen *et al.*, 2003). For example, certain tribes in Africa have been traditionally specialized in collecting and trading wild mushrooms for meat and agricultural crops. Unfortunately, forest management is typically carried out by government or large international logging companies that are not aware of, or do not care about, the alternative forest revenues such as wild mushrooms, fruits and medicinal plants. Habitat conservation and careful management planning, along with establishment of plantations, are promising for securing sustainable use of these forest products in the long term. So far, there is no evidence that intensive collection of fruit bodies reduces the reproductive output in fungi, because a few fruit bodies not found by mushroom hunters always remain for sporulation. Transportation of collected mature fruit bodies within the forest may facilitate distribution of spores. Mycelium of these prized mushrooms may live in the soil for decades.

Mycorrhizal associations have received negligible attention in conservation planning of forests. Both rare and common plants are associated with the same mycorrhizal networks, except certain orchids. Germination and early growth of all orchids depends on energy provided by specific fungi. Mycoheterotrophic orchids obtain carbon and mineral nutrients from fungi throughout all stages of their life. Mycoheterotrophic orchids are usually more rare, at least partly because they associate with EcM fungi rather than saprotrophic fungi and their mycorrhizal associations tend to be more specific compared with green orchids (Dearnaley *et al.*, 2012). EcM fungi exhibit habitat and host preferences and therefore have patchy distribution. Thus, the sites for successful establishment and growth for orchids is limited as well. Because we know too little about the autecology of orchid-associated fungal species, habitat conservation offers the best possible means for protection of fungi and plants with specific mycorrhizal associations.

Conclusions

Mycorrhizal associations play an integral role in the mineral nutrition of plants and nutrient cycling in forest ecosystems. Disruption of mycorrhizal connections reduces primary production, soil microbial biomass and biodiversity. Fungi involved in mycorrhiza types differ in their capacity to degrade soil organic compounds and transfer nutrients. Climate change, particularly in temperature and precipitation, causes shifts in vegetation and prevailing mycorrhiza types, which may dramatically alter ecosystem nutrient cycling. These changes are also important to consider in management of exotic forest plantations and understanding the invasion biology of plants. In addition to improved timber and fruit production, mycorrhizal systems offer alternative benefits for cultivation and commercial harvesting of highly prized edible mushrooms.

References

Allen, E.B., Allen, M.F., Helm, D.J., Trappe, J.M., Molina, R.M. and Rincon, M. (1995) 'Patterns and regulation of mycorrhizal plant and fungal diversity', *Plant and Soil*, vol 170, pp. 47–62.

Babikova, Z., Gilbert, L., Bruce, T.J.A., Birkett, M.A., Caulfield, J.C. and others (2013) 'Underground signals carried through common mycelial networks warn neighbouring plants of aphid attack', *Ecology Letters*, vol 16, pp. 835–843.

Bahram, M., Põlme, S., Kõljalg, U., Zarre, S. and Tedersoo, L. (2012) 'Regional and local patterns of ectomycorrhizal fungal diversity and community structure along an altitudinal gradient in the Hyrcanian forests of northern Iran', *New Phytologist*, vol 193, pp. 465–473.

Beaumont, L.J., Pitman, A., Perkins, S., Zimmermann, N.E., Yoccoz, N.G. and Thuiller, W. (2011) 'Impacts of climate change on the world's most exceptional ecoregions', *Proceedings of the National Academy of Sciences USA*, vol 108, pp. 2306–2311.

Brundrett, M.C. (2004) 'Diversity and classification of mycorrhizal associations', *Biological Reviews*, vol 79, pp. 473–495.

Brundrett, M.C. (2009) 'Mycorrhizal associations and other means of nutrition of vascular plants: Understanding global diversity of host plants by resolving conflicting information and developing reliable means of diagnosis', *Plant and Soil*, vol 320, pp. 37–77.

Brundrett, M.C., Bougher, N., Dell, B., Grove, T. and Malajczuk, N. (1996) 'Working with mycorrhizas in forestry and agriculture', *ACIAR Monographs*, vol 21, pp. 1–374.

Brundrett, M.C., Malajczuk, N., Mingqin, G., Daping, X., Snelling, S. and Dell, B. (2005) 'Nursery inoculation of *Eucalyptus* seedlings in western Australia and Southern China using spores and mycelial inoculum of diverse ectomycorrhizal fungi from different climatic regions', *Forest Ecology and Management*, vol 209, pp. 193–205.

Bueé, M., De Boer, W., Martin, F., van Overbeek, L. and Jurkevich, E. (2009) 'The rhizosphere zoo: An overview of plant-associated communities of microorganisms, including phages, bacteria, archaea, and fungi, and some of their structuring factors', *Plant and Soil*, vol 321, pp. 189–212.

Cairney, J.W.G. and Bastias, B. (2007) 'Influences of fire on soil fungal communities', *Canadian Journal of Forest Research*, vol 37, pp. 207–215.

Cairney, J.W.G. and Meharg, A.A. (2003) 'Ericoid mycorrhiza: A partnership that exploits harsh edaphic conditions', *European Journal of Soil Science*, vol 54, pp. 735–740.

Claridge, A.W., and Trappe, J.M. (2005) 'Sporocarp mycophagy: Nutritional, behavioral, evolutionary, and physiological aspects', in J. Dighton, J.F. White and P. Oudemans (eds) *The Fungal Community. Its Organization and Role in the Ecosystem*, CRC Press, Boca Rayton, USA.

Clemmensen, K.E., Bahr, A., Ovaskainen, O., Dahlberg, A., Ekblad, A. and others (2013) 'Roots and associated fungi drive long-term carbon sequestration in boreal forest', *Science*, vol 339, pp. 1615–1618.

Courty, P.-E., Bueé, M., Diedhiou, A.G., Frey-Klett, P., Le Tacon, F. and others (2010) 'The role of ectomycorrhizal communities in forest ecosystem processes: New perspectives and emerging concepts', *Soil Biology and Biochemistry*, vol 42, pp. 679–698.

Dearnaley, J.D.W., Martos, F. and Selosse, M.-A. (2012) 'Orchid mycorrhizas: Molecular ecology, physiology, evolution and conservation aspects', in B. Hock (ed.) *Fungal associations, The Mycota IX*, Springer-Verlag, Berlin, Germany.

Durall, D.M., Jones, M.D. and Lewis, K.J. (2005) 'Effects of forest management on fungal communities', in J. Dighton, J.F. White and P. Oudemans (eds) *The Fungal Community. Its Organization and Role in the Ecosystem*, CRC Press, Boca Rayton, USA.

Gadd, G.M. (2007) 'Geomycology: Biogeochemical transformations of rocks, minerals, metals and radionuclides by fungi, bioweathering and bioremediation', *Mycological Research*, vol 111, pp. 3–49.

Härkönen, M., Niemelä, T. and Mwasumbi, L. (2003) 'Tanzanian mushrooms: Edible, harmful and other fungi', *Norrlinia*, vol 10, pp. 1–200.

Hawkes, C.V., Kivlin, S.N., Rocca, J.D., Huguet, V., Thomsen, M.A., and Blake Suttle, K. (2011) 'Fungal community responses to precipitation', *Global Change Biology*, vol 17, pp. 1637–1645.

Hawksworth, D.L. (2001) 'The magnitude of fungal diversity: The 1.5 million species estimate revisited', *Mycological Research*, vol 105, pp. 1422–1432.

Heinemeyer, A., Hartley, I.P., Evans, S.P., Carreira de la Fuente, J.A. and Ineson, P. (2007) 'Forest soil CO_2 flux: Uncovering the contribution and environmental responses of ectomycorrhizas', *Global Change Biology*, vol 13, pp. 1786–1797.

Hobbie, E.A. (2006) 'Carbon allocation to ectomycorrhizal fungi correlates with belowground allocation in culture studies', *Ecology*, vol 87, pp. 563–569.

Hobbie, E.A., van Diepen, L.T., Lilleskov, E.A., Ouimette, A.P., Finzi, A.C. and Hofmockel, K.S. (2014) 'Fungal functioning in a pine forest: Evidence from a ^{15}N-labeled global change experiment', *New Phytologist*, vol 201, pp. 1431–1439.

Högberg, M.N., Briones, M.J.I., Keel, S.G., Metcalfe, D.B., Campbell, C. and others (2010) 'Quantification of effects of season and nitrogen supply on tree below-ground carbon transfer to ectomycorrhizal fungi and other soil organisms in a boreal pine forest', *New Phytologist*, vol 187, pp. 485–493.

Johnson, N.C., Angelard, C., Sanders, I.R. and Kiers, E.T. (2013) 'Predicting community and ecosystem outcomes of mycorrhizal responses to global change', *Ecology Letters*, vol 16, pp. 140–153.

Joner, E.J. and Leyval, C. (2003) 'Phytoremediation of organic pollutants using mycorrhizal plants: A new aspect of rhizosphere interactions', *Agronomie*, vol 23, pp. 495–502.

Jones, M.D. and Smith, S.E. (2004) 'Exploring functional definitions of mycorrhizas: Are mycorrhizas always mutualisms?', *Canadian Journal of Botany*, vol 82, pp. 1089–1109.

Kiers, R.T., Duhamel, M., Beesetty, Y., Mensah, J.A., Franken, O. and others (2011) 'Reciprocal rewards stabilize cooperation in the mycorrhizal symbiosis', *Science*, vol 333, pp. 880–882.

Kõljalg, U., Nilsson, R.H., Abarenkov, K., Tedersoo, L., Taylor, A.F.S. and Bahram, M. (2013) 'Towards a unified paradigm for sequence-based identification of Fungi', *Molecular Ecology*, vol 22, pp. 5271–5277.

Kuikka, K., Härmä, E., Markkola, A., Rautio, P., Roitto, M. and others (2003) 'Severe defoliation of Scots pine reduces reproductive investment by ectomycorrhizal symbionts', *Ecology*, vol 84, pp. 2051–2061.

Lambers, H., Raven, J.A., Shaver, G.A. and Smith, S.E. (2008) 'Plant nutrient-acquisition strategies change with soil age', *Trends in Ecology and Evolution*, vol 23, pp. 95–103.

Leake, J., Johnson, D., Donnelly, D., Muckle, G., Boddy, L. and Read, D.J. (2004) 'Networks of power and influence: The role of mycorrhizal mycelium in controlling plant communities and agroecosystem functioning', *Canadian Journal of Botany*, vol 82, pp. 1016–1045.

Leake, J.R. (2004) 'Myco-heterotroph/epiparasitic plant interactions with ectomycorrhizal and arbuscular mycorrhizal fungi', *Current Opinion in Plant Biology*, vol 7, pp. 422–428.

Li, D.-W. (2005) 'Release and dispersal of basidiospores from *Amanita muscaria* var. *alba* and their infiltration into a residence', *Mycological Research*, vol 109, pp. 1235–1242.

Linder, H.P. (2001) 'Plant diversity and endemism in sub-Saharan tropical Africa', *Journal of Biogeography*, vol 28, pp. 169–182.

Liu, X., Liang, M., Etienne, R.S., Wang, Y., Staehelin, C. and Yu, S. (2012) 'Experimental evidence for a phylogenetic Janzen-Connell effect in a subtropical forest', *Ecology Letters*, vol 15, pp. 111–118.

McGuire, K.L. (2007) 'Common ectomycorrhizal networks may maintain monodominance in a tropical rain forest', *Ecology*, vol 88, pp. 567–574.

Nara, K. (2006) 'Ectomycorrhizal networks and seedling establishment during early primary succession', *New Phytologist*, vol 169, pp. 169–178.

Newbery, D.M., Alexander, I.J. and Rother, J.A. (1997) 'Phosphorus dynamics in a lowland African rain forest: The influence of ectomycorrhizal trees', *Ecological Monographs*, vol 67, pp. 367–409.

Okuda, A., Yamato, M. and Iwase, K. (2011) 'The mycorrhiza of *Schizocodon soldanelloides* var. *magnus* (Diapensiaceae) is regarded as ericoid mycorrhiza from its structure and fungal identities', *Mycoscience*, vol 52, pp. 425–430.

Öpik, M., Zobel, M., Cantero, J.J., Davison, M., Facelli, J.M. and Hiiesalu, I. (2013) 'Global sampling of plant roots expands the described molecular diversity of arbuscular mycorrhizal fungi', *Mycorrhiza*, vol 23, pp. 411–430.

Peay, K., Garbelotto, M. and Bruns, T.D. (2009) 'Spore heat resistance plays an important role in disturbance-mediated assemblage shift of ectomycorrhizal fungi colonizing *Pinus muricata* seedlings', *Journal of Ecology*, vol 97, pp. 537–547.

Põlme, S., Bahram, M., Yamanaka, T., Nara, K., Dai, Y.C. and Grebenc, T. (2013) 'Biogeography of ectomycorrhizal fungi associated with alders *(Alnus* spp.) in relation to biotic and abiotic variables at the global scale', *New Phytologist*, vol 198, pp. 1239–1249.

Pringle, A., Bever, J.D., Gardes, M., Parrent, J.L., Rillig, M.C. and Klironomos, J.N. (2009) 'Mycorrhizal symbioses and plant invasions', *Annual Reviews in Ecology, Evolution and Systematics*, vol 40, pp. 699–715.

Querejeta, J.I., Egerton-Warburton, L.M. and Allen, M.F. (2003) 'Direct nocturnal water transfer from oaks to their mycorrhizal symbionts during severe soil drying', *Oecologia*, vol 134, pp. 55–64.

Read, D., Leake, J.R. and Perez-Moreno, J. (2004) 'Mycorrhizal fungi as drivers of ecosystem processes in heathland and boreal forest biomes', *Canadian Journal of Botany*, vol 82, pp. 1243–1263.

Richard, F., Roy, M., Shahin, O., Sthultz, C., Duchemin, M. and others (2011) 'Ectomycorrhizal communities in a Mediterranean forest ecosystem dominated by *Quercus ilex*: Seasonal dynamics and response to drought in the surface organic horizon', *Annals of Forest Science*, vol 68, pp. 57–68.

Richardson, D.M., Allsopp, N., D'Antonio, C., Milton, S.J. and Rejmanek, M. (2000) 'Plant invasions – The role of mutualisms', *Biological Reviews*, vol 75, pp. 65–93.

Rinaldi, A.C., Comadini, O. and Kuyper, T.W. (2008) 'Ectomycorrhizal fungal diversity: Separating the wheat from the chaff', *Fungal Diversity*, vol 33, pp. 1–45.

Ryberg, M. and Matheny, P.B. (2012) 'Asynchronous origins of ectomycorrhizal clades of Agaricales', *Proceedings of the Royal Society, series B*, vol 279, pp. 2003–2011.

Schwartz, M.W., Hoeksema, J.D., Gehring, C.A., Johnson, N.C., Klironomos, J.N. and others (2006) 'The promise and the potential consequences of the global transport of mycorrhizal fungal inoculum', *Ecology Letters*, vol 9, pp. 501–515.

Simard, S.W., Beiler, K.J., Bingham, M.A., Deslippe, J.R., Philip, L.J. and Teste, F.P. (2012) 'Mycorrhizal networks: Mechanisms, ecology and modelling', *Fungal Biology Reviews*, vol 26, pp. 39–60.

Simard, S.W., Martin, K., Vyse, A. and Larson, B. (2013) 'Meta-networks of fungi, fauna and flora as agents of complex adaptive systems', in C. Messier, K.J. Puettman and K.D. Coates (eds) *Managing Forests as Complex Adaptive Systems*, Routledge, Oxon, UK.

Smith, S.E. and Read, D.J. (2008) *Mycorrhizal Symbiosis*, Academic Press, London, UK.

Štursová, M., Šnajdr, J., Cajthaml, T., Barta, J., Šantručková, H. and Baldrian, P. (2014) 'When the forest dies: The response of forest soil fungi to a bark beetle-induced tree dieback', *The ISME Journal*, vol 8, pp. 1920–1931.

Swaty, R.L., Deckert, R.J., Whitham, T.G. and Gehring, C.A. (2004) 'Ectomycorrhizal abundance and community structure shifts with drought: Predictions from tree rings', *Ecology*, vol 85, pp. 1072–1084.

Taylor, A.F.S. and Alexander, I.J. (2005) The ectomycorrhizal symbiosis: Life in the real world. *Mycologist*, vol 19, pp. 102–112.

Tedersoo, L. and Smith, M.E. (2013) 'Lineages of ectomycorrhizal fungi revisited: foraging strategies and novel lineages revealed by sequences from belowground', *Fungal Biology Reviews*, vol 27, pp. 83–99.

Tedersoo, L., Bahram, M., Toots, M., Diédhiou, A.G., Henkel, T.W. and others (2012) 'Towards global patterns in the diversity and community structure of ectomycorrhizal fungi', *Molecular Ecology*, vol 21, pp. 4160–4170.

Tedersoo, L., Gates, G., Dunk, C., Lebel, T., May, T.W. and others (2009) 'Establishment of ectomycorrhizal fungal community on isolated *Nothofagus cunninghamii* seedlings regenerating on dead wood in Australian wet temperate forests: Does fruit-body type matter?' *Mycorrhiza*, vol 19, pp. 403–416.

Torti, S.D., Coley, P.D. and Kursar, T. (2001) 'Causes and consequences of monodominance in tropical lowland forests', *American Naturalist*, vol 157, pp. 141–153.

Trocha, L.K., Mucha, J., Eissenstat, D.M., Reich, P.B. and Oleksyn, J. (2010) 'Ectomycorrhizal identity determines respiration and concentrations of nitrogen and non-structural carbohydrates in root tips: A test using *Pinus sylvestris* and *Quercus robur* saplings', *Tree Physiology*, vol 30, pp. 648–654.

van der Heijden, E.W. and Kuyper, T.W. (2003) 'Ecological strategies of ectomycorrhizal fungi of *Salix repens*: Root manipulation versus root replacement', *Oikos*, vol 103, pp. 668–680.

Whipps, J.M. (2004) 'Prospects and limitations of mycorrhizas in biocontrol of root pathogens', *Canadian Journal of Botany*, vol 82, pp. 1198–1227.

Wilkinson, D.M. (1998) 'Mycorrhizal fungi and Quaternary plant migrations', *Global Ecology and Biogeography Letters*, vol 7, pp. 137–140.

23
BIOGEOCHEMICAL CYCLING

David Paré, Daniel Markewitz and Håkan Wallander

Forest nutrient cycles

Of the 34 elements known or believed to be essential to plants, animals and microorganisms, six are known as plant macro nutrients – nitrogen (N), phosphorus (P), potassium (K), calcium (Ca), magnesium (Mg) and sulfur (S). N and P are the ones that are the most studied and that most often limit plant growth (Sterner and Elser 2002). These elements are found in relatively small amounts in the soil in a form that is available to plants and are usually efficiently recycled and conserved within the ecosystem. The study of forest nutrient cycling uses three interlinked cycles as proposed by Switzer and Nelson (1972):

The biochemical cycle (or cycling within organisms)

This cycle includes fluxes of elements that are recovered from dying tissues. Nutrient resorption is a fundamental process through which plants withdraw nutrients from leaves before abscission. A large portion of mobile elements such as N and P are recovered prior to leaf shedding but other elements such as K could also be substantially recycled within the plant (Vergutz *et al.* 2012). Typically, resorption rates are around 50 percent but can be as high as 80 percent (Aerts 1996). Resorption from roots may be important but is difficult to quantify because so little information is available. Nutrients that play a structural role in the plant such as Ca are not recycled biochemically by plants. In infertile soil conditions, a consistent strategy for conserving elements is through long leaf lifespan. This allows for a greater photosynthetic C return per unit of nutrient used (high nutrient use efficiency) (Chapin 1980). For example, black spruce (*Picea mariana* Mill.) can maintain its needles for up to 13 years.

The biogeochemical cycle (or cycling within the ecosystem)

The biogeochemical cycle involves fluxes of nutrients from the trees to the soil in solid form (leaves, dead roots, wood, other plants such as mosses) or in soluble form (stemflow, leaching). It also includes the transformation of nutrients that accumulate in soil organic matter and litter into soluble forms through the processes of decomposition and mineralization. Finally, the uptake of nutrients from the soil completes this cycle. Quantifying or describing these fluxes necessitates a definition of system boundaries. It can be as large as a biome, but usually a small area is used, such as a forest stand. The plant rooting depth is often used as the lower boundary for biogeochemical nutrient cycling. There is also a chemical boundary as nutrients that are present in recalcitrant organic matter or in rock or clay minerals that have limited solubility can be considered as being outside of the ecosystem boundaries while still being in the rooting zone. The solubilization of these elements can be considered as geochemical input.

Biogeochemical fluxes can be affected by numerous factors at various temporal and spatial scales. When comparing across biomes, climate strongly controls nutrient fluxes and these fluxes closely correlate with ecosystem productivity. Cold temperatures, and excessive humidity or dryness can reduce plant productivity and nutrient uptake, the flow of nutrients in organic forms to the soil, and decomposition or mineralization of organic nutrients in the soil. Suboptimal temperatures and water content limit organic matter decomposition and release of nutrients. The decomposition process follows an exponential relationship with temperature that is often represented by a Q_{10} relationship where the rate of the process increases every 10°C rise by a given factor. A Q_{10} of 2, indicating a doubling of the rate for every 10°C rise, is widely used (Davidson and Janssens 2006). The relationship between soil water availability and decomposition (and concomitant nutrient mineralization) follows a bell-shaped curve with a sharp rise and decline at both ends and a large plateau in-between. Apart from microclimatic conditions, major factors affecting biogeochemical cycles are soil type and forest community composition (see this chapter, third section: Plant–soil interactions), all these factors being to a certain degree interrelated.

Several soil properties affect nutrient conservation and facilitate efficient nutrient cycling. The accumulation of soil organic matter enhances nutrient retention in two ways: first it contains nutrients in proportions that are relatively similar to that of plants thus promoting efficient uptake upon decay, and second, organic matter possesses both negatively and positively charged functional groups that provide cation and anion exchange sites where nutrients that are in the form of cations (K^+, Ca^{++}, Mg^{++}, NH_4^+) or anions (NO_3^-, $H_2PO_4^-$) can be maintained in a form easily accessible to plant roots. Soil properties such as drainage and the abundance of small particles (clay and silt) have great influence on the accumulation of organic matter. Other soil properties such as the presence and type of clay also strongly influence the soil exchange capacity. Very acidic and very basic soils have the potential to sequester P anions into recalcitrant forms that are poorly available to plants or microbes. Nutrients that are not retained may be lost from the soil by leaching. Trees develop a root system that together with mycorrhizal fungi is extremely efficient in preventing these losses. While tree roots can extend to great depths, the bulk of fine roots are located within the top 30 cm of soil, especially in boreal forests (Jackson *et al.* 1997). A very high density of fine roots in the top soil horizon is often observed in tropical, boreal and temperate forests. The term *direct cycling* is used when roots and their associated mycorrhizal fungi are observed growing directly on decomposing organic matter and on freshly fallen leaves. This direct transfer ensures a very efficient nutrient cycling by preventing leaching out of the system or P fixation on mineral surfaces.

The geochemical cycle (or ecosystem gains and losses)

This cycle involves both inputs of nutrients to ecosystems of atmospheric, geological and biological origin as well as outputs or losses from ecosystems through solid, gaseous or liquid form. The same issues of ecosystem boundaries apply as for other cycles. Gains and losses of elements can vary rapidly over short distances as they are affected by soil types, the biota (tree composition for example) or forest stand development stage. While plots of a few square metres are often considered, the boundaries of a watershed (i.e., drainage basin) are often used to constrain the limit of these studies, especially when the focus is on ecosystem scale nutrient balance. A watershed is beneficial in defining rainfall inputs and streamwater outputs but may include a mosaic of conditions within the basin with various rates of geochemical cycles.

The main ecosystem input fluxes (often defined in $kg.ha^{-1}.yr^{-1}$) are from mineral weathering, particularly for base cations and P, and from atmospheric inputs in rainfall, fine particles and aerosols, and N fixation. The highest N fixing rates are observed for the few species of plants that carry symbiotic N fixing bacteria with rates that can reach over 100 $kg.ha^{-1}$ of N annually. Non-obligatory symbiotic organisms such as the blue-green algae *Nostoc*, are present in the soil, on decaying wood and in certain mosses and have rates that are in general lower than 3 $kg.ha^{-1}.yr^{-1}$. As N fixation is a highly energy demanding process, there is some evidence that N fixing activity is regulated by the amount of N available in the soil in a negative feedback loop. When levels of available N are high, N fixation rates are often low. This is observed in N fixing algae in moss communities (Gundale *et al.* 2011) as well as in natural successional pattern from early N fixing plants that performed well in N poor soil to non-nitrogen fixing communities. Ecosystems lose small amounts of nutrients continuously through processes of leaching and denitrification but episodic disturbances can cause high losses through biomass removal, fire (mostly through oxidation when material is combusted) and erosion.

Overview of the three cycles

The nutrients circulating in a forest ecosystem originate from geochemical inputs. However, on a yearly basis, biochemical and biogeochemical cycles strongly dominate the fluxes of nutrients (Table 23.1). The bulk of the amounts of nutrients required by plants to meet growth comes from these two fluxes. The pool of nutrients that is cycling quickly in trees and within the tree–soil system makes it possible to produce high amounts of biomass in relatively poor soils. Because the maintenance of a foliar cover creates a major nutrient requirement within forest ecosystems on a yearly basis, having a large portion of the requirement satisfied by foliar resorption alleviates the need to prospect the soil for additional nutrients and thus compete with other plants or microbes. Greater geochemical inputs such as fertilization or soil richness can reduce rates of resorption but this is not always observed (Aerts 1996). Plants adapted to nutrient-poor environments generally have slow growth rates, slow growth response to nutrient addition and greater leaf longevity, which is associated with a greater production of anti-herbivore products (Chapin 1980). These properties are conducive to a positive retroaction between soil fertility and nutrient cycling. Table 23.1 illustrates some nutrient fluxes in a tropical, a temperate and a boreal forest ecosystem. A universal property of forest ecosystems is that the quantity of nutrients lost in leachates, in the absence of severe disturbance, represents a small fraction of the amounts cycling within these systems on an annual basis, especially for N and P. Climate, topography, soil drainage, and parent material define potential limits to nutrient cycling rates. Nevertheless, these rates are very dynamic in space and time and are dependent on time since disturbance (see the following section) and on the biological composition of the ecosystem (see this chapter, third section: Plant–soil interactions). Forest management activities influence nutrient cycling at the scale of all three interlinked cycles. They have direct impacts by enhancing nutrient inputs or outputs and indirect impacts by influencing the biological composition of the ecosystem or the soil properties that control the rates of nutrient cycling.

Table 23.1 Nutrient fluxes in a tropical rain forest, a deciduous temperate forest and a boreal forest. Fluxes are in kg.ha^{-1}.yr^{-1}. Leaching losses as a percentage of uptake give an approximation of the intensity of recycling within the ecosystems: 100 minus this estimate gives as a percentage the proportion of the plant uptake that is met by biogeochemical recycling.

	N	P	K	Ca	Mg
Tropical (rain forest, Brazil); from: Markewitz et al. (2004)					
Uptake	149	3.3	65	115	27
Precipitation inputs	4	0.03	5	3	1
Leaching	12	0.1	1.4	9.9	5.2
Leaching loss as % of uptake	**8**	**3**	**2**	**9**	**19**
Biogeochemical recycling efficiency (%)	**91**	**97**	**98**	**91**	**81**
Temperate (Northern hardwood, USA); from: Whittaker et al. (1979)					
Uptake	114	13	47	50.6	8.5
Precipitation inputs	6	0.1	1.1	2.6	0.7
Stream leaching	2.3	0.02	1.7	11.7	2.8
Leaching loss as % of uptake	**2**	**0**	**4**	**23**	**33**
Biogeochemical recycling efficiency (%)	**98**	**100**	**96**	**77**	**67**
Boreal (mixed, Alaska); from: Van Cleve et al. (1983)					
Uptake	42.9	4.7	23.9	30.1	7.3
Precipitation inputs (n.a.: not available)	n.a.	n.a	n.a.	n.a.	n.a.
Stream leaching	0.7	0.2	0.3	3.1	0.6
Leaching loss as % of uptake	**2**	**4**	**1**	**10**	**8**
Biogeochemical recycling efficiency (%)	**98**	**96**	**99**	**90**	**92**

Nutrient cycling: change over time

Stand development

Biogeochemical cycling of nutrients through forests is defined by the balance of inputs and outputs from the ecosystem and drives changes in ecosystem storage (Figure 23.1). Nutrient cycling in forest ecosystems changes during stand development and species succession. These changes in nutrient cycles differ depending on the initial conditions of the site, ideas that developed originally during studies of primary or secondary forest succession. In primary succession, forest development occurs on recently exposed or deposited parent materials (e.g., deglaciation, sand dunes, lava flows). In these scenarios, the parent material is often nutrient rich (but see Olson 1958 for consideration of sand dunes) and can contribute substantially to the nutrients accumulated by plant biomass. In these cases, however, there is limited organic matter or actively cycling nutrients in the soil because the sites have not been previously vegetated. Studies have been conducted to examine primary succession using a chronosequence approach at the edge of retreating glaciers in Alaska (Chapin et al. 1994). In these ecosystems that extended over ~200 years, a successional sequence of species from blue-green algae to Sitka spruce (*Picea sitchensis*) was described and the accumulation of plant biomass was associated with a pattern of decreasing soil pH, increasing contents of soil C and N, and an increase in soil exchangeable cations. In a chronosequence of primary succession on sand dunes along the shores of Lake Michigan, Olson (1958) described a plant succession over ~300 years from

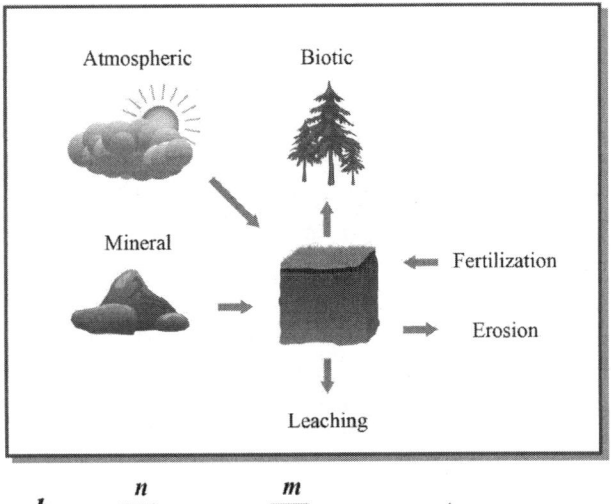

$$\frac{dm_c}{dt} = \sum_{i=1}^{n} F_c \, in - \sum_{o=1}^{m} F_c \, out \pm \triangle F_c \, storage$$

Figure 23.1 Generalized model of biogeochemical cycling of elements in ecosystems demonstrating a balance of inputs, outputs, and change in storage. The equation represents the change in the mass of an element c with time ($\frac{dm_c}{dt}$) as a sum of all the fluxes in from 1 to n of element c ($\sum_{i=1}^{n} F_c\,in$) minus all the fluxes out from 1 to m of element c ($\sum_{o=1}^{m} F_c\,out$) plus or minus any change in storage of element c ($\triangle F_c\,storage$).

bunch grasses (*Andropogen* sp.) to jack pine (i.e., *Pinus banksiana*) and black oak (*Quercus velutina*) but a similar pattern of accumulation of soil C and N with an increase in exchange capacity and exchangeable acidity. The development of nutrient cycles in these ecosystems includes an accumulation of nutrients in living biomass and soil organic matter pools (see Figure 23.2 for relative rates of soil N accumulation) as well as an increase in the intrasystem cycling of nutrients (see biogeochemical cycles defined above). The case of N is particularly interesting as N content in primary soil substrates is often quite limited such that N fixation, as noted above, can be a critical process for N inputs. Thus whether N fixers are part of the plant succession can dramatically alter N accumulation and dynamics.

These nutrient accumulations during stand development on newly exposed substrates contrast with secondary succession where sites have previously been vegetated but have been reset to an earlier stage of plant succession through some form of forest disturbance (e.g., harvest, fire, landslide, or hurricane). An important difference in stand development after disturbance is that there may be substantial organic matter and nutrients already cycling in the ecosystem. In studies of stand development in old agricultural fields, previous fertilizer inputs that have accumulated in soil may also be used by plants and thus transfer soil stocks to plant biomass (Richter *et al.* 2000). Furthermore, rates of plant regeneration may be slower in primary succession that requires the development of soil and the import of plant stock (i.e., seeds) while in secondary succession, rates may be quite rapid due to a large seed bank and prolific root or stump sprouting. There have been a substantial number of studies that have investigated secondary succession after a range of disturbances: Agriculture – (Richter and Markewitz 2001); Fire – (Nave *et al.* 2011); Harvest – (Nave *et al.* 2010). In many of these studies, there is

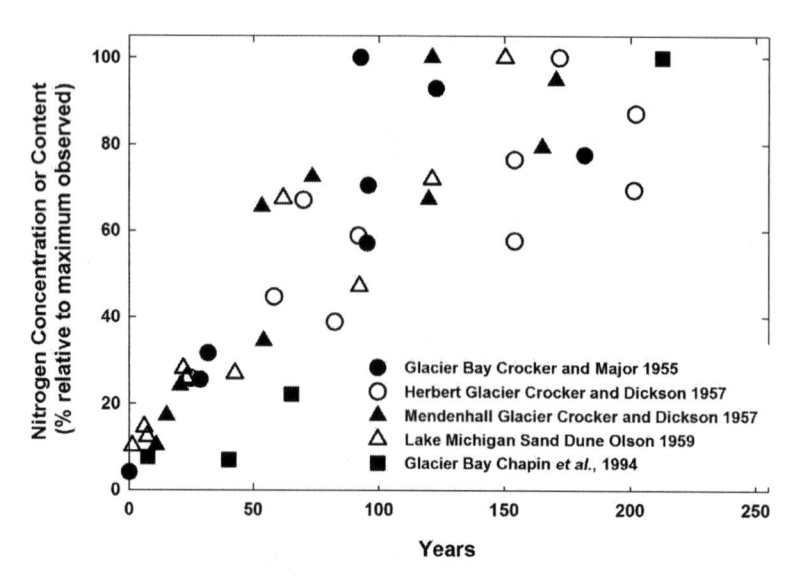

Figure 23.2 Patterns of soil N accumulation in primary successional sequences. Data at each location are presented as relative to the maximum observed at that location

a period shortly after disturbance where nutrient availability exceeds plant demand such that losses of elements from the ecosystem in solution (groundwater or streams) are apparent. These processes were amply demonstrated by Bormann and Likens (1979) in the watersheds of the Hubbard Brook ecosystems of New Hampshire. As plants begin to once again re-vegetate the site and increase plant demand, losses of elements from the ecosystem can decline.

Supply and demand

During ecosystem development, both supply of and demand for nutrients change. This balance was described by Vitousek and Reiners (1975) as a critical framework for understanding when ecosystems should be expected to tightly retain nutrients as compared with periods when the ecosystem appears leaky, in other words, outputs from the ecosystem in drainage waters or gases would be large and nearly equal to inputs (see Figure 23.3 for conceptual model). Very shortly after substrate exposure or disturbance, the demand for nutrients is low due to an absence of plant demand. Nutrients coming into the ecosystem with wet deposition, for example, should go through the ecosystem rapidly. In the case of forest disturbance, we might observe an additional release of nutrients that had accumulated in forest floor or soil organic matter such that losses might exceed that being contributed in rainfall.

In a young and vigorously growing forest, the demand by plants for essential elements will be high and thus retention of these elements in growing biomass can be great (Figure 23.3). During this early phase supply can be high for some elements (for example, rock-derived elements such as P or Ca) or low (i.e. N). The situation for N is dynamic and can be highly affected by plant succession or the previous history of the site (Richter *et al.* 2000). N mineralization of organic matter on site can increase N availability after disturbance and such increases have been demonstrated in many studies after forest harvest (Nave *et al.* 2010). In contrast, newly exposed substrates can have a limited N supply (Chadwick *et al.* 1999) as can sites after disturbance from severe fire. As stands continue to develop and plant growth slows,

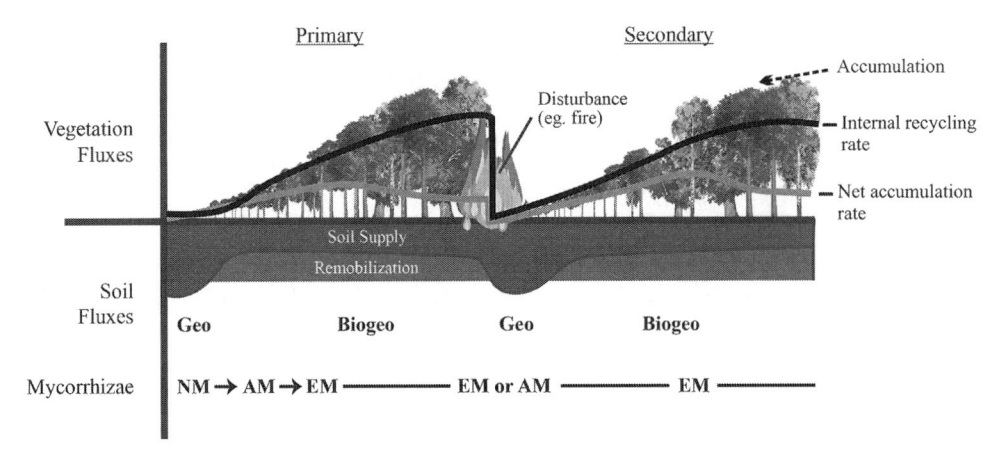

Figure 23.3 Conceptualized patterns of changes with time in rates of nutrient fluxes along primary and secondary successions. The height of stands represents the accumulating stocks of carbon and nutrients in the vegetation. Lines represent the yearly fluxes of nutrients accumulating in the vegetation and the requirement that is met by internal recycling. Soil supplies and the portion that comes from remobilization are shown in the opposite direction. Geochemical N fixation is frequent in early primary succession while disturbance can enhance nutrient availability in secondary succession. The dominance of biogeochemical versus geochemical cycles is indicated as well as that of non-, arbuscular- and ecto-mycorrhizal fungi (NM, AM and EM, respectively) as observed in temperate and boreal climates

so will plant demand and thus there is increased potential for removal of nutrients outside the defined ecosystem boundaries. This pattern of supply versus demand during ecosystem development is not observed for non-essential plant elements since there is no plant demand (e.g., Na), and such elements demonstrate a relatively constant balance of inputs and outputs. For those elements that are plant essential but are not perceived as limiting to growth, an intermediate pattern is expected to develop (i.e., some decline in outputs relative to inputs but some output flux is still measurable). The dichotomy of tight versus leaky relates most to the geochemical cycle of ecosystem inputs and outputs as described above and should not overshadow the efficient and persistent internal biogeochemical cycles that are sustained even in well-developed forest ecosystems. These efficient cycles are particularly relevant for plant-limiting nutrients of N and P where individual atoms of the element may cycle through the ecosystem many times despite a relative geochemical balance of inputs and outputs.

Change in limiting element

During any phase of forest growth, there is presumed to be some element that limits the growth of the forest or species of interest (i.e., Liebig's law of the minimum). Over time there has been an increasing recognition of co-limitations to forest growth that has been demonstrated by a range of irrigation × fertilizer studies (Albaugh *et al.* 1998) and CO_2 × nutrient studies (McCarthy *et al.* 2010). Despite this recognition, much work has adopted the paradigm of the limiting element with a particular focus on N and P. In a millennial-scale chronosequence study in Hawaii, Chadwick *et al.* (1999) proposed a developmental sequence in which ecosystems are most limited by N early in development because P is available from rock weathering at that time. Over time, P is slowly converted from available to unavailable forms as originally hypothesized by Walker and Syers (1976). Changes in P availability are described

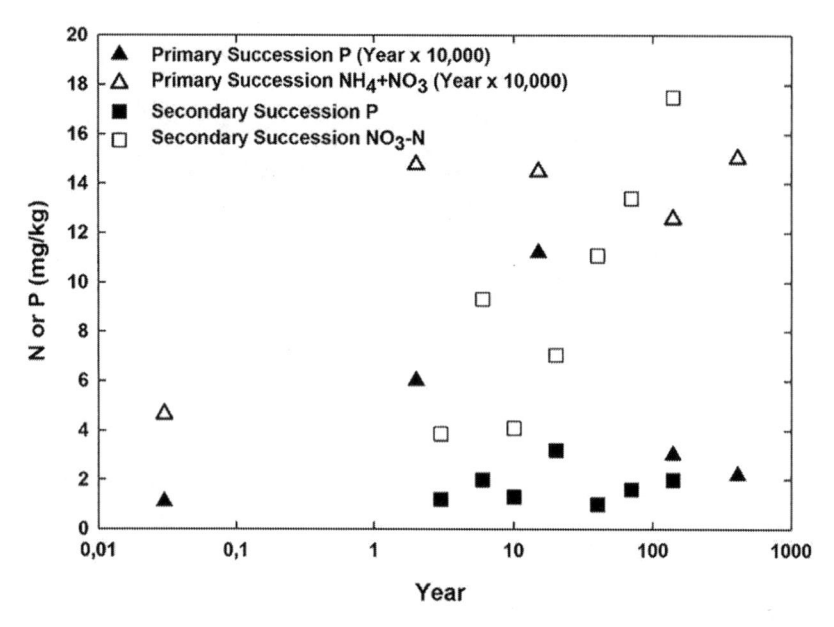

Figure 23.4 Patterns of soil N and P availability throughout a primary chronosequence over a millennial timescale from Hawaii (Chadwick *et al.*, 1999) and a secondary chronosequence over a decadal timescale from the Brazilian Amazon (Davidson *et al.*, 2007)

as progressing from a condition of high availability due to weathering of the primary mineral apatite to a state of low availability as weathered and biogeochemical cycling P is slowly adsorbed on Fe and Al oxides and hydroxides, as predominate in highly weathered soils, such that P is no longer plant available (Peltzer *et al.*, 2010). During this developmental sequence in the Hawaii chronosequence, P became increasingly more limiting to forest growth such that atmospheric dust inputs of P displaced soil derived inputs as the predominant ecosystem supply. During this same millennial-scale development period, N accumulated in the ecosystem and became less of a constraint to growth. Similar patterns in nutrient limitation from N to P have also been proposed during secondary stand development (see Figure 23.4 for data comparison). Davidson *et al.* (2007) measured increasing N capital in a chronosequence of secondary tropical forests while P availability and capital remained constant, suggesting increasing P limitation.

In all these cases of biogeochemical cycling through time, both inputs and outputs can be dynamic and other macro (i.e., K) and micro (i.e., B, Mo) elements can constrain productivity and cycling in the ecosystem. These changes can have effects on vegetation composition with feedbacks on nutrient cycling (see case study in the next section).

Plant–soil interactions

Plant and soil organisms play important roles in regulating biogeochemical cycling. As indicated above, forest stand composition is influenced by succession, disturbance, management activities and the presence of invasive species. Forest tree species composition, together with associated organisms including plants of different strata, soil fauna and microorganisms, influence nutrient capture by the ecosystem as well as nutrient retention and biogeochemical nutrient cycling rates. Of particular importance are plant associations with mycorrhizal fungi. These associations are almost ubiquitous among trees, as well as other plants, and large amounts of carbon are invested in exchange for the nutrients, especially under nutrient-poor conditions. The fine mycelia formed by mycorrhizal fungi have an outstanding capacity to explore the soil substrate in the search for nutrients. They extend beyond the depletion zones around roots and enter micropores and cavities in the soil where roots have no access (Smith and Read 2008). Arbuscular mycorrhiza, the most ancient form of mycorrhizal association, evolved with the first land plants that colonized the planet around 400 million years ago. This was considered a necessary step in order to obtain enough P when moving from the water to the riparian zones where these plants flourished. Arbuscular mycorrhizal fungi colonize around 80 percent of all plants and are the most common association in tropical forest trees where P is usually the growth limiting element (Smith and Read 2008). Ectomycorrhizal (EM) associations evolved much later and are the dominant associations in temperate and boreal forests. In these ecosystems N is usually the growth limiting element because of slow decomposition rates, which results in N retention in the organic material. This effect is expressed more fully further north since lower temperatures retard decomposition. For this reason, forest trees are usually more strongly dependent on their EM symbionts in northern ecosystems (Smith and Read 2008).

Plant–soil interaction and biogeochemical cycling

The chemical and physical characteristics of plant litter, which varies with species and plant components, affect rates of decay and the accumulation of soil organic matter. In boreal forests, ericaceous shrubs and mosses, typically associated with conifers, slow down rates of cycling of N through production of recalcitrant organic molecules and through the cooling of the soil surface (Thiffault *et al.* 2013). The impact of plant litter can be complex since the physical and chemical properties will influence the community of decomposers which in turn will affect decomposition rates. While the decomposition process is largely conducted by soil fungi and bacteria, larger organisms, such as earthworms, can influence the process dramatically both by physically reducing the size of organic matter and by altering its chemical composition. In a controlled experiment, Laganière *et al.* (2010) observed that deciduous litter decayed much faster than conifer litter only when lying on its own soil, which hosted earthworms. This example shows that litter properties were only part of the explanation. Tree species influenced the litter decay rates in great part by harboring a distinct soil fauna.

It is generally considered that EM fungi evolved from saprophytic fungi (Wolfe *et al.* 2012) and many species have retained the oxidative or enzymatic machineries to attack organic matter while mining for nutrients, especially N (Rineau *et al.* 2013). EM fungi produce large amounts of external mycelium when exploring the soil for nutrients, and residues of this mycelium are important for soil formation and carbon sequestration. Recent research revealed that EM associations contributed to more than 50 percent of the soil organic matter in northern boreal forests (Clemmensen *et al.* 2013), probably an effect of the large production and slow decomposition rates of roots and external EM mycelium in nutrient-poor forest soils. When carbon is sequestrated through this mechanism, N will be retained in N-rich fungal residues. Ectomycorrhizal fungi may thus enhance C sequestration and N retention through their residues but at the same time reduce C sequestration and N retention when they mine the organic matter for N. The outcome of these opposing processes (Figure 23.5) is critical for the C and N balance of forest ecosystems. At the global scale, EM ecosystems have been shown to enhance C sequestration by 1.7 times per unit of N relative to AM ecosystems (Averill *et al.* 2014).

The production of recalcitrant N-rich litter by EM fungi has been suggested as a way to reduce competition from more N-demanding species that lack the N mining capacity of EM associations. In support for this hypothesis, Näsholm *et al.* (2013) demonstrated strong N

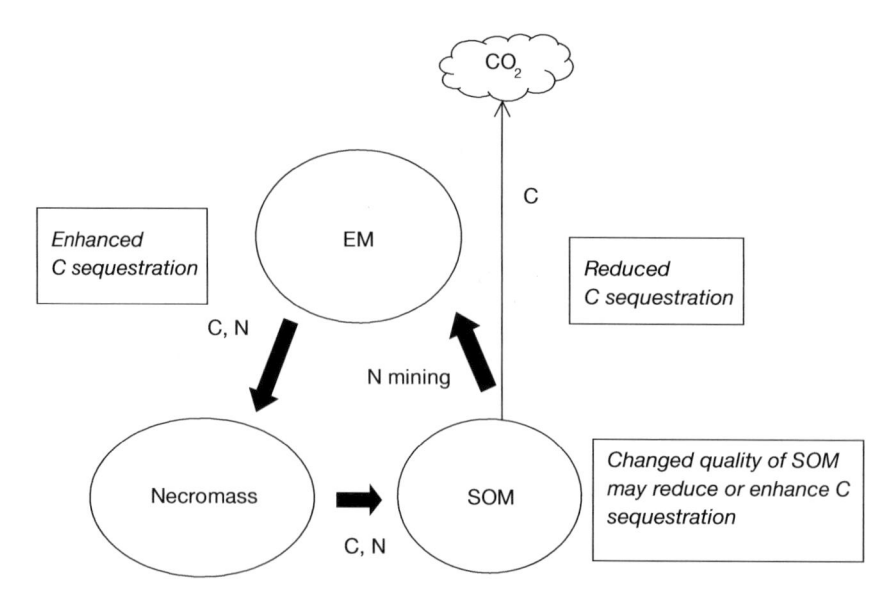

Figure 23.5 Ectomycorrhizal fungi (EM) effects on C and N cycling in forest soil. Carbon is released when EM mine for N in soil organic matter. But C and N are also sequestered when recalcitrant residues of EM accumulate in the soil. The quality of soil organic matter also changes after fungal colonization which will affect its decomposability

retention in EM mycelia in forests in northern Sweden, which could explain exclusion of N-demanding understory, which proliferate on clear cuts or after forest fires when the dominance of EM symbionts has diminished (Figure 23.3).

Biotic influence on geochemical inputs

Symbiotic N fixation activities contribute to building the ecosystem N stocks especially in primary successions (see this chapter, second section: Nutrient cycling: change over time). However, such a process is found only in the root systems of a few tree and plant species. The architectural structure of trees influences the capture of nutrients. Atmospheric depositions will accumulate to a larger extent in conifers, which have foliage the whole year around, compared with deciduous trees, which drop their foliage for winter, especially under conditions of high atmospheric pollution (Berger *et al.* 2009). Overall this will lead to a higher atmospheric input in coniferous compared with deciduous forests and potentially to greater soil acidification and leaching losses. Mycorrhizal fungi also play a role in nutrient capture by accelerating mineral weathering. Many EM species together with associated bacteria produce acids that can release nutrients such as P from primary minerals (Hoffland *et al.* 2004).

Regulation of nutrient cycling by EM fungi

N deposition case study

Growth of EM mycelia depends on below ground allocation of carbon, which is regulated by the nutrient status of the trees (Ericsson 1995). Deficiency of N and P enhances, while deficiency of K and Mg reduces, below ground carbon allocation and EM growth (Ericsson

1995). Fertilization or elevated N deposition usually results in retarded EM growth since the carbon allocated below ground will be used instead to incorporate inorganic N into amino acids to enhance above ground shoot growth (Wallander 1995). The composition of the EM communities is also influenced by N input and this will have consequences for uptake and leaching of nutrients from the soil. For instance, EM species with a high capacity to mine organic matter for N usually decline when availability of inorganic N increases (Taylor et al. 2000). These species are carbon demanding, produce large external mycelium and are classified as long-distance or medium-distance types, according to Agerer (2001). In contrast, species that proliferate in N-rich areas are less carbon demanding, produce low amounts of mycelia, and are classified as short distance or contact types (Agerer 2001). Kjøller et al. (2012) found dominance of contact types, low EM growth and high N leaching at a forest edge exposed to elevated N levels from a nearby poultry farm, while EM growth increased and N leaching declined when moving towards the forest interior where N deposition declined. Similar changes in EM communities were found along a N deposition gradient in Alaska (Lilleskov et al. 2002).

It can be assumed that reduced EM mycelial production at elevated N levels reduces the overall capacity to take up other nutrients such as P, K and Mg from the soil. But shifts in EM communities may also enhance species that are more efficient in uptake of nutrients other than N. For instance, mycorrhizal roots formed by *Paxillus involutus* increased in abundance along the N gradient studied in Alaska by Lilleskov et al. (2002) and this species is known to be efficient in P uptake (Colpaert et al. 1999). Although most species that are enhanced by N deposition are contact types with low mycelial production and low capacity to degrade organic matter, *P. involutus* is an exception. It has a high capacity to degrade organic matter but in contrast to other species this capacity does not decline under conditions of high inorganic N levels (Rineau et al. 2013).

In contrast to N and P, deficiency of K and Mg results in reduced below ground carbon allocation due to a shortage of carbon in the leaves and needles, and impaired carbon loading into the phloem when Mg is in short supply (Ericsson 1995). For this reason, Mg deficiency was suggested to be one of the major causes for the forest decline in central Europe that occurred during the second half of the last century (Schulze et al. 2005). The poor Mg status in these sites was a result of low bedrock Mg levels, low atmospheric inputs and severe soil acidification due to extensive SO_2 emissions.

The differential response of carbon allocation and EM growth to different nutrients may stem from the fact that K and Mg are relatively abundant in most ecosystems and specific mechanisms to increase the availability of these nutrients have thus not evolved. On the other hand, N and P deficiency is common in many ecosystems and plants have evolved a number of traits to cope with these stresses. For these reasons, directions for sustainable forest management should make sure that the removal of nutrients through harvesting and leaching does not result in K and Mg deficiencies unless these nutrients are added as fertilizers.

Succession–retrogression case study

Chronosequence studies often show continued weathering of soil minerals during progression (Peltzer et al. 2010) and a peak in ecosystem biomass will eventually be reached, followed by a retrogressive phase when biomass declines due to deficiency of P. The reason for this is that P-containing minerals are relatively rare and P released during weathering is often transformed into occluded forms that are unavailable for plant uptake. The community of plants will change when the ecosystem undergoes these changes from a progressive to a retrogressive stage. Plants producing dauciform roots, with high capacity to release P through exudation of citric acid, will become more common, and plants forming strong associations with mycorrhizal fungi, especially arbuscular mycorrhizal fungi which are efficient in taking up P at low concentrations, will increase in abundance during the retrogressive stage (Peltzer et al. 2010). Turpault et al. (2009) found a small but significant enhancement of apatite weathering at 20 cm soil depth when exposed to EM fungi, and Wallander and Thelin (2009) found stimulated colonization of apatite by EM fungi in forests with low P status. It can thus be expected that EM-induced weathering of P-containing minerals increases when forests undergo a transition from N to P deficiency, but a similar mechanism is much more uncertain when forests approach Mg or K deficiency.

Biogeochemical services

The biogeochemical cycling of nutrients through forest ecosystems provides a critical service in sustaining productivity. The biochemical, biogeochemical, and geochemical cycles of nutrients facilitate recycling of elements that allows for the continued photosynthetic capture of CO_2. As ecosystems progress through primary or secondary succession the opened or closed nature of the cycle will change relative to plant demand and one element might develop predominance over another as the limiting element in the ecosystem changes. Forest ecosystems have evolved plant–microbe interactions to facilitate the capture of these critical elements that over the long sweep of time have sustained these ecosystems over cycles of disturbance and regeneration. Managing and maintaining these fundamental biogeochemical cycles will require continued research and enlightened application of this knowledge.

References

Aerts, R. (1996) Nutrient resorption from senescing leaves of perennials: Are there general patterns? *Journal of Ecology*, vol. 84: 597–608

Agerer, R. (2001) Exploration types of ectomycorrhizae – A proposal to classify ectomycorrhizal mycelial systems according to their patterns of differentiation and putative ecological importance. *Mycorrhiza*, vol. 11, pp. 107–114

Albaugh, T., Allen, H.L., Dougherty, P.M., Kress, L., and King, J.S. (1998) Leaf area and above- and belowground responses of loblolly pine to nutrient and water additions. *Forest Science*, vol. 44, pp. 317–327

Averill, C., Turner, B.L., and Finzi, A.C. (2014) Mycorrhiza-mediated competition between plants and decomposers drives soil carbon storage. *Nature*, vol. 505, pp. 543–545

Berger, T.W., Inselbacher, E., Mutsch, F., and Pfeffer, M. (2009) Nutrient cycling and soil leaching in eighteen pure and mixed stands of beech (*Fagus sylvatica*) and spruce (*Picea abies*). *Forest Ecology and Management*, vol. 258, pp. 2578–2592

Bormann, F.H., and Likens, G.E. (1979) *Pattern and Process in a Forested Ecosystem*. Springer-Verlag, New York. pp. xi-253.

Chadwick, O.A., Derry, L., Vitousek, P.M., Huebert, B.J., and Hedin, L.O. (1999) Changing sources of nutrients during four million years of ecosystem development. *Nature*, vol. 397, pp. 491–497

Chapin, F.S. III (1980) The mineral nutrition of wild plants. *Annual Review of Ecology and Systematics*, vol. 11, pp. 233–260

Chapin, F.S. III, Walker, L.R., Fastie, C.L., and Sharman, L.C. (1994) Mechanisms of primary succession following deglaciation at Glacier Bay, Alaska. *Ecological Monographs*, vol. 64, pp. 149–175

Clemmensen, K.E., Bahr, A., Ovaskainen, O., Dahlberg, A., Ekblad, A., and others (2013) Roots and associated fungi drive long-term carbon sequestration in boreal forest. *Science*, vol. 339, pp. 1615–1618

Colpaert, J.V., van Tichelen, K.K., van Assche, J.A., and Van Laere, A. (1999) Short-term phosphorus uptake rates in mycorrhizal and non-mycorrhizal roots of intact *Pinus sylvestris* seedlings. *New Phytologist*, vol. 143 (3), pp. 589–597

Davidson, E.A., and Janssens, I.A. (2006) Temperature sensitivity of soil carbon decomposition and feedbacks to climate change. *Nature*, vol. 440, pp. 165–173

Davidson, E.A., de Carvalho, C.J.R., Figueira, A.M., Ishida, F.Y., Ometto, J.P.B., and others (2007) Recuperation of nitrogen cycling in Amazonian forests following agricultural abandonment. *Nature*, vol. 447, pp. 995–998

Ericsson, T. (1995) Growth and shoot: Root ratio of seedlings in relation to nutrient availability. *Plant and Soil*, vols. 168–169 (1), pp. 205–214.

Gundale, M.J., Deluca, T.H., and Nordin, A. (2011) Bryophytes attenuate anthropogenic nitrogen inputs in boreal forests. *Global Change Biology*, 17 (8), pp. 2743–2753

Hoffland, E., Kuyper, T.W., Wallander, H., Plassard, C., Gorbushina, A., and others (2004) The role of fungi in weathering. *Frontiers of Ecology and the Environment*, vol. 2, pp. 258–264

Jackson, R.B., Mooney, H.A., and Sculze, E.D. (1997) A global budget for fine root biomass, surface area and nutrient contents. *Proceedings of the National Academy of Sciences*, vol. 94, pp. 7362–7366

Kjøller, R., Nilsson, L.-O., Hansen, K., Schmidt, I.K., Vesterdal, L., and Gundersen, P. (2012) Dramatic changes in ectomycorrhizal community composition, root tip abundance and mycelial production along a stand-scale nitrogen deposition gradient. *New Phytologist*, vol. 194, pp. 278–286

Laganière, J., Paré, D., and Bradley, R.L. (2010) How does a tree species influence litter decomposition? Separating the relative contribution of litter quality, litter mixing, and forest floor conditions. *Canadian Journal of Forest Research*, vol. 40, pp. 465–475

Lilleskov, E.A., Fahey, T.J., Horton, T.R., and Lovett, G.M. (2002) Belowground ectomycorrhizal fungal community change over a nitrogen deposition gradient in Alaska. *Ecology*, vol. 83, pp. 104–115

Markewitz, D., Davidson, E., Moutinho, P., and Nepstad, D. (2004) Nutrient loss and redistribution after forest clearing on a highly weathered soil in Amazonia. *Ecological Applications*, vol. 14, pp. S177–S199

McCarthy, H.R., Oren, R., Johnsen, K.H., Gallet-Budynek, A., Pritchard, S.G., and others (2010) Re-assessment of plant carbon dynamics at the Duke free-air CO2 enrichment site: Interactions of atmospheric [CO2] with nitrogen and water availability over stand development. *New Phytologist*, vol. 185, pp. 514–528

Näsholm, T., Högberg, P., Franklin, O., Metcalfe, D., Keel, S.G., and others (2013) Are ectomycorrhizal fungi alleviating or aggravating nitrogen limitation of tree growth in boreal forests? *New Phytologist*, vol. 198, pp. 213–221

Nave, L.E., Vance, E.D., Swanston, C.W., and Curtis, P.S. (2010) Harvest impacts on soil carbon storage in temperate forests. *Forest Ecology and Management*, vol. 259, pp. 857–866

Nave, L.E., Vance, E.D., Swanston, C.W., and Curtis, P.S. (2011) Fire effects on temperate forest soil C and N storage. *Ecological Applications*, vol. 21, pp. 1189–1201

Olson, J.S. (1958) Rates of succession and soil changes on southern Lake Michigan sand dunes. *Botanical Gazette*, vol. 119, pp. 125–170

Peltzer, D.A., Wardle, D.A., Allison, V.J., Baisden, W.T., Bardgett, R.D., and others (2010) Understanding ecosystem retrogression. *Ecological Monographs*, vol. 80, pp. 509–529

Richter, D.D., and Markewitz, D. (2001) *Understanding Soil Change*. Cambridge University Press, Cambridge, UK. 255 pp.

Richter, D.D., Markewitz, D., Heine, P.R., Jin, V., Raikes, J., and others (2000) Legacies of agriculture and forest regrowth in the nitrogen of old-field soils. *Forest Ecology and Management*, vol. 138, pp. 233–248

Rineau, F., Shah, F., Smits, M.M., Persson, P., Johansson, T., and others (2013) Carbon availability triggers the decomposition of plant litter and assimilation of nitrogen by an ectomycorrhizal fungus. *The ISME Journal*, vol. 7, pp. 2010–2022

Schulze, E.D., Beck, E., and Muller-Hohenstein, K. 2005. *Plant Ecology*. Springer, Berlin

Smith, S.E., and Read, D.J. (2008) *Mycorrhizal Symbiosis*, third edition. Academic Press, San Diego, US. 56 pp.

Sterner, R.W., and Elser, J.J. (2002) *Ecological Stoichiometry, The biology of elements from molecules to the biosphere*. Princeton University Press. Princeton, US and Oxford, UK. 439 pp.

Switzer, G.L. and Nelson, L.E. (1972) Nutrient accumulation and cycling in loblolly pine (*Pinus taeda* L.) plantation ecosystems: The first twenty years. *Soil Science Society of America Journal Proceedings*, vol. 36, pp. 143–147

Taylor, A.F.S., Martin, F., and Read, D.J. (2000) Fungal diversity in ectomycorrhizal communities of Norway spruce (*Picea abies* (L.) Karst.) and beech (*Fagus sylvatica* L.) along north-south transects in Europe. *Ecological Studies*, vol. 142, pp. 343–365

Thiffault, N., Fenton, N.J., Munson, A.D., Hébert, F., Fournier, R.A., and others (2013) Managing understory vegetation for maintaining productivity in black spruce forests: A synthesis within a multi-scale research model. *Forests,* vol. 4, pp. 613–631

Turpault, M.-P., Nys, C., and Calvarusco, C. (2009) Rhizophere impact on the dissolution of test minerals in a forest ecosystem. *Geoderma,* vol. 153, pp. 147–154

Van Cleve, K., Oliver, L., Schlentner, R., Viereck, L.A., and Dyrness, C.T. (1983) Productivity and nutrient cycling in taiga forest ecosystems. *Canadian Journal of Forest Research,* vol. 13, pp. 747–766

Vergutz, L., Manzoni, S., Porporato, A., Novais, R.F., and Jackson, R.B. (2012) Global resorption efficiencies and concentrations of carbon and nutrients in leaves of terrestrial plants. *Ecological Monographs,* 82 (2), pp. 205–220

Vitousek, P.M., and Reiners, W.A. (1975) Ecosystem succession and nutrient retentions: A hypothesis. *BioScience,* vol. 25, pp. 376–381

Walker, T.W., and Syers, J.K. (1976) The fate of phosphorus during pedogenesis. *Geoderma,* vol. 15, pp. 1–19

Wallander, H. (1995) A new hypothesis to explain allocation of dry matter between ectomycorrhizal fungi and pine seedlings. *Plant and Soil,* vols. 168–169, pp. 243–248

Wallander, H., and Thelin, G. (2009) Influence of phosphorus and potassium fertilization on weathering of apatite in Norway spruce forest with low phosphorus status. *Soil Biology and Biochemistry,* vol. 40, pp. 2517–2522

Whittaker, R.H., Likens, G.E., Bormann, F.H., Easton, J.S. and Siccama, T.G. (1979) The Hubbard Brook ecosystem study: Forest nutrient cycling and element behavior. *Ecology,* vol. 60, pp. 203–220

Wolfe, B.E., Tulloss, R.E., and Pringle, A. (2012) The irreversible loss of a decomposition pathway marks the single origin of an ectomycorrhizal symbiosis. *PLoS ONE,* vol. 7, pp. 1–9

24
HYDROLOGICAL CYCLING

Michael Bredemeier and Shabtai Cohen

Introduction

The hydrological cycle (HC) is central to life on Earth. Forests interact extensively with that cycle, with interdependencies going both ways. Thus the HC may influence the structure and function of the forest, while forest properties can feed back and influence the cycle. Understanding these interactions is important to us because of the extensive services that the forest HC provides, e.g. pure drinking water for our cities and flood mitigation.

Solar radiation is the ultimate energy source of most of the significant processes on Earth's surface. One energy intensive process is the HC. Based on estimates of the total annual evaporation rates on Earth's surface (Bonan, 2008) it can be shown that more than 30 per cent of the total solar radiation reaching Earth's surface is dissipated by evapotranspiration resulting in increased latent energy in the lower atmosphere. Of course most of this is latent heat from the oceans but a significant part is from land surfaces including forests. Forests determine land surface characteristics including surface albedo, energy fluxes and resistances to the transport of heat of different types. Thus they modify the atmosphere and weather as well as surface energy fluxes, soil properties and surface hydrology. How solar energy is absorbed and dissipated in different ecological systems is central to the HC, and changes in these aspects will in part determine whether, as climate changes, the HC "spins up" leading to increased rainfall and a wetter globe or "spins down" leading to increased aridity due to global warming and drying, drought and desertification. Changes in potential evapotranspiration and rainfall will also determine the extent, character and distribution of future forests.

Several books, review papers and book chapters cover aspects of forest interactions with the HC and the current short chapter cannot deal with all aspects. Bonan (2008) discusses in detail the influence of different ecological systems including forests on surface properties, with special interest in how these features influence climate and how the physics of these interacts in the current generation of climate models. Waring and Running (2007) discuss the water cycle in a similar quantitative approach to that of Bonan (2008), but with the final objective that the reader understands forest water balance modelling. Bond *et al.* (2008) review our understanding of water movement in trees and tree canopies and how tree structure and function modulates and limits water use. Many of the aspects covered by their review are results of more recent research, which has not yet been integrated into the models used for predicting forest water balance and forest influences on climate. Bredemeier *et al.* (2011) present a compendium of review chapters and case studies on forest water interactions with particular attention to climate change influences in European forestry. The first section of that book deals with tree to small catchment processes, including below ground and above ground processes, the second with forest structures, management and water fluxes, and the third with the problem of scaling knowledge of processes at different spatial and temporal scales to deal with real-life issues. In this chapter we will follow the latter approach and give a brief overview of forest HC processes at the tree process, stand and regional scales.

Tree level

Water can enter the forest as precipitation, stream flow, or as condensation of fog or dewfall. Capture of fog by "cloud forests" is unusual, but can be a major source of water for forests in humid montane regions (García-Santos *et al.*, 2004). Similarly, dewfall (Ben-Asher *et al.*, 2010) and condensation of water from warm humid air on cold forest surfaces (Waring and Running, 2007) can be significant, but normally play only a minor role in the HC (e.g. Monteith, 1957). Precipitation can follow several routes before it returns to the atmosphere at a later time. Some is intercepted by the forest canopy (Gash, 1979), from where it may return to the atmosphere within hours or days as evaporation, and some passes through the canopy and reaches the forest floor as throughfall or stemflow (Crockford and Richardson, 2000). Water reaching the forest floor can evaporate, become surface runoff or enter the soil where it can drain into groundwater or aquifers, or be taken up by plants. All of these processes depend on climate, surface geography like slope and aspect, forest canopy and root structure and soil and bedrock properties. Entry of precipitation into the soil depends on soil infiltration rate and when throughfall exceeds the infiltration rate water runs off. In some soils a seal can form as soon as the surface becomes wet, which makes the infiltration rate negligible, leading to floods even when the sub-surface soil is dry. In general, forests reduce seal formation.

Below ground

Hydrological effects on below ground processes were reviewed by Rewald *et al.* (2011). Water movement in the soil depends on soil structure, water content and hydraulic conductance. Models of flow in the soil usually consider the gradient of water potential in the soil and the hydraulic conductance as described by Darcy's law of flow in porous media (Darcy, 1856). Hydraulic conductance of soils is large for sandy and coarse textured soils and decreases as soil clay content increases. It usually decreases exponentially as soils dry and the pathway for water flow becomes sparse. In addition to flow described this way, flow can be through preferential pathways, e.g. cracks in soil and between rocks. Sometimes the "Darcy" flow is only a small part of the total, while most flow is via preferential pathways.

Forests contribute organic matter to soils, which improves soil structure and aeration and increases hydraulic conductance. Roots grow easily in the upper soil layers where water and air are more abundant. However, roots can grow down to 10 m or more in deep soils or in cracks in the bedrock in order to access deep groundwater in dry regions. Great diversity exists in root morphology, depth and hydraulic properties, which is the result of the importance of these organs to the plant for accessing water and nutrients from the soil in a large range of climates and soil types. Mycorrhizal fungi live in symbiotic association with many roots. Their mycelia can improve root access to water and nutrients both physically (due to their spatial extent) and chemically through their ability to alter rhizosphere chemistry (Read and Perez-Moreno, 2003; Simard *et al.*, 2012).

Water uptake by simple plants (e.g. mosses and liverworts) is through osmosis. However, for higher plants, root water uptake is an active and complex process. Root hairs grow into appropriate moist soil and active production of osmotic gradients in the root membranes causes osmotic powered flow into the root, across the membranes and radially into the central conductive xylem of the root. Membranes also contain aquaporins which are active in modulating hydraulic conductance according to the plant's needs. This is the likely explanation for circadian rhythms in hydraulic conductance of root systems, which have been observed even when roots are disconnected from the plant (Henzler et al., 1999; Javot and Maurel, 2002).

Water uptake, transport and evapotranspiration

Plant life depends on photosynthetic fixation of light energy as sugars. But to do that, plants must expose the wet surfaces of the sub-stomatal cavity to the environment, which inevitably results in transpirational water loss. Competition for light has driven plant evolution to climb towards the sky, develop arboreal forms and given us forests. The unsuccessful climbers have either moved into less favourable climates where for a number of reasons a tall habit is disadvantageous, or have made do with the low light (and photosynthetic productivity) in the understory. As plant canopies climb higher and farther from the soil, structural support and access to water become critical for their survival. Arboreal plants have solved the problem of water transport with one general scheme – flow along a pressure gradient in capillary sized conduits of the xylem. In gymnosperms and primitive plants structural support is also provided by the xylem, but in angiosperms supportive tissue can be separate and specialized. The xylem has been the focus of much research in recent years and we will elaborate on a few xylem properties and limits, which in many cases are the limiting factors for canopy height and water use of forests. More detail can be found in the many papers published each year on this topic, and Tyree and Zimmermann (2002) provide an overview of xylem structure and function.

The xylem of higher plants, including monocot and dicot angiosperms and gymnosperms, is made up of thousands of conduits whose diameters can range from ten to several hundred microns. The cohesion–tension theory describes water flow in these capillaries where the flow is powered by a negative pressure gradient between the water in the roots and leaves. Surface tension of the conduit walls maintains continuity of the water columns at negative tension far below that at which wider water columns would break. Hydraulic efficiency increases with the square of conduit diameter, but conduit diameter is directly related to the negative tension at which columns will break. These two properties define the trade-off between hydraulic efficiency and hydraulic safety. Thus in dry environments where the xylem must withstand large pressure gradients conduits are narrow, while in wet environments conduits can be large and transport water more efficiently. Even so, in dry climates trees live in danger of losing xylem conductivity through cavitation of xylem conduits. Although it is clear that xylem can refill in certain conditions, extensive cavitation is an avenue leading to tree death. Recent reports of widespread die-offs of forests that may be related to climate change (Allen et al., 2010; Breshears et al., 2005) have drawn attention to this (Choat et al., 2012) and other mechanisms of drought related tree mortality (McDowell et al., 2008). Better understanding of the mechanisms of hydraulic safety and recovery from cavitation are necessary in order to select and breed new tree varieties for managed forests in future climate conditions. For unmanaged forests, shifts in temperature, potential evapotranspiration and rainfall will lead to shifts in forest composition and distribution and changes in the forest HC.

Once water reaches the leaves it is transpired mostly from the sub-stomatal cavities when stomata are open. Stomata are the valves that allow gas exchange and photosynthesis when conditions are appropriate (i.e. when light is sufficient for photosynthesis), but can close and reduce leaf conductance to water vapour in order to maintain hydraulic integrity by preventing leaf water potential from dropping below the level that would lead to cavitation. It can be shown that this type of control may be achieved by a reduction in leaf conductance (i.e. stomatal closure) when the air's water vapour deficit (VPD) increases (Jones, 1992). Reductions in leaf conductance on a whole tree scale (as canopy conductance) in response to increased VPD have been shown in many studies (e.g. Granier et al., 1996).

Transpiration from leaves and canopies is best described with the help of two types of physical models – flux models that describe mass (i.e. water vapour) and heat transfer, and energy balance models. A combination of the two modelling approaches leads to the well-known Penman–Monteith (P-M) equation (Monteith, 1965) and further analysis to the idea of climate coupling and decoupling (McNaughton and Jarvis, 1983) which determines for a particular climate the extent to which stomata and canopy conductance can control water loss.

Evapotranspiration from forests and the HC

One application of the P-M model is to determine the potential or reference evaporation for a particular climate, which is essentially the water that would be evaporated from a free water surface or a well-irrigated grass crop (Allen et al., 1998). Potential evaporation (E_p) is useful in defining aridity by comparison with precipitation (P). For arid cases $E_p >> P$.

Evaporation from soils depends on an energy source like solar radiation, but also depends on whether the soil surface is wet. When it is not, evaporation is limited by the ability of the soil to transport water to the surface, i.e. soil hydraulic conductivity, which is usually very low. Trees set down root systems which are much more efficient than soil in transporting water. Thus, water availability for transpiration is usually much higher than for soil evaporation. In many cases forests can draw more water from the soil than would be lost without a forest and the price of forest growth may be reduced recharge of aquifers, while the increased humidity from transpiration and evaporation can increase the intensity of the HC (Ellison et al., 2012). While it is clear that land use changes, e.g. the extent of forests, influence climate (Pepin et al., 2010), empirical evidence which would enable quantification of the influence of forest hydrology on the HC is rare, and this topic should be given more attention in future research. The need for this quantification is amplified by the prediction of significant climate changes in many parts of the globe in the near future which will surely influence forest processes, distribution of forests (see Figure 1 in Khatun et al., 2013) and their contribution to the HC. Finally, economic evaluation of forest water use must be included in a full evaluation of ecosystem services and environmental costs (Cohen and Bredemeier, 2011).

The mesoscale

The stand level is essentially the interface between the individual tree and spatial mesoscales of hydrological cycling in forests. A stand integrates a larger number (~ 10^2–10^5) of trees and their ecological interactions (competition, mutualism, etc.). It is also typically the smallest unit for which forest management measures are planned and performed, which is of practical importance for active management of forest water relations. The mesoscale includes catchment and landscape scales. Above that comes the macro-scale, comprising regional and (sub)continental scales and the global scale, which are not considered here. On the mesoscale, we shall examine

– on the basis of recent case studies – the issues of drought, followed by aspects of catchment hydrology and controls on water flows, and finally flood control by forests, considering the latter's potentials and limitations.

Drought

Availability of water determines forest vitality and productivity to a large extent, particularly where potential evapotranspiration is equal to or exceeds precipitation. Water limitation may occur permanently or in temporal (seasonal) cycles. Permanent limitation frequently defines the limits of forest growth and transition to other less productive and more drought tolerant ecosystems types. European areas where seasonal water limitation is prevalent are, for instance, the Mediterranean and parts of the south-east. There, mechanisms for efficient drought adaptation of forest tree species can be observed.

Moreno *et al.* (2011) and Schiller *et al.* (2010) studied Mediterranean oak species (deciduous *Quercus pyrenaica* and *Q. Itaburensis,* and evergreen *Q. Caleprinos*), which are forest tree species well adapted to summer drought. Soil water content in these cases declines significantly over the rainless summer season. The observed summer drought, however, had no discernible effects on vitality and growth of the forest stand. This demonstrates that Mediterranean tree species are apparently well adapted to surviving summer drought periods while maintaining photosynthetic activity. They have a tendency to increasingly exploit soil water from the deeper and moister layers as the dry season proceeds, as documented for *Pinus halepensis* (Klein *et al.*, 2014). Thus, these Mediterranean oak species, probably similar to many other species, are only slightly water-limited during summer drought.

It is interesting to inquire where the limits of forests' drought tolerance lie, and how forest growth under drought conditions interacts with an important and much valued service of forest ecosystems, i.e. the transfer and supply of water to aquifers.

Schiller (2011) reports a case study under very arid conditions in a forest plantation of Aleppo pine (*P. halepensis* Mill.) in the northern Israeli Negev desert, the Yatir Forest. There, long-term average rainfall is only about 300 mm/a, which is concentrated in the winter months, but often with an irregular pattern. In three consecutive years of study, transpiration of the canopy amounted on average to 155 mm/a. This represented on average 58 per cent of the annual rainfall in the same period. Under those low rainfall conditions the proportion of transpiration is very high, while seepage and discharge approach zero. Under the arid conditions of the Yatir Forest, cumulative daily canopy transpiration constituted only 15 to 20 per cent of potential evapotranspiration. Still, the Aleppo pine forest can survive and be maintained as a recreational area under the arid conditions of the Negev desert. Transpiration rates were found to be highly dependent on slope aspect, particularly in very dry years. Trees growing on north-facing slopes transpired less than those on the plateau or south-facing slopes. Thus stocking density should be adjusted on the different slope aspects by careful forest management aimed at alleviating competition for water by wider spacing of the trees under drier conditions.

Intuition would suggest that there is a connection between leaf area index (LAI) and transpirational demand such that transpiration increases with LAI, or conversely, LAI would be reduced if transpirational demand exceeds the available supply of water. Naithani *et al.* (2013) describe the interaction of spatial tree species distribution and spatio-temporal co-dynamics of LAI and soil moisture across a forested landscape. It appears that the spatial distribution of tree species across the landscape creates unique spatio-temporal patterns of LAI, which in turn generate patterns of water demand reflected in varying soil moisture across space and time. The authors found a lag of about 11 days between the increase in LAI and decline in soil water content at the onset of the vegetative period.

Figure 24.1 shows the temporal dynamics of LAI and the negative correlation of LAI with soil water content. This relationship supports Schiller's idea that drought stress could be alleviated by a targeted reduction of LAI through forest management. However, as reduction of LAI through a reduction of evapotranspiration (ET) also entails reduced evaporative cooling and less insertion of water vapour into the air, one has to be aware of another trade-off emerging here for the regional scale.

Establishment of a *mixed forest* may be another option for alleviating drought effects, as reported, in particular, for forests in more humid environments. A general hypothesis in this context is that different tree species tap different portions of the water stored in the porous soil. For instance, one species would extract more water from the upper layers, and the other from the deeper ones. Compared to a monoculture stand where individuals compete for the same water supply locations, competition would be alleviated and the mixed forest could be more productive. This form of mutual furtherance (or relief from competition) has been termed "facilitation" in the literature.

For example, Lebourgeois et al. (2013) observed that mixed stands reduce sensitivity of silver fir (*Abies alba*) to summer drought. The effect was particularly pronounced under the driest conditions. Mixture effects, such as diversification in rooting depth and water input pathways (e.g. stemflow and rainfall interception), may constitute different facilitation processes. This robust functioning of mixed forests highlights their importance for adapting forest management to climate change.

Similarly, Pretzsch et al. (2013) found indications of alleviation of drought stress by interspecific facilitation in mixed versus pure forests. Their study shows that tree resistance and resilience to drought stress can be modified noticeably through species mixing. Moreover, the extent of the effect is species-dependent. Norway spruce (*Picea abies*) and sessile oak (*Quercus petraea*) in mixture performed similarly to pure stands, but beech (*Fagus silvatica*) was significantly more resistant and resilient in mixture than in monoculture. Beech is facilitated significantly when mixed with oak. The authors hypothesize that the discovered water stress release of beech emerges in mixture because of the asynchronous stress reaction pattern of beech and oak

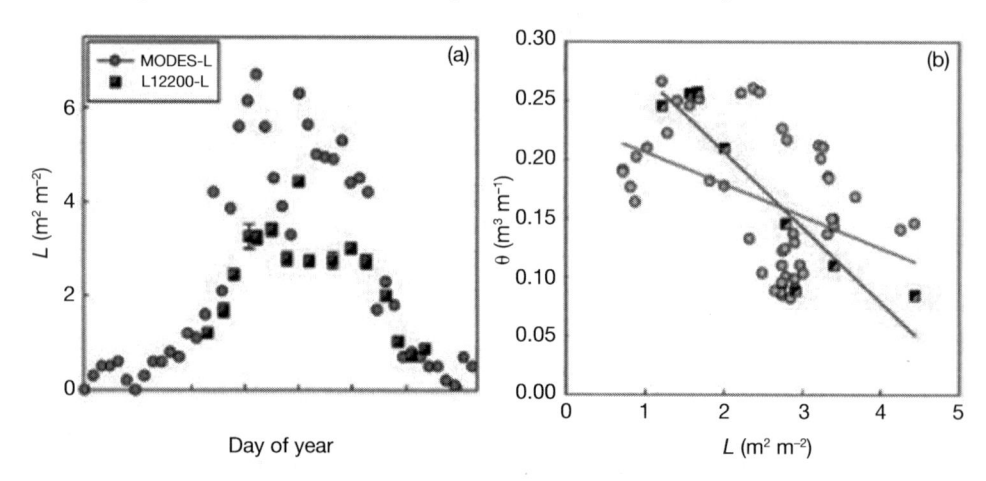

Figure 24.1 (a) Temporal development of leaf area index (L) and (b) regressions of L on soil water content (θ) in two approaches employed by Naithani et al. (2013); figures reproduced with permission from PLoS ONE

and a facilitation of beech by hydraulic lift of water by oak. Hydraulic lift is a phenomenon whereby trees with limited deep root systems (i.e. sinker roots) lift water at night from deep soil layers by root-assisted redistribution of soil water, thus providing water in the day to the abundant roots in the upper soil layers.

This facilitation of beech in mixture with oak probably contributes to the frequently reported "over-yield" of beech in mixed versus pure stands. The differences in stress response open up some options for adapted forest management under climatic change.

Mesoscale aspects of drought can thus be summarized by two major inferences: (1) an option to alleviate drought stress and support forest vitality in drier environments is the reduction of stand density (and thereby LAI and root density), particularly on the most drought-susceptible sites such as south-facing slopes; (2) an option for the more humid environments may be establishment and support of mixed stands (at the appropriate sites), based on facilitation effects which occur between some species when they are mixed.

Catchment hydrology and controls on water flows

The scale of catchments is relevant for describing and assessing the HC in forests. Forested catchments occur to different spatial extents, although forested catchments are usually on the smaller side of the size distribution and located in the headwater areas uphill, whereas downhill in the landscape catchments with mixed land use tend to prevail, and in the fertile and broad floodplains of large rivers there may be a predominance of agricultural, urban and industrial lands. At any rate, catchments are integrators of water flow; collecting precipitation and storing it transiently or longer term, modifying it chemically, and finally releasing it as stream and aquifer flow. The chemical modification is usually favourable, i.e. pollution retention and cleansing, which comprises the forest ecosystem service of water purification. The capacity for transient storage of water in the porous soil space within the catchments is the basis of their flood control capacities (to which we shall turn later).

For our present consideration of hydrological cycling in forests, it is particularly interesting to examine how forests modify and control water flows through catchments.

A review by Schleppi (2011) shows the difference in flow characteristics between small headwater catchments and larger (integrated) catchment areas, catchments of similar size but different vegetation cover, and temporal dynamics due to seasonal influences. Forested catchments show gentler flow duration curves while grassland shows steeper ones, i.e. higher discharge rates occur at lower frequency in the forested catchments, indicating more infiltration and transient retention of water under forest, and a damping of the peak and speed of discharge.

There are numerous reports that forest cover reduces stream flow (and groundwater recharge) – or that cutting forests in catchments enhances stream flow, e.g. Surfleet and Skaugset (2013). The latter studied the effects of timber harvest on base flows and low flows in summer. In a paired watershed study in the Pacific north-western USA, summer low stream flows increased on average by 45 per cent for the three summers following forest harvest (13 per cent of cover). Following a second harvest of an additional 13 per cent of forest cover, August stream flow increased by 106 per cent in the first summer and 47 per cent in the second summer. However, stream flow in small watersheds was not distinguishable from pre-harvest levels within five years for all but one watershed, which had the highest proportion of watershed area harvested. This shows that the reduction of forest cover by cutting has a *transient* effect on water efflux: as long as (forest) vegetation regenerates, it will soon compensate for the decrease in water uptake and transpiration.

Afforestation in a catchment should bring about opposite effects to forest cutting. The relationships of *active* afforestation with the water cycle were the focus of a study by Raftoyannis *et al.* (2011). Large scale afforestation usually reduces ground water yields and runoff. Energy wood plantations are presently increasing due to the fossil energy crisis and related political processes such as the Kyoto protocol. Plantation yield is sufficient to be financially attractive in the more humid environments of Europe. Figure 24.2 (reprinted from Raftoyannis *et al.*, 2011; data from Lindroth and Bath, 2013) shows the increase in wood yield of willow short rotation coppices along a precipitation gradient in southern Sweden. Where rainfall amounts limit plantation growth there is an increasing risk of depleting recharge of aquifers, groundwater and aquatic ecosystems.

Similarly, it has been found that in other afforestation efforts in Europe, benefits are often counterbalanced by drawbacks and problems. In south-eastern Europe, major afforestation efforts were undertaken, mostly motivated by the desire for soil protection, e.g. in torrential precipitation environments. While the desired increased soil protection can be achieved in most instances, more or less severe reductions in runoff and water yields have to be anticipated and considered.

Overall, the following general picture emerges: in the moister environments of Europe with ample precipitation, afforestation offers attractive ecological services, such as better peak flow control combined with biodiversity support, along with attractive economic returns, for instance from energy wood grown in short rotations. However, where water supply is more limited, a distinct *trade-off* emerges: while afforestation may still provide desired ecological services, including soil protection and good habitats, the enhanced water use will result in decreases in water yield, groundwater recharge rates, surface water flow and recharge of reservoirs. Careful consideration of the trade-offs at catchment, landscape and regional scales should be made in these circumstances.

There is a basic disagreement between these two positions in the literature – one that forests withdraw water from downstream systems by their own consumption (via transpiration) and the other one that forests support the water cycle and thus eventually "provide" or "secure" water. Ellison *et al.* (2012) investigated this argument, and showed that the role of forests in the water cycle is scale dependent and depends on local conditions. Thus, evidence can be found to support both sides of the argument, since trees can reduce runoff at the small catchment scale, but at larger scales forests increase precipitation and water availability, owing simply to their function of enhancing the HC. Forest-driven evapotranspiration removed from a particular catchment will contribute to the availability of atmospheric water vapour elsewhere, particularly in continental interiors more distant from oceans. This elucidates the generally beneficial relationship between forest cover and the intensity of the HC at larger spatial scales.

Flood control by forests – potentials and limitations

We finally turn to the question of the forest's contribution to flood attenuation and prevention of destructive flooding. The notion that forests mitigate flooding is passed on like a popular belief – but what is the quantitative reality behind it?

A comprehensive study by Wahren and Feger (2011) reveals potentials and limitations of forest impact on water balance and floods at the mesoscale, i.e. from small catchments (6.8 km^2) up to larger, third order river basins (129 km^2). Based on the hydrological characteristics of an area in eastern Germany which was badly flooded in 2002, a model was applied to estimate the effects of different land uses on water flow with two climate scenarios: frequent very wet events (ca. 60 mm d^{-1}) and the highly infrequent "catastrophic" event of August 2002 (ca. 230 mm

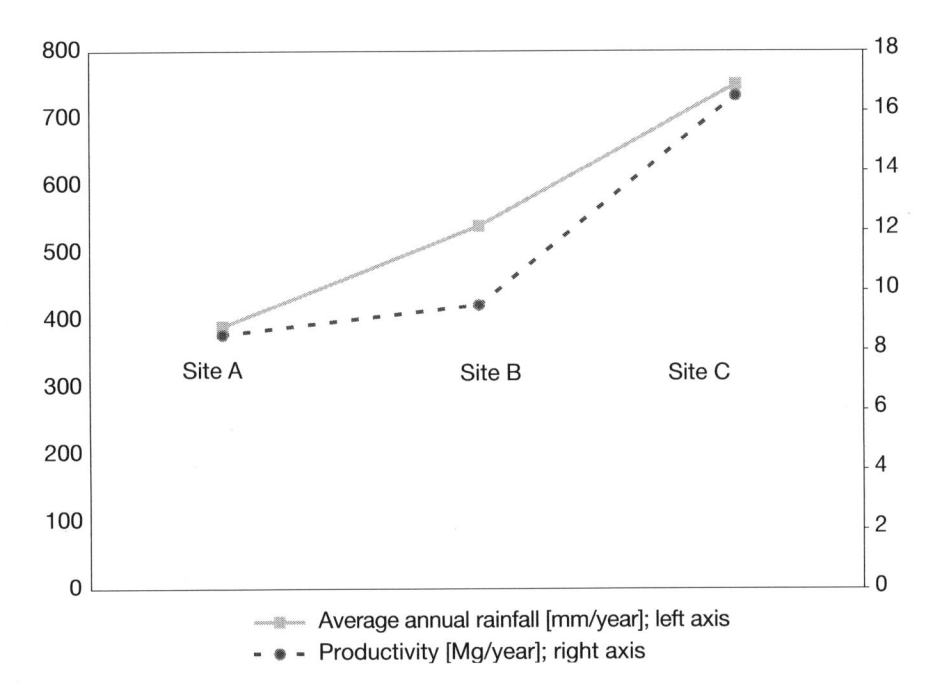

Figure 24.2 Stemwood production of willow short rotation coppices along a precipitation gradient in southern Sweden (based on data from Lindroth and Bath, 2013)

within 2 d). Figure 24.3 (reprinted from Wahren and Feger, 2011) shows the change in flood effective rainfall for these two scenarios. The flood effective rainfall is the fraction of the storm rainfall which is neither retained on the land surface (e.g., in basins) nor infiltrated and retained in the soil, but directly transformed into fast runoff components. While forest land can still absorb a share of the runoff, the ratio of fast runoff to the absorption potential of the land is so large that the impact is marginalized. Thus, the influence of land use during such heavy rain events is negligible. However, for more frequent events the forested areas do significantly decrease flood effective rainfall. The previously cited case study is one model quantification of the general direct relationship between vegetation cover, in particular forest, and reduced flooding intensity, which is more important under "normal" precipitation conditions than under "exceptional" ones. This demonstrates the potential for flood control by forests, which depends on management options, although it is clear that the potential is limited under "catastrophic" flooding events.

In a further study, Wahren *et al.* (2012) demonstrate that the peak reduction of flooding during flood events varies over a wide range, i.e., from 3 per cent to 70 per cent, and that it is highly related to the event characteristics, and especially pre-event soil moisture, which determines the residual water storage capacity in soils during the event.

Bathurst *et al.* (2011) analysed flood peak data for focus catchments in Costa Rica, Ecuador, Chile and Argentina to examine the hypothesis that, as the size of the hydrological event increases, the influence of land use and vegetation cover on peak discharge declines. In all cases they found relative or absolute convergence of the responses as discharge increases. The point of convergence, expressed as magnitude of the event, seemed to be at the point of flood return periods of around ten years. They conclude (in agreement with Wahren and Feger, 2011; Wahren *et al.*, 2012) that in general forest cover is unlikely to significantly reduce peak

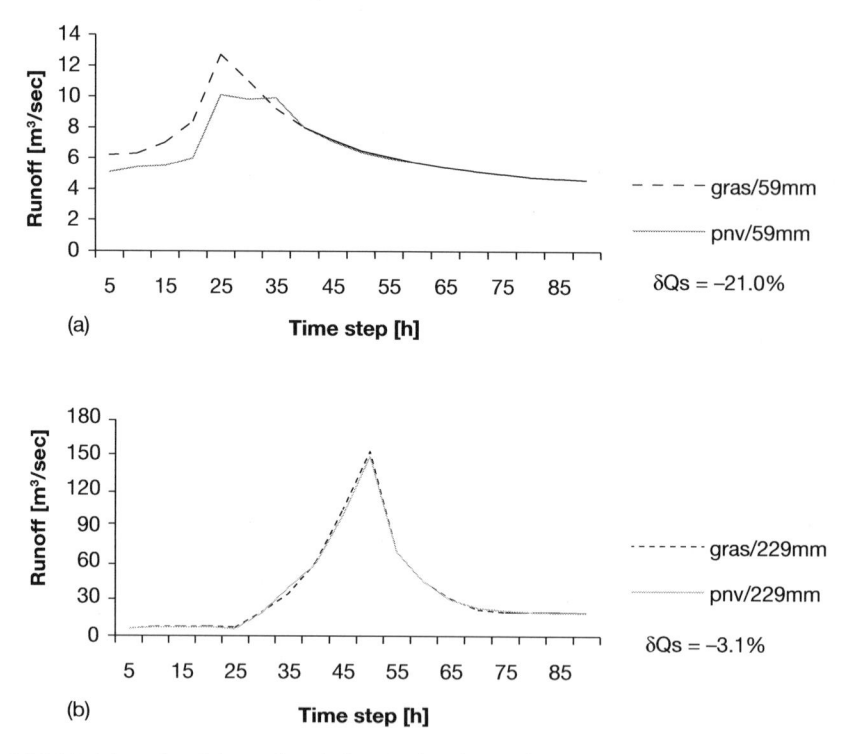

Figure 24.3 Model results of changed peak flow (Qs) and runoff hydrograph due to changed land use (black dashed = low vegetation (grassland); grey solid = potential natural vegetation, forest (PNV) for two different simulated rainfall events: (a) strong but "normal" precipitation event with 59 mm/d and (b) extreme event with 229 mm/d. Note different scaling between parts (a) and (b) (re-drawn based on Wahren and Feger, 2011)

discharges generated by extreme rainfall, but may still offer considerable mitigation benefits for moderate (and hence more frequent) rainfall events.

So there is some truth to the general proposition that forest cover can offer significant services with respect to flood attenuation and control in comparison with other land use forms. However, this effect is limited, and at its limit approaches nil in the situations of strongest rainfall and flooding.

Concluding remarks

This brief overview of hydrological cycling in forests demonstrates that a relatively broad and consolidated knowledge base exists on the processes involved at various relevant scales from stomata to large river basins. At the lowest scale that we discussed, i.e. that of plant structure and function, a distinct trade-off between hydraulic efficiency and hydraulic safety can be discerned, with important implications for tree adaptation to drought. It is interesting to note that at the largest scale, that of forests and landscape, hydraulic flows and safety, especially as related to the ability of forests to store rainwater and mitigate floods, are major issues.

The ability of forests to increase the access of the atmosphere to soil water via the hydraulically efficient root systems and improved soil structure means that the forest is particularly efficient in generating ET and recycling water into the atmosphere as water vapour. The increased ET is sometimes at the expense of aquifer recharge or water that might be used for other purposes and therefore forest water use comes at a finite price. Thus, quantification of forest ecosystem and other services is important in order to determine the economics of forestry, where water use is one of the expenditures.

Increased hydrological cycling is important for the generation of rainfall on extended spatial scales, and can be important in regions which would otherwise remain much dryer and less productive. A positive feedback cycle might result, i.e. ET, rainfall and "greening", as recently discussed by Ellison et al. (2012). However, the increases in precipitation due to forestry may only occur downwind, leading to trans-boundary issues that might be positive in the case of increased precipitation in dry regions or negative where increased humidity and precipitation might lead to flooding and other damage.

These and other issues are likely to come into better focus as our understanding of forest hydrological cycling and its interaction with forest management moves from the current conceptual and modelling stages into a future quantitative deterministic stage allowing more precise economic analysis.

References

Allen, C.D., Macalady, A.K., Chenchouni, H., Bachelet, D., McDowell, N., and others 2010. A global overview of drought and heat-induced tree mortality reveals emerging climate change risks for forests. *For. Ecol. Manage.* 259: 660–684.

Allen, R.G., Pereira, L.S., Raes, D., Smith, M. 1998. *Crop Evapotranspiration*, Guidelines for computing crop water requirements. FAO Irrigation and drainage paper 56. FAO, Rome.

Bathurst, J.C. Iroumé, A., Cisneros, F., Fallas, J., Iturraspe, R. and others 2011. Forest impact on floods due to extreme rainfall and snowmelt in four Latin American environments 1: Field data analysis. *Journal of Hydrology*, 400(3–4): S.281–291.

Ben-Asher, Y., Alpert, P., and Ben-Zvi, A. 2010. Dew is a major factor affecting vegetation water use efficiency rather than a source of water in the eastern Mediterranean area. *Water Resources Research* 46 (10) Article number W10532.

Bonan, G. 2008. *Ecological Climatology. Concepts and Applications*, second edition. Cambridge, UK: Cambridge University Press.

Bond, B.J., Meinzer, F.C. and Brooks, J.R. 2008. How trees influence the hydrological cycle in forest ecosystems. In: *Hydroecology and Ecohydrology: Past, Present and Future*. P.J. Wood, D.M. Hannah and J.P. Sadler, (Eds). John Wiley & Sons pp. 7–35.

Bredemeier, M., Cohen, S., Godbold, D.L., Lode, E., Pichler, V. and Schleppi, P. (Eds) 2011. *Forest Management and the Water Cycle – An Ecosystem-Based Approach*, Heidelberg, Germany: Springer.

Breshears, D.D., Cobb, N.S., Rich, P.M., Price, K.P. and Allen, C.D. (2005) Regional vegetation die-off in response to global-change-type drought. *Proc. Natl. Acad. Sci. U.S.A.* 102: 15144–15148.

Choat, B., Jansen, S., Brodribb, T.J., Cochard, H., Delzon, S., and others 2012. Global convergence in the vulnerability of forests to drought. *Nature* 491: 752–755.

Cohen, S. and Bredemeier, M. 2011. Synthesis and outlook. In *Forest Management and the Water Cycle – An Ecosystem-Based Approach*. Bredemeier, M., Cohen, S., Godbold, D.L., Lode, E., Pichler, V. and Schleppi, P. (Eds) Ecological Studies. Heidelberg, Germany: Springer, S. 507–512.

Crockford, R.H. and Richardson, D.P. 2000. Partitioning of rainfall into throughfall, stemflow and interception: Effect of forest type, ground cover and climate. *Hydrological Processes* 14: 2903–2920.

Darcy, H. (1856) *Les Fontaines Publiques de la Ville de Dijon*, Paris: Dalmont.

Ellison, D., Futter, M.N. and Bishop, K., 2012. On the forest cover-water yield debate: from demand- to supply-side thinking. *Global Change Biology*, 18(3): S.806–820.

García-Santos, G., Marzol, M.V. and Aschan, G. 2004. Water dynamics in a laurel montane cloud forest in the Garajonay National Park (Canary Islands, Spain), *Hydrol. Earth Syst. Sci.*, 8: 1065–1075.

Gash, J.H.C. 1979. Analytical model of rainfall interception by forests. *Quarterly Journal of the Royal Meteorological Society* 105: 43–55.

Granier, A., Huc, R. and Barigah, S.T. 1996. Transpiration of natural rain forest and its dependence on climatic factors. *Agric. For. Meteorol.* 78: 19–29.

Henzler, T., Waterhouse, R.N., Smyth, A.J., Carvajal, M., Cooke, D.T. and others 1999. Diurnal variations in hydraulic conductivity and root pressure can be correlated with the expression of putative aquaporins in the roots of lotus japonicas. *Planta* 210: 50–60.

Javot, H. and Maurel, C. 2002. The role of aquaporins in root water uptake. *Annals of Botany* 90: 301–313.

Jones, H.G. 1992. *Plants and Microclimate*, second edition. Cambridge, UK: Cambridge University Press.

Khatun, K., Imbach, P. and Zamora, J.C. 2013. The implications of climate change impacts on conservation strategies for Central America using the Holdridge Life Zone (HLZ) land classification. *iForest* 6: 183–189. See http://www.sisef.it/iforest/contents/?id=ifor0743-006

Klein, T., Rotenberg, E., Cohen-Hilaleh, E., Raz-Yaseef, N., Tatarinov, F., and others 2014. Quantifying transpirable soil water and its relations to tree water use dynamics in a water-limited pine forest. *Ecohydrology* 7: 409–419.

Lebourgeois, F., Gomez, N., Pinto, P., Mérian, P. 2013. Mixed stands reduce Abies alba tree-ring sensitivity to summer drought in the Vosges mountains, western Europe. *Forest Ecology and Management*, 303: S.61–71.

Lindroth, A. and Bath, A. 2013. Assessment of regional willow coppice yield in Sweden on basis of water availability. *Forest Ecology and Management*, 121: S.57–65.

McDowell, N.G., Pockman, W.T., Allen, C.D., Breshears, D.D., Cobb, N. and others 2008. Mechanisms of plant survival and mortality during drought: Why do some plants survive while others succumb? *New Phytology* 178: 719–739.

McNaughton, K.G. and Jarvis, P.G. 1983. Predicting effects of vegetation changes on transpiration and evaporation. In: Kozlowski, T.T. (Ed.), *Water Deficits and Plant Growth*, vol. VII. Academic Press, pp. 1–47.

Monteith, J.L. 1957. Dew. *Quarterly Journal of the Royal Meteorological Society* 83: 322–341.

Monteith, J.L. 1965. Evaporation and environment. *Symposia of the Society for Experimental Biology* 19: 205–224.

Moreno, G., Gallardo, J.F. and Angeles Vicente, M. 2011. How Mediterranean deciduous trees cope with long summer drought? The case of Quercus pyrenaica forests in Western Spain. In *Forest Management and the Water Cycle – An Ecosytem-Based Approach*. Bredemeier, M., Cohen, S., Godbold, D.L., Lode, E., Pichler, V. and Schleppi, P. (Eds) Ecological Studies. Heidelberg, Germany: Springer, S. 187–201.

Naithani, K.J., Baldwin, D.C., Gaines, K.P., Lin, H. and Eissenstat, D.M. 2013. Spatial distribution of tree species governs the spatio-temporal interaction of leaf area index and soil moisture across a forested landscape. T. Wang (ed.), hrsg. *PLoS ONE*, 8(3): S.e58704.

Pepin, N.C., Duane, W.J. and Hardy, D.R. 2010. The montane circulation on Kilimanjaro, Tanzania and its relevance for the summit ice fields: Comparison of surface mountain climate with equivalent reanalysis parameters. *Global and Planetary Change* 74: 61–75.

Pretzsch, H., Schütze, G. and Uhl, E. 2013. Resistance of European tree species to drought stress in mixed versus pure forests: evidence of stress release by inter-specific facilitation: Drought stress release by inter-specific facilitation. *Plant Biology*, 15(3): S.483–495.

Raftoyannis, Y., Bredemeier, M., Buozyte, R., Lamersdorf, N., Mavrogiakoumos, A. and others 2011. Afforestation strategies with respect to forest–water interactions. In *Forest Management and the Water Cycle – An Ecosystem-Based Approach*. Bredemeier, M., Cohen, S., Godbold, D.L., Lode, E., Pichler, V. and Schleppi, P. (Eds) Ecological Studies. Heidelberg, Germany: Springer, S. 225–245.

Read, D. J. and Perez-Moreno, J., 2003. Mycorrhizas and nutrient cycling in ecosystems – A journey towards relevance? *New Phytologist* 157(3): 475–492.

Rewald, B., Michopoulos, P., Dalsgaard, L., Jones, D.L., Godbold, D.L. 2011. Hydrological effects on below ground processes in temperate and Mediterranean forests. In *Forest Management and the Water Cycle – An Ecosystem-Based Approach*. Bredemeier, M., Cohen, S., Godbold, D.L., Lode, E., Pichler, V. and Schleppi, P. (Eds) Ecological Studies. Heidelberg, Germany: Springer, S. 5–30.

Schiller, G., 2011. The case of Yatir Forest. In *Forest Management and the Water Cycle – An Ecosystem-Based Approach*. Bredemeier, M., Cohen, S., Godbold, D.L., Lode, E., Pichler, V. and Schleppi, P. (Eds) Ecological Studies. Heidelberg, Germany: Springer, S. 163–186.

Schiller, G., Ungar, E. D., Cohen, S., Herr, N., 2010. Water use by Tabor and Kermes oaks growing in their respective habitats in the lower Galilee region of Israel. *Forest Ecology and Management* 259: 1018–1024.

Schleppi, P., 2011 Forested water catchments in a changing environment. In *Forest Management and the Water Cycle – An Ecosystem-Based Approach*. Bredemeier, M., Cohen, S., Godbold, D.L., Lode, E., Pichler, V. and Schleppi, P. (Eds) Ecological Studies. Heidelberg, Germany: Springer, S. 89–110.

Simard, S.W., Beiler, K.J., Bingham, M.A., Deslippe, J.R., Philip, L.J. and Teste, F.P. 2012. Mycorrhizal networks: Mechanisms, ecology and modeling. *Fungal Biology Review* 26: 39–60.

Surfleet, C.G. and Skaugset, A.E., 2013. The effect of timber harvest on summer low flows, Hinkle Creek, Oregon. *Western Journal of Applied Forestry*, 28(1): S.13–21.

Tyree, M.T. and Zimmermann, M.H. 2002. *Xylem Structure and the Ascent of Sap*. Second edition. Berlin: Springer.

Wahren, A. and Feger, K.H., 2011. Model-Based assessment of forest land management on water dynamics at various hydrological scales – A case study. In *Forest Management and the Water Cycle – An Ecosystem-Based Approach*. Bredemeier, M., Cohen, S., Godbold, D.L., Lode, E., Pichler, V. and Schleppi, P. (Eds) Ecological Studies. Heidelberg: Springer S. 453–470.

Wahren, A., Schwärzel, K. and Feger, K.-H. 2012. Potentials and limitations of natural flood retention by forested land in headwater catchments: Evidence from experimental and model studies: Potentials and limitations of natural flood retention. *Journal of Flood Risk Management*, 5(4), S.321–335.

Waring, R.H. and Running, S.W. 2007. *Forest Ecosystems. Analysis at Multiple Scales*. Third Edition. New York: Academic Press.

25
PRIMARY PRODUCTION AND ALLOCATION

Frank Berninger, Kelvin S.-H. Peh and Hazel K. Smith

The products of photosynthesis constitute almost the entirety of plant dry biomass and contribute critically to global primary production. It is for this reason that increasing photosynthesis, both in terms of efficiency and level is so desirable for stakeholders such as farmers and foresters. However, some systems, such as tropical rain forests, produce relatively low levels of woody biomass despite high photosynthetic production levels. Scenarios such as this complicate the relationship between photosynthesis and dry biomass yield. Another example of the ambiguous relationship between these two interlinked traits is that of inter-annual variation in photosynthetic production, which shows only a moderate correlation with stem biomass growth. This suggests that the percentage of photosynthetic production used for stem growth varies from year to year as well as between sites (Rocha et al. 2006, Gea-Izquierdo et al. 2014). In this chapter we will explore why the relationship between photosynthetic production and wood growth is not straightforward. We will consider which environmental factors determine the photosynthetic production of forests and how the products of photosynthesis are allocated within the plant for biomass production.

An introduction to primary production

Primary production is the synthesis of organic compounds from atmospheric carbon dioxide and occurs primarily through photosynthesis. The organisms which are responsible for primary production are termed primary producers. They include plants and algae, the former of which will be the focus in this chapter. Generally, the photosynthetic production of an ecosystem is termed *gross primary production* (GPP) and quantifies the total amount of carbon which is fixed as carbohydrates by plants in photosynthesis. Not all of GPP is able to contribute to plant growth since a portion of it is lost as plant respiration, which is also known as autotrophic respiration (H_A). The remainder of GPP is used for biomass accumulation and is referred to as *net primary production* (NPP), and is the rate of production of new organic matter in the plant. The actual net biomass increment is smaller than NPP because a fraction is lost through senescence and plant mortality. This relationship can be explained as:

$$\text{Net biomass increment} = \text{GPP} - H_A - \text{senescence} - \text{mortality}$$
$$= \text{NPP} - \text{senescence} - \text{mortality}$$

In the example of the biomass accumulation of tropical rain forests, we have considered only woody biomass, however, biomass production is also allocated to non-woody tissues such as plant foliage and roots and this uses some of the available resources. Therefore, wood production is less than total (biomass) production. Another complicating factor when considering forest level production is that of carbon balance where, in addition to autotrophic respiration, large pools of carbon exist in the soil, where they can be lost through heterotrophic (H_R) respiration. Carbon loss also occurs when dissolved organic carbon is lost through run-off, although these losses are typically not large. On the other hand, although important when calculating plant GPP, senescence and mortality have no immediate effect on the ecosystem carbon balance of the forest since dead biomass remains in the soil until it eventually decomposes. The exception to this is when whole trees or parts are harvested which has the immediate effect of removing carbon from the ecosystem. The carbon balance of an ecosystem is therefore GPP minus H_R and H_A minus losses by carbon in run-off or:

$$\text{Ecosystem carbon balance} = GPP - H_A - H_R - \text{carbon lost through run-off}$$

Figure 25.1 describes the different components of photosynthetic production, growth and carbon balances of three contrasting forest types and much of the terminology is defined in Table 25.1. All forests displayed in Figure 25.1 have a high rate of photosynthetic production, which would lead to a high growth rate if all photosynthetic products were used for stem growth. For typical GPP and NPP values of different biomes, see Chapter 37.

Table 25.1 Definition of different productivity terms

GPP	*Gross primary production*	Photosynthetic production of all organisms in an ecosystem excluding losses by respiration.
H_A	*Autotrophic respiration*	Respiration of photosynthesising organisms.
H_R	*Heterotrophic respiration*	Respiration of non-photosynthesising organisms.
NPP	*Net primary production*	$NPP=GPP-H_A$ Growth of biomass of all plants excluding losses of biomass through litter production and herbivory.
NEE	*Net ecosystem exchange*	$NEE=H_R+H_A-GPP=H_R-NPP$ Difference of respiration and the gross primary production. Net carbon balance of an ecosystem. Negative numbers usually indicate a carbon sink and positive numbers a carbon source to the atmosphere (see NEP below).
NEP	*Net ecosystem production*	$NEP= -NEE=GPP-H_A-H_R$ usually $-1 \times NEE$
NBP	*Net biome productivity*	Carbon balance for large regions. Average NEP minus losses from disturbance.
I_B	*Net Biomass growth*	$NPP-L-M$ Biomass growth in an ecosystem including losses from litter and tree mortality. Can be measured as the difference of biomass at two points in time.
L	*Litter production*	Death of biomass. Usually excluding sapwood turnover and litter production due to tree mortality.
M	*Tree mortality*	Biomass loss due to the mortality of whole trees.

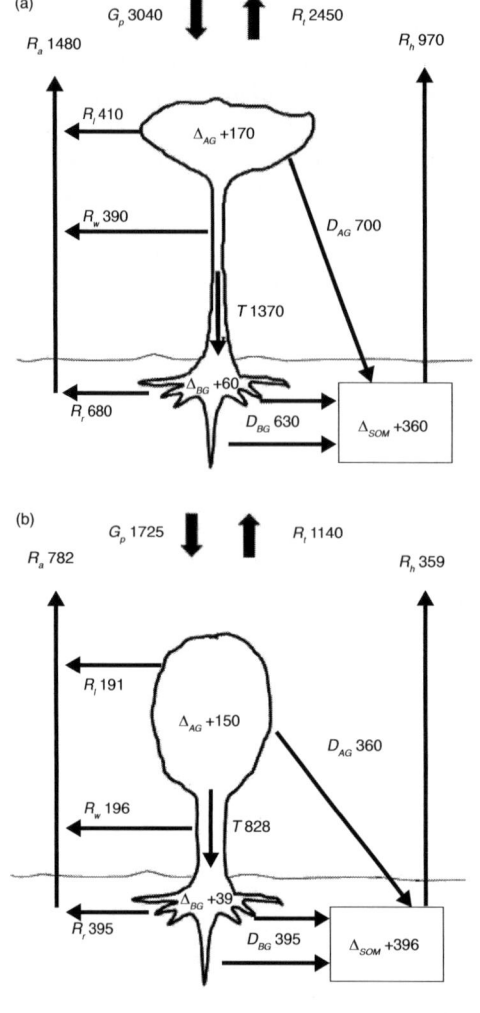

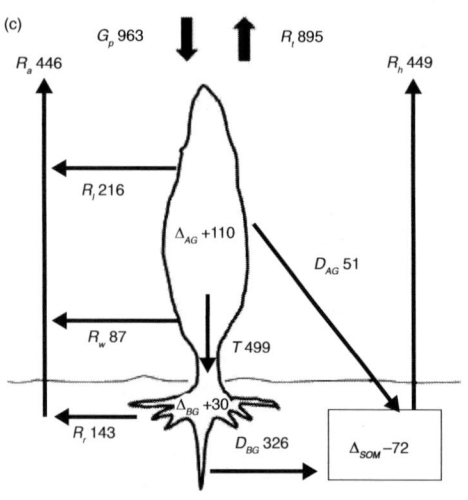

Figure 25.1 Productivity of a boreal, a temperate and a humid tropical ecosystem (all in g C m^{-2} yr^{-1}; Malhi *et al.* 1999). Forests are: (a) tropical evergreen rainforest from the Amazon (Brazil); (b) a deciduous broadleaf forest from Tennessee (USA); and (c) a northern boreal black spruce forest from Canada. Gp = gross primary productivity (GPP); R_t = ecosystem respiration (H$_A$); R_a = autotrophic respiration; R_w = above-ground wood respiration; DAG = above-ground senescence and mortality; DBG = below-ground senescence and mortality; ΔAG = above-ground net biomass carbon increment; ΔBG = below-ground net biomass carbon increment; ΔSOM = net increment in soil organic carbon (from Malhi *et al.* 1999)

Environmental effects on photosynthetic production

Land availability is a critical limiting resource when we consider plant productivity at a site-scale (Running 2012) but other factors are also important. Primary production is affected, both positively and negatively, by environmental factors such as temperature, solar radiation and water. A very rudimental example of this is evident when one examines the different productivity levels of plants which are grown in warm and wet environments compared to those from colder regions of the world. This chapter will outline how some of these environmental factors affect photosynthetic production in more detail.

Light and temperature

The first step for photosynthetic production is the absorption of light by chlorophyll in the leaves, with unabsorbed photons not contributing to photosynthesis. Consequently, plants and stands with a lower leaf area tend to absorb less light and therefore demonstrate lower photosynthetic production. In addition to the leaf area index (LAI), which is the projected leaf area of all leaves per ground area, the spatial arrangement of leaves greatly influences light absorption and photosynthesis. There are numerous imperfect ways to estimate light interception of forest stands based on empirical measurements and models. For example, hemispherical photographs taken using a fisheye lens can be used to calculate how much free sky can be seen, while estimating the quantity of light which originates from each direction indicates the light interception of the canopy above the sampling point. Other methods rely on direct measurements of the reflection of light, which can be measured using satellite images. Although different methods are favoured in particular scenarios, each is based on non-trivial assumptions regarding the distribution of foliage and interception of light by non-photosynthetic organs, such as tree stems. Different methods tend to give different results of intercepted radiance because the assumptions differ between methods (Chen *et al.* 2006).

The relationship between photosynthetic production and the amount of light intercepted by a single leaf at a moment in time is curvilinear (Monteith 1994) and saturates at high light intensities. Therefore, the instantaneous light use efficiency of a leaf (LUE_I), defined as the rate of photosynthesis divided by the quantity of light absorbed by the leaf, declines with increasing light intensity. However, there are several definitions of light use efficiency in the literature and readers of scientific papers must be careful to understand the definition used in each case. At low light levels, initial light use efficiency, or the light use efficiency of photosynthesis, is relatively constant across plant species and is slightly higher than the theoretical rate that can be derived from stoichiometry of photosynthesis (Long *et al.* 1993). A canopy photosynthesising at medium and high light intensities will exhibit a higher light use efficiency than that of a single leaf as some leaves in the canopy will be shaded and operate at higher light use efficiencies. Conversely, sunlit leaves will be exposed to high light and therefore operate at lower light use efficiencies (Long *et al.* 1993).

Figure 25.2 shows the instantaneous and long term dependence of photosynthetic production on light levels in leaves and canopies of Scots pine at Hyytiälä. For a single leaf at a single moment, the relationship between intercepted irradiance and photosynthetic production is highly nonlinear, but when we look at canopy production over longer time periods a linear relationship evolves. This linear relationship was described by John Monteith about 20 years ago (Monteith 1994, Turner *et al.* 2005) and is still used by most of the remote sensing based products to estimate plant production (see Running *et al.* 2004 for a description of the MODIS GPP algorithm). However, it is widely recognised that a linear relationship between absorbed

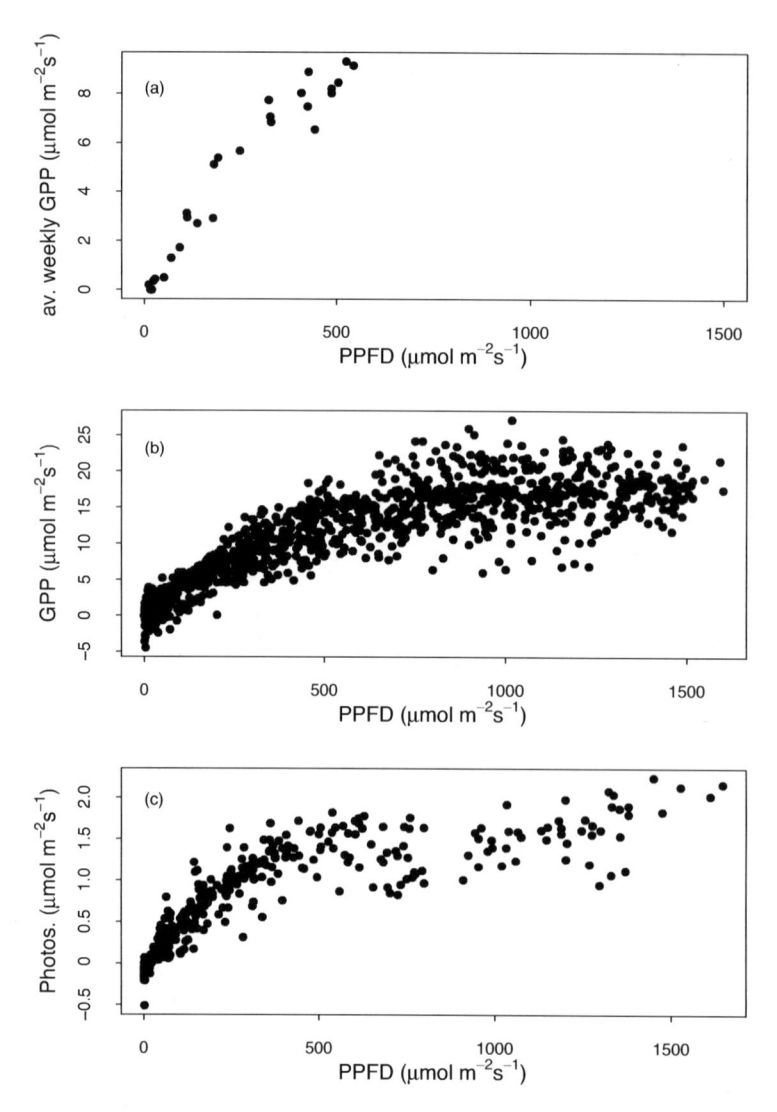

Figure 25.2 Rates of photosynthesis (photos.) of Scots pine at different light levels measured in terms of photosynthetic photon flux density (PPFD): (a) displays the long term average (av.) weekly responses of gross primary production to incoming photosynthetically active radiation (July–November 2010); (b) the half hourly response of gross primary production (GPP) to incoming photosynthetically active radiation (July 2010); and (c) the response of a single Scots pine shoot to photosynthetically active radiation (July 2008). Note that the relationship becomes more linear from the bottom to the top of the graphs. Data is from the Hyytiälä SMEAR II station

irradiance and production holds only approximately. As we elaborate below, different environmental factors can alter plant light use efficiency and need to be accounted for when considering plant productivity.

An example of an environmental factor which affects the potential linear relationship between absorbed irradiance and productivity is temperature, which limits photosynthetic production in several ways. Although photosynthetic rates are lower at very high and low temperatures, photosynthetic production has a broad temperature optimum and the temperature response can adapt to changing environmental conditions (Hikosaka *et al.* 2006). However, long lasting effects due to low temperatures can be observed in boreal forests that have low rates of photosynthesis during the winter months. These lower rates of photosynthesis are partially due to suboptimal temperatures but can also be attributed to a deactivation of the photosynthetic apparatus over the winter months. Temperature is the main driver of the recovery process of these photosynthetic systems (Mäkelä *et al.* 2004, Suni *et al.* 2003) and recovery is relatively slow, often taking several weeks to complete. Gea-Izquierdo *et al.* (2010) showed that the recovery process is slower in more northern boreal ecosystems than in southern ecosystems. In the case of a maritime evergreen Douglas fir system, differences between winter and summer states of photosynthesis were minimal due to the relatively high winter temperatures in this maritime environment. In contrast, continental boreal coniferous forests showed an important, yet slow, recovery process which lasted several weeks (Gea-Izquierdo *et al.* 2010).

Water

The survival and growth of plants is reliant on water availability and it is widely recognised that water availability is the most limiting environmental factor for plant productivity. Plants need water directly for photosynthesis as well as for a range of other functions, such as transpiration. Transpiration acts to transport water from the roots to the rest of the plant, which, in turn, facilitates the movement of nutrients from the soil to the growing tissues of the plant. Furthermore, the driving force for transpiration is diffusion from the stomatal pores in the foliage and stomatal aperture also controls CO_2 uptake. When plants experience drought, stomatal aperture is reduced, this not only reduces the transport of nutrients from the roots to the growing regions of the plant, but also limits CO_2 uptake for photosynthesis (Chaves *et al.* 2003). These mechanisms underlie the reduction in productivity, which is caused by water deficits. The effect of water availability can be seen through experimental evidence that NPP is strongly affected by mean annual precipitation across a range of ecosystems (Sala *et al.* 2012).

Wood productivity has been investigated in northern hardwood forests alongside a range of other environmental factors (Baribault *et al.* 2010). Water availability was shown to positively correlate with wood productivity, although other factors, such as nutrient level, were found to be more related. Productivity has previously been linked to nutrient level in studies which examined multiple environmental factors. For example, Patagonian steppe systems span a precipitation gradient and it has been shown that, although productivity is affected by both nitrogen content and soil water, they are not independent of one another as soil water controls nitrogen levels (Austin and Sala 2002). In the future, it is predicted that the world will experience an increase in droughts, both in terms of frequency and severity and this is likely to negatively affect productivity in many regions.

Nutrients

Nitrogen is essential for photosynthetic production and approximately half of the nitrogen which exists in leaves is in the main photosynthetic enzyme Ribulose-1,5-bisphosphate carboxylase/oxygenase (RuBisCO). Therefore, nitrogen concentration in the leaves affects the GPP of trees in several ways. First, the photosynthetic capacity of leaves is strongly correlated with the nitrogen concentration in leaves due to its integral role in the enzyme RuBisCO (Reich *et al.* 1995). This effect is pronounced for sunlit leaves and species with low LAI since nitrogen affects photosynthetic capacity, while there is no effect on light use efficiency at low light intensities.

Second, there seems to be a larger leaf area per leaf dry mass in leaves that have higher nitrogen concentrations (Coll et al. 2011), increasing the interception of light per unit of leaf mass. The effect of nitrogen on the photosynthetic capacity is higher in deciduous broadleaf trees than for conifers (Reich et al. 1995; Figure 25.3) and might explain the dominance of deciduous trees on fertile sites. One reason for this is possibly the dual role of foliage in nitrogen storage and metabolism, which is pronounced in conifers (Vapaavuori et al., 1995).

While nitrogen is generally accepted to be the nutrient most often limiting plant productivity in the temperate and boreal forests, phosphorous availability is the main factor influencing tropical forest productivity (Quesada et al. 2012). A deficiency of phosphorous could potentially limit photosynthesis and thus community level primary production because this nutrient is essential for adenosine triphosphate and other sugar phosphates for metabolism. Nevertheless, productivity of tropical montane forests and some lowland tropical forests may still be limited by nitrogen availability (Quesada et al. 2012).

In temperate and boreal forests, base cations could enter the systems through atmospheric deposition. A drastic increase in base cations, especially extractable calcium and magnesium, due to a high level of air pollution could negatively affect nutrient cycling and tree productivity (see Chapter 31). Nevertheless, extractable calcium content could also have a positive influence on productivity at varying levels in these forest systems. Fertilisation with calcium in field experiments promotes growth in seedlings (Kobe et al. 2002), saplings (Juice et al. 2006) and mature trees (Gradowski and Thomas 2008). In northern hardwood forests, it has been shown that this fertilisation effect is finite for above-ground productivity, with the effect of exchangeable calcium saturating at a concentration of approximately 1 mol/kg of soil (Baribault et al. 2010, Figure 25.4). In contrast, calcium deficiency limits productivity in plants, which demonstrates the effect that changing soil nutrition can have on plant productivity (McLaughlin and Wimmer 1999). A deficiency in other nutrients may also limit tree productivity, especially when grown in habitats that are anthropologically modified such as old fields and drained peatland.

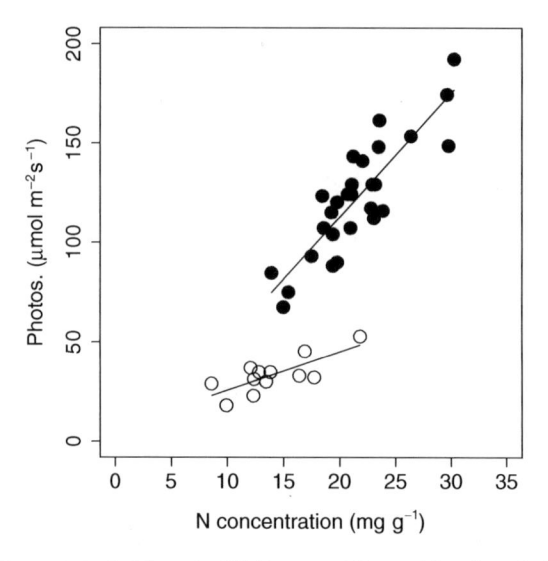

Figure 25.3 Rates of photosynthesis (photo.) of different coniferous (closed symbols) and deciduous (open symbols) species in Wisconsin as a function of the leaf nitrogen (N) concentration. Redrawn from Reich et al. (1995)

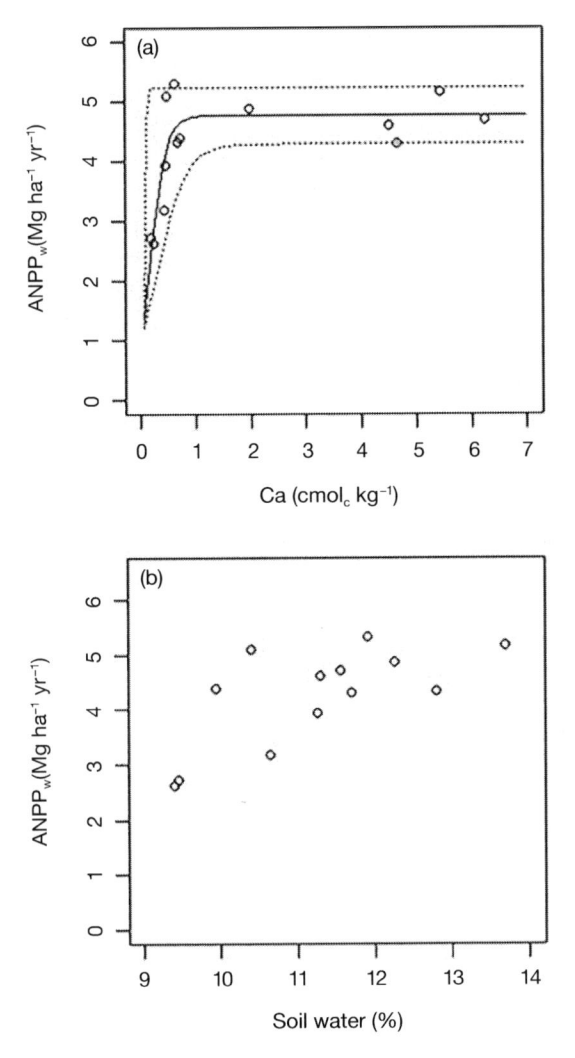

Figure 25.4 Wood above-ground net primary production (ANPP) expressed as functions of soil calcium (a) and soil water content (b) (Baribault *et al.* 2010)

Soil resources are crucial for plant productivity and have been shown to account for a high proportion of wood NPP and leaf NPP when considering calcium, nitrogen and water availability. It is likely that these factors act together to jointly control productivity with input from other environmental factors (Baribault *et al.* 2010).

Respiration

The products of photosynthesis are not all used for growth since about half of fixed carbon is lost through respiration. Respiration increases exponentially with temperature and initially researchers thought that trees in warm regions had higher losses of carbon in respiration than trees in cold regions (Berninger and Nikinmaa 1997). The traditional framework which was used to explain respiration rates was proposed by Warren Wilson (1967) and divides respiration into three components. These components are: maintenance respiration which depends on metabolic enzyme turnover; growth respiration which is when sugars are converted to more energy rich substances during growth; and substrate induced respiration. Maintenance respiration is temperature dependent, while growth and substrate induced respiration depend on growth rates and the chemical composition of new biomass. Specifically, the amount of carbon used for growth respiration is relatively well constrained by the chemical composition of the tissue and directly reflects the metabolic costs of producing new biomass. These are usually higher for energy rich tissues such as protein rich leaves rather than for cellulose rich plant stems (Amthor 2000).

Despite the short term dependence of respiration on temperature, it has been shown that respiration of trees does not increase globally from high to low latitudes. Furthermore, respiration is able to acclimate to higher temperatures and so respiration rates decline when plants are exposed to higher temperatures for extended periods of time. Therefore, Waring *et al.* (1998) proposed that when large geographical and environmental ranges were considered, respiration approximated half of all photosynthetic production. The idea of a constant ratio between photosynthesis and respiration has been further investigated by Mäkelä and Valentine (2001), who concluded that the ratio of GPP to respiration should decline with tree size but that this decrease is possibly not very important. As a generality, we can conclude that under present climate conditions, plant respiration accounts for approximately half the GPP. However, it is important to remember that the ratio is not exact or constant and might be affected by climate change (see Chapter 37).

Species diversity and composition

Plant biomass production provides the 'fuel' for heterotrophs to survive. Understanding the consequences of declining diversity on plant production is becoming increasingly important given the high rate of species loss in tree-dominated ecosystems due to human disturbance. 'Natural' experiments in Mediterranean forests (Vilà *et al.* 2007), temperate forests (Caspersen and Pacala 2001), tropical forests (Peh 2009) and tropical tree plantations (Erskine *et al.* 2006) have provided evidence of a positive association between species richness and wood production at the landscape scale. It has been hypothesised that species diversity may influence net primary productivity in two ways: (a) areas with more species may have increased plant productivity caused by mechanisms such as resource-use complementarity and facilitation; or (b) production heavily depends upon the traits of individual species and their presence or absence.

Productivity can be affected by species composition whereby some species are more adapted to certain resource availability conditions and therefore grow at a faster rate in relation to their neighbours. In the case of northern hardwood forests, species such as *Quercus rubra* grows more quickly than *Quercus alba* under equivalent resources (Comas and Eissenstat 2004). Thus, the species composition of a stand can mediate growth response to resource availability. In addition, species composition would also affect resource levels both through resource use (Fujinuma *et al.* 2005) and litter cycling rates (Dijkstra 2003). For example, abundance of *Tilia americana* and *Acer saccharum* has been shown to be strongly correlated to calcium availability, while the latter also correlates with increased nitrate levels (Baribault *et al.* 2010). However, NH_4^+ does not correlate with species abundance in this northern hardwood forest.

Plant allocation

The resources which are obtained from the environment and produced by the plant must be allocated to a number of plant tissues and functions so that the resource uptake and use is balanced. Processes such as growth, reproduction and defence are all important and allocation between them is regulated by a number of factors, as discussed below. When considering allocation from a productivity perspective, it is necessary to consider your purpose. For example, for those in agriculture, productivity is measured in terms of the harvest index (HI) of a crop. In this case, allocation to reproduction is critical since the fruit is the harvest product. In contrast, when considering tree species, stem growth is often considered to be the most important allocation aspect since it is the site of wood production.

Productivity and allocation are intricately linked with growth being the product of NPP and the proportion of NPP the tree allocates to a given tissue. In reality, increasing yield in agriculture and forestry is more easily achieved by increasing allocation to the harvestable tissues, rather than by increasing NPP. For example, although agriculture has seen large increases in grain yield over the last 100 years, studies have shown that genetic gains in NPP in this time have been negligible. The yield increase has therefore been attained by changes in allocation within the plant (Richards 2000). In addition to allocation changes, it is also important to note that advances in agricultural management during this time, such as the use of fertilisers, has also aided this increase in primary productivity of wheat fields. It is not yet clear how productivity and allocation are related in woody species (Pulkkinen et al. 1989).

Physiologically, allocation is based on the transport of carbohydrates from the leaves to the sites of growth. While water is transported in the xylem down a water potential gradient which runs from the roots to the leaves, the translocation of carbohydrates in the phloem is more complicated. Translocation is an active process, which allows the plant to transport sugars and other solutes from sites of photosynthesis to tissues where growth is occurring. In translocation, solutes move from mature leaves, which are also known as 'sources', to 'sinks', which can be any other type of organ, such as roots, stems, young leaves or seeds. In contrast to transpiration in the xylem, translocation in the phloem can occur in any direction and can change depending on plant allocation. For example, early in the season translocation can primarily occur downwards from the leaves to the roots, before shifting later in the season when allocation is redirected to the flowers and seeds of the plant. Allocation is strongly influenced by hormonal regulation with auxins and gibberellins increasing shoot growth, and abscisic acid promoting allocation to roots. For a more detailed review of the physiology of translocation, see Taiz and Zeiger (2010).

In practice we do not yet have a good quantitative mechanistic understanding of how the translocation of carbohydrates and the related plant growth are regulated in a tree. The process of translocation is complex and quantitative analysis of it usually relies on mathematical models. Recent research has emphasised the importance of water relations for translocation (Nikinmaa et al. 2013), but these models have not yet been applied to the long term development of trees. A simplified approach has previously been implemented by Thornley (1991) with some promise. This approach modelled a tree as a series of inter-connected labile carbon and labile nitrogen reservoirs. Transport of carbon and nitrogen depends on the concentration differences of these elements between these reservoirs, and growth is proportional to the product of carbon and nitrogen concentrations. Models based on these extremely simplified versions of the transport and growth processes simulate a realistic development of forest stands but they have not yet been used extensively.

Allocation between fine roots and leaves

A functional balance is usually assumed to exist between fine root and leaf allocation. The origins of this concept can be traced back to Pearsall (1927) who recognised the close connection between shoot and root growth in plants. Later it was found that the root–shoot ratio depends on nutrient supply, with nitrogen being most important. As can be seen in Figure 25.5 the shoot fraction of birch trees increases when trees are approaching optimal nutrition (Ingestadt and Ågren 1991). From an evolutionary perspective, these shifts in allocation have been explained as an adaptation that maximises plant productivity under variable nutrient availability. Under high nutrient supply, a low root volume can take up sufficient nutrients to maintain foliar growth and photosynthetic production. In this scenario, a shift in allocation from fine roots to leaves allows trees to invest in vegetative growth, thereby achieving a higher net growth rate and the ability to compete successfully for above-ground resources, such as light. In contrast, under low nutrient levels, plants must shift allocation to root production, in order to acquire the nutrients which are necessary for basic plant maintenance and thus foliar investment is limited compared to that under high nutrient conditions. This is consistent with the 'mechanistic' Thornley model which is described above. Furthermore, optimal models of allocation give similar results to the functional balance of allocation between root and shoots (Mäkelä and Sievänen 1987). The role of nutrition as a determinant of the allocation between roots and shoots has been challenged by Coyle and Coleman (2005), who argue that the higher root–shoot ratio observed with fertiliser application in young *Populus deltoides* is an ontogenetic effect since they found root–shoot ratios depended on tree size but not on nutrient availability experimentally.

Interestingly, Malhi *et al.* (2011) found that – by analysing a global dataset of NPP allocation in tropical forests – the main variation in allocation among the different sites was a shifting allocation between wood and fine roots, while allocation to foliage remaining relatively constant. Instead, this observation implies a trade-off between allocation to fine roots and woods (Malhi *et al.* 2011). Also, there is evidence that tropical forests on fertile soils may in some cases have greater NPP and allocate more of that NPP to fine roots, contradicting the predictions of resource-allocation theory (Doughty *et al.* 2013).

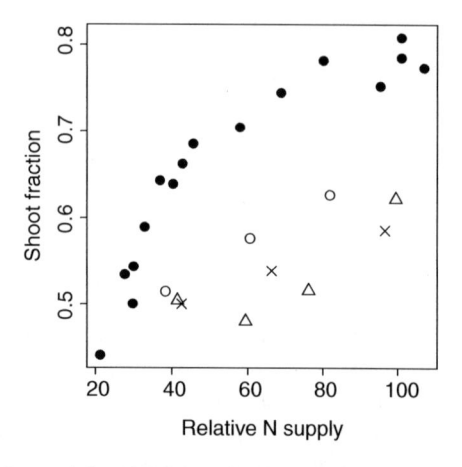

Figure 25.5 Shoot fraction for pendulous birch (*Betula pendula*) (closed circles), Scots pine (*Pinus sylvestris*) (triangles), Contorta pine (*Pinus contorta*) (open circles) and Norway spruce (*Picea abies*) (crosses) as a function of relative N supply. Redrawn from Ingestadt and Ågren (1991)

Drought

The effect of drought on root–shoot allocation is less well understood when compared to nutrient effects. Moderate droughts decrease the uptake of water and nutrients from the soil, which, according to the functional balance theory, should increase root allocation. However, low soil water potentials can increase fine root mortality and severe droughts limit root growth directly. Nevertheless, many experimental studies indicate that root–shoot ratios do increase under drought (Zhang *et al.* 2004). Allocation to fine roots along a precipitation gradient in European beech was described by Meier and Leuschner (2008). While fine root biomass did not differ between stands with different precipitation levels, it was observed that drier stands had thinner roots and a higher root turnover resulting in more resources being used below-ground.

Water supply is not spatially or temporally homogeneous and generally, even in arid regions, plants are able to employ different water use strategies which enable them to access sufficient water resources for life. Li *et al.* (2000) compared *Eucalyptus microtheca* F. Muell. from different regions under drought in a glasshouse experiment. They showed that the water use strategy of seedlings depended on annual rainfall distribution. In plants which originated from a climate with evenly distributed rainfall, the root–foliage ratio was not strongly affected by the drought treatment, while for seedlings originating from climates with a pronounced dry season, the effect of drought on root–foliage ratio was large. However, the root–foliage ratio was always higher in the drought treatment than in the control. Li *et al.* (2000) linked these changes in allocation to plant strategies, which allowed them to withstand periods of drought. Plants with a diffuse dry season but low average rainfall showed a conservative water use strategy with high allocation to roots. This strategy allows a steady growth rate when water is limiting and should be more successful when there is a low but consistent water supply (Li *et al.* 2000). Conversely, plants which originated from regions with a pronounced drought period generally exhibited an aggressive water use strategy with a high water use and comparatively low allocation to roots. This allows these plants to effectively exploit short pulses of water and to achieve a high growth rate when water is available.

This experimental evidence supports the theoretical simulation study of Schwinning and Ehleringer (2001). They used a simple model for desert plants and simulated optimal allocation, water use and rooting depth. Optimisation was then used to find two optimal evolutionary stable allocation strategies for plant growth and resource use. In the first strategy, plants grow deep root systems, have relatively low stomatal conductance and shift their allocation towards the roots. This allows a long lasting water supply at the initiation of drought and helps plants to maintain a positive carbon balance for a relatively long period of drought. For example, O'Grady *et al.* (1999) measured water fluxes in *Eucalyptus* trees in northern Australia and found low stomatal conductance and high transpiration rates – which indicates an ability to access water in the subsoil via deep roots – during the dry season. The second strategy is to grow shallow root systems that can harvest rainfall water efficiently. Water is rapidly consumed before it evaporates from the surface or percolates below-ground. This means that when water supply is high, root–shoot ratios are low and stomatal conductance is elevated. Stomatal conductance, water use and photosynthetic production decline rapidly in the latter strategy as the surface soil water store is exhausted. Indeed, some Mediterranean species like holm oak (*Quercus ilex* L.) have shallow root systems, high stomatal conductance and seem to follow the second water use strategy. However, the success of the strategy depends on how trees can survive long drought periods and mortality in Mediterranean trees seems to increase as the root–shoot ratio decreases (Padilla and Pugnaire 2007). Different plant adaptive traits such as drought-deciduousness, rooting depth or sclerophyllous leaves may improve drought tolerance or resistance. If plants avoid drought by shedding their leaves or persist during drought at a reduced level of physiological activity, plants will not necessarily change allocation significantly despite large productivity effects. This is due to allocation occurring primarily during the productive wet periods. Taken together, this shows that allocation strategies and drought tolerance and resistance are interlinked and that numerous strategies are viable.

In summary, allocation between fine roots and foliage is sensitive to the environment as we have discussed regarding nutrient and water availability. Generally, allocation to the roots will increase if the acquisition of resources from the root system is becoming limited.

Allocation to woody plant tissue

Although the distribution of biomass between physiologically active plant tissues, such as foliage and fine roots is a large research area as we have discussed above, woody plant tissue is also an important part of tree biomass. Since woody tree stems do not photosynthesise or take up significant quantities of water or nutrients they are mainly considered as support structures for foliage in the canopy. Since the late nineteenth century, two main theories for the importance of stems in trees have evolved. Pressler (1864) emphasised the role of the stem to provide mechanical support for the branches and leaves while Jaccard (1913) put forward the idea that the stem serves as a water pipeline from the roots to the foliage. Both theories mean that, in order to grow a unit of new foliage, trees must also provide 'sufficient infrastructure' in terms of stem, coarse roots and branches to mechanically and hydraulically support the foliage.

Analysis of allocation to the stem based on the mechanical theory of stem function originates from observations which were made on the stem taper of trees. Tree diameter decreases with height up the stem and this allows the amount of additional load, which would lead to the breakage of the stem, to be maintained at an approximately constant size (Domec and Gartner 2002). Also, it has been observed that during early development young stands maintain similar safety margins against stem breakage over time and with different stand densities (King 1990). Furthermore, tree diameter growth responds to mechanical stimuli indicating plasticity with respect to allocation towards woody plant tissues (Valinger 1992).

Hydraulic theories of stem form are more frequently used than mechanical theories in the modelling of plant allocation. Tree water transport is known to function near to the limit of operation and that, during dry and hot periods, small increases in transpiration could lead to catastrophic xylem failure. Therefore, it could be assumed that under drier or more evaporative conditions, a higher proportion of carbohydrates would be allocated to woody parts of the tree that conduct water from the roots to the foliage. The risk of cavitation and xylem failure could be increased by a longer hydraulic pathway and Yoder and colleagues (1994) proposed that water transport might effectively limit productivity and height growth in old trees. Mencuccini and Grace (1996) showed that indeed water transport from roots to the foliage in *Pinus sylvestris* (Scots pine) is more difficult in taller trees than in shorter trees. An influential theoretical equation that relates hydraulic conductance, climate, stomatal conductance and foliage mass was proposed by Whitehead *et al.* (1984). Here, the authors concluded that stomatal conductance, transport-resistance of wood and evaporative demand are closely interlinked, and that they are able to maintain consistent water potential in the leaves.

Most quantitative approaches addressing allocation in trees are based on the so called 'pipe model theory' formulated by Shinozaki *et al.* (1964). Shinozaki *et al.* (1964) proposed simply that foliage is connected to the roots by a set of pipes. This leads to a linear relationship between the foliage mass or foliage area and the cross-sectional area of the stem. Although not initially proposed by Shinozaki, the relationship was later interpreted as a hydraulic model of tree form, where foliage mass and sapwood area were related. According to the pipe model theory, the production of new foliage is assumed to require the production of sufficient support in terms of the pipe system in the branches and stem. Based on modelling studies (Berninger and Nikinmaa 1997), it seems that allocation to the stem increases with tree height since 'the length of the pipes' required to support the foliage is greater. Also, allocation to the stem is high when foliage biomass increases rapidly. Processes like sapwood turnover and height growth, as well as the ratio of foliage biomass to sapwood area become important determinants of tree growth and stand dynamics.

More recently West et al. (1997, 1999) tried to reconcile the mechanical and hydraulic theories of tree growth using geometric scaling laws and considered the scaling of tubes between organs. The model is derived from a consideration of optimal architecture and aims to explain both the size of branching network (i.e. trunks, branches and petioles) as well as vessel diameter in the wood. These models are well suited to explaining large scale differences between plants of very different sizes. However, the model does not necessarily work as well for small scale differences, such as differences between trees of similar size (Nygren and Pallardy 2008). Summarily, the approach of West et al. (1997, 1999), can be used to understand differences between trees of very different sizes and different ecosystems but may fail to adequately explain differences in allocation at finer scales.

What determines stem growth of trees: changes in allocation or changes in production?

Both tree production per unit leaf area and the amount of foliage increase as a response to a higher resource availability. A higher allocation of carbohydrate to the stem, because of, for example, decreased root growth or different pipe model ratios, may contribute to higher growth of stands under irrigation or fertilisation. The main drivers of changes in stemwood production in trees from sites with differing fertility are not yet certain but ideas are now emerging. Coll et al. (2011) used a process based model to analyse the response of *Populus* spp. (poplar) to nitrogen fertilisation and found that changes in photosynthetic capacity and increases in GPP dominate the response to nitrogen fertilisation. In conifers, Mäkelä (1997) focused on allocation responses and reported increases in stand productivity which occurred uniquely through increased foliage biomass when less carbon is invested in root growth.

In summary, it seems that primary production and allocation jointly determine the production of stemwood and other biomass in forests. While our understanding of photosynthetic production in forest ecosystems is advanced, much less is known about plant allocation. However, allocation is closely interlinked with production via various feedback cycles and it is therefore important to gain a better understanding of the long and short term dynamics of these processes.

References

Amthor, J. S. (2000) 'The McCree-de Wit-Penning de Vries-Thornley respiration paradigms: 30 years later', *Annals of Botany,* vol 86, pp. 1–20

Austin, A. T. and Sala, O. E. (2002) 'Carbon and nitrogen dynamics across a natural precipitation gradient in Patagonia, Argentina', *Journal of Vegetation Science*, vol 13, no 3, pp. 351–360

Baribault, T. W., Kobe, R. K. and Rothstein, D. E. (2010) 'Soil calcium, nitrogen, and water are correlated with aboveground net primary production in northern hardwood forests', *Forest Ecology and Management*, vol 260, pp. 723–733

Berninger, F. and Nikinmaa, E. (1997) 'Differences in pipe model parameters determine differences in growth and affect response of Scots pine to climate change', *Functional Ecology*, vol 11, pp. 146–156

Caspersen, J. P. and Pacala, S. W. (2001) 'Successional diversity and forest ecosystem function', *Ecological Research*, vol 16, pp. 895–903

Chaves, M. M., Maroco, J. P. and Pereira, J. S. (2003) 'Understanding plant responses to drought – from genes to the whole plant', *Functional Plant Biology*, vol 30, pp. 239–264

Chen, J. M., Govind, A., Sonnentag, O., Zhang, Y., Barr, A. and Amiro, B. (2006) 'Leaf area index measurements at Fluxnet-Canada forest sites', *Agricultural and Forest Meteorology*, vol 140, pp. 257–268

Coll, L., Schneider, R., Domenicano, S., Messier, C. and Berninger, F. (2011) 'Quantifying the effect of nitrogen induced physiological and structural changes on poplar growth using a carbon-balance model', *Tree Physiology*, vol 31, pp. 381–390

Comas, L. H. and Eissenstat, D. M. (2004) 'Linking fine root traits to maximum potential growth rate among 11 mature temperate tree species', *Functional Ecology*, vol 18, pp. 388–397

Coyle, D. R. and Coleman, M. D. (2005) 'Forest production responses to irrigation and fertilization are not explained by shifts in allocation', *Forest Ecology and Management*, vol 208, pp. 137–152

Dijkstra, F. A. (2003) 'Calcium mineralization in the forest floor and surface soil beneath different tree species in the northeastern US', *Forest Ecology and Management*, vol 175, pp. 185–194

Domec, J. C. and Gartner, B. (2002) 'Age- and position-related changes in hydraulic versus mechanical dysfunction of xylem: Inferring the design criteria for Douglas-fir wood structure', *Tree Physiology*, vol 22, pp. 91–104

Doughty, C. E., Metcalfe, D. B., da Costa, M. C., de Oliveira, A. A. R., Neto, G. F. C. and others (2013) 'The production, allocation and cycling of carbon in a forest on fertile terra preta soil in eastern Amazonia compared with a forest on adjacent infertile soil', *Plant Ecology and Diversity*, vol 7, nos 1–2, pp. 41–53

Erskine, P. D., Lamb, D. and Bristow, M. (2006) 'Tree species diversity and ecosystem function: can tropical multi-species plantations generate greater productivity?', *Forest Ecology and Management*, vol 233, pp. 205–210

Fujinuma, R., Bockheim, J. and Balster, N. (2005) 'Base-cation cycling by individual tree species in old-growth forests of Upper Michigan, USA', *Biogeochemistry*, vol 74, pp. 357–376

Gea-Izquierdo, G., Mäkelä, A., Margolis, H., Bergeron, Y., Black, A. T. and others (2010) Modelling acclimation of photosynthesis to temperature in evergreen boreal conifer forests. *New Phytologist*, vol 188, pp. 175–186

Gea-Izquierdo, G., Bergeron, Y., Huang, J. G., Lapointe-Garant, M. P., Grace, J. and Berninger, F. (2014) 'The relationship between productivity and tree-ring growth in boreal coniferous forests', *Boreal Environmental Research*, vol 19, 363–378.

Gradowski, T. and Thomas, S. C. (2008) 'Responses of *Acer saccharum* canopy trees and saplings to P, K and lime additions under high N deposition', *Tree Physiology*, vol 28, no 2, pp. 173–185

Hikosaka, K., Ishikawa, K., Borjigidai, A., Muller, O. and Onoda, Y. (2006) 'Temperature acclimation of photosynthesis: Mechanisms involved in the changes in temperature dependence of photosynthetic rate', *Journal of Experimental Botany*, vol 57, pp. 291–302

Ingestadt, T. and Ågren, G. I. (1991) 'The influence of plant nutrition on biomass allocation', *Ecological Applications*, vol 1, pp. 169–174

Jaccard, P. (1913) 'Eine neue Auffassung über die Ursachen des Dickenwachstums', *Naturwissentschaftliche Zeitung Forst- und Landwirtschaft*, vol 11, pp. 241–279

Juice, S. M., Fahey, T. J., Siccama, T. G., Driscoll, C. T., Denny, E. G. and others (2006) 'Response of sugar maple to calcium addition to northern hardwood forest', *Ecology*, vol 87, no 5, pp. 1267–1280

King, D. A. (1990) 'The adaptive significance of tree height', *American Naturalist*, vol 135, pp. 809–828

Kobe, R. K., Likens, G. E. and Eagar, C. (2002) 'Tree seedling growth and mortality responses to manipulations of calcium and aluminium in a northern hardwood forest', *Canadian Journal of Forest Research*, vol 32, no 6, pp. 954–966

Li, C., Berninger, F., Koskela, J. and Sonninen, E. (2000) 'Different origins of *Eucalyptus microtheca* differ in their ability to acclimate to drought', *Australian Journal of Plant Physiology*, vol 27, pp. 231–238

Long, S. P., Postl, W. F. and Bolhár-Nordenkampf, H. R. (1993) 'Quantum yields for uptake of carbon dioxide in C3 vascular plants of contrasting habitats and taxonomic groupings', *Planta*, vol 189, pp. 226–234

Malhi, Y., Baldocchi, D. D. and Jarvis, P. G. (1999) 'The carbon balance of tropical, temperate and boreal forests', *Plant, Cell and Environment*, vol 22, pp. 715–740

Malhi, Y., Doughty, C. and Galbraith, D. (2011) 'The allocation of ecosystem net primary productivity in tropical forests', *Philosophical Transactions of the Royal Society B*, vol 366, pp. 3225–3245

Mäkelä, A. (1997) 'A carbon balance model of growth and self-pruning in trees based on structural relationships', *Forest Science*, vol 43, no. 1, pp. 7–24

Mäkelä, A. and Sievänen, R. (1987) 'Comparison of two shoot-root partitioning models with respect to substrate utilization and functional balance', *Annals of Botany*, vol 59, pp. 129–140

Mäkelä, A. and Valentine, H. T. (2001) 'The ratio of NPP to GPP: Evidence of change over the course of stand development', *Tree Physiology*, vol 21, pp. 1015–1030

Mäkelä, A., Hari, P., Berninger, F., Hänninen, H., Nikinmaa, E. (2004) 'Acclimation of photosynthetic capacity in Scots pine to the annual cycle of temperature', *Tree Physiology*, vol 24, pp. 369–376

McLaughlin, S. B. and Wimmer, R. (1999) 'Tansley review no. 104 calcium physiology and terrestrial ecosystem processes', *New Phytologist*, vol 142, no 3, pp. 373–417

Meier, I. C. and Leuschner, C. (2008) 'Below-ground drought response of European beech: Fine root biomass and carbon partitioning in 14 mature stands across a precipitation gradient', *Global Change Biology*, vol 14, pp. 2081–2095

Mencuccini, M. and Grace, J. (1996) 'Developmental patterns of above-ground hydraulic conductance in a scots pine (*Pinus sylvestris* L.) age sequence', *Plant, Cell and Environment*, vol 19, pp. 939–948

Monteith, J. L. (1994) 'Validity of the correlation between intercepted radiation and biomass', *Agricultural and Forest Meteorology*, vol 68, nos 3–4, pp. 213–220

Nikinmaa, E., Hölttä, T., Hari, P., Kolari, P., Mäkelä, A. and others (2013) 'Assimilate transport in phloem sets conditions for leaf gas exchange', *Plant, Cell and Environment*, vol 36, pp. 655–669

Nygren, P. and Pallardy, S. G. (2008) 'Applying a universal scaling model to vascular allometry in a single-stemmed, monopodially branching deciduous tree (Attim's model)', *Tree Physiology*, vol 28, pp. 1–10

O'Grady, A. P., Eamus, D. and Hutley, L. B. (1999) 'Transpiration increases during the dry season: Patterns of tree water use in eucalypt open-forests of northern Australia', *Tree Physiology*, vol 19, pp. 591–597.

Padilla, F. M. and Pugnaire, F. I. (2007) 'Rooting depth and soil moisture control Mediterranean woody seedling survival during drought', *Functional Ecology*, vol 21, pp. 489–495

Pearsall, W. H. (1927) 'Growth Studies VI, On the relative size of organs in plants', *Annals of Botany*, vol 41, pp. 549–556

Peh, K. S.-H. (2009) 'The relationship between species diversity and ecosystem function in low- and high-diversity tropical African forest', PhD thesis, University of Leeds, UK

Pressler, M. R. (1864) 'Das Gesetz der Stammbildung. Arnoldische Buchhandlung', Leipzig, no 153

Pulkkinen, P., Pöykkö, T., Tigerstedt, P. M. A. and Velling, P. (1989) 'Harvest index in northern temperate cultivated conifers', *Tree Physiology*, vol 5, pp. 83–98

Quesada, C. A., Phillips, O. L., Schwarz, M., Czimczik, C. I., Baker, T. R. and others (2012) 'Basin-wide variations in Amazon forest structure and function are mediated by both soils and climate', *Biogeosciences*, vol 9, pp. 2203–2246

Reich, P. B., Kloeppel, B. D., Ellsworth, D. S. and Waiters, M. S. (1995) 'Different photosynthesis-nitrogen relations in deciduous hardwood and evergreen coniferous tree species', *Oecologia*, vol 104, pp. 24–30

Richards, R.A. (2000) 'Selectable traits to increase crop photosynthesis and yield of grain crops', *Journal of Experimental Botany*, vol 51, pp. 447–458

Rocha, A. V., Goulden, M. L., Dunn, A. L. and Wofsy, S. C. (2006) 'On linking interannual tree ring variability with observations of whole-forest CO_2 flux', *Global Change Biology*, vol 12, no 8, pp. 1378–1389

Running, S. W. (2012) 'A measurable planetary boundary for the biosphere', *Science*, vol 337, no 6101, pp. 1458–1459

Running, S. W., Nemani, R. R., Heinsch, F. A., Zhao, M., Reeves, M. and Hashimoto, H. (2004) 'A continuous satellite-derived measure of global terrestrial production', *BioScience*, vol 54, pp. 547–560

Sala, O. E., Gherardi, L. A., Reichmann, L., Jobbágy, E. and Peters, D. (2012) 'Legacies of precipitation fluctuations on primary production: Theory and data synthesis', *Philosophical Transactions of the Royal Society of Biological Sciences*, vol 367, pp. 3135–3144

Schwinning, S. and Ehleringer, J. D. (2001) 'Water use trade-offs and optimal adaptations to pulse-driven arid ecosystems', *Journal of Ecology*, vol 89, pp. 464–480

Shinozaki, K., Yoda, K., Hozumi, K. and Kira, T. (1964) 'A quantitative analysis of plant form–the pipe model theory. I. Basic analyse', *Japanese Journal of Ecology*, vol 14, pp. 133–139

Suni, T., Berninger, F., Vesala, T., Markkanen, T., Hari, P. and others (2003) 'Air temperature triggers the commencement of evergreen boreal forest photosynthesis in spring', *Global Change Biology*, vol 9, no 10, pp. 1410–1426

Taiz, L. and Zeiger, E. (2010) *Plant Physiology*, Sinauer Associates, Sunderland, MA, USA

Thornley, J. H. M. (1991) 'A transport-resistance model of forest growth and partitioning', *Annals of Botany*, vol 68, pp. 211–226

Turner, D. P., Ritts, W. D., Cohen, W. B., Maeirsperger, T. K., Gower, S. T. and others (2005) 'Site-level evaluation of satellite-based global terrestrial gross primary production and net primary production monitoring', *Global Change Biology*, vol 11, pp. 666–684

Valinger, E. (1992) 'Effects of wind sway on stem form and crown development of Scots pine (*Pinus sylvestris* L.)', *Australian Forestry*, vol 55, pp. 15–21

Vapaavuori, E. M., Vuorinen, A. H., Aphalo, P. J. and Smolander, H. (1995) 'Relationship between net photosynthesis and nitrogen in Scots pine: Seasonal variation in seedlings and shoots', *Plant and Soil*, vols 168–169, pp. 263–270

Vilà, M., Vayreda, J., Comas, L., Ibáñez, J. J., Mata, T. and Obón, B. (2007) 'Species richness and wood production: A positive association in Mediterranean forests', *Ecology Letters*, vol 10, pp. 241–250

Waring, R. H., Landsberg, J. J. and Williams, M. (1998) 'Net primary production of forests: A constant fraction of gross primary production?', *Tree Physiology*, vol 18, no 2, pp. 129–134

Warren Wilson, J. (1967) 'Ecological data on dry-matter production by plants and plant communities' in E. F. Bradley and O. T. Denmead (eds) *The Collection and Processing of Field Data*, John Wiley & Sons, New York

West, G. B., Brown, J. H. and Enquist, B. J. (1997) 'A general model for the origin of allometric scaling laws in biology', *Science*, vol 276, pp. 122–126

West, G. B., Brown, J. H. and Enquist, B. J. (1999) 'A general model for the structure and allometry of plant vascular systems', *Nature*, vol 400, pp. 664–667

Whitehead, D., Edwards, W. R. N. and Jarvis, P. G. (1984) 'Conducting sapwood area, foliage area, and permeability in mature trees of *Picea sitchensis* and *Pinus contorta*', *Canadian Journal of Forest Research*, vol 14, pp. 940–947

Yoder, B. J., Ryan, M. G., Waring, R. H., Schoettle, A. W. and Kaufmann, M. R. (1994) 'Evidence of reduced photosynthetic rates in old trees', *Forest Science*, vol 40, no 3, pp. 513–527

Zhang, Y., Zang, R. and Li, C. (2004) 'Population differences in physiological and morphological adaptations of *Populus davidiana* seedlings in response to progressive drought stress', *Plant Science*, vol 166, pp. 791–797

PART V

Forest conservation and management

26
NATURAL REGENERATION AFTER HARVESTING

Nelson Thiffault, Lluís Coll and Douglass F. Jacobs

Introduction

Natural regeneration of forest stands refers to the replacement of mature individual trees that have died or have been removed, by recruits of sexual or vegetative origin, many of which will eventually form the new canopy. Disturbances, either natural or anthropogenic, are events that modify elements of the stand structure and affect resources availability by killing (or removing) vegetation and releasing growing space that becomes available for other plants and individuals to occupy (Oliver and Larson 1996). Depending on their severity, frequency, spatial pattern and duration, disturbances associated with harvesting interact with the stand characteristics to influence the ecological niches that become available for new individuals. Thus, the natural regeneration of forest stands is intimately linked with the disturbances affecting the stands.

In the context of forest management, *natural* regeneration is distinguished from *artificial* regeneration, which implies planting or direct seeding to establish the new stand. In many forestry contexts, natural regeneration is the most economical means to re-establish desirable species following harvesting activities (Greene *et al.* 2002). However, the outcomes of natural regeneration are less predictable than those of artificial regeneration as they are driven by complex interactions between many factors including seed and propagule availability, management history, stand and soil characteristics, light, water and nutrient levels, species autecology, plant–plant interactions, predation and browsing. When natural regeneration is successful, stands can be overstocked and must be thinned to minimize loss in productivity.

In this chapter, we summarize the fundamental concepts related to natural forest regeneration dynamics after harvesting. We explore how harvesting disturbs the ecosystem and how it affects and interacts with stand characteristics, soil and environmental properties to influence regeneration success. We then exemplify these concepts by examining the impacts of harvesting and its interaction with key ecosystem properties on natural forest regeneration of three contrasting ecosystems, namely the boreal forest, the temperate deciduous forest, and the dry Mediterranean ecosystems.

Harvesting as a disturbance

Forest harvesting can serve many purposes as a tool for silviculture. The main objective is usually to extract commercially valuable wood products and alter the characteristics and density of a forest stand. Timber harvesting can also serve to alter the structure of the tree component of the ecosystem, develop specific visual characteristics in a stand, provide a particular kind of cover for wildlife, or promote or dampen various understory responses (Nyland 2002).

Such a large array of potential objectives implies that harvesting approaches can vary greatly in terms of their severity, spatial scale, frequency, and hence, their impacts and interactions with key ecosystems characteristics and functions. Figure 26.1 illustrates how various harvesting regeneration methods are distributed along gradients of relative severity, clearing size, and time between disturbances. Also represented along the severity gradient is the relative level of available light, as a proportion of full radiation, along with the expected seed supply remaining in the standing trees after harvesting. Figure 26.1 also includes examples of natural disturbances along the severity scale, which can be considered as being of low, moderate, or high severity.

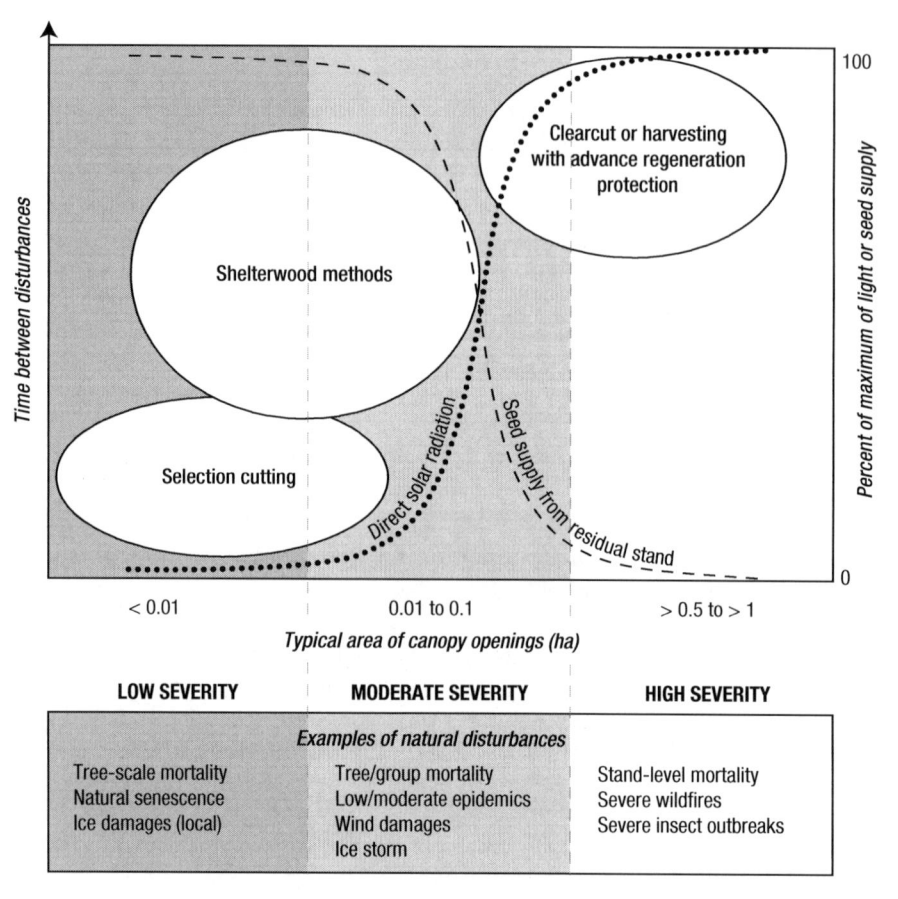

Figure 26.1 Conceptual representation of regeneration methods within a gradient of severity, clearing size, and frequency. Adapted from Raymond et al. (2013), Coates and Burton (1997) and Smith (1962). Severity is defined by the degree to which the canopy is removed, microclimate and soils are altered, the extent of organic layer consumption or removal and the degree to which living organisms are killed (Haeussler and Kneeshaw 2003)

For example, harvesting activities conducted in the context of selection cutting are generally characterized by frequent (every 10–15 years), low intensity cuttings, in which a small proportion of the total stand volume is harvested. The characteristics of the trees that are harvested and the volume extracted depend on the initial properties of the stands and the specific objectives of the cut. Globally, such a treatment mimics, to a certain point, small gap natural mortality that could be caused by individual tree senescence. This harvesting approach is generally realized through the creation of small gaps in which advance regeneration is already established. Partial mechanical soil scarification is sometimes used to favour the germination of buried or incoming seeds. Light levels in the understory remain low, as the canopy is never fully harvested. At the other end of the severity spectrum, clearcut harvesting mimics stand-level mortality that can be caused by high severity, infrequent disturbances such as wildfires or major insect outbreaks. In clearcutting systems the canopy is completely removed, which drastically changes the environmental conditions compared to pre-cut conditions. By contrast, fire or windstorms leave behind boles that cast shade and thus moderate the tendency for disturbance to create hotter, drier conditions near the ground. Complete removal of the canopy typically favours the establishment of light-demanding species that are either wind-dispersed (from adjacent stands) or have persistent seed banks. Variants of the clearcutting method protect advance regeneration by restricting machinery operations to specific skidding trails, and thus are somewhat like windstorms or ice storms where most damage is limited to the crown.

Depending on its severity, harvesting can have a limited or dramatic effect on the availability of key environmental resources, which in turn impact seed germination, germinant survival, as well as the growth and survival of any advance regeneration (see next section). The main impacts of canopy removal include increased light levels in the understory, changes in the seedbeds (e.g. reduced broadleaf litter accumulation), and changes in soil temperature and moisture. For example, light levels measured at 0.8 m were 27.5 per cent, 17.5 per cent and 7.5 per cent of full sun following high, medium, and low intensity harvesting, respectively, in northern hardwood forests of central Ontario, Canada (Domke *et al.* 2007). In addition, soil temperature at 10 cm was increased by 2.4°C between 6 and 22 months after harvesting an east Texas bottomland hardwood forest in the United States (Londo *et al.* 1999). Soil moisture can also be affected by harvesting: for example, Moroni *et al.* (2009) reported that the surface layer of harvested areas tended to dry faster than uncut stands in the boreal forest in Newfoundland in eastern Canada, with evidence of soil drying as deep as 3.5 cm. Operation of machinery during harvesting and skidding also disturbs the humus, and can alter the bulk density, as well as the temperature/moisture regime of the upper mineral soil (Dickerson 1976). Any soil disturbances that remove the bulk of the fibrous material from the humus layer almost invariably have a positive effect on seedbed quality, as most species of commercial interest have small seeds, with the necessarily small germinants arising from those seeds needing access to a low-porosity upper layer such as humus or mineral soil for a high probability of successful establishment (Greene and Johnson 1998). For example, removal of the organic layer has proved beneficial to yellow birch (*Betula alleghaniensis*) germination in boreal mixedwood in eastern Canada (Elie *et al.* 2009) and to Spanish black pine (*Pinus nigra*) in the Mediterranean forest (Lucas-Borja *et al.* 2012). By removing a portion or the entire canopy, harvesting also influences the aerial seed bank, that is, seed contained in fruits or ovulate cones on standing trees (see next section).

The process of regeneration

Natural regeneration of forest stands after harvesting depends on the availability of an abundant and viable seed supply or perennating organs at the time of harvest, as well as the presence of an adequate substrate for the establishment of seeds after harvest (Kozlowski 2002). Figure 26.2 summarizes this process. The original stand is generally the main supplier of seeds or propagules, although in some forests certain species may have a long-lived soil seed bank; further, a large fraction of the seeds can arrive from non-harvested trees at the edges of the cut if it is sufficiently small that dispersal is not greatly constrained (Greene and Johnson 1995). As an example of a soil seed bank, abundant red maple (*Acer rubrum*) seedlings recruit from persistent seed banks in temperate forests (Hille Ris Lambers and Clark 2005). The dispersal capacity of seeds moved by the wind is generally inversely proportional to seed size; for animal dispersal of seeds, there is no clear relationship with propagule mass. Further, small animals will tend to avoid recent clearings; for example, the relatively big (>1 g) fruits of walnut (*Juglans*) are generally dispersed by rodents no farther than 50 m from the seed tree (Hewitt and Kellman 2002) because these caching animals tend to avoid open spaces. On the other hand, birch species (*Betula* spp.) produce small (around 1 mg) seeds that can travel large distances as they are blown by the wind along the surface of the snow (Houle 1998). Some boreal species, such as black spruce (*Picea mariana*) or jack pine (*Pinus banksiana*), accumulate aerial seed banks that remain on the branches and serve as a reliable seed source until a natural disturbance occurs. Importantly, for this process to happen, harvesting should maintain the closed cones on site rather than allow them to accumulate at a landing. Finally, tree regeneration can occur from vegetative reproduction through stump sprouting, root suckering, rhizomes, or layering from parent trees. This capacity

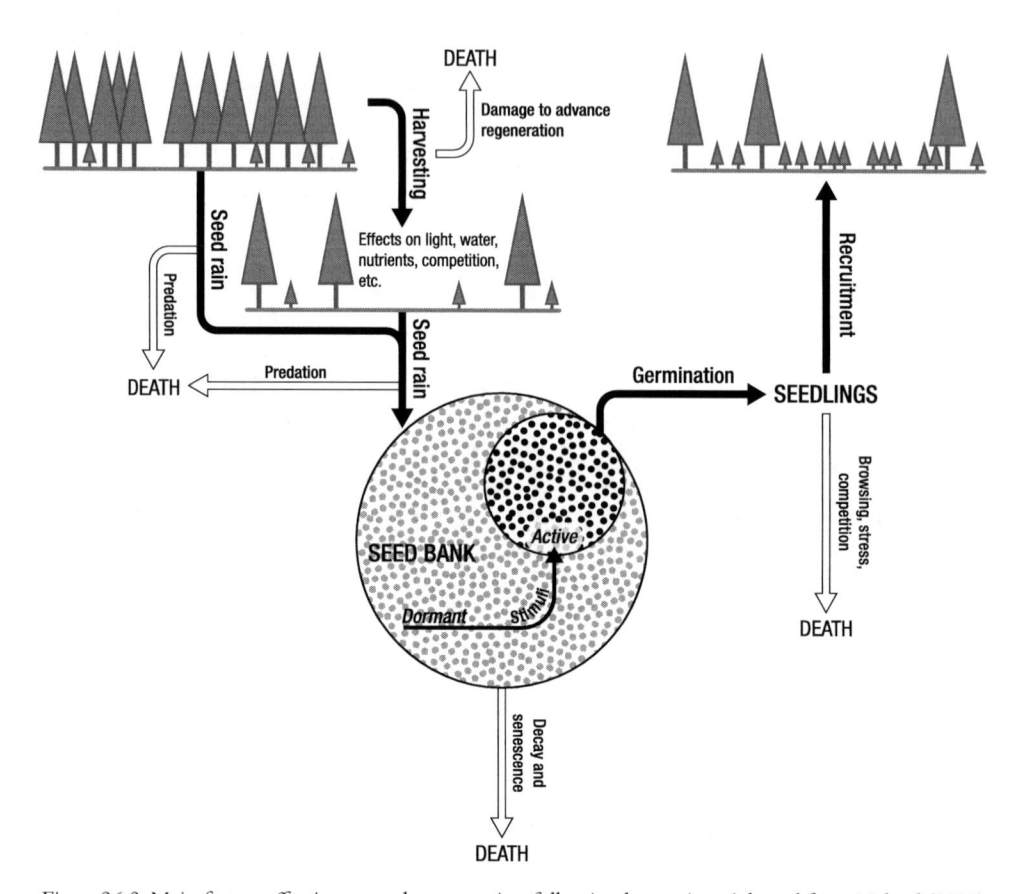

Figure 26.2 Main factors affecting natural regeneration following harvesting. Adapted from Nyland (2002)

is far more common among angiosperms (e.g. *Populus tremuloides*); within the conifers, the commercially valuable *Pinaceae* have very few examples of species with the ability to sprout.

Seed and vegetative sources can be depleted by a number of abiotic and biotic factors such as predation, desiccation, flooding or diseases, occurring either before or after harvest. Once dispersed, germination is influenced by habitat filters (such as seedbed characteristics) and biotic filters (such as above ground biomass of dominant plants). The harvesting technique that is used can promote or hinder those filters, depending on its severity and timing (Figure 26.1). For asexual recruits, the ratio of auxin to cytokinin is a key factor in promoting the development of shoots from dormant buds. Moreover, asexual stems invariably grow more rapidly than sexually derived stems, as they can rely on the starch stored in root systems prior to harvest.

The survival and growth of the newly established seedlings is dependent upon environmental conditions found in the harvested area (light availability, air and soil temperature and moisture, soil nutrient availability, soil compaction, etc.), inter- and intra-competitive interactions, facilitation interactions and other biotic factors such as insect attacks, diseases, or browsing (Kozlowski 2002). Certain species are more sensitive than others to these factors, which influences regeneration dynamics. For example, Box 26.1 illustrates how preferential browsing can affect forest succession to a point where alternative succession trajectories are induced. The severity of the cut has a direct influence on these factors. Surviving seedlings exhibiting sufficient growth are then recruited in the regenerating cohort.

The following sections illustrate how the general process of natural regeneration following harvesting takes place in contrasting ecosystems. Although the mechanisms are mainly exemplified using North American and European ecosystems, the principles generally apply to other forests and species. Transposition to other contexts and ecosystems can be made using, for example, key functional traits characterizing the vegetative and regeneration stages of the tree life cycle (Kattge *et al.* 2011).

Natural regeneration after harvesting in boreal ecosystems

Boreal ecosystems share a subarctic climate that constrains growing conditions by impacting precipitation, air temperature, soils, and the length of the growing season (among other factors). For example, the boreal forest of North America is characterized by a cold, continental climate with severe winters, short growing seasons, cold soils, slow decomposition rates, and, thus, strong limitations to plant productivity. It is dominated by relatively few tree species, mostly conifers, and is subject to a cycle of natural disturbances driven primarily by insect outbreaks and wildfires (Burton *et al.* 2003). The understory of nutrient-poor, moist to dry sites of boreal forest stands is mainly composed of mosses (e.g. *Pleurozium* and *Ptillium*), lichens (e.g. *Cladina*), and ericaceous shrubs (e.g. *Kalmia*, *Rhododendron* and *Vaccinium*).

The two dominant conifer species of the northern American boreal forest (*Picea mariana*, *Pinus banksiana*) possess long-lived aerial seed banks contained in serotinous cones; high-intensity fires or increased heat at soil surface are required for the cones of some species to open and seed dispersal to occur (Greene *et al.* 1999). *In situ* aerial seed banks can also ensure the recolonization of sites for non-serotinous species, such as *Picea glauca* (Michaletz *et al.* 2013). Regeneration through natural seeding generally occurs within five years following harvesting in these northern ecosystems as the seedbeds soon deteriorate due to leaf litter accumulation. Black spruce also reproduces from vegetative growth through the formation of layers around trees, a capacity described long ago that is shared by many conifer species in America and elsewhere (Cooper 1911). But, this can only occur for recruits long after the harvest when aggrading moss encourages adventitious rooting. On the other hand, advance black spruce regeneration layers take five to seven years to adapt to the new light conditions created by canopy removal; a delay is thus observed before increased growth of the already-present asexual stems. A high percent of live crown and the absence of stem wounds on advance regeneration are good predictors of their future survival in these ecosystems (Ruel *et al.* 1995).

The dominant harvesting approach used in boreal stands, in North America and elsewhere, is clearcutting or variants of this treatment that protect advance regeneration (Figure 26.3). In the context of ecosystem-based management, which promotes the use of close-to-nature approaches, clearcutting is intended to emulate the impacts of the dominant natural disturbances that drive these ecosystems. Indeed, as fire does, clearcut harvesting removes most of the canopy trees and creates open sky conditions with high light levels (Figure 26.1). However, clearcut harvesting does not necessarily reproduce all the effects that fire has on the thick organic layer typical of boreal stands; it leaves an intact layer of undisturbed thick moss, the worst possible organic seedbed on upland sites (Greene and Johnson 1998). Moreover, it can promote cover of competing species, which may necessitate mechanical soil preparation to expose patches of mineral soil and humus, and create appropriate seedbeds that reduce the cover of competitors. For example, by exposing mineral soil after harvesting, mechanical scarification increased black spruce stocking from natural seeding compared with clearcut only, five years following the treatment (Prévost 1997). Similarly, Nilsson *et al.* (2006) have reported enhanced density and height of naturally regenerated Scots pine (*Pinus sylvestris*) following scarification in Sweden, compared to control conditions.

The southern part of the North American boreal forest is characterized by a few more tree species not present in its northern part. With the potential exception of paper birch (*Betula*

Figure 26.3 Careful logging around advance growth restricts circulation of harvesting, forwarding and skidding machinery to evenly spaced, parallel trails and thus protects existing forest regeneration between the trails in northern Québec (Canada). Photo courtesy of the Ministère des Forêts, de la Faune et des Parcs du Québec, Canada

> **Box 26.1 Heavy browsing as a filter to stand regeneration after harvesting in boreal ecosystems – the case of Anticosti Island (Québec, Canada)**
>
> High densities of large herbivores can have strong top-down effects on tree regeneration in some ecosystems. Anticosti Island, located in the Gulf of St. Lawrence, is a ~8,000 km^2 territory naturally dominated by balsam fir, white spruce, black spruce and paper birch. White-tailed deer were introduced 100 years ago for sport hunting. Since then, low hunting harvest rates and the absence of large predators have favoured the development of unusually high densities of deer. As balsam fir is preferably browsed by deer, compared to white spruce, this browsing pressure has led to a drastic reduction in the representation of balsam fir-dominated stands on the island (Potvin *et al.* 2003). Forest managers face significant challenges as they implement restoration scenarios based on innovative forest harvesting techniques, enrichment planting, and sport hunting that could avoid the alternative succession trajectories forced by browsing on this ecosystem (Hidding *et al.* 2013). Similar cases have been documented in other ecosystems around the world.

papyrifera), the dominant tree species do not have seed dormancy within the soil extending beyond one year (Greene *et al.* 1999). Balsam fir (*Abies balsamea*), often a dominant conifer species in this ecosystem, establishes dense understory seedling banks that can survive for decades under low light conditions and positively react to the opening of the canopy (Parent and Ruel 2002). Similarly, Norway spruce (*Picea abies*), a dominant conifer of the Scandinavian boreal forest, depends largely on seedling banks that can survive in an extremely stunted condition for many years, retaining the ability to respond to canopy opening (Jonsson and Esseen 1990).

Trembling aspen (*Populus tremuloides*), one of the few northern hardwood species, reproduces through aggressive root suckering when above ground portions of the tree are removed or killed (Frey *et al.* 2003). Clearcut harvesting is also the preferred treatment in these ecosystems. However, clearcut approaches generally do not leave large amounts of coarse woody debris on sites, as would insect outbreaks or wildfires. The lack of coarse woody debris can affect the ability of some species to regenerate. For example, dead wood is a preferential germination substrate for white spruce in intact forests; repeated cycles of clearcut harvests tend to limit the ability of this species to regenerate. In some ecosystems, harvesting must thus be followed by enrichment planting to compensate for the low stocking and density of this species.

In certain contexts, boreal forest stands dominated by shade tolerant species can also be harvested and naturally regenerated using shelterwood approaches, usually with a combined treatment that exposes the mineral substrate to favour seed germination (Raymond *et al.* 2000). For example, regular shelterwood cutting removes part of the canopy and leaves selected trees to act as seed sources. The small increase in understory light levels can be sufficient to promote germination and establish new regeneration, or stimulate the growth of advance regeneration. Moreover, the limited light levels due to the presence of seed trees can control the growth of fast growing, opportunistic species that would otherwise compete with the preferred species.

Natural regeneration after harvesting in temperate deciduous ecosystems

Temperate deciduous forests occupy large areas in America, Asia and Europe. They are also found in the southern hemisphere, specifically in South America, Africa, and Australia. Temperate deciduous ecosystems are characterized by seasonal shedding of leaves of the dominant species, which impacts ecosystem processes, including regeneration. As they are typically associated with fertile soil conditions, deciduous forests have historically been subjected to extensive logging activities as well as land clearing for agriculture and urbanization (Bergmeier et al. 2010). As an example of one temperate deciduous forest, the Central Hardwood Forest Region of North America comprises more than 89 million ha, with about 50 per cent of this area in forests (Johnson et al. 2009). These are mainly secondary forests that established from the late nineteenth to middle twentieth century due to land abandonment driven by competition with new farms in the prairies (Parker 1989). Because these forests have been exposed to intensive anthropogenic disturbances including extensive land clearing through clearcutting, fire and grazing, they are dominated largely by shade-intolerant species such as oaks (*Quercus* spp.) and yellow poplar (*Liriodendron tulipifera*). In recent decades, however, the trend for intensive disturbance in these forests systems has been replaced with management favouring fire suppression and selective cutting usually by means of single-tree selection (Abrams 2003). Simultaneously, the region has experienced dramatic increases in white-tailed deer populations (*Odocoileus virginianus*), which have led to greater herbivory pressures and selective browsing of certain species (McEwan et al. 2011) that ultimately favours browse-tolerant species (Horsley et al. 2003; and see Box 26.1 for an example in a boreal ecosystem). This has resulted in a shift toward a very different forest composition, characterized by decreased regeneration of oaks and an increase in the importance of maples (*Acer* spp.) (Abrams 2003; Fei et al. 2011).

The shift in the natural dominance of temperate forests is an important issue worldwide, as naturally regenerating stands are considered of low quality and biodiversity (Dudley et al. 2005). For example, decline of natural oak regeneration in north-eastern North America has been a source of concern for forest managers for decades because oaks are economically important in this region for timber and also comprise critical components of wildlife habitat, aesthetics, and culture (Johnson et al. 2009). Even-aged silviculture (i.e. via clearcutting, see Figure 26.1) has been recommended as the most reliable means of establishing oak regeneration in this region (Roach and Gingrich 1968). However, success of natural oak regeneration under even-aged regeneration harvesting systems depends principally on the presence of advanced regeneration or likelihood of stump sprouting; regeneration from seed contributes only nominally to the oak regeneration pool (Morrissey et al. 2008). Regeneration is further dependent on site characteristics; higher quality sites are commonly dominated by sugar maple (*Acer saccharum*), red maple, yellow poplar, and white ash (*Fraxinus americana*), while oak species compete best on poorer sites (Johnson et al. 2009; Morrissey et al. 2008). For example, Morrissey et al. (2008) reported regeneration responses on 70 clearcut sites, ranging in age from 21–35 years, in southern Indiana, United States. They found that oak regeneration success was mainly dependent on site edaphic characteristics as well as the pre-harvest abundance of oaks. Nearly half of the dominant oak stems originated from stump sprouts. While oak abundance declined greatly during the early stages of regeneration (i.e. stand initiation), dominant stems of oak increased during the subsequent stem exclusion stage of stand development. This was largely attributed to the relatively greater drought tolerance of oak, which helped to enhance competitiveness against associated species during several consecutive extreme drought years (Figure 26.4). The authors concluded that to ensure oak regeneration success under even-aged regeneration systems, intermittent silvicultural manipulation (i.e. selective thinning of less desirable competitors) would be necessary.

Figure 26.4 Regeneration development during the stem exclusion stage following clearcutting in southern Indiana (USA), showing an oak and yellow poplar stem (a). While the oak stem was alive, the yellow poplar exhibited canker associated with exposure to drought (b). Photos: R. C. Morrissey

Natural regeneration after harvesting in dry ecosystems

Particular challenges arise in natural regeneration after harvesting of dry ecosystems. These forests are usually exposed to climates with highly seasonal rainfall. The abiotic conditions under such a climatic pattern strongly impact tree phenology, seed production and germination, as well as seedling establishment and survival (Ray and Brown 1994; Swaine 1992). For example, the northern Mediterranean basin is characterized by a climate with a pronounced bi-seasonality due to the migration of the subtropical high pressure system leading to dry, hot summers, and relatively moist and cool autumns and winters. Forests are generally characterized by a complex structure and composition resulting from the inherent environmental and topographic heterogeneity of the Mediterranean region and its very long history of human alteration (Nocentini and Coll 2013). This produces a number of forest types organized along elevational and drought gradients each with its particular constraints regarding natural regeneration.

The Thermo- and Meso-Mediterranean forests (at the lower altitudes) are dominated by thermophilous *Pinus* species (e.g. *Pinus pinaster*, *P. halepensis*, *P. brutia*) and evergreen sclerophyllous oak species (*Quercus suber*, *Q. ilex*, *Q. rotundifolia*, *Q. coccifera*). These species share a high drought tolerance and have developed a number of mechanisms to adapt and rapidly respond to the occurrence of large disturbances such as fire or overgrazing (e.g. serotinous cones, resprouting capacity, long-lived seed banks). Although there is little information about how these species react to different harvesting methods (Rodríguez-García et al. 2010), in general Thermo- and Meso-Mediterranean pines are known to require high light regimes to develop (Thanos 2000) and thus they need large openings to adequately regenerate, a trait consistent with the disturbance regimes characteristic of their distributional area.

In the case of Aleppo pine (*Pinus halepensis*), the most widespread coniferous species in the Mediterranean area, adequate natural regeneration in fire-free systems is not common. This is mainly due to a number of factors acting at the microsite level (e.g. seed predation, competition) that cause complex spatiotemporal patterns (Nathan and Ne'eman 2004). However, recent experiences with this species indicate that recruitment could be enhanced by applying soil and ground vegetation treatments after the cuttings such as mechanical soil scarification or the high-intensity burning of slash, which reduces both the soil seed bank and perennating tissues of the competing herbaceous species (Prévosto and Ripert 2008). Post-harvest natural regeneration of Maritime pine (*Pinus pinaster*) has indeed been found to be highly dependent on the seedbed characteristics (texture, litter) through the effects they have on the water-holding capacity of soils (Rodríguez-García et al. 2011). In addition to soil and microsite characteristics, natural regeneration success of these pine species growing in dry environments may undoubtedly depend on post-harvest micro-climate conditions and in particular on the amount of spring and summer rainfall reaching the soil (Rodríguez-García et al. 2010).

Evergreen sclerophyllous oak species in the Mediterranean area have traditionally been managed as coppice systems consisting of total or partial removal of the standing biomass of trees. These actions favour asexual regeneration of the species by stump or root sprouting (Retana et al. 1992). In the last decades, basal coppicing is applied less frequently due to the progressive decrease in use of firewood and charcoal as energy source. This has led to an increased density and growth stagnation of the stands with some of them already showing signs of decay (Cañellas et al. 2004). Converting coppice stands into more conventional forest systems (with regeneration mainly originating from seeds) by progressively cutting the weakest stems while leaving the best ones for acorn production has been proposed as an alternative. However, this process can be seriously constrained by the high pre- and post-dispersal seed predation that oak species suffer in the area (Espelta et al. 2009). In addition, the predicted increasing drought associated with global warming is expected to reduce acorn production and seedling survival (Sánchez-Humanes and Espelta 2011), which adds a further difficulty to the successful establishment of the seedlings under the coppices. Finally, there is a recruitment bottleneck occurring with evergreen oaks at sapling stage: seedlings need a certain degree of cover to establish (to be protected from high evaporative demand) but then require increasing light levels to reach larger sizes (Espelta et al. 1995).

With increasing elevation, vegetation communities change from the Thermo- and Meso-Mediterranean to Supra- and Montane-Mediterranean. These forests are mainly composed of conifers such as *Pinus nigra* or *P. sylvestris* and deciduous species (e.g. *Quercus faginea*, *Q. pubescens*). In general post-harvest natural regeneration of these species does require a certain degree of protection from increased evaporation (and associated moisture stress) to survive. This is one reason (in addition to easy implementation) behind the wide use of shelterwood systems or progressive cuttings in the region (Montes and Cañellas 2007). In pine-dominated forests, group selection cuttings could also be envisaged in stands tending to form uneven age structures (Piqué et al. 2001) (Figure 26.5). Finally, soil preparation could be needed in some cases to reduce herbaceous competition and to break thick organic layers (del Cerro Barja et al. 2009). Caution should be taken not to apply these treatments at the driest periods of the year (which would cause desiccation) but when the availability of seeds is at their highest, typically at the start of the winter rains (Barbeito et al. 2011).

Figure 26.5 Experimental group selection cuttings of 0.2 and 0.4 ha applied in a Supra-Mediterranean Scots pine stand near Solsona, Lleida (Spain). Photo: Santiago Martín Alcón

Conclusion

Forest ecosystems are dynamic. Their species composition and distribution, vertical and horizontal spatial patterns, leaf area, volume and other physical characteristics change over time, as a result of the inherent growth of the individual plants that compose them and of their interactions (Oliver and Larson 1996). Energy and nutrient fluxes within the ecosystems are changing as well. The natural regeneration of forest stands after harvesting is part of this dynamic process. Harvesting is a perturbation of the ecosystem that triggers the onset of the secondary succession. Depending on its severity and the characteristics of the ecosystems in which it is applied (including the macro-scale climatic and biogeographic context), harvesting influences the direction of the succession by affecting seed and propagule sources, the germination and survival substrate, and the level of environmental resources. The three examples presented here (boreal, temperate deciduous, and European dry ecosystems) illustrate how particular abiotic or biotic factors can influence forest regeneration. The success of natural regeneration after harvesting depends on complex interactions between these factors.

Acknowledgements

We thank Dr Patricia Raymond for insightful discussions regarding the gradient of harvesting severity illustrated in Figure 26.1. Mario Beltrán and Pau Vericat provided helpful comments on the section devoted to Mediterranean forests. We are indebted to two anonymous reviewers for commenting on an earlier version of this chapter.

References

Abrams M. D. (2003) 'Where has all the white oak gone?', *Bioscience*, vol 53, no 10, pp. 927–938

Barbeito I., LeMay V., Calama R. and Cañellas I. (2011) 'Regeneration of Mediterranean *Pinus sylvestris* under two alternative shelterwood systems within a multiscale framework', *Canadian Journal of Forest Research*, vol 41, no 2, pp. 341–351

Bergmeier E., Petermann J. and Schröder E. (2010) 'Geobotanical survey of wood-pasture habitats in Europe: diversity, threats and conservation', *Biodiversity and Conservation*, vol 19, no 11, pp. 2995–3014

Burton P. J., Messier C., Weetman G. F., Prepas E. E., Adamowicz W. L. and Tittler R. (2003) 'The current state and boreal forestry and the drive for change', in P. J. Burton, C. Messier, D. W. Smith and W. L. Adamowicz (eds) *Towards Sustainable Management of the Boreal Forest*, NRC Research Press, Ottawa, ON

Cañellas I., Del Rio M., Roig S. and Montero G. (2004) 'Growth response to thinning in *Quercus pyrenaica* Willd. coppice stands in Spanish central mountain', *Annals of Forest Science*, vol 61, no 3, pp. 243–250

Coates K. D. and Burton P. J. (1997) 'A gap-based approach for development of silvicultural systems to address ecosystem management objectives', *Forest Ecology and Management*, vol 99, no 3, pp. 337–354

Cooper W. S. (1911) 'Reproduction by layering among conifers', *Botanical Gazette*, vol 52, no 5, pp. 369–379

del Cerro Barja A., Lucas Borja M. E., Martínez García E., López Serrano F. R., Andrés Abellán M., Garcia Morote F. A. and Navarro López R. (2009) 'Influence of stand density and soil treatment on the Spanish Black Pine (*Pinus nigra* Arn. ssp. Salzmannii) regeneration in Spain', *INIA: Sistemas y Recursos Forestales*, vol 18, no 2, pp. 167–180

Dickerson B. P. (1976) 'Soil compaction after tree-length skidding in northern Mississippi', *Soil Science Society of America Journal*, vol 40, no 6, pp. 965–966

Domke G. M., Caspersen J. P. and Jones T. A. (2007) 'Light attenuation following selection harvesting in northern hardwood forests', *Forest Ecology and Management*, vol 239, nos 1–3, pp. 182–190

Dudley N., Mansourian S. and Vallauri D. (2005) 'Forest landscape restoration in context', in S. Mansourian, D. Vallauri and N. Dudley (eds) *Forest Restoration in Landscapes: Beyond Planting Trees*, Springer, New York, NY

Elie J.-G., Ruel J.-C. and Lussier J.-M. (2009) 'Effect of browsing, seedbed, and competition on the development of yellow birch seedlings in high-graded stands', *Northern Journal of Applied Forestry*, vol 26, no 3, pp. 99–105

Espelta J. M., Riba M. and Retana J. (1995) 'Patterns of seedling recruitment in West-Mediterranean *Quercus ilex* forests influenced by canopy development', *Journal of Vegetation Science*, vol 6, no 4, pp. 465–472

Espelta J. M., Cortés P., Molowny-Horas R. and Retana J. (2009) 'Acorn crop size and pre-dispersal predation determine inter-specific differences in the recruitment of co-occurring oaks', *Oecologia*, vol 161, no 3, pp. 559–568

Fei S., Kong N., Steiner K. C., Moser W. K. and Steiner E. B. (2011) 'Change in oak abundance in the eastern United States from 1980 to 2008', *Forest Ecology and Management*, vol 262, no 8, pp. 1370–1377

Frey B. R., Lieffers V. J., Landhäusser S. M., Comeau P. G. and Greenway K. J. (2003) 'An analysis of sucker regeneration of trembling aspen', *Canadian Journal of Forest Research*, vol 33, no 7, pp. 1169–1179

Greene D. F. and Johnson E. A. (1995) 'Long-distance wind dispersal of tree seeds', *Canadian Journal of Forest Research*, vol 73, no 7, pp. 1036–1045

Greene D. F. and Johnson E. A. (1998) 'Seed mass and early survivorship of tree species in upland clearings and shelterwoods', *Canadian Journal of Forest Research*, vol 28, no 9, pp. 1307–1316

Greene D. F., Zasada J. C., Sirois L., Kneeshaw D. D., Morin H. and others (1999) 'A review of the regeneration dynamics of North American boreal forest tree species', *Canadian Journal of Forest Research*, vol 29, no 6, pp. 824–839

Greene D. F., Kneeshaw D. D., Messier C., Lieffers V. J., Cormier D. and others. (2002) 'Modelling silvicultural alternatives for conifer regeneration in boreal mixedwood stands (aspen/white spruce/balsam fir)', *The Forestry Chronicle*, vol 78, no 2, pp. 281–295

Haeussler S. and Kneeshaw D. D. (2003) 'Comparing forest management to natural processes', in P. J. Burton, C. Messier, D. W. Smith and W. L. Adamowicz (eds) *Towards Sustainable Management of the Boreal Forest*, NRC Research Press, Ottawa, ON

Hewitt N. and Kellman M. (2002) 'Tree seed dispersal among forest fragments: II. Dispersal abilities and biogeographical controls', *Journal of Biogeography*, vol 29, no 3, pp. 351–363

Hidding B., Tremblay J.-P. and Côté S. D. (2013) 'A large herbivore triggers alternative successional trajectories in the boreal forest', *Ecology*, vol 94, no 12, pp. 2852–2860

Hille Ris Lambers J. and Clark J. S. (2005) 'The benefits of seed banking for red maple (*Acer rubrum*): maximizing seedling recruitment', *Canadian Journal of Forest Research*, vol 35, no 4, pp. 806–813

Horsley S. B., Stout, S. L. and deCalesta D. S. (2003) 'White-tailed deer impact on the vegetation dynamics of a northern hardwood forest', *Ecological Applications*, vol 13, no 1, pp. 98–118

Houle G. (1998) 'Seed dispersal and seedling recruitment of *Betula alleghaniensis*: spatial inconsistency in time', *Ecology*, vol 79, no 3, pp. 807–818

Johnson P. S., Shifley S. R. and Rogers R. (2009) *The Ecology and Silviculture of Oaks. Second Edition.* CABI Publishing, CAB International, Wallingford, UK

Jonsson B. G. and Esseen P.-A. (1990) 'Treefall disturbance maintains high bryophyte diversity in a boreal spruce forest', *Journal of Ecology*, vol 78, no 4, pp. 924–936

Kattge J., Diaz S., Lavorel S., Prentice I. C., Leadley P. and others (2011) 'TRY - a global database of plant traits', *Global Change Biology*, vol 17, no 9, pp. 2905–2935

Kozlowski T. T. (2002) 'Physiological ecology of natural regeneration of harvested and disturbed forest stands: implications for forest management', *Forest Ecology and Management*, vol 158, nos 1–3, pp. 195–221

Londo A. J., Messina M. G. and Schoenholtz S. H. (1999) 'Forest harvesting effects on soil temperature, moisture, and respiration in a bottomland hardwood forest', *Soil Science Society of America Journal*, vol 63, no 3, pp. 637–644

Lucas-Borja M. E., Fonseca T. F., Lousada J. L., Silva-Santos P., Garcia E. M. and Andrés Abellán M. A. (2012) 'Natural regeneration of Spanish black pine [*Pinus nigra* Arn. ssp. *salzmannii* (Dunal) Franco] at contrasting altitudes in a Mediterranean mountain area', *Ecological Research*, vol 27, no 5, pp. 913–921

McEwan R. W., Dyer J. M. and Pederson N. (2011) 'Multiple interacting ecosystem drivers: Toward an encompassing hypothesis of oak forest dynamics across eastern North America', *Ecography*, vol 34, no 2, pp. 244–265

Michaletz S. T., Johnson E. A., Mell W. E. and Greene D. F. (2013) 'Timing of fire relative to seed development may enable non-serotinous species to recolonize from the aerial seed banks of fire-killed trees', *Biogeosciences*, vol 10, no 7, pp. 5061–5078

Montes F. and Cañellas I. (2007) 'Analysis of the spatial relationship between the remaining trees from the previous crop and the establishment and development of saplings in *Pinus sylvestris* L. stands in Spain', *Applied Vegetation Science*, vol 10, no 2, pp. 151–160

Moroni M. T., Carter P. Q. and Ryan D. A. J. (2009) 'Harvesting and slash piling affect soil respiration, soil temperature, and soil moisture regimes in Newfoundland boreal forests', *Canadian Journal of Soil Science*, vol 89, no 3, pp. 343–355

Morrissey R. C., Jacobs D. F., Seifert J. R., Fischer B. C. and Kershaw J. A. (2008) 'Competitive success of natural oak regeneration in clearcuts during the stem exclusion stage', *Canadian Journal of Forest Research*, vol 38, no 6, pp. 1419–1430

Nathan R. and Ne'eman G. (2004) 'Spatiotemporal dynamics of recruitment in Aleppo pine (*Pinus halepensis* Miller)', *Plant Ecology*, vol 171, nos 1–2, pp. 123–127

Nilsson U., Örlander G. and Karlsson M. (2006) 'Establishing mixed forests in Sweden by combining planting and natural regeneration—Effects of shelterwoods and scarification', *Forest Ecology and Management*, vol 237, nos 1–3, pp. 301–311

Nocentini S. and Coll L. (2013) 'Mediterranean forests: human use and complex adaptive systems', in C. Messier, K. J. Puettmann and K. D. Coates (eds) *Managing Forests as Complex Adaptive Systems. Building Resilience to the Challenge of Global Change*, Routledge, New York, NY

Nyland R. D. (2002) *Silviculture: Concepts and Applications. Second Edition*, Waveland Press, Inc., Long Grove, IL

Oliver C. D. and Larson B. C. (1996) *Forest Stand Dynamics. Updated Edition*, John Wiley & Sons, Inc., New York, NY, USA

Parent S. and Ruel J.-C. (2002) 'Chronologie de la croissance chez des semis de sapin baumier (*Abies balsamea* (L.) Mill.) après une coupe à blanc avec protection de la régénération', *The Forestry Chronicle*, vol 78, no 6, pp. 876–885

Parker G. R. (1989) 'Old-growth forests of the central hardwood region', *Natural Areas Journal*, vol 9, no 1, pp. 5–11

Piqué M., Beltrán M., Vericat P., Cervera T., Farriol R. and Baiges T. (2001) *Models de gestió per als boscos de pi roig (Pinus sylvestris L.): producció de fusta i prevenció d'incendis forestals. Sèrie: Orientacions de gestió forestal sostenible per a Catalunya (ORGEST)*. Centre de la Propietat Forestal. Departament d'Agricultura, Ramaderia, Pesca, Alimentació i Medi Natural. Generalitat de Catalunya, Spain

Potvin F., Beaupré P. and Laprise G. (2003) 'The eradication of balsam fir stands by white-tailed deer on Anticosti Island, Québec: A 150-year process', *Écoscience*, vol 10, no 4, pp. 487–495

Prévost M. (1997) 'Effects of scarification on seedbed coverage and natural regeneration after a group seed-tree cutting in a black spruce (*Picea mariana*) stand', *Forest Ecology and Management*, vol 94, pp. 219–231

Prévosto B. and Ripert C. (2008) 'Regeneration of *Pinus halepensis* stands after partial cutting in southern France: Impacts of different ground vegetation, soil and logging slash treatments', *Forest Ecology and Management*, vol 256, no 12, pp. 2058–2064

Ray G. J. and Brown B. J. (1994) 'Seed ecology of woody species in a Caribbean dry forest', *Restoration Ecology*, vol 2, no 3, pp. 156–163

Raymond P., Ruel J.-C. and Pineau M. (2000) 'Effet d'une coupe d'ensemencement et du milieu de germination sur la régénération des sapinières boréales riches de seconde venue du Québec', *The Forestry Chronicle*, vol 76, no 4, pp. 643–652

Raymond P., Guillemette F. and Larouche C. (2013) 'Les grands types de couvert et les groupements d'essences principales', in C. Larouche, F. Guillemette, P. Raymond and J.-P. Saucier (eds) *Le guide sylvicole du Québec - Tome 2. Les concepts et l'application de la sylviculture*, Les Publications du Québec, Québec, QC, Canada

Retana J., Riba M., Castell C. and Espelta J. M. (1992) 'Regeneration by sprouting of holm-oak (*Quercus ilex*) stands exploited by selection thinning', *Vegetatio*, vols 99–100, no 1, pp. 355–364

Roach B. A. and Gingrich S. F. (1968) *Even-Aged Silviculture for Upland Central Hardwoods*. USDA Forest Service, Upper Darby, PA, USA

Rodríguez-García E., Juez L. and Bravo F. (2010) 'Environmental influences on post-harvest natural regeneration of *Pinus pinaster* Ait. in Mediterranean forest stands submitted to seed-tree selection method', *European Journal of Forest Research*, vol 129, no 6, pp. 1119–1128

Rodríguez-García E., Gratzer G. and Bravo F. (2011) 'Climatic variability and other site factors influences on natural regeneration of *Pinus pinaster* Ait. in Mediterranean forests', *Annals of Forest Science*, vol 68, no 4, pp. 811–823

Ruel J.-C., Doucet R. and Boily J. (1995) 'Mortality of balsam fir and black spruce advance growth 3 years after clear-cutting', *Canadian Journal of Forest Research*, vol 25, no 9, pp. 1528–1537

Sánchez-Humanes B. and Espelta J. M. (2011) 'Increased drought reduces acorn production in *Quercus ilex* coppices: thinning mitigates this effect but only in the short term', *Forestry*, vol 84, no 1, pp. 73–82

Smith D. M. (1962) *The Practice of Silviculture. Seventh Edition*, John Wiley & Sons, Inc., New York, NY

Swaine M. D. (1992) 'Characteristics of dry forest in West Africa and the influence of fire', *Journal of Vegetation Science*, vol 3, no 3, pp. 365–374

Thanos C. A. (2000) 'Ecophysiology of seed germination in *Pinus halepensis* and *P. brutia*', in G. Ne'eman and L. Trabaud (eds) *Ecology, Biogeography and Management of Pinus halepensis and P. brutia Forest Ecosystems in the Mediterranean Basin*, Backhuys Publishers, Leiden, The Netherlands

27
TROPICAL DEFORESTATION, FOREST DEGRADATION AND REDD+

John A. Parrotta

Introduction: Tropical forest loss and its climate change implications

Forests today cover an estimated 31 percent of Earth's land surface (4.03 billion hectares), of which 93 percent are natural forest and 7 percent are planted (FAO, 2010). They contain a substantial proportion of the world's terrestrial biodiversity (SCBD, 2010) and play a major role in the global carbon cycle, removing carbon dioxide (CO_2) from the atmosphere (carbon sequestration) and storing carbon for extended periods of time in biomass, dead organic matter and soil carbon pools. Of the global forest carbon stocks (including soils to 1 m depth), an estimated 55 percent (471 Pg C) is stored in tropical and subtropical forests, of which more than half is stored in biomass (Pan *et al.*, 2011). Between 1990 and 2007, forests globally were estimated to have contributed a net sink of 1.1 Pg C yr^{-1}. The absorption of atmospheric carbon by intact forests in tropical and subtropical regions was 1.2 Pg C yr^{-1}, but this was offset by net emissions of 1.3 Pg C yr^{-1} resulting from land-use changes, i.e., deforestation and clearing emissions minus storage in regrowth (secondary and planted) forests (Pan *et al.*, 2011).

Deforestation and forest degradation – the result of, for example, agricultural expansion and conversion to pastureland, infrastructure development, destructive logging, and fires – account for approximately 12–22 percent of global greenhouse gas emissions, more than the entire global transportation sector and second only to the energy sector (van der Werf *et al.*, 2009; Pan *et al.*, 2011; Le Quéré *et al.*, 2013). In recent years the potential to mitigate climate change caused by increased concentrations of atmospheric CO_2 and other greenhouse gases (principally from the burning of fossil fuels) by enhancing the carbon sequestration by forests and forested landscapes has attracted increasing interest worldwide.

What is REDD+?

REDD+ is a mechanism for climate change mitigation developed in the United Nations Framework Convention on Climate Change (UNFCCC) and being implemented through a growing number of international organizations, institutions and governments. REDD+ aims to reduce CO_2 and other greenhouse gas emissions from deforestation and forest degradation, and enhance forest carbon stocks in 'developing' countries (principally in tropical and subtropical regions) by providing financial and other incentives to governments, landholders and/or communities for managing their forest lands towards this end.

The topic of reducing emissions from deforestation and forest degradation was first introduced in the 2005 meeting of the UNFCCC's Conference of the Parties, and further elaborated two years later in the UNFCCC's 'Bali Action Plan.' In 2010, UNFCCC Conference of the Parties in Cancún, Mexico, reached an agreement on policy approaches and positive incentives for reducing greenhouse gas emissions from forests. The Cancún decision on REDD+ specifically encourages developing countries to pursue climate change mitigation actions in the forest sector by: (a) reducing emissions from deforestation; (b) reducing emissions from forest degradation; (c) conservation of forest carbon stocks; (d) sustainable management of forests; (e) enhancement of forest carbon stocks. The Cancún decision also specified that these five REDD+ activities should, among other things: be country-driven; consistent with the objective of environmental integrity and take into account the multiple functions of forests and other ecosystems; be implemented in the context of sustainable development and reducing poverty; be consistent with the climate change adaptation needs of the country; be results-based; and promote sustainable management of forests. The Warsaw REDD+ agreements (2013) finalized the details for funding, implementation and monitoring of projects under the UN-REDD program.

The development of REDD+ has raised hopes and expectations in many quarters, particularly among those who see the potential for significant environmental and socio-economic 'co-benefits.' These anticipated co-benefits include conservation of forest biodiversity, water regulation, soil conservation, timber, forest foods and other non-timber forest products, and direct social benefits – jobs, livelihoods, land tenure clarification, carbon payments, enhanced participation in decision making and improved governance. Critics of REDD+ emphasize the lack of clarity regarding the eventual architecture of the international REDD+ regime and the international financial mechanisms that will underpin it, the environmental and social risks and inequity associated with various aspects of REDD+ policy development, planning and implementation (e.g., issues of sovereignty, risk of 'land grabs'), the high likelihood of 'leakage' (displacement of deforestation and forest degradation from a REDD+ project area to another location), and long-standing difficulties in addressing the underlying causes of deforestation and forest degradation.

The following sections examine deforestation and forest degradation in tropical and subtropical regions of the world where REDD+ activities are under way or may be carried out in the future. The proximal and underlying causes of forest loss and degradation are discussed, as well as their impacts on the provision of ecosystem services[1] with a particular focus on forest carbon sequestration (explored in depth in Thompson et al., 2012). Forest and landscape management activities that may be implemented in countries to meet REDD+ objectives will be considered, as will the likely environmental impacts and their social and economic impacts (both positive and negative) on people most likely to be affected by REDD-related changes in agricultural and forest management policies, incentives and governance structures.

---

1 Forest ecosystem services include *supporting services* such as nutrient cycling, soil formation and primary productivity; *provisioning services* such as food, water, timber and medicine; *regulating services* such as erosion control, climate regulation, flood mitigation, purification of water and air, pollination and pest and disease control; and *cultural services* such as recreation, ecotourism, educational and spiritual values (MA, 2005)

Deforestation and forest degradation in tropical and subtropical regions

Deforestation

Deforestation is the conversion of forest land to another land use, generally the result of transformation of forested lands to other land uses that are maintained by a continued human-induced or natural perturbation. Deforestation includes areas of forest converted to agriculture, pasture, water reservoirs and urban areas, but excludes areas where trees have been removed as a result of harvesting or logging, where the forest is expected to regenerate naturally or with the aid of silvicultural measures. Deforestation also includes areas where, for example, the impact of disturbance, overuse or changing environmental conditions affects the forest to an extent that it cannot sustain a tree cover.

At the global level, deforestation has been estimated at between 13 to 16 million hectares (Mha) per year between 1990 and 2010 (FAO, 2010). However, as a result of large-scale forest planting efforts, natural expansion of forests, and successes in slowing deforestation rates in some countries, the net global loss in forest area has slowed from 8.3 Mha (from 1990 to 2000) to 5.2 Mha (an area the size of Costa Rica) from 2000 to 2010 (FAO, 2010).

Rates of deforestation are particularly high in tropical regions, with an estimated net forest loss of 8.0 Mha yr^{-1} between 2000 and 2005 (FAO, 2011). South America and Africa continue to have the largest net loss of forest. Forest area in Central America was estimated as almost the same in 2010 as in 2000. Asia, which had a net loss in the 1990s, reported a net gain of forest in the period 2000–2010, primarily due to the large-scale afforestation reported by China and despite continued high rates of loss in many countries in South and Southeast Asia. In both Brazil and Indonesia, which had the highest net losses of forest in the 1990s, rates of forest loss have been significantly reduced in recent years, while in Australia, severe drought and forest fires since 2000 have contributed significantly to loss of forest cover.

Although recent deforestation rates have fallen in some countries, continued pressure on forests would suggest that rates of forest loss in tropical and subtropical countries are likely to remain high in the foreseeable future (e.g., Rudel *et al.*, 2009; FAO, 2011). Recent human impacts across global forest ecosystems have not been equal, with some forest types now under severe threat. Both tropical and subtropical dry and montane forests have been converted to a large extent because they are located in climates highly suitable for agriculture and grazing. Mangrove forest area declined by 19 percent from 1980 to 2005 (FAO, 2007) due to land clearing, aquaculture, changes to hydrological regimes and coastal development. Extensive areas of freshwater swamp and peat forests in Southeast Asia, which store vast amounts of carbon in their soils, have been lost or severely degraded in recent decades by unsustainable logging and agricultural expansion, including oil palm plantations.

Agricultural expansion has been the most important direct cause of global forest loss, accounting for 80 percent of deforestation worldwide, with the majority occurring during the 1980s and 1990s through conversion of tropical forests (Gibbs *et al.*, 2010). The conversion of forest lands for commercial agriculture and pasture has been responsible for approximately two-thirds of deforestation in Latin America, while in Africa and tropical and subtropical Asia forest clearing for subsistence farming is the major driver of land-use change (Kissinger *et al.*, 2012).

The underlying drivers of forest loss – which cannot be ignored – include rapid population growth, increasing global natural resource consumption, and the often over-riding effects of economic globalization and global land scarcity. These are exacerbated by problems of weak governance, inadequate policies, lack of cross-sectoral coordination, perverse incentives and illegal activities (Kissinger *et al.*, 2012).

Forest degradation

Forest degradation may be broadly defined as a reduction in the capacity of a forest to produce ecosystem services, such as carbon storage and wood products as a result of anthropogenic and environmental changes. Forest degradation is widespread and has become an important consideration in global policy processes that deal with biodiversity, climate change, and forest management. The International Tropical Timber Organization estimated that up to 850 Mha of tropical forest could already be characterized as degraded (ITTO, 2002).

Forest may be degraded from several perspectives, depending on the cause, the particular goods or services of interest, and the temporal and spatial scales considered. Forest degradation must therefore be measured against a desired baseline condition, and the types of degradation can be represented using five criteria that relate to the drivers of degradation, loss of ecosystem services and sustainable management, including: productivity, biodiversity, unusual disturbances, protective functions, and carbon storage (Thompson et al., 2013).

The proximate drivers of forest degradation include unsustainable logging, over-harvest of fuelwood and non-timber forest products (NTFPs), over-grazing, human-induced fires (or fire suppression in dry forests) and poor management of shifting cultivation (Chazdon, 2008; Kissinger et al., 2012). For example, unsustainable timber extraction accounts for more than 70 percent of tropical forest degradation in Latin America and Asia (Kissinger et al., 2012). Unsustainable logging has resulted in forests being degraded by removal of high-value trees, the collateral damage associated with timber extraction, and subsequent burning and clearing (Asner et al., 2006). Fuelwood collection and charcoal production, which account for an estimated 40 percent of global removal of wood from forests (FAO, 2006), as well as forest grazing, are major causes of forest degradation, particularly across Africa (Kissinger et al., 2012).

Although fire is a natural element in many forest ecosystems, humans have altered fire regimes across an estimated 60 percent of terrestrial habitats (Shlisky et al., 2009). Fires have spread in increasing extent and frequency across tropical rainforests with the expansion of agriculture, forest fragmentation, unsustainable shifting cultivation and logging.

In addition to human-induced and disturbance-related forest degradation, climate change poses an additional and growing threat to global forest ecosystems, in particular through an increase in the frequency of severe droughts that can result in a long-term reduction of forest cover. Tropical regions that appear particularly vulnerable to warming and drought include Central America, Southeastern Amazonia and West Africa.

Deforestation and forest degradation can act synergistically. Deforestation fragments forest landscapes, which often results in degradation of remaining forests due to edge effects (e.g., drying of the forest floor, increased fire frequency, increased tree mortality and shifts in tree species composition). Poorly planned logging activities increase road access to remaining forest interiors, further facilitating shifting cultivation and other land clearing, hunting, illegal logging, blowdown and fire (Griscom et al., 2009).

Degradation can lead to subsequent deforestation. In the Brazilian Amazon basin, for example, Asner et al. (2006) estimated that 16 percent of unsustainably logged areas were deforested during the following year, and 32 percent in the following three years. Degraded forests can often remain in a degraded state for long periods of time if degradation drivers (e.g., fire, human and livestock pressures) remain, or if ecological thresholds have been passed beyond which forests cannot recover their structure and composition through natural successional processes.

Impacts of deforestation and forest degradation on carbon sequestration and storage

CO_2 emissions associated with forest conversion

Tropical and subtropical forests store an estimated 247 Gt C in biomass (both above and below ground) (Saatchi *et al.*, 2011). When forests are converted to croplands, often through burning, a large portion of carbon stored in above-ground vegetation is immediately released to the atmosphere as CO_2 (and other greenhouse gases), or over time through the decomposition of debris (Figure 27.1). Carbon in soils following deforestation can also become a large source of emissions because of increased soil respiration with warmer ambient temperatures. There is increased soil loss with higher flooding and erosion rates, with carbon being transported downstream where a large fraction of the decayed organic matter is released as CO_2.

CO_2 emissions associated with forest degradation

Carbon emissions from forest degradation are difficult to assess because of a lack of consistent data. Forest degradation is often pooled with deforestation to estimate emissions from land-use change (e.g., Houghton, 2003), or is estimated as less than 10 percent of tropical carbon emissions (e.g., Nabuurs *et al.*, 2007). Emissions from degradation, however, are likely to be more substantial (Lambin *et al.*, 2003).

Poor logging practices create large canopy openings and cause collateral damage to remaining trees, sub-canopy vegetation and soils (Asner *et al.*, 2006). During timber harvest, a substantial

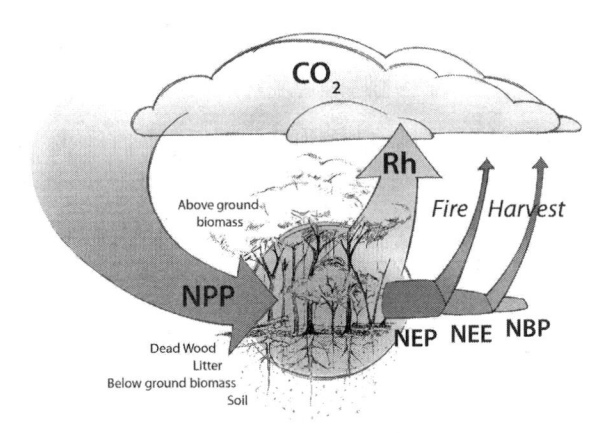

Figure 27.1 The major carbon fluxes in forest ecosystems. Net Primary Production (NPP) quantifies the amount of organic matter produced annually. Most of this carbon uptake is offset through losses from the decomposition of litter, dead wood and soil C pools (Rh = heterotrophic respiration). The net balance (Net Ecosystem Production, NEP) is further reduced through direct fire emissions to yield Net Ecosystem Exchange (NEE), from which harvest losses are subtracted to estimate the annual C stock change in forest ecosystems (Net Biome Production, NBP). Positive NBP indicates increasing forest carbon stocks, a sink from the atmosphere, while negative NBP indicates a carbon source. NEE is reported from the perspective of the atmosphere and has the opposite sign convention

Source: Figure by Avril Goodall (Canadian Forest Service), reprinted from Thompson *et al.* (2012), in Parrotta *et al.* (2012)

portion of biomass carbon (approximately 50 percent) can be left as logging residues, and about 20 percent of harvested wood biomass is further lost in the process of manufacturing wood products (Pan *et al.*, 2011). There is a continuing loss of carbon from decomposition or combustion of wood products from the significant proportion of harvested wood that is used for fuelwood and paper production.

CO_2 emissions associated with forest fires

The impact of forest fires on carbon emissions is particularly significant (van der Werf *et al.*, 2009). In extreme drought years, carbon emissions from tropical forest fires can exceed those from deforestation (Houghton, 2003). For example, total estimated carbon emissions from tropical forest fires during the 1997–1998 El Niño event were 0.83 to 2.8 Pg C yr^{-1} (Cochrane, 2003).

In recent decades, the frequency and size of forest fires have increased in many (sub-)tropical regions, including areas where fires have not been known to occur commonly, and where forest ecosystems are not well adapted to fire effects. Such fires are often associated with deforestation and land-use practices (Cochrane, 2003). Fire frequency may be intensified in forests that have been degraded by logging or previously burned, because these areas become more flammable and fire is more likely with human encroachment. Even low to medium severity fires in undisturbed or lightly degraded intact forest can kill over 50 percent of all trees (Barlow *et al.*, 2003). Trees in tropical humid forests are particularly susceptible to fire damage because fires are historically rare.

Shifting cultivation ('slash-and-burn agriculture'), practiced throughout the tropics and subtropics, also contributes to overall greenhouse gas emissions. However, it should be noted that carbon sequestration by secondary forest regeneration during the fallow phase may offset a significant proportion of these losses.

REDD+ activities and their potential for climate change mitigation

A number of forest and associated land management options offer potential to meet REDD+ climate change mitigation objectives while serving biodiversity conservation and broader societal needs (Kapos *et al.*, 2012, summarized below).

These management actions may be applied at varying spatial scales both within and outside of the forest sector to address the proximate as well as underlying drivers of deforestation and forest degradation in tropical and subtropical regions as well as approaches for increasing forest cover on degraded forest lands (Table 27.1). Actions that seek to maintain existing carbon and biodiversity through effectively reducing deforestation and forest degradation are more likely to have the greatest and most immediate benefits compared to those that seek to restore them.

Actions that address drivers of deforestation and forest degradation are essential to reduce greenhouse gas emissions from forest landscapes. Improvements in agricultural practices on lands currently in use for food and livestock production have a particularly important role to play in REDD+ strategies. Sustainable agricultural intensification and other improvements in existing production systems, including agroforestry and shifting cultivation may help both to limit the increase in demand for new land (and associated deforestation and forest degradation) and to reduce direct impacts such as those from unsustainable shifting cultivation, the use of fire in land preparation and management, and the application of agrochemicals. Such improvements can also help to enhance carbon stocks across agro-forest landscapes.

Table 27.1 Relevance of management interventions to the five REDD+ activities. Some interventions have a strong and direct role to play in a given REDD+ activity (••), while others may have less immediate relevance but may still play a role (•). Source: Kapos et al. 2012

Forest Management Type Management actions likely to be used in REDD	Relevance to REDD+ activities				
	Reduction of emissions from deforestation	Reduction of emissions from forest degradation	Carbon stock enhancement	Sustainable management of forests	Conservation of carbon stocks
Improving agricultural practice					
Sustainable agricultural intensification	••	•	•		•
Agroforestry		•	•		
Sustainable shifting cultivation	•	••	•		•
Fire management	•	••	••	••	•
Protection measures	••	••	•	•	••
Reducing impacts of extractive use					
Reduced Impact Logging		••		••	
Efficiencies, alternative production, or substitution of fuelwood and NTFPs		••		••	
Hunting regulation		••		•	
Restoration/Reforestation					
Assisted natural regeneration	•	•	••	•	
Afforestation and reforestation primarily for wood/fiber production			••		
Reforestation primarily for biodiversity and ecosystem services		•	••		
Landscape scale planning and coordination	••	••	•	••	•

One of the most effective means for protecting forests from degradation and loss – and thereby conserving forest carbon stocks and reducing greenhouse gas emissions – is the establishment of formally protected areas or other conservation units. Management approaches for such areas may range from strict protection of biodiversity to allowing multiple uses, including limited extractive activities. Protected area governance may also vary, with some managed by government authorities and others by private landowners or by local communities. Establishing protected areas may not always result in the net reduction of carbon emissions, if other unprotected areas become more degraded as a result.

Reducing emissions from forest degradation will also require in many regions actions to reduce the impacts of extraction of forest products including timber, fuelwood and other NTFPs. Such actions may include improved timber harvesting practices such as reduced impact logging (Putz et al., 2008), as well as promoting the sustainable harvesting and use of NTFPs. Depending on national and local circumstances, such actions and the policies used to promote them can play an important role in REDD+ through their contribution to reducing forest degradation and promoting sustainable management of forests.

The enhancement of forest carbon stocks may include a broad set of management actions involving various forms of forest restoration, reforestation and afforestation. All have the potential to yield positive results in terms of carbon sequestration as well as the conservation of biodiversity and provision of forest ecosystem services. In this context, it should be noted that there is increasing evidence that biodiversity plays a critical role in the provision of a wide range of forest ecosystem services, including carbon sequestration and storage (Thompson et al., 2012). Assisted natural regeneration in deforested or degraded areas can be used in many locales to accelerate natural secondary forest development by removing stressors or barriers to natural regeneration (such as recurrent fire, grazing or dominant invasive grasses). Reforestation may also be achieved through establishment of planted forests, using either plantation monocultures or mixed species plantings (Lamb et al., 2005). In some cases enrichment planting may be used to modify the composition of existing degraded forests.

Factors influencing the environmental impacts of REDD+ activities

The forest and land management activities outlined above can have highly variable impacts on carbon, biodiversity, and provision of forest ecosystem services, depending on location, scale of implementation, initial conditions, historical impacts, forest type and the wider landscape context. While any or all of these actions may potentially form part of REDD+ programs and strategies, coordination and planning at landscape and broader scales are key to minimizing negative impacts, and ensuring positive outcomes for both carbon and biodiversity (SCBD, 2011).

Different management actions require different time periods to deliver benefits for carbon, biodiversity and other ecosystem services (Kapos et al., 2012). In some cases, actions that yield positive carbon sequestration benefits in the short term may fail to deliver biodiversity conservation benefits and in some cases may cause negative impacts. Trade-offs between carbon and biodiversity outcomes can occur both locally and at wider spatial scales. For example, plantations of introduced species may provide large and rapid – though short-lived – carbon benefits but contribute little to local biodiversity and provision of valued ecosystem services. Depending on factors such as their management and prior land uses, such plantations may actually have detrimental impacts. At landscape scales, efforts to alleviate deforestation pressure on natural forests through agricultural intensification and associated inputs of agrochemicals can lead both to detrimental impacts on biodiversity and to increased greenhouse gas emissions.

Not all impacts on carbon and biodiversity are easily anticipated or measured. Impacts can occur outside the area of management and/or in the future (Kapos et al., 2012). Indirect impacts resulting from displacement of land-use pressures or extractive activities (e.g. following the creation of protected areas) are particularly problematic. Unintended increases in net greenhouse gas emissions may result if, for example, constraints on timber harvesting lead to the replacement of wood products with more emissions-intensive alternatives such as concrete, steel or plastics.

Both the magnitude and the direction of the environmental impacts can change over time. For example, fire suppression in naturally fire-dependent forest ecosystems can lead to increased carbon stocks in the short term, but can be severely detrimental in the long term for both carbon and biodiversity if the accumulation of fuel (flammable biomass) leads to catastrophic fires.

It is generally accepted that of the five REDD+ activities, reducing deforestation and forest degradation have by far the greatest potential to yield positive results in terms of carbon sequestration and biodiversity conservation. As a means to enhance forest carbon stocks, forest restoration practices to restore riparian forests in degraded watersheds or to create corridors and improve forest connectivity in fragmented landscapes can provide substantial benefits for biodiversity while at the same time enhance the provision of many forest ecosystem services valued by people (Mansourian et al., 2005). There is much uncertainty, however, about the potential impacts on biodiversity of other activities to enhance forest carbon stocks and those related to the sustainable management of forests. Further, there is both uncertainty and concern about how all REDD+ activities may directly and indirectly affect the well-being of people, especially indigenous and local communities who may depend on a variety of ecosystem services other than carbon.

Social and economic impacts of REDD+ activities

While REDD+ has the potential to generate substantial positive impacts for climate mitigation and biodiversity conservation, the way in which it is implemented will determine its social and economic effects (Strassburg et al., 2012), summarized in Figure 27.2. The primary objectives of REDD+, avoiding deforestation and forest degradation, can greatly benefit landless or otherwise disadvantaged people in rural areas who typically have a greater dependence on products derived from forests for their subsistence and livelihood needs. Such people are often disproportionately impacted by the loss of forest cover and forest biodiversity and the environmental services they provide (Chomitz, 2007). If a significant fraction of expected financial resources associated with REDD+ reach the rural poor, considerable benefits may be realized.

On the other hand, the poor are also most vulnerable to changes in resource management systems and rights of access to forest resources that may be associated with REDD+, with severe negative consequences for their already marginal livelihoods. There is abundant evidence that security of tenure and associated authority for local decision support better environmental management, as well as the realization of livelihood benefits (Strassburg et al., 2012, Angelsen et al., 2009). Tenure security includes recognition of all forms of ownership and control, especially communal tenure. Poor recognition of such rights excludes the rural poor from decision making, and usually denies them access to potential benefits from market-based interventions, such as payments for environmental services and REDD+. Weak tenure security also facilitates 'land grabbing' and other irregularities related to land ownership and transfer, which often result in expropriation of lands from the most vulnerable segments of rural society.

If REDD+ programs and activities are to address the social, political and economic factors that produce inequitable outcomes, socio-economic objectives need to be given high priority in REDD+ planning and implementation. The most adverse social and economic consequences can be avoided through the adoption of strong environmental and social safeguards, which are sensitive to, and include monitoring systems for, tracking social impacts. Evidence from past experience, including other PES programs, strongly suggests that pursuing these social objectives alongside REDD+ is likely to not only make the process more equitable but also increase the likelihood of achieving carbon and biodiversity goals (Strassburg et al., 2012). However, it is important to recognize that 'win–win' outcomes may not always be possible and that difficult trade-offs may need to be negotiated between carbon, biodiversity and social objectives. Integrated landscape management is a powerful tool for addressing and reconciling the many environmental, social and economic aspects relevant to REDD+ inside and outside forests (DeFries and Rosenzweig, 2010). Careful and inclusive (participatory) spatial planning can positively influence the distribution of winners and losers across the landscape so that REDD+ serves the interests of the most vulnerable groups, thereby increasing the likelihood of positive impacts on both equity and environmental effectiveness (Strassburg et al., 2012).

John A. Parrotta

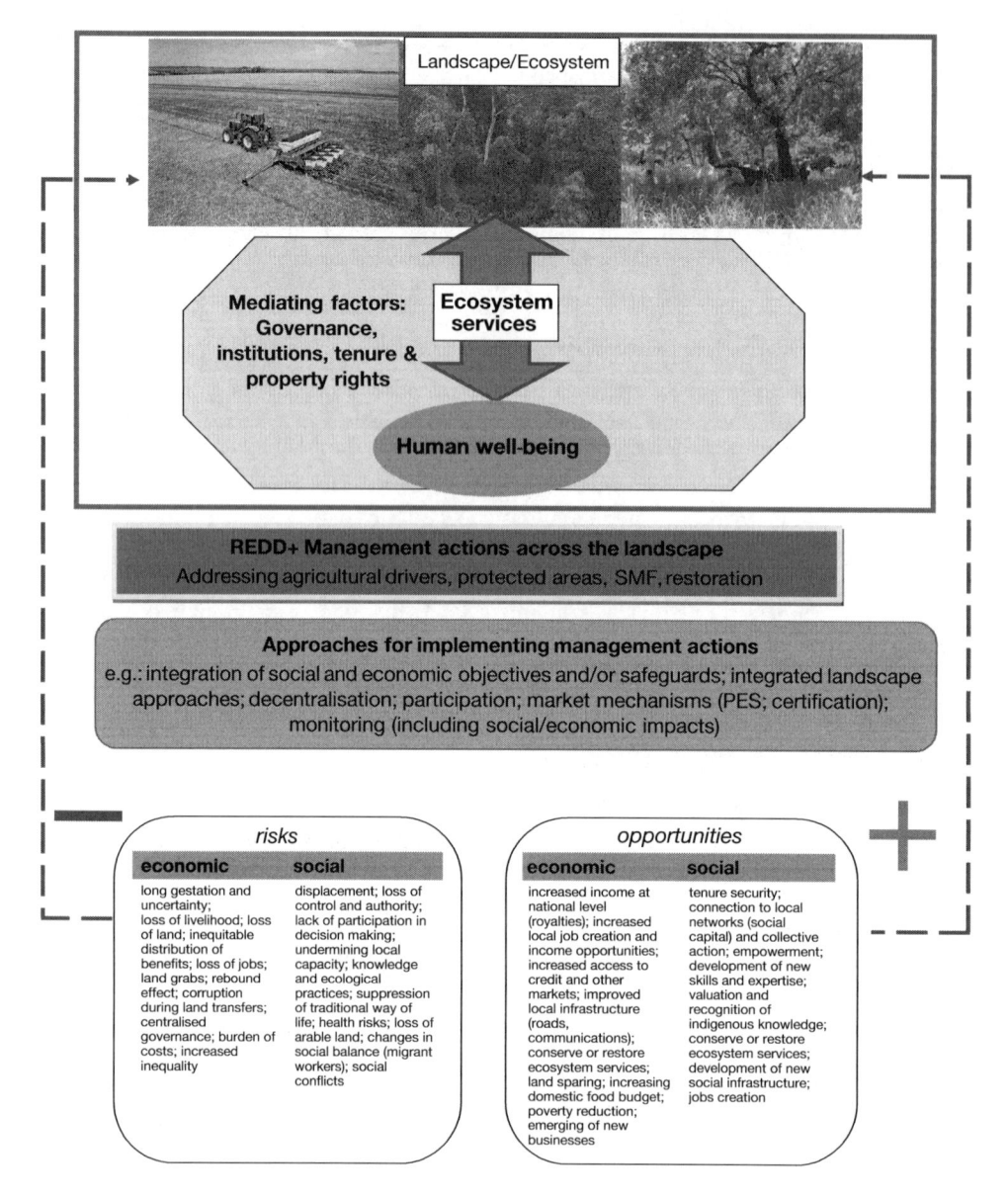

Figure 27.2 Economic and social impacts of REDD+ management actions on different stakeholders within a landscape

Source: Reprinted from Strassburg *et al.* (2012), in Parrotta *et al.* (2012)

Conclusion

REDD+ was conceived as climate change mitigation strategy, focusing almost exclusively on measures to enhance the role of forests in sequestering carbon and reducing CO_2 and other greenhouse gas emissions associated with deforestation and forest degradation in tropical and subtropical countries. The conservation, sustainable management and restoration of forests are critical not only to efforts to help mitigate climate change, but to stem the loss of biodiversity which to a great extent underpins the capacity of forests to provide the ecosystem services that sustain the livelihoods of people worldwide. The success of REDD+ initiatives will depend on the degree to which REDD+ planning and implementation of specific activities recognize – and address – the complexity of historical and contemporary environmental, social, economic and political factors that drive deforestation and forest degradation. It must also learn from decades of experience why efforts to date to slow these processes have often failed, and recognize the importance (and more immediate value) of the broad range of other ecosystem services (beyond carbon sequestration and storage) that forests provide to people, particularly to those segments of rural society who depend most directly on forest goods and services to sustain their livelihoods.

References

Angelsen, A., Brockhaus, M., Kanninen, M., Sills, E., Sunderlin, W. D. and Wertz-Kanounnikoff, S. (eds.) (2009) *Realising REDD+: National Strategy and Policy Options*, CIFOR, Bogor, Indonesia

Asner, G. P., Broadbent, E. N., Oliveira, P. J. C., Keller, M., Knapp, D. and Silva, J. N. M. (2006) 'Condition and fate of logged forests in the Brazilian Amazon', *Proceedings of the National Academy of Sciences of the USA*, vol 103, pp. 12947–12950

Barlow, J., Peres, C. A., Lagan, B. O. and Haugaasen, T. (2003) 'Large tree mortality and the decline of forest biomass following Amazonian wildfires', *Ecology Letters*, vol. 6, pp. 6–8

Chazdon, R. L. (2008) 'Beyond deforestation: restoring forests and ecosystem services on degraded lands' *Science*, vol. 320, pp. 1458–1460

Chomitz, K. M. (2007) *At Loggerheads?: Agricultural Expansion, Poverty Reduction, and Environment in the Tropical Forests*, World Bank, Washington DC

Cochrane, M. A. (2003) 'Fire science for rainforests', *Nature*, vol. 421, pp. 913–919

DeFries, R. and Rosenzweig, C. (2010) 'Toward a whole-landscape approach for sustainable land use in the tropics', *Proceedings of the National Academy of Sciences*, vol. 107, pp. 19627–19632

FAO (2006) *Global Forest Resources Assessment 2005 (Forestry Paper 147)*, Food and Agriculture Organization of the United Nations, Rome

FAO (2007) 'The world's mangroves 1980–2005. A thematic study prepared in the framework of the Global Forest Resources Assessment 2005', *FAO Forestry Paper 163*, Food and Agriculture Organization of the United Nations, Rome

FAO (2010) 'Global Forest Resources Assessment', *Forestry Paper 163*, Food and Agriculture Organization of the United Nations, Rome

FAO (2011) *Global Forest Land Use Change from 1990 to 2005*, Food and Agriculture Organization of the United Nations, Rome

Gibbs, H. K, Ruesch, A. S., Achard, F., Clayton, M. K., Holmgren, P. and others (2010) 'Tropical forests were the primary sources of new agricultural land in the 1980s and 1990s', *Proceedings of the National Academy of Sciences*, vol. 107, pp. 16732–16737

Griscom, B., Ganz, D., Virgilio, N., Price, F., Hayward, J. and others (2009) *The Hidden Frontier of Forest Degradation: A Review of the Science, Policy and Practice of Reducing Degradation Emissions*, The Nature Conservancy, Arlington, VA, US

Houghton, R. A. (2003) 'Revised estimates of the annual net flux of carbon to the atmosphere from changes in land use and land management 1850–2000', *Tellus*, vol. 55B, pp. 378–390

ITTO (2002) 'ITTO guidelines for the restoration, management and rehabilitation of degraded and secondary tropical forests', *ITTO Policy Development Series No. 13*, International Tropical Timber Organization, Yokohama, Japan

Kapos, V., Kurz, W. A., Gardner, T., Ferreira, J., Guariguata, M. and others (2012) 'Impacts of forest and land management on biodiversity and carbon', in J. A. Parrotta, C. Wildburger, and S. Mansourian (eds) *Understanding Relationships between Biodiversity, Carbon, Forests and People: The Key to Achieving REDD+ Objectives*, IUFRO World Series Volume 31, International Union of Forest Research Organizations, Vienna

Kissinger, G., Herold, M. and De Sy, V. (2012) *Drivers of Deforestation and Forest Degradation: A Synthesis Report for REDD+ Policymakers*, Lexeme Consulting, Vancouver, Canada

Lamb, D., Erskine, P. D. and Parrotta, J. A. (2005) 'Restoration of degraded tropical forest landscapes', *Science*, vol. 310, no. 5754, pp. 1628–1632

Lambin, E. F., Geist, H. J. and Lepers, E. (2003) 'Dynamics of land-use and land-cover change in tropical regions', *Annu. Rev. Environ. Resour*, vol. 28, pp. 205–224

Le Quéré, C., Peters, G. P., Andres, R. J., Andrew, R. M., Boden, T. and others (2013) *Earth System Science Data* Discussion. doi: 10.5194/essdd-6-689-2013

MA (2005) 'Ecosystems and human well-being: Synthesis'. *Millennium Ecosystem Assessment*, Island Press, Washington DC

Mansourian, S., Vallauri, D. and Dudley, N. (2005) *Forest Restoration in Landscapes: Beyond Planting Trees*, Springer, New York

Nabuurs, G. J., Masera, O., Andrasko, K., Benitez-Ponce, P., Boer, R. and others (2007) 'Forestry', in: Metz, B., Davidson, O. R., Bosch, P. R., Dave, R. and Meyer, L. A. (eds). *Climate Change 2007: Mitigation. Contribution of Working Group III to the Fourth Assessment Report of the Intergovernmental Panel on Climate Change*, Cambridge University Press, Cambridge, UK and New York

Pan, Y., Birdsey, R. A., Fang, J., Houghton, R., Kauppi, P. E. and others (2011) 'A large and persistent carbon sink in the world's forests', *Science Express*, vol. 333, pp. 988–993

Parrotta, J. A., Wildburger, C. and Mansourian, S. (eds) (2012) *Understanding Relationships between Biodiversity, Carbon, Forests and People: The Key to Achieving REDD+ Objectives. A Global Assessment Report Prepared by the Global Forest Expert Panel on Biodiversity, Forest Management, and REDD+*, IUFRO World Series Volume 31. International Union of Forest Research Organizations, Vienna, http://www.iufro.org/science/gfep/

Putz, F. E., Sist, P., Fredericksen, T. and Dykstra, D. (2008) 'Reduced-impact logging: challenges and opportunities' *Forest Ecology and Management*, vol. 256, pp. 1427–1433

Rudel, T. K., DeFries, R., Asner, G. P. and Laurance, W. F. (2009) 'Changing drivers of deforestation and new opportunities for conservation', *Conserv. Biol.*, vol. 23, pp. 1396–1405

Saatchi, S. S., Harris, N. L., Brown, S., Lefsky, M., Mitchard, E. T. A. and others (2011) 'Benchmark map of forest carbon stocks in tropical regions across three continents', *Proceedings of the National Academy of Sciences*, vol. 108, no. 24, pp. 9899–9904

SCBD (2010) *Global Biodiversity Outlook 3*. Secretariat of the Convention on Biological Diversity, Montreal, Canada

SCBD (2011) 'REDD-plus and Biodiversity', *CBD Technical Series No. 59*. Secretariat of the Convention on Biological Diversity, Montreal, Canada

Shlisky, A., Alencar, A. A. C., Nolasco, M. M. and Curran, L. M. (2009) 'Global fire regime conditions, threats, and opportunities for fire management in the tropics', in M. A. Cochrane *Tropical Fire Ecology: Climate Change, Land Use, and Ecosystem Dynamics*, Praxis (ebook)

Strassburg, B. B. N., Vira, B., Mahanty, S., Mansourian, S., Martin, A. and others (2012) 'Social and economic considerations relevant to REDD+', in J. A. Parrotta, C. Wildburger, and S. Mansourian (eds) *Understanding Relationships between Biodiversity, Carbon, Forests and People: The Key to Achieving REDD+ Objectives*, IUFRO World Series Volume 31, International Union of Forest Research Organizations, Vienna

Thompson, I. D., Ferreira, J., Gardner, T., Guariguata, M., Koh, L. P. and others (2012) 'Forest biodiversity, carbon and other ecosystem services: relationships and impacts of deforestation and forest degradation', in J. A. Parrotta, C. Wildburger, and S. Mansourian (eds) *Understanding Relationships between Biodiversity, Carbon, Forests and People: The Key to Achieving REDD+ Objectives*, IUFRO World Series Volume 31, International Union of Forest Research Organizations, Vienna

Thompson, I. D., Guariguata, M. R., Okabe, K., Bahamondez, C., Nasi, R. and others (2013) 'An operational framework for defining and monitoring forest degradation', *Ecology and Society*, vol. 18, no. 2, p.20

van der Werf, G. R., Morton, D. C., DeFries, R. S., Olivier, J. G. J., Kasibhatla, P. S. and others (2009) 'CO_2 emissions from forest loss', *Nature Geoscience*, vol. 2, pp. 737–738

28
RESTORATION OF FOREST ECOSYSTEMS

David Lamb

Introduction

The world's forests have been declining for many years but especially over the last 100 years. Some former forest lands are used for agriculture but many now lie abandoned in a degraded state. There is much debate over how to define and assess land degradation but one estimate suggests there could be as much as two billion ha of lands that have been degraded and may now be available for some form of reforestation (Laestadius *et al.* 2011). A number of international agencies are now calling for this to happen. For example, the Convention on Biological Diversity has suggested a target of 15 percent of all degraded lands be restored and many national governments have adopted national reforestation policies to overcome degradation (Suding 2011). The question is, how should this be done? Reforestation using simple monocultures of fast-growing exotic species has been carried out on a large scale in many countries but is it possible to actually restore the original forests?

The first studies of ecological successions were carried out over 100 years ago but it is only recently that ecologists have begun to think about how successional theories might be used to restore degraded ecosystems. The Society for Ecological Restoration (www.ser.org) refers to the process of initiating or accelerating the recovery of an ecosystem and returning the ecosystem to its historical trajectory as Ecological Restoration. It defines Ecological Restoration as 'the process of assisting the recovery of an ecosystem that has been degraded, damaged or destroyed' (SER 2004). The attributes of a restored ecosystem are seen to be: (i) the re-assembly of a characteristic suite of indigenous species that can be found in undisturbed natural reference or target ecosystems; (ii) the restoration of key ecological processes and functioning; and (iii) the capacity of the new system to be self-sustaining.

Some doubt this can be done because of uncertainties about the identity of the target itself (especially in heavily modified landscapes where no undisturbed forest remains). Others have raised concerns about the presence of exotic species that have been naturalised or because of the extinction of some former native species (Aronson *et al.* 2007). Nonetheless, there are many places where relatively undisturbed sites remain that can act as reference areas and as a source of seeds or colonists, leaving open the possibility of recreating the historic ecosystem.

However, an additional set of concerns have recently emerged about the feasibility of restoration. One is that future global climate change is likely to mean that many natural ecosystems will change as some species go extinct while others flourish. The new ecosystems that evolve will be unlike those now present although it is presently difficult to specify their composition or how different they will be in the way that they function. From a restoration viewpoint, this means that historic ecosystems may not always be useful points of reference and novel ecosystems will have to be created to suit the changed environmental conditions of the future (Suding 2011). A second concern is that Ecological Restoration may be too costly. The sheer scale of the area of degraded lands now present mean that Ecological Restoration, as originally conceived, may not be feasible simply because it is too expensive to implement on a large scale. This is especially the case in the many degraded landscapes occupied by poor rural populations.

In short, new forms of reforestation will be needed to restore forests to the large areas of degraded landscapes that have now accumulated. These will have to be able to accommodate future environmental conditions, be resilient, self-sustaining and be capable of supplying the variety of services and goods needed by human communities. These new forests may, or may not, be identical to the historic ecosystems formerly present. If not, some might see this as being a case of 'lowering the bar' or reducing standards and another step along the way towards a more biologically impoverished world. Others will see it as a pragmatic and more realistic alternative means of counteracting a trend towards simplification in the face of both environmental uncertainty and the pressing need to improve the well-being of large numbers of impoverished people.

There are a number of terms to describe the ways in which forests might be re-established at cleared or degraded lands (Carle and Holmgren 2003, Aronson et al. 2007). Reforestation shall be used here to include the establishment of forests or woodlands on deforested land by deliberately planting or sowing seeds as well as through natural regeneration (i.e. excluding situations where production forests are being reforested as part of the normal silvicultural cycle). Reforestation can be done in a variety of ways and restoration is sometimes used to describe forms of reforestation where the emphasis is on multi-species plantings or on some degree of 'naturalness' but the distinction is not always clear and the two terms (reforestation and restoration) will be used here interchangeably. Ecological Restoration is a particular form of reforestation and will be retained to describe attempts to restore the original forest ecosystems to degraded lands.

This chapter will discuss the approaches that might be used and the challenges involved in implementing these across large, deforested landscape areas. It begins by discussing the main methods used to initiate the restoration process.

Methods of forest restoration

There are three main ways in which plants can be regrown at deforested sites. These include: through natural regeneration, by direct seeding or by planting seedlings. In some cases restoration will involve just one of these approaches although it may sometimes be necessary to use all three methods. The advantages and disadvantages are summarised in Table 28.1.

Natural regeneration

Natural regeneration can occur if a site is protected from further disturbances (e.g. wildfire, grazing) and if sufficient trees remain on the area as old roots, stumps or in the soil seed pool. In such cases trees can resprout from the remnant stumps or the seeds can germinate and re-occupy the site (Chazdon 2014). Natural regeneration can also occur if there is a (protected) source of seeds nearby (e.g. in a patch of residual natural forest) and these seeds can be carried to the site by dispersal agents such as birds or bats, or taken there by wind or water. Dispersal is most effective when the distances are not great and dispersal vectors are able to move easily across the landscape but dispersal can be substantially reduced when these factors are limiting. Once trees become established these can become perch sites and, possibly, food sources for birds and act as attractants. This obviously accelerates the dispersal process.

The greatest advantage of natural regeneration is its low cost (although it may still be relatively expensive to protect regenerating sites from disturbances). This means that very large areas may be restored with limited resources. Examples can be seen in many parts of the world where farmland has been abandoned and forests have begun to recover (Carter and Gilmour 1989, Janzen 2002, Kuemmerle et al. 2011). Some of the most extensive cases where natural regeneration has occurred are in Africa, where very large areas of woody regrowth have developed (Sendzimir et al. 2011). An example is shown in Figure 28.1.

However, there are also some disadvantages associated with natural regeneration (Table 28.1). One disadvantage is that natural regeneration may be unable to restore the original ecosystem within time frames desired by humans. Much depends on the proportion of the original biota able to persist at the site following the original clearing process and it is likely this proportion will decline over time, especially if the site is repeatedly disturbed. This is because successive disturbances are likely to diminish the pool of stumps or old roots. Natural regeneration will then be increasingly dependent on seed dispersed into the site from external sources. Species with small seeds are likely to be more readily dispersed than those with larger

Figure 28.1 Natural regeneration from residual stumps and roots in old fields in Niger (Photo: T. Rinaudo)

seeds and some of these may be exotic species. Likewise, more seeds are likely to be dispersed if seed sources are nearby than if they are distant (McConkey et al. 2012). Both factors mean only a sub-set of the original species may be represented in the new forest.

A second disadvantage is that natural regrowth can be patchy, with some areas well stocked with trees and other areas remaining bare. This means overall recovery will be slow which means, in turn, that it will be risky; the longer that grasses and shrubs are dominant then the more likely it is that wildfires will occur and destroy the new forest. Of course, undisturbed forests can also burn but the presence of grasses increases the risk of this happening. Likewise, the longer it takes for trees to become established then the more likely it is that someone will declare the site 'unused wastelands' and try to use it for some other purpose.

All this means that a key question facing those wishing to initiate restoration at a deforested site is whether to rely on natural regeneration alone or whether to actively intervene and sow seeds or plant seedlings. Just what density of seedling regrowth must be present and how big must these seedlings be to be assessed? Much depends on the resources available and on the landowner's objectives (Holl and Aide 2011). Working in the humid tropics, Elliott et al. (2013) suggest that a high seedling density might be needed to achieve rapid canopy closure but much lower densities may be satisfactory in drier locations. This means there are no universal standards and guidelines prescribing minimal acceptable seedling densities should be developed to suit local circumstances.

Table 28.1 Some advantages and disadvantages of alternative methods of initiating forest restoration

	Advantages	*Disadvantages*
Natural regrowth	• Low cost. • Can treat large areas. • Need little ecological knowledge of biota.	• Prone to further disturbances because it may not be obvious that a site is owned and being restored. • Difficult to restore all flora; may be dominated by only a few species (some of which may be exotic weeds). • May result in sparse or patchy distribution of trees. • Rate of regrowth may be slow.
Direct seeding	• Relatively low cost. • Seed can be dispersed from the air, meaning inaccessible (e.g. steep) sites can be treated.	• Not always effective. • Need large quantities of seed since establishment rates often low. • Hence, may only be possible to sow some of the original biota because of limited seed availability. • Seeds of some species may need to be buried to achieve germination success. • Best used at sites that have been entirely cleared (e.g. by fire or ploughing) because subsequent weed control is difficult.
Planting seedlings	• More reliable. • Makes efficient use of scarce seed resources. • Can ensure preferred species are established. • Can plant at required locations and densities. • Weed control easier to manage.	• More costly to raise seedlings in nurseries and plant in field. • May be difficult to raise seedlings of all species in species-rich forests.

Direct seeding

A second approach to initiating forest restoration bypasses the need to await natural seed dispersal and directly sows seeds at the site. Direct seeding is likely to be much cheaper than planting seedlings because it overcomes the need for a costly nursery stage and, unlike relying on natural regeneration, allows specific locations to be targeted. Direct seeding is likely to be most effective if seeds are spread over bare ground. This can be managed at former mine sites by carrying out seeding immediately after the mining has ceased and before weeds are able to colonise a site. Otherwise weeds must be removed using fire, herbicides or ploughing. In this case the bare ground provides temporary respite from weed competition which is difficult to manage once the tree seeds germinate and grow.

A variety of techniques can be used to undertake direct seeding. Many direct seeding operations have been carried out using modified agricultural implements towed behind tractors which remove weeds, prepare a seedbed and bury seeds at appropriate depths (Figure 28.2). Accounts of such field operations are provided by Stanturf *et al.* (2009) and Jonson (2010). But seeds can also be dispersed from aircraft which means sites difficult to access can be treated (Hodgson and McGhee 1992). In these cases fire might be used to temporarily remove weeds before the seed is dispersed although this may be hazardous in many situations.

Despite its lower initial cost, direct seeding has some significant disadvantages. One is that it requires large amounts of seed of each species. Some of the dispersed seed will be consumed by predators and some will fail to germinate or the new seedlings will die from drought. These risks mean the technique may only be appropriate for those species regularly producing large amounts of seed and other techniques will be needed to restore other species. Much depends on the attributes of the particular species and their seed (Tunjal and Elliott 2012). Success also

Figure 28.2 Five-year-old multi-species eucalypt forest established using direct seeding on former agricultural lands in Western Australia (Photo: J. Jonson)

depends on favourable climatic conditions being experienced immediately following seed dispersal so that the timing of this is critical and there may be large year-to-year variations in the effectiveness of direct seeding. Germination success can be substantially improved by burying seed but different species may require different burial depths so that application methods using a standard sowing depth for a mixture of seeds may be sub-optimal for some.

The key issue confronting those wishing to use direct sowing is to develop methods that allow multiple species to be sown at once in ways that minimise predation losses and improve seedling establishment.

Planted tree seedlings

Despite the advantages of using natural regeneration and direct seeding, many new restoration activities are undertaken by planting seedlings because this is a more reliable method of initiating a new succession. Seedlings are raised in the sheltered conditions of nurseries so that seeds are used effectively. This means even species producing few seeds can be included in restoration programs. Infrequently seeding species can also be raised from cuttings or by collecting wildlings (i.e. naturally occurring seedlings) from natural forest. These seedlings can then be planted in desired locations and at preferred densities. It is relatively easy to apply fertilisers at the time of planting and subsequently undertake weed control. This ensures that community composition can be controlled and that growth can be enhanced, leading to rapid canopy closure and further successional development.

Techniques for raising seedlings and establishing these in the field are widely known so that the main research questions involve developing methods for raising seedlings of specific species and clarifying the precise site requirements of these species.

Methods of ecosystem assembly

Each of these methods can be used to initiate a new succession but each has some limitations. This means that, by themselves, they are unlikely to be able to restore the original flora to degraded sites, especially in species-rich ecosystems. These constraints point to the need to use a combination of several approaches. But just how should this be done? Suppose, for example, it is possible to identify all the plant species that should be present in a particular ecosystem and raise seedlings of these, might the seedlings then all be planted at the same time? Or are there more complex 'rules of assembly' that govern the way in which species are combined and new ecosystems are re-established? There has been much debate over the process of ecosystem assembly (Temperton *et al*. 2004). But it is still not clear that there are widely applicable 'rules' that must be followed. However, there appear to be three key issues. These are the numbers and types of species that must be initially established, the sequence in which these species are established and the relative abundance of the various species being sown or planted.

The numbers and types of species initially established

In some relatively simple ecosystems it may be possible to establish all plant species by direct seeding or planting seedlings. But this is difficult in most forests, especially if an attempt is being made to restore woody understorey species as well as the tree species over large areas. In many tropical forests, for example, a large number of species are represented by trees present at very low densities. In such cases a pragmatic solution may be to only plant seedlings of those species having readily available seed and hope that natural regeneration and further seed dispersal from external sources will bring in other species (most restoration projects necessarily assume that wildlife, including key mutualists, will also recolonise sites in this way). The extent to which this will happen depends on there being patches of natural forest nearby and on the availability of wildlife able to carry seeds across the intervening landscape.

The issue is further complicated, however, in situations where environmental degradation has been such as to preclude the possibility of some of the original species re-occupying the site. For example, drainage conditions may have changed, topsoil may have been lost or the site may now be too exposed to allow certain of the original flora to grow. In these situations it may be necessary to include exotic species in the initial plantings either to facilitate the establishment of other species (see further below) or because they are functional analogues of the species now unable to regenerate at the site. There has been debate over the use of exotics because of the perceived risk that they may become invasive (e.g. Richardson and Rejmanek 2011, Dodet and Collet 2012). Native species should be used when possible but there may be little alternative to the use of exotics in some ecological situations.

The choice of number and type of species to use is also being influenced by a growing appreciation of the importance of fostering resilience (Walker and Salt 2006). Resilience is the capacity to tolerate and adapt to change and can be enhanced not only by fostering species diversity, but also by including representatives of a variety of functional types. It will also be improved by including, wherever possible, more than one species of each of these types. This is because functionally diverse communities are more likely to be able to withstand future environmental disturbances than more simple communities (Mori *et al.* 2013). In short, it is preferable to seek to restore the original tree flora but functionally diverse species mixtures may be all that is possible at some sites.

The establishment sequence

Studies of plant successions often identify so-called pioneer species that occupy sites at the commencement of successions. These are species that are easily dispersed and shade intolerant but otherwise able to tolerate a variety of ecological conditions. Must these pioneer species always be planted first and other species only planted or sown several years later once the pioneers are established? The empirical evidence suggests that in most cases the answer is 'no' and that successional development can often be initiated by planting all species at the same time. That is, the pioneers have no unique facilitatory role and are simply early colonists because of their dispersal abilities. But there are exceptions to this generalisation and some species that do appear to require pioneers to facilitate their establishment by providing some with early shelter or protection. Examples are some of the dipterocarp species from Asia which require at least some initial shade to become established (Ashton *et al.* 2001, McNamara *et al.* 2006). Similar examples can be found in other biomes. In such cases a two-step approach may be needed whereby cover is provided by a nurse species and the less tolerant species are directly seeded or under-planted once this canopy has been established.

The relative abundance of each species

The final issue concerns the relative proportions of each species that should be introduced to the new succession as seedlings (or seed). Perhaps one might use the relative abundance of each species in a reference ecosystem as a guide. But species differ in their competitive abilities, growth rates and longevities so that the relative proportions observable at a late successional stage are not necessarily the same as those needed at an earlier stage. Sometimes certain species are favoured because of their functional importance, their conservation status or their value for human uses. In such cases it might seem useful to increase the relative proportion of these. But abnormally high proportions of some species can disrupt successional trajectories (Siddique *et al.* 2008). An indication of the likely impact of particular species may be possible by assessing attributes such as their growth rates, final stature and longevities as well as crown sizes and densities (Noble and Slatyer 1980). The problem is one where adaptive management may be the only way of minimising unexpected successional diversions.

Restoration in practice

The way in which restoration is actually undertaken will depend on the resources available as well as on the site conditions and landscape context. Some of the options available for deliberately intervening to accelerate successional development (rather than simply relying on natural regrowth) are outlined in Table 28.2. Several of these just initiate successional development and then depend on subsequent colonisation to achieve their objectives. Others involve substantially greater efforts to include all species because remnant vegetation is too distant or the vectors usually able to disperse seed are too limited.

Perch trees

Natural regrowth may not always occur, but even when it does, it can be slow. Several studies have shown that perch trees scattered about the site will accelerate the process by acting as attractants for seed-dispersing birds (Elliott *et al.* 2013). Planting small clumps rather than single trees increases the effectiveness of the technique (Zahawi *et al.* 2013). The cost obviously varies with the planting density but scattered clumps of trees may be enough to accelerate the process while minimising costs. The approach obviously depends on there being sufficient wildlife to actively disperse seed. The chief disadvantage of the technique is that, like natural regeneration

Table 28.2 Methods of restoring forests using planted seedlings

Method	Approach
Perch trees	Plant scattered individuals or clumps of indigenous trees across old fields to act as attractants for birds able to disperse seed into the site.
Nurse trees	Establish simple monocultures of a species able to tolerate site conditions. These are then under-planted or seeded with the preferred species once the nurse trees have closed the canopy. Alternatively, the site can be left to enable natural colonisation to take place beneath the nurse trees.
Framework species method	Establish a variety of short-lived and longer-lived species including species that are poorly dispersed (e.g. because of their large seed size). Again, these trees grow to form a canopy and enable colonisation from external seed sources to take place. The inclusion of short-lived species means canopy gaps are created when these die, allowing seedlings of the colonists to flourish and grow into the new canopy layer.
High diversity plantings	Plant and/or sow as many species as possible. Give preference to indigenous species but use long-lived exotic species if it is necessary to compensate for the absence of functionally important native species.
Ecological Restoration	Plant and sow only indigenous species.

more generally, it can be risky because the site may be seen as abandoned wasteland rather than as a regenerating forest and, consequently, may be at risk of further disturbances, especially wildfires or being cleared for agriculture.

Nurse trees

Grasslands and shrublands are difficult places for trees to colonise because of the competition experienced by the new tree seedlings. Sometimes only a restricted variety of tree species can colonise such sites. Stressful environmental conditions like these can be transformed once a simple tree cover is established. These more tolerant trees facilitate the subsequent establishment of a wider variety of native species by reducing weed competition and ameliorating environmental conditions. This ameliorative role is often seen in commercial timber plantations where diverse understories containing a variety of other tree species develop over time when natural forests are nearby (e.g. Keenan *et al.* 1997). In situations where colonisation from external sources is limited, the preferred species might be under-planted or sown beneath the nurse tree canopy (Löf *et al.* 2010). The density of the nurse trees will determine how rapidly the seedlings grow and it may be necessary to eventually remove some of the nurse trees and create canopy gaps to allow more light to reach the forest floor. Just when this should be done and how much canopy cover should be removed obviously depends on the attributes of the particular species involved as well as geographic location (larger gaps may be needed at higher latitudes with different solar elevations than in the tropics). An example of the process is provided by McNamara *et al.* (2006) who describe a case in Vietnam where the nurse tree species had a commercial value and the harvesting and sale of these to create canopy gaps was used to fund further expansion of the restoration process.

Framework species method

The nurse tree method described above has the advantage of simplicity since only a single species must be raised in a nursery. On the other hand, the method requires subsequent canopy manipulation to create gaps and relies entirely on further plantings, sowings or natural colonisation to increase diversity. The latter may bring a diverse variety of species to the site but most of these will be species with small seeds or fruit while poorly dispersed and large-fruited species will not normally be represented. The framework species method overcomes both these limitations and was developed to suit species-rich tropical forests (Figure 28.3). The need for future interventions is reduced by planting a mixture of 15–20 species which include short-lived as well as long-lived species and include poorly dispersed species as well as those likely to attract seed-dispersing wildlife (Goosem and Tucker 2013, Lamb 2011). These are all planted at the same time and establish a closed forest canopy. Subsequent deaths of short-lived species allow species that have subsequently colonised the site to grow up in the canopy gaps created by the tree deaths. The method facilitates successional recovery by quickly establishing a forest cover, overcoming the dispersal limitations of certain species and ensuring the progressive development of canopy gaps needed to drive the successional process.

High diversity plantings

In some situations there may be limited scope for natural regeneration or subsequent seed dispersal to enrich the initial plantings. This may be because there are no remnants of natural vegetation nearby or because there is simply not enough wildlife still able to disperse seeds from

Figure 28.3 Forest established using the framework species method at a former degraded site in Thailand. The site is now 16 years old and the planted trees are now being supplemented by additional species brought to the site from nearby natural forest by wildlife

remnant areas to the sites being restored. Alternatively, the species are those that are normally poorly dispersed. In such cases all the required species must be deliberately brought to the site. If the sites have not been severely degraded then Ecological Restoration may be possible (see further below). But if environmental conditions such as soil fertility or hydrological flows have been too greatly modified then a more appropriate objective may be to forgo restoring the original biodiversity and, instead, use indigenous and exotic species representing a variety of functional types to restore ecological processes and functions. The former ecosystem may provide some guidance here but the changed environmental conditions may mean some judgement is needed about the species to use. In most cases it would be useful to include species differing in their longevities, rates of growth and capacities to access and cycle nutrients. Preference should also be given to those able to provide resources for wildlife. Novel ecosystems of this type are necessarily experimental although experience from rehabilitating former mine sites suggests sustainable forest covers can be achieved (Cooke and Johnson 2002).

Ecological Restoration

There are a variety of ways in which Ecological Restoration might be implemented depending on the type of forest and on the environmental conditions present. These may involve relying on natural regeneration as well as on sowing seed or planting seedlings. In most cases only indigenous species are used, although there may be occasions where a short-lived exotic species may have a role to play as a nurse species or facilitator. The numbers of species initially planted or sown, the sequence in which these are introduced to the site and the relative proportions of each are all issues likely to be unique to a particular restoration project. The approach successfully used in one temperate forest restoration project is described in Box 28.1.

Box 28.1 Restoration in *Eucalyptus* forest after mining, Western Australia

Bauxite mining is being carried out in an area of south-western Australia occupied by *Eucalyptus marginata* forest (Koch and Hobbs 2007). This area is known as one of the world's biodiversity hotspots. The forests have only a small number of tree species but a very large diversity of understorey shrubs. Prior to open-cut mining, the top 15 cm of topsoil containing soil seed banks and mycorrhiza is removed and stockpiled. Following mining, overburden material is returned, the sites are recontoured and topsoil is respread over the soil surface. Plants are regenerated using direct seeding and by planting seedlings. Around 78–113 tree and understorey shrub species are directly sown into the topsoil at a single time depending on the particular location. Sowing is carried out in late summer or autumn by a seeder attached to a ripping bulldozer. These species are supplemented by planting a further 28 species that are not in the soil seed bank or cannot be directly sown, largely because of limited seed supplies. The identity of the species used is based on those found in pre-mining surveys. Experience has shown that few species are able to recolonise these sites from external areas. This has meant that considerable effort is needed to ensure all species are restored to the site at the commencement of the restoration program in densities matching those of natural forests. Evidence suggests there are no special assembly rules in these ecosystems and that these restoration techniques have enabled most of the original flora and the key ecological functions to be restored.

Need for monitoring

Re-assembling complex systems is difficult when so little is usually known about the component species or how they interact. This emphasises the need to monitor the successional trajectory and to practice adaptive management. Ideally this should be based on a series of specific questions so that answers can be linked to management actions. For example, are relative population densities of each species being maintained? Are seedlings planted beneath nurse trees growing or stagnating? Are exotic weeds invading the site? Are the newly established plants flowering and fruiting? Are seedlings of these species present on the forest floor? Are these seedlings increasing in height?

Depending on the outcome of monitoring it might be necessary to carry out supplementary enrichment plantings or seed sowing in the new forest or to adjust the relative abundances of species at the time of their introduction to the system. The failure of new seedlings on the forest floor to grow beyond the understorey layer may also point to the need to reduce overall tree density in order to create a more open canopy and allow more light to reach the forest floor. Alternatively, it may point to a need to (temporarily) control certain wildlife populations. Over time this process of monitoring and adaptive management should generate more knowledge of how the original assembly process can be improved.

Discussion and conclusions

All forms of restoration are long-term enterprises and it is difficult to predict how a successional trajectory will develop over periods of several decades or more. Reviews of the outcomes of various restoration projects have shown encouraging early signs of success (Benayas *et al.* 2009, Suding 2011). Of course 'success' can be measured in various ways – restoration of structure, biomass or of the original species diversity. Likewise, it may be expressed as the re-establishment of ecological functioning or the restoration of the capacity to supply ecosystem services. In many of the projects reviewed by Benayas *et al.* (2009) and Suding (2011) the restoration standards that had to be met were not specified. However, their findings are heartening results, particularly given the complexity of the systems involved and the fact that most of the projects reviewed were, ecologically speaking, comparatively young.

While promising, surveys like these show that restoration remains a risky business and more needs to be known in order to reduce risks and improve success rates. Certain problems can probably be solved fairly easily by simply applying existing silvicultural knowledge. But there is a large variety of ecological questions that need to be addressed if we are to acquire a predictive understanding of the restoration process sufficient to improve the reliability of the restoration process. Some examples are outlined in Box 28.2.

Box 28.2 Questions deserving further study

Regrowth: How to decide when to rely on natural regrowth alone to restore forest cover or when some form of intervention (e.g. sowing seeds, planting seedlings) is necessary? What are the minimum species numbers or seedling densities that must be present before natural regeneration alone is deemed reliable enough to restore degraded sites?

Dispersal: How does the distance of a site from natural forest affect the variety of plant species able to be transported (by wind? by wildlife?) and the rate at which their seeds reach the site? How might this knowledge influence the design of landscape restoration programs?

Direct seeding: How can the effectiveness of direct seeding be increased? How can the effectiveness of aerial applications be improved?

Facilitation: Nurse trees are sometimes used to facilitate the establishment of tree species with more specialised habitat requirements. How much overstorey cover is needed and when should it be removed?

Relative abundances: How can the most appropriate relative abundances of species to sow as seed or plant as seedlings be judged when commencing restoration?

Successional transitions: Most newly restored forests are even-aged because the trees were established at the same time. On the other hand, most natural forests have a variety of age classes. How (and when) can the transition from even-aged to uneven-aged forest be facilitated?

Disturbance regimes: All forms of restoration require that the site be protected from disturbances while forest establishment takes place. But natural forests are subject to disturbance regimes of various kinds. At what point can protection be reduced (or particular kinds of disturbances such as fires even be re-introduced)?

> *Wildlife*: Restorationists commonly assume wildlife will colonise sites once appropriate habitats develop. Is this the case in fact? Carnivores requiring large habitat areas can play an important role in structuring ecosystems (Ripple *et al.* 2014). How can these be included in restoration programs?
>
> *Relationship with agriculture*: How can restoration be undertaken on a landscape scale in ways that make it more attractive to landholders and help improve future food security within the broader landscape mosaic?

Finally, some reference should be made to the socio-economic and political context in which restoration is carried out. Whatever ecologists or conservationists might wish, restoration will only occur on a large scale if the opportunity costs are low and landowners believe it is in their interests to adopt it as a new land use practice (note that self-interest is not necessarily defined solely in financial terms). This means the future task will be to find forms of restoration that balance conservation needs as well as improve livelihoods (Lamb 2014). This may be difficult to achieve at particular sites but may be more feasible at a landscape scale across which a mosaic of land uses and restoration methods can be used.

References

Aronson, J., Milton, S. J. and Blignaut, J. N. 2007. Restoring natural capital: Definitions and rationale. Pages 3–8 *in* J. Aronson, S. J. Milton, and J. N. Blignaut, editors. *Restoring Natural Capital: Science, Business, and Practice*. Island Press, Washington, Covelo, US and London.

Ashton, M. S., Gunatilleke, C. V. S., Singhakumara, B .M. P. and Gunatilleke, I. A. U. 2001. Restoration pathways for rain forest in southwest Sri Lanka: A review of concepts and models. *Forest Ecology and Management* 154:409–430.

Benayas, J. M. R., Newton, A., Diaz, A. and Bullock, J. M. 2009. Enhancement of biodiversity and ecosystem services by ecological restoration: A meta-analysis. *Science* 325:1121–1124.

Carle, J. and Holmgren, P. 2003. *Definitions Related to Planted Forests. Working Paper 79*, Forestry Department, UN Food and Agriculture Department, Rome.

Carter, A. S., and Gilmour, D. 1989. Increase in tree cover on private farm land in Central Nepal. *Mountain Research and Development* 9:381–391.

Chazdon, R. 2014. *Second Growth: The Promise of Tropical Forest Regeneration in an Age of Deforestation*. University of Chicago Press, Chicago, US.

Cooke, J. A., and Johnson, M. 2002. Ecological restoration of land with particular reference to the mining of metals and industrial minerals: A review of theory and practice. *Environmental Reviews* 10:41–71.

Dodet, M., and Collet, C. 2012. When should exotic forest plantation tree species be considered as an invasive threat and how should we treat them? *Biological Invasions* 14:1765–1788.

Elliott, S., Blakesley, D.and Hardwick, K. 2013. *Restoring Tropical Forests: A Practical Guide*. Royal Botanic Gardens, Kew, UK.

Goosem, S. and Tucker, N. 2013. *Repairing the Rainforest: Theory and Practice of Rainforest Re-establishment in North Queensland's Wet Tropics*. Wet Tropics Management Authority and Biotropica Australia, Cairns, Australia.

Hodgson, B. and McGhee, P. 1992. Development of aerial seeding for the regeneration of Tasmanian eucalypt forests. *Tasforests* 4:77–85.

Holl, K. and Aide, M. 2011. When and where to actively restore ecosystems? *Forest Ecology and Management* 261:1558–1563.

Janzen, D. 2002. Tropical dry forest: Area de Conservacion Guanacaste, northwest Costa Rica. Pages 559–583 *in* M. Perrow and A. Davy, editors. *Handbook of Ecological Restoration*. Cambridge University Press, Cambridge, UK.

Jonson, J. 2010. Ecological restoration of cleared agricultural land in Gondwana Link: Lifting the bar at 'Peniup'. *Ecological Management and Restoration* 11:16–26.

Keenan, R., Lamb, D., Woldring, O., Irvine, T. and Jensen, R. 1997. Restoration of plant diversity beneath tropical tree plantations in northern Australia. *Forest Ecology and Management* 99:117–132.

Koch, J. and Hobbs, R. 2007. Synthesis: Is Alcoa successfully restoring a Jarrah forest ecosystem after bauxite mining in Western Australia? *Restoration Ecology* 15:S137–S154.

Kuemmerle, T., Olofsson, P., Chaskovskyy, O., Baumann, M., Ostapowicz, K. and others. 2011. Post-Soviet farmland abandonment, forest recovery, and carbon sequestration in western Ukraine. *Global Change Biology* 17:1335–1349.

Laestadius, l., Maginnis, S. Minnemeyer, S., Potapov, P., Saint-Laurent, C. and Sizer, N. 2011. Mapping opportunities for forest landscape restoration. *Unasylva* 62:47.

Lamb, D. 2011. *Regreening the Bare Hills: Tropical Forest Restoration in the Asia-Pacific Region*. Springer, Dordrecht, Netherlands.

Lamb, D. 2014. *Large-Scale Forest Restoration*. Earthscan Forest Library, Routledge, Abingdon, UK.

Löf, M., Berquist, J., Brunet, J., Karlsson, M. and Welander, N. 2010. Conversion of Norway spruce stands to broadleaved woodland – regeneration systems, fencing and performance of planted seedlings. *Ecological Bulletins* 53:165–173.

McConkey, K. R., Prasad, S., Corlett, R. T., Campos-Arceiz, A., Brodie, J. F., Rogers, H. and Santamaria, L. 2012. Seed dispersal in changing landscapes. *Biological Conservation* 146:1–13.

McNamara, S., Tinh, D. V., Erskine, P. D., Lamb, D., Yates, D. and Brown, S. 2006. Rehabilitating degraded forest land in central Vietnam with mixed native species plantings. *Forest Ecology and Management* 233:358–365.

Mori, A. S., Furukawa, T. and Sasaki, T. 2013. Response diversity determines the resilience of ecosystems to environmental change. *Biological Reviews* 88:349–364.

Noble, I. R., and Slatyer, R. 1980. The use of vital attributes to predict successional changes in plant communities subject to recurrent disturbances. *Plant Ecology* 43:5–21.

Richardson, D. M., and Rejmanek, M. 2011. Trees and shrubs as invasive alien species: A global review. *Diversity and Distributions* 17:788–809.

Ripple, W. J., Estes, J. A., Beschta, R. L., Wilmers, C. C., Ritchie, E .G. and others. 2014. Status and ecological effects of the world's largest carnivores. *Science* 343:1241484

Sendzimir, J., Reij, C. P. and Magnuszewski, P. 2011. Rebuilding resilience in the Sahel: regreening in the Maradi and Zinder regions of Niger. *Ecology and Society* 16:1.

SER. 2004. *The SER International Primer on Ecological Restoration*. Society for Ecological Restoration International, Tucson, US.

Siddique, I., Engel, V. L., Parrotta, J., Lamb, D., Nardoto, G., and others. 2008. Dominance of legume trees alters nutrient relations in mixed species forest restoration plantings within seven years. *Biogeochemistry* 88:89–101.

Stanturf, J. A., Gardiner, E. S., Shepard, J. P., Schweitzer, C. J., Portwood, C. J. and Dorris Jr., L. C. 2009. Restoration of bottomland hardwood forests across a treatment intensity gradient. *Forest Ecology and Management* 257:1803–1814.

Suding, K. N. 2011. Toward an era of restoration in ecology: Successes, failures, and opportunities ahead. *Annual Review of Ecology, Evolution, and Systematics* 42:465–487.

Temperton, V., Hobbs, R. J., Nuttle, T. and Halle, S. 2004. *Assembly Rules and Restoration Ecology*. Island Press, Washington.

Tunjal, P. and Elliott, S. 2012. Effects of seed traits on the success of direct seeding for restoring southern Thailand's lowland evergreen forest ecosystem. *New Forests* 43:319–333.

Walker, B., and Salt, D. 2006. *Resilience Thinking: Sustaining Ecosystems and People in a Changing World*. Island Press, Washington.

Zahawi, R., Holl, K., Cole, R., and Leighton, J. 2013. Testing applied nucleation as a strategy to facilitate tropical forest recovery. *Journal of Applied Ecology* 50:88–96.

29
FOREST FRAGMENTATION

Edgar C. Turner and Jake L. Snaddon

Introduction

The majority of the world's forests are now degraded and fragmented, often as a direct result of agricultural expansion (Foley *et al.* 2005). For example, in the tropics over 80 per cent of agricultural expansion that took place between 1980 and 2000 replaced forest (Gibbs *et al.* 2010). Other key drivers of forest loss and fragmentation include urbanization and expansion of road networks, which can open up forest areas for further change (Laurance *in press*). Continued demand for agricultural land and timber is likely to exacerbate these pressures on remaining forest areas still further. As a result of reduced habitat and population sizes and edge effects within fragments, forest fragmentation frequently leads to changes in animal and plant communities and shifts in ecosystem functioning.

In this chapter we will introduce the impacts of forest fragmentation on biodiversity and the underlying causes for changes in communities of plants and animals that result from forest fragmentation. We will investigate the role of habitat area, edge effects and the habitat matrix in determining changes that occur in forests following fragmentation. We will also see how the functional traits of species can influence their susceptibility to fragmentation. We will go on to consider some of the ways in which multiple drivers of ecosystem change can act together to influence communities within fragmented habitats. We will finish by investigating the impacts of fragmentation on community interactions and processes; particularly how changes in one component of a community can have cascading effects and impact on important ecosystem functions.

From the Species–Area Relationship to landscape-scale fragmentation studies

The positive relationship between habitat area and the number of species an ecosystem can support (the 'Species–Area Relationship') has been a central topic in ecological research for over 40 years (Connor and McCoy 1979) and comes close to being a universal theory in ecology. This relationship was key to the development of MacArthur and Wilson's equilibrium Theory of Island Biogeography (MacArthur and Wilson 1967), which modelled the number of species islands can support. Focussing on islands of homogenous habitat, they predicted that larger islands closer to the mainland would contain a higher number of species, as larger islands would suffer lower rates of extinction and closer islands would have higher rates of immigration. These findings have been supported by numerous observational studies, although whether area and isolation alone drives this relationship is less clear (Connor and McCoy 1979; Ricklefs and Lovette 1999).

In the wake of accelerating forest loss and fragmentation worldwide (Foley *et al.* 2005), theories relating to the number of species different sized areas can support have been expanded and adapted to apply to habitat islands (Diamond 1975): areas of forest or natural habitat isolated as a result of anthropogenic habitat conversion. Expanding research in this area has demonstrated that the impact of forest fragmentation on biological communities is driven by multiple interconnected factors that surpass those directly addressed within Island Biogeography Theory (Laurance 2008) and which occur simultaneously as forest areas become fragmented. The interdependence of landscape effects and species responses to habitat fragmentation have inspired considerable debate, stimulating the emergence of new conceptual models based on multiple and interconnected drivers (Didham *et al.* 2012).

Forest fragmentation studies

Over the last 30 years the number of studies focussing on the biological impacts of forest fragmentation has expanded dramatically. For example, from a search using the scientific search engine 'Web of Science', no publications were listed for the search terms 'forest AND fragmentation' pre 1970, only 2 for 1971–1980, 27 for 1981–1990, 1323 for 1991–2000, and 5228 for 2001–2010. The majority of these studies are observational, cataloguing and comparing biological communities and ecosystem processes between existing forest fragments. However, a growing number of studies have taken an experimental approach, allowing confounding factors, such as underlying differences in communities, to be taken into account.

A key development in fragmentation research has been the design and implementation of large-scale experimental forest fragmentation projects, which allow the impacts of forest fragmentation to be tested at the landscape scale. These include some of the longest-running and most influential experimental ecological studies worldwide (Ewers *et al.* 2011). Originally designed to answer key questions relating to minimum habitat size, the role of these studies has expanded to address a much broader range of questions, such as the impact of regeneration and matrix connectivity in determining species patterns. Key large-scale experimental forest fragmentation studies include: the Biological Dynamics of Forest Fragments Project (BDFFP), located in the tropical rainforest of the Brazilian Amazon; the Calling Lake Fragmentation Experiment, located in the boreal forest of Alberta, Canada; the Savannah River Site Corridor Experiment, located in the pine forests of South Carolina; the Wog Wog Habitat Fragmentation Experiment, located in eucalypt forest in Australia (Debinski and Holt 2000); and, most recently, the Stability of Altered Forest Ecosystems Project (SAFE), located in the lowland rainforest of Malaysian Borneo (Ewers *et al.* 2011). Most of these studies have focussed on habitat size at the core of their experimental design. For example, the BDFFP study experimentally isolated fragments of varying sizes (five 1-hectare, four 10-hectare, and two 100-hectare fragments) (Laurance *et al.* 2002) to observe the effects of changes in habitat area on species numbers and ecosystem functions. Now in its thirty-sixth year (initiated in 1979), the BDFFP study provides detailed and wide-ranging insights into the long-term drivers and effects of forest fragmentation (Laurance *et al.* 2011).

The value of these experiments is enormous: by taking an experimental approach, they allow direct testing of the impacts of fragmentation. This is harder to achieve in observational

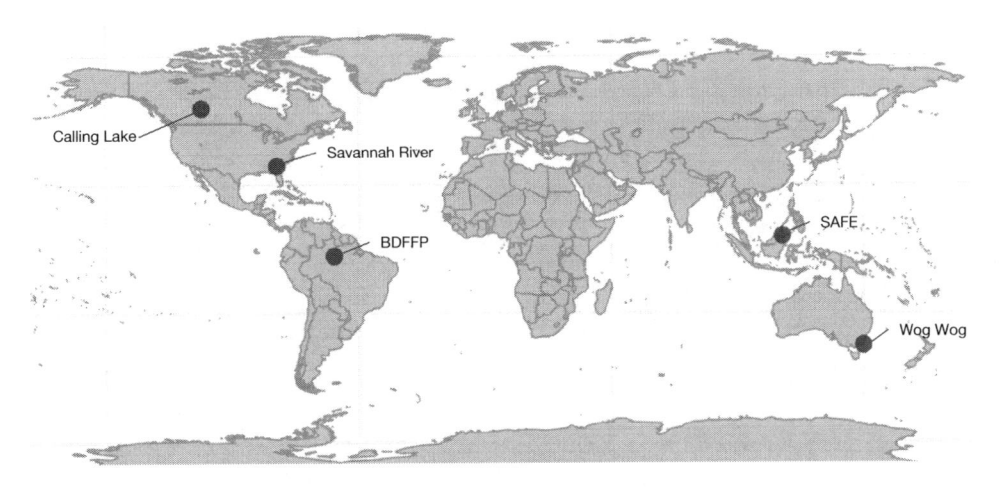

Figure 29.1 Location of some of the largest forest fragmentation studies globally

studies, which are confounded by the patchy distribution of species in the original continuous habitat and non-random forest clearance. As a result, those patches of forest remaining after non-experimental clearance are often in difficult-to-access areas, such as steep slopes, or areas prone to flooding, which may never have contained representative communities of the wider intact forest. The sheer scale that is necessary for fragmentation experiments to provide meaningful information means that they are fairly few in number, with Africa and large areas of Asia not being represented (Figure 29.1).

Fragmentation impacts

Biological communities in habitat fragments differ from those in continuous forest for a number of inter-dependent reasons:

- Fragments only contain a subset of the original community, especially where habitats were varied and species patchily distributed in the original habitat.
- Population sizes in fragments are lower, increasing the likelihood of stochastic extinction events.
- Fragments are isolated from other forest areas, reducing immigration and the chance of recolonization once local extinctions have occurred, and potentially increasing levels of inbreeding and loss of genetic diversity.
- Fragments are impacted by edge effects and changing conditions as a result of exposure to the matrix, potentially reducing the area still further for forest specialists.
- Fragments may be more vulnerable to other perturbations such as invasion by non-native species, increased hunting pressure or climate change impacts.
- Community changes as a result of fragmentation may precipitate further shifts in biotic interactions, altering functional links and processes within fragments.

Separating these different factors and the causal mechanisms behind these alterations is problematic as most changes occur simultaneously as a result of fragmentation and interact to impact on the biological community (Figure 29.2).

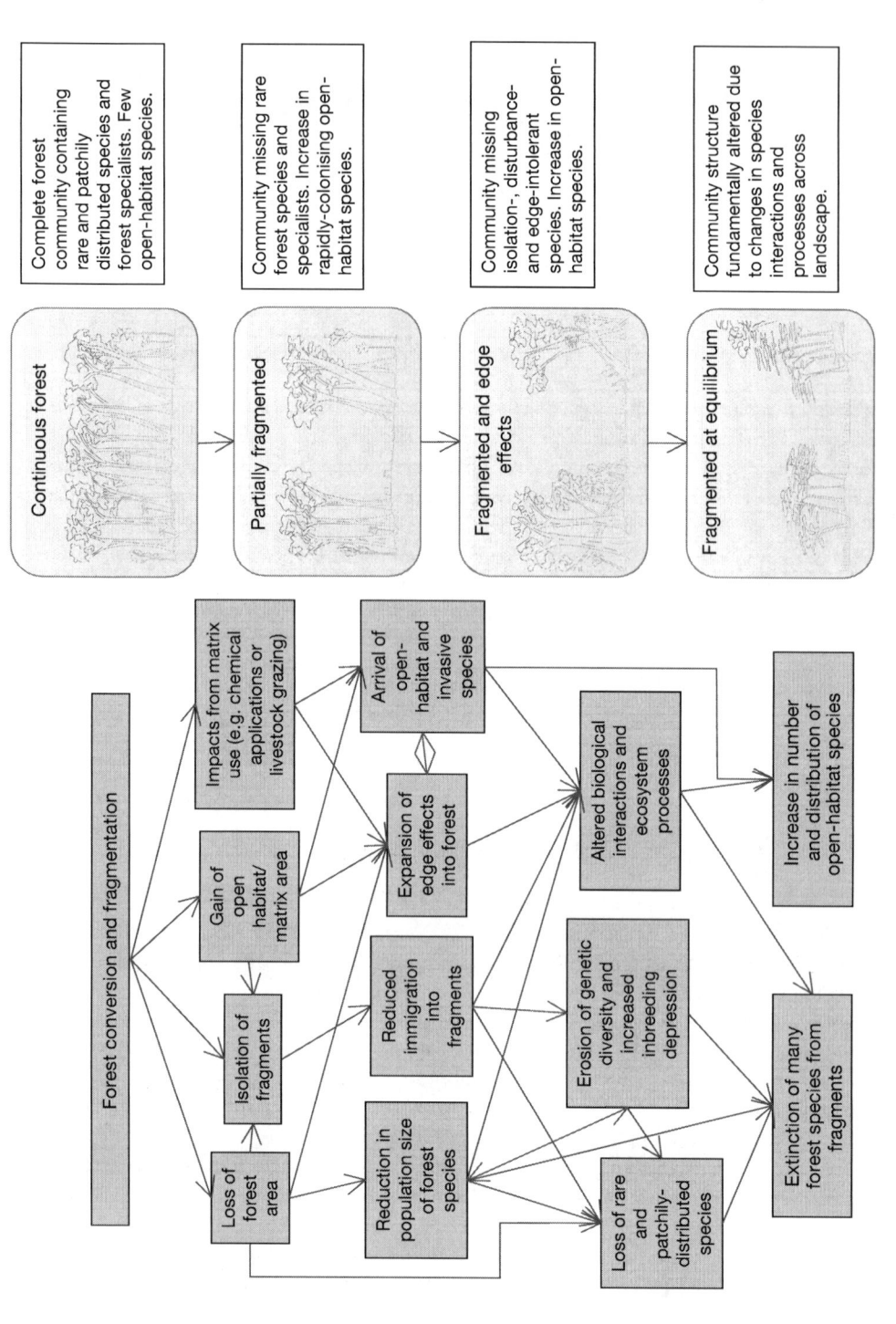

Figure 29.2 Conceptual diagram showing some of the interacting factors that lead to changes in a forest community following fragmentation

Factors associated with fragmentation impacts

Habitat patchiness and size

The lower number of species found in fragments is, in part, owing to the patchy nature of species distributions in natural habitats, meaning that some species may not be represented in forest fragments, simply because they did not occur in that area in the first place (termed 'sample effects' [Connor and McCoy 1979]). This may be exacerbated by underlying habitat heterogeneity, which is often correlated with habitat size. For example, in a study of a range of taxonomic groups on real islands in the Lesser Antilles, species diversity increased with island size. However, for several of the taxa this was related to habitat diversity increases, rather than habitat size itself (Ricklefs and Lovette 1999). Related to this point, loss of forest habitat is itself non-random, with flatter areas generally being cleared sooner than hilly areas. In addition, the clearance of forest in one area makes further clearance in adjacent areas much more likely (Boakes *et al.* 2010). As a result of these factors, forest fragments are likely to only include a subset of the original habitat types and, therefore, species represented.

A second factor is the intrinsic rarity of many species in even pristine forest (Novotny and Basset 2000), which means that, although a species may initially exist in a newly fragmented habitat, it is unlikely to persist in the long term. This loss of species over time, which may not occur immediately (and is therefore sometimes termed 'the extinction debt'), is in-line with Island Biogeography Theory, in that losses tend to be more marked and rapid in smaller fragments (Laurance *et al.* 2011), which contain smaller populations of any resident species. For example, in the Amazonian rainforest, Stouffer *et al.* (2011) found 44–84 per cent of the original bird fauna had gone extinct in 1-ha fragments 25 years after isolation, but only 31–45 per cent in 10-ha fragments, and 8–16 per cent in 100-ha fragments over this time period. Extinction debt may also take a very long time to be fully repaid: continued loss of mammal species in forest fragments in NE Australia were still being detected 70 years after the initial fragmentation event (Laurance *et al.* 2008).

Edge effects

Habitat islands differ from real islands in several ways that fundamentally affect the number of species they can support. One key factor is the rapid change in abiotic and biotic factors (such as increased wind-speed, changing microclimate or the invasion of open-habitat species) that occur during fragmentation events in forest fragment edges (called 'edge effects'), which can precipitate further changes in the fragments and increase species losses (Laurance *et al.* 2011).

Edge effects have been shown to be surprisingly extensive and important in determining the communities of plants and animals and the ecosystem functioning within forest fragments. Some of the most extensive influences of edge effects can be seen from the results of the BDFFP study (Laurance *et al.* 2002). Increased disturbance from higher wind speeds at forest edges have been measured up to 400 m into forest fragments. Effectively this means that the whole area of fragments less than 50 hectares in size may be influenced by edge effects. A wide range of other biotic and abiotic changes have also been reported within 100 m of forest edges. These include lower canopy foliage density, increased incidence of tree-fall gaps, higher leaf litter inputs, lower relative humidity, lower soil moisture, and increased air temperature (Laurance *et al.* 2002). Because forest fragments are frequently irregular in shape, the impacts of edge effects may be even more severe – with irregular-shaped forest fragments there is a greater edge to area ratio, so a larger proportion of the fragment will be impacted by edge effects.

Factors associated with edge effects can also interact to exacerbate further changes. Forest trees experience higher mortality closer to forest edges, partly owing to wind damage and partly to physiological stress caused by the rapidly changing abiotic conditions. This changes the gap dynamics of forest edges and can lead to an increase in the prevalence of early-successional tree species and a much higher abundance of climbers and lianas (Laurance et al. 2002). In some cases, these influences can be even more marked: partly owing to drier conditions and a larger volume of dead wood and partly to increased access by people, forest edges are vulnerable to fire. Once a fire has occurred the likelihood and severity of future burns increases (Cochrane et al. 1999), resulting in even higher mortality of remaining trees and a reduction in the seed bank, which can hinder forest regeneration.

Habitat isolation

Increasing isolation in smaller forest fragments can leave populations vulnerable to stochastic processes such as birth and death rates, loss of genetic variation, and environmental fluctuations. This can lead to a reduction in species richness in fragments, as once species are lost from isolated fragments there is little chance of recolonization. In a study of dung beetles on forest islands in Venezuela, more isolated islands were found to contain a lower diversity of beetles (Larsen et al. 2008). More isolated and smaller populations are also likely to suffer increased levels of inbreeding, inbreeding depression and loss of genetic diversity (Young et al. 1996; Finger et al. in press), which may also interact with environmental change in fragments to reduce populations still further (e.g. Ismail et al. 2014).

The matrix matters

Another factor that differs between real islands and habitat islands is the nature of the surrounding habitat or 'matrix' (Ricketts 2001). In real islands this is a true barrier to dispersal, but effects can be much more variable for habitat islands. Some types of matrix, such as plantation forest, may be relatively benign and help to buffer forest edges from the detrimental impacts of wind or changes in microclimate. However, other matrix types, such as pasture or arable areas, provide little buffering capacity and can lead to dramatic and expanding impacts into remaining habitat islands. In some cases the land use in the matrix can directly affect nearby forest areas. Forest areas next to cultivated land may, for example, experience drift of pesticides, herbicides and fertilizers or may trap airborne pollutants resulting in higher rates of deposition in forest edges (Weathers et al. 2001).

The matrix type can also act to support or reduce species migration between fragments. A matrix that is more favourable for forest species can help to reconnect isolated populations, increasing gene-flow and reducing the chance of local extinctions. For example, in an experimental study on butterflies in meadow patches surrounded by willow or conifer thickets in Colorado, Ricketts (2001) showed that the degree of matrix permeability for different species can influence the effective isolation of a fragment, with matrix dominated by willow being easier for a range of species to disperse through than that dominated by conifers. The permeability of the matrix can also change through time. If pasture surrounding a forest fragment begins to regrow, this can reconnect forest fragments, leading to recolonization by forest species (Stouffer and Bierregaard 1995).

Species-specific effects

The exact effect of fragmentation depends on the characteristics of the species involved (Henle et al. 2004), such as their population size and fluctuation, competitive ability, habitat specialization, sensitivity to disturbance and dispersal ability. Fragments therefore tend to contain a non-random subset of the full forest assemblage (Sekercioglu et al. 2002). In Costa Rica, more dispersive insectivorous birds are more likely to persist in forest fragments (Sekercioglu et al. 2002). In Venezuela, large-bodied, forest-specialist, and rare dung beetles are more vulnerable than other groups to fragmentation (Larsen et al. 2008). Species with close biotic interactions may also be particularly vulnerable to fragmentation, as the disappearance of one species will result in the loss of the other. For example, birds which follow foraging army ant colonies to catch fleeing insects are generally lost from forest fragments (Stouffer and Bierregaard 1995) as the army ants themselves require extensive forest areas to persist.

Synergistic impacts with other environmental change

Habitat fragmentation is not the only environmental change that is affecting forest communities; it is clear that multiple drivers are responsible for losses of biodiversity and changing ecosystem functions globally (Brook et al. 2008). Other drivers of habitat change such as the introduction of invasive species, hunting, pollution and the use of pesticides, and climate change can act synergistically to further influence forest fragment communities. For example, edge effects can facilitate the colonization of invasive species into disturbed environments (Hobbs and Yates 2003), potentially outcompeting native species. Fragmented habitats are easier to access for hunting and may be in closer proximity to human populations, leading to greater hunting pressure (Peres 2001). The ability of species to cope with changes in climate as a result of global warming may be influenced by fragmentation (Corlett in press) and be dependent on the level of structural complexity and habitat fragmentation (Brodie et al. 2012). A fragmented habitat may hinder dispersal, effectively isolating species in unfavourable climatic conditions and precipitating higher rates of extinction.

Community interactions and ecosystem functions

Forest fragmentation and the resulting loss of species and changing community interactions can have a measurable effect on a wide range of ecosystem processes, although the severity and direction of these impacts can vary (Schleuning et al. 2011). The predictability of changes generally depends on the range of factors that determine particular ecosystem functions (Peh et al. in press). Impacts of fragmentation on ecosystem functions such as herbivore control or long-distance seed dispersal, which may be dependent on a few large species, are generally predictable. In contrast, fragmentation impacts on functions such as decomposition, which are dependent on a large number of species as well as environmental conditions, are more variable. For example, Didham (1998) found higher decomposition rates in Amazonian forest fragment edges compared to the forest interior, but lower rates of decomposition in smaller forest fragments.

The loss of top predators from smaller fragments has been widely reported and may fundamentally alter and shorten the food web of an area, impacting on lower trophic levels, reducing predation rates and precipitating further species losses. Smaller forest islands in Venezuela without top predators were also found to have an increased number of seed predators and herbivores and a resultant higher herbivory and reduced seedling and sapling recruitment (Terborgh et al. 2001).

The loss of pollinators in fragmented habitats has also been reported to have an impact on plant pollination and reproduction, with animal-pollinated plant species being more severely impacted than other groups (Aguilar et al. 2006). Forest fragments in the Brazilian Atlantic Forest contain a lower functional diversity of pollination systems compared to continuous forest. As a result, trees dependent on pollination by birds, flies and non-flying mammals, and self-incompatible species, decline in fragments, while hermaphroditic tree species increase (Girão et al. 2007).

Forest fragmentation can also lead to reduced fruit removal and dispersal both within and between fragments, with the severity of this effect being influenced by multiple factors including the size of the fragment, the permeability of the matrix and characteristics of the seed plants and their dispersers (McConkey et al. 2012). The loss and reduced abundance of large fruit feeders, such as howler monkeys, that occurs alongside fragmentation can also affect the abundance and species richness of dung beetles (Feer and Hingrat 2005). This in turn can reduce the percentage of seeds buried by the beetles (Andresen 2013) with potential impacts on plant communities, as seeds buried by dung beetles are much less likely to be predated by rats than non-buried seeds.

Predicting the impacts of forest fragmentation on ecosystem functioning remains a key challenge for ecologists. Central to this is careful quantification of the exact relationship between factors that predict species loss ('species disassembly rules') and the role that species perform in ecosystems. This depends on the relationship between the response of species to habitat change (response traits) and the effect of species on ecosystem functioning (effect traits). There is evidence that larger bodied dung beetle species may be more sensitive to habitat fragmentation and also more effective in supporting ecosystem functioning (Larsen et al. 2005). Early loss of these species will therefore have a non-random and disproportionately severe impact on ecosystem functioning.

Applications of forest fragmentation studies

Understanding the changes in communities of animals and plants and the consequences for ecosystem functions as a result of forest fragmentation is important for both theoretical and practical reasons. With pressure mounting on remaining forest areas worldwide, the number of large, intact forest habitats is dwindling. It is therefore important to minimize the losses that occur in fragmented landscapes, as these are becoming, or are, the norm in large parts of the world. Tailoring management and conservation practices based on scientific evidence is central to this. The results of fragmentation studies have long been applied to reserve design, with recommendations focussed on favouring larger reserves or increasing the size of small areas, linking together isolated reserves, favouring the location of new reserves so they are close to areas of existing habitat and creating circular and regularly-shaped reserves, which minimize edge effects (Diamond 1975).

The value of corridors in linking together forest fragments and reconnecting isolated populations has received considerable attention. There is strong evidence that corridors are effective in promoting movement of species between isolated fragments and therefore in reducing some of the negative impacts of habitat isolation (Gilbert-Norton et al. 2010). Such forest corridors are likely to be particularly valuable in extensive and intensive agricultural systems, where the matrix is unsuitable for most forest species and migrations between forest fragments is therefore minimal. More recently, the relationship between fragmented, but interconnected populations in neighbouring habitat fragments ('metapopulations' [Hanski 1999]) has been highlighted as important in allowing species to persist at the landscape scale. In these systems, recurrent immigrations from neighbouring patches allow recolonization of areas where species have recently become extinct, stabilizing the population as a whole.

An understanding of habitat fragmentation and its effect on species diversity can also be used to model and predict future biodiversity loss (e.g. Wearn *et al.* 2012), informing and forewarning decision makers. Understanding the landscape history of an area may also be important for this. Recent work indicates that the historical fragmentation process and shared history of forest fragments may be important in determining the identity of species found in different fragments and could be used to identify priority patches for conservation (Ewers *et al.* 2013).

Finally, the maintenance of forest fragments in a matrix of agriculture can also have beneficial impacts on crop production (Klein *et al. in press*), by supporting beneficial species which carry out important ecosystem functions, such as pollination and pest control (Shackelford *et al.* 2013). In this context, the maintenance of forest fragments is not only a conservation issue, but also one of sustainable agricultural productivity.

Conclusion

Fragmentation research has come a long way since the development of the Theory of Island Biogeography. It has moved from a focus on fragment area to a more sophisticated approach that aims to understand the underlying ecological processes determining and resulting from community changes across fragmented landscapes. It is clear that the effects of forest fragmentation are far from straightforward with area, edge, matrix, community interactions and time all playing a part in determining the changes observed in biological communities once forest becomes fragmented. Major gaps still exist in our knowledge of fragmentation effects, including a detailed understanding of the long-term, whole-ecosystem impacts of fragmentation or the interacting impacts of environmental change on communities of plants and animals in fragmented habitats. It is also unclear how these changes will effect earth–atmosphere interactions and how these may scale-up to effect processes at larger spatial scales (Klein *et al. in press*).

Despite these gaps, the wealth of information gleaned from forest fragmentation studies already has huge potential to inform conservation management strategies, to predict future ecosystem changes and to prioritize conservation actions. Much of the world's forests have already been cleared and fragmented or logged, and it is becoming clear that these changes have been ongoing for hundreds if not thousands of years (Ellis *et al.* 2013). What is now important is to make the best decisions to manage what is left.

Acknowledgements

We would like to thank Richard Corlett, William Foster, Chris Kettle and Sarah Luke for their helpful comments on drafts of this manuscript. ECT is funded by the Isaac Newton Trust, Cambridge and PT SMART Research Institute.

References

Aguilar, R., Ashworth, L., Galetto, L., and Aizen M.A. (2006) 'Plant reproductive susceptibility to habitat fragmentation: review and synthesis through a meta-analysis', *Ecology Letters*, vol 9, pp. 968–980

Andresen, E. (2013) 'Effect of forest fragmentation on dung beetle communities and functional consequences for plant regeneration', *Ecography*, vol 26, no 1, pp. 87–97

Boakes, E.H., Mace, G.M., McGowan, P.J.K. and Fuller, R.A. (2010) 'Extreme contagion in global habitat clearance', *Proceedings of the Royal Society B: Biological Sciences*, vol 277, no 1684, pp. 1081–1085

Brodie, J., Post, E. and Laurance, W.F. (2012) 'Climate change and tropical biodiversity: a new focus', *Trends in Ecology and Evolution*, vol 27, no 3, pp. 145–150

Brook, B.W., Sodhi, N.S. and Bradshaw, C.J.A. (2008) 'Synergies among extinction drivers under global change', *Trends in Ecology and Evolution*, vol 23, no 8, pp. 453–460

Cochrane, M., Alencar, A., Schulze, M.D., Souza, C.M. Jr., Nepstad, D.C. and others (1999) 'Positive feedbacks in the fire dynamic of closed canopy tropical forests', *Science*, vol 284, no 5421, pp. 1832–1835

Connor, E.F. and McCoy, E.D. (1979) 'The statistics and biology of the species-area relationship', *The American Naturalist*, vol 113, no 6, pp. 791–833

Corlett, R.T. (*in press*) 'Forest fragmentation and climate change', in C.J. Kettle and L.P. Koh (eds) *Global Forest Fragmentation*, CABI, Wallingford, UK

Debinski, D. and Holt, R. (2000) 'A survey and overview of habitat fragmentation experiments', *Conservation Biology*, vol 14, pp. 342–355

Diamond, J. (1975) 'The island dilemma: lessons of modern biogeographic studies for the design of natural reserves', *Biological Conservation*, vol 7, pp. 129–146

Didham, R.K. (1998) 'Altered leaf-litter decomposition rates in tropical forest fragments', *Oecologia*, vol 116, pp. 397–406

Didham, R.K., Kapos, V. and Ewers, R.M. (2012) 'Rethinking the conceptual foundations of habitat fragmentation research', *Oikos*, vol 121, pp. 161–170

Ellis, E.C., Kaplan, J.O., Fuller, D.Q., Vavrus, S., Goldewijk, K.K. and Verburg, P.H. (2013) 'Used planet: a global history', *Proceedings of the National Academy of Sciences of the United States of America*, vol 110, no 20, pp. 7978–7985

Ewers, R.M., Didham, R.K., Fahrig, L., Gonçalo, F., Hector, A. and others (2011) 'A large-scale forest fragmentation experiment: the Stability of Altered Forest Ecosystems Project', *Philosophical Transactions of the Royal Society B: Biological Sciences*, vol 366, pp. 3292–3302

Ewers, R.M., Didham, R.K., Pearse, W.D., Lefebvre, V., Rosa, I.M.D. and others (2013) 'Using landscape history to predict biodiversity patterns in fragmented landscapes', *Ecology Ltters*, vol 16, pp. 1221–1233

Feer, F. and Hingrat, Y. (2005) 'Effects of forest fragmentation on a dung beetle community in French Guiana', *Conservation Biology*, vol 19, no 4, pp. 1103–1112

Finger, A., Radespiel, U., Habel, J.C. and Kettle, C.J. (*in press*) 'Forest fragmentation genetics: What can genetics tell us about forest fragmentation?', in C.J. Kettle and L.P. Koh (eds) *Global Forest Fragmentation*, CABI, Wallingford, UK

Foley, J.A., DeFries, R., Asner, G.P., Barford, C., Bonan, G. and others (2005) 'Global consequences of land use', *Science*, vol 309, pp. 507–574

Gibbs, H.K., Ruesch, A.S., Achard, F., Clayton, M.K., Holmgren, P. and others (2010) 'Tropical forests were the primary sources of new agricultural land in the 1980s and 1990s', *Proceedings of the National Academy of Sciences of the United States of America*, vol 107, no 38, pp. 16732–16737

Gilbert-Norton, L., Wilson, R., Stevens, J.R. and Beard, K.H. (2010) 'A meta-analytic review of corridor effectiveness', *Conservation Biology*, vol 24, no 3, pp. 660–668

Girão, L.C., Lopes, A.V., Tabarelli, M. and Brunaet, E.M. (2007) 'Changes in tree reproductive traits reduce functional diversity in a fragmented Atlantic forest landscape', *PLoS ONE*, vol 2, no 9, p. e908

Hanski, I. (1999) *Metapopulation Ecology*, Oxford University Press, Oxford, UK

Henle, K., Davies, K.F., Kleyer, M., Margules, C. and Settele, J. (2004) 'Predictors of species sensitivity to fragmentation', *Biodiversity and Conservation*, vol 13, pp. 207–251

Hobbs, R. and Yates, C.J. (2003) 'Turner Review No. 7 Impacts of ecosystem fragmentation on plant populations: generalising the idiosyncratic', *Australian Journal of Botany*, vol 51, pp. 471–488

Ismail, S.A., Ghazoul, J., Ravikanth, G., Kushalappa, C.G., Uma Shaanker, R., and Kettle, C.J. (2014) 'Forest trees in human modified landscapes: ecological and genetic drivers of recruitment failure in *Dysoxylum malabaricum* (Meliaceae)', *PLoS One*, vol 9, no. 2, pp. e89437

Klein, A-M., Boreux, V., Bauhus, J., Jahi Chappell, M., Fischer, J. and Philpott, S.M. (*in press*) 'Forest islands in an agricultural sea', in C.J. Kettle and L.P. Koh (eds) *Global Forest Fragmentation*, CABI, Wallingford, UK

Larsen, T.H., Williams, N.M. and Kremen, C. (2005) 'Extinction order and altered community structure rapidly disrupt ecosystem functioning', *Ecology Letters*, vol 8, no 5, pp. 538–547

Larsen, T.H., Lopera, A. and Forsyth, A. (2008) 'Understanding trait-dependent community disassembly: dung beetles, density functions, and forest fragmentation', *Conservation Biology*, vol 22, no. 5, pp. 1288–1298

Laurance, W.F. (2008) 'Theory meets reality: How habitat fragmentation research has transcended island biogeographic theory', *Biological Conservation*, vol 141, no 7, pp. 1731–1744

Laurance, W.F. (*in press*) 'Contemporary drivers of habitat fragmentation', in C.J. Kettle and L.P. Koh (eds) *Global Forest Fragmentation*, CABI, Wallingford, UK

Laurance, W.F., Lovejoy, T.E., Vasconcelos, H.L., Bruna, E.M., Didham, R.K. and others (2002) 'Ecosystem decay of Amazonian forest fragments: a 22-year investigation', *Conservation Biology*, vol 16, no 3, pp. 605–618

Laurance, W.F., Camargo, JL.C., Luizão, R.C.C., Laurance, S.G., Pimm, S.L. and others (2011) 'The fate of Amazonian forest fragments: a 32-year investigation', *Biological Conservation*, vol 144, no 1, pp. 56–67

Laurance, W.F., Laurance, S.G. and Hilbert, D.W. (2008) 'Long-term dynamics of a fragmented rainforest mammal assemblage', *Conservation Biology*, vol 22, no 5, pp. 1154–1164

MacArthur, R. and Wilson, E.O. (1967) *The Theory of Island Biogeography*, Princeton University Press, Princeton, USA

McConkey, K.R., Prasad, S., Corlett, R.T., Campos-Arceiz, A., Brodie, J.F. and others (2012) 'Seed dispersal in changing landscapes', *Biological Conservation*, vol 146, no 1, pp. 1–13

Novotny, V. and Basset, Y. (2000) 'Rare species in communities of tropical insect herbivores: pondering the mystery of singletons', *Oikos*, vol 3, pp. 564–572

Peh, K.S.-H., Lin, Y., Luke, S.H., Foster, W.A. and Turner, E.C. (*in press*) 'Forest fragmentation and ecosystem function', in C.J. Kettle and L.P. Koh (eds) *Global Forest Fragmentation*, CABI, Wallingford, UK

Peres, C.A. (2001) 'Synergistic effects of subsistence hunting and habitat fragmentation on Amazonian forest vertebrates', *Conservation Biology*, vol 15, no 6, pp. 1490–1505

Ricketts, T.H. (2001) 'The matrix matters: effective isolation in fragmented landscapes', *The American Naturalist*, vol 158, no 1, pp. 87–99

Ricklefs, R.E. and Lovette, I.J. (1999) 'The roles of island area per se and habitat diversity in the species-area relationships of four Lesser Antillean faunal groups', *Journal of Animal Ecology*, vol 68, pp. 1142–1160

Schleuning, M., Farwig, N., Peters, M.K., Bergsdorf, T., Bleher, B. and others (2011) 'Forest fragmentation and selective logging have inconsistent effects on multiple animal-mediated ecosystem processes in a tropical forest', *PloS One*, vol 6, no 11, p. e27785

Sekercioglu, C.H., Ehrlich, P.R., Daily, G.C., Aygen, D., Goehring, D. and Sandi, R.F. (2002) 'Disappearance of insectivorous birds from tropical forest fragments', *Proceedings of the National Academy of Sciences of the United States of America*, vol 99, no 1, pp. 263–267

Shackelford, G., Steward, P.R., Benton, T.G., Kunin, W.E., Potts, S.G. and others (2013) 'Comparison of pollinators and natural enemies: a meta-analysis of landscape and local effects on abundance and richness in crops', *Biological Reviews of the Cambridge Philosophical Society*, vol 88, no 4, pp. 1002–1021

Stouffer, P. and Bierregaard, R. (1995) 'Use of Amazonian forest fragments by understory insectivorous birds', *Ecology*, vol 76, no 8, pp. 2429–2445

Stouffer, P.C., Johnson, E.I., Bierregaard, R.O. Jr and Lovejoy, T.E. (2011) 'Understory bird communities in Amazonian rainforest fragments: species turnover through 25 years post-isolation in recovering landscapes', *PLoS ONE*, vol 6, no 6, p. e20543

Terborgh, J., Lopez, L., Nunez, V.P., Rao, M., Shahabuddin, G. and others (2001) 'Ecological meltdown in predator-free forest fragments', *Science*, vol 294, no 5548, pp. 1923–1926

Wearn, O.R., Reuman, D.C. and Ewers, R.M. (2012) 'Extinction debt and windows of conservation opportunity in the Brazilian Amazon', *Science*, vol 337, no 6091, pp. 228–232

Weathers, K.C., Cadenasso, M.L. and Pickett, S.T.A. (2001) 'Forest edges as nutrient and pollutant concentrators: potential synergisms between fragmentation, forest canopies, and the atmosphere', *Conservation Biology*, vol 15, no 6, pp. 1506–1514

Young, A., Boyle, T. and Brown, T. (1996) 'The population genetic consequences of habitat fragmentation for plants', *Trends in Ecology and Evolution*, vol 11, pp. 413–418

30
ECOLOGICAL EFFECTS OF LOGGING AND APPROACHES TO MITIGATING IMPACTS

Paul Woodcock, Panu Halme and David P. Edwards

History and significance of logging

Globally, approximately 31 per cent (1.2 billion hectares) of forests are primarily managed for wood production, with a further 25 per cent designated for multiple uses including wood (FAO 2010). Wood removals are valued at around $100 billion annually and forest management thus represents an important source of employment (Figure 30.1; FAO 2010). Historically, logging was motivated by demand for firewood, housing materials and agricultural land, although the extent of logging and current trends differ substantially amongst biomes.

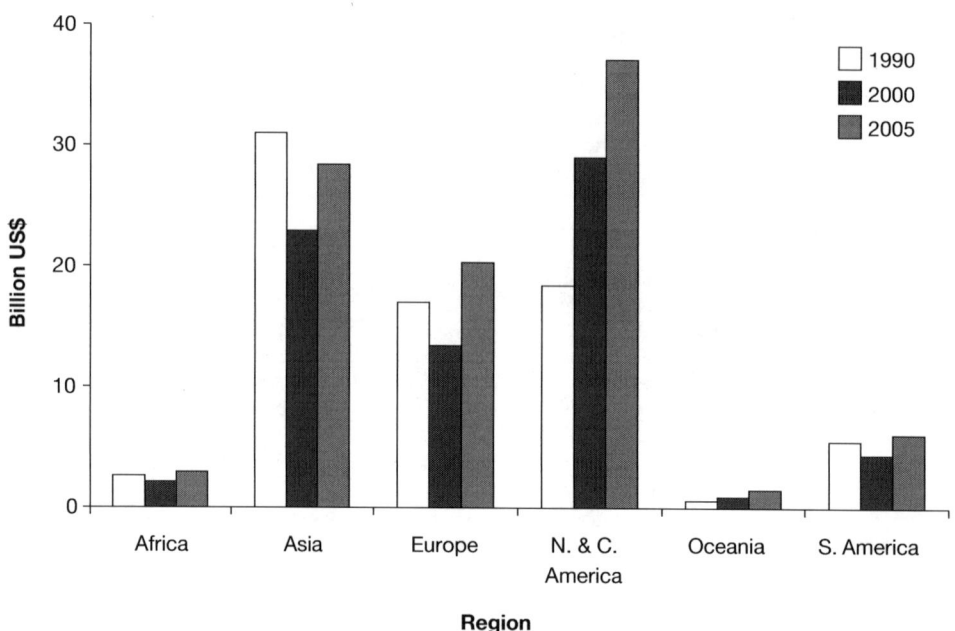

Figure 30.1 Value of wood removals (billion US$) by region in 1990, 2000, and 2005

Source: FAO (2010), *Global Forest Resources Assessment. Main Report.* http://www.fao.org/docrep/013/i1757e/i1757e00.htm (adapted with permission).

In Europe, logging of temperate forest dates back at least 5,000 years and from the fifteenth century became increasingly important for providing timber for shipbuilding, with demand for tar and charcoal also driving logging in boreal regions (Wallenius *et al.* 2010). Two centuries later, logging in North America followed a similar pattern. Consequently, although remote regions of Europe and North America retain significant areas of natural forest, logging in boreal and temperate biomes now occurs primarily in semi-natural and plantation forest (Brunet *et al.* 2010; FAO 2010). In contrast, until the 1950s, logging in the tropics was geographically restricted and either motivated by demand for fuel and housing, or focused on selectively harvesting precious woods such as mahogany and ebony. Improved transportation and increased availability of machinery has resulted in a rapid expansion of tropical selective logging during the past 50 years with a wider range of species commercially exploited (Edwards *et al.* 2014). Now encompassing around half the permanent tropical forest estate (Blaser *et al.* 2011), selective logging is amongst the most widespread forms of anthropogenic disturbance in the tropics.

Conventional approaches to logging

Clear-cutting

Clear-cutting is probably the most common form of logging in boreal and temperate zones (Rosenvald and Lõhmus 2008; Kuuluvainen *et al.* 2012), and involves harvesting almost all tree cover in a single operation. Clear-cutting may be uniform across an area or carried out in smaller strips or patches, with operations ranging in extent from <1 hectare (ha) up to around 50 ha. Although small numbers of trees are often retained, the regenerating stand is predominantly even-aged, with silvicultural treatments (e.g. soil preparation, planting seeds or saplings) sometimes applied to facilitate regrowth. Clear-cutting also occurs in the tropics, but generally precedes conversion to different land uses (e.g. agriculture). Even tropical timber plantations are highly artificial systems relative to natural forest, and we therefore consider clear-cutting of tropical forest as land-use conversion and do not discuss this topic further.

Shelterwood cutting

Unlike clear-cutting, shelterwood forestry aims to maintain a continuous (partial) canopy cover throughout harvesting. Practised in some temperate deciduous forests (Brunet *et al.* 2010), this approach favours the regeneration of shade-tolerant species and involves an initial harvest of selected trees to thin the canopy, with further thinning guided by the light requirements of the desired regeneration. Some mature seed trees are retained through these thinnings, and when sufficient regeneration is achieved the remaining canopy is felled, resulting in a young, even-aged stand.

Selective logging

Selective logging is the most common form of timber extraction in the tropics, and involves harvesting all trees of commercially valuable species above specified diameter limits within a logging coupe. The volume of wood harvested varies substantially across the tropics depending on the density of marketable trees and the accessibility of the timber concession, with logging intensities typically highest in Southeast Asia and lower in South America and Africa (Putz *et al.* 2012).

Selective logging has been largely replaced by clear-cutting and shelterwood forestry in Eurasia and North America, although the potential for broader ecosystem benefits (e.g. aesthetic value, reduced soil erosion) has rekindled interest (Kuuluvainen *et al.* 2012). Operations can involve harvesting individual trees across the stand, or 'group selection' in which all trees from particular size classes are harvested from patches of forest and lower intensity thinning of the surrounding matrix takes place. Logging is often planned to retain trees from a range of size classes, with small-diameter trees dominating (Kuuluvainen *et al.* 2012). Unlike shelterwood systems, selective logging thus results in an uneven-aged stand, although both practices maintain a continuous canopy cover throughout.

Logging cycles

Forests managed for long-term wood production are intended to follow logging cycles – also termed rotations – in which trees are harvested and the forest is then left to regenerate for several decades before being relogged. These cycles ostensibly aim to maximise the volume of timber from each harvest without compromising future yields. In clear-cutting and shelterwood systems, regeneration must allow for growth from seed or sapling to optimal harvest age and so logging cycles are typically long (50–140 years, Brunet *et al.* 2010; Kuuluvainen *et al.* 2012). By contrast, selective logging in temperate and boreal forests involves much shorter cutting cycles, and may be viable with 20–30 year rotations (Kuuluvainen *et al.* 2012).

In the tropics, selective logging cycles of 30–60 years are typically proposed. However, the relatively recent advent of large-scale logging means that the long-term sustainability of these operations is uncertain. Furthermore, relogging sometimes takes place considerably earlier than originally intended (Edwards *et al.* 2014). To make early relogging economically viable, harvests of smaller trees are allowed, but despite removing large numbers of stems, timber yields are low and long-term harvest potential may be severely compromised (Putz *et al.* 2012). As the extent of unharvested tropical forest declines, repeated logging is likely to become increasingly common, and so an improved understanding of relogging impacts is central to successful long-term forest management and conservation.

Ecological impacts of logging

Forest structure and abiotic conditions

The immediate physical impacts of logging are most pronounced in clear-cutting: the loss of tree cover causes major changes to abiotic conditions (e.g. large increases in light levels, greater variability in temperature), and intensive machine use compacts the soil and breaks dead wood into small pieces (Brunet *et al.* 2010). Contemporary clear-cutting operations are sometimes accompanied by additional mechanical removal of dead wood and plant debris for safety reasons, to facilitate stand regeneration or, increasingly, as biofuel (Bouget *et al.* 2012), causing further damage to soil and residual vegetation. Without appropriate mitigation (see section on mitigating the impacts of logging), regenerating stands can thus become highly homogenous in composition, age, and structure.

Shelterwood and selective logging systems have less pronounced physical impacts than clear-cutting, but changes can still be substantial. Conventional shelterwood management progressively opens up the canopy and ultimately results in relatively homogeneous even-aged stands. Whilst selectively logged forest can maintain greater heterogeneity, forest structure nonetheless deteriorates as the largest, oldest trees are progressively harvested (Kuuluvainen *et al.* 2012). Shelterwood harvesting and selective logging also cause substantial mortality to non-target vegetation. In the tropics, felling and the construction of skid trails (corridors in the forest in which large vegetation is flattened to allow extraction of logs), logging roads, and logging decks can result in severe damage or mortality of 10–20 trees for every stem harvested (Feldpausch *et al.* 2005; Putz *et al.* 2008a), whilst in boreal and temperate forests, post-harvest mortality of residual trees from windthrow is problematic (Thorpe *et al.* 2008). Although shelterwood and selective logging systems aim to maintain continuous canopy cover, large gaps and canopy damage can thus result, leading to greater variability in soil and understory temperatures and increased light levels at the forest floor. Fast-growing understory plants and/or pioneer tree species are then favoured, often accompanied by dense tangles of lianas in tropical regions (Perry and Thill 2013; Edwards *et al.* 2014). Forest regrowth closes most canopy gaps within a few years, but recovery of mature forest structure can take far longer (Asner *et al.* 2004; Josefsson *et al.* 2010). Furthermore, in the most heavily impacted areas (e.g. frequently used skid trails), severe soil compaction combines with changes in microclimate to restrict regrowth to grassy, scrub-like vegetation for several years or even decades.

In addition to the smaller-scale changes in forest structure outlined above, at larger scales most selectively logged forests contain areas in which felling is limited by a lack of marketable trees or steepness, as well as parts subjected to very intensive harvesting. Large selective logging concessions thus encompass considerable variation in disturbance levels, sometimes including pockets of relatively intact forest. At wider spatial scales, differences in the timing of logging operations (for clear-cutting or for continuous cover forestry) usually also result in a mosaic of patches, each in a distinct stage of recovery from disturbance.

Biodiversity

The effects of logging on biodiversity are driven by shifts in forest structure and abiotic conditions and by the associated changes in resource availability. Impacts are most immediate following clear-cutting, in which short-term changes strongly favour species tolerant of open habitats (Graham-Sauvé *et al.* 2013). Where the conservation of early successional species is required, regenerating clear-cuts can thus provide valuable habitat – particularly if key structures typical of old-growth forest are retained (Perry and Thill 2013). Equally, however, the removal of most of the large, old trees and woody debris during clear-cutting causes declines amongst many birds and dead wood-dependent invertebrates, bryophytes, and fungi, whilst the major changes in microclimate regimes following canopy removal are detrimental to a range of forest specialists. In the absence of mitigation (see section on mitigating the impacts of logging), the structurally homogeneous single-species stand typical of regenerating clear-cuts thus has limited value for many forest taxa, with several species groups taking decades to recover the composition found in mature forest (Paillet *et al.* 2010).

Through gradual harvesting and the maintenance of continuous canopy cover, shelterwood management has fewer immediate effects on forest taxa than clear-cutting (Graham-Sauvé *et al.* 2013), and has been proposed for situations where early successional forest is required but clear-cutting is considered problematic (Perry and Thill 2013). Nonetheless, conventional shelterwood systems still cause declines across a range of taxa (Brunet *et al.* 2010), particularly following the final harvest of seed trees. In the case of bryophytes and fungi, adverse effects of shelterwood management are caused by the removal of dead wood and by the hotter, drier microclimate that results from increased canopy openness. Similarly, many bird species typical of mature forest require structures for nesting and foraging that are rare in conventional shelterwoods (e.g. large, old trees and standing dead wood).

Total species richness in selectively logged forest is often very similar to that in unlogged forest, and in some instances may be higher in the former (Gibson et al. 2011; Graham-Sauvé et al. 2013; Figure 30.2). As with other logging practices, however, species richness patterns can mask significant changes in community composition, with generalist and disturbance-tolerant species increasing in abundance following logging whilst old-growth forest specialists decline. In tropical forests, for example, the dense post-logging understory vegetation adversely affects insectivorous bird and bat species that require an open understory for effective hunting (Peters et al. 2006; Edwards et al. 2011), whilst the combination of soil compaction and changes in temperature and humidity cause declines amongst soil-dwelling invertebrates (Vasconcelos et al. 2000; Jones et al. 2003). These changes in composition highlight the vulnerability of many tropical forest species to selective logging, although there is also evidence that the majority of species found in unlogged forest can persist despite intensive logging disturbance (Edwards et al. 2011; Woodcock et al. 2011), perhaps partly by surviving at lower densities in pockets of less degraded habitat.

In boreal and temperate forests, significant impacts of selective logging result particularly from the removal of large, old trees, which reduces nesting and foraging habitats in the short term and limits the accumulation of dead wood in the longer term (Abrego and Salcedo 2013). The latter changes can have detrimental impacts on dead wood-dependent taxa that persist for several decades (Josefsson et al. 2010). Equally, however, some forest taxa that decline following selective logging may recolonise relatively rapidly (Vanderkerkhove et al. 2011), highlighting the potential for selectively logged forest to contribute to biodiversity conservation.

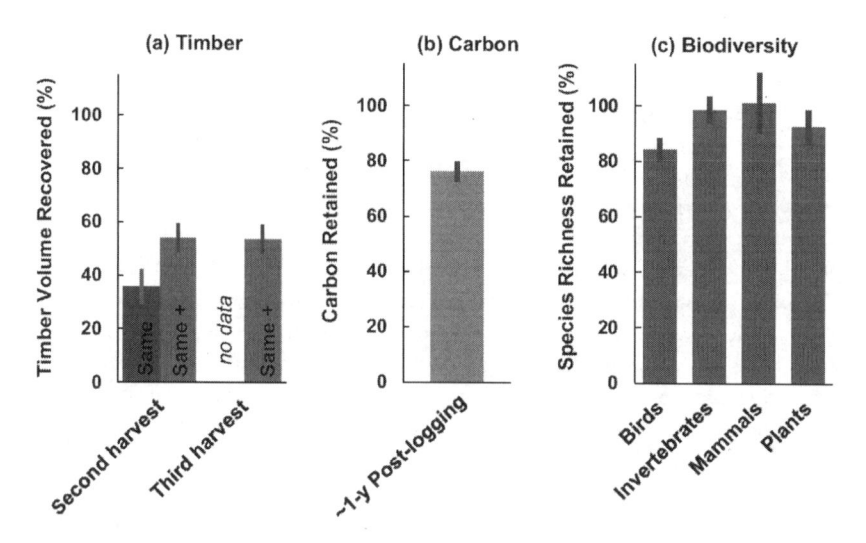

Figure 30.2 Results from meta-analysis examining the effects of tropical selective logging on (a) merchantable timber volume, (b) carbon stocks in living tree biomass approximately one year after logging, and (c) species richness in selectively logged forest as a percentage of that in old-growth forest. In (a), the volume harvested during the second and third cut is shown (cutting cycles 20–40 years). 'Same' refers to logging in which the same species are harvested, 'Same+' refers to logging in which additional species are harvested. Means and standard errors for (a), (b) and (c) are based on 59, 22 and 109 studies, respectively.

Source: Reproduced from Putz, F. E. et al. (2012), Sustaining conservation values in selectively logged tropical forests: the attained and the attainable, Conservation Letters, 5, 296–303, with permission from John Wiley and Sons. Copyright and Photocopying © 2012 Wiley Periodicals Inc.

Biotic interactions and ecological processes

Research on the ecological impacts of logging has generally focused on biodiversity metrics (e.g. species richness and composition), but logging disturbance can also alter key intra- and inter-specific interactions with implications for ecosystem functioning (Edwards *et al.* 2014). Perhaps the most extreme instances are major pest outbreaks in managed stands of boreal and temperate forests. Such outbreaks occur naturally in the relatively low diversity forests of these regions, and so may be exacerbated in systems managed for a small number of tree species with a homogenous age structure and reduced genetic variation (Rönnberg 2000; Wermelinger 2004). Similar factors may underlie increased infection rates by root rot fungi in trees damaged by logging (Rönnberg 2000).

Other changes in biotic interactions are sometimes inferred from shifts in the relative abundances of functional guilds (e.g. herbivores, frugivores, etc.). For example, meta-analysis of avian community responses to tropical selective logging identified insectivores and frugivores as being particularly vulnerable, suggesting that ecological processes such as seed dispersal and insect control may be affected by logging (Gray *et al.* 2007). Although relatively few studies have quantitatively compared ecological processes between unlogged and logged forest, there is evidence for reduced seed removal rates and shorter dispersal distances in logged tropical forest (Markl *et al.* 2012 and references therein, see also later section on the indirect effects of logging). Moreover, changes in biotic interactions can occur independently of shifts in species composition, as the ecology of individual species shifts in response to disturbance (Edwards *et al.* 2013; Woodcock *et al.* 2013).

Carbon stocks

Forests are increasingly recognised as an important global carbon sink, and so understanding and mitigating the effects of logging on forest carbon stocks represents a key aspect of strategies to combat climate change. This section focuses on changes in forest carbon, but note that from a carbon accounting perspective, the fate of timber products will differ amongst forest types and influence net CO_2 emissions from logging.

Clear-cutting operations remove almost all above ground vegetation, and therefore cause a major reduction in carbon stocks. Some carbon is retained in remaining dead wood, although, because fallen branches decay quickly, this pool is rapidly depleted. Accumulation of carbon occurs slowly as the forest regenerates, and can take up to 200 years or longer to reach pre-logging levels (Holtsmark 2012). Planting multi-species mixtures of deciduous species rather than single-species coniferous stands can accelerate the recovery of carbon stocks, because the former has faster rates of biomass production (Figure 30.3; Gamfeldt *et al.* 2013).

Selective logging reduces tropical forest carbon stocks through the direct removal of timber. However, decomposition of the large volumes of non-target vegetation damaged during felling combined with vegetation clearance for logging infrastructure (skid trails, logging decks) generally has the larger impact (Pearson *et al.* 2014). On average, carbon stocks are 24 per cent lower in selectively logged forest, with considerable variation reflecting differences in logging intensity and harvest methods (Figure 30.4; Putz *et al.* 2008a; Putz *et al.* 2012). Post-logging carbon accumulation can be rapid (Gourlet-Fleury *et al.* 2013; Edwards *et al.* 2014), and although recovery of carbon stocks to primary forest levels may take several decades, regenerating selectively logged tropical forests provide a substantial carbon sink – particularly when contrasted with alternative land uses such as agricultural plantations (Edwards *et al.* 2014). Repeated logging without appropriate regeneration periods

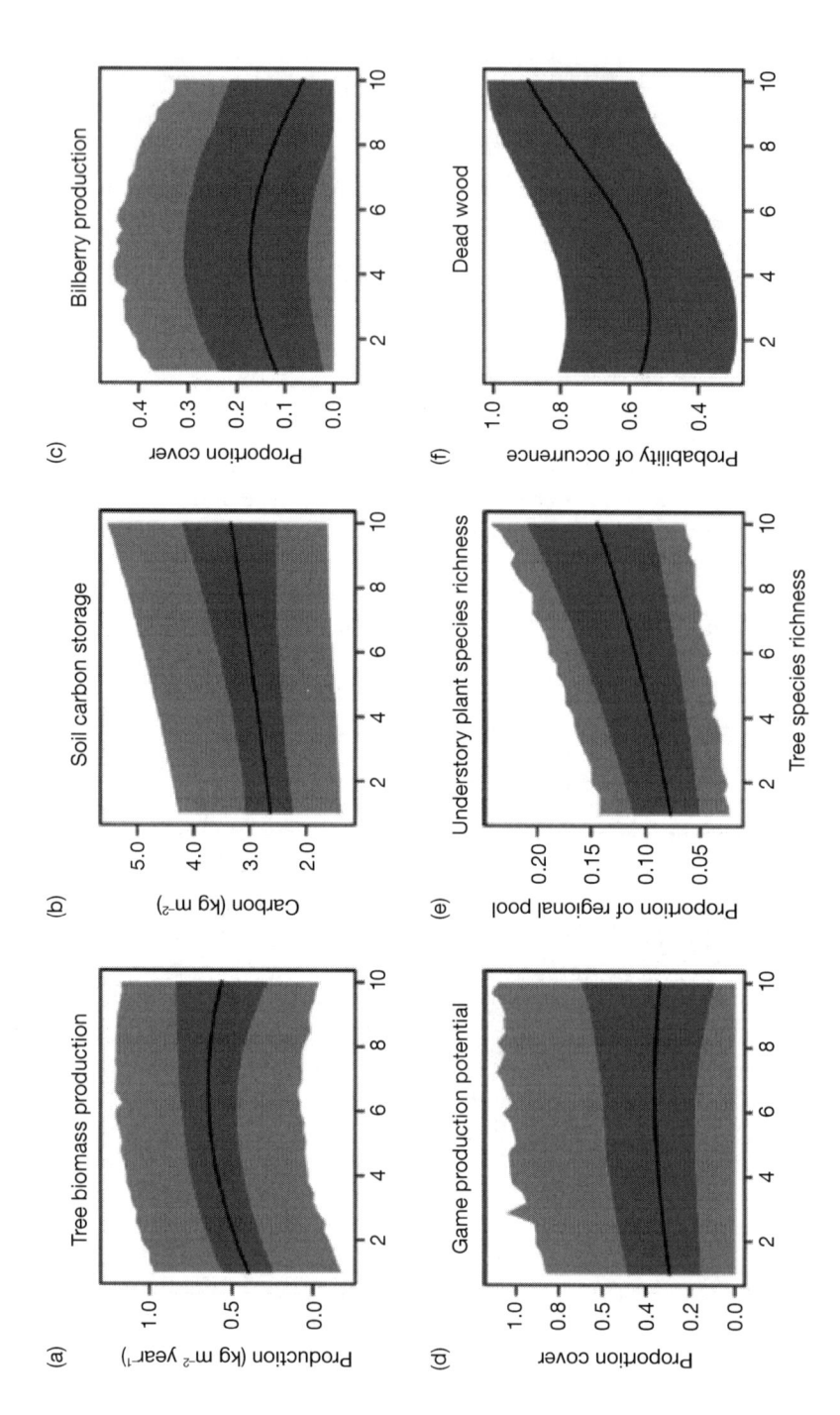

Figure 30.3 Relationships between tree species richness and ecosystem services in boreal and temperate production forests. Ecosystem services are: (a) tree biomass production, (b) soil carbon storage, (c) bilberry production, (d) game production potential, (e) understory plant species richness, (f) occurrence of dead wood. Mean relationships (black) are shown, with 95 per cent confidence intervals for the relationships excluding (dark grey) and including (light grey) the residual variation

Source: Figure and legend text reproduced from Gamfeldt et al. (2013) *Nature Communications*, 4, 1340. doi:10.1038/ncomms2328 (with permission).

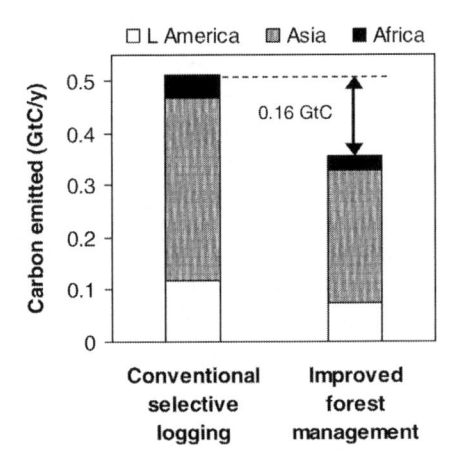

Figure 30.4 Estimated annual reduction in global carbon emissions that would result from improved forest management

Source: Figure and legend text reproduced from Putz et al. (2008a) *PLoS Biology*, 6(7), e166. doi:10.1371/journal.pbio.0060166 (with permission).

may erode carbon stocks, however, and recovery within relogged forest has received relatively little attention (Putz et al. 2012).

Indirect effects of logging

Logging operations are associated with a number of generic anthropogenic impacts. Perhaps the most pervasive of these indirect effects stems from the construction of networks of logging roads for transporting timber. In Central Africa, for example, Laporte et al. (2007) documented rapid expansion in road-building within timber concessions, and conservatively estimated >50,000 km of logging roads in the region, whilst an estimated 250,000 km of logging roads were constructed in Borneo from 1973–2013 (Gaveau et al. 2014). Roads make previously remote areas accessible, resulting in increased hunting pressure and greater risk of complete forest clearance (Asner et al. 2006; Edwards et al. 2014). Logged tropical forests can also be more susceptible to further degradation through fires, as a consequence of the hotter, drier conditions and large volumes of debris from non-target vegetation (Asner et al. 2006).

Mitigating the impacts of logging

Approaches to mitigating the impacts of logging by modifying forestry practices can be divided into actions that take place before or during logging, and those that occur post-logging.

Mitigation before and during logging

Close-to-nature forestry

Developed and applied in Europe and North America (and also known variously as 'back-to-nature forestry', 'near-to-nature forestry', 'nature-based silviculture', 'natural disturbance-based forest management', 'emulating natural disturbance regimes' or 'ecological silviculture'), close-to-nature forestry seeks to maintain wood production with less detrimental impacts on biodiversity and wider ecosystem values than conventional logging. 'Nature' in this context is defined either as restoring forest structure and composition to a reference state (e.g. unlogged forest) or using forestry interventions that mimic processes occurring within natural forest (Gamborg and Larsen 2003; Long 2009): both approaches therefore move away from intensively managed monocultures and conventional clear-cutting.

A suite of practices can be considered aspects of close-to-nature forestry, including maintaining a continuous canopy cover, maintaining mixed-species stands of native trees, using natural regeneration, retaining dead wood and standing dead trees, and minimising the use of fertilisers and pesticides (Gamborg and Larsen 2003). These practices buffer forests against climate extremes and help species typical of old-growth forest to persist. Equally, however, maintaining wood production necessitates harvesting commercially valuable large, old trees together with some dead wood removal (Gossner et al. 2013). The ecological impacts of close-to-nature forestry therefore fall on a continuum between more conventional logging and strict forest protection (Gamborg and Larsen 2003).

Retention forestry

Retention forestry focuses on maintaining structures that are important for biodiversity and ecological processes but that are typically lost during conventional logging. Originally developed in North America, the practice is most commonly applied in boreal and temperate forests managed using clear-cutting, although the relevance of retention forestry for other systems is increasingly recognised (Gustafsson et al. 2012; Lindenmayer et al. 2012). The structures retained encompass features that are rare and/or take a long time to develop, and may include pieces of dead wood, individual trees (often above a minimum diameter), or small patches of unharvested forest (Figure 30.5). Retention of these structures may be aggregated or dispersed across the landscape, with a minimum of 5–10 per cent of the total volume of the stand recommended (Gustafsson et al. 2012).

Retention forestry can buffer against more extreme changes in microclimate and provide a 'lifeboat' to allow old-growth forest species to persist following clear-cutting, whilst areas of intact forest also act as stepping stones or corridors for dispersal between mature stands. These effects increase species richness and abundance for some taxa (e.g. birds, ectomycorrhizal fungi, beetles). Benefits for other groups (e.g. bryophytes) are less clear, however, and syntheses are complicated by variation amongst studies in the volume, type and distribution of structures retained (Rosenvald and Lõhmus 2008).

The concept of Woodland Key Habitats (WKH) is closely related to retention forestry, and applied most prominently in northern Europe. WKH are selected based on edaphic conditions, hydrology or geomorphology, or following the identification of key structural features such as late successional forest, riparian forest, aggregations of logs, and forest springs. These habitat patches are then either excluded from logging or logged at lower intensities (Timonen et al. 2012). Similar networks of key habitats are also retained in other regions – e.g. in North America, relatively wide riparian buffer strips are often protected to conserve aquatic and terrestrial biota (Marczak et al. 2010). Where WKH or similar patches are incorporated strategically into a reserve network, connectivity between habitats can be enhanced, although small and isolated WKH have more limited biodiversity benefits, particularly for taxa with short dispersal capabilities (Laita et al. 2010).

Figure 30.5 (a) Group retention in coastal Canada (photograph: William Beese), (b) Tree and habitat retention in a gap release treatment in Western Australia (photograph: Deirdre Maher), (c) Small aggregate and created dead wood in boreal Sweden (photograph: Lena Gustafsson), (d) Dispersed retention in Washington State (photograph: Cassandra Koerner)

Source: Reproduced from Gustafsson *et al.* (2012), Retention forestry to maintain multifunctional forests: a world perspective, *BioScience*, 62, 633–645, with permission from Oxford University Press on behalf of the American Institute of Biological Sciences.

Reduced Impact Logging

Reduced Impact Logging (RIL) is applied in tropical forests as an alternative to conventional selective logging (CL), and encompasses several practices that are intended primarily to improve the long-term sustainability of timber harvesting (Putz *et al.* 2008a). The practices employed vary but include one or more of the following: (1) a full inventory and mapping prior to logging; (2) planning and restrictions on skid trails, logging roads, and log decks to avoid unnecessary construction; and (3) pre-harvest liana cutting and directional felling to minimise damage to non-target vegetation (Putz *et al.* 2008b).

RIL is most effective at mitigating logging impacts that are related to forest regeneration. By combining planned extraction routes with directional felling, residual damage can be decreased by 50 per cent relative to CL (Putz *et al.* 2008a). Pre-harvest liana cutting further reduces damage to non-target trees, as well as benefiting regenerating forest by restricting liana densities post-logging (Alvira *et al.* 2004). Appropriately practised, RIL thus allows logged forest to maintain a more similar forest structure to unlogged forest, with significantly less canopy damage (Asner *et al.* 2004). RIL forest may also retain higher carbon stocks than CL forest (Figure 30.4; Putz *et al.* 2008a), although the benefits of RIL for minimising CO_2 emissions are highly dependent on the particular suite of measures applied (Griscom *et al.* 2014).

Relatively few studies have examined the extent to which RIL mitigates the direct effects of logging on biodiversity, and whilst there is some evidence for higher species richness in RIL versus CL forest, other studies found no difference between forest types (see Edwards et al. 2012 and references therein). Moreover, the benefits of RIL for biodiversity reflect the intensity and type of practices employed, and there is a need for more targeted research in this respect. For example, pre-felling planning to reduce the number of skid trails may be particularly important in limiting access by poachers, and also results in less vegetation clearance. Similarly, practices that maintain a more intact canopy structure may reduce fire risk, whilst improved long-term timber yields from more sustainable management could provide greater economic and political incentives for forest retention over conversion.

Mitigation post-logging

Post-logging mitigation measures generally seek to: (1) maintain future timber yields; and/or (2) assist the recovery of logged forest back towards the natural state ('ecological restoration'; SER 2004). Whilst these objectives overlap, they are not necessarily identical. Silvicultural treatments designed to maintain timber yields involve the removal of lianas and understory vegetation (liberation cutting), combined with planting saplings of commercial species (enrichment planting). These treatments enhance forest regeneration and carbon sequestration relative to naturally regenerating forest (Gourlet-Fleury et al. 2013), and can have positive effects on biodiversity. For example, insectivorous bird species that decline following logging benefit from the more open understory that results from liberation cutting and enrichment planting (Ansell et al. 2011).

Ecological restoration can include the silvicultural techniques described above, and may involve major ecosystem modifications such as reforesting cleared areas and utilising prescribed fires (Suding 2011; Halme et al. 2013). More subtle measures are also sometimes employed, such as creating dead wood and small gaps, and blocking ditches (Halme et al. 2013). In the short term, ecological restoration can benefit target groups (e.g. species dependent on fires or on dead wood), provided these groups are able to colonise the restored sites. More permanent benefits of restoration are less certain, however, reflecting the temporary nature of some techniques and the need for longer-term research (Kouki et al. 2012; Komonen et al. 2014).

Conclusion

Logging operations encompass a wide range of practices and mitigation strategies, and cannot be regarded as having a single, globally consistent set of effects. The picture is further complicated by the lack of long-term (i.e. decadal) data on the impacts of repeated logging, and by inherent differences between ecosystems. Nonetheless, some general patterns emerge. The impacts of clear-cutting on habitat structure, carbon stocks, and biodiversity are more severe than the impacts of selective logging, and for both practices, mitigation measures can alleviate negative effects to some extent. There is also increasing recognition of the potential conservation significance of permanent production forest, with selective logging (in particular) apparently less detrimental to biodiversity than other forms of anthropogenic disturbance (e.g. complete forest clearance for agriculture; Gibson et al. 2011; Edwards et al. 2014) and mitigation measures such as retention forestry proposed to allow more sustainable management (Lindenmayer et al. 2012). Whilst production forests should not be viewed as a substitute for unlogged, old-growth forest, the value of well-managed logged forest as a complement to strictly protected areas should not be ignored, particularly in regions where the conservation of extensive areas of pristine habitat is no longer possible, or is economically or socially unfeasible.

References

Abrego, N. and Salcedo, L. (2013) 'Variety of woody debris as the factor influencing wood inhabiting fungal richness and assemblages: Is it a question of quantity or quality?', *Forest Ecology and Management*, vol 291, 377–385

Alvira, D., Putz, F. E. and Frederiksen, T. S. (2004) 'Liana loads and post-logging liana densities after liana cutting in a lowland forest in Bolivia', *Forest Ecology and Management*, vol 190, 73–86

Ansell, F. A., Edwards, D. P. and Hamer, K. C. (2011) 'Rehabilitation of logged rain forests: avifaunal composition, habitat structure, and implications for biodiversity-friendly REDD+', *Biotropica*, vol 43, 504–511

Asner, G. P., Keller, M., Pereira Jr, R., Zweede, J. C. and Silva, J. N. M. (2004) 'Canopy damage and recovery after selective logging in Amazonia: field and satellite studies', *Ecological Applications*, vol 14, 280–298

Asner, G. P., Broadbent, E. N., Oliveira, P. J. C., Keller, M., Knapp, D. E. and Silva, J. N. M. (2006) 'Condition and fate of logged forests in the Brazilian Amazon', *Proceedings of the National Academy of Sciences USA*, vol 103, 12947–12950

Blaser, J., Sarre, A., Poore, D. and Johnson, S. (2011) 'Status of tropical forest management', *ITTO Technical Series No. 38*. International Tropical Timber Organisation, Yokohama, Japan

Bouget, C., Lassauce, A. and Jonsell, M. (2012) 'Effects of fuelwood harvesting on biodiversity – a review focused on the situation in Europe', *Canadian Journal of Forest Research*, vol 42, 1421–1432

Brunet, J., Fritz, Ö. and Richnau, G. (2010) 'Biodiversity in European beech forests – a review with recommendations for sustainable forest management', *Ecological Bulletins*, vol 10, 1093–1101

Edwards, D. P., Larsen, T. H., Docherty, T. D. S., Ansell, F. A., Hsu, W. W. and others (2011) 'Degraded lands worth protecting: the biological importance of Southeast Asia's repeatedly logged forests', *Proceedings of the Royal Society: Biological Sciences*, vol 278, 82–90

Edwards, D. P., Woodcock, P., Edwards, F. A., Larsen, T. H., Hsu, W. W. and others (2012) 'Reduced-impact logging and biodiversity conservation: a case study from Borneo', *Ecological Applications*, vol 22, 561–571

Edwards, D. P, Woodcock, P., Newton, R. J., Edwards, F. A., Andrews, D. J. R. and others (2013) 'Trophic flexibility and the persistence of understory birds in intensively logged rainforest', *Conservation Biology*, vol 27, 1079–1086

Edwards, D. P., Tobias, J. A., Sheil, D., Meijaard, E. and Laurance, W. F. (2014) 'Maintaining ecosystem function and services in logged tropical forests', *Trends in Ecology and Evolution*, vol 29, 511–520

FAO (2010) 'Global Forest Resources Assessment – Main Report'. *FAO Forestry Paper No. 163*. Rome, Italy

Feldpausch, T. R., Jirka, S., Passos, C. A. M., Jasper, F. and Riha, S. J. (2005) 'When big trees fall: damage and carbon export by reduced impact logging in southern Amazonia', *Forest Ecology and Management*, vol 219, 199–215

Gamborg, C. and Larsen, B. J. (2003) '"Back-to-nature" – a sustainable future for forestry', *Forest Ecology and Management*, vol 179, 559–571

Gamfeldt, L., Snäll, T., Bagchi, R., Jonsson, M., Gustafsson, L. and others (2013) 'Higher levels of multiple ecosystem services are found in forests with more tree species', *Nature Communications*, vol 4, 1340. doi:10.1038/ncomms2328

Gaveau, D. L. A., Sloan, S., Molidena, E., Yaen, H., Sheil, D. and others (2014) 'Four decades of forest persistence, clearance and logging on Borneo'. *PLoS ONE*, vol 9(7), e101654. doi:10.1371/journal.pone.0101654

Gibson, L., Lee, T. M., Koh, L. P., Brooke, B. W., Gardner, T. A. and others (2011) 'Primary forests are irreplaceable for sustaining tropical biodiversity', *Nature*, vol 478, 378–382

Gossner, M. M., Lachat, T., Brunet, J., Isacsson, G., Bouget, C. and others (2013) 'Current near-to-nature forest management effects on functional trait composition of saproxylic beetles in beech forests', *Conservation Biology*, vol 27, 605–614

Gourlet-Fleury, S., Mortier, F., Baya, F. A., Quédraogo, D., Bénédet, F. and Picard, N. (2013) 'Tropical forest recovery from logging: a 24 year silvicultural experiment from Central Africa', *Philosophical*

Transactions of the Royal Society: Biological Sciences, vol 368, 20120302. http://dx.doi.org/10.1098/rstb.2012.0302

Graham-Sauvé, L., Work, T. T., Kneeshaw, D. and Messier, C. (2013) 'Shelterwood and multicohort management have similar initial effects on ground beetle assemblages in boreal forests', *Forest Ecology and Management*, vol 306, 266–274

Gray, M. A., Baldauf, S. L., Mayhew, P. J. and Hill, J. K. (2007) 'The response of avian feeding guilds to tropical forest disturbance', *Conservation Biology*, vol 21, 133–141

Griscom, B., Ellis, P. and Putz, F. E. (2014) 'Carbon emissions performance of commercial logging in East Kalimantan, Indonesia', *Global Change Biology*, vol 20, 923–937

Gustafsson, L., Baker, S. C., Baugus, J., Breese, W. J., Brodie, A. and others (2012) 'Retention forestry to maintain multifunctional forests: a world perspective', *BioScience*, vol 62, 633–645

Halme, P., Allen, K. A., Auniņš, A., Bradshaw, R. H. W., Brumelis, G. and others (2013) 'Challenges of ecological restoration: lessons from forests in Northern Europe', *Biological Conservation*, vol 167, 248–256

Holtsmark, B. (2012) 'Harvesting in boreal forests and the biofuel carbon debt', *Climatic Change*, vol 112, 415–428

Jones, D. T., Susilo, F. X., Bignell, D. E., Hardiwinoto, S., Gillison, A. N. and Eggleton, P. (2003) 'Assemblage collapse along a land-use intensification gradient in lowland central Sulawesi, Indonesia', *Journal of Applied Ecology*, vol 40, 380–391

Josefsson, T., Olsson, J. and Östlund, L. (2010) 'Linking forest history and conservation efforts: long-term impact of low-intensity timber harvest on forest structure and wood-inhabiting fungi in northern Sweden', *Biological Conservation*, vol 143, 1803–1811

Komonen, A., Kuntsi, S., Toivanen, T. and Kotiaho, J. S. (2014) 'Fast but ephemeral effects of ecological restoration on forest beetle community', *Biodiversity and Conservation*. doi 10.1007/s10531–014–0678–6.

Kouki, J., Hyvärinen, E., Lappalainen, H., Martikainen, P. and Similä, M. (2012) 'Landscape context affects the success of habitat restoration: large-scale colonization patterns of saproxylic and fire-associated species in boreal forests', *Diversity and Distributions*, vol 18, 348–355

Kuuluvainen, T., Tahvonen, O. and Aakala, T. (2012) 'Even-aged and uneven-aged forest management in boreal Fennoscandia: a review', *Ambio*, vol 41, 720–737

Laita, A., Mönkkönen, M. and Kotiaho, J. S. (2010) 'Woodland key habitats evaluated as part of a functional reserve network', *Biological Conservation*, vol 143, 1212–1227

Laporte, N. T., Stabach, J. A., Grosch, R., Lin, T. S. and Goetz, S. J. (2007) 'Expansion of industrial logging in Central Africa', *Science*, vol 316, 1451

Lindenmayer, D. B., Franklin, J. F., Lõhmus, A., Baker, S. C., Bauhus, J. and others (2012) 'A major shift to the retention approach for forestry can help resolve some global forest sustainability issues', *Conservation Letters*, vol 5, 421–431

Long, J. N. (2009) 'Emulating natural disturbance regimes as a basis for forest management: a North American view', *Forest Ecology and Management*, vol 257, 1868–1873

Marczak, L. B., Sakamaki, T., Turvey, S. L., Deguise, I., Wood, S. L. R. and Richardson, J. S. (2010) 'Are forested buffers an effective conservation strategy for riparian fauna? An assessment using meta-analysis', *Ecological Applications*, vol 20, 126–134

Markl, J., Schleuning, M., Forget, P. M., Jordano, P., Lambert, J. E. and others (2012) 'Meta-analysis of the effects of human disturbance on seed dispersal by animals', *Conservation Biology*, vol 26, 1072–1081

Paillet, Y., Bergès, L., Hjältén, J., Odor, P., Avon, C. and others (2010) 'Biodiversity differences between managed and unmanaged forests: meta-analysis of species richness in Europe', *Conservation Biology*, vol 24, 101–112

Pearson, T. R. H., Brown, S. and Cas, F. M. (2014) 'Carbon emissions from tropical forest degradation caused by logging', *Environmental Research Letters*, vol 9, 034017, doi:10.1088/1748–9326/9/3/034017

Perry, R. W. and Thill, R. E. (2013) 'Long-term responses of disturbance-associated birds after different timber harvests', *Forest Ecology and Management*, vol 307, 274–283

Peters, S. L., Malcolm, J. R. and Zimmerman, B. L. (2006) 'Effects of selective logging on bat communities in the Southeastern Amazon', *Conservation Biology*, vol 20, 1410–1421

Putz, F. E., Zuidema, P. A., Pinard, M. A., Boot, R. G. A., Sayer, J. A. and others (2008a) 'Improved tropical forest management for carbon retention', *PLoS Biology*, vol 6(7), e166. doi:10.1371/journal.pbio.0060166

Putz, F. E., Sist, P., Fredericksen, T. and Dykstra, D. (2008b) 'Reduced-impact logging: challenges and opportunities', *Forest Ecology and Management*, vol 256, 1427–1433

Putz, F. E., Zuidema, P. A., Synnott, T., Peña-Claros, M., Pinard, M. A. and others (2012) 'Sustaining conservation values in selectively logged tropical forest: the attained and the attainable', *Conservation Letters*, vol 5, 296–303

Rönnberg, J. (2000) 'Logging operation damage to roots of clear-felled *Picea abies* and subsequent spore infection by *Heterobasidium annosum*', *Silva Fennica*, vol 34, 29–36

Rosenvald, R. and Lõhmus, A. (2008) 'For what, when, and where is green-tree retention better than clear-cutting? A review of the biodiversity aspects', *Forest Ecology and Management*, vol 255, 1–15

SER (2004) *The SER International Primer on Ecological Restoration*. Society for Ecological Restoration, Science and Policy Working Group.

Suding, K. N. (2011) 'Toward an era of restoration in ecology: successes, failures, and opportunities ahead', *Annual Review of Ecology, Evolution, and Systematics*, vol 42, 465–487

Thorpe, H. C., Thomas, S. C. and Caspersen, J. P. (2008) 'Tree mortality following partial harvests is determined by skidding proximity', *Ecological Applications*, vol 18, 1652–1663

Timonen, J., Gustafsson, L., Kotiaho, J. S. and Mönkkönen, M. (2012) 'Hotspots in cold climate: conservation value of woodland key habitats in boreal forests', *Biological Conservation*, vol 144, 2061–2067

Vanderkerkhove, K., Keersmaeker, L. D., Walleyn, R., Köhler, F., Crevecoeur, L. and others (2011) 'Reappearance of old-growth elements in lowland woodlands in northern Belgium: Do the associated species follow?', *Silva Fennica*, vol 45, 909–935

Vasconcelos, H. L., Vilhena, J. M. S., Caliri, G. J. A. (2000) 'Responses of ants to selective logging of a central Amazonian forest', *Journal of Applied Ecology*, vol 37, 502–514

Wallenius, T., Niskanen, L., Virtanen, T., Hottole, J., Brumelis, G. and others (2010) 'Loss of habitats, naturalness and species diversity in Eurasian forest landscapes', *Ecological Indicators*, vol 10, 1093–1101

Wermelinger, B. (2004) 'Ecology and management of the spruce bark beetle *Ips typographus* – a review of recent research', *Forest Ecology and Management*, vol 202, 67–82

Woodcock, P., Edwards, D. P., Fayle, T. M., Newton, R. J., Chey, V.-K. and others (2011) 'The conservation value of South East Asia's highly degraded forests: evidence from leaf-litter ants', *Philosophical Transactions of the Royal Society: Biological Sciences*, vol 366, 3256–3264

Woodcock, P., Edwards, D. P., Newton, R. J., Chey, V.-K., Bottrell, S. H. and Hamer, K. C. (2013) 'Impacts of intensive logging on the trophic organization of ant communities in a biodiversity hotspot', *PLoS ONE*, vol 8(4), e60756. doi:10.1371/journal.pone.0060756

31
POLLUTION IN FORESTS

Mikhail V. Kozlov and Elena L. Zvereva

Introduction

The influence of industrial pollution on forests has long been recognised as a serious social and scientific problem. We inherited large areas of forests disturbed or destroyed by pollution, and in many countries the destruction continues to the present. Disturbance-induced changes in ecosystems are of central concern in fundamental ecology. From this perspective, the impacts of pollutants on ecosystems can be seen as unintentional large-scale disturbance experiments, the results of which can be studied by ecologists in order to understand the factors that affect the resilience of community structures and ecosystem functions. Habitats transformed by pollution, including unusual landscapes of denuded barren land, can be considered opportunistic macrocosms (unique laboratories) for integrated research on the effects of harsh environmental conditions on terrestrial biota.

Several thousand case studies have documented various biotic effects occurring in contaminated areas. Still large uncertainties concern the responses of forest ecosystems to nitrogen and ozone and ecosystem recovery following reduction in pollution load. At what level of pollutant deposition can we expect the start of forest recovery? How do different forms of nitrogen deposition affect the forest ecosystems? How will polluted forests respond to climate change and what new synergetic effects may occur under rapidly changing environmental conditions? These and many other problems still await solutions, and these solutions can only be based on a deep understanding of the combined effects of pollution and other environmental changes on structure and functions of forest ecosystems.

Pollution, polluters and pollutants

Although every scientist understands the meaning of the term 'pollution', no straightforward definition is available. Any foreign substance, primarily waste from human activities, that renders the air, soil, water or other natural resource harmful or unsuitable for a specific purpose can be classified as a pollutant. The innocuous and even useful substances become pollutants when they appear where they should not be, such as gasoline in drinking water. In this chapter, we consider only aerial pollution, leaving out the impacts of contaminated water and of solid waste on forests. The pollutants in the ambient air occur as gases (e.g., SO_2, NO_2, O_3), as aerosol particles (mainly 0.1 to 3.0 μm in diameter) containing e.g. heavy metals and as cloud droplets (3 to 20 μm in diameter) containing the ions (e.g. SO_4^{2-}, Cl^-).

Historically, SO_2 was the very first pollutant that caused local but severe forest degradation near large cities and industrial enterprises. This gas is emitted into the atmosphere by many industries, with the largest contribution from non-ferrous smelters and power plants. This is the best-studied pollutant in terms of its effects on forests.

Fluorine pollution (both gaseous HF and fluorine-containing dust) is primarily associated with the aluminium industry, but it also originates from glass, brick, enamel and ceramic tile production. In high concentrations, both SO_2 and fluorine cause yellowing and chlorosis of leaves and needle, followed by growth retardation and sometimes by the death of trees. Effective control strongly decreased global fluorine emissions by the 1980s and the global emissions of SO_2 by the late 1990s.

Heavy metals (e.g. Pb, Ni, Cu and Zn) have been very common pollutants since ancient times; they are mostly emitted by smelting industries. Fortunately, they generally do not spread far away from the smelters: only a few of the largest polluters caused detectable heavy metal contamination of soils and vegetation beyond 50 km from the emission source. Most heavy metals are extremely toxic, but they were rarely reported as proximate reasons of forest damage.

Another pollutant that usually does not spread far from the emission source is an alkaline dust originating from cement and magnesite factories, as well as from power plants (in a form of fly ash). The alkaline dust and organic (mostly hydrocarbon) pollutants emitted by chemical industries generally impose less harmful impacts on ecosystems than SO_2, fluorine and heavy metals.

Increased deposition of nitrogen, associated primarily with the formation of nitrogen oxides during fossil fuel combustion, started to play an important role in European and North American forests several decades ago. Although forest death was sometimes observed near factories producing nitrogen fertilisers, nitrogen deposition does not usually kill trees, but drastically changes the structure and productivity of forests, decreases plant diversity (Bobbink et al., 2010) and predisposes forests to increased risk of damage by other environmental stressors. Effects of nitrogen deposition are harmful and long lasting, in particular due to changes in soil acidity and nutritional quality.

Ozone has been recently identified as a probable contributor to the observed forest degradation in Europe and North America. Ozone is not emitted directly by car engines or by industrial operations, but is formed by the atmospheric reaction of sunlight with hydrocarbons and nitrogen oxides. Given the recent decrease in industrial emissions, ozone is regarded at present as the air pollutant potentially most detrimental to vegetation. Consequently, during the past decades pollution ecology shifted from the exploration of the acute damage imposed by large industrial enterprises (emitting SO_2, fluorine and metals) towards investigation of regional effects that are caused primarily by ozone and, to a lesser extent, nitrogen deposition.

Extent and severity of impacts

Pollution may affect the components of an ecosystem in different ways (Kozlov and Zvereva, 2011), and the effects of pollution may interact with the effects of other stressors, such as climate, pests and pathogens. Furthermore, most polluters emit several classes of pollutants which may impose different, sometimes opposite, effects on biota. These effects may be both direct and indirect. Direct effects are caused by the toxicity of pollutants, whereas indirect effects are mediated by habitat loss, changes in microclimate and alteration in ecosystem structure and species interactions, such as competition, facilitation, predation and mutualism.

For decades, ecologists performed studies of pollution effects on forests mostly in the vicinities of large industrial areas. Therefore, acute local effects of pollution on forests are best described in scientific literature (reviewed by Kozlov et al., 2009). In contrast, ecological studies in forests suffering from nitrogen deposition or ozone are relatively scarce (but see Allen et al., 2007; Bobbink et al., 2010).

The most striking examples of severe local pollution impacts on forests, documented in scientific literature, have long been associated with the Canadian smelters (e.g. Sudbury). After implementation of strong emission controls and rehabilitation programs in Canada, the most extensive forest damage by industrial pollution is observed in Russia and China. The Norilsk smelter in north-east Siberia imposes the largest environmental problems at the world scale: forest damage has been observed up to 100–150 km from this smelter. The biggest sources of fluorine-containing emissions are also situated in Siberia (e.g. the aluminium plant in Bratsk). The most extensive ecological information concerning the consequences of severe pollution impact on forest ecosystems has been collected around the Monchegorsk nickel-copper smelter located in the Kola Peninsula, north-west Russia (Kozlov et al., 2009).

Regional effects of pollution attracted considerable attention in the 1970s, when large areas of forest required rehabilitation to mitigate the impacts of acidic deposition and other forms of pollution. The problem was particularly apparent in Central Europe, with the most striking example being the 'Black Triangle', an area along the German–Czech–Polish border, which has long been heavily affected by industrial pollution (Kozlov et al., 2000). In North America, intense episodes of photochemical smog have been known in the Los Angeles Basin since the early 1950s. Contaminated air masses move inland with the westerly onshore winds and are pushed against the San Bernardino Mountains, damaging and even killing sensitive ponderosa and Jeffrey pines (Bytnerowicz et al., 2008).

Model calculations demonstrate that by 2050 severe regional problems will also occur in South East Asia, South Africa, Central America, and along the Atlantic coast of South America (Fowler et al., 1999). Forests of temperate and northern parts of Europe and North America and of the Mediterranean ecoregion are especially vulnerable to nitrogen deposition.

At the global scale, 8 per cent of the forests of the world received annually >1 kg H^+ ha^{-1} as sulphur in 1985, i.e. the load at which the forests are at risk on acid sensitive soils; and it is estimated that 17 per cent of the forests will receive this pollution load by 2050. Similarly, 24 per cent of global forest was exposed to ozone concentrations >60 ppb (i.e. much higher than the threshold value of 40 ppb, above which direct adverse effects on plants and ecosystems may occur) by 1990, and this proportion is expected to increase to 50 per cent by 2100 (Fowler et al., 1999). These changes will obviously have global consequences for carbon cycles, primary productivity and other characteristics of forest ecosystems. At present, the concern is related to the long-term resilience of forests to cope with the interacting impacts of air pollution and climate change at large spatial scales (de Vries et al., 2014).

Effects of pollution on forests

Landscape-level effects

Soils near industrial enterprises often contain increased amounts of phytotoxic pollutants, such as fluorine and heavy metals. The large fraction of metals is deposited in insoluble forms, such as oxides, which are slowly transformed to bioavailable forms. Pollution may cause both increases and decreases in soil pH, which lead to changes in the bioavailability of nutrients and in the toxicity of pollutants for plants and animals. In particular, soil acidification promotes nutrient leaching from upper soil horizons, making nutritional deficiency one of the key factors behind forest deterioration in polluted regions (de Vries et al., 2014). Both extreme acidification and extreme alkalisation have drastic consequences for forest ecosystems.

Large industrial polluters are often surrounded by extensive areas of disturbed land. Forest death changes visual perception of landscapes, making them unattractive – especially when dead trees, that remain standing for decades, cut the skyline (Figure 31.1a). However, standing trees, even when they are dead, maintain climatic and biotic stability of contaminated habitats, in particular by ameliorating microclimate (Wołk, 1977) and preventing soil erosion. Therefore, it is not recommended to cut dead trees in severely polluted areas prior to the beginning of vegetation recovery.

Extreme deposition of airborne pollutants can transform forests to industrial barrens (Figure 31.1a) – bleak open landscapes, with only small patches of vegetation surrounded by bare land. These landscapes first appeared about a century ago in North America. Currently, about 40 industrial barrens exist worldwide, and most of them are located next to non-ferrous smelters. The extent of industrial barrens varies from a few hectares (e.g. Harjavalta in Finland) to several hundreds of square kilometres (Norilsk and Monchegorsk in Russia). In spite of the overall impression of 'dead land' and general reduction in biodiversity, industrial barrens still support a variety of life, including regionally rare and endangered species, as well as populations that have evolved specific adaptations to the harsh and toxic environment (Kozlov and Zvereva, 2007).

Ecosystem-level effects

Decrease in net primary productivity (NPP) is one of the most important ecosystem-level responses to different kinds of disturbances (Odum, 1985). The overall effect of industrial pollutants on plant abundance is negative, but the degree of damage varies depending on the type of vegetation and emitted chemicals. The non-ferrous smelters impose the strongest

Figure 31.1 (a) Industrial barren located 4 km south of the nickel–copper smelter at Monchegorsk, north-west Russia. This barren evolved from dense Norway spruce forest. Note dead trees, the absence of field layer vegetation and extensive soil erosion. (b) Modifications of crown structure of Norway spruce (foreground) and Scots pine (background) in this barren site. Note dense branches near the ground: they are protected by snow during the winter time. (Photos by V. Zverev)

detrimental impact on forest productivity. The density of a stand decreases around polluters emitting both acidifying (SO_2, NO_x, NH_3) and alkalifying (fly ash, magnesite and cement dust) pollutants, whereas the density of shrubs and field layer vegetation decreases only near acidifying polluters. The response of bryophyte cover to pollution mirrors that of tree cover, but does not correlate with changes in the vascular field layer. Ground lichens also decline with pollution, although less rapidly than epiphytic lichens which are extremely sensitive to gaseous pollutants and whose disappearance indicates pollution impact well before other organisms are affected.

Information on the effects of regional pollution on NPP is scarce. The analysis of field data did not reveal the effects of ozone on forest NPP in Italy (De Marco et al., 2013), although meta-analysis of experimental data suggests that ozone is currently decreasing net tree photosynthesis by 11 per cent and tree biomass production by 7 per cent. However, these estimates are based largely on individual young trees growing in a non-competitive environment, and extrapolation of these results may not be appropriate for predicting the response of mature trees and forests to ozone (Ainsworth et al., 2012).

Pollution-induced decrease in NPP affects nutrient cycling of forest ecosystems (see Chapter 23). Lower density and height of a stand, as well as reduced cover of filled layer vegetation, result in a smaller litter input and, consequently, in smaller forest floor thickness and topsoil depth. At the same time, pollution adversely affects soil microbiota and detritophagous arthropods, thus significantly reducing the litter decomposition rate and leading to an accumulation of non-decomposed plant material, especially near non-ferrous smelters (Kozlov et al., 2009).

The reductions in biomass production and litter decomposition affect the carbon and nutrient budget of forest ecosystems, although the direction of the change depends on both pollutants and the affected ecosystems. Nitrogen deposition, ozone exposure and aerosols/fine particulates can reduce the carbon storage capacity of the forests (Matyssek et al., 2012). However, on nutrient-poor soils, nitrogen deposition may stimulate forest growth and thus cause a net carbon sink (Chapter 37).

Community-level effects

Clements and Newman (2002) provide a broad perspective for bridging ecological and toxicological approaches in studying community-level effects of pollution on biota, in particular attracting attention to indirect effects of contaminants, mediated by changes in species interactions. Biotic interactions, in particular the ways in which they determine tolerance to and avoidance of environmental stress, were recently identified as one of the focal points in studying forest responses to pollution and climate change (Matyssek et al., 2012).

Pollution effects on the diversity of different groups of forest biota are not uniform (Figure 31.2). While the diversity of soil microfungi, bryophytes, vascular plants, soil/ground living arthropods and birds generally decreases with industrial pollution, the species richness of insect herbivores often increases. Nitrogen deposition reduces the diversity of forest vegetation (Bobbink et al., 2010), while the presumed adverse impacts of ozone on biodiversity are mostly projected from fumigation experiments, with limited supporting evidence from the affected forests (Allen et al., 2007). The observed changes in diversity can be caused not only by the direct impacts of pollutants, but also by habitat loss and fragmentation.

Gordon and Gorham (1963) describe the deterioration of polluted forests as a peeling off of the vegetation layers. The trees decline first, followed by tall shrubs and, finally, under the severest conditions, the short shrubs and herbs are affected. The generality of this 'downward' pattern of ecosystem destruction is confirmed by a decrease in the magnitude of the effects of

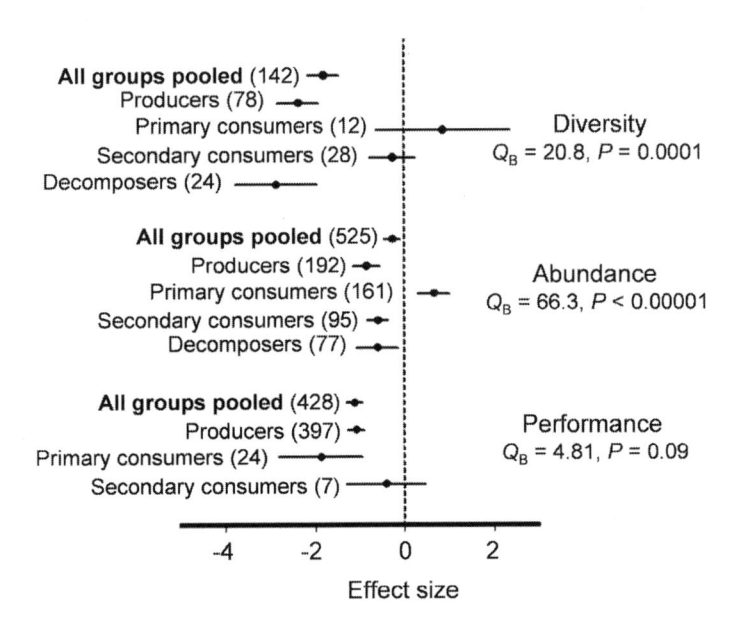

```
All groups pooled (142) ──●──
      Producers (78)  ──●──
Primary consumers (12) ──┼──●──        Diversity
Secondary consumers (28) ──●┼──    $Q_B = 20.8$, $P = 0.0001$
    Decomposers (24)  ──●──┤

All groups pooled (525) ──●┤
      Producers (192) ──●──┤
Primary consumers (161) ┤   ──●──      Abundance
Secondary consumers (95) ──●──┤    $Q_B = 66.3$, $P < 0.00001$
    Decomposers (77)   ──●──┤

All groups pooled (428) ──●┤
      Producers (397) ──●──┤
Primary consumers (24) ────●────      Performance
Secondary consumers (7)    ──●──    $Q_B = 4.81$, $P = 0.09$

         -4    -2    0    2
              Effect size
```

Figure 31.2 Results of meta-analysis of published data on the effects of industrial pollution on different characteristics and different trophic levels of biota. Producers – vascular plants and bryophytes; primary consumers – herbivorous insects; secondary consumers – predatory and parasitic arthropods; decomposers – soil microfungi and some arthropods. Mean effect sizes (dots), 95 per cent confidence intervals (horizontal lines) and sample sizes (in parentheses). Effect size is the difference between the mean values of the parameter in polluted and control site(s), divided by the pooled standard deviation and adjusted for sample sizes. Thus, a negative effect size indicates that the parameter under study has lower values in the polluted site(s) than in the control site(s). The effect is significant if the 95 per cent confidence interval does not overlap the zero value. Q_B values indicate the difference among the trophic levels. (After Kozlov and Zvereva, 2011)

pollution on plant abundance, from trees to shrubs and, then, to dwarf shrubs and herbs (Zvereva and Kozlov, 2012). However, under some circumstances, dwarf shrubs and herbs decline earlier than top-canopy trees (Kozlov et al., 2009). The disappearance of some functional groups of plants from polluted forests contributed to the development of the hypothesis (Odum, 1985) on simplification of ecosystem structure under pollution impact.

Differential, and sometimes opposite, pollution effects on organisms that belong to different trophic levels (Figure 31.2) may seriously disrupt ecosystem structure and function (Kozlov and Zvereva, 2011). In particular, pollution generally imposes larger detrimental effects on predatory than on herbivorous arthropods, creating enemy-free space for herbivores and thus contributing to the increase in their abundance.

Population-level effects

Pollution effects on a suite of parameters driving population dynamics, such as death rate, life expectancy, survival and fecundity, are properly documented only for a handful of forest species. The list includes white birch, *Betula pubescens*; Scots pine, *Pinus sylvestris*; bilberry, *Vaccinium myrtillus*; willow feeding leaf beetle, *Chrysomela lapponica*; great tit, *Parus major*; pied flycatcher, *Ficedula hypoleuca*; grey-sided vole, *Clethrionomys rufocanus* and bank vole, *C. glareolus* (Kozlov et al., 2009).

Pollution may affect the genetic structure of populations, often leading to the development of pollution tolerance. These processes, referred to as microevolution due to pollution (Medina et al., 2007), are best documented for microbiota, and they have been reported more frequently for plants than for animals. The heavy metal tolerance of plants, especially grasses, provides a well-documented example of rapid evolutionary adaptation. The data on long-lived trees are scarce, although development of pollution tolerance has been demonstrated for birches and willows. However, the establishment of birch seedlings in industrial barrens near non-ferrous smelters is hampered even when they belong to the resistant genotypes. The laboratory experiments show that microevolutionary processes in plant populations could also occur under long-term ozone exposure, but field evidence is still lacking. Among animals, genetic adaptations to metal pollutants, such as Cu, Cd and Zn, have been described in earthworms, terrestrial isopods, springtails and insects.

The acute shortage of information on wildlife mortality in pollution gradients (except for bird nestlings) could be due to practical difficulties in direct measurements of death rates. Although observations on mortality of cattle and wildlife attributable to air pollution have repeatedly been summarised (e.g., Newman et al., 1992), they almost exclusively concern extreme levels of exposure or catastrophic local events. So far, only indirect data, derived from measurements of abundance of necrophagous beetles, suggest that mortality of vertebrates increases near big polluters (Kozlov et al., 2005).

The adverse effects of industrial pollutants on the abundance of organisms generally become stronger with time from the beginning of pollution, while the effects on individual fitness become weaker (Figure 31.3). The latter result can be seen as evidence of the development of pollution resistance in the affected populations, while the decline in abundance can either be the mechanism behind this process (survival selection) or the consequence of adaptation, which may involve some costs. Because the detected patterns are uniform across the investigated taxa, the evolutionary adaptation to the impact of industrial pollution is likely to be a general

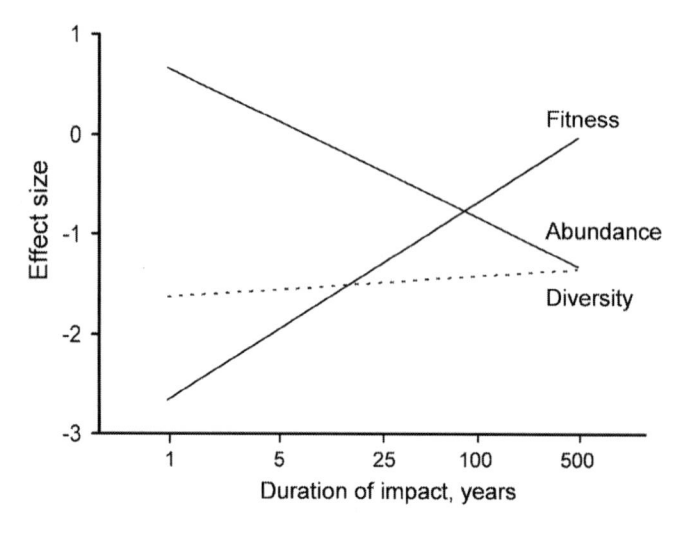

Figure 31.3 Relationships between the duration of pollution impacts and the magnitudes of the pollution effects on individual fitness, abundance and diversity of organisms. The analysis is based on the data on the following taxa: non-mycorrhizal soil microfungi, vascular plants, bryophytes and arthropods. The dotted line indicates the non-significant regression model. For explanation of effect size consult Figure 31.2. (After Kozlov and Zvereva, 2011)

phenomenon. The adaptation potential of individual organisms and entire ecosystems should be accounted for in models predicting pollution effects at regional and global scales.

A number of studies compared densities of organisms between polluted and unpolluted sites. However, not only can the same species demonstrate different changes in density around different polluters; there is also temporal variation in these responses. Hence, idiosyncratic relationships between population density and pollution load were observed for the same species during different study years, being the result of asynchronous density fluctuations in spatially isolated populations. Several case studies suggest that outbreaks of some forest pests are favoured by pollution. However, the biases in published data may have led to overestimation of the generality of this effect (Zvereva and Kozlov, 2010).

Organism-level effects

Many studies conducted around industrial polluters focus on organism-level effects. However, a critical problem, which remains to the present day, is the common absence of linkages between the recorded individual responses and processes occurring in affected populations and communities.

A great number of publications reported visible injury of plants attributable to pollution. However, variety of symptoms, subjectivity in estimation of damage, absence of statistical analysis, especially in older publications, and absence of links between visible injury and changes in plant fitness made interpretation of these data rather difficult. This conclusion also concerns various indices of tree health and visual estimates of crown defoliation and discoloration that serve as the basis of large-scale forest monitoring programmes. In contrast, a decrease in needle life span in conifers is a reliable index of the severity of pollution impact.

Pollution frequently modifies crown structure of trees. The most common symptoms of pollution damage in mature coniferous trees are the increase in crown transparency and the death of the tops of canopies. At the borders of industrial barrens, trees of Norway spruce, *Picea abies*, often have vital lower twigs creeping over the ground, whereas the mid-crown is completely dead; the uppermost branches, if alive, have very short needle longevity and prominent discoloration (Figure 31.1b). Individuals of mountain birch, *Betula pubescens* ssp. *czerepanovii*, growing in industrial barrens, have a shrubby growth habit and highly branching compact crowns resulting from increased numbers of long shoots and more densely spaced buds, in contrast to tall slender individuals in unpolluted forests. These changes increase the crown density; nearly spherical and compact crowns could minimise the impacts of unfavourable environmental conditions on trees and are therefore considered as adaptive features (Zverev *et al.*, 2013).

Pollution negatively affects fitness of both vascular and cryptogamic plants. Growth and reproduction of vascular plants near industrial polluters are reduced by about 10 per cent, while growth and reproduction of bryophytes are reduced by about 50 per cent relative to unpolluted forests (Zvereva *et al.*, 2010; Zvereva and Kozlov, 2011). The decrease in radial growth of trees is often used for the retrospective analysis of pollution impacts on forests. The effects of ozone on tree growth, measured in fumigation experiments, are of about the same magnitude as the effects of SO_2 and fluorine (Wittig *et al.*, 2009), while the field evidence for impacts of ambient ozone exposure on tree growth is weak (de Vries *et al.*, 2014).

Animals living in polluted areas may have a smaller size and lower fecundity. For example, mean body size (or weight) of the investigated species of terrestrial arthropods significantly decreased near industrial polluters (Zvereva and Kozlov, 2010); decreased clutch sizes have been reported for some bird species breeding in polluted environments (Eeva and Lehikoinen, 2013).

Causality of responses to pollution

Pollution ecology suffers from generally poor understanding of cause-and-effect relationships. The 'presumption of guilt' is commonly accepted, and changes observed around industrial polluters are often attributed to the toxicity of pollutants without proper justification. At the regional scale, it is even more difficult to identify the immediate cause(s) of forest damage than near industrial polluters, not only because of generally weaker effects on forests, but also because other stressors may cause effects that are similar to the effects of pollution. For example, the forest damage observed in Finnish Lapland in the late 1980s, which was assumed to be due to emissions from the smelters located in the Kola Peninsula, north-west Russia, after thorough examination was attributed to the combination of nutrient deficiency caused by the exceptional weather conditions of winter 1986–1987 with the epidemics of scleroderris canker (Tikkanen and Niemelä, 1995).

In many situations the causal links behind the changes observed in polluted forests are far from being transparent due to both natural variation in study systems and strong indirect impacts, which may enhance primary disturbance in a positive feedback fashion. For example, pollution-induced forest thinning and dieback results in higher wind speed that increases climatic stress both directly and by changes in snowpack structure, and also imposes mechanical damage on plants (Kozlov, 2001). A thin and compact snow layer explains lower soil temperatures recorded in industrial barrens during the wintertime. Changes in microclimate in combination with a pollution-induced decrease in cold hardiness of conifers increase the probability of death of extant trees from freezing injury. Forest death results in accumulation of woody debris, which, in combination with proximity to large human settlements, increases the risk of occasional fires. Changes in forest structure cause the disappearance of some animal species due to loss of nesting and feeding habitats. Forest decline reduces recovery of topsoil and facilitates soil erosion, with adverse consequences for all components of biota. All these effects are apparent only at the landscape scale and cannot be reproduced in small-scale experiments with industrial pollutants (Kozlov et al., 2009).

To prove the causal relationships between pollution and changes observed in forest ecosystems, the researchers should follow the criteria of causation (Hill, 1965). These criteria include: strength and consistency of observed association, its specificity and temporality (a logical time sequence of events), existence of the relationship between the dose (the putative cause) and the response, coherence with known biological facts, plausibility, experimental support and analogy with similar, better known, situations. Importantly, none of the criteria are taken as indicative by themselves, but equally, none are seen as absolutely necessary to evaluate causal significance of associations. The more criteria that are satisfied and the stronger the association, the more confidence we have in our judgement that the association is causal.

Variation in biotic responses to pollution

The first studies of the effects of industrial pollution on terrestrial biota were driven by economic losses in agriculture and forestry. As a result, the most polluted areas, where forests have been severely damaged or even killed by industrial emissions, have attracted the greatest attention from scientists, who carefully described lifeless industrial barrens around several non-ferrous smelters. The similarity between the first documented examples of extreme forest deterioration created a deceptive impression that all types of pollution caused uniform changes in different types of ecosystems.

Further accumulation of data, which was especially intensive in the 1980s, demonstrated that the responses of biota to pollution are far from being uniform. It soon became clear that, because each impacted area developed in its own way due to a unique history of events, the experiences from one system are rarely directly applicable to predict the fate of another system. The need to reduce uncertainties in predicting the consequences of the pollution impact called for the quantitative exploration of factors influencing the responses of organisms, communities and ecosystems to pollution. Accomplishing this task became possible with the development of meta-analysis, which allowed statistical exploration of an unlimited amount of diverse information.

Several meta-analyses (Zvereva *et al.*, 2008, 2010; Zvereva and Kozlov, 2010, 2011, 2012) consistently demonstrated the diversity in responses of different components of terrestrial ecosystems to industrial pollution, and this discovery is one of the most interesting and important findings of these meta-analyses. The magnitude and even the direction of the effects of pollution on biota depend on characteristics of the polluter (type, amount of emissions, duration of the impact), the affected organisms (trophic level, life history), the measured character (individual fitness, abundance, diversity), and the environment (biome, climate). In particular, acidifying pollutants generally cause stronger effects on biota than alkaline pollutants; the effects of non-ferrous smelters, emitting both SO_2 and heavy metals, are generally more detrimental than the effects of polluters only emitting SO_2. Furthermore, the effect of one factor is often modified by other factors. These findings stress, in particular, the need to record (in observational studies) or to control (in manipulative studies) not only the concentration(s) of principal pollutant(s), but also other environmental factors that may substantially modify the responses of organisms and ecosystems to pollution.

Although it became evident long ago that the same pollution loads may cause different effects in different ecosystems and/or regions, little is known about the ecosystem properties that determine their sensitivity to pollution. Differential sensitivities to pollutants were formalised through the concept of critical loads. The critical load for a pollutant defines a deposition level below which sensitive parts of ecosystems are not affected. A critical load approach has been widely and successfully used to predict the effects of acid precipitation on ecosystems in Europe (Posch, 2002); it has shaped air pollution control policies since the 1980s.

Another approach to quantifying the sensitivity of ecosystems to disturbances is based on theoretical concepts linking sensitivity with ecosystem structure and functional organisation. In particular, relationships between species diversity and ecosystem stability, that have been and still are widely debated, may have direct implications for understanding how communities respond to pollution and other anthropogenic stressors (Clements and Newman, 2002). However, the data collected from polluted ecosystems do not indicate that more diverse ecosystems are more resilient to pollution impacts. The effects of industrial pollution on terrestrial biota were absent in tundra, lowest in boreal and temperate forests and highest in the desert scrub (Figure 31.4), and this pattern appeared to be related to climate rather than to the diversity of impacted communities (Kozlov and Zvereva, 2011).

Pollution and climate

The research addressing the biotic effects of global change has overlooked the pollution issue until very recently. Only a few attempts have been made to predict the consequences of climate change for polluted forests. At the organism level, a simulation showed that the sensitivity of trees to air pollution would increase in large areas of Europe, primarily in boreal forests, as the

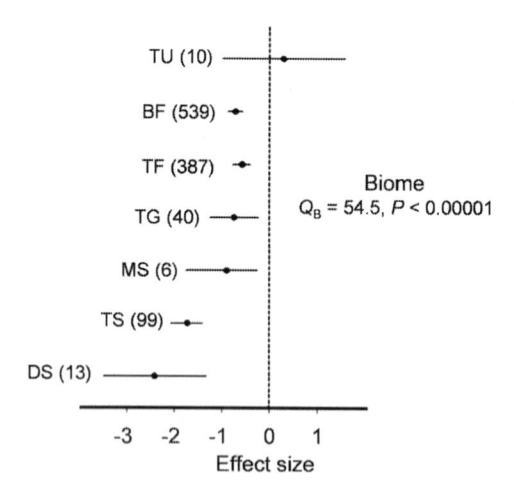

Figure 31.4 Results of meta-analysis of published data on the effects of industrial pollution on organisms in different biomes. The meta-analysis is based on the data on the following taxa: non-mycorrhizal soil microfungi, vascular plants, bryophytes and arthropods; BF – boreal forest or taiga; DS – desert scrub; MS – Mediterranean scrub; TF – temperate broadleaf deciduous forest; TG – temperate grassland; TS – tropical savannah; TU – tundra. For further explanations consult Figure 31.2. (After Kozlov and Zvereva, 2011)

temperature increases (Guardans, 2002). At the community level, pollution effects on plant diversity are stronger in warmer climates (Zvereva et al., 2008).

On the other hand, the researchers exploring the distribution of pollutants and the biotic effects of pollution have long neglected possible effects of climate, even though changes in the toxicity of pollutants as a result of temperature changes were already known several decades ago (discussed by Kozlov et al., 2009). Furthermore, climate differentially modifies responses of functional groups of terrestrial biota to pollution. Increases in temperature enhance the adverse effects of pollution on producers and decomposers but mitigate the adverse effects on primary and secondary consumers (Figure 31.5). Thus, warming is likely to increase harmful impacts of pollution on forests also through differential responses of trophic groups to temperature increases in polluted environments, which may distort ecosystem functioning (Kozlov and Zvereva, 2011).

In general, high-latitude ecosystems are less sensitive to the impact of industrial pollution than low-latitude ecosystems (Figure 31.4). This conclusion, which contradicts a widespread opinion on the fragility of arctic ecosystems, can be explained by several reasons. First, lower temperatures may result in lower stomata opening, decreasing the impact of gaseous pollutants on plants. Second, the shorter vegetation period naturally decreases the exposure to pollutants. Third, high-latitude soils generally contain fewer nutrients, and therefore the adverse effects of toxic pollutants may be counterbalanced to some extent by the fertilising effect of nitrogen oxides. Finally, higher temperatures and increased precipitation may enhance mobility and increase toxicity of pollutants, contributing to the generally higher adverse effects seen in a warmer climate. Thus, the existing levels of pollution may become more detrimental for biota as the climate warms (Zvereva et al., 2008).

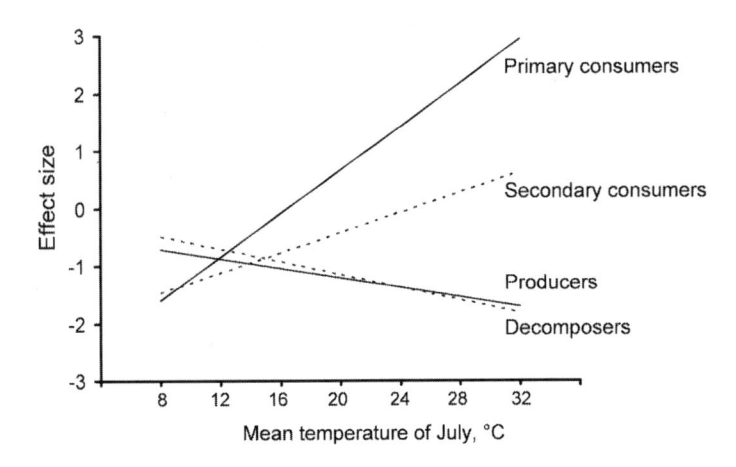

Figure 31.5 Relationships between mean temperatures in July at the polluter's locality and the magnitudes of the observed pollution effects on organisms from four trophic levels. Producers – vascular plants and bryophytes; primary consumers – herbivorous insects; secondary consumers – predatory and parasitic arthropods; decomposers – soil microfungi and some arthropods. Dotted lines indicate non-significant regression models. For explanation of effect size consult Figure 31.2. (After Kozlov and Zvereva, 2011)

Forest restoration in polluted habitats

Decrease in pollutant deposition due to implementation of the emission control or decommissioning of the polluter creates prerequisites for natural recovery of forests that have been severely damaged or killed by industrial pollution. The regeneration of heavily polluted ecosystems sometimes begins as early as 5–15 years after emission decline. For example, obvious signs of forest ecosystem recovery in the Krusne Hory Mountains, Czech Republic, were observed 16 years after considerable reduction of SO_2 emissions in the region (Cerny, 1995), and plant species richness in an industrial barren near Karabash, Russia, proved to increase significantly within five years after copper smelter shutdown (Chernenkova et al., 2001). Insectivorous birds returned to their former nesting sites in forests near the copper–nickel smelter in Harjavalta, Finland, seven years after emissions from this plant were reduced (Eeva and Lehikoinen, 2000). However, in many situations, the beginning of natural recovery is delayed due to legacy effects such as extremely high concentrations of heavy metals, the complete leaching of which from contaminated soils will take centuries after the cease of emissions, and shortage of nutrients in the thin derelict soil layer. Long-term monitoring showed that the time lag between the decrease in emissions of SO_2 and heavy metals and the beginning of recovery in industrial barrens surrounding non-ferrous smelters may last for several decades (Zverev, 2009; Vorobeichik et al., 2014). At the regional scale, deterioration of Norway spruce stands in the Ardennes has continued until recently in spite of the sharp decline in acid deposition since the beginning of the 1980s (Jonard et al., 2012).

The slow rate or even absence of natural forest regeneration has prompted many rehabilitation projects. Restoration of terrestrial ecosystems after the reduction of emissions is based on the suggestion that some successional stages are not essential and can be bypassed by proper intervention. Still, the restoration of a forest on a heavily disturbed site requires a long time, and there is no known example of a successfully completed restoration of forest ecosystems on former industrial barrens. The land reclamation programme in Sudbury, Canada (Gunn, 1995), is the best known practical example of a large-scale restoration of an area that had long been impacted by extreme pollution from smelting industries. This programme showed that it is possible to convert heavily contaminated barrens into forests (albeit depauperated in terms of biodiversity) over a period of about two decades, assuming that there is the social demand for such a conversion and sufficient financial support is granted.

The methods used to restore forests in heavily polluted sites (Kozlov et al., 2000) differ in many details from the restoration measures applied in unpolluted sites (see Chapter 28). Forest restoration around Sudbury started with liming of barren land and seeding of grass, followed by planting tree seedlings, mostly of species that were common in the previous forest. The original motivation was to improve Sudbury's image as a treeless wasteland, but gradually the revegetation philosophy turned into the need to restore the landscapes and ecosystems that were destroyed by pollution (Gunn, 1995).

Restoration measures applied at Sudbury ranged from simple abiotic enhancements (like liming and fertilisation) to complex, multistage revegetation treatments. These measures yielded different results, with the largest success achieved in heavily polluted sites that received complex treatments. Twenty-five years after the beginning of restoration, the native understory vascular species richness in these sites recovered to about the same extent as in naturally recovering, mildly disturbed sites (Rayfield et al., 2005). Thus, in practical terms, it appeared possible to partially substitute time needed for ecosystem recovery with money.

The restoration efforts at Sudbury and at some other heavily polluted sites have been successful in turning industrial barrens into forested landscapes, i.e. in increasing the vegetation cover and plant diversity. However, restoration of vegetation does not always imply restoration of the full range of fauna and, consequently, of ecosystem structure and functions. In particular, the diversity of the arthropod community lags behind the diversity of the restored plant community (Babin-Fenske and Anand, 2010). Similarly, birds and mammals associated with mature forests can only re-colonise the formerly barren areas after the recovery of their habitats. As a result, newly restored sites have sometimes been called 'depauperate imitations', as they superficially resemble the undisturbed sites in some ways but do not have the same diversity or composition of flora and fauna. Therefore the measures that assist natural recovery should have priority over the placement of agricultural soils or soil substitutes rich in nutrients over the polluted soils (as had been done at some sites near the non-ferrous smelter in Monchegorsk, Russia), because the latter method, although it ensures fast re-greening of barren lands and thus improves the visual perception of the landscape, is unlikely to assure long-term success in forest recovery.

Future research directions

Several recent reviews stressed the need to advance the understanding of pollution impacts on forest ecosystems through the concerted effort of scientists monitoring the effects of pollution, exploring them by means of field and laboratory experiments, and modelling both distribution of pollutants and their impacts on ecosystem processes in the recent and possible future climate. A starting point for integrative research is the introduction of a new kind of research site in forest ecosystems, so-called 'supersites'. These 'supersites' are specifically designed to explore effects of relatively low levels of regional pollution, unify air pollution and climate change research for forest ecosystems, and on this basis to promote predictive modelling for reliable risk assessment (Matyssek et al., 2012). It was also suggested (Kozlov et al., 2009) that pollution ecologists expand the use of the following research approaches: comparisons between polluters, pollutants, organisms, communities, and ecosystems; long-term studies, especially those addressing recovery of polluted ecosystems; exploration of dose-and-effect and cause-and-effect relationships and meta-analysis. In terms of geography, more studies should be conducted in the forests of Africa, South East Asia, South America and Oceania. Especially urgent is exploration of pollution impacts on tropical forests. Among research topics, the most critical knowledge gaps are associated with pollution impacts on below ground processes, spatial structure and demography of populations, functional diversity of communities and biotic interactions.

Conclusion

The adverse consequences of pollution's impacts on forest ecosystems have been under careful investigation for more than a century. The earlier studies attempted large-scale generalisations, but accumulation of knowledge resulted in the discovery of a great diversity in the responses of organisms, populations, communities and ecosystems to different pollutants. Identification of factors explaining this diversity is crucial for understanding causal relationships between pollution and its effects on the structure and functions of forest ecosystems. Although the development of evolutionary adaptations to pollution is likely to be a common phenomenon, still the evolved pollution tolerance is insufficient to assure the survival of forests in severely polluted areas. Polluted forests may require specific management, in particular because the harmful effects of pollution are likely to increase with global warming. The need for intelligent management decisions requires further investigation of the combined effects of relatively low but chronic levels of pollution and other environmental stressors, in particular climate, on ecological processes in forest ecosystems and the development of the theoretical background of pollution ecology.

References

Ainsworth, E. A., Yendrek, C. R., Sitch, S., Collins, W. J. and Emberson, L. D. (2012) 'The effects of tropospheric ozone on net primary productivity and implications for climate change', *Annual Review of Plant Biology*, vol 63, pp. 637–661

Allen, E. B., Temple, P. J., Bytnerowicz, A., Arbaugh, M. J., Sirulnik, A. G. and Rao, L. E. (2007) 'Patterns of understory diversity in mixed coniferous forests of southern California impacted by air pollution', *The Scientific World Journal*, vol 7 suppl 1, pp. 247–263

Babin-Fenske, J. and Anand, M. (2010) 'Terrestrial insect communities and the restoration of an industrially perturbed landscape: assessing success and surrogacy', *Restoration Ecology*, vol 18 suppl 1, pp. 73–84

Bobbink, R., Hicks, K., Galloway, J., Spranger, T., Alkemade, R. and others (2010) 'Global assessment of nitrogen deposition effects on terrestrial plant diversity: a synthesis', *Ecological Applications*, vol 20, pp. 30–59

Bytnerowicz, A., Arbaugh, M., Schilling, S., Frączek, W. and Alexander, D. (2008) 'Ozone distribution and phytotoxic potential in mixed conifer forests of the San Bernardino Mountains, southern California', *Environmental Pollution*, vol 155, pp. 398–408

Cerny, J. (1995) 'Recovery of acidified catchments in the extremely polluted Krusne Hory Mountains, Czech Republic', *Water, Air, and Soil Pollution*, vol 85, pp. 589–594

Chernenkova, T. V., Kabirov, R. R., Mekhanikova, E. V., Stepanov, A. M. and Gusarova, A. Yu. (2001) 'Demutation of vegetation after copper smelter shutdown', *Lesovedenie [Forestry, Moscow]*, vol 0(6), pp. 31–37

Clements, W. H. and Newman, M. C. (2002) *Community Ecotoxicology*, Wiley, New York

De Marco, A., Screpanti, A., Attorre, F., Proietti, C. and Vitale, M. (2013) 'Assessing ozone and nitrogen impact on net primary productivity with a generalised non-linear model', *Environmental Pollution*, vol 172, pp. 250–263

de Vries, W., Dobbertin, M. H., Solberg, S., van Dobben, H. F. and Schaub, M. (2014) 'Impacts of acid deposition, ozone exposure and weather conditions on forest ecosystems in Europe: an overview', *Plant and Soil*, vol 380, pp. 1–45

Eeva, T. and Lehikoinen, E. (2000) 'Pollution: recovery of breeding success in wild birds', *Nature*, vol 403, pp. 851–852

Eeva, T. and Lehikoinen, E. (2013) 'Density effects on great tit (*Parus major*) clutch size intensifies in a polluted environment', *Oecologia*, vol 173, pp. 1661–1668

Fowler, D., Cape, J. N., Coyle, M., Flechard, C., Kuylenstierna, J. and others (1999) 'The global exposure of forests to air pollutants', *Water, Air, and Soil Pollution*, vol 116, pp. 5–32

Gordon, A. G. and Gorham, E. (1963) 'Ecological aspects of air pollution from an iron-sintering plant at Wawa, Ontario', *Canadian Journal of Botany*, vol 41, pp. 1063–1078

Guardans, R. (2002) 'Estimation of climate change influence on the sensitivity of trees in Europe to air pollution concentrations', *Environmental Science and Policy*, vol 5, pp. 319–333

Gunn, J. M., (ed.) (1995) *Restoration and Recovery of an Industrial Region: Progress in Restoring the Smelter-Damaged Landscape near Sudbury, Canada*, Springer, New York

Hill, A. B. (1965) 'Environment and disease: association or causation', *Proceedings of the Royal Society of Medicine*, vol 58, pp. 295–300

Jonard, M., Legout, A., Nicolas, M., Dambrine, E., Nys, C. and others (2012) 'Deterioration of Norway spruce vitality despite a sharp decline in acid deposition: a long-term integrated perspective', *Global Change Biology*, vol 18, pp. 711–725

Kozlov, M. V. (2001) 'Snowpack changes around a nickel-copper smelter at Monchegorsk, northwestern Russia', *Canadian Journal of Forest Research*, vol 31, pp. 1684–1690

Kozlov, M. V. and Zvereva, E. L. (2007) 'Industrial barrens: extreme habitats created by non-ferrous metallurgy', *Reviews in Environmental Science and Biotechnology*, vol 6, pp. 231–259

Kozlov, M. V. and Zvereva, E. L. (2011) 'A second life for old data: global patterns in pollution ecology revealed from published observational studies', *Environmental Pollution*, vol 159, pp. 1067–1075

Kozlov, M. V., Haukioja, E., Niemelä, P., Zvereva, E. and Kytö, M. (2000) 'Revitalization and restoration of boreal and temperate forests damaged by aerial pollution', in J. L. Innes and J. Oleksyn, eds, *Forest Dynamics in Heavily Polluted Regions*, CAB International, Wallingford, UK, pp. 193–218

Kozlov, M. V., Zvereva, E. L., Gilyazov, A. S. and Kataev, G. D. (2005) 'Contaminated zone around a nickel-copper smelter: a death trap for birds and mammals?', in A. R. Burk (ed), *Trends in Biodiversity Research*, Nova Science, New York, pp. 81–99

Kozlov, M. V., Zvereva, E. L. and Zverev, V. E. (2009) *Impacts of Point Polluters on Terrestrial Biota: Comparative Analysis of 18 Contaminated Areas*, Springer, Dordrecht, Netherlands

Matyssek, R., Wieser, G., Calfapietra, C., de Vries, W., Dizengremel, P. and others (2012) 'Forests under climate change and air pollution: gaps in understanding and future directions for research', *Environmental Pollution*, vol 160, pp. 57–65

Medina, M. H., Correa, J. A. and Barata, C. (2007) 'Micro-evolution due to pollution: possible consequences for ecosystem responses to toxic stress', *Chemosphere*, vol. 67, pp. 2105–2114

Newman, J. R., Schreiber, R. K. and Novakova, E. (1992) 'Air pollution effects on terrestrial and aquatic animals', in J. R. Barker and D. T. Tingey (eds), *Air Pollution Effects on Biodiversity*, Van Nostrand Reinhold, New York, pp. 177–233

Odum, E. P. (1985) 'Trends expected in stressed ecosystems', *BioScience*, vol 35, pp. 419–422

Posch, M. (2002) 'Impacts of climate change on critical loads and exceedances in Europe', *Environmental Science and Policy*, vol 5, pp. 307–317

Rayfield, B., Anand, M. and Laurence, S. (2005) 'Assessing simple versus complex restoration strategies for industrially disturbed forests', *Restoration Ecology*, vol 13, pp. 639–650

Tikkanen, E. and Niemelä, I. (eds) (1995) *Kola Peninsula Pollutants and Forest Ecosystems in Lapland*, Finnish Forest Research Institute, Rovaniemi, Finland

Vorobeichik, E. L., Trubina, M. R., Khantemirova, E. V. and Bergman, I. E. (2014) 'Long-term dynamic of forest vegetation after reduction of copper smelter emissions', *Russian Journal of Ecology*, vol 45, pp. 498–507

Wittig, V. E., Ainsworth, E. A., Naidu, S. L., Karnosky, D. F. and Long, S. P. (2009) 'Quantifying the impact of current and future tropospheric ozone on tree biomass, growth, physiology and biochemistry: a quantitative meta-analysis', *Global Change Biology*, vol 15, pp. 396–424

Wołk, A. (1977) 'Microclimate alterations in relation to the extent of forest destruction in vicinity of the nitrogen plant at Puławy', *Sylwan*, vol 121, pp. 33–46 (in Polish)

Zverev, V. E. (2009) 'Mortality and recruitment of mountain birch (*Betula pubescens* ssp. *czerepanovii*) in the impact zone of a copper-nickel smelter in the period of significant reduction of emissions: the results of 15-year monitoring', *Russian Journal of Ecology*, vol 40, pp. 254–260

Zverev, V. E., Kozlov, M. V. and Zvereva, E. L. (2013) 'Changes in crown architecture as a strategy of mountain birch for survival in habitats disturbed by pollution', *Science of the Total Environment*, vol 444, pp. 212–223

Zvereva, E. L. and Kozlov, M. V. (2010) 'Responses of terrestrial arthropods to air pollution: a meta-analysis', *Environmental Science and Pollution Research*, vol 17, pp. 297–311

Zvereva, E. L. and Kozlov, M. V. (2011) 'Mosses under impact of industrial polluters: a meta-analysis of observational studies', *Water, Air, and Soil Pollution*, vol 218, pp. 573–586

Zvereva, E. L. and Kozlov, M. V. (2012) 'Changes in the abundance of vascular plants under the impact of industrial air pollution: a meta-analysis', *Water, Air, and Soil Pollution*, vol 223, pp. 2589–2599

Zvereva, E. L., Toivonen, E. and Kozlov, M. V. (2008) 'Changes in species richness of vascular plants under the impact of air pollution: a global perspective', *Global Ecology and Biogeography*, vol 17, pp. 305–319

Zvereva, E. L., Roitto, M. and Kozlov, M. V. (2010) 'Growth and reproduction of vascular plants under pollution impact: a synthesis of existing knowledge', *Environmental Reviews*, vol 18, 355–367

32
BIOLOGICAL INVASIONS IN FORESTS AND FOREST PLANTATIONS

Marcel Rejmánek

Biological invasions (spontaneous spread of non-native human-introduced taxa) have become a global phenomenon (Figure 32.1). Among the best-known examples are the spread of Africanized bees in the Neotropics, kudzu (*Pueraria montana* var. *lobata*) in the southeastern US, opuntias (Cactaceae) in the Old World, pines in the Southern Hemisphere, Gypsy moth (*Lymantria dispar*) in North America, feral pigs in Hawaii and many other islands, brushtail possum in New Zealand, beavers in Tierra del Fuego, and brown tree snakes in Guam. Some invasions – e.g., Eurasian earthworms in North America or Argentine ants in southern Africa – are not so visible, but may have profound ecosystem effects. The economic and/or environmental impact of many invasions may be essentially inconsequential, but only time will tell.

It is usually expected that unless logged, burned, or disturbed by storms or floods, forests are, in general, resistant to biological invasions, i.e., penetrations of non-native taxa. Indeed, most natural forests host fewer nonindigenous organisms than do surrounding human-modified habitats. Nevertheless, alien plants and animals have invaded even some old-growth undisturbed forests. So far, only specific studies or reviews on invasions of particular groups of organisms into particular types of forests have been provided (e.g., Denslow and DeWalt 2008, Hendrix et al. 2008, Kohli et al. 2009, Liebhold and McCullough 2011, McKenzie et al. 2005, Rejmánek 1996, Sanderson et al. 2012).

This chapter sketches a global overview of biological invasions in forests. Special attention is paid to invasions of alien plants, nematodes, earthworms, snails, slugs, insects, reptiles, birds, and mammals. For each group of organisms, lists of representative invaders are provided, and their environmental impacts in boreal, temperate and tropical forests are discussed.

Plants

For any non-native species to invade there must be suitable climatic conditions and resources. Essential resources for plants are light, water, and nutrients. Light and nutrients are more likely available in disturbed ecosystems, and this is the reason why relatively few plant invasions are known from closed, undisturbed forests. Some examples of such species are in Table 32.1. More examples may be found in Dawson *et al.* (2009), Foxcroft *et al.* (2013), Kohli *et al.* (2009), Martin *et al.* (2009), Rejmánek (1996), Rejmánek and Richardson (2013) and Woodward and

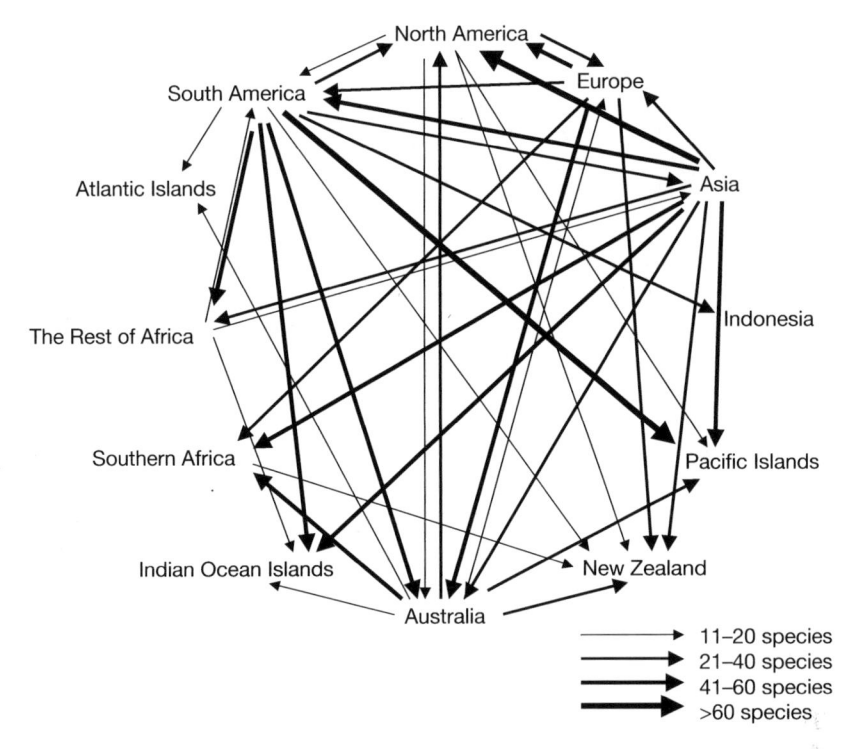

Figure 32.1 The global donor–acceptor network of invasive alien trees and shrubs (derived from Tables 3 and 4 in Rejmánek, 2014). Arrow thickness corresponds to numbers of species. "South America" also includes the Caribbean islands and Central America. Some regions play a major role as donors (Asia, South America), whereas others are mostly acceptors (Southern Africa, New Zealand, Pacific Islands)

Quinn (2011). Besides straight competition for resources with resident species, it is very often a layer of slow-decomposing litter on the ground that prevents establishment of exotic seedlings in forests. While plants invading disturbed areas are usually fast-growing, light-demanding species that rapidly produce many small wind-dispersed seeds (Rejmánek *et al.* 2013), plants invading closed forests have very different attributes. They are shade-tolerant, and often vertebrate- (mostly bird) dispersed. Extended leaf phenology (earlier leaf emergence and/or later leaf senescence) is another attribute that can promote invasiveness of some non-native woody species (Rejmánek 2013a). There are also some unexpected combinations of attributes. For example, Chinese tallow (*Triadica sebifera*) is a fast-growing species with a very short juvenile period, but shade-tolerant. Even typical disturbed-area invaders can be important in forest dynamics. After any disturbance, they can block vegetation succession for a long time, and they can compete with native pioneer species in forest gaps.

From the healthy forest regeneration point of view, the most important invaders are clonal species, which form dense compact stands, often with high litter biomass (e.g., *Microstegium vimineum, Lantana camara, Fallopia* spp., *Rubus* spp., *Stenotaphrum secundum, Ailanthus altissima*). The second most important group of plant invaders is the climbers (*Pueraria montana, Hedera* spp., *Lygodium* spp., *Celastrus orbiculatus, Lonicera japonica, Cryptostegia grandiflora, Passiflora* spp.). They can form dense blankets preventing any forest regeneration after disturbances. Fire-promoting grasses (e.g., jaragua grass *Hyparrhenia rufa*, molasses grass *Melinis minutiflora*, Guinea grass *Panicum maximum*) are particularly important in the tropics where forest trees are not

Table 32.1 Some prominent examples of non-native plant species invading forests and forest plantations

Latin name (Family)	Common name	Life form	Native range	Invaded range	Habitats and impacts
Acacia spp. (Fabaceae, Mimosoideae)	Acacia, Wattle	Trees, with nitrogen-fixing rhizobia and ant or bird dispersed seeds	Australia (Africa, Asia)	All temperate and tropical continents, many islands	Open forests Displace native plant species, enrich soil by nitrogen
Acer negundo (Sapindaceae)	Box elder, Ash-leaf maple, Manitoba maple	Dioecious shade-tolerant, fast growing tree	N & C America	Europe, Australia (New South Wales)	Riparian forests Displaces native plant species
Acer platanoides (Sapindaceae)	Norway maple	Deciduous shade-tolerant wind-dispersed tree	Europe & W Asia	Eastern N America & British Isles	Broadleaved forests Displaces native plant species
Ageratina adenophora (Asteraceae)	Crofton weed, Eupatory, Snakeroot	Perennial herbaceous shrub	C America	Asia, Australia, S Africa, many islands	Forests and open areas Forms monospecific understory
Alliaria petiolata (Brassicaceae)	Garlic mustard	Shade-tolerant biennial, not palatable to herbivores	Eurasia	N America	Produces allelochemicals suppressing mycorrhizal fungi of native plants
Archontophoenix cunninghamiana (Arecaceae)	Bangalow palm, King palm	Shade-tolerant, bird-dispersed palm	Australia	S America	Atlantic rainforest in Brazil Forms compact carpets of seedlings and saplings
Berberis thunbergii (Berberidaceae)	Japanese barberry, Thunberg's barberry	Thorny fleshy-fruiting shrub, veget. reproduction	Japan, E Asia	N America	Forest gaps, floodplains Forms dense thickets
Chromolaena odorata (Asteraceae)	Siam weed, Christmas bush, Devil weed	Evergreen wind-dispersed shrub, climber	N & C America	Africa, Asia, Australia, Indonesia, many islands	Forest margins and gaps Dense thickets prevent establishment of native species
Clidemia hirta (Melastomataceae)	Soap bush, Koster's curse	Densely branching shade-tolerant bird-dispersed shrub	C & S America	Asia, many islands	Forest edges and gaps Competes with native plants
Cytisus scoparius (Fabaceae)	Scotch broom	Perennial leguminous nitrogen-fixing shrub	W Europe	N & S America, Australia	Open forests and forest plantations Outcompetes tree seedlings
Fallopia japonica (Polygonaceae)	Japanese knotweed	Tall herbaceous perennial	E Asia	Europe, N America, New Zealand	Forests and open wet habitats Outcompetes >80% of native plants

Species (Family)	Common names	Description	Native range	Invasive range	Habitat / Impact
Hedera helix s.l. (Araliaceae)	Ivy, Common ivy, English ivy	Shade-tolerant bird-dispersed root climber	Europe & Asia	Australia, USA, New Zealand	Forests and forest plantations; Outcompetes native vegetation
Hovenia dulcis (Rhamnaceae)	Japanese raisin tree	Deciduous tree, vertebrate-dispersed	E Asia	S America, E Africa	Subtropical and tropical forests; Competes with native plants
Lantana camara (Verbenaceae)	Lantana, Big sage, Wild sage	Fleshy-fruiting, spiny, shrub	C & S America	Africa, Asia, Australia, many tropical islands	Open areas, forests and plantations; Outcompetes all seedlings
Lygodium japonicum (Lygodiaceae)	Japanese climbing fern	Fern with creeping stem and long leaves	E Asia	SE USA	Swamp forests; Forms dense mats
Microstegium vimineum (Poaceae)	Japanese stiltgrass, Nepalese browntop	Shade-tolerant C4 grass	Asia	SE USA	Broadleaved forests; Displaces native plants
Morella (Myrica) faya (Myricaceae)	Fire tree	Evergreen nitrogen fixing shrub or small tree	Macronesia	Hawaii	Metrosideros polymorpha forests; Displaces native plants
Oeceoclades maculata (Orchidaceae)	Monk orchid, African spotted orchid	Terrestrial orchid, self-pollinated	Africa, Madagascar	The Neotropics and Florida	Disturbed and old-growth forests; Impact not detected
Pinus radiata (Pinaceae)	Monterey pine, Radiata pine	Fast growing, medium-density softwood	California	Australia, S Africa, Chile	Native eucalyptus forests, fynbos; Competes with native plants
Prunus serotina (Rosaceae)	Wild black cherry, Rum cherry	Moderately shade-tolerant, bird-dispersed tree	E North America	W & C Europe	Disturbed and undisturbed forests; Competes with native plants
Psidium cattleianum (Myrtaceae)	Strawberry guava, Cattley guava	Small fleshy-fruiting, vertebrate-dispersed tree	Tropical S America	Many tropical islands	Disturbed and undisturbed forests; Forms dense monocultures
Rubus armeniacus (Rosaceae)	Himalayan blackberry (R. discolor misapplied)	Fleshy-fruiting shrub, vertebrate-dispersed tree	Armenia, Iran	N America, Europe	Disturbed and riparian forests; Outcompetes native plants
Stenotaphrum secundum (Poaceae)	St. Augustine grass	Perennial, C4, stoloniferous grass	Africa & Americas	Australia	Swamp forests; Local extinction of native plants
Syzigium jambos (Myrtaceae)	Rose apple	Fleshy-fruiting tree	SE Asia	C America, Africa, many tropical islands	Forests (all successional stages); Decline in plant species diversity
Triadica sebifera (Euphorbiaceae)	Chinese tallow	Shade-tolerant, fast growing tree	E Asia, Japan	SE US, India, Pacific Islands	Successional forests; Blocking vegetation succession

adapted to fires. Invasive woody species associated with nitrogen fixing microorganisms (legumes, *Morella faya*) can completely change successional trajectories (Gaertner *et al.* 2014). Finally, hybridization with native congeners may cause permanent genetic changes in native populations (e.g., species in the genera *Eucalyptus, Populus, Rosa, Rubus,* and *Ulmus*).

Some impacts of invasive plants are less obvious. Here are three examples. Invasive *Rhamnus cathartica* changes the vegetation structure of the forest, affecting nest predation of birds. *Berberis thunbergii* provides questing sites for the blacklegged ticks that carry the spirochete *Borrelia burgdorferi*, the causative agent of Lyme disease. Dense exotic *Lantana camara* understory in Australian *Eucalyptus saligna* forests seems to be the most critical in the development of forest dieback caused by psyllids (*Glycaspis spp.*). *Lantana* provides cover for the bell miner (*Manorina melanophrys*), a highly territorial bird that drives off other bird species that predate on psyllids. Impacts of invasive plants on forest biodiversity are context dependent and often only careful observations and long-term studies in permanent plots may bring conclusive results (Brown *et al.* 2006, Norghauer *et al.* 2011). Also, environmentally safe experiments should be encouraged (Green *et al.* 2004, Rejmánek 2013b).

Tropical and boreal forests seem to be less invaded by exotic plants than are temperate forests (Rejmánek 1996, Sanderson *et al.* 2012). However, the degree to which this is a result of differences in propagule pressure is difficult to judge. The quantity of shade and fast post-disturbance recovery of wet tropical forests may be the major reasons that these forests are relatively resistant to plant invasions. Riparian forests are usually more invaded than other types of forests. Frequent disturbances and high input of propagules of invasive plant species are the two underlying causes.

The vast majority of plants invading forests or forest plantations around the world are the result of intentional introductions. Exotic tree plantations, nurseries, gardens and erosion control projects are major local sources of invasive plant species (Reichard 2011, Richardson 2011). Surprisingly, many species, while recognized as influential invaders, are still widely planted. As with many other major groups of invasive organisms, forest fragmentation, roads, and movement of propagules in soil attached to vehicles and forest industry machinery are the major reasons that forests are invaded by non-native taxa. Besides prevention, early detection of harmful invaders is the key to successful eradication. In general, eradication of exotic pest plant infestations in excess of 1,000 ha is very unlikely, given a realistic amount of resources (Rejmánek and Pitcairn 2002).

Nematodes, earthworms, snails and slugs

Nematodes (phylum Nematoda) are wormlike in appearance but quite distinct taxonomically from the true worms. Plant parasitic nematodes (>1,100 species) are small, less than 2.5 mm in length. Because of their size, invasiveness of nematodes is poorly studied and their impacts are not well known (Singh *et al.* 2013). The most important invasive nematode in forests is the pine wilt or pine wood nematode (*Bursaphelenchus xylophilus*; Mota and Vieira 2008). This species infects pine trees and causes the pine wilt disease. It affects, with different severity, more than 30 pine species and several other conifers. It is probably native to North America and is invasive in Japan, eastern Asia, and Portugal. This and related species are vectored by a number of bark beetles and wood borers, most often the cerambycid longhorn beetles in the genus *Monochamus*. Both *Monochamus* and *Bursaphelenchus* may be found in pine chips, unseasoned lumber, and logs. Consequently, they are easily transported. For this reason, the European Plant Protection Organization has banned import of softwood products except kiln-dried lumber from North America.

Semiaquatic and terrestrial earthworms are usually classified into 17 families that belong to the subclass Oligochaeta. Over 3,000 species of earthworms have been described, which may be just about a half of all existing species. Since Darwin's classic *The Formation of Vegetable Mould through the Action of Worms*, published in 1881, there has been increasing appreciation of ecosystem consequences of earthworm activities. So far, about 120 species of earthworms can be classified as invasive species. Representative examples of earthworm species invading forests and forest plantations are found in Table 32.2. Additional examples can be found in Edwards (2004), Hendrix (2006), Hendrix et al. (2008), James (2011), and Reynolds (2012). Success of invasive earthworms depends to a large extent on a resistant stage (diapause), ability to use diverse food resources, human-mediated transport, and parthenogenesis in many species. Because earthworms are small and their egg capsules even smaller, they travel with soil and on the roots of living plants without our knowledge. The horticulture industry, topsoil movement by landscapers and other machinery, and use of earthworms as fish bait are major contributors to their long-distance dispersal.

Impacts of earthworm invasions are most pronounced in forests north of the southern limit of the last glaciation in North America and Europe. Such areas have no native earthworm populations, due to slow natural range expansions. In the absence of earthworms, a thick, slowly decomposing layer of leaf litter develops on the surface of mineral soil. When earthworms are introduced, they consume the litter and mix the mineral soil with the finely divided organic

Table 32.2 Some prominent examples of non-native earthworm species invading forests and forest plantations

Taxa	Life form	Native range	Invaded range
Lumbricideae			
Aporrectodea rosea and *A. tuberculata*	Parthenogenetic, endogeic	Western Palearctic[a]	All other realms
Dendrobaena octaedra	Parthenogenetic, epigeic	Western Palearctic	All other realms except Ethiopian
Lumbricus rubellus	Sexually reproducing, epi-endogeic	Western Palearctic	All other realms
Lumbricus terrestris	Cross-fertilization hermaphrodite, aneic[e]	Western Palearctic	All other realms
Octolasion tyrtaeum	Parthenogenetic, endogeic	Palearctic	Nearctic, Neotropical
Megascolecidae			
Amynthas spp., *Metaphire* spp.	Parthenogenetic, epigeic	Eastern Palearctic[b]	All other realms
Glossoscolecidae			
Pontoscolex corethrurus	Sexual & parthenogenetic, endogeic	Neotropical[c]	All other realms
Eudrilidae			
Eudrilus eugeniae	Sexually reproducing, epigeic	Ethiopian[d]	Neotropical

[a] Europe, western Asia, and northern Africa. [b] Eastern Asia. [c] Central and South America. [d] Sub-Saharan Africa. [e] Litter-feeding.

matter of their fecal material. Reduction of the forest-floor litter layer has many consequences. Nutrient mineralization is accelerated and soil C:N ratio is greatly reduced. Abundance of soil organisms depending on the litter dominated by fungi is probably reduced as well. In some cases, the impacted soil is less cohesive, and herbs and tree seedlings may be uprooted by deer feeding, rather than just deprived of upper parts. This may lead to reductions of forest-floor herb layers and the influx of some invasive plant species. In northern hardwood forests, invasions of exotic earthworms may be facilitated by the production of high quality leaf litter produced by exotic woody invaders like *Rhamnus cathartica* (buckthorn) or *Lonicera* spp. (honeysuckle). Removal of such species may result in reductions of exotic earthworm abundance by 50 percent. Earthworm burrowing can disrupt mycorrhizae of native species and promote invasive plant species that are either nonobligate or amycorrhizal (*Alliaria petiolata, Berberis thunbergii, Microstegium vimineum*). Invasive plants are conspicuous and are often assumed to be the drivers of environmental changes; however, much of the change may occur below ground, and invasive plants are the beneficiaries, rather than drivers of such changes (Nuzzo *et al.* 2009). With increased human activity and global warming, invasions by earthworms are likely to increase in northern forests in the future.

Snails (gastropods with a coiled shell) and slugs (shell-less gastropods) are also important invasive invertebrates, particularly in forests on islands. For example, about 100 species of land snails are naturalized on the islands of the Pacific (Cowie 2005). Among the best-known examples are *Achatina fulica* (giant African snail) and *Euglandina rosea* (rosy wolfsnail or cannibal snail). *A. fulica* is a macrophytophagous herbivore introduced to Hawaii from Japan. By the 1950s it had become so abundant that 15 predatory snails were introduced as biological control agents; a small number of them became established, the most notable of which is *E. rosea* (Cowie 2005). There is no reliable evidence that *E. rosea* has controlled populations of *A. fulica*, but *E. rosea* was very likely responsible for the decline and possible extinction of Hawaiian tree snails and possibly of other endemic snails. The Hawaiian Islands have no native slugs, but at least 12 introduced slug species are now established. They can reduce seedling survival of endangered plant species by 50 percent, without significant effects on the survival of invasive plant species (Joe and Daehler 2008). Similarly, the European slug *Deroceras reticulatum* (the gray field slug) significantly damages native herbaceous plants in North American forests, while the invasive *Alliaria petiolata* is avoided (Hahn *et al.* 2011).

Insects

As far as we know, insects are, besides vascular plants, the most numerous group of invasive organisms. In the continental United States alone, more than 450 exotic herbivorous insect species have colonized forests and urban trees since European settlement. Only about 15 percent of these insects have caused notable damage to trees (Aukema *et al.* 2010). However, some impacts on forests are much more drastic than the impacts of any invasive plant species. Interestingly, the situation in New Zealand is very different: the total number of exotic insect species that affect woody plants is about 200 and only about 5 percent of them are serious pests, mostly on introduced tree species. Many New Zealand native woody plants appear to be resistant to exotic herbivorous insects, perhaps because of their phylogenetic uniqueness (Brockerhoff *et al.* 2010). Also, forests in European countries seem to be much less invaded and influenced by invasive herbivorous insects. For example, in the Czech Republic, among 90 exotic forest herbivores, only up to 10 cause any substantial damage (Šefrová and Laštůvka 2005, and personal communication). Several prominent examples of insect species invading forests and forest plantations worldwide are in Table 32.3. More examples may be found in

Table 32.3 Some prominent examples of non-native insect species invading forests and forest plantations

Latin name (Order)	Common name	Native range	Invaded range	Host species	Tissue attacked/damage
Adelges tsugae (Hemiptera)	Hemlock wooly adelgid	Eastern Asia	Eastern N. America	Tsuga canadensis & T. caroliniana	Ray parenchyma of twigs/needle loss, mortality
Agrilus planipennis (Coleoptera)	Emerald ash borer	Asia	N. America, eastern Europe	Fraxinus spp. (≥16 species)	Xylem, phloem, bark/dieback, mortality
Anoplophora glabripennis (Coleoptera)	Asian long-horned beetle	Eastern Asia	N. America, Europe, Trinidad	Acer spp., Populus spp., Ulmus spp., Betula spp., Aesculus spp., Salix spp.	Xylem, phloem/dieback, breakage, mortality
Apis mellifera hybrid (Hymenoptera)	Africanized honey bees, killer bees	Produced by cross-breeding	S. & C. America, southern N. America	N/A	Aggressively protective behavior
Cameraria ohridella (Lepidoptera)	Horse chestnut leaf miner	Southern Europe	Central and western Europe	Aesculus hippocastanum	Leaves/defoliation, reduction of growth and reproduction
Coptotermes formosanus (Blattodea)	Formosan subterranean termite	Southern China, Taiwan	Southern U.S., Hawaii, S. Africa	Quercus spp., Fraxinus spp., Liquidambar styraciflua, etc.	Wood and bark/mortality, damage on woody structures
Cryptococcus fagisuga (Hemiptera)	Beech scale	The Caucasus Mts., Greece	Eastern N. America	Fagus grandifolia	Parenchyma of bark/fungal infection, mortality
Dendroctonus valens (Coleoptera)	Red turpentine beetle	N. America	China	Pinus spp., occasionally Picea spp.	Phloem/mortality
Dreyfusia nordmannianae (Hemiptera)	Fir woolly adelgid	The Caucasus region	Central and northern Europe	Abies spp.	Ray parenchyma of twigs/needle loss, mortality
Eriococcus orariensis (Hemiptera)	Manuka blight scale	Australia	New Zealand	Leptospermum scoparium, Kunzea ericoides	Twigs/sap-feeding, dieback, mortality
Gonipterus scutellatus (Coleoptera)	Eucalyptus weevil	Australia	Africa, N. & S. America, Europe, Asia, New Zealand	Eucalyptus spp.	Leaves & soft bark of twigs/defoliation, dieback of twigs
Heteropsylla cubana (Hemiptera)	Leucaena psyllid	C. & S. America	Tropical Asia & Africa, Australia, Pacific islands	Leucaena★ spp., Albizia spp.	Leaves, buds/sap-feeding, defoliation, dieback
Hyphantria cunea (Lepidoptera)	Fall webworm	N. America	Eurasia	Broadleaf trees (>600 species)	Leaves/defoliation

Table 32.3 continued

Latin name (Order)	Common name	Native range	Invaded range	Host species	Tissue attacked/damage
Linepithema humile (Hymenoptera)	Argentine ant (syn.: *Iridomymex humilis*)	S. America	N. America, S. Africa Europe, Australia, etc.	N/A	Displaces most or all native ants
Lymantria dispar (Lepidoptera)	Gypsy moth	Eurasia	N. America	*Quercus* spp., *Populus* spp., and about 200 other spp.	Leaves/defoliation
Neodiprion sertifer (Hymenoptera)	European pine sawfly	Europe	N. America	*Pinus* spp.	Needles/defoliation
Operophtera brumata (Lepidoptera)	Winter moth	Europe & Near East	N. America	Many broadleaf trees	Buds & leaves/defoliation
Pheidole megacephala (Hymenoptera)	Big-headed ant, Coastal brown ant	Africa	N. America, Australia, etc.	N/A	Displaces most or all native ants
Scolytus multistriatus (Coleoptera)	Smaller European elm bark beetle	Europe	N. America	*Ulmus* spp., *Zelkova* spp.	Bark/vector of Dutch elm disease (fungus *Ophiostoma*)
Sirex noctilio (Hymenoptera)	Sirex woodwasp	Eurasia, N. Africa	N. & S. America, S. Africa Australia, New Zealand	*Pinus* spp.	Xylem, phloem/phytotoxic secretion, mortality
Vespula vulgaris (Hymenoptera)	Common wasp, common yellowjacket	Eurasia	Australia, New Zealand, N. America, many islands	N/A	Predation on native invertebrates and nesting birds
Xylosandrus crassiusculus (Coleoptera)	Asian ambrosia beetle, granulate ambrosia beetle	Tropical Asia	Africa, N. America	>120 host species	Twigs, branches/mortality of some species in plantations

*Because *Leucaena leucocephala* is very often an invasive species itself, *H. cubana* may be considered more of a biological control agent than a forest pest.

Cartwell (2007), Ciesla (2011), Holway *et al.* (2002), Lach and Hooper-Bui (2010), Liebhold *et al.* (1995), Liebhold and Tobin (2008), Liebhold and McCullough (2011), Paine (2008), Paine *et al.* (2011), and Wylie and Speight (2012).

In contrast to vascular plants, reptiles, birds, and mammals, most exotic insect introductions are unintentional and are typically by-products of economic activities. Most problems from forest insect invasions can be largely attributed to global trade. Plants imported for nurseries or landscapes have always been an important pathway by which exotic herbivorous insects have arrived. Recently, solid wood packing material has become another important pathway for invasive forest insects. The International Plant Protection Convention recently enacted an international phytosanitary measure (ISPM 15) that limits the movement of raw wood. This should help, but expanding global trade and travel will very likely result in more introductions of exotic forest insects in the near future.

Established exotic insect species usually spread much faster than plants. Species like *Adelges tsugae* (hemlock wooly adelgid), *Cryptococcus fagisuga* (beech scale), or *Lymantria dispar* (gypsy moth) exhibit a radial rate of spread 3 to 30 km/year. Probably the fastest rate of spread ever recorded was that of the Africanized honeybees (resulting from hybridization of *Apis mellifera scutellata* with other subspecific taxa of the same species) in South and Central America: 300 to 500 km/year (Liebhold and Tobin 2008). Spread of exotic species is usually not a continuous process but a combination of more or less continuous spread due to dispersal abilities of individual species and long-distance dispersal events, usually human- or bird-mediated transport.

Most of the invasive forest herbivorous insect species in the continental United States are either sap feeders (192 species) or foliage feeders (155 species). However, phloem and wood borers (71 species) and foliage feeders often exert greater impact than expected (Table 32.4). Detection of insects that feed on phloem or wood has increased markedly in recent years. If species feeding on trees belonging to only one genus are classified as monophagous, species feeding on trees belonging to only one family as oligophagous and species feeding on trees from multiple families as polyphagous, then the majority of invasive forest herbivorous insects in the continental United States are either monophagous (49 percent) or polyphagous (33 percent) (Aukema *et al.* 2010). In contrast, the majority of successful exotic herbivorous forest insects in New Zealand seem to be rather polyphagous (Brockerhoff *et al.* 2010).

Table 32.4 Numbers of non-native forest insect species established in the continental United States (from Aukema *et al.* 2010)

Order	Number of species	Phloem and wood borers	Foliage feeders	Sap feeders	Other	Number of high-impact species	Percentage of high-impact species
Blattodea (Isoptera)	1	1	0	0	0	0	0
Orthoptera	1	0	0	0	1	0	0
Thysanoptera	4	0	0	4	0	2	50.0
Diptera	13	0	6	0	7	1	7.7
Hymenoptera	41	2	35	0	4	11	26.8
Lepidoptera	87	5	76	0	6	10	11.5
Coleoptera	119	63	38	0	18	20	16.8
Hemiptera	189	0	0	188	1	18	9.5
Total	455	71	155	192	37	62	13.6

The biology and impact of many forest insects are closely connected with microorganisms such as fungi. For example, the very well-known Dutch elm disease (DED) is caused by three species of the genus *Ophiostoma* (Ascomycota) that are dispersed in North America by native (*Hylurgopinus rufipes*) and exotic (*Scolytus multistriatus* and *S. schevyrewi*) bark beetles. In an attempt to block the fungus from spreading, the tree reacts by plugging its own xylem tissue with gum and tyloses. As the xylem delivers water and nutrients to the rest of the plant, these plugs prevent them from travelling up the trunk of the tree. As the disease progresses, major limbs die and eventually the entire tree is killed. Another example of insect–fungus association impacts is the considerable damage to pine plantations in the Southern Hemisphere following invasion of the wood-wasp *Sirex noctilio*, along with its fungus symbiont *Amylostereum areolatum*. Some exotic insect–fungus relationships are even more complicated. In New Zealand, the Australian scale insect *Eriococcus orariensis* caused significant damage on native manuka (*Leptospermum scoparium*). The insects suck sap and excrete large amounts of excess sap sugars as fine droplets of solution called honeydew. Where the honeydew falls on stems and foliage it provides food for a soot fungus *Capnodium walteri*, that blackens infested plants. Although the photosynthetic activity of leaves covered with the fungus is reduced, this alone does not cause the death of the plant. Debilitation and death are believed to be the direct result of massive withdrawal of sap nutrients by nymphs and adult females. Initially, heavy attack often resulted in the death of manuka, but later *E. orariensis* suffered from an entomopathogenic fungus *Angatia thwaitesii*, probably also introduced, and due to this spontaneous "biological control" the impact became less serious (Brockerhoff *et al.* 2010).

A special group of forest insect invaders are generalist predators. The most prominent cases are wasps. For example, the Eurasian *Vespula germanica* (German wasp) and *V. vulgaris* (common yellowjacket) predate on native forest invertebrates and nesting birds in many countries where they have been introduced. Some ant species are particularly important as omnivores and generalist predators (Holway *et al.* 2002). Out of about 150 ant species that have been introduced by humans beyond their native range, only a few are really widespread and damaging. The best examples are *Linepithema humile* (Argentine ant), *Pheidole megacephala* (big-headed ant), and *Solenopsis invicta* (red fire ant). These species often displace native ants and disrupt existing ant–plant mutualistic relations. Fortunately, invasive ants mostly favor open and disturbed habitats and only rarely invade undisturbed forests.

Preventing the arrival and potential establishment of exotic insects (and any other potential invaders) is ideal. Early detection of newly established exotic species is also critical. A localized population of an invader can sometimes be successfully eradicated. The availability of taxonomic experts is an important component of early detection. Containment, or slowing the spread of established species, is a management approach increasingly considered for certain forest insect invasions. Suppression of nascent populations along the invasion front prevents their growth and coalescence, ultimately resulting in slower spread into new regions (Liebhold and McCullough 2011).

Reptiles, birds and mammals

Over 200 species of reptiles are established or spreading outside their natural range, mainly because of inadvertent introductions via the pet trade (Kraus 2009). Two prominent examples of species invading forests are the brown tree snake (*Boiga irregularis*) and Burmese python (*Python molurus bivittatus*). Both are predators, extremely dangerous to native wildlife. The brown tree snake is responsible for the extinction of 10 of 13 native bird species, 6 of 12 native lizards and 2 of 3 bat species on the island of Guam. Some of the lost species were important pollinators or seed dispersers. The Burmese python is a giant constrictor that has invaded swamp forests of southern Florida. It preys mostly on native warm-blooded animals (deer, bobcats, raccoons, herons, etc.), although reptiles including iguanas and crocodilians are also taken (Reed and Rodda 2011). Eradication of snakes is extremely difficult.

Almost all established non-native birds are results of intentional introductions. Extreme cases are religious, mainly Buddhist, releases of birds (over 500,000 individuals every year in Hong Kong or Taiwan). Many of the released birds are exotic. Birds seem to be somewhat less important as invaders in forests, particularly on continents. Some examples of non-native bird species invading forests and forest plantations are in Table 32.5. More complete lists can be derived from Duncan *et al.* (2006), Lever (1994, 2005), Long (1981), and Scott *et al.* (1986).

Native and exotic bird species are usually segregated along gradients related to anthropogenic disturbance (Barnagaud *et al.* 2014). Most introduced birds are restricted to urban habitats, rural areas, forest edges, and exotic plantations. Even on islands, only a tiny fraction of introduced birds penetrates into undisturbed forests. For example, in Hawaii where more exotic bird species have been established than anywhere else, out of 58 naturalized bird species, only 7 have invaded native forests: *Cardinalis cardinalis*, *Garrulax caerulatus*, *G. canorus*, *G. pectoralis*, *Leiothrix lutea*, *Lophura leucomelana*, and *Zosterops japonicus* (Scott *et al.* 1986). Similarly, out of 34 established exotic bird species in New Zealand, only 2 have successfully invaded intact forests: *Fringilla coelebs* and *Turdus merula* (Duncan *et al.* 2006, and personal communication). In contrast, among 19 bird species introduced by people in Hong Kong, 10 invaded secondary forests (Leven and Corlett 2004). In this case, however, over half of these species were very likely present in the area prior to extensive deforestation.

Negative impacts of invasive birds are usually limited to competition with native birds (only rarely conclusively documented) and to dispersal of exotic plant species. However, in some cases the most important impact is the spread of avian pathogens. Via introduced birds infected with *Plasmodium relictum* and the introduced African mosquito *Culex quinquefasciatus*, avian malaria was introduced to and spread throughout the Hawaiian Islands. Avian malaria and avian pox together caused the extinction of many native bird species. The only positive impact of exotic birds in some forests is that they can partially substitute for extinct native pollinators and seed dispersers (Aslan *et al.* 2013, Foster and Robinson 2007).

Across continents and islands, introduced mammals are more successful invaders than birds (Jeschke 2008). Mammals are also more successful as invaders of undisturbed forests, and their direct impacts on native biota are much larger. Some examples of non-native mammal species invading forests and forest plantations are in Table 32.6. More examples can be found in Lever (1994), Long (2003), Nentwig *et al.* (2009), and Woodward and Quinn (2011). Different groups of mammals are important as soil disturbers (wild boar and feral pigs), drastic habitat modifiers (beaver), competitors (gray squirrel vs. native red squirrel, North American mink vs. European otter), herbivores (nutria, several deer taxa, brushtail possum), seed predators (rats, wild boar and feral pigs), and animal predators (mink, feral cats, foxes, rats). Impacts are particularly strong on islands where animal predators and/or large herbivores were previously absent.

There are many indirect impacts of invasive mammals. For example, non-native wild boar and deer are dispersing non-native ectomycorrhizal fungi that support invasive pine species in northern Patagonia. Moose (*Alces alces*), a non-native herbivore in Gros Monte National Park (Canada), acts as the primary conduit for non-native plant invasions by dispersing propagules and creating disturbance by trampling and browsing native vegetation. Some introduced mammals have turned out to be vectors of infectious diseases (invasive gray squirrels do not die from the parapox virus but can infect European red squirrels).

Table 32.5 Some prominent examples of non-native bird species invading forests and forest plantations

Latin name (Family)	Common name	Diet	Native range	Invaded range	Impacts
Acridotheres tristis (Sturnidae)	Common myna (mynah), Indian myna bird	Omnivorous	Asia	Australia, N. America, S. Africa, many islands	Competition with many hollow-nesters and killing of native young birds
Cardinalis cardinalis (Cardinalidae)	Northern cardinal	Mainly granivorous	N. America	Hawaii, Atlantic islands	?
Fringilla coelebs (Fringillidae)	Chaffinch (many subspecies)	Omnivorous, adults mostly granivorous	Eurasia, N. Africa	New Zealand, rare in S. Africa	?
Lophura leucomelanos (Phasianidae)	Kalij pheasant	Herbivorous	Asia	Hawaii, Argentina	Dispersal of invasive plant species (e.g., *Passiflora* and *Clidemia*)
Myiopsitta monachus (Psittacidae)	Monk or quaker parakeet	Herbivorous	S. America	USA, Mexico, Spain	Eating buds and fruits of native trees
Pycnonotus cafer (Pycnonotidae)	Red-vented bulbul	Omnivorous	Asia	Hawaii, Fiji, probably eradicated in New Zealand	Dispersal of invasive plant species (*Lantana camara*, *Miconia clavescens*)
Sturnus vulgaris (Sturnidae)	European starling	Omnivorous	Europe, W. Asia, N. Africa	Australia, N. America, S. Africa, New Zealand	Seed dispersal, competition with native birds for nesting sites
Turdus philomelos (Turdidae)	Song thrush	Omnivorous	Eurasia	Australia, New Zealand	?
Turdus merula (Turdidae)	Common blackbird, Eurasian blackbird	Omnivorous	Eurasia, N. Africa	Australia, New Zealand, rare in N. America	Competition with native birds, dispersal of invasive plant species
Tyto alba (Tytonidae)	Barn owl	Predator	All continents except Antarctica	Seychelles, Hawaii	Competes with and predates on native birds
Zosterops japonicus (Zosteropidae)	Japanese white-eye	Omnivorous	East Asia	Hawaii	Competition with native birds, vector for avian parasites

Table 32.6 Some prominent examples of non-native mammal species invading forests and forest plantations

Latin name (Order)	Common name	Life form	Native range	Invaded range	Impacts
Castor canadensis (Rodentia)	North American beaver, Canadian beaver	Large semi-aquatic rodent	N. America	Tierra del Fuego, Finland	Conversion of forests into wetlands
Cervus spp. (Artiodactyla)	Deer	Grazers/browsers	Many continents	Many continents and islands	Prevention of forest regeneration
Felis catus (Carnivora)	Feral cat	Predator	Eurasia	Many continents and islands	Preying on small mammals, birds, reptiles, and amphibians
Myocastor coypus (Rodentia)	Nutria, coypu	Semi-aquatic rodent	S. America	N. America, Europe, Africa, Japan	Destruction of wetlands, prevention of bottomland forest regeneration
Neovison vison (Carnivora)	America mink	Predator	N. America	Continental Europe, Iceland, S. America	Preying on small mammals and birds
Rattus spp. (Rodentia)	Rats	Omnivorous rodent	Asia	All continents except Antarctica, many islands	Elimination of native plants, birds, reptiles, insects, and snails
Sciurus carolinensis (Rodentia)	Eastern gray squirrel	Herbivorous and fungivorous rodent	N. America	Europe, S. Africa	Stripping bark from trees, outcompeting native squirrels*
Sus scrofa (Artiodactyla)	Wild boar and feral pigs	Fast reproducing, omnivorous	Eurasia, N. & E.	Americas, N. Africa, Australia, many islands	Extensive rooting, killing plants and ground-nesting birds
Vulpes vulpes (Carnivora)	Red fox	Carnivore/frugivore	Eurasia, N. Africa, boreal N. America	Australia, eastern N. America	Predation on mammals and ground-nesting birds
Trichosurus vulpecula (Diprotodontia)	Common brushtail possum	Nocturnal, semi-arboreal herbivorous marsupial	Australia	New Zealand	Defoliation of native trees, predation on eggs and chicks

*Also transmitting deadly parapox virus to native squirrels.

One of the major problems in European forests is sika deer (*Cervus nippon*) introduced from East Asia. Sika is a serious pest, causing significant damage in forests and forest plantations. It also plays a role in the epidemiology of *Asworthius sidemi*, a nematode affecting native deer (DAISIE 2009). Moreover, hybrids with native deer (*C. elaphus*) are fertile and further hybridization or back-crossing is rapidly threatening genetic integrity of the native species.

What does the future hold?

Rates of almost all human activities are accelerating and so are the rates of intentional and unintentional introductions of non-native biota. A majority of introduced species are not established. However, with global climate change, many new species will be established in areas where they would not have survived before (Dukes *et al.* 2009, Zizka and Dukes 2011). With increasing numbers of introduced species the probability of establishment of some of them increases (Hulme 2012) and with increasing numbers of established non-native species the probability that some of them will have an environmental or economic impact increases as well (Rejmánek and Randall 2004). Therefore, prevention, if we have such option, is usually cheaper than control (Finnoff *et al.* 2007, Roy *et al.* 2014). Early detection of pest species increases the chance of successful eradication (Rejmánek and Pitcairn 2002). However, once a harmful invader is widespread, eradication is unlikely and several management options, including species-specific biological control, have to be considered (Clout and Williams 2009, Van Driesche *et al.* 2008, Miller *et al.* 2010, Wylie and Speight 2012). Ecologically sound forestry practices (e.g., preventing of dispersal of alien propagules by machinery, reduced impact of logging, fast reforestation of disturbed areas with native species, or creation of refugia for endangered species) can certainly alleviate problems caused by biological invasions.

Acknowledgements

Clare Aslan, Robert Cowie, Richard Corlett, Richard Duncan, Paul Hendrix, Jonathan Jeschke, Zdeněk Laštůvka, Andrew Liebhold, Dane Panetta, John Reynolds, and Toni Withers contributed in many ways to the final version of this chapter. This project was supported by the University of California Agricultural Experiment Station.

References

Aslan, C.E., Zavaleta, E.S., Tershy, B., Croll, D. and R.H. Robichaux (2013) 'Imperfect replacement of native species by non-native species as pollinators of endemic Hawaiian plants' *Conservation Biology*, vol. 28, pp. 478–488

Aukema, J.E., McCullough, D.G., Von Holle, B., Liebhold, A.M., Britton, K. and S.J. Frankel (2010) 'Historical accumulation of nonindigenous forest pests in the continental United States' *BioScience*, vol. 60, pp. 886–897

Barnagaud, J-Y., Barbaro, L., Papaix, J., Deconchat, M. and E.G. Brockerhoff (2014) 'Habitat filtering by landscape and local forest composition in native and exotic New Zealand birds' *Ecology*, vol. 95, pp. 78–87

Brockerhoff, E.G., Barratt, B.I.P., Beggs, J.R., Fagan, L.L., Kay, M.K. and others (2010) 'Impacts of exotic invertebrates on New Zealand's indigenous species and ecosystems' *New Zealand Journal of Ecology*, vol. 34, pp. 158–174

Brown, K.A., Scatena, F.N. and J. Gurevitch (2006) 'Effects of an invasive tree on community structure and diversity in a tropical forest in Puerto Rico' *Forest Ecology and Management*, vol. 226, pp. 145–152

Cartwell, C.G. (2007) *Invasive Forest Pests*, Nova Science Publishers, New York

Ciesla, W.M. (2011) *Forest Entomology: A global Perspective*, Willey-Blackwell, Chichester, UK

Clout, M.N. and P.A. Williams (2009) *Invasive Species Management – A Handbook of Principles and Techniques*, Oxford University Press, Oxford, UK

Cowie, R.H. (2005) 'Alien non-marine molluscs in the islands of the tropical and subtropical Pacific: a review' *Amer. Malac. Bull.*, vol. 20, pp. 95–103

DAISIE (2009) *Handbook of Alien Species in Europe*, Springer, Berlin

Dawson, W., Burslem, D.F.R.P. and P.E. Hulme (2009) 'Factors explaining alien plant invasion success in a tropical ecosystem differ at each stage of invasion' *Journal of Ecology*, vol. 97, pp. 657–665

Denslow, J.S. and S. DeWalt (2008) 'Exotic plant invasions in tropical forests: patterns and hypotheses', in W.P. Carson and S.A. Schnitzer (eds.) *Tropical Forest Community Ecology*, University of Chicago Press, Chicago, US

Dukes, J.S., Pontius, J., Orwig, D., Garnas, J.R., Rodgers, V.L. and others (2009) 'Responses of insect pests, pathogens, and invasive plant species to climate change in the forests of northeastern North America: what can we predict?' *Canadian Journal of Forest Research*, vol. 39, pp. 231–248

Duncan, R.P., Blackburn, T.M. and P. Cassey (2006) 'Factors affecting the release, establishment and spread of introduced birds in New Zealand', in R.B. Allen and W.G. Lee (eds.) *Biological Invasions in New Zealand*, Springer-Verlag, Berlin, pp. 137–154

Edwards, C.A. (ed.) (2004) *Earthworm Ecology*, 2nd ed., CRC Press, Boca Raton, US

Finnoff, D., Shogren, J.F., Leung, B. and D. Lodge (2007) 'Take a risk: preferring prevention over control of biological invaders' *Ecological Economics*, vol. 62, pp. 216–222

Foster, J.T. and S.K. Robinson (2007) 'Introduced birds and the fate of Hawaiian rainforests' *Conservation Biology*, vol. 21, pp. 1248–1257

Foxcroft, L.C., Pyšek, P., Richardson, D.M. and P. Genovesi (eds.) (2013) *Plant Invasions in Protected Areas*, Springer, Dordrecht, Netherlands

Gaertner, M., Biggs, R., Te Beest, M., Hui, C., Molofsky, J. and D.M. Richardson (2014) 'Invasive plants as drivers of regime shifts: identifying high-priority invaders that alter feedback relationships' *Diversity and Distributions*, vol. 20, pp. 733–744

Green, P.T., Lake, P.S. and D.J. O'Dowd (2004) 'Resistance of island rainforest to invasion by alien plants: influence of microhabitat and herbivory on seedling performance' *Biological Invasions*, vol. 6, pp. 1–9

Hahn, P.G., Draney, M.L. and M.E. Dornbush (2011) 'Exotic slugs pose previously unrecognized threat to the herbaceous layer in a Midwestern woodland' *Restoration Ecology*, vol. 19, pp. 786–794

Hendrix, P.F. (ed.) (2006) *Biological Invasions Belowground: Earthworms as Invasive Species*, Springer, Dordrecht, Netherlands

Hendrix, P.F., Callaham, M.A., Drake, J.M., Huang, C-Y., James, S.W. and others (2008) 'Pandora's box contained bait: the global problem of introduced earthworms' *Annual Review of Ecology, Evolution, and Systematics*, vol. 39, pp. 593–613

Holway, D.A., Lach, L., Suarez, A.V., Tsutsui, N.D. and T.J. Case (2002) 'The causes and consequences of ant invasions' *Annual Review of Ecology and Systematics*, vol. 33, pp. 181–233

Hulme, P.E. (2012) 'Weed risk assessment: a way forward or a waste of time?' *Journal of Applied Ecology*, vol. 49, pp. 10–19

James, S.W. (2011) 'Earthworms', in D. Simberloff and M. Rejmánek (eds.) *Encyclopedia of Biological Invasions*, University of California Press, Berkeley, US, pp. 177–183

Jeschke, J.M. (2008) 'Across islands and continents, mammals are more successful invaders than birds' *Diversity and Distributions*, vol. 14, pp. 913–916

Joe, S.M. and C.C. Daehler (2008) 'Invasive slugs as under-appreciated obstacles to rare plant restoration: evidence from Hawaiian Islands' *Biological Invasions*, vol. 10, pp. 245–255

Kohli, R.K., Jose, S., Singh, H.P. and D.R. Batish (2009) (eds.) *Invasive Plants and Forest Ecosystems*, CRC Press, Boca Raton, Florida, US

Kraus, F. (2009) *Alien Reptiles and Amphibians: A Scientific Compendium and Analysis*. Springer, Dordrecht, Netherlands

Lach, L. and L.M. Hooper-Bui (2010) 'Consequences of ant invasions', in L. Lach, C.L. Parr and K.L. Abbott (eds.) *Ant Ecology*, Oxford University Press, Oxford, UK, pp. 261–286

Leven, M.R. and Richard T. Corlett (2004) 'Invasive birds in Hong Kong, China' *Ornithological Science*, vol. 3, pp. 43–55

Lever, C. (1994) *Naturalized Animals: The Ecology of Successfully Introduced Species*, T and A D Poyser, London

Lever, C. (2005) *Naturalized Birds of the World*, T and A D Poyser, London

Liebhold, A.M. and P.C. Tobin (2008) 'Population ecology of insect invasions and their management' *Annual Review of Entomology*, vol. 53, pp. 387–408

Liebhold, A.M. and D.G. McCullough (2011) 'Forest insects', in D. Simberloff and M. Rejmánek (eds.) *Encyclopedia of Biological Invasions*, University of California Press, Berkeley, US, pp. 238–241

Liebhold, A.M., Macdonald, W.L., Bergdahl, D. and V.C. Mastro (1995) 'Invasion by exotic forest pests: A threat to forest ecosystems' *Forest Science Monographs*, vol. 30, pp. 1–49

Long, J.L. (1981) *Introduced Birds of the World*, David and Charles, London

Long, J.L. (2003) *Introduced Mammals of the World*, CSIRO Publishing, Collingwood, Australia

Martin, P.H., Canham, C.D. and P.L. Marks (2009) 'Why forests appear resistant to exotic plant invasions: intentional introductions, stand dynamics, and the role of shade tolerance' *Frontiers of Ecology and Environment*, vol. 7, pp. 142–149

McKenzie, P., Brown, C., Jianghua, S. and W. Jian (2005) (eds.) *The Unwelcome Guests, Proceedings of the Asia-Pacific Forest Invasive Species Conference*, Food and Agriculture Organization of the United Nations, Regional Office, Bangkok

Miller, J.H., Manning, S.T., and S.F. Enloe (2010) *A Management Guide for Invasive Plants in Southern Forests*, USDA, Forest Service, General Technical Report SRS-131, Southern Research Station, Asheville, North Carolina, US

Mota, M.M. and P. Vieira (2008) (eds.) *Pine Wilt Disease: A Worldwide Threat to Forest Ecosystems*, Springer, Dordrecht, Netherlands

Nentwig, W., Kühnel, E. and S. Bacher (2009) 'A generic impact-scoring system applied to alien mammals of Europe' *Conservation Biology*, vol. 24, pp. 302–311

Norghauer, J.M., Martin, A.R., Mycroft, E.E., James, A. and S.C. Thomas (2011) 'Island invasion by a threatened tree species: evidence for natural enemy release of mahogany (*Swietenia macrophylla*) on Dominica, Less Antiles', *PLos ONE*, vol. 6, e18790

Nuzzo, V.A., Maerz, J.C. and B. Blossey (2009) 'Earthworm invasion as a driving force behind plant invasion and community change in northeastern North American forests' *Conservation Biology*, vol. 23, pp. 966–974

Paine, T.D. (2008) (ed.) *Invasive Forest Insects, Introduced Forest Trees, and Altered Ecosystems*, Springer, Dordrecht, Netherlands

Paine, T.D., Steinbauer, M.J. and S.A. Lawson (2011) 'Native and exotic pests of *Eucalyptus*: A worldwide perspective' *Annual Review of Entomology*, vol. 56, pp. 181–201

Reed, R.N. and G.H. Rodda (2011) 'Burmese python and other giant constrictors', in D. Simberloff and M. Rejmánek (eds.) *Encyclopedia of Biological Invasions*, University of California Press, Berkeley, US, pp. 85–91

Reichard, S. (2011) 'Horticulture', in D. Simberloff and M. Rejmánek (eds.) *Encyclopedia of Biological Invasions*, University of California Press, Berkeley, US, pp. 336–342

Rejmánek, M. (1996) 'Species richness and resistance to invasions', in G. Orians, R. Dirzo and J.H. Cushman (eds.) *Ecosystem Functions of Biodiversity in Tropical Forests*, Springer, Berlin, pp. 153–172

Rejmánek, M. (2013a) 'Extended leaf phenology: a secret of successful invaders?' *Journal of Vegetation Science*, vol. 24, pp. 975–976

Rejmánek, M. (2013b) 'Assessing the impacts of plant invaders on native plant species diversity' in CONABIO (ed.) *Proceedings of the 2012 Weeds Across Borders Conference*, CONABIO, Mexico City, pp. 63–69, www.weedcenter.org/wab/2012/docs/MemoriaWAB_2012_final.pdf

Rejmánek, M. (2014) 'Invasive trees and shrubs – where do they come from and what we should expect in the future?' *Biological Invasions*, vol. 16, pp. 483–498

Rejmánek, M. and M.J. Pitcairn (2002) 'When is eradication of exotic pest plants a realistic goal?' in C.R. Vietch and M.N. Clout (eds.) *Turning the Tide: The Eradication of Island Invasives*, IUCN, Gland, Switzerland, pp. 249–253, http://www.issg.org/database/species/reference_files/onoaca/Rejmanek.pdf

Rejmánek, M. and J.M. Randall (2004) 'The total number of naturalized species can be a reliable predictor of the number of alien pest species' *Diversity and Distributions*, vol. 10, pp. 367–369

Rejmánek, M. and D.M. Richardson (2013) 'Trees and shrubs as invasive alien species – 2013 update of the global database' *Diversity and Distributions*, vol. 19, pp. 1093–1094

Rejmánek, M., Richardson, D.M. and P. Pyšek (2013) 'Plant invasions and invasibility of plant communities', in E. van der Maarel and J. Franklin (eds.) *Vegetation Ecology*, 2nd ed., Wiley-Blackwell, Chichester, UK, pp. 387–424

Reynolds, J.W. (2012) 'The status of terrestrial earthworm (Oligochaeta) surveys in North America and some Caribbean countries' *Megadrilogica*, vol. 15, pp. 227–247

Richardson, D.M. (2011) 'Forestry and agroforestry', in D. Simberloff and M. Rejmánek (eds.) *Encyclopedia of Biological Invasions*, University of California Press, Berkeley, US, pp. 241–248

Roy, B.A., Alexander, H.M., Davidson, J., Campbell, F.T., Burdon, J. and others (2014) 'Increasing forest loss worldwide from invasive pests requires new trade regulations' *Frontiers in Ecology and Environment*, vol. 12, pp. 457–465

Sanderson, L.A., McLaughlin, J.A. and P.M. Antunes (2012) 'The last great forest: a review of the status of invasive species in the North American boreal forest' *Forestry*, vol. 85, pp. 329–340

Scott, J.M., Mountainspring, S., Ramsey, F.L. and C.B. Kepler (1986) *Forest Bird Communities of the Hawaiian Islands*. Allen Press, Lawrence, Kansas, US

Šefrová, H. and Z. Laštůvka (2005) 'Catalogue of alien animal species in the Czech Republic' *Acta Universitatis Agriculturae et Silviculturae Mendelianae Brunensis*, vol. 53, pp. 151–170

Singh, S.K., Hodda, M., Ash, G.J. and N.C. Banks (2013) 'Plant-parasitic nematodes as invasive species: characteristics, uncertainty and biosecurity implications' *Annals of Applied Biology*, vol. 163, pp. 323–350

Van Driesche, R., Hoddle, M. and T. Center (2008) *Control of Pests and Weeds by Natural Enemies*, Blackwell, Oxford, UK

Woodward, S.L. and J.A. Quinn (2011) *Encyclopedia of Invasive Species, Vols. 1 and 2*, Greenwood, Santa Barbara, California, US

Wylie, F.R. and M.R. Speight (2012) *Insect Pests in Tropical Forestry*, 2nd ed., CABI, Wallingford, UK

Zizka, L.H. and J.S. Dukes (2011) (eds.) *Weed Biology and Climate Change*, Wiley-Blackwell, Ames, Iowa, US

PART VI

Forest and climate change

33
FIRE AND CLIMATE
Using the past to predict the future

Justin Waito, Martin P. Girardin, Jacques C. Tardif, Christelle Hély, Olivier Blarquez and Adam A. Ali

Fire as a major ecosystem process

The unique distribution and stand characteristics of global forests are the result of interactions between vegetation and the environment over millennial time spans. The spatial distribution of forests is governed in part by physical processes, such as incoming solar radiation, growth season length, and moisture and nutrient availability, which impose limitations on the establishment and growth of vegetation (Bonan and Shugart 1989). Additionally, vegetation distribution is influenced by disturbance events such as fire, damage from wind and ice storms, insect outbreaks, and rockfalls and avalanches in mountain regions (Attiwill 1994). These are important for regeneration and maintenance of the heterogeneous distribution of species and stand structure.

Fire has been an important component of ecosystems since the appearance of the first terrestrial plants, with records of fire existing from as far back as 420 million years ago (Bowman *et al.* 2013). It is present in most of Earth's ecosystems except in areas of sparse vegetation and near the poles (Bowman *et al.* 2013; Flannigan *et al.* 2013). It has the potential to affect hundreds of millions of hectares annually, with the majority being located in grasslands and savannas. Woodland fires, however, contribute larger amounts of biomass burned. Therein, fire is capable of temporarily and rapidly reducing previously vegetated areas to mineral soil.

Human manipulation of fire for land-clearing and recreational activities, as well as fire suppression practices, has disrupted the natural pattern of fire activity in many forested regions (Bowman *et al.* 2013). In addition, anticipated human-caused climate change is expected to further alter global fire occurrence, with important consequences for species distribution, ecosystem integrity and function, atmospheric greenhouse gas balance, and human safety (Flannigan *et al.* 2013). To date, characterizing the range of fire regime modifications remains a major challenge owing to the large interannual variability in the area burned (Figure 33.1) that tends to mask long-term and subtle changes.

The importance of fire in Earth's ecosystems necessitates an understanding of historic variability in order to place current observations and future projections in context. In this chapter, we outline the physical processes of fire in relation to the forest environment and climate. We also review the methods used in fire history reconstructions as well as results from

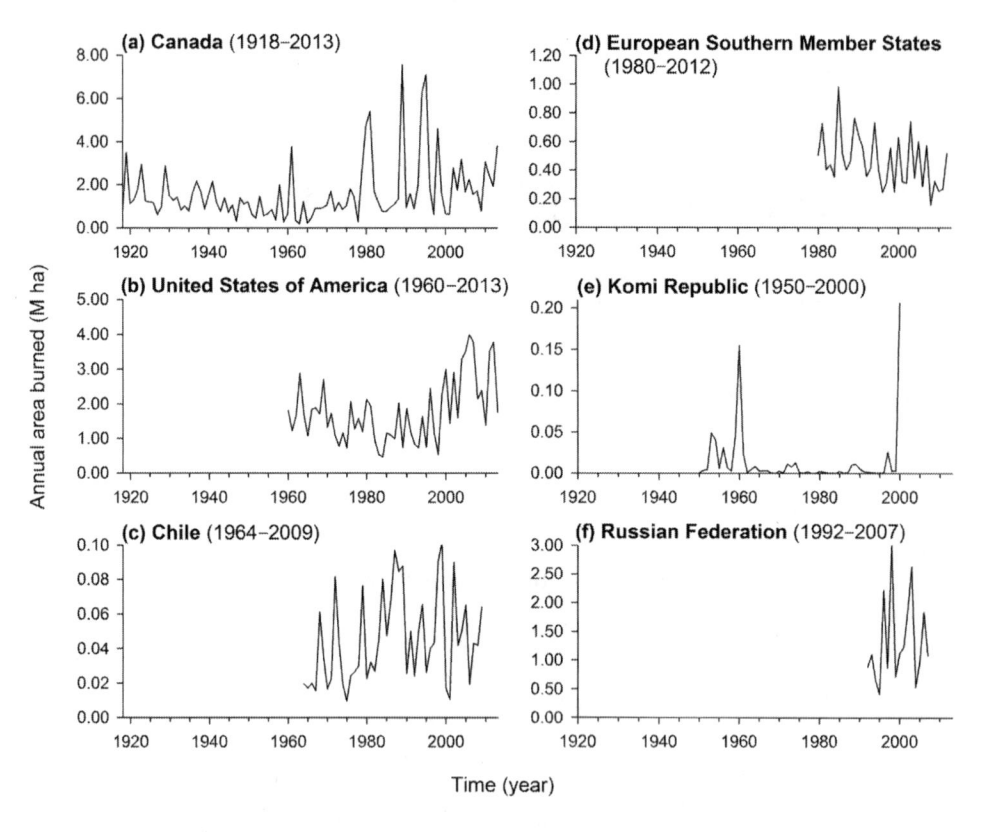

Figure 33.1 Examples of annual area burned time series for various countries: (a) Canada (van Wagner 1988); (b) United States of America (https://www.nifc.gov/fireInfo/fireInfo_stats_totalFires.html; last accessed on August 1, 2014); (c) Chile (http://www.fire.uni-freiburg.de/; last accessed on August 1, 2014); (d) European Southern Member States (Portugal, Spain, France, Italy, and Greece (Schmuck *et al.* 2013); (e) Komi Republic (Drobyshev and Niklasson 2004); and (f) Russian Federation (Goldammer *et al.* 2007). Parentheses indicate the periods covered by each fire statistic

fire history reconstructions and modelling. The best proxy[1] records for longer term fire reconstructions are tree-ring and lake-sediment records. Our discussion will thus focus on fire history originating from these two proxy types within North American forests.

Climate, fuel and forest fires

To begin, it is worth introducing some concepts and processes associated with forest fires as they relate to landscape processes. Fire in a forested landscape is described through the fire regime concept that encompasses fire ignition (human or lightning caused), fire type (surface or crown fire), fire intensity and severity as well as spatial (size) and temporal (frequency, seasonality) characteristics (Keeley 2009). Generally speaking, a small number of fires determine the total area burned each year in temperate and boreal forests. For instance, annual area burned

---

1 Proxies are preserved physical, chemical and biological characteristics of the past that stand in for direct measurements.

in boreal forests of North America and Russia can exceed 7 million and 3 million hectares, respectively. The majority of these are infrequent lightning-caused large fires exceeding 200 hectares in size (Stocks *et al.* 2003; Goldammer *et al.* 2007). Conversely, in tropical regions (e.g. southern Africa), the cumulative annual area burned by numerous but small anthropogenic fires is equivalent to the lightning-caused area burned (Archibald *et al.* 2012).

The occurrence of fire is largely determined by the combination of factors that facilitate ignition and spread. Large fires are predominantly influenced by climate, often associated with prolonged blocking high-pressure systems in the upper atmosphere over or upstream from the affected regions (e.g. Pereira *et al.* 2005; Macias-Fauria and Johnson 2008). The blocking high-pressure systems cause air subsidence in the upper atmosphere, obstructing the normal west-to-east progress of migratory storms, resulting in typically sunny, warm days that create dry fuel conditions extending over several hundred kilometres. Such dry fuel conditions facilitate the rapid spread of fires. Ignition occurs due to lightning strikes associated with the penetration of short-wave atmospheric troughs or a cold front along the west side of the ridges (Macias-Fauria and Johnson 2008). Additionally, fire spread is enhanced by the strong and gusty surface winds that are commonly associated with the breakdown of the blocking ridges.

In addition to climate, vegetation and fuel types also exert important influences on fire ignition and propagation. For example, fire size and intensity are generally higher in forests dominated by coniferous species compared with those dominated by broadleaf species (Terrier *et al.* 2013). This difference is largely due to the quantity and type of organic matter that is available to be burned. The fuel conditions in coniferous forests lead to fire susceptibility because of an abundance of small-diameter, dry organic material (Hély *et al.* 2000). In contrast, broadleaf forests are less likely to burn because of higher leaf moisture loading and decreased flammability (Päätalo 1998; Campbell and Flannigan 2000). In some areas of the world, such as tropical grasslands and savannas, precipitation is an additional prerequisite for fire ignition and spread as it is required to initiate the plant growth (during the wet season) and subsequent fuel build-up (during the dry season) that allows continuous fire spread on the landscape (Flannigan *et al.* 2013). In such situations, the interval between successive fire events is dependent on the rate of fuel build-up and so ignition is a combination of the annual precipitation amount, time-since-last-fire and favourable climate or anthropic activities.

Landscape characteristics are also important drivers of fire through features such as slope and fire barriers (Turner and Romme 1994). The influence of slope on fire is through preferential fire spread in an upslope direction. In addition, slope aspect influences fire spread as south-facing slopes are more prone to moisture deficits. However, this influence is limited as severe fires can be relatively impervious to landscape features. Fire occurrence and extent is also affected by natural fire breaks, such as water bodies, that can alter the direction of fire by restricting fire spread in certain directions.

The behaviour of individual fires is highly variable and is determined by the interaction between climate, fuels and landscape characteristics. Fires are typically ground fires, surface fires or crown fires, each situated along an increasing burn intensity gradient (with intensity being the energy released during a fire event) (Bowman *et al.* 2013). Ground fires are slow-moving low-intensity fires that burn through soils with deep organic matter. Despite the low intensity and slow rate of spread, ground fires can burn during several weeks or months and therefore can consume more biomass per hectare than the other fire types. Surface fires represent low-intensity fast-moving fires where thin soil organic matter below the litter layer and vegetation close to the surface can burn over large areas. Crown fires are the most intense and severe of the three and occur when a fire spreads through the canopy layer of the forest (van Wagner 1977; Turner and Romme 1994). The rate of spread through the canopy can be described as being passive, active or independent (van Wagner 1977). A passive crown fire generally occurs during low wind conditions, which limits the spread of fire through the forest canopy and requires the release of surface fire energy to sustain the crown fire spread (van Wagner 1977). Active crown fires represent the intermediate class where wind speed is more conducive to spread of fire through the forest canopy. Independent crown fires are the most severe type and are characterized by separation of the surface and crown portions of the fire. The type of fire is an important component contributing to the spatial extent of individual fires as well as to the total forested area burned each year.

Historical fire records during the instrumental period

Records used in fire history research include written accounts, aerial photographs, and remote sensing (van Wagner 1988; Stocks *et al.* 2003; Macias-Fauria and Johnson 2008; Bowman *et al.* 2013). Intuitively, written records provide the longest account of fire data but the resolution is dependent on whether an area was settled, events were recorded and that the records survived to the present day (Figure 33.1). Generally speaking, the earliest written records of fire events are sporadic and infrequent. However, records became more spatially explicit and continuous with the advent of systematic record keeping by national institutions. Record keeping has further improved with the advent of active fire suppression that allowed collection of first-hand written accounts and boundary maps of fires (Stocks *et al.* 2003). Since their inception, fire records have been steadily improving as new technologies and monitoring networks have been developed. In many regions of the world, very accurate and reliable records have been available since the 1970s (Figure 33.1).

Following written records, the longest available fire-related records can be found in aerial photography coverage of forested landscapes associated with the development of forest inventories. In the southern boreal forest of North America, records obtained from aerial surveying generally extend back to the early twentieth century (van Wagner 1988). These records can be used in fire history reconstructions to locate boundaries between forest types and age classes that are indicative of past disturbance events. A more recent development in fire history mapping is the application of remote sensing techniques involving satellite imagery coupled with GIS analyses (e.g. Lehsten *et al.* 2009). This method is still being developed through ongoing classification and validation of interpretations from field studies.

Fire history reconstruction

Research into past fire activity relies on several indicators, one of the main being the direct analyses of damage caused to trees by fire. Fire as a physical process is the result of a chemical reaction leading to combustion of organic material (Michaletz and Johnson 2007). Combustion is a four-fold process whereby an endothermic reaction first completely evaporates the fuel water content followed by an exothermic oxidation reaction that breaks down complex molecules (pyrolysis). This process creates volatile organic compounds (Simpson *et al.* 2011) whose increasing concentration in the surrounding air, combined with increasing temperature, leads to an explosive reaction in the presence of oxygen, which results in flame occurring. This third, flaming phase is the most efficient in terms of energy release and proceeds as long as the breakdown of the particle molecules is exothermic enough to produce volatile gas to feed the flame. When the emission of volatile compounds stops, the flame disappears and the exothermic oxidation process, now less efficient, only occurs at the surface of the particle, thereby moving the combustion into the fourth phase, known as the glowing phase, which will ultimately cease once the decreasing energy release cannot sustain the high temperatures needed for rapid oxidation. In trees, the exothermic reaction can result in complete combustion or, more often, transitions to a lower energy reaction that leaves some material unconsumed. Burned material left behind after fires (Figure 33.2a) can be used to determine the ages of previous forest cohorts (Heinselman 1973). In a crown fire ecosystem such as the boreal forest, all standing trees are killed and pioneer species regenerate following fire events. Therein, comparison of stand ages across the landscape reveals the succession of fire events that have occurred on a landscape over time (Figure 33.2). In addition, charred material produced during the flaming and smouldering phases of a fire can become incorporated into soil, lake sediments and peat bogs, thereby preserving a long-term record of fire events.

In a non-lethal surface fire, cambial injury to the stems of living trees can lead to the formation of fire scars on the leeward side (Figure 33.2) (Gutsell and Johnson 1996). Fire scar formation is explained through fluid dynamic processes where the physical behaviour of fire in relation to a tree is governed by differential airflow (Gutsell and Johnson 1996). Airflow patterns around the tree result in the formation of vortices on the leeward side that contribute to fire scar formation by drawing the flame up the stem of the tree, which increases heat exposure time. The size and shape of the scar is dependent on the position of the vortices. Accumulation of organic matter around the stem is also a prerequisite for fire scar formation as it will act as the available fuel build-up and fuel bed. The analysis of fire scars is one of the most common methods of inferring past fire activity (e.g. Swetnam 1993). Properly dated through their position along the tree-ring sequence, fire scars can provide information on the year of fire scar formation and seasonality of burn. Comparison of samples with fire scars of the same year provides the spatial extent of individual fires whereas comparison of multiple fire scar years provides the frequency of fires in a given area (Swetnam 1993). However, analysis of fire scars is not without limitations, such as limited temporal depth, formation restricted to surface or mixed fire regimes and the potential for confusion between fire scars and scars produced by other means.

The theoretical basis for reconstructing past fires from sedimentation in lakes, wetlands and peatlands lies in the fact that, following a fire, the depositional processes of material (i.e. charcoal particles) into the sediment are altered, thus producing a record of the events (Patterson et al. 1987; Higuera et al. 2007). Fire events are identified when the charcoal peak component exceeds the background component of the record (Figure 33.3). For example, analysis of macro-charcoal fragments (particles 0.3–0.5 mg in weight and with a diameter >0.2 mm) extracted from the soil profile can yield direct information of past fires in forest stands that underwent incomplete wood combustion (Payette et al. 2012). That said, there is a high probability that successive fires burn the charcoal deposited in the soil by a previous fire; thus, numerous events may not be recorded using this method (Ohlson 2012). On the other hand, the finest particles emitted from fires and that are air- and runoff-transported over the landscape can be extracted from lake sediments and used to reconstruct long-term fire history at the landscape and regional scales. Note that it is common practice in such fire history studies to also examine plant macro-remains and pollen assemblages as fire events alter the influx of pollen and macro-remains into sediments by burning the vegetation that produces it (Whitlock and Bartlein 2003; Girardin et al. 2013a). The careful examination of pollen and macro-remains allows us to determine vegetation assemblages before and after fire events.

The use of particles from soil profiles and lake, peatland and wetland sediments for fire history reconstruction requires accurate determination of the age of the fire. In fire reconstructions based on charred particles extracted from soils, age determination is undertaken by the dating of the wood that produced the charcoal (Payette et al. 2012). Of note with this

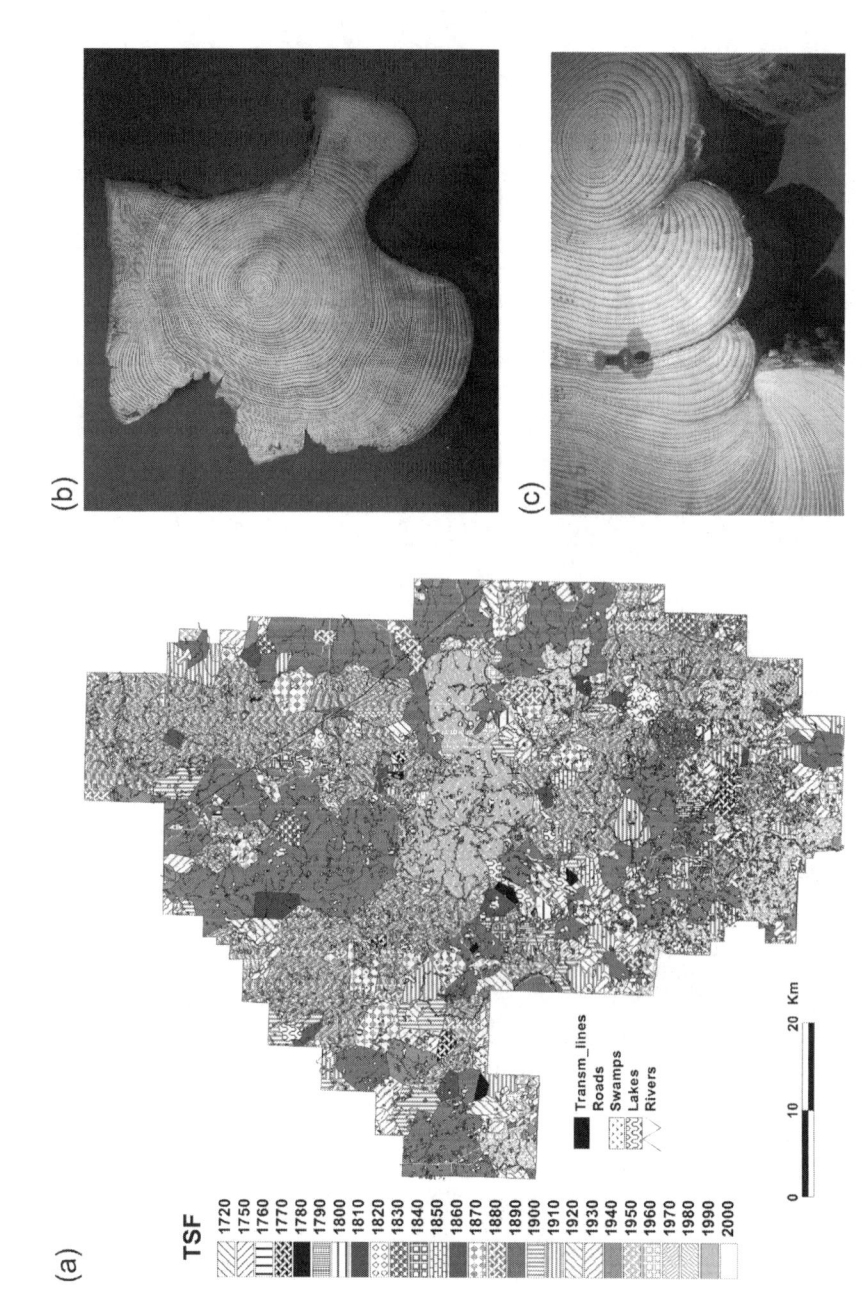

Figure 33.2 (a) Time-since-fire (TSF) map for the Duck Mountain Provincial Forest, Manitoba, Canada, as of 2002. The TSF date (i.e. approximate year of burn) for each of the sites was either provided by the oldest trees in the overstory cohort, by sampling of burned material left on the ground (b), and/or by fire scar date (c). The TSF map indicated that a large portion of the landscape originated from fires between 1880 and 1899. The TSF map also showed that during the early twentieth century, numerous small fires occurred at the southwest periphery. The largest fire of the twentieth century was that of 1961 with about 6 per cent of the study area burned. For more details, see Tardif (2004)

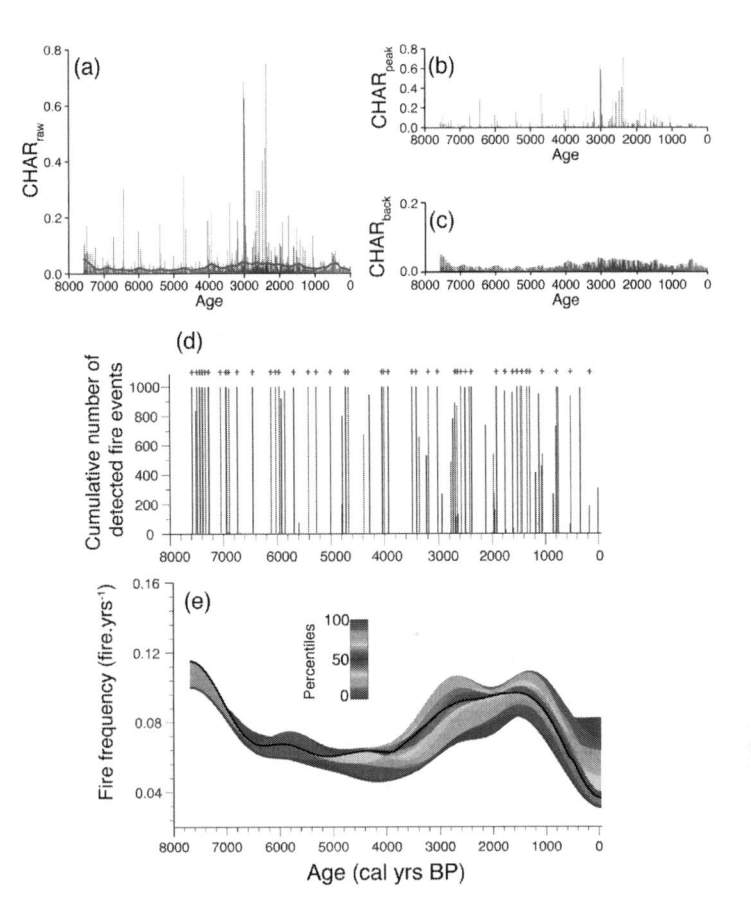

Figure 33.3 Summary of the main analytical steps required for reconstructing past fire frequency from sedimentary charcoal records. (a) Raw charcoal values expressed in Charcoal Accumulation Rate (CHAR mm2.cm-2.yr-1) were smoothed using a locally weighted regression (LOWESS) with a 500-year window width, which enabled us to discriminate the CHARpeak (b) and CHARback (c) components. CHARback (c) represented long-distance fires, charcoal redeposition and noise while the CHARpeak (b) represented local fires. Noise is also present in the remaining CHARpeak component and a Gaussian mixture model is generally used to discriminate the two CHARpeak subpopulations: CHARnoise and CHARfire. The CHARfire component represented the occurrence of one or more fires locally around lakes (<10 km). The Gaussian mixture could be applied globally to the entire records (such as in this example) or locally in order to better fit the centennial to millennial variations in CHAR. Analytical steps (CHARraw filtering technique and window width) that are able to provide well-discriminated fire events are not unique (e.g. Higuera *et al.* 2007) and techniques that aimed at replicating the analysis of a record using ensembles have been proposed (Blarquez *et al.* 2013). Therein, 1,000 replicates of the steps (a–c) are made using multiple smoothing techniques and window width, and the cumulative number of fires by the ensemble is calculated for each fire event date (d). From this ensemble of fire event dates, it is possible to assess fire frequency (or fire return intervals). In the example illustrated in (e), the tinted bands represent the distribution of the fire frequencies obtained from the 1,000 iterations of the model parameters. From these, one can be confident that the changes from low to high fire frequency ca. 3,000 cal BP are not a consequence of the selection of model parameters but, rather, reflect important changes that have occurred in burning activity (Blarquez *et al.* 2013)

method is that a lag, which may range from decades to centuries, separates the date of the wood and the date of the actual fire. In contrast to soils, dating of fire events in lake sediment is undertaken through age determination of plant macro-remains or bulk sediments. This procedure ensures continuity of sediments and minimal lag between the effective incorporation of radioactive elements in the organic material and its sedimentation. This methodology also has limitations that mostly involve data analysis methods that require a thorough statistical framework and the source origin of the deposited charred particles.

The process of dating sediments is through analysis of isotopes including lead (^{210}Pb) and carbon (^{14}C). The theoretical underpinning to ^{210}Pb dating is that lead is a product of the uranium decay series, the proportion of which can be used to accurately date the uppermost 150–200 years of sediment (Appleby 2001). Lead^{210} analysis is made possible because it is continuously added to water bodies from the atmosphere where it becomes incorporated into the sediment. Analysis of the unsupported (atmospheric) ^{210}Pb fraction against the supported (in situ) fraction in the sediment is based on its half-life (22.3 years) and provides an approximation of when the isotope was incorporated into the sediment. In order to obtain dates from older sediments, the ^{14}C isotope is often used as it has a half-life of 5,568 years (Björk and Wohlfarth 2001). Carbon^{14} is typically used to date material as it is a heavy isotope that is less likely to be preferentially selected by plants (fractionation) and so provides the most accurate date. Accurate dating of material is made possible because known ratios of ^{13}C and ^{12}C have been established, and calibration curves developed, from proxy records such as wood and sedimentary rocks. In addition, the ratio of atmospheric ^{14}C to ^{13}C has been established from wood and so an age determination of material can be made through comparison of the ratio of ^{14}C to the calibration curve developed for other carbon isotopes (Reimer et al. 2013).

Modelling of fire behaviour

The impact of ongoing climatic changes on forests has become a growing concern, with many studies projecting increasing fire activity worldwide as a result of global warming. Particularly vulnerable to climatic warming are the length of the fire season and the seasonal distribution of fire occurrences (Flannigan et al. 2013). In addition, the intensity of drought is expected to increase in many areas of the boreal forests, which could cause larger, more intense, severe forest fires (Turetsky et al. 2011; Flannigan et al. 2013). Even so, other ecosystem processes may play a role in determining future fire trajectories. Notably, vegetation dynamics and changes in fuel conditions could modify fire behaviour. For example, favouring broadleaf species in management planning could provide negative feedback on fire (Terrier et al. 2013). A high frequency of successive forest fires could also cause regeneration failures in forest stands, thereby reducing fuel loads and again inducing negative feedback on fire. An important challenge in fire science is the quantification of such effects over the short period covered by fire statistics.

Fire history research may allow for the documentation and identification of the long-term fire trends that, ultimately, can serve to improve the ability to project future fire impacts. Modifications of fire activity that serve as a marker between past periods provide a valuable setting for analysis of the sensitivity of fire to various ecosystem and climatic processes. Taking advantage of this opportunity, recent advances include the use of global climate models (GCMs; also referred to as general circulation models) to develop a mechanistic understanding of the evolution of fire in ecosystems at different temporal and spatial scales. A GCM is a three-dimensional time-dependent numerical representation of the atmosphere, oceans and sea ice Earth compartments, using the equations of motion and including radiation, photochemistry, and the transfer of heat and water vapour. Such climatic simulations are available for past millennia and future centuries (e.g. Singarayer and Valdes 2010). The millennial reconstructions of fire events extracted from lake sediments can be compared with past climatic conditions as simulated using GCMs. One can, for instance, look at whether or not there could be a correlation between variability in fire frequency inferred from lake-sediment records during past millennia and drought as simulated by a GCM over the same horizon (Ali *et al.* 2012). In the case of no correlation, one can test alternative hypotheses such as the influence of human activities or changes in fuel conditions in relation to changes in fire frequency.

More advanced modelling experiments include the coupling of a GCM to an ecosystem simulator to assess the effect of climate change and changing vegetation structure and composition on fire (e.g. Smith *et al.* 2001; Lehsten *et al.* 2009). Generally, the processes included in the ecosystem simulator cover photosynthesis, respiration, evapotranspiration, organic matter decomposition in soil, post-fire tree mortality, succession, and biomass burned (e.g. Pfeiffer *et al.* 2013). In these experiments, vegetation composition is often summarized with the presence and abundance of Plant Functional Types (PFTs) that grow and develop within a given geographical location characterized by its soil type and a given climate. PFTs are assumed to represent all main species within a plant form (tree or grass) and a given bioclimatic limit (boreal, temperate or tropical) that have the same photosynthetic pathway, as well as ecological affinities and tolerance for light and water. Alternatively, the processes of forest growth may also be simpler and be dictated by yield curves determined from the analyses of forest inventories (de Groot *et al.* 2013). Through cross-comparison with the proxy fire and vegetation records, these modelling experiments allow testing of different temporal trajectories of fire activity in association with changes in forest types and climatic conditions.

Post-glacial fire history of North America

Fire history in North America has been documented using a large array of methods and is used here as an example. The syntheses, presented in Figure 33.4, made use of ratios (δ_g) of current to past burn rates or fire frequencies (depending on the type of historical fire records used) calculated as follows:

$$\delta_g = \frac{b_{CURRENTg}}{b_{PASTg}}$$

where $b_{CURRENTg}$ and b_{PASTg} are current and historical fire activities, respectively, at a given location g. Values of δ_g above 1.0 are indicative of a higher incidence of fire activity in recent decades as opposed to the historical period, and vice versa. The metric δ_g is essentially qualitative. Detailed information on the data used here may be obtained by consulting the work by Bergeron *et al.* (2004) in relation to stand-replacing fire history studies, and Daniau *et al.* (2012) for lake-sedimentary charcoal syntheses. Additionally, we present an example in Figure 33.5 of a modelling experiment in which GCM simulations of the past 11,000 years were coupled with vegetation information and a fire history reconstruction to highlight the importance of vegetation and climate interactions in driving long-term fire trajectories in eastern North America.

Contextually, the boreal forest of North America evolved during a geological period known as the Holocene that began with the end of the most recent glaciation approximately 11,700 years BP (Björk 1985; Viau *et al.* 2006). During that period, the margins of the glaciers retreated northward followed closely by the formation of proglacial lakes (Björk 1985). Fire history records indicate that arrival of fire in the area currently occupied by the boreal forest

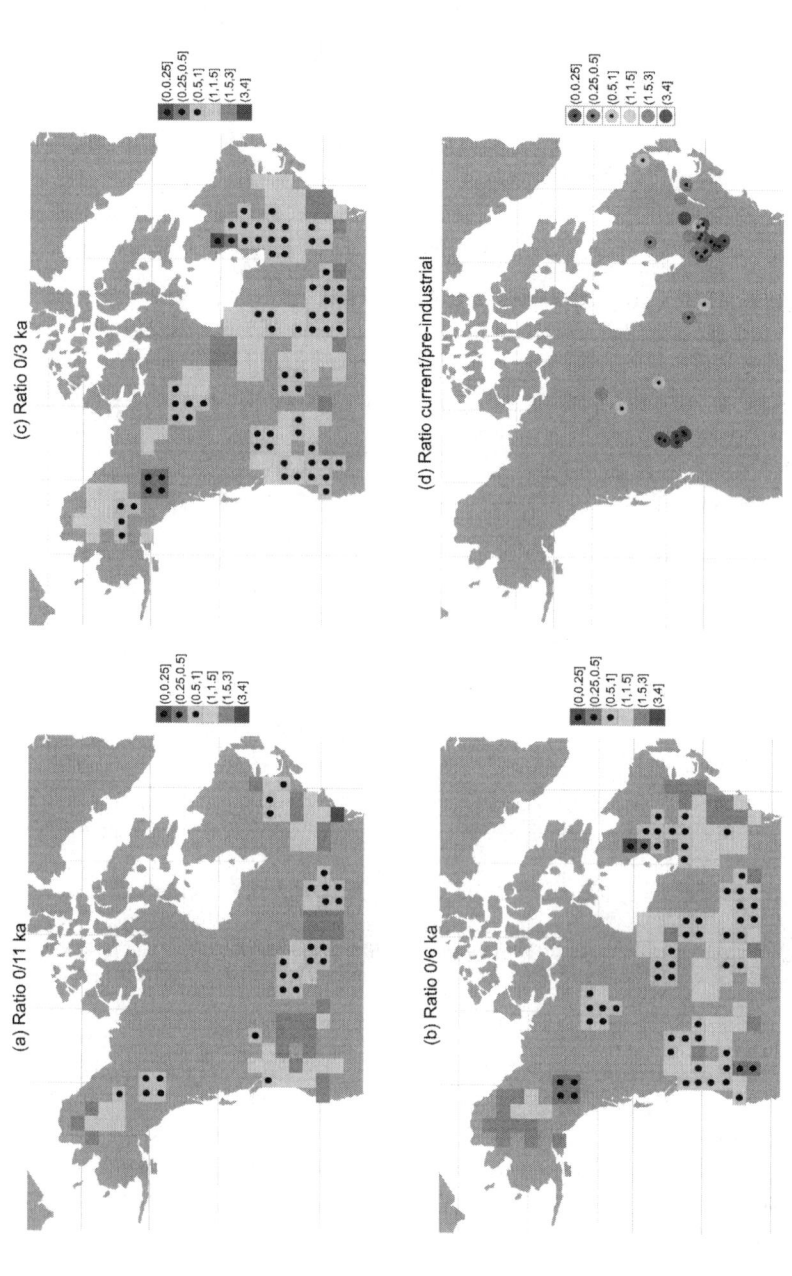

Figure 33.4 (a–c) Ratios δ_g of current (0 calibrated years before present, BP) to past (11,000, 6,000 and 3,000 BP) burning activity for North America above 40°N inferred from sedimentary charcoal records. Sedimentary charcoal records from the Global Charcoal database were used (Daniau et al. 2012). A 200 × 200 km regular spatial grid was used to interpolate transformed charcoal values for each key period (11,000, 6,000, 3,000 and 0 cal BP). Treatments of charcoal records were done using the R paleofire package (http://cran.r-project.org/web/packages/paleofire/). (d) Ratios δ_g of current (1959–1999) to past (pre-industrial era) proportions of annual burned areas for boreal Canada inferred from stand-replacing fire history studies (data from Girardin et al. 2013b). The colour corresponds to the magnitude of the ratio δ_g at each location; values of δ_g above 1.0 are indicative of a higher incidence of fire activity in the present as opposed to the past, and vice versa

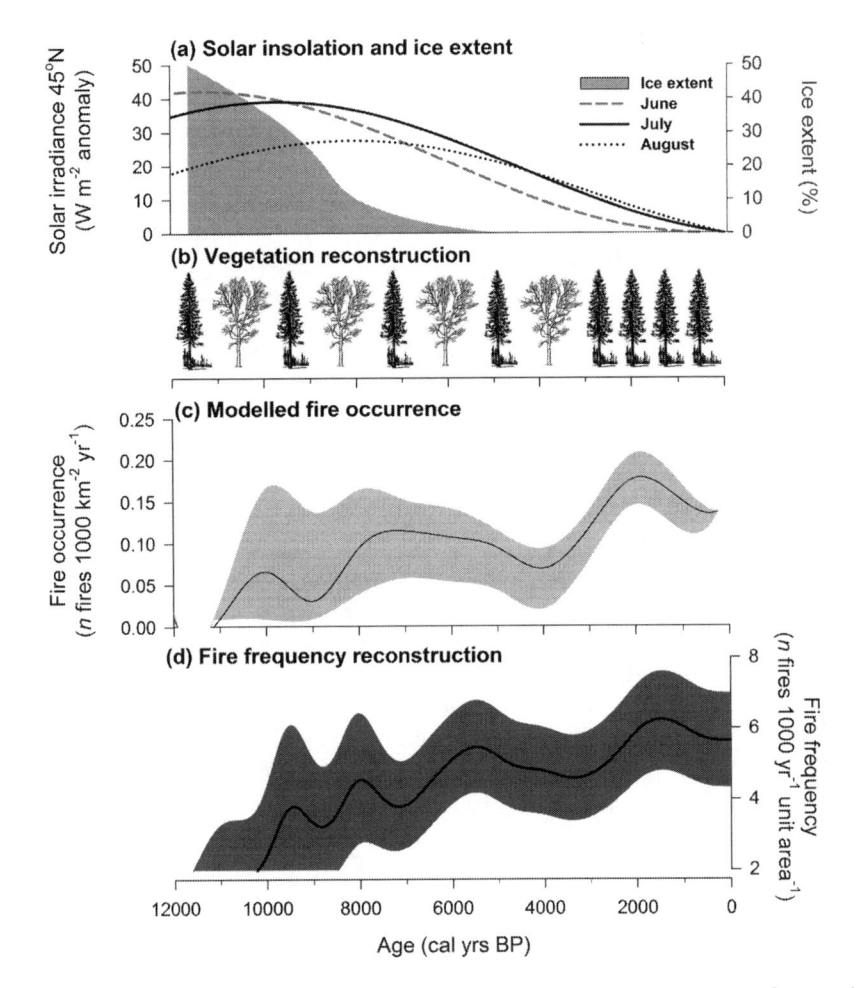

Figure 33.5 Comparison between projected versus reconstructed fire-history trajectories for mixedwood forests in eastern Canada. Multi-millennial forcings such as those arising from (a) solar insolation (solid lines, computed for 45°N) and Laurentide Ice Sheet extent (shaded area) were assimilated by a global climate model (GCM) to produce climatic outputs. These climatic outputs were combined with information extracted from the pollen-based vegetation reconstructions shown in (b) to project past fire occurrence in mixedwood forests (shown in [c]). The projected fire activity was then verified against the reconstructed fire frequency obtained by the analysis of charcoal records. The shaded bands in (c) and (d) denote 90 per cent confidence intervals for uncertainty in mean regional histories. From the comparison of (c) and (d), Girardin *et al.* (2013a) concluded that in spite of a warmer summer climate 6,000 to 3,000 calibrated years before present (cal yrs BP), the fire activity (c) and (d) was not significantly higher than today's level (0 cal yrs BP) owing to the lower proportion of flammable conifers in landscapes. Adapted from Girardin *et al.* (2013a).

closely follows the northward retreat of the glaciers *ca.* 13,000–8,000 BP and associated migration of vegetation. Fire in northern boreal regions was, nonetheless, restricted by a lack of sedimentation during this transitory period. In the eastern and western United States regions, fire at 11,000 BP was lower than current levels (Figure 33.4a). The reverse holds true in the United States central regions, albeit there is spatial heterogeneity. At 6,000 BP, fire was relatively low compared with today in the southeastern United States regions (Figure 33.4b). It was higher than today in northern boreal regions, with the exception of the southwestern Hudson Bay area and Alaska. The dipole between northeastern and southeastern North America is particularly noticeable. These patterns remained relatively stable up to 3,000 BP (Figure 33.4c). The decline in fire activity seen in many areas of the boreal forest persisted until very recently according to stand-replacing fire history reconstructions (Figure 33.4d). Indeed, most of the sampled regions have a current proportion of burned areas below that of past levels. The synchronicity of the decline in sedimentary charcoal and stand-replacing fire history reconstructions is striking and highlights the importance of climate as a dominant driver of fire (Bergeron et al. 2004; Daniau et al. 2012). While heterogeneity in climatic drivers can explain some of the spatial differences seen in these records, feedback effects arising from local vegetation composition changes, or human impacts on fire, likely also played a role. Below we present an example in which these effects are clarified.

Modelling is a useful tool for testing hypotheses about the sensitivity of ecosystem processes to different types of forcing and interactions among them. The material presented in Figure 33.5 focusses on an important region of North America in which a southerly displacement of the transition zone of the mixedwood and needleleaf forests, in association with cooler climatic conditions, marked the transition from the mid- to the late-Holocene about 3,000 years ago. Having this information in hand, Girardin et al. (2013a) postulated that declining risks brought about by climate cooling may well have been offset by a more fire-prone landscape. They combined data from analyses of charcoal particles and pollen from sediment in boreal forest lakes in eastern Canada with climate models and the modelling of fire frequency. Their fire simulations included forcing from the prolonged presence of the residual Laurentide Ice Sheet in eastern North America and changing solar radiation (Figure 33.5a) and vegetation (Figure 33.5b). The simulations projected higher levels of fire frequency during the last 2,000 years relative to the mid-Holocene period (Figure 33.5c), which is consistent with the fire trajectory deduced from the fire history reconstruction (Figure 33.5d). The changes in fire frequency were due to the fact that the elevated risk of fire in the south 6,000 to 3,000 years ago was counteracted by a larger number of broadleaf trees in landscapes. Such a historical perspective of fire, vegetation and climate interactions provides significant insight into how fire activity may evolve in the future with climatic changes and modification of vegetation composition and structure across landscapes (Terrier et al. 2013).

Conclusions

The latitudinal distribution of vegetation on Earth is largely imposed by climatic limitations, but the spatio-temporal distribution of individual species is often the result of physical limitations and of repeated disturbances. In woodlands for instance, fire leads to species regeneration and therefore contributes to maintaining the health and diversity of the forests. Accordingly, management strategies are now increasingly being focussed on replicating fire as a disturbance in order to maintain ecosystem integrity. Additionally, many studies predict that future fire activities will be higher than present ones owing to increasing global temperatures. Yet, fire activity is the result of complex feedback processes between temperature, precipitation, wind, humidity, and vegetation composition and structure; the influence of each of these factors in a given landscape setting is often difficult to tackle when studying short time series such as those provided by fire agencies. These two examples illustrate the need for more work to determine the natural patterns of fire activity and factors contributing to its variability. As outlined here, archival records and fire proxies are suitable for determining past fire, notwithstanding the fact that each type of proxy has its own limitations. Proxy records stored in lake sediments are particularly suited to regional fire history reconstructions, while information at the stand level can be better obtained from tree-fire scars and charcoal particles preserved in soils.

References

Ali, A. A., Blarquez, O., Girardin, M. P., Hély, C., Tinquaut, F., and others (2012) 'Control of the multimillennial wildfire size in boreal North America by spring climatic conditions', *Proceedings of the National Academy of Sciences of the United States of America*, vol 109, no 51, pp. 20966–20970

Appleby, P. G. (2001) 'Chronostratigraphic techniques in recent sediments', in W. M. Last and J. P. Smol (eds.) *Tracking Environmental Change Using Lake Sediments Volume 1: Basin Analysis, Coring, and Chronological Techniques*, Kluwer Academic, New York

Archibald, S., Staver, C., and Levin, S. A. (2012) 'Evolution of human-driven fire regimes in Africa', *Proceedings of the National Academy of Sciences of the United States of America*, vol 109, no 3, pp. 847–852

Attiwill, P. M. (1994) 'The disturbance of forest ecosystems – the ecological basis for conservative management', *Forest Ecology and Management*, vol 63, nos 2–3, pp. 247–300

Bergeron, Y., Flannigan, M., Gauthier, S., Leduc, A., and Lefort, P. (2004) 'Past, current and future fire frequency in the Canadian boreal forest: Implications for sustainable forest management', *Ambio*, vol 33, no 6, pp. 356–360

Björk, S. (1985) 'Deglacial chronology and revegetation in Northwestern Ontario', *Canadian Journal of Earth Science*, vol 22, no 6, pp. 850–871

Björk, S., and Wohlfarth, B. (2001) '^{14}C chronostratigraphic techniques in paleolimnology', in W. M. Last and J. P. Smol (eds.) *Tracking Environmental Change Using Lake Sediments Volume 1: Basin Analysis, Coring, and Chronological Techniques*, Kluwer Academic, New York

Blarquez, O., Girardin, M. P., Leys, B., Ali, A. A., Aleman, A. C., and others (2013) 'Paleofire reconstruction based on an ensemble-member strategy applied to sedimentary charcoal', *Geophysical Research Letters*, vol 40, pp. 2667–2672

Bonan, G. B., and Shugart, H. H. (1989) 'Environmental factors and ecological processes in boreal forests', *Annual Review of Ecology and Systematics*, vol 20, pp. 1–28

Bowman, D. M. J. S., O'Brien, J. A., and Goldammer, J. G. (2013) 'Pyrogeography and the global quest for sustainable fire management', *Annual Review of Environment and Resources*, vol 38, pp. 57–80

Campbell, I. D. and Flannigan, M. D. (2000) 'Long-term perspectives on fire-climate-vegetation relationships in the North American boreal forest'. In E. Kasischke and B. J. Stocks (eds.), *Fire, Climate Change and Carbon Cycling in North American Boreal Forests*, Springer-Verlag, Berlin

Daniau, A. L., Bartlein, P. J., Harrison, S. P., Prentice, I. C., Brewer, S., and others (2012) 'Predictability of biomass burning in response to climate changes', *Global Biogeochemical Cycles*, vol 26, no 4, pp. 1–12

de Groot, W. J., Flannigan, M. D., and Cantin A. S. (2013) 'Climate change impacts on future boreal fire regimes', *Forest Ecology and Management*, vol 294, pp. 35–44

Drobyshev, I., and Niklasson, M. (2004) 'Linking tree rings, summer aridity, and regional fire data: An example from the boreal forests of the Komi Republic, East European Russia', *Canadian Journal of Forest Research*, vol 34, pp. 2327–2339

Flannigan, M., Cantin, A. S., de Groot, W. J., Wotton, M., Newberry, A., and Gowman, L. M. (2013) 'Global wildland fire season severity in the 21st century', *Forest Ecology and Management*, vol 294, pp. 54–61

Girardin, M. P., Ali, A. A., Carcaillet, C., Blarquez, O., Hély, C., and others (2013a) 'Vegetation limits the impact of a warm climate on boreal wildfires', *New Phytologist*, vol 199, no 4, pp. 1001–1011

Girardin, M. P., Ali, A. A., Carcaillet, C., Gauthier, S., Hély, C., and others (2013b) 'Fire in managed forests of eastern Canada: Risks and options', *Forest Ecology and Management*, vol 294, pp. 238–249

Goldammer, J. G., Sukhinin, A., and Davidenko, E. P. (2007) 'Advance publication of wildland fire statistics for Russia 1992–2007', *International Forest Fire News*, vol 37, http://www.fire.uni-freiburg.de/inventory/Russia-1996-2007.pdf (last accessed May 2015)

Gutsell, S. L., and Johnson, E. A. (1996) 'How fire scars are formed: Coupling a disturbance process to its ecological effect', *Canadian Journal of Forest Research*, vol 26, no 2, pp. 166–174

Heinselman, M. L. (1973) 'Fire in the virgin forests of the Boundary Waters Canoe Area, Minnesota', *Quaternary Research*, vol 3, no 3, pp. 329–382

Hély, C., Bergeron, Y., and Flannigan, M. D. (2000) 'Effects of stand composition on fire hazard in mixed-wood Canadian boreal forest', *Journal of Vegetation Science*, vol 11, no 6, pp. 813–824

Higuera, P. E., Peters, M. E., Brubaker, L. B., and Gavin, D. G. (2007) 'Understanding the origin and analysis of sediment-charcoal records with a simulation model', *Quaternary Science Review*, vol 26, nos 13–14, pp. 1790–1809

Keeley. J. E. (2009) 'Fire intensity, fire severity and burn severity: A brief review and suggested usage', *International Journal of Wildland Fire*, vol 18, no 1, pp. 116–126

Lehsten, V., Tansey, K., Balzter, H., Thonicke, K., Spessa, A., and others (2009) 'Estimating carbon emissions from African wildfires', *Biogeosciences*, vol 6, no 3, pp. 249–360

Macias-Fauria, M. and Johnson, E. A. (2008) 'Climate and wildfires in the North American boreal forest', *Philosophical Transactions of the Royal Society B.*, vol 363, no 1501, pp. 2317–2329

Michaletz, S. T., and Johnson, E. A. (2007) 'How forest fires kill trees: A review of the fundamental biophysical processes', *Scandinavian Journal of Forest Research*, vol 22, no 6, pp. 500–515

Ohlson, M. (2012) 'Soil charcoal stability over the Holocene – Comment to the paper published by de Lafontaine and Asselin, *Quaternary Research*, v.76, 2011, pp. 196–200'. *Quaternary Research* 2012, vol 78, no 1, p. 154

Päätalo, M.-L. (1998) 'Factors influencing occurrence and impacts of fires in northern European forests', *Silva Fennica*, vol 32, no 2, pp. 185–202

Patterson, W. A. III., Edwards, K. J., and Maguire, D. J. (1987) 'Microscopic charcoal as a fossil indicator of fire', *Quaternary Science Review*, vol 6, no 1, pp. 3–23

Payette, S., Delwaide, A., Schaffhauser, A., and Magnan, G. (2012) 'Calculating long-term fire frequency at the stand scale from charcoal data', *Ecosphere*, vol 3, no 7, article 59, pp. 1–16

Pereira, M. G., Trigo, R. M., da Camara, C. C., Pereira, J. M. C., and Leite, S. M. (2005) 'Synoptic patterns associated with large summer forest fires in Portugal', *Agricultural and Forest Meteorology*, vol 129, nos 1–2, pp. 11–25

Pfeiffer, M., Spessa, A., and Kaplan, J. O. (2013) 'A model for global biomass burning in preindustrial time: LPJ-LMfire (v1.0)', *Geoscientific Model Development*, vol 6, no 3, pp. 643–685

Reimer, P. J., Bard, E., Bayliss, A., Beck, J. W., Blackwell, P. G., and others (2013) 'IntCal[13] and Marine[13] radiocarbon age calibration curves 0–50,000 years cal BP, *Radiocarbon*, vol 55, no 4, pp. 1869–1887

Schmuck, G., San-Miguel-Ayanz, J., Camia, A., Durrant, T., Boca, R., and Libertá, G. (2013) *Forest Fires in Europe, Middle East and North Africa 2012.* Joint Research Centre, European Commission, Institute for Environment and Sustainability, Land Management and Natural Hazards Unit, Luxembourg

Simpson, I. J., Akagi, S. K., Barletta, B., Blake, N. J., Choi, Y., and others (2011) 'Boreal forest fire emissions in fresh Canadian smoke plumes: C1-C10 volatile organic compounds (VOCs), CO_2, CO, NO_2, NO, HCN and CH_3CN', *Atmospheric Chemistry and Physics*, vol 11, no 13, pp. 6445–6463

Singarayer, J. S., and Valdes, P. J. (2010) 'High-latitude climate sensitivity to ice-sheet forcing over the last 120 kyr', *Quaternary Science Reviews*, vol 29, pp. 43–55

Smith, J. B., Schellnhuber, H. J., Mirza, M. Q., Fankhauser, S., Leemans, R., and others (2001) 'Vulnerability to climate change and reasons for concern: A synthesis', in McCarthy, J., Canziana, O., Leary, N., Dokken, D., White, K. (eds.), *Climate Change 2001: Impacts, Adaptation, and Vulnerability*, Cambridge University Press, New York

Stocks, B. J., Mason, J. A., Todd, J. B., Bosch, E. M., Wotton, B. M., and others (2003) 'Large forest fires in Canada, 1959–1997', *Journal of Geophysical Research*, vol 107, no D1, pp. 1–12

Swetnam, T. W. (1993) 'Fire history and climate change in giant sequoia groves', *Science*, vol 262, no 5135, pp. 885–889

Tardif, J. (2004) *Fire History in the Duck Mountain Provincial Forest, Western Manitoba.* Sustainable Forest Management Network Projects report 2003/2004. Edmonton, Alberta, Canada

Terrier, A., Girardin, M. P., Périé, C., Legendre, P., and Bergeron, Y. (2013) 'Potential changes in forest composition could reduce impacts of climate change on boreal wildfires', *Ecological Applications*, vol 23, no 1, pp. 21–35

Turetsky, M. R., Kane, E. S., Harden, J. W., Ottmar, R. D., Manies, K. L., and others (2011) 'Recent acceleration of biomass burning and carbon losses in Alaskan forests and peatlands', *Nature Geoscience*, vol 4, pp. 27–31

Turner, M. G., and Romme, W. H. (1994) 'Landscape dynamics in crown fire ecosystems', *Landscape Ecology*, vol 9, no 1, pp. 59–77

van Wagner, C. E. (1977) 'Conditions for the start and spread of crown fire', *Canadian Journal of Forest Research*, vol 7, no 1, pp. 23–34

van Wagner, C. E. (1988) 'The historical pattern of annual burned area in Canada', *Forestry Chronicle*, vol 64, no 3, pp. 182–185

Viau, A. E., Gajewski, K., Sawada, M. C., and Fines, P. (2006) 'Millennial-scale temperature variations in North America during the Holocene', *Journal of Geophysical Research*, vol 111, no D09102, pp. 1–12

Whitlock, C., and Bartlein, P. J. (2003) 'Holocene fire activity as a record of past environmental change', in A. Gillespie, S. C. Porter, and B. Atwater (eds.) *Developments in Quaternary Science*, vol 1, Elsevier, Amsterdam

34
ECOLOGICAL CONSEQUENCES OF DROUGHTS IN BOREAL FORESTS

Changhui Peng

Introduction

Climatic warming during the past century has led to a variety of responses by terrestrial ecosystems, including changes in net primary productivity (Ciais *et al.*, 2005; Zhao and Running, 2010), forest growth (Barber *et al.*, 2000), carbon balances (Piao *et al.*, 2008; Arnone *et al.*, 2008), plant phenology (Cleland *et al.*, 2007), and species distributions towards the poles (Parmesan *et al.*, 1999). These changes have been accompanied by increases in forest dieback and mortality around the world (Allen *et al.*, 2010; Phillips *et al.*, 2009; van Mantgem *et al.*, 2009; Carnicer *et al.*, 2011; Peng *et al.*, 2011). Ongoing climate change has resulted in increases in climate extremes, such as droughts, heat waves, heavy rainfall, and frosts (AchutaRao *et al.*, 2013). These unprecedented climate extremes can trigger a suite of ecological processes to regulate plant and ecosystem responses and could affect plant growth, community structure, and ecosystem functions and services in fundamentally different ways from that the normal climatic variability does (Figure 34.1). Several recent reviews synthesized plant phonological and physiological processes (Reyer *et al.*, 2013; Niu *et al.*, 2014) and ecosystem carbon cycles (Reichstein *et al.*, 2013; van der Molen *et al.*, 2011) in response to extreme climate, which greatly advance our understanding of those subjects.

One of the greatest uncertainties in global climate change is predicting changes in the feedbacks between the biosphere and the atmosphere (AchutaRao *et al.*, 2013). Forests exert strong controls on the global carbon cycle and influence regional hydrology and climatology directly through their effects on water and surface energy budgets (Bonan, 2008). Recent studies have indicated that forest mortality caused by rising temperatures and increased drought have rapidly increased around the world during the past decade (Allen *et al.*, 2010; Phillips *et al.*, 2009; van Mantgem *et al.*, 2009; Carnicer *et al.*, 2011; Peng *et al.*, 2011). To date, much progress has been made on quantifying the impacts of drought and increased water stress on forest mortality in tropical rainforests (Phillips *et al.*, 2009), in temperate forests of the southwestern and western United States (van Mantgem *et al.*, 2009) and southern Europe (Carnicer *et al.*, 2011) as well as in Canadian boreal forest (Peng *et al.*, 2011; Ma *et al.*, 2012).

The boreal forest of Canada encompasses approximately 30 per cent of global boreal forests and 77 per cent of forested land in Canada. Accordingly, its impact on the albedo of the planet's surface is significant (Bonan 2008). Climate change throughout the last century has had a

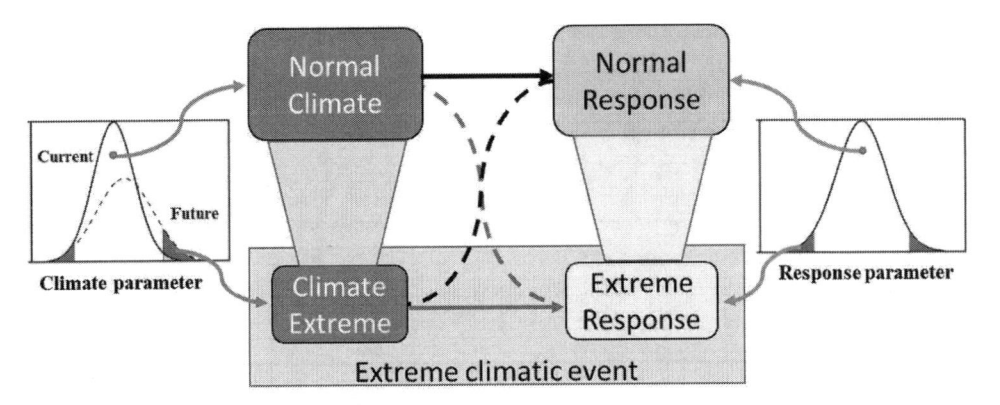

Figure 34.1 Conceptual representation of ecological response to an extreme climatic event that is defined synthetically as involving extremeness in both the climate driver and the ecological response (modified from Smith, 2011). Climate variability can trigger a range of ecological responses (small to extreme, distribution on the right). Changes in climate means or variability may lead to a response that is well within the range of variability for a system (solid black arrow) or one that is extreme (i.e., exceeds this range, dashed grey arrow). Similarly, climate extremes (represented by tails of the distribution on the left) may (solid grey arrow) or may not (dashed black arrow) result in an ecological response that is outside the typical or normal range of variability for a system (modified from Niu et al., 2014)

serious effect on the boreal forest of Canada. The possibility of increasing tree mortality in boreal forests is a particular concern because boreal forests have been identified as a critical "tipping element" of Earth's climate system and are believed to be more sensitive to drought than other forests (Lenton et al., 2008). Recent progress has been made in investigations of the impacts of severe drought on trembling aspen (*Populus tremuloides*) mortality in western Canada (Hogg et al., 2008; Michaelian et al., 2011), suggesting that Canadian boreal forests may be vulnerable to rapid increases in tree mortality due to a combination of warmer temperatures (i.e., increased evapotranspiration) and more severe drought (i.e., decreased water supply).

Drought-induced tree mortality, which rapidly alters forest ecosystem composition, structure, and function, as well as the feedbacks between the biosphere and climate, has occurred worldwide over the past few decades, and is expected to increase pervasively as climate change progresses. The objectives of this chapter are to: (1) highlight the likely ecological consequences of drought-induced tree mortality; (2) synthesize and quantify the impact of drought on growth, mortality and biomass carbon of boreal forests in Canada; (3) discuss the causes of forest mortality and biomass carbon change; and (4) highlight global prospects and future research directions.

Methods and data sources

Although many permanent sample plots (PSPs) have been established in each province (580 in Alberta [AB], 2,426 in Saskatchewan [SK], 368 in Manitoba [MA], greater than 4,000 in Ontario [ON], and approximately 12,000 in Quebec [QC]), most plots did not meet the criteria of this study (Peng et al., 2011). In this study, the 96 PSPs located within the boreal forest of Canada were selected from locations in AB, SK, MB, ON, and QC (Peng et al., 2011). Because the purpose of this study was to explore whether tree growth and mortality rates have changed and whether this change is related to recent environmental changes, such as global warming, drought, etc., the 96 PSPs were selected based on several strict criteria. They had to be plots from natural regenerated forest stands that qualify under at least three consecutive measurements. They also needed to have undergone at least a ten year census interval without experiencing any natural disturbances, such as wildfire, floods, and storms or experiencing any insect or anthropogenic disturbances, such as thinning, harvesting, and fertilization. Additionally, they also had to meet the criterion of being mature forest plots with a stand age greater than or equal to 80 years. Other conditions also had to be met, such as mandatory spatial location data for each selected plot. Detailed information regarding the data selection criteria can be found in Peng et al. (2011).

For the 96 plots chosen, the primary tree species consisted of black spruce (*Picea mariana*), jack pine (*Pinus banksiana*), trembling aspen (*Populus tremuloides*), and white spruce (*Picea glauca*). Minor species consisted of white birch (*Betula papyrifera*), balsam fir (*Abies balsamifera*), lodgepole pine (*Pinus contorta*), Engelmann spruce (*Picea engelmanni*), etc. Measurement periods varied between 10 and 38 years, with the first measurements taking place between 1963 and 1994 and the last measurements taking place between 1990 and 2008. Censuses were carried out every three to five years on all plots. The sizes of plots ranged from 0.04 to 0.82 ha ($\bar{x}$ = 0.12 ha). The plots contained a total of 14,300 trees that had survived the entire census period. Refer to Peng et al. (2011) for more detailed information concerning plot information.

For this study, the basal area growth rate (hereafter $BAGR_{tree}$) was calculated at individual tree and plot levels (hereafter $BAGR_{plot}$) to reflect tree growth rates in the boreal forest of Canada. BAGR was calculated as $(BA_f - BA_i)/t$, where BA_f is the final basal area and BA_i is the basal area at the start of the measurement interval. Thus, the same formula was used to calculate BAGR at the individual tree level ($BAGR_{tree}$) and the plot level ($BAGR_{plot}$) as shown in Equations [1] and [2]:

$$BAGR_{tree} = (BA_f - BA_i)/t \quad [1]$$
$$BAGR_{plot} = (BA_f - BA_i)/t \quad [2]$$

where t is the time in years between the two census dates. It should be emphasized that $BAGR_{tree}$ is based on individual trees that survived the entire measuring period and, thus, $BAGR_{plot}$ for this study only includes the effects of growth of surviving trees. In other words, $BAGR_{plot}$ does not reflect effects of tree recruitment and mortality.

Climatic data from the Daily 10 km Raster-Gridded Climate Dataset for Canada from 1961 to 2003 (the Daily 10 km Gridded Climate Dataset: 1961–2003, Agri-Geomatics, a division of Agriculture and Agri-Food Canada, 2007) was used to obtain climatic variables associated with individual plots. The dataset included daily maximum temperature (°C, T_{max}), minimum temperature (°C, T_{min}), and precipitation (mm, PCP) for Canadian landmasses south of 60°N. The 10 × 10 km grids were interpolated from daily Environment Canada weather station observations, applying a thin plate smoothing spline surface fitting method implemented with the help of ANUSPLIN V4.3 software (Hutchinson, 2004). Since plots with census years after 2003 were used, climate data was downloaded from the nearest weather stations and adjusted for use with interpolation by ANUSPLIN V4.3 (Hutchinson, 2004). Peng et al. (2011) provide detailed information on these weather stations. For this study, the annual climate moisture index (cm, CMI) (Hogg, 1997) was used to provide the annual climatic water deficit. Monthly CMI values were calculated as monthly PCP minus potential evapotranspiration (PET), estimated from T_{max}, T_{min}, and elevation (Hogg 1997). Annual CMI was calculated by summing monthly CMI values from January through December. Positive CMI values indicated relatively moist conditions and negative CMI values indicated relatively dry conditions. Annual mean temperature and annual precipitation were both calculated for this study. To model changes in tree growth rates as a function of climatic variables, annual climatic variables were averaged across all years for each census interval for a given plot.

Results

Impact of drought on forest growth

Much progress has been made in quantifying the dynamics of tree growth rates in boreal forests using tree-ring analysis on relatively small spatial scales. Yet spatially extensive analyses of climate warming effects on boreal tree growth are very limited. Recently, Ma et al. (2014) used long-term PSPs to investigate long-term changes in tree growth rates across Canadian boreal forests. Results indicate that tree growth across the boreal forest of Canada is changing, but these changes are not spatially constant across regions. Growth rates for about 60 per cent of individual trees in western Canada (AB, SK, and MB) were found to be decreasing. Conversely, growth rates for approximately 70 per cent of individual trees in eastern Canada (ON and QC) were found to be increasing. There is little doubt that the recent climate warming contributed to changes in tree growth rates, but it is not the sole factor driving this change (Figure 34.2). Moreover, results indicate that the decrease observed in surviving tree growth rates has contributed to the decrease in above ground biomass in western Canada. However, corresponding increases of growth rates of surviving trees may offset the flux out of the biomass pool from the increase of tree mortality in eastern Canada.

Impact of drought on forest mortality

The results showed (Figure 34.3) that mortality rates increased significantly ($P < 0.0001$, two-tailed binomial test) in 83 per cent of the PSPs, including 91 per cent of these plots (64/70) in western Canada and 62 per cent of the plots (16/26) in eastern Canada. Mortality rates increased significantly for all plots combined and in both regions analysed separately (Figure 34.2, Table

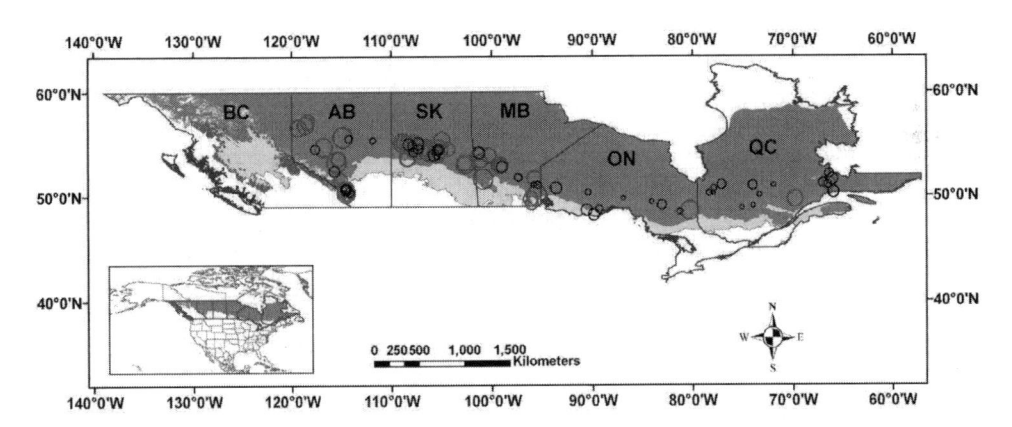

Figure 34.2 Locations of the 96 forest permanent sample plots in Canada's boreal forest. Percentage of trees with decreasing growth rates for the 96 forest plots selected. Grey and black circles represent plots with > 50 per cent and < 50 per cent trees, respectively, with decreasing growth rates. The size of the circle is proportional to the percentage of trees with decreasing growth rates for each plot. Background colours in grey and light grey represent boreal and hemiboreal regions, respectively

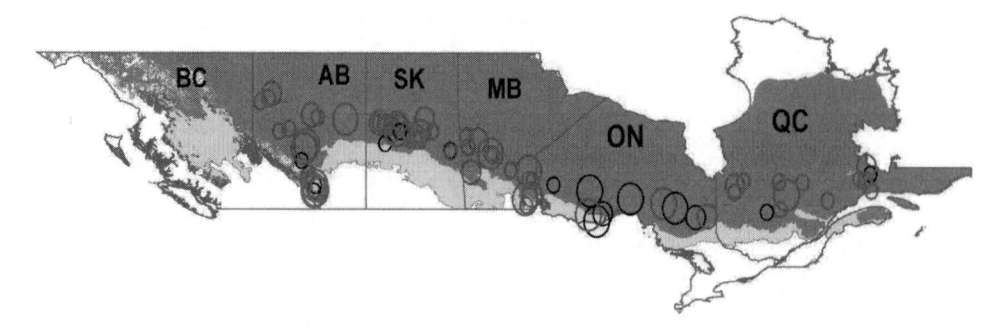

Figure 34.3 The black and grey circles represent plots with decreasing and increasing mortality rates, respectively. Circle size corresponds to the magnitude of the annual change in mortality rates (smallest symbols, < 0.05 per cent year^{-1}; largest symbols, > 0.10 per cent year^{-1}; medium symbols, 0.05 to 0.10 per cent year^{-1}), calculated using a generalized nonlinear model. The background colours of grey and light grey represent, respectively, Canada's boreal and hemiboreal regions. Of these plots, 70 were located in western Canada including AB, SK, and MB and 26 were located in eastern Canada including ON and QC

1 in Peng *et al.*, 2011). Mortality rates also increased for small, medium, and large trees (DBH < 10, 10 to 15, and > 15 cm, respectively) and at low, middle, and high elevations (< 500, 500 to 1,200, and > 1,200 m asl, respectively) (Figure 34.2, Table 1 in Peng *et al.*, 2011). The four most abundant tree species in our plots (comprising 69 per cent of all trees) were trembling aspen (*Populus tremuloides*), jack pine (*Pinus banksiana*), black spruce (*Picea mariana*), and white spruce (*Picea glauca*). All four showed increasing mortality rates, as did trees of all the remaining species (31 per cent of all trees).

In contrast to mortality rates, recruitment rates increased in only 42 per cent of the plots. This proportion was not significantly different from a random result ($P = 0.13$, two-tailed binomial test). There was no detectable trend in recruitment for all plots combined ($P = 0.076$), but recruitment decreased significantly in western Canada ($P = 0.002$, generalized nonlinear mixed model [GNMM]) and increased significantly in eastern Canada ($P = 0.006$, GNMM) when the two regions were analysed separately.

We also found that mean annual precipitation showed no directional trend over the study period ($P = 0.11$, linear mixed model [LMM]), whereas both mean annual temperature and the annual moisture index (AMI) increased significantly ($P < 0.0001$, LMM), and the CMI decreased significantly ($P < 0.0001$, LMM). At our study sites, temperature and water deficits that are represented using the parameters CMI and AMI were both significantly positively correlated with tree mortality rates for all plots combined and for western Canada, whereas the correlation was only significant for temperature in eastern Canada ($P < 0.0001$, GNMM). Our results are consistent with recent findings of a widespread moisture-driven increase in tree mortality in tropical forests in the Amazon basin (Phillips *et al.*, 2009), temperate forests in the western United States (van Mantgen *et al.*, 2009), and trembling aspen stands in western Canada (Hogg *et al.*, 2008; Michaelian *et al.*, 2011).

Impact of drought on forest biomass carbon

The analysis (Figure 34.4) showed that the rate of biomass change for all plots combined and for the western region showed significant decreasing trends, but there was no significant change for the eastern region. In addition to the statistical analysis, we compared the average rates of

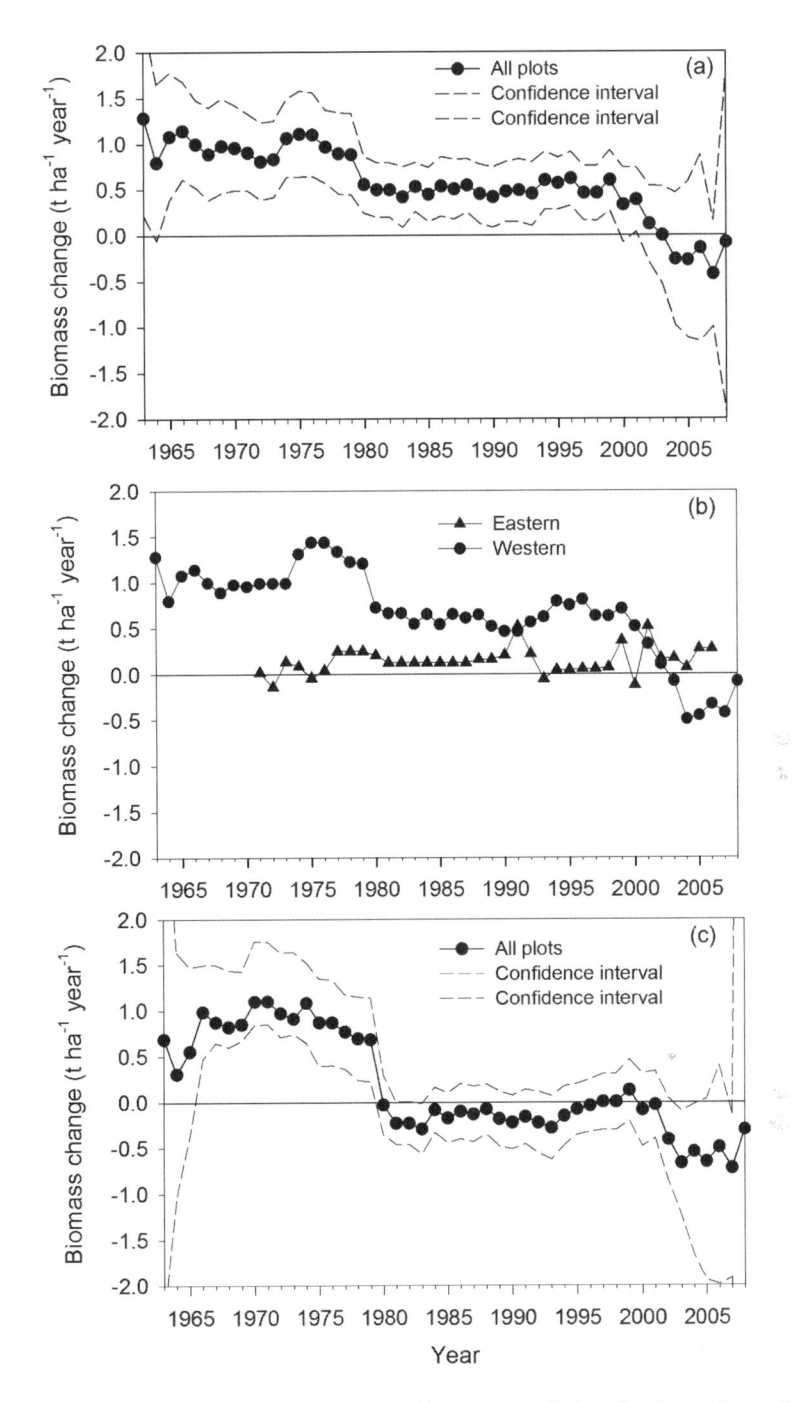

Figure 34.4 (a) Annual rate of change in above ground biomass for all Canadian boreal forest plots combined from 1963 to 2008. The dotted lines represent 95 per cent confidence intervals. (b) Annual rate of change of above ground biomass for the western and eastern regions. (c) Annual rate of change in stand-age corrected above ground biomass for all Canadian boreal forest plots combined from 1963 to 2008. The dotted lines represent 95 per cent confidence intervals (from Ma *et al.*, 2012)

biomass change for the western and eastern regions between the first and final censuses (Ma et al., 2012). Consistent with our model results, the biomass in the western region during the last census interval was significantly lower than that during the first interval ($p < 0.0001$, paired two-sample t-test); for the eastern region, there was no significant difference ($p = 0.1339$, paired two-sample t-test). Moreover, the biomass increment decreased significantly during three periods (before 1980, from 1980 to 2000, and after 2000) in the western region, but there was no significant change in the eastern region (Ma et al., 2012).

Discussion

Changes in tree growth rate between western and eastern Canada

Climate change is one of the dominant external forces controlling the dynamics of northern boreal forest ecosystems. These forests are especially vulnerable to climate change due to tree longevity and the extent of expected climate change within their region and lifespan (AchutaRao et al., 2013). One of the main drivers of uncertainty for future forest management is the potential effect of climate change on forest dynamics. Peng et al., (2011) showed that western Canada (AB, SK, and MB) and eastern Canada (ON and QC) undergo different climatic conditions. Annual mean temperatures increased significantly for both regions; however, annual precipitation and CMI decreased significantly for the western region but increased significantly for the eastern region. For more details concerning climatic variables analysis, refer to Table S4 in Peng et al. (2011).

The basal area increment (BAI) was used in place of tree diameter width growth dendrometric measurements to study changes in tree growth rates because, for mature forests, the width of trees may biologically decline with age or size, which would likely lead to uncertainties when using diameter width alone as a growth decline indicator (Phipps and Whiton, 1988). This study found that $BAGR_{tree}$ for most individual trees in western Canada (AB, SK, and MB) decreased during 1963–2008. Most individual trees in eastern Canada (ON and QC) experienced an increase in $BAGR_{tree}$. In addition, statistically significant decreasing trends in $BAGR_{plot}$ were found in AB and SK and statistically significant increasing trends in $BAGR_{plot}$ in QC. Although MB showed large numbers of trees with increasing individual tree level trends in BAGR, plot level BAGR was random. This may be due to the decreasing rates in BAGR for large numbers of small trees while bigger trees showed increasing rates in BAGR. Thus, the decrease in growth rates for smaller trees and the increase in growth rates for bigger trees may result in the random plot level trend in BAGR. This decreasing BAGR trend in western Canada was consistent with previous studies (e.g., Barber et al., 2000; Beck and Goetz, 2011; Berner et al., 2011); however, this study found no evidence of declining tree growth in ON as previous studies had reported (e.g., Silva et al., 2010).

Regional differences in tree mortality between western and eastern Canada

To understand the different responses of tree mortality in western and eastern Canada to a warming climate, we examined two key variables that govern boreal forest growth and mortality: temperature and moisture. The changes in tree mortality rates differed significantly between boreal regions of western and eastern Canada (Figure 34.3). The increase of tree mortality in western Canada ($a = 0.0488 = 4.9$ per cent) was 2.6 times the rate in eastern Canada ($a = 0.0193 = 1.9$ per cent). This is probably because western Canada has experienced more severe and extensive droughts, such as the one in 2001–2002, than eastern Canada

(Bonsal and Wheaton, 2005). The average annual temperature (°C) was relatively higher in western Canada than in eastern Canada while CMI in western Canada has been much lower than that in eastern Canada since 1970. Moreover, the summer water deficit has been considered to be the dominant factor that controls tree growth in western boreal Canada (Hogg *et al.*, 2008), but this may not be the case for eastern Canada (Girardin *et al.*, 2004). Although most areas of the Canadian boreal forest experienced some degree of drought in 2001–2002, trees in western Canada were most susceptible to mortality because these provinces experienced more serious and extensive droughts than eastern Canada (Bonsal and Wheaton, 2005).

In western Canada, most available moisture for precipitation comes from the Pacific Ocean during the winter (November–March) and from the Gulf of Mexico during the summer (May–August) (Liu *et al.*, 2004). Thus, regional precipitation may be strongly connected to global sea surface temperature patterns (Shabbar and Skinner, 2004). For eastern Canada, Girardin *et al.*, (2004) showed that climate warming and increases in the amount and frequency of precipitation during the last century had no significant impact on the severity of summer drought. In addition, a long-term reduction in the amount of solar radiation in the Canadian Prairies between 1951 and 2005 (Cutforth and Judiesch, 2007) may also have increased the tree mortality rate due to a decline in forest productivity (i.e., net photosynthesis) in western Canada (Hogg *et al.*, 2008; Michaelian *et al.*, 2011). Combined with the other aforementioned factors, this may explain the regional differences in tree mortality between western and eastern Canada.

Causes of forest biomass carbon change

The observed patterns of the rate of biomass change for western and eastern regions may have resulted from changes in tree mortality, growth of surviving trees, or a combination of both factors. To further explore the ecological causes of the rate of biomass change, we analysed the trends of mortality, stand density, and biomass of the surviving trees (see Ma *et al.*, 2012). The analysis revealed that the biomass increment associated with mortality for both regions showed significant increasing trends. These trends were further confirmed by the changes in stand density, which indicated that the number of surviving trees was decreasing significantly for both the western and the eastern regions. However, the rate of biomass change for surviving trees in the western region showed a significant decreasing trend, whereas there was a significant increasing trend for surviving trees in the eastern region (Ma *et al.*, 2012, 2014). Thus, our analysis indicated that the trend of decreasing biomass change in the western region resulted from the cumulative effects of increased mortality and decreased growth of surviving trees, but mortality had a larger effect ($\beta = 0.0289$) than growth of surviving trees ($\beta = -0.0116$) on the rate of biomass change. For the eastern region, the simultaneous increase in mortality and growth of surviving trees represented offsetting factors that concealed any significant trend in biomass change (Ma *et al.*, 2012, 2014).

Global prospects

Alongside global warming, droughts are expected to increase in frequency, severity, and extent in the near future, which will likely result in significant impacts on forest growth, production, structure, composition, and ecosystem services. However, due to spatial and temporal characteristics, it is difficult to monitor and assess the effects of droughts. Remote sensing can provide an effective way to obtain real-time conditions of forests affected by drought and offer a range of spatial and temporal insights into drought-induced changes to forest ecosystem structure, function, and services (Zhang *et al.*, 2013). Remote sensing is rapidly developing as more satellites are launched. In situ and remotely sensed data fusion techniques have achieved notable success in assessing drought-induced damage to forests and carbon cycles. Even so, constraints still exist when using satellite data. Long-term observations will be an addition to improve our understanding of the ecological consequences of tree mortality and are expected to improve our modelling ability (Wang et al., 2012).

Summary and conclusions

In summary, for this study, plot level growth rates of surviving trees were found to be decreasing in AB and SK while no equivalent trends were detected in MB and ON. An increasing trend, however, was detected in QC. Results from this study provided more detailed information on the forest dynamics of the boreal forest of Canada than previous reports had, such as an increase in tree mortality, a decrease in recruitment rates for western Canada (Peng et al., 2011), and an increase in tree mortality alongside an increase in tree recruitment for eastern Canada (Peng et al., 2011). Thus, decreasing rates observed in above ground biomass for western Canada (AB, SK, and MB) (Ma et al., 2012) were due to a decrease in surviving tree growth, an increase in tree mortality, and a decrease in tree recruitment. This simultaneous increase in growth rates of surviving trees, tree mortality, and recruitment rates have led to a random rate of above ground biomass in eastern Canada (ON and QC) (Ma et al., 2012). Results from this study indicate that the decline in surviving tree growth worsened decreasing rates in above ground biomass accumulation for western Canada, but the increase in growth rates of surviving trees may offset the effects on the biomass pool of an increase in tree mortality in eastern Canada. This study provides the first plot-based evidence for ecological variability and regional differences in tree growth detected by satellite-based remote-sensing and tree-ring studies in Canada's boreal forests. Findings from this study could provide valuable insight into forest dynamics and regional carbon budgets and could be useful in developing strategies for adaptive forest management under a changing climate.

Acknowledgments

Funding for this study was provided by the NSERC strategic Network (ForValueNet), NSERC discovery grant, and China QianRen program. We thank the Forest Management Branch of Alberta Ministry of Sustainable Resource Development, Saskatchewan Renewable Resources Forestry Branch, Forestry Branch of Manitoba, Ontario Terrestrial Assessment Program, and Ministère des Ressources Naturelles et de la Faune du Québec, and our colleagues (P. Comeau, J. Liu, V. LeMay, S. Huang, J. Parton, K. Zhou) for providing detailed data. C. Peng acknowledges the support he received during his sabbatical leave at Northwest A&F University, China. We also thank T. Hogg for his help with our calculations of CMI, P.J. van Mantgem for his assistance with our statistical models, W. Zhang for her technical assistance, and G. Hart for editorial help on the earlier draft of chapter.

References

AchutaRao, K.M., Allen, M.R., Gillett, N., Gutzler, D., Hansingo, K. and others. Detection and attribution of climate change: From global to regional. In: *Climate Change 2013: The Physical Science Basis. Contribution of Working Group I to the Fifth Assessment Report of the Intergovernmental Panel on Climate Change* (eds. Stocker, T.F., Qin, D., Plattner, G.-K., Tignor, M., Allen, S.K. and others) Cambridge University Press, Cambridge, UK, pp. 867–952 (2013).

Allen, C.D., Macalady, A.K., Chenchouni, H., Bachelet, D., McDowell, N. and others. A global overview of drought and heat-induced tree mortality reveals emerging climate change risks for forests. *Forest Ecology and Management,* 259, 660–684 (2010).

Arnone III, J.A., Verburg, P.S.J., Johnson, D.W., Larsen, J.D., Jasoni, R.L. and others. Prolonged suppression of ecosystem carbon dioxide uptake after an anomalously warm year. *Nature,* 455, 383–386 (2008).

Barber, V.A., Juday, G.P. and Finney, B.P. Reduced growth of Alaskan white spruce in the twentieth century from temperature-induced drought stress. *Nature,* 405, 668–673 (2000).

Beck, P.S.A. and Goetz, S.J. Satellite observations of high northern latitude vegetation productivity changes between 1982 and 2008: Ecological variability and regional differences. *Environ. Res. Lett.* 6 045501 (10pp) (2011).

Berner, L.T., Beck, P.S.A., Bunn, A.G., Lloyd, A.H., and Goetz, S.J. High-latitude tree growth and satellite vegetation indices: Correlations and trends in Russia and Canada (1982–2008). *Journal of Geophysical Research,* 116, G01015 (2011)

Bonan, G.B. Forests and climate change: Forcings, feedbacks, and the climate benefits of forests. *Science,* 320, 1444–1449 (2008).

Bonsal, B.R. and Wheaton, E.E. Atmospheric circulation comparisons between the 2001 and 2002 and the 1961 and 1988 Canadian prairie droughts. *Atmosphere-Ocean,* 43, 163–172 (2005).

Carnicer, J., Coll, M., Ninyerola, M., Pons, X., Sanchez, G. and Penuelas, J. Widespread crown condition decline, food web disruption, and amplified tree mortality with increased climate change-type drought. *Proceedings of the National Academy of Sciences USA,* 108, 1474 (2011).

Ciais, P., Reichstein, M., Vivoy, N., Granier, A., Ogee, J. and others. Europe-wide reduction in primary productivity caused by the heat and drought in 2003. *Nature,* 437, 529–533 (2005).

Cleland, E.E., Chuine, I., Menzel, A., Mooney, H.A. and Schwartz, M.D. Shifting plant phenology in response to global change. *Trends in Ecology and Evolution,* 22, 357–365 (2007).

Cutforth, H.W. and Judiesch, D. Long-term changes to incoming solar energy on the Canadian Prairie. *Agricultural and Forest Meteorology,* 145, 167–175 (2007).

Girardin, M.P., Tardif, J., Flannigan, M.D., Wotton, B.M. and Bergeron, Y. Trends and periodicities in the Canadian Drought Code and their relationships with atmospheric circulation for the southern Canadian boreal forest. *Canadian Journal of Forest Research,* 34, 103–119 (2004).

Hogg, E.H. Temporal scaling of moisture and the forest grassland boundary in western Canada. *Agricultural and Forest Meteorology* 84, 115–122 (1997).

Hogg, E.H., Brandt, J.P. and Michaelian, M. Impacts of a regional drought on the productivity, dieback, and biomass of western Canadian aspen forests. *Canadian Journal of Forest Research,* 38, 1373–1384 (2008).

Hutchinson, M.F. *ANUSPLIN Version 4.36.* Centre for Resource and Environmental Studies, Australian National University (2004).

Lenton, T.M., Held, H., Kriegler, E., Hall, J.W., Lucht, W. and others. Tipping elements in the Earth's climate system. *Proceedings of the National Academy of Sciences,* 105, 1786–1793 (2008).

Liu, J., Stewart, R.E. and Szeto, K.K. Moisture transport and other hydrometeorological features associated with the severe 2000/01 drought over the Western and Central Canadian Prairies. *Journal of Climate,* 17, 305–319 (2004).

Ma, Z., Peng, C., Zhu, Q., Chen, H., Yu, G., and others. Regional drought-induced reduction in the biomass carbon sink of Canada's boreal forests since 1963. *Proceedings of National Academy of Sciences of United States of America (PNAS),* 109 (7), 2423–2427 (2012).

Ma, Z., Peng, C., Zhu, Q., Liu, J., Xu, X. and Zhou, X. Contrasted response of tree growth to recent climate warming in the boreal forest of Canada. *Ecoscience* (in review), (2014).

Michaelian, M., Hogg, E.H., Hall, R.J. and Arsenault, E. Massive mortality of aspen following severe drought along the southern edge of the Canadian boreal forest. *Global Change Biology,* 17, 2087–2094 (2011).

Niu, S., Luo, Y., Li, D., Cao, S., Xia, J. and others. Plant growth and mortality under climatic extremes: An overview. *Environmental and Experimental Botany* 98, 13–19 (2014).

Parmesan, C., Ryrholm, N., Stefanescu, C., Hill, J.K., Thomas, C.D. and others. Poleward shifts in geographical ranges of butterfly species associated with regional warming, *Nature,* 399, 579–583 (1999).

Peng, C.H., Ma, Z.H., Lei, X.D., Zhu, Q., Chen, H. and others. A drought-induced pervasive increase in tree mortality across Canada's boreal forests. *Nature Climate Change* 1, 467–471 (2011).

Phillips, O.L., Aragao, L.E.O.C., Lewis, S.L., Fisher, J.B., Lloyd, J. and others. Drought Sensitivity of the Amazon Rainforest. *Science,* 323, 1344–1347 (2009).

Phipps, R.L., and Whiton, J.C. Decline in long-term growth trends of white oak. *Canadian Journal of Forest Research* 18, 24–32 (1988).

Piao, S., Ciais, P., Friedlingstein, P., Peylin, P., Reichstein, M. and others. Net carbon dioxide losses of northern ecosystems in response to autumn warming. *Nature,* 451, 49–53 (2008).

Reichstein, M., Bahn, M., Ciais, P., Frank, D., Mahecha, M.D. and others. Climate extremes and the carbon cycle. *Nature* 500, 287–295 (2013).

Reyer, C.P.O., Leuzinger, S., Rammig, A., Wolf, A., Bartholomeus, R.P. and others. A plant's perspective of extremes: Terrestrial plant responses to changing climatic variability. *Global Change Biology* 19, 75–89 (2013).

Shabbar, A. and Skinner, W. Summer drought patterns in Canada and the relationship to global sea surface temperatures. *Journal of Climate,* 17, 2866–2880 (2004).

Silva, L.C.R., Anand, M. and Leithead, M.D. Recent widespread tree growth decline despite increasing atmospheric CO_2. *PLoS ONE* 5(7): e11543. doi:10.1371/journal.pone.0011543 (2010).

Smith, M.D. An ecological perspective on extreme climatic events: A synthetic definition and framework to guide future research. *Journal of Ecology* 99, 656–663 (2011).

Van der Molen, M.K., Dolman, A.J., Ciais, P., Eglin, T., Gobron, N. and others. Drought and ecosystem carbon cycling. *Agricultural and Forest Meteorology* 151, 765–773 (2011).

Van Mantgem, P., Stephenson, N.L., Byrne, J.C., Daniels, L.D., Frankin, J.K. and others. Widespread increase of tree mortality rates in the western United States. *Science* 323, 521 (2009).

Wang, W., Peng, C., Kneeshaw, D.D., Larocque, G.R., and Luo, Z. Drought-induced tree mortality: Ecological consequences, causes, and modeling. *Environmental Reviews* 20, 109–121 (2012).

Zhang, Y., Peng, C., Li, W., Fang, X., Zhang, T. and others. Monitoring and estimating drought-induced impacts on forest structure, growth, function, and ecosystem services using remote-sensing data: Recent progress and future challenges. *Environmental Reviews* 21, 103–115 (2013).

Zhao, M. and Running, S.W. Drought-induced reduction in global terrestrial net primary production from 2000 through 2009. *Science* 329, 940–943 (2010).

35
ASSESSING RESPONSES OF TREE GROWTH TO CLIMATE CHANGE AT INTER- AND INTRA-ANNUAL TEMPORAL SCALE

Sergio Rossi, Jian-Guo Huang and Hubert Morin

Introduction

In trees, the length of the growing season, which influences the resulting volume of wood produced by the stem, is determined by the phenology of primary and secondary meristems and the growth rate, which are in turn determined by climate (Rathgeber *et al.* 2011; Rossi *et al.* 2014). In addition, the phenology of the secondary meristems, a much less studied trait, has an impact on the quality of wood, including its density. The wide temporal scale of forest production, covering from a few decades to a century, requires the adaptation of the trees to both current and future environmental conditions. Thus, knowledge of how and when the meristems are active and trees grow could support forest management strategies. If well planned, stands in most productive areas could achieve the desired growth characteristics and adaptation to local climate for ensuring a constant and sustainable wood production. The accuracy of growth estimations of trees and their responses to the environment should be considered especially carefully within a context of climate change. In its report, the Intergovernmental Panel on Climate Change (IPCC) has confirmed that climate change is a reality (IPCC 2013). Over the next 50 years, the increase in atmospheric CO_2 will be accompanied by a substantial warming, mainly at the highest latitudes and altitudes. Such changes will have significant impacts on the growth of most tree species, particularly on production and quality of wood (O'Neill and Nigh 2011).

The productivity of a stand is represented by the amount of wood produced in a predetermined period of time. Wood is the result of gradual accumulation of xylem cells by trees in order to renew their transport system, store substances and ensure mechanical support to branches and leaves. At the basis of wood formation, and, in the broad sense, of plant growth, are the meristems. Cambium is a lateral meristem designed to produced xylem and phloem cells. Through a cyclical alternation of periods of activity and rest, the cambium produces the conducting elements during the more favourable periods, leaving indelible traces in the form of tree rings. A tree ring is the product of an entire season of cambial activity that, in the boreal and temperate–cold zones of the planet, corresponds to the period between spring and autumn. In these environments, air temperature is the key factor that influences the resumption of activity of the meristems and the water balance in plants during growth. It is therefore a priority to understand the effects of warming, and of climate change more generally, on forest species to predict the potential future evolution and the ecological and economic consequences of modifications of tree growth over time.

In this chapter, we describe how tree growth can be analysed at different temporal scales, between years and within a year, and how tree rings can be used to assess the responses of trees to a changing environment by giving examples of growth under natural and controlled environmental conditions. Black spruce is taken as case study because of its widespread distribution in the boreal forest of North America, and the availability of long-term growth data in the form of inter- and intra-annual chronologies collected over a wide geographical area. The different scales of analysis show the importance of investigating tree responses to climate at several levels of detail, and how some results from inter- and intra-annual study may at first glance seem in partial disagreement. These divergent interpretations are discussed and clarified in the second part of the chapter.

Wood growth at inter-annual scale

Long-term records of inter-annual tree-ring growth or inter-annual xylem formation are a natural archive for documenting environmental signals. Tree-ring data have a high resolution (annual or seasonal), absolute accuracy, and can cover wide spatial distributions compared to other proxy data such as ice cores, lake sediments and historical documents. Hence, tree rings have long been used as a valid tool to explore the long-term growth reactions to past climate variations and to assess the impacts of future climate warming on tree growth and forest ecosystems within the context of increased temperatures and atmospheric CO_2.

Dendrochronology is a science that uses inter-annual variability in tree-ring width, wood density or isotope concentration to understand the effects of environmental factors (e.g., climate, insect or fire disturbances, human activities) on tree growth. In dendrochronology, the growth series of individual trees can be translated into an aggregation of environmental factors that affect growth over time (Fritts 1976; Schweingruber 1996). Cook (1987) suggested that radial growth (in terms of tree-ring width) is an aggregate function of a number of factors: tree-ring width at year t can be split into several components according to the following equation:

$$R_t = A_t + C_t + \delta D1_t + \delta D2_t + \varepsilon_t$$

where R_t is the observed tree-ring width, A_t is the age- or size-related growth trend due to normal physiological aging processes, C_t is the effect of climate on tree growth, $D1_t$ is the occurrence of disturbance factors within the forest stand (e.g., a blow down of trees), $D2_t$ is the occurrence of disturbance factors from outside the forest stand (e.g., an insect outbreak that defoliates the trees), and ε_t represents the random processes not accounted for by the other components, δ is either 0 or 1, for absence or presence of the disturbance signal, respectively (Cook 1987).

According to the specific purposes of a study, the age- or size-related growth trend has to be removed by appropriate detrending approaches (called *standardization*, Fritts 1976) in order to attain the objective factors (called *signals*) from the above equation. To understand the climatic effects on growth, for example, it is necessary to highlight the climatic factor C_t of the equation. Alternatively, the disturbance factor $D2_t$ has to be preserved to analyse the effects of insect outbreaks on growth. These effects can also be observed by comparing tree-ring width of host and non-host trees, the latter reacting to the other variables in the same way as host trees (Swetnam et al. 1985; Morin et al. 2010). This allows the insect signal on growth to be detected and chronologies to be compiled of insect outbreaks over time. Insects are thermo-dependent organisms. The impact, recurrence and location of outbreaks in space and time gives important information on the present and past climate, and provides some valuable clues about what can occur in the context of climate change (Morin et al. 2010; Régnière et al. 2012).

Dendrochronology has been widely used in different fields such as climatology, ecology, hydrology and geomorphology. Here, we describe two cases in which tree-ring width was used to understand the effects of climate change on radial growth of black spruce over a latitudinal gradient in the boreal forest of eastern Canada and on the spatial and temporal dynamics of the spruce budworm (*Choristoneura fumiferana* Clem.).

Effects of climate change on black spruce

Trees growing at high latitudes or elevations are very sensitive to environmental changes, such as temperature and precipitation. Black spruce grows in a broad transcontinental band from Alaska to Newfoundland, but only forms extended closed forests in the boreal region of north-eastern North America. It has proved useful in dendroclimatology to understand the effects of climate change on growth (Hofgaard et al. 1999; Tardif et al. 2001). Hofgaard et al. (1999) investigated the growth–climate relationships for spruce over a latitudinal gradient from 48 to 50°N and found that previous summer moisture and current growing season temperature are the main climatic factors affecting radial growth. Girardin et al. (2014) reconstructed the net primary productivity using tree-ring chronologies from black spruce and reported that the unusual forest growth decline in boreal North America covaries with the retreat of Arctic sea ice.

Huang et al. (2010) conducted a dendroclimatological study over a broad latitudinal gradient from 46–54°N in the eastern Canadian boreal forest. They found that previous summer temperature (June–August) and current June–July temperature were negatively associated with radial growth of black spruce at most latitudes (Figure 35.1). This negative effect of temperature on growth was also reflected by negative correlations between radial growth and the previous summer monthly drought code (MDC), which is a numerical parameter representing the average moisture content of deep and compact organic layers and was previously shown to correlate well with growth in boreal tree species (Girardin and Wotton 2009). In addition, June precipitation was shown to be positively correlated with radial growth. These results were consistent with other growth–climate studies on black spruce (Dang and Lieffers 1989; Hofgaard et al. 1999; Tardif et al. 2001; Nishimura and Laroque 2011). Altogether, these effects indicate that moisture stress during the previous and current summer is the main factor limiting radial growth over the studied gradient. The negative effect of previous summer temperature and positive effect of previous June precipitation on growth might also be explained by a climatic driven induction of masting trees belonging to the genus *Picea*, because these climatic factors were found to be critical for predicting seed production of Norway spruce (*Picea abies* (L.) Karst.) in southern Norway (Selås et al. 2002). Late-summer climate conditions in the previous year are believed to generally influence the size of the buds and number of leaf primordials produced within them for growth of boreal conifers, which then affects the amount of leaf area for the most photosynthetically efficient class of needles produced in the current year of bud expansion and ring formation (Kozlowski et al. 1991).

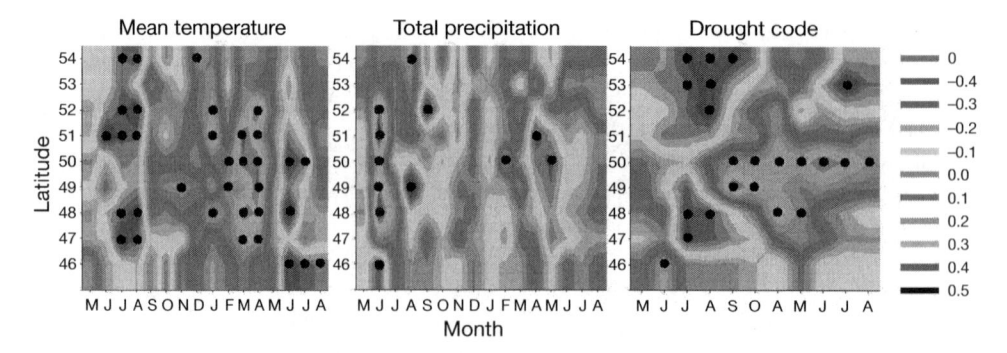

Figure 35.1 Growth–climate response function for black spruce over the latitudinal gradient from 46–54 °N in the boreal forest of eastern Canada. Note: climate variables included mean monthly temperature, total monthly precipitation and monthly drought code from the previous May to the current August; Drought code available from May to October only. Solid dots indicate significance level at P < 0.05 (modified from Huang *et al.* 2010)

In addition to the aforementioned effect of summer moisture on growth, temperatures during winter and early spring (December–April) showed positive correlations with radial growth over most latitudes. A warm winter may reduce risk of damage and foliage losses induced by snow and wind in a cold winter, which could allow an earlier resumption of tree growth during warmer springs. Warm temperature may also supply more moisture by hastening snowmelt, thus providing enough moisture and warmth to favour earlier onsets of cambial activity and subsequent xylem growth. Consequently, a wider tree ring is formed in a prolonged growing season during warmer years (Huang *et al.* 2011). These growth–climate responses over the broad spatial gradient can be used to approximate climate warming over time and thus allow further predicting of forest growth and productivity under a warming climate, as shown in Huang *et al.* (2013).

Spruce radial growth was also reported to react to climate extremes such as drought. Drobyshev *et al.* (2013) investigated non-anomalous growth years with climate and found that the non-anomalous growth was negatively responded to June precipitation, indicating a strong signal of drought stress on growth. Drought was also reported to induce spruce mortality in the Canadian boreal forest ecosystem (Peng *et al.* 2011). Therefore with the expected increase in climate extremes under future warming, the effect of climate extremes on growth should also be well assessed and included in forest ecosystem management.

Effect of climate change on the spatial and temporal dynamics of spruce budworm

Insects are very sensitive to climate. They require a minimum amount of heat to complete their life cycle. The spatial and temporal dynamics of erupting insects, such as the spruce budworm, which can reach epidemic levels and have a tremendous impact on forest dynamics, are affected by a changing climate. The impact of recent climate warming has already been reported for some erupting insects. For example, in the Alps, the cycle of *Zeiraphera diniana* Gn. that has been very regular for centuries has been interrupted in the last decades (Esper *et al.* 2007; Büntgen *et al.* 2009). In Western America, the mountain pine beetle, *Dendroctonus ponderosae* Hopkins, formerly restricted to some valleys of the Rocky Mountains, has expanded its range due to a recent favourable climate (Cudmore *et al.* 2010). In the northern part of their range, the population of erupting insects may fluctuate in relation to good climatic conditions.

Spruce budworm is a *Lepidopteran* that feeds on conifer needles during its larval stage. Its primary hosts are balsam fir (*Abies balsamea*) and white spruce (*Picea glauca*) but it is also found on black spruce and red spruce (*Picea rubens*). It is an insect that primarily affects mixed stands in the southern part of the boreal forest, in the balsam fir–white birch zone, although its range can extend to the black spruce–feather moss forest at a latitude of 54°N (Levasseur 2000). Dendroecological investigations on old spruces (probably white spruce) used for building churches in southern Quebec, where the Europeans first settled, has permitted the longest chronology to be reconstructed for the southern-central part of spruce budworm distribution (Boulanger *et al.* 2012) (Figure 35.2). There is a remarkable regularity in the recurrence of insect outbreaks in the southern part of its range. However, chronologies from the boreal forest show an increase in regularity, impact and synchronism over large areas of the boreal region, including the southern black spruce zone, during the twentieth century (Figure 35.3). Wood subfossils buried in peat show the same trend: in the boreal zone, it is possible to locate the

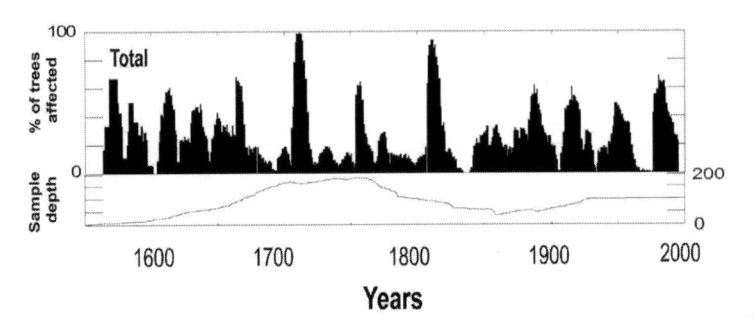

Figure 35.2 Percentage of trees (probably white spruce) sampled in old churches in southern Quebec that had been affected by spruce budworm in a given year (modified from Boulanger *et al.* 2012)

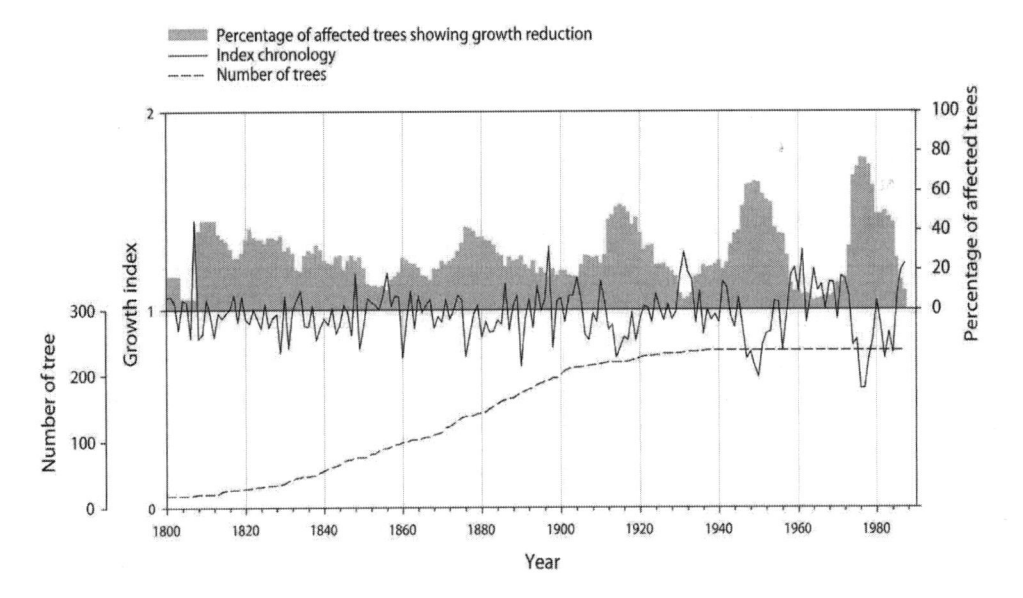

Figure 35.3 Percentage of living white spruces affected by spruce budworm in a given year in the boreal zone of Quebec

major outbreaks of the twentieth century but we have to go back many thousands of years to obtain equivalent signals of outbreaks in the past (Simard et al. 2011; Lapointe 2013). Many hypotheses have been put forward to explain this increase in the impact of outbreaks during the twentieth century: (i) the augmentation of balsam fir, its primary host, by cutting practices; (ii) the control of fire that preserves the old-growth forest, which is more susceptible; and (iii) the changing landscape caused by a diminution in the frequency of fires after the Little Ice Age (Blais 1983; Jardon and Morin 2003; Morin et al. 2008). A hypothesis that reconciles dendroecologists and entomologists has recently emerged: climate change. Spruce budworm would expand its range in the boreal zone when the climate is favourable (Régnière et al. 2012), resulting in significant defoliation and hence growth reductions in the host species that could allow outbreaks to be located in the chronologies. It would recede towards the southern boreal and mixed stands when the climate is cooler, such as during the Little Ice Age, so that growth reductions caused by defoliation would disappear. Although confirmation is required by additional studies, the hypothesis is supported by independent paleoecological analyses using macrofossils of the insects, which have shown a trend similar to the expected one (Simard et al. 2006; Morin et al. 2008).

Wood growth at intra-annual scale

Wood growth occurs in a well-defined time window. During this period, trees are directly affected by environmental signals at different timescales: long-term general patterns (climate) or punctual events (weather). Investigations at intra-annual scale, on time resolutions shorter than one year, can allow the effects of short-term or punctual environmental signals on wood production to be detected and the consequences of environmental changes on radial growth to be assessed (Seo et al. 2008; Rossi et al. 2012).

The analysis of wood formation divides the growing season into short periods by means of sampling at time intervals of between a few days and two weeks. Such monitoring provides information on different developmental stages of the xylem cells and timings of their occurrence, thus identifying the phases of xylem phenology. Intra-annual observations of wood formation, also called xylem phenology, take into account growth dynamics and allow several key questions to be answered: when and how does wood production occur? What environmental signals trigger timings and dynamics of xylogenesis? Does xylogenesis vary between individuals and how do such differences affect the response of trees to climate?

Observations of wood formation began a century ago (Knudson 1913) with histological measurements using light microscopy. Since then, cambium and its derivatives, which are at the origin of wood, have been the subject of investigation by plant physiologists and ecologists. According to Wolter (1968), cambium was pierced with a thin needle, and the wood samples analysed at the end of the year to reconstruct the growing season. This technique was later applied in several studies (Nobuchi et al. 1995; Schmitt et al. 2004) and is nowadays termed pinning (see Seo et al. 2007 for a review). During the 1970s, several studies regarding wood formation were published with an emphasis on data analysis and the relationship with climatic factors (Wodzicki 1971; Denne 1974).

In the last decade, the ecological and economic importance of forest ecosystems has given a new impetus to intra-annual investigations. The study of wood formation covers the phenology of xylem formation (Camarero et al. 1998; Deslauriers et al. 2003; Mäkinen et al. 2003; Rossi et al. 2003; Schmitt et al. 2004) or focuses on precise events such as the onset (Oribe et al. 2001; Gričar et al. 2006; Vieira et al. 2014a) or end of growth (Gindl et al. 2000; Gričar et al. 2005). Investigations have attempted to link xylogenesis with endogenous (Rossi et al. 2008a;

Rathgeber et al. 2011; Anfodillo et al. 2012; Vieira et al. 2014b) or environmental factors (Deslauriers and Morin 2005; Moser et al. 2010; Rossi et al. 2011; Vieira et al. 2014a). Because of the relatively recent development of this technique, some details are provided in the next paragraphs to better understand how the intra-annual analysis of wood formation works.

Periodic micro-samplings are based on repeated observations on wood samples of the developing tree ring during the growing season (Bäucker et al. 1998; Camarero et al. 1998; Horacek et al. 1999; Jones et al. 2004). In order to be the least invasive possible, wood samples are repeatedly extracted around the stem as blocks (Gričar et al. 2005) or small cores (microcores, Forster et al. 2000). These samples contain the dead outer bark, phloem, cambium, developing xylem and recently-formed tree rings. Microcoring is performed by means of small cutting tubes inserted into the stem and then extracted. The microcores are stored in an ethanol solution at 5°C. They consist of three parts (wood, cambium and phloem-bark) with different consistency and density, requiring specific procedures for sample preparation. Recent preparations of microcores involve dehydration in ethanol and D-limonene, and embedding in paraffin (Rossi et al. 2006a) or glycol methacrylate (Gruber et al. 2010). Although embedding is not essential (Moser et al. 2010), paraffin penetrates deeply into the wood, fills cell lumens of the tissues and improves the quality of sections. Transverse sections of 10–30 mm thickness are cut from the samples with a rotary or sledge microtome, stained with cresyl violet acetate (Rossi et al. 2006b) or safranin and astrablue (Moser et al. 2010), and examined under bright-field and polarized light at 400–500 magnifications to differentiate the developing and mature xylem cells.

Xylem formation can be assessed either by counting cells (Wodzicki 1971; Antonova and Stasova 1997; Deslauriers et al. 2003; Rossi et al. 2006b) or by measuring the radial distance covered by the cells (Marion et al. 2007; Seo et al. 2007, 2008). According to Savidge (2000), counting of cell numbers in a radial file, for each developmental zone and mature state, gives more useful information about current activity of cambium as cells represent the most important unit to have quantitative data on. As the cambial activity varies with time, counting the radial number of cells in the cambial zone, post-cambial growth and mature cells on several different dates highlights the tree-ring developmental pattern. Counts of cell number have been intensively applied in conifers, which have a xylem with quite homogeneous cells. Because of the presence of vessels and wide variations in cell size, cell count has been less frequently applied in broadleaves, being substituted by measurements of radial distance (van der Werf et al. 2007; Marion et al. 2007; Zhai et al. 2012). However, in addition to cambial activity, measurements of radial distance take post-cambial growth into account, i.e. the increase in size of the cells.

During development, the cambial derivatives alter both morphologically and physiologically, gradually assuming definite features and differentiating into the specific elements of the stem tissues (Figure 34.4). In cross section, cambial cells are characterized by thin cell walls and small radial diameters. During cell enlargement, the derivatives consist of a protoplast still enclosed in the thin and elastic primary wall. Following positive turgor increase by water movement into the vacuoles, the cell wall stretches, increasing the radial diameter of the tracheid and consequently, the lumen area. In cross section, observations under polarized light discriminate between enlarging and cell-wall thickening cells. Because of the arrangement of the cellulose microfibrils, the developing secondary walls shine under polarized light. Instead, there is no glistening in enlargement zones. Once their final size has been reached, the cells begin maturing through cell-wall thickening and lignification. The progress of lignification is detected with cresyl violet acetate that reacts with lignin and shows a colour change from violet (unlignified secondary cell walls) to blue (lignified cell walls). A blue colour over the whole cell wall indicates the end of lignification and attainment of the mature stage for the tracheid.

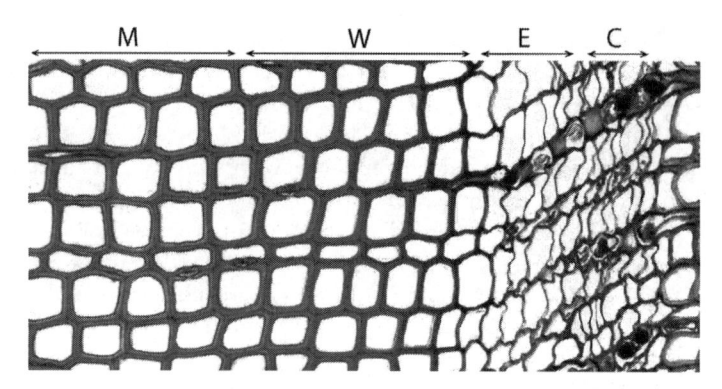

Figure 35.4 Cross section of developing xylem in black spruce with mature cells (M), cells in phases of wall thickening and lignification (W), enlargement (E), and cambial zone (C)

The temporal and spatial zones of cambium, differentiation and mature cells allow the dynamics of xylem growth to be described and the phases of xylem phenology to be identified. All species of temperate and boreal ecosystems show the same trend of xylem formation and the radial files of differentiating cells have an evident pattern of variation during the year (Rossi *et al.* 2006b). The observed patterns are connected with the number of cells passing through each differentiation phase and with the gradual accumulation of mature cells in the tree ring. Analysis of the dynamics of xylogenesis requires the dates of each phenophase to be summarized in a concise but informative way. The selection and use of the descriptive statistics are particularly important with the phenophases of wood formation because a substantial variability exists both within and between trees, and has to be carefully quantified and analysed (Lupi *et al.* 2012b).

The growth patterns in Mediterranean climates or in ecosystems with marked summer droughts partially differ from those observed in temperate and boreal regions. In early summer, water stress can lead to an earlier interruption of wood formation, possibly associated with the formation of tracheids with small lumen areas. Precipitations occurring during the mild autumn months allow xylem cell differentiation to resume (enlargement and cell-wall thickening), and, in some cases, also cambial activity (Deslauriers *et al.* 2009; Vieira *et al.* 2014a). This double interruption of wood formation in winter and summer produces a bimodal pattern of stem radial increment, which is distinctly recorded by changes in stem size with two periods of increment: one in spring and another in autumn. As a result, radial profiles of tree rings frequently exhibit intra-annual density fluctuations (Novak *et al.* 2013). In some Mediterranean species, this bimodal pattern of xylem growth may correspond to a double flushing of buds and branches. However, accurate investigations of xylem anatomy suggest that the autumnal increment period mostly involves a rehydration of the stem tissues rather than effective growth (i.e. new formed cells) (Vieira *et al.* 2014a).

The intra-annual analyses of xylem phenology in black spruce began 15 years ago in order to identify the factors involved in wood production and understand how their modifications could affect the productivity of the boreal forest. The monitoring has been performed at three different levels that are described in detail in the next sections: natural environment, manipulated natural environment and greenhouse. Each level corresponds to a specific intensity of control of the sources of variation and naturalness, and clearly affects the representativeness of the results.

Natural environment

Monitoring xylogenesis in untreated trees was first used to define the growing season of species and its variability among years and sites according to specific gradients in climatic factors (Deslauriers et al. 2003; Rossi et al. 2006b; Marion et al. 2007; Schmitz et al. 2007; van der Werf et al. 2007; Moser et al. 2010).

It is not surprising that temperature is one of the key ecological factors controlling the growing season of black spruce. However, other related and unrelated climatic factors have been observed to play a role in defining the timings of wood formation in this species. Is the temperature sufficient to comprehensively explain the pattern of wood production? If so, xylem cell production should closely follow the seasonal thermal trend, gradually increasing in spring, attaining a maximum close to the second half of July, and reducing during late summer and autumn. Cell production was expected to follow the same pattern. Rossi et al. (2006c) compared the rates of xylem cell production in black spruce and other Canadian and European species and observed that the maximum occurs during the summer solstice, when day length, and not temperature, culminates. After 21 June, cell production gradually decreases until ceasing. These findings suggest that the influence of temperature on cambium depends on photoperiod, which is a factor independent of temperature and constant in time. Trees would have evolved by synchronizing their growth rates with day length, which represents a reliable natural calendar. The advantages of connecting cambial activity with photoperiod are explained by the need to avoid the period thermally unfavourable to growth. So, in boreal forests, increases in temperature can induce cell division only in certain periods of the year, in spring, when the conditions are and will probably remain favourable for completing the maturation of the newly produced cells.

There is evidence of threshold temperatures controlling the boundaries of the growing season in conifers of cold environments, characterized by an alternation of thermally favourable and unfavourable climate conditions during the year (Deslauriers et al. 2008; Swidrak et al. 2011). These values were demonstrated to be similar among all studied conifer species, across a wide geographical range of monitored sites and temperature regimes (Rossi et al. 2008b). How does the growing season change along a thermal gradient, and how does this change affect cell production? Xylem phenology and cell production were analysed along a latitudinal gradient, from the 48th to 53rd parallels, the range covering the whole closed black spruce forest in Quebec, Canada (Rossi et al. 2014). As logically expected, the southern and warmer site showed the highest radial growth, which corresponded to both the highest rates and longest durations of cell production. More surprisingly, the differences between the other sites in terms of xylem phenology and growth were marginal. Within the range analysed, the relationship between temperature and most phenological phases of xylogenesis was linear. On the contrary, temperature was related to cell production according to a threshold effect (Figure 35.5). Small increases in the periods of xylogenesis corresponded to substantial and disproportionate increases in cell production, approximately occurring at a May–September average temperature of 14°C (Figure 35.5). As a consequence, it was deduced that at the lower boundary of the distribution of black spruce, small environmental changes allowing marginal lengthening of the period of cell division could potentially lead to disproportionate increases in xylem cell production, with substantial consequences for the productivity of this boreal species.

The effect of extreme events was analysed by Deslauriers et al. (2008), who monitored xylem formation in *Pinus leucodermis* at the treeline in southern Italy during two contrasting years, 2003 and 2004. Spring 2003 was hot, with a temperature up to 2.6°C above historical means. Warming and prolonged drought affected several forest areas all over Europe and only

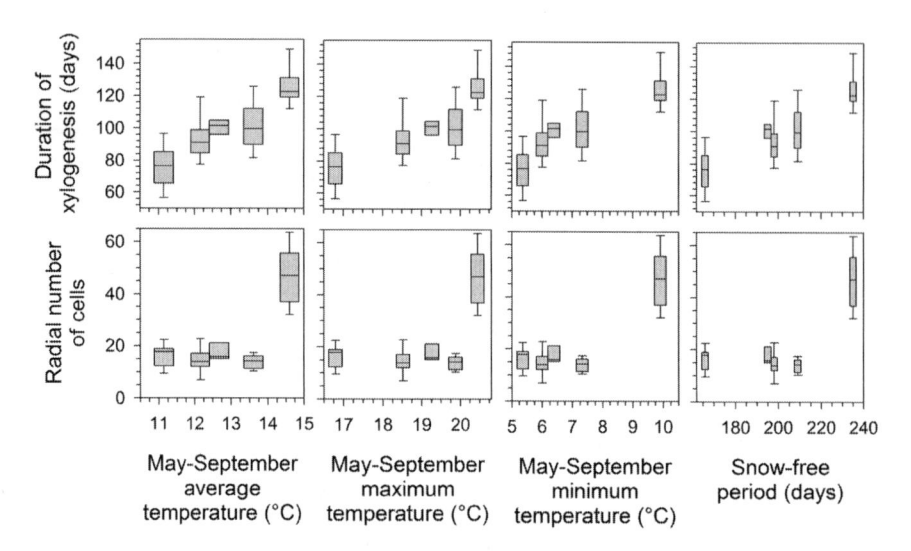

Figure 35.5 Duration of xylogenesis and amount of cell production versus the May–September temperatures and the snow-free period recorded in five study sites in the boreal forest of Quebec, Canada. Boxes represent upper and lower quartiles, whiskers achieve the 10th and 90th percentiles, and the median is drawn as a horizontal solid line (modified from Rossi et al. 2014)

trees growing above a certain altitude benefited. These conditions led to an earlier onset of about 20 days in cambial activity and all differentiation phases, resulting in a substantial increase in duration of xylogenesis of about 23 days.

Wood is the result of xylem cell production during a given period (duration) and with a given intensity (rate). Both factors, duration and rate of growth, play a role in xylem cell production of conifers, although the importance of these factors has been observed to change with species. In black spruce, the effect of the rate was marginal in respect to duration (Lupi et al. 2010; Rossi et al. 2014). However, contrasting results were observed in another species under different environmental conditions, i.e. silver fir growing in a French temperate forest (Rathgeber et al. 2011; Cuny et al. 2012). To date, the scarcity of information on the timings and mechanisms explaining the variability in cell production prevents the question being resolved of whether rate or duration is the most important factor in wood formation.

Manipulated natural environment

Some experiments were conducted by manipulating the growing conditions in natural environments in order to verify whether and how phenology and tree growth were affected. Such manipulations allow hypotheses to be tested on systems growing under natural conditions. However, intra-annual analyses are expensive and can be performed on just a few study trees. The wide variability in the responses observed among trees reduces the power of tests and modest differences can be interpreted as not statistically significant.

Plants of the boreal forest are generally considered to be limited by low temperatures and low nitrogen availability, both of these being projected to rise in the future due to climate warming and increased human activities. For three years, Lupi et al. (2012a, 2012b) increased soil temperature by 4°C during the first part of the growing season, and added precipitation containing three times the current nitrogen concentration by means of frequent canopy applications. Soil warming resulted in earlier onsets of xylogenesis and interacted with nitrogen addition, producing longer durations of xylogenesis in black spruce. The effect of warming was evidently more important for root phenology, while wood production, in terms of number of tracheids, was not significantly affected by the treatment. However, after three years of soil warming, the gap between treated and control trees increased substantially in both sites (Figure 35.6). This trend revealed the occurrence of long-term potential effects of soil temperature on nutrient cycles and, indirectly, on tree growth.

Belien et al. (2012) installed plastic under-canopy roofs on black spruce trees in four sites of the boreal forest of Quebec, Canada, thus excluding them from precipitation during the period June–September. The rain exclusion treatment significantly reduced the lumens of xylem cells, but no effect was observed in cell-wall thickness, cell production or xylem phenology. After removal of the under-canopy roofs, the treated trees and soil quickly recovered a water status similar to the control. The results obtained by Lupi et al. (2012a, 2012b) and Belien et al. (2012) demonstrated the resistance of this species and/or its environment to changes lasting one or a few growing seasons.

Greenhouse

The wide variability observed in tree growth, in terms of tree-ring width, and in the response of trees to their environment is expected to be partly related to site heterogeneity or to specific microclimates. Thus, some hypotheses need to be tested or validated under more controlled growing conditions. Experiments in greenhouses allow one or a few factors to be modified while maintaining all other factors constant among individuals. Plant selection can be more rigorous, allowing the use of plant material homogeneous in age and size. Nevertheless, the destructive sampling of wood phenology requires a very large number of individuals, thus even greenhouses of medium dimensions can only accommodate plants in the form of seedlings, or, at best, small saplings. Young plants exhibit higher resistance to stress and lower sensitivity to climatic factors than adult or old trees (Carrer and Urbinati 2004; Rossi et al. 2009a, 2009b), probably because of their higher metabolism, lower proportion of non-photosynthetic biomass, and greater capacity for enhancing photosynthesis (Bond 2000; Boege 2005; Mencuccini et al.

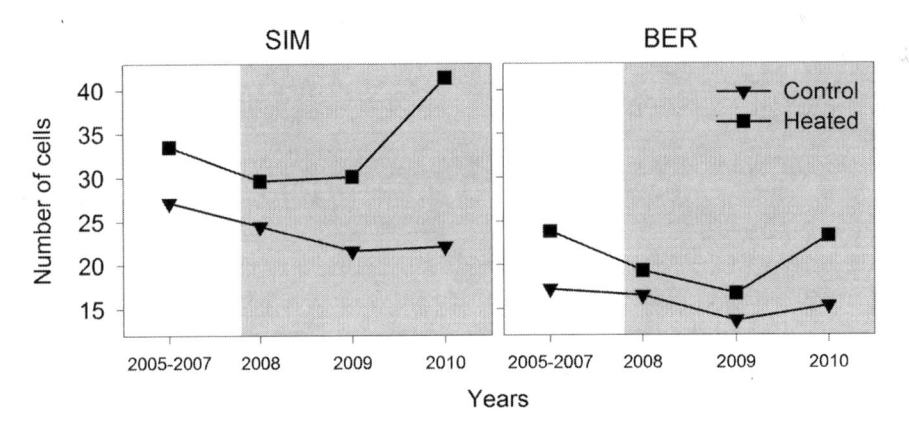

Figure 35.6 Number of cells produced before (white background) and during (grey background) the heating treatment in control and heated trees in two sites of the boreal forest of Quebec, Canada. The values represent LS means (modified from Lupi et al. 2012b)

2005). Thus, any direct application in the field of results from greenhouse experiments should be evaluated with caution.

Balducci *et al.* (2013) investigated cambial activity and wood formation during the May–October period in four-year-old saplings of black spruce subjected to both water deficit and warming. The saplings growing in greenhouses were submitted to temperatures 2 and 5°C higher than a control set at the same temperature as the external one. Saplings were also submitted to a water deficit of one month in late spring. Water stress produced a reduction or an interruption of cell division in cambium, which resulted in fewer cells in differentiation and thinner tree rings (i.e. tree rings with fewer cells along the radius) (Figure 35.7). Under warmer conditions, the recovery of non-irrigated saplings was slow and they needed 2–4 weeks to completely restore a cambial activity similar to the control. Despite a substantial mortality of between 5 and 12 per cent, the survival of black spruce saplings demonstrated an interesting resistance and plastic response to intense water deficit occurring during the period of maximum cambial activity and under substantial warming.

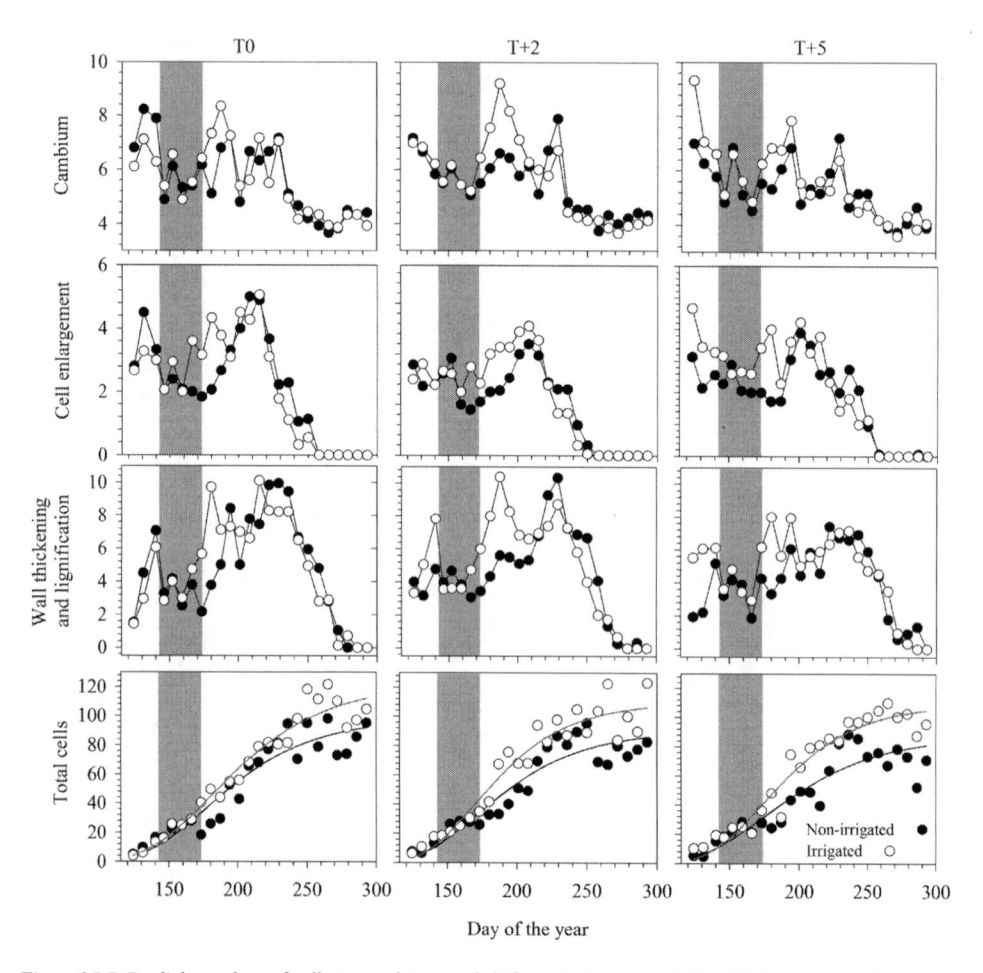

Figure 35.7 Radial number of cells in cambium and differentiation recorded in black spruce saplings before, during and after a water-deficit period (grey background) at three thermal conditions (T0, temperature equal to the external air temperature; T+2 and T+5, temperatures of 2 and 5°C higher than T0, respectively) (modified from Balducci *et al.* 2013)

Scales of analysis and interpretation

Forest productivity is based on wood production. An understanding of the mechanisms of xylem production and maturation and the environmental factors driving wood formation is essential for predicting how trees will respond to climate change in the future. It is obvious that intra-annual analyses of wood formation require that samplings, observations or measurements are performed when growth occurs. This constraint limits the temporal depth of the investigation, which is closely connected to the starting date of the monitoring. Generally, the cost of the analyses and short time span of research projects allow intra-annual sampling to be performed during a few growing seasons. Chronologies from the same site and/or the same trees are thus rarely longer than three years, and more frequently last just one or two years (Marion *et al.* 2007; Deslauriers *et al.* 2008; Rathgeber *et al.* 2011). On the contrary, dendrochronology is based on the detection of the environmental signals engraved on tree rings of living or dead trees, whose growth represents a reliable proxy of climate. As a consequence, chronologies lasting from hundreds to thousands of years can be compiled, as long as overlapped sequences of tree rings are found. The different length of the chronologies, longer at inter-annual scale and shorter at intra-annual scale, entails two contrasting depths of analysis and different interpretations of the results. The former chronologies can include large datasets and represent long-term responses of tree growth to climate, but fail to distinguish the effects of short-term events on tree rings. As also shown elsewhere in this chapter, dendrochronology is able to detect the inter-annual influences of climate, such as the effect of the previous year's temperature on current growth. The latter chronologies are mostly very short, but provide details of variation in wood production as a response to punctual events occurring during the growing season.

Several dendrochronological studies demonstrated that black spruce growth is affected by reductions in moisture due to drought events in the previous and current growing season (Huang *et al.* 2010 and other aforementioned papers) or by temperature-driven drought stress (Girardin *et al.* 2014). These results differ in part with those observed at short timescale, when the sensitivity to water stress was tested under controlled conditions or manipulated environments. A water stress occurring in late June, when cambial activity culminates, reduced the amount of cells produced by black spruce saplings in a greenhouse by approximately 20 per cent (Balducci *et al.* 2013). However, irrigation was completely withheld for 32 days, an unusually long drought period for Quebec. Three years of rain exclusion in the field did not affect cell production of black spruce, even in the warmer sites of its distribution (Belien *et al.* 2012, 2013). Results from inter- and intra-annual investigations are not necessarily in contrast: they are probably connected to, and explained by, the different temporal scales of the analysis. Forests are complex ecosystems whose persistence is related to a structured network among components and functions, allowing the organisms to attain optimal levels of resistance or resilience to natural disturbances. Short-term events or treatments applied for a few years have been demonstrated to produce only minor effects on xylem cell production (Houle and Moore 2008; Belien *et al.* 2012; Lupi *et al.* 2012a, 2012b). This is particularly true when a dry period occurs in sites with wet soils or superficial groundwater levels, where an episodic reduction in precipitation could be considered favourable for root respiration and development. On the contrary, in the long term, even small variations in environmental factors could lead to substantial modifications of the growing conditions and affect tree growth. This hypothesis will be better explained in the following paragraph with an example on the timings of snowmelt and their effect on tree growth.

The presence of snow is related to temperature, but timings of snowmelt have been indicated as a crucial moment affecting tree growth in boreal environments. Despite the marked increasing trend of temperatures detected at northern latitudes, a tree-ring and climate analysis in subarctic Eurasia showed a tendency towards reduced growth in conifers. The observed narrower tree rings were linked by Vaganov *et al.* (1999) to a significant increase in snowfall that would have delayed the date of complete snowmelt and, consequently, growth resumption in spring. Rossi *et al.* (2011) therefore tested the hypothesis that late snowmelts delay the onset of xylogenesis and reduce xylem production. The spatial and temporal variability of the climatic (snowfall) and biological (xylem phenology) chronologies was analysed taking into account both the long- and short-term variation. The highest cell productions were detected in the warmer sites, where the longest growth and snow-free periods were also observed. In the long term, the effects of snowmelt were significant for both duration and amount of growth. However, in the short term, the date of complete snowmelt did not affect cell production. The results obtained by Rossi *et al.* (2011) demonstrated the multi-scale influence of snowfall on tree growth and the determinant role of other factors (most probably nutrient cycling) in the productivity of boreal ecosystems. It is likely that a similar effect of scale is present with other environmental factors, which could explain the divergent responses of trees to temperature and water availability that have been observed using inter- and intra-annual analyses.

Most of the approaches described here are based on statistical tools, with results that are often empirical and specific to particular species or environments. In the last decade, there has been increasing interest in process-oriented approaches that develop explanatory models based on ecophysiological processes, and these should help the components of the complex environment–tree-growth system to be understood more deeply and exhaustively (Misson 2004; Vaganov *et al.* 2006). These innovative mechanistic approaches: (i) involve various spatial and time scales of analyses, from intra- to inter-annual; (ii) consider the main ecological factors that are supposed to affect growth; and (iii) are able to predict a number of growth parameters such as variations in bud phenology, tree-ring width, and xylem cell sizes.

Conclusions

Inter- and intra-annual studies of wood formation allow analysis of how wood production changes in time and according to the environmental factors by dividing the growth occurring during the overall lifespan of trees into its basic structural units, the tree ring and xylem cell. A long-term strategy of forest management requires knowledge of the natural development dynamics of tree species, which also includes the study of the mechanisms of growth and wood formation and the responses of trees to their environment at several temporal scales. In this sense, tree-based growth models built on the study of tree-ring formation and growth could be valuable decision-making tools in forestry and forest ecology, and for the timber industry. If and how tree growth, in terms of tree-ring production, adapts to a changing environment allows scenarios of forest productivity in the near future to be estimated using precise and reliable predictions. At cell level, information on the period and rate of lignification of woody tissues could be useful in tree crops for the production of wood pulp with low lignin content for the paper industry, or for the study of the origin of internal tensions in logs, which cause deformation and twisting of timber during seasoning (Donaldson 1991; Plomion *et al.* 2001). Climate–growth relationships have opened up interesting fields of exploration and discussion in predicting how a forest, and its components, the trees, are responding or will evolve under current, or potentially different, growing conditions. Ecologists and physiologists are now called upon to face a new challenge, which requires the results of climate–growth relationships from the different scales of analysis to be combined in a convincing, conclusive and inclusive final interpretation of tree growth.

References

Anfodillo, T., Deslauriers, A., Menardi, R., Tedoldi, L., Petit, G. and Rossi, S. (2012) 'Widening of xylem conduits in a conifer tree depends on the longer time of cell expansion downwards along the stem', *Journal of Experimental Botany*, vol 63, pp. 837–845

Antonova, G. F. and Stasova, V. V. (1997) 'Effects of environmental factors on wood formation in larch (*Larix sibirica* Ldb.) stems', *Trees*, vol 11, pp. 462–468

Balducci, L., Deslauriers, A., Giovannelli, A., Rossi, S. and Rathgeber, C. B. K. (2013) 'Effects of temperature and water deficit on cambial activity and woody ring features in *Picea mariana* saplings', *Tree Physiology*, vol 33, pp. 1006–1017

Bäucker, E., Bues, C. T. and Vogel, M. (1998) 'Radial growth dynamics of spruce (*Picea abies*) measured by micro-cores', *IAWA Journal*, vol 19, pp. 301–309

Belien, E., Rossi, S., Morin, H. and Deslauriers, A. (2012) 'Xylogenesis in black spruce subjected to a rain exclusion in the field', *Canadian Journal of Forest Research*, vol 42, pp. 1306–1315

Belien, E., Morin, H., Rossi, S. and Deslauriers, A. (2013) '*Wood anatomy of mature black spruce subjected to repeated artificial drought*', Poster presented to the International Symposium on Wood Structure in Plant Biology and Ecology WSE, 17–20 April, Napoli, Italy

Blais, R. (1983) 'Trends in the frequency, extent, and severity of spruce budworm outbreaks in eastern Canada', *Canadian Journal of Forest Research*, vol 13, pp. 539–547

Boege, K. (2005) 'Influence of plant ontogeny on compensation to leaf damage', *American Journal of Botany*, vol 92, pp. 1632–1640

Bond, B. J. (2000) 'Age-related changes in photosynthesis of woody plants', *Trends in Plant Science*, vol 5, pp. 349–355

Boulanger, Y., Arseneault, D., Morin, H., Jardon, Y., Bertrand, P. and Dagneau, C. (2012) 'Dendrochronological reconstruction of spruce budworm (*Choristoneura fumiferana* Clem.) outbreaks in southern Québec for the last 400 years', *Canadian Journal of Forest Research*, vol 42, pp. 1264–1276

Büntgen, U., Frank, D., Liebhold, A., Johnson, D., Carrer, M. and others (2009) 'Three centuries of insect outbreaks across the European Alps', *New Phytologist*, vol 182, pp. 929–941

Camarero, J. J., Guerrero-Campo, J. and Gutiérrez, E. (1998) 'Tree-ring growth and structure of *Pinus uncinata* and *Pinus sylvestris* in the Central Spanish Pyrenees', *Arctic and Alpine Research*, vol 30, pp. 1–10

Carrer, M. and Urbinati, C. (2004) 'Age-dependent tree-ring growth responses to climate in *Larix decidua* and *Pinus cembra*', *Ecology*, vol 85, pp. 730–740

Cook, E. R. (1987) 'The decomposition of tree-ring series for environmental studies', *Tree-Ring Bulletin*, vol 47, pp. 37–59

Cudmore, T. J., Björklund, N., Carroll, A. L. and Lindgren, B. S. (2010) 'Climate change and range expansion of an aggressive bark beetle: evidence of higher reproductive success in naïve host tree populations', *Journal of Applied Ecology*, vol 47, pp. 1036–1043

Cuny, H. E., Rathgeber, C. B. K., Lebourgeois, F., Fortin, M. and Fournier, M. (2012) 'Life strategies in intra-annual dynamics of wood formation: example of three conifer species in a temperate forest in north-east France', *Tree Physiology*, vol 32, pp. 612–625

Dang, Q. L. and Lieffers, V. J. (1989) 'Climate and annual ring growth of black spruce in some Alberta peatlands', *Canadian Journal of Botany*, vol 67, pp. 1885–1889

Denne, M. P. (1974) 'Effects of light intensity on tracheid dimensions in *Picea sitchensis*', *Annals of Botany*, vol 38, pp. 337–345

Deslauriers, A. and Morin, H. (2005) 'Intra-annual tracheid production in balsam fir stems and the effect of meteorological variables', *Trees*, vol 19, pp. 402–408

Deslauriers, A., Morin, H. and Begin, Y. (2003) 'Cellular phenology of annual ring formation of *Abies balsamea* in the Quebec boreal forest (Canada)', *Canadian Journal of Forest Research*, vol 33, pp. 190–200

Deslauriers, A., Rossi, S., Anfodillo, T. and Saracino, A. (2008) 'Cambium phenology, wood formation and temperature thresholds in two contrasting years at high altitude in Southern Italy', *Tree Physiology*, vol 28, pp. 863–871

Deslauriers, A., Giovannelli, A., Rossi, S., Castro, G., Fragnelli, G. and Traversi, L. (2009) 'Intra-annual cambial activity and carbon availability in stem of poplar', *Tree Physiology*, vol 29, pp. 1223–1235

Donaldson, L. A. (1991) 'Seasonal changes in lignin distribution during tracheid development in *Pinus radiata* D. Don', *Wood Science and Technology*, vol 25, pp. 15–24

Drobyshev, I., Gewehr, S., Berninger, F. and Bergeron, Y. (2013) 'Species specific growth responses of black spruce and trembling aspen may enhance resilience of boreal forest to climate change', *Journal of Ecology*, vol 101, pp. 231–242

Esper, J., Büntgen, U., Frank, D. C., Nievergelt, D. and Liebhold, A. (2007) '1200 years of regular outbreaks in alpine insects', *Proceedings of the Royal Society of London, Series B: Biological Sciences,* vol 274, pp. 671–679

Forster, T., Schweingruber, F. H. and Denneler, B. (2000) 'Increment puncher: a tool for extracting small cores of wood and bark from living trees', *IAWA Journal,* vol 21, pp. 169–180

Fritts, H. C. (1976) *Tree-rings and Climate,* Academic Press, New York

Gindl, W., Grabner, M. and Wimmer, R. (2000) 'The influence of temperature on latewood lignin content in treeline Norway spruce compared with maximum density and ring width', *Trees,* vol 14, pp. 409–414

Girardin, M. P. and Wotton, B. M. (2009) 'Summer moisture and wildfire risks across Canada', *Journal of Applied Meteorology and Climatology,* vol 48, pp. 517–533

Girardin, M. P., Jing Guo, X., De Jong, R., Kinnard, C., Bernier, P. and Raulier, F. (2014) 'Unusual forest growth decline in boreal North America covaries with the retreat of Arctic sea ice', *Global Change Biology,* vol 20, pp. 851–866

Gričar, J., Čufar, K., Oven, P. and Schmitt, U. (2005) 'Differentiation of terminal latewood tracheids in silver fir trees during autumn', *Annals of Botany,* vol 95, pp. 959–965

Gričar, J., Zupancic, M., Čufar, K., Koch, G., Schmitt, U. and Oven, P. (2006) 'Effect of local heating and cooling on cambial activity and cell differentiation in the stem of Norway spruce (*Picea abies*)', *Annals of Botany,* vol 97, pp. 943–951

Gruber, A., Strobl, S., Veit, B. and Oberhuber, W. (2010) 'Impact of drought on the temporal dynamics of wood formation in *Pinus sylvestris*', *Tree Physiology,* vol 30, pp. 490–501

Hofgaard, A., Tardif, J. and Bergeron, Y. (1999) 'Dendroclimatic response of *Picea mariana* and *Pinus banksiana* along a latitudinal gradient in the eastern Canadian boreal forest', *Canadian Journal of Forest Research,* vol 29, pp. 1333–1346

Horacek, P., Slezingerova, J. and Gandelova, L. (1999) 'Effects of environment on the xylogenesis of Norway spruce (*Picea abies* (L.) Karst.)', in R. Wimmer and R. Vetter (eds) *Tree-Ring Analysis: Biological, Methodological and Environmental Aspects,* CABI, Wallingford, UK

Houle, D. and Moore, J. D. (2008) 'Soil solution, foliar concentrations and tree growth response to three-year of ammonium-nitrate addition in two boreal forests of Québec, Canada', *Forest Ecology and Management,* vol 255, pp. 2049–2060

Huang, J.-G., Tardif, J., Bergeron, Y., Denneler, B., Berninger, F. and Girardin, M. (2010) 'Radial growth response of four dominant boreal tree species to climate along a latitudinal gradient in the eastern Canadian boreal forest', *Global Change Biology,* vol 16, pp. 711–731

Huang, J.-G., Bergeron, Y., Zhai, L. H. and Denneler, B. (2011) 'Variation in intra-annual radial growth (xylem formation) of *Picea mariana* (Pinaceae) along a latitudinal gradient in western Quebec, Canada', *American Journal of Botany,* vol 98, pp. 792–800

Huang, J.-G., Bergeron, Y., Berninger, F., Zhai, L. H., Tardif, J. and Denneler, B. (2013) 'Impact of future climate on radial growth of four major boreal tree species in the eastern Canadian boreal forest', *PLoS ONE,* vol 8, p. e56758

IPCC (2013) *Climate Change 2013: The Physical Science Basis. Contribution of Working Group I to the Fifth Assessment Report of the Intergovernmental Panel on Climate Change,* Cambridge University Press, Cambridge, UK

Jardon, Y. and Morin, H. (2003) 'Périodicité des épidémies de la tordeuse des bourgeons de l'épinette au cours des deux derniers siècle', *Canadian Journal of Forest Research,* vol 33, pp. 1947–1961

Jones, B., Tardif, J. and Westwood, R. (2004) 'Weekly xylem production in trembling aspen (*Populus tremuloides*) in response to artificial defoliation', *Canadian Journal of Botany,* vol 82, pp. 590–597

Knudson, L. (1913) 'Observations on the inception, season, and duration of cambium development in the American larch [*Larix laricina* (Du Roi) Koch.]', *Bulletin of the Torrey Botanical Club,* vol 40, pp. 271–293

Kozlowski, T. T., Kramer, P. J. and Pallardy, S. G. (1991) *The Physiological Ecology of Woody Plants,* Academic Press, San Diego, US

Lapointe, P. (2013) 'Épidémies de la tordeuse des bourgeons de l'épinette au nord du Lac Saint-Jean à travers les arbres subfossiles', Master's thesis, Université du Québec à Chicoutimi, Chicoutimi, Canada

Levasseur, V. (2000) 'Analyse dendroécologique de l'impact de la tordeuse des bourgeons de l'épinette (*Choristoneura fumiferana*) suivant un gradient latitudinal en zone boréale au Québec'. Master's thesis, Université du Québec à Chicoutimi, Chicoutimi, Canada

Lupi, C., Morin, H., Deslauriers, A. and Rossi, S. (2010) 'Xylem phenology and wood production: resolving the chicken-or-egg dilemma', *Plant Cell Environment*, vol 33, pp. 1721–1730

Lupi, C., Morin, H., Deslauriers, A., Rossi, S. and Houle, D. (2012a) 'Increasing nitrogen availability and soil temperature: effects on xylem phenology and anatomy of mature black spruce', *Canadian Journal of Forest Research*, vol 42, pp. 1277–1288

Lupi, C., Morin, H., Deslauriers, A. and Rossi, S. (2012b) 'Xylogenesis in black spruce: does soil temperature matter?', *Tree Physiology*, vol 32, pp. 74–82

Mäkinen, H., Nöjd, P. and Saranpää, P. (2003) 'Seasonal changes in stem radius and production of new tracheids in Norway spruce', *Tree Physiology*, vol 23, pp. 959–968

Marion, L., Gričar, J. and Oven, P. (2007) 'Wood formation in urban Norway maple trees studied by the micro-coring method', *Dendrochronologia*, vol 25, pp. 97–102

Mencuccini, M., Martinez-Vilalta, J., Vanderklein, D., Hamid, H. A., Korakaki, E., Lee, S. and Michiels, B. (2005) 'Size-mediated ageing reduces vigour in trees', *Ecology Letters*, vol 8, pp. 1183–1190

Misson, L. (2004) 'MAIDEN: a model for analyzing ecosystem processes in dendroecology', *Canadian Journal of Forest Research*, vol 34, pp. 874–887

Morin, H., Laprise, D., Simard, A. A. and Amouch, S. (2008) 'Le régime des épidémies de la TBE', in S. Gauthier, M.-A. Vaillancourt, A. Leduc, L. De Grandpré, D. Kneeshaw and others (eds) *L'aménagement écosystémique en forêt boréale*, Les presses de l'Université du Québec, Québec, Canada

Morin, H., Jardon, Y., Simard, S. and Simard, I. (2010) 'Détection et reconstitution des épidémies de la tordeuse des bourgeons de l'épinette (*Choristoneura fumiferana* Clem.) à l'aide de la dendrochronologie', in S. Payette and L. Filion (eds) *La dendroécologie - Principes, methodes et applications*, Les Presses de l'Université Laval, Québec, Canada

Moser, L., Fonti, P., Büntgen, U., Franzen, J., Esper, J. and others (2010) 'Timing and duration of European larch growing season along altitudinal gradients in the Swiss Alps', *Tree Physiology*, vol 30, pp. 225–233

Nishimura, P. H. and Laroque, C. P. (2011) 'Observed continentality in radial growth-climate relationships in a twelve site network in western Labrador, Canada', *Dendrochronologia*, vol 29, pp. 17–23

Nobuchi, T., Ogata, Y. and Siripatanadilok, S. (1995) 'Seasonal characteristics of wood formation in *Hopea odorata* and *Shorea henryana*', *IAWA Journal*, vol 16, pp. 361–369

Novak, K., Saz Sánchez, M. A., Čufar, K., Raventós, J. and de Luis, M. (2013) 'Age, climate and intra-annual density fluctuations in *Pinus halepensis* in Spain', *IAWA Journal*, vol 34, pp. 459–474

O'Neill, G. A. and Nigh, G. (2011) 'Linking population genetics and tree height growth models to predict impacts of climate change on forest production', *Global Change Biology*, vol 17, pp. 3208–3217

Oribe, Y., Funada, R., Shibagaki, M. and Kubo, T. (2001) 'Cambial reactivation in locally heated stems of the evergreen conifer *Abies sachalinensis* (Schmidt) Masters', *Planta*, vol 212, pp. 684–691

Peng, C., Ma, Z., Lei, X., Zhu, Q., Chen, H. and others (2011) 'A drought-induced pervasive increase in tree mortality across Canada's boreal forests', *Nature Climate Change*, vol 1, 467–471

Plomion, C., Leprovost, G. and Stokes, A. (2001) 'Wood formation in trees', *Plant Physiology*, vol 127, pp. 1513–1523

Rathgeber, C. B. K., Rossi, S. and Bontemps, J.-D. (2011) 'Tree size influences cambial activity in a mature silver fir plantation', *Annals of Botany*, vol 108, pp. 429–438

Régnière, J., St-Amant, R. and Duval, P. (2012) 'Predicting insect distributions under climate change from physiological responses: spruce budworm as an example', *Biological Invasions*, vol 14, pp. 1571–1586

Rossi, S., Deslauriers, A. and Morin, H. (2003) 'Application of the Gompertz equation for the study of xylem cell development', *Dendrochronologia*, vol 21, pp. 1–7

Rossi, S., Anfodillo, T. and Menardi, R. (2006a) 'Trephor: a new tool for sampling microcores from tree stems', *IAWA Journal*, vol 27, pp. 89–97

Rossi, S., Deslauriers, A. and Anfodillo, T. (2006b) 'Assessment of cambial activity and xylogenesis by microsampling tree species: an example at the Alpine timberline', *IAWA Journal*, vol 27, pp. 383–394

Rossi, S., Deslauriers, A., Anfodillo, T., Morin, H., Saracino, A. and others (2006c) 'Conifers in cold environments synchronize maximum growth rate of tree-ring formation with day length', *New Phytologist*, vol 170, pp. 301–310

Rossi, S., Deslauriers, A., Anfodillo, T. and Carrer, M. (2008a) 'Age-dependent xylogenesis in timberline conifers', *New Phytologist*, vol 177, pp. 199–208

Rossi, S., Deslauriers, A., Gričar, J., Seo, J.-W., Rathgeber, C. B. K. and others (2008b) 'Critical temperatures for xylogenesis in conifers of cold climates', *Global Ecology and Biogeography*, vol 17, pp. 696–707

Rossi, S., Simard, S., Deslauriers, A. and Morin, H. (2009a) 'Wood formation in *Abies balsamea* seedlings subjected to artificial defoliation', *Tree Physiology*, vol 29, pp. 551–558

Rossi, S., Simard, S., Rathgeber, C. B. K., Deslauriers, A. and De Zan, C. (2009b) 'Effects of a 20-day-long dry period on cambial and apical meristem growth in *Abies balsamea* seedlings', *Trees*, vol 23, pp. 85–93

Rossi, S., Morin, H. and Deslauriers, A. (2011) 'Multi-scale influence of snowmelt on xylogenesis of black spruce', *Arctic, Antarctic, and Alpine Research*, vol 43, pp. 457–464

Rossi, S., Morin, H. and Deslauriers, A. (2012) 'Causes and correlations in cambium phenology: towards an integrated framework of xylogenesis', *Journal of Experimental Botany*, vol 63, pp. 2117–2126

Rossi, S., Girard, M.-J. and Morin, H. (2014) 'Lengthening of the duration of xylogenesis engenders disproportionate increases in xylem production', *Global Change Biology*, vol 20, pp. 2261–2271

Savidge, R. A. (2000) 'Biochemistry of seasonal cambial growth', in R. Savidge, J. Barnett and R. Napier (eds) *Cell and Molecular Biology of Wood Formation*, BIOS Scientific Publishers Ltd, Oxford, UK

Schmitt, U., Jalkanen, R. and Eckstein, D. (2004) 'Cambium dynamics of *Pinus sylvestris* and *Betula* spp. in the northern boreal forest in Finland', *Silva Fennica*, vol 38, pp. 167–178

Schmitz, N., Verheyden, A., Kairo, J. G., Beeckman, H. and Koedam, N. (2007) 'Successive cambia development in *Avicennia marina* (Forssk.) Vierh. is not climatically driven in the seasonal climate at Gazi Bay, Kenya', *Dendrochronologia*, vol 25, pp. 87–96

Schweingruber, F. H. (1996) *Tree-rings and Environment: Dendroecology*, Swiss Federal Institute for Forest, Snow and Landscape Research, WSL/FNP, Brimensdorf, Switzerland

Selås, V., Piovesan, G., Adams, J. M. and Bernabei, M. (2002) 'Climatic factors controlling reproduction and growth of Norway spruce in southern Norway', *Canadian Journal of Forest Research*, vol 32, pp. 217–225

Seo, J.-W., Eckstein, D. and Schmitt, U. (2007) 'The pinning method: from pinning to data preparation', *Dendrochronologia*, vol 25, pp. 79–86

Seo, J.-W., Eckstein, D., Jalkanen, R., Rickebusch, S. and Schmitt, U. (2008) 'Estimating the onset of cambial activity in Scots pine in northern Finland by means of the heat-sum approach', *Tree Physiology*, vol 28, pp. 105–112

Simard, I., Morin, H. and Lavoie, C. (2006) 'A millennial-scale reconstruction of spruce budworm abundance in Saguenay, Québec, Canada', *The Holocene*, vol 16, pp. 31–37

Simard, S., Morin, H. and Krause, C. (2011) 'Long-term spruce budworm outbreak dynamics reconstructed from subfossil trees', *Journal of Quaternary Science*, vol 26, pp. 734–738

Swetnam, T. W., Thompson, M. A. and Sutherland, E. K. (1985) *Using Dendrochronology to Measure Radial Growth of Defoliated Trees*, US Department of Agriculture, Forest Service, Washington

Swidrak, I., Gruber, A., Kofler, W. and Oberhuber, W. (2011) 'Effects of environmental conditions on onset of xylem growth in *Pinus sylvestris* under drought', *Tree Physiology*, vol 31, pp. 483–493

Tardif, J., Brisson, J. and Bergeron, Y. (2001) 'Dendroclimatic analysis of *Acer saccharum*, *Fagus grandifolia*, and *Tsuga canadensis* from an old-growth forest, southwestern Quebec', *Canadian Journal of Forest Research*, vol 31, pp. 1491–1501

Vaganov, E. A., Hughes, M. K., Kirdyanov, A. V., Schweingruber, F. H. and Silkin, P. P. (1999) 'Influence of snowfall and melt timing on tree growth in subarctic Eurasia', *Nature*, vol 400, pp. 149–151

Vaganov, E. A., Hughes, M. K. and Shashkin, A. V. (2006) *Growth Dynamics of Conifer Tree Rings*, Springer-Verlag, Berlin, Heidelberg, Germany

van der Werf, G. W., Sass-Klaassen, U. G. W. and Mohren, G. M. J. (2007) 'The impact of the 2003 summer drought on the intra-annual growth pattern of beech (*Fagus sylvatica* L.) and oak (*Quercus robur* L.) on a dry site in the Netherlands', *Dendrochronologia*, vol 25, pp. 103–112

Vieira, J., Rossi, S., Campelo, F., Freitas, H. and Nabais, C. (2014a) 'Xylogenesis of *Pinus pinaster* under a Mediterranean climate', *Annals of Forest Science*, vol 71, pp. 71–80

Vieira, J., Rossi, S., Campelo, F. and Nabais, C. (2014b) 'Are neighbouring trees in tune? Wood formation in *Pinus pinaster*', *European Journal of Forest Research*, vol 133, pp. 41–50

Wodzicki, T. J. (1971) 'Mechanism of xylem differentiation in *Pinus silvestris* L.', *Journal of Experimental Botany*, vol 22, pp. 670–687

Wolter, E. K. (1968) 'A new method for marking xylem growth', *Forest Science*, vol 14, pp. 102–104

Zhai, L. H., Bergeron, Y., Huang, J.-G. and Berninger, F. (2012) 'Variation in intra-annual wood formation, and foliage and shoot development of three major Canadian boreal species', *American Journal of Botany*, vol 99, pp. 827–837

36
PLANT MOVEMENTS IN RESPONSE TO RAPID CLIMATE CHANGE

Richard T. Corlett

Introduction

The Fifth Assessment Report of the Intergovernmental Panel on Climate Change (IPCC, 2013) makes grim reading. Atmospheric concentrations of CO_2 have increased from 278 ppm in 1750, before the industrial revolution, to 400 ppm in 2013: an increase of 43 percent. The other major greenhouse gases, CH_4 and N_2O, have risen by more than 150 percent and by 20 percent, respectively. Present-day concentrations of these greenhouse gases are higher than at any time in the last 800,000 years. As a consequence, Earth has been getting warmer: in most regions, 0.5–1.0°C warmer over the last hundred years. Although the rate of warming slowed in the first decade of the twenty-first century (Fyfe and Gillett, 2014), each of the past three decades has been warmer than the previous one and also warmer than all previous decades since records began. Earth has been warmer in the past, but in most regions the last 30 years seems to have been the warmest for at least 800 years and probably longer. Other climatic variables are also changing, but trends in rainfall are less clear and more varied.

The global climate models agree fairly well with each other for the next 20 years or so, with a predicted additional increase of 0.3–0.7°C in global mean surface air temperature, giving a total warming of 1.0–1.5°C (somewhat higher over land) (IPCC, 2013). Globally, rainfall is expected to increase, but some regions (such as most of the Mediterranean) will get drier and confidence in detailed rainfall predictions in most parts of the world is low. In the longer term, predictions vary greatly even for temperatures, depending on the choice of climate model and the assumptions made about future greenhouse gas emissions and carbon-cycle feedbacks. Temperatures over land are expected to increase by 3–6°C by 2100, compared with 2000, but the range of plausible values is considerably wider.

There is abundant evidence from a variety of sources that the rising temperatures in recent decades, as well as changes in other climatic variables, are already impacting wild species and ecosystems (IPCC, 2014). Plant responses to climate change are of particular interest to climate modellers because of the known and potential feedbacks between vegetation—particularly forests—and climate. The global significance of the carbon fluxes through forests is shown by the annual decline in global atmospheric CO_2 concentrations (of 3–9 ppm) at the late-summer peak of photosynthesis in the northern hemisphere, where there is much more forest than in the southern hemisphere. Forests, particularly tropical forests, are also the major control on the interannual variability in the global carbon dioxide balance, with the high temperatures and droughts associated with strong El Niño events leading to large increases in the atmospheric CO_2 growth rate (Wang et al., 2014). Moreover, in the last few decades global forests have absorbed as much as 30 percent of annual global anthropogenic CO_2 emissions, partly in forests recovering from past disturbance but also in mature forests (Bellassen and Luyssaert, 2014). The long-term stability of this crucial sink depends on how plants respond to climate change. Plant responses will also have a major influence on the persistence of animals, particularly the hyperdiverse and relatively plant-dependent invertebrates (Stork and Habel, 2014).

In the face of climate change, plant species must either acclimate (plastic change within the lifetime of an individual), adapt (genetic change over > 1 generation), or move so as to stay within their climate envelope (Corlett and Westcott, 2013). Evidence for genetic change is limited so far, but the widely reported changes in plant phenology (IPCC, 2014) represent a plastic response that appears in many cases to be adaptive (e.g. Cleland et al., 2012). Other plastic responses are less well understood, but studies on tree species from boreal to tropical forests suggest that significant thermal acclimation of leaf respiration can occur within a few days, reducing concerns that warming will reduce the capacity of forests to fix and store carbon (Slot et al., 2014). Movement is more difficult for plants than most animals because dispersed seeds must grow to reproductive maturity before a new generation of seeds can move again, and most plants take much longer to mature than most animals. Movement potential for plants is considered in more detail below.

The paleoecological record

Movement was a near universal response of plant populations to climate change in the paleoecological record (Harrison and Goñi, 2010; Pardi and Smith, 2012). Species did not necessarily track climate change closely, however, and there is evidence that many plant species in Europe are not in equilibrium with the current climate as a result of post-glacial migration lag (Normand et al., 2011). On the other hand, only one plant species, *Picea critchfieldii*, which was previously widespread in eastern North America, is known to have become globally extinct from climate change during the Late Pleistocene (Jackson and Weng, 1999). Overall, the paleoecological record strongly suggests that plant species are more likely to move than acclimate or adapt, even in response to gradual climate change (Wiens et al., 2010).

The need for speed

Although climates have changed in the past and have sometimes been warmer than today, there are good reasons to think that twenty-first century climate change will be a problem. Evidence from the paleoecological record suggests that the *rates* of warming and other climate changes over the twenty-first century will be exceptional, since changes in well-studied periods in the past were 10–100 times slower (Wing and Currano, 2013). A useful concept is the 'velocity of climate change', which is the speed and direction that something needs to move to keep within its current 'climate envelope', i.e. to stay within the range of climate conditions it can tolerate. Velocities of temperature change are expected to exceed 1 km/year over much of Earth's land surface during the twenty-first century (Loarie et al., 2009; Diffenbaugh and Field, 2013), although steep temperature gradients will result in much lower velocities in areas of rugged topography (< 10 m/year on the steepest slopes). Velocities of rainfall change are projected to be similar. Few, if any, plant species have their distributions determined by only one or two climatic factors, such as mean annual temperature and rainfall. Instead, plant species respond individualistically to multivariate climate change. Species distribution models (see below) can be used to estimate species-specific 'bioclimatic velocities' (Serra-Diaz et al., 2014). These different approaches, however, have resulted in similar estimates of the velocities needed for organisms to track twenty-first century climate change (e.g. Cunze et al., 2013).

For comparison, the estimated global mean velocity of temperature change between the last glacial maximum and the present day was only 5.9 m year^{-1} (Sandel et al., 2011), although more detailed studies have shown that there were short periods with much higher velocities (400–700 m year^{-1}), between 20,000–10,000 years ago (Correa-Metrio et al., 2013). Moreover, the resolution of the paleoecological record is insufficient to rule out even more rapid changes on a century scale. However, these past changes differ from those projected for the twenty-first century in another important way: the extremes were maintained for relatively short periods, before swinging back. This makes it more likely that plant populations could persist through periods of unfavourable climate as suppressed, non-reproductive, individuals, whereas the forecast monotonic trends for the twenty-first century and beyond are likely to favour extinction (Correa-Metrio et al., 2013).

The velocity of climate change is a measure of the exposure of a plant population to climate change, not necessarily the impact. Current ranges are not necessarily limited by climate and a species restricted by non-climatic factors well within its potential climate envelope may not need to move. Moreover, as discussed above, the capacity of plant populations for acclimation and adaptation are still poorly understood. However, the plant responses exhibited in the paleoecological record suggest that climate niches, in general, are highly conservative (Wiens et al., 2010).

The potential for plant movement

Plants are immobile as adults, but plant populations can move if seeds are dispersed and establish beyond the current range at the leading edge, while the plants at the trailing edge fail to regenerate. Most studies have focused on the dispersal phase of the colonization process, but for movement to occur the dispersed propagules must then establish in conditions that are likely to be increasingly dissimilar to the home environment the further a population moves outside its original range (De Frenne et al., 2014). Except at the treeline, forest plants will normally be moving into established forests, which may have been weakened by climate change, but still have the advantages of numbers and local adaptation to non-climatic factors (Bai et al., 2011; Zhu et al., 2012). Currently, the role of establishment in promoting or reducing plant movements is poorly understood, but negative impacts are most likely for slow-growing, mature-phase forest taxa moving into existing forests, while pioneer and open-country species are less likely to suffer problems (Corlett and Westcott, 2013).

There have been major advances in our understanding of the mechanisms of seed dispersal over the last decade or so, both for wind-dispersed taxa (e.g. Trakhtenbrot et al., 2014) and for dispersal inside the guts of frugivorous animals (e.g. Dennis and Westcott, 2006; Morales et al., 2013). However, while this mechanistic understanding can help predict median dispersal distances for various combinations of plants and vectors, we are much further from being able to predict the rare long-distance dispersal events—the 99th percentile dispersal distances—which models suggest will have a major influence on plant movement velocities (Nathan et al., 2011). We can therefore say with some confidence that most seeds will be dispersed < 1,500 m from the parent plant (Corlett and Westcott, 2013), with the upper end of this range occurring mostly outside forests, but cannot rule out a significant long-distance 'tail' to the right of this distribution. Routine dispersal distances > 1,500 m are most common in species that have small, wind-dispersed seeds, such as epiphytic orchids, and those moved by the largest birds, bats and terrestrial mammals. At the other extreme, ants are the commonest dispersal agents for herb species in temperate deciduous forests, resulting in maximum dispersal distances < 5 m (Whigham, 2004; Kelemen et al., 2014). In between these extremes are the largely rodent-dispersed Fagaceae, which dominate many forests in the northern subtropics and temperate zone and are typically dispersed tens of metres at most.

Forests have been reduced and fragmented worldwide. In general, this is likely to greatly reduce dispersal distances, particularly for plants with vectors that cannot move between fragments (Corlett and Westcott, 2013). However, the movement of seeds within fragmented forest landscapes is currently a major research gap, so it is unwise to generalize too widely (McConkey et al., 2012). Restoring connectivity across such landscapes is one of the most frequent suggestions for climate change adaptation, but again the effectiveness of various suggested options—from creating forest corridors to ameliorating conditions in the non-forest matrix—is unclear and needs experimental study.

In tropical forests, most seeds are adapted for dispersal by vertebrate frugivores and all the larger frugivore species have declined—often to local extinction—as a result of hunting (McConkey et al., 2012). In contrast, wind and other abiotic dispersal agents are still present, as are large numbers of small, disturbance-tolerant, frugivorous birds and bats. The result of this size-biased defaunation has been a selective loss of dispersal services for large, large-seeded fleshy fruits across huge areas of the tropics (Markl et al., 2012; McConkey et al., 2012). The consequences for plant movements in response to climate change are clear: plant species that depended on large frugivores for dispersal—at least for long-distance dispersal—will be far less mobile in the future than they were in the past.

Converting dispersal distances into potential velocities also requires knowledge of the number of dispersal events in the time period under consideration, which depends on the time from dispersed seed to reproductive adult. Assuming most plant species are dispersing 50–1,500 m every 1–30 years, the range of plant velocities in unfragmented habitats with no barriers to movement would be 1.7–1,500 m/year (Corlett and Westcott, 2013). However, the longer times to plant maturity in forests and the positive relationship between dispersal distance and plant height (Thomson et al., 2011) suggest that most forest plant species will be capable of moving less than 1,000 m per year and many less than 100 m. Projected velocities of climate change (often > 1 km/year) and potential velocities of plant movement (< 1.5 km/year) broadly overlap, but most plant velocities are likely to be at the lower end of the range and will be exceeded by climate velocities except in steep topography.

Recent movements

For animals, there are numerous reports of both poleward and upward movements in response to recent warming (Parmesan and Yohe, 2003; Chen et al., 2011; Freeman and Freeman, 2014; IPCC, 2014). For plants, altitudinal range shifts have been reported across a range of environments from the temperate zone to the tropics (Kelly and Goulden, 2008; Lenoir et al., 2008; Jump et al., 2009; Feeley et al., 2011, 2013; Telwala et al., 2013), but evidence for large-scale poleward movements in response to recent warming is weak (Groom, 2013; Zhu et al., 2014), although some individual species appear to be moving (e.g. Delzon et al., 2013). It is possible that the warming experienced so far has simply not been enough to require movement by most species, but it is also likely that the barriers to latitudinal movements over hundreds of kilometres are usually much greater than those for short-distance movements upslope.

Modelling movements of species and communities

Species distribution models (SDMs) are currently the most widely used method to assess the potential impacts of climate change on the distributions of individual plant species (Bateman *et al.*, 2013). SDMs typically correlate current spatial distributions with current climatic and other (soil, topography etc.) variables and then predict the future distribution from the future climate. Such correlative models are easy to run and are generally fairly accurate in reproducing the current distributions of plants. However, they have a number of limitations when used for prediction, including the inherent dangers of extrapolation from correlations and their inability to deal with novel future climates (or novel combinations of climate and other factors). However, they can still be useful as a way of determining the exposure of a species to climate change. More sophisticated, process-based (or mechanistic) models incorporate information on observed relationships between environmental conditions and the performance of the species, although these require more information than is currently available for most plant species. Process-based models tend to reproduce current distributions fairly poorly (e.g. Kramer *et al.*, 2010), since our understanding of the relevant processes is incomplete, but should do better than correlative models when predicting future distributions, particularly in novel climates.

Both correlative and process-based SDMs are usually implemented with two extreme dispersal scenarios, full (unlimited) dispersal and no dispersal, on the assumption that the truth will lie somewhere between these extremes. Recently, there have been many attempts to incorporate realistic dispersal into SDMs, with varying degrees of success (e.g. Bateman *et al.*, 2013; Saltré *et al.*, 2013). Other recent refinements have included attempts to incorporate biotic interactions into SDMs (Wisz *et al.*, 2013), since such interactions could inhibit or, less often, facilitate the movement of a species through a landscape (Corlett and Westcott, 2013). Note, however, that in most cases the major limitation is not the modelling process itself, but the availability of the data needed to parameterize these more sophisticated models. Less progress has been made in modelling species abundances (SAMs), rather than simple occurrences, yet in many potential applications—including the assessment of population viability and the prediction of ecosystem processes—abundances are crucial information (Chambers *et al.*, 2013).

SDMs of individual species can be 'stacked' to predict future plant species assemblages (e.g. Pottier *et al.*, 2013). Large-scale vegetation models—dynamic global vegetation models (DGVMs)—take a different approach, however, simulating the response of a limited range of plant functional types (PFTs), based largely on physiology. DGVMs usually assume everywhere is accessible to all PFTs, i.e. full migration. This is unfortunate because, in practice, 'migration lag' by late-successional species is likely to lead to carbon storage being reduced below the potential modelled by physiology alone (see below). Moreover, DGVMs are often coupled with global climate models (GCMs) in order to simulate the bi-directional carbon-cycle feedbacks between the biosphere and atmosphere, so systematic errors in vegetation predictions could affect predictions for future climates. It is possible to include migration rates where these are known (Snell, 2014), but where many species are involved this may require partitioning physiology-based PFTs into multiple additional types on the basis of their potential movement velocities (Corlett and Westcott, 2013).

Climate change refugia

In all but the flattest topography, local variations in temperature, moisture and other climate variables create potential 'refugia', where plant populations may persist in favourable conditions outside their broadly defined climatic envelope. SDMs are usually run at too coarse a spatial resolution to detect them, but the use of finer scale climate projections can capture these topographically controlled variations, revealing refugia (and dispersal corridors) that were not apparent at a coarser scale (Franklin et al., 2013). Cryptic refugia north of the main treeline have been invoked to explain improbably high post-glacial tree migration rates (Vegas-Vilarrúbia et al., 2012; Tzedakis et al., 2013; Cheddadi et al., 2014) and could similarly reduce the movement velocities needed to track future climate change (Corlett and Westcott, 2013). Identifying and protecting these refugia is one of the more plausible suggestions for adaptation to climate change (Keppel et al., 2012).

The potential consequences of migration lag

Model-based projections of future plant species extinction risks due to climate change can be catastrophic (e.g. Thomas et al., 2004; Warren et al., 2013), contrasting strongly with the rarity of extinctions in the paleoecological record and the absence of well-documented recent extinctions. The models may be overestimating extinction risk, because they largely ignore acclimation capacity and are typically run at a coarse spatial resolution, but it is also possible that the combination of unprecedented rates of climate change with massively fragmented landscapes and a suite of other adverse human impacts really will be catastrophic. Extinction, however, is likely to be a slow and untidy process, since plant species can typically persist as individuals well outside the range of conditions that will support a self-sustaining population (Sax et al., 2013). Such individuals may be 'committed to extinction' if conditions do not change, but as long as they persist there is the possibility of rescue by a future reversal in climate change, or by active human intervention.

The impacts of migration lags on the global carbon sink are more easily predicted, since the species that fail to track climate change are likely to grow more slowly and fix less carbon. Moreover, the forests of the future, whether they replace existing forests or occupy deforested areas, will be dominated by those species that can move fast enough, which are more likely to be relatively short-lived, short-stature pioneers with low-density wood, rather than long-lived, carbon-rich, late-successional species (Meier et al., 2012).

Assisted migration

Assisted migration (also called managed translocation and assisted colonization) is the deliberate establishment of a population of organisms *outside* their known historical range to save the species from extinction. First applied to the movement of vulnerable species to predator-free offshore islands, these terms are now most often used in relation to climate change (Schwartz et al., 2012). If a species is vulnerable to extinction because it is isolated on a mountain top or edaphic island, or its habitat has become fragmented, or it simply cannot move fast enough, then why not move it artificially to a suitable site? Opponents highlight the risk of introduced species becoming invasive (Mueller and Hellmann, 2008), but the plant traits that make a species likely to need human assistance—including poor dispersal, long life-cycle, low competitive ability—are the opposite of those that favour invasiveness. Moreover, most species translocations are likely to be into similar communities near the edge of the natural range, reducing the risk of unpredictable impacts. More realistic are concerns for the practicality of assisted migration on anything approaching the scale that is likely to be needed. In addition, climate change is a moving target and long-lived species will experience significant changes during their lifetime, so the risks of present or future maladaptation of translocated plants are high.

Poleward translocations of suitable genotypes *within* the overall native range are likely to be much less controversial than translocations outside the range, as is the use of warmth-adapted genotypes of native species in ecological restoration, reforestation and native plantations. However, in the longer term these options are unlikely to be enough and we will have to learn how to carry out assisted migration of species in a way that minimizes the risk/reward ratio (Corlett and Westcott, 2013). Given the large uncertainties, there is an urgent need for experimental studies in a wide variety of situations, from using translocated species in reforestation of degraded and open habitats to the more challenging task of trying to establish populations in intact natural communities. There are legal barriers to such experiments in many areas (Schwartz et al., 2012) and these will need to be modified in a way that does not open the flood gates to uncontrolled plant movements.

Conclusions

Rapid climate change coupled with massive fragmentation and a host of other adverse impacts could make the remainder of the twenty-first century a very bad time for plants. Just how bad is currently unclear, with massive uncertainties in both climate projections and climate change impacts on plant persistence. These uncertainties favour uncontroversial, 'no regrets' adaptation measures, including reducing non-climate impacts, restoring forests, maximizing landscape connectivity, protecting and, where necessary, re-introducing long-distance seed dispersal agents and protecting refugia where climate change is expected to be less than the regional mean. However, if current climate projections are at all realistic, these measures will not be enough, since many plants could not move fast enough even in intact landscapes and refugia cannot hold everything.

Areas of low projected climate change velocity are likely to be the best places for *in situ* conservation (Serra-Diaz et al., 2014). Most of these will be in regions of steep and rugged topography, which are also likely to provide climate change refugia for species that cannot track even low-velocity changes. However, sharp declines in area with altitude will limit the long-term potential of many such areas, with well-vegetated foothills perhaps the best long-term prospect. Conversely, areas with high projected climate change velocities, which are usually in landscapes with low-relief topography, may be most appropriate for managed translocations and other human interventions. Indeed, the whole concept of 'natural vegetation' in such landscapes is probably meaningless in times of rapid change. Large-scale experiments involving multiple sites and species are urgently needed, but we will not have time to wait for the results so some degree of risk is unavoidable, even with detailed monitoring and adaptive management. There are no easy options.

References

Bai, F., Sang, W. and Axmacher, J. C. (2011) 'Forest vegetation responses to climate and environmental change: A case study from Changbai Mountain, NE China', *Forest Ecology and Management,* vol 262, pp. 2052–2060.

Bateman, B. L., Murphy, H. T., Reside, A. E., Mokany, K. and VanDerWal, J. (2013) 'Appropriateness of full-, partial- and no-dispersal scenarios in climate change impact modelling', *Diversity and Distributions,* vol 19, pp. 1224–1234.

Bellassen, V. and Luyssaert, S. (2014) 'Carbon sequestration: Managing forests in uncertain times', *Nature,* vol 506, pp. 153–155.

Chambers, D., Périé, C., Casajus, N. and de Blois, S. (2013) 'Challenges in modelling the abundance of 105 tree species in eastern North America using climate, edaphic, and topographic variables', *Forest Ecology and Management,* vol 291, pp. 20–29.

Cheddadi, R., Birks, H. J. B., Tarroso, P., Liepelt, S., Gömöry, D. and others (2014) 'Revisiting tree-migration rates: *Abies alba* (Mill.), a case study', *Vegetation History and Archaeobotany*, vol 23, pp. 113–122.

Chen, I. C., Hill, J. K., Ohlemüller, R., Roy, D. B. and Thomas, C. D. (2011) 'Rapid range shifts of species associated with high levels of climate warming', *Science*, vol 333, pp. 1024–1026.

Cleland, E. E., Allen, J. M., Crimmins, T. M., Dunne, J. A., Pau, S. and others (2012) 'Phenological tracking enables positive species responses to climate change', *Ecology*, vol 93, pp. 1765–1771.

Corlett, R. T. and Westcott, D. A. (2013) 'Will plant movements keep up with climate change?', *Trends in Ecology and Evolution*, vol 28, pp. 482–488.

Correa-Metrio, A., Bush, M., Lozano-García, S. and Sosa-Nájera, S. (2013) 'Millennial-scale temperature change velocity in the continental northern Neotropics', *PLoS ONE*, vol 8, e81958.

Cunze, S., Heydel, F. and Tackenberg, O. (2013) 'Are plant species able to keep pace with the rapidly changing climate?', *PLoS ONE*, vol 8.

De Frenne, P., Coomes, D. A., De Schrijver, A., Staelens, J., Alexander, J. M. and others (2014) 'Plant movements and climate warming: intraspecific variation in growth responses to nonlocal soils', *New Phytologist*, vol 202, pp. 431–441.

Delzon, S., Urli, M., Samalens, J.-C., Lamy, J.-B., Lischke, H. and others (2013) 'Field evidence of colonisation by holm oak, at the northern margin of its distribution range, during the Anthropocene period', *PLoS ONE*, vol 8, e80443.

Dennis, A. J. and Westcott, D. A. (2006) 'Reducing complexity when studying seed dispersal at community scales: a functional classification of vertebrate seed dispersers in tropical forests', *Oecologia*, vol 149, pp. 620–634.

Diffenbaugh, N. S. and Field, C. B. (2013) 'Changes in ecologically critical terrestrial climate conditions', *Science*, vol 341, pp. 486–492.

Feeley, K. J., Silman, M. R., Bush, M. B., Farfan, W., Cabrera, K. G. and others (2011) 'Upslope migration of Andean trees', *Journal of Biogeography*, vol 38, pp. 783–791.

Feeley, K. J., Hurtado, J., Saatchi, S., Silman, M. R. and Clark, D. B. (2013) 'Compositional shifts in Costa Rican forests due to climate-driven species migrations', *Global Change Biology*, vol 19, pp. 3472–3480.

Franklin, J., Davis, F. W., Ikegami, M., Syphard, A. D., Flint, L. E. and others (2013) 'Modeling plant species distributions under future climates: how fine scale do climate projections need to be?', *Global Change Biology*, vol 19, pp. 473–483.

Freeman, B. G. and Freeman, A. M. (2014) 'Rapid upslope shifts in New Guinean birds illustrate strong distributional responses of tropical montane species to global warming', *Proceedings of the National Academy of Sciences*, vol 111, pp. 4490–4494.

Fyfe, J. C. and Gillett, N. P. (2014) 'Recent observed and simulated warming', *Nature Climate Change*, vol 4, pp. 150–151.

Groom, Q. J. (2013) 'Some poleward movement of British native vascular plants is occurring, but the fingerprint of climate change is not evident', *PeerJ*, vol 1, e77.

Harrison, S. P. and Goñi, M. F. S. (2010) 'Global patterns of vegetation response to millennial-scale variability and rapid climate change during the last glacial period', *Quaternary Science Reviews*, vol 29, pp. 2957–2980.

IPCC (2013) *Climate Change 2013: The Physical Science Basis*, Cambridge University Press, Cambridge, UK.

IPCC (2014) *Climate Change 2014: Impacts, Adaptation, and Vulnerability*, Cambridge University Press, Cambridge, UK.

Jackson, S. T. and Weng, C. (1999) 'Late Quaternary extinction of a tree species in eastern North America', *Proceedings of the National Academy of Sciences*, vol 96, pp. 13847–13852.

Jump, A. S., Mátyás, C. and Peñuelas, J. (2009) 'The altitude-for-latitude disparity in the range retractions of woody species', *Trends in Ecology and Evolution*, vol 24, pp. 694–701.

Kelemen, K., Krivàn, A. and Standovár, T. (2014) 'Effects of land-use history and current management on ancient woodland herbs in Western Hungary', *Journal of Vegetation Science*, vol 25, pp. 172–183.

Kelly, A. E. and Goulden, M. L. (2008) 'Rapid shifts in plant distribution with recent climate change', *Proceedings of the National Academy of Sciences*, vol 105, 11823–11826.

Keppel, G., Van Niel, K. P., Wardell-Johnson, G. W., Yates, C. J., Byrne, M. and others (2012) 'Refugia: identifying and understanding safe havens for biodiversity under climate change', *Global Ecology and Biogeography*, vol 21, pp. 393–404.

Kramer, K., Degen, B., Buschbom, J., Hickler, T., Thuiller, W. and others (2010) 'Modelling exploration of the future of European beech (*Fagus sylvatica* L.) under climate change—range, abundance, genetic diversity and adaptive response', *Forest Ecology and Management*, vol 259, pp. 2213–2222.

Lenoir, J., Gégout, J. C., Marquet, P. A., de Ruffray, P. and Brisse, H. (2008) 'A significant upward shift in plant species optimum elevation during the 20th century', *Science*, vol 320, pp. 1768–1771.

Loarie, S. R., Duffy, P. B., Hamilton, H., Asner, G. P., Field, C. B. and Ackerly, D. D. (2009) 'The velocity of climate change', *Nature*, vol 462, pp. 1052–1055.

Markl, J. S., Schleuning, M., Forget, P. M., Jordano, P., Lambert, J. E. and others (2012) 'Meta-analysis of the effects of human disturbance on seed dispersal by animals', *Conservation Biology*, vol 26, pp. 1072–1081.

McConkey, K. R., Prasad, S., Corlett, R. T., Campos-Arceiz, A., Brodie, J. F. and others (2012) 'Seed dispersal in changing landscapes', *Biological Conservation*, vol 146, pp. 1–13.

Meier, E. S., Lischke, H., Schmatz, D. R. and Zimmermann, N. E. (2012) 'Climate, competition and connectivity affect future migration and ranges of European trees', *Global Ecology and Biogeography*, vol 21, 164–178.

Morales, J. M, García, D., Martínez, D., Rodriguez-Pérez, J. and Manuel Herrera, J. (2013) 'Frugivore behavioural details matter for seed dispersal: A multi-species model for Cantabrian thrushes and trees', *PLoS ONE*, vol 8.

Mueller, J. M. and Hellmann, J. J. (2008) 'An assessment of invasion risk from assisted migration', *Conservation Biology*, vol 22, pp. 562–567.

Nathan, R., Katul, G. G., Bohrer, G., Kuparinen, A., Soons, M. B. and others (2011) 'Mechanistic models of seed dispersal by wind', *Theoretical Ecology*, vol 4, pp. 113–132.

Normand, S., Ricklefs, R. E., Skov, F., Bladt, J., Tackenberg, O. and Svenning, J.-C. (2011) 'Postglacial migration supplements climate in determining plant species ranges in Europe', *Proceedings of the Royal Society B-Biological Sciences*, vol 278, pp. 3644–3653.

Pardi, M. I. and Smith, F. A. (2012) 'Paleoecology in an era of climate change: how the past can provide insights into the future' in J. Louys (ed.) *Paleontology in Ecology and Conservation*, pp. 93–116, Springer-Verlag, Berlin.

Parmesan, C. and Yohe, G. (2003) 'A globally coherent fingerprint of climate change impacts across natural systems', *Nature*, vol 421, pp. 37–42.

Pottier, J., Dubuis, A., Pellissier, L., Maiorano, L., Rossier, L. and others (2013) 'The accuracy of plant assemblage prediction from species distribution models varies along environmental gradients', *Global Ecology and Biogeography*, vol 22, pp. 52–63.

Saltré, F., Saint-Amant, R., Gritti, E. S., Brewer, S., Gaucherel, C. and others (2013) 'Climate or migration: what limited European beech post-glacial colonization?', *Global Ecology and Biogeography*, vol 22, pp. 1217–1227.

Sandel, B., Arge, L., Dalsgaard, B., Davies, R. G., Gaston, K. J. and others (2011) 'The influence of Late Quaternary climate-change velocity on species endemism', *Science*, vol 334, pp. 660–664.

Sax, D. F., Early, R. and Bellemare, J. (2013) Niche syndromes, species extinction risks, and management under climate change. *Trends in Ecology and Evolution*, 28, 517–523.

Schwartz, M. W., Hellmann, J. J., McLachlan, J. M., Sax, D. F., Borevitz, J. O. and others (2012) 'Managed relocation: integrating the scientific, regulatory, and ethical challenges', *Bioscience*, vol 62, pp. 732–743.

Serra-Diaz, J. M., Franklin, J., Ninyerola, M., Davis, F. W., Syphard, A. D. and others (2014) 'Bioclimatic velocity: the pace of species exposure to climate change', *Diversity and Distributions*, vol 20, pp. 169–180.

Slot, M., Rey-Sánchez, C., Gerber, S., Lichstein, J. W., Winter, K. and Kitajima, K. (2014) 'Thermal acclimation of leaf respiration of tropical trees and lianas: response to experimental canopy warming, and consequences for tropical forest carbon balance', *Global Change Biology*, vol 20, pp. 2915–2926.

Snell, R. S. (2014) 'Simulating long-distance seed dispersal in a dynamic vegetation model', *Global Ecology and Biogeography*, vol 23, pp. 89–98.

Stork, N. E. and Habel, J. C. (2014) 'Can biodiversity hotspots protect more than tropical forest plants and vertebrates?', *Journal of Biogeography*, vol 41, pp. 421–428.

Telwala, Y., Brook, B. W., Manish, K. and Pandit, M. K. (2013) 'Climate-induced elevational range shifts and increase in plant species richness in a Himalayan biodiversity epicentre', *PLoS ONE*, vol 8.

Thomas, C. D., Cameron, A., Green, R. E., Bakkenes, M., Beaumont, L. J. and others (2004) 'Extinction risk from climate change', *Nature*, vol 427, pp. 145–148.

Thomson, F. J., Moles, A. T., Auld, T. D. and Kingsford, R. T. (2011) 'Seed dispersal distance is more strongly correlated with plant height than with seed mass', *Journal of Ecology*, vol 99, pp. 1299–1307.

Trakhtenbrot, A., Katul, G. G. and Nathan, R. (2014) 'Mechanistic modeling of seed dispersal by wind over hilly terrain', *Ecological Modelling*, vol 274, 29–40.

Tzedakis, P. C., Emerson, B. C. and Hewitt, G. M. (2013) 'Cryptic or mystic? Glacial tree refugia in northern Europe', *Trends in Ecology and Evolution*, vol 28, pp. 696–704.

Vegas-Vilarrúbia, T., Nogué, S. and Rull, V. (2012) 'Global warming, habitat shifts and potential refugia for biodiversity conservation in the Neotropical Guayana Highlands', *Biological Conservation*, vol 152, pp. 159–168.

Wang, X., Piao, S., Ciais, P., Friedlingstein, P., Myneni, R. B. and others (2014) 'A two-fold increase of carbon cycle sensitivity to tropical temperature variations', *Nature*, vol 506, pp. 212–215.

Warren, R., VanDerWal, J., Price, J., Welbergen, J. A., Atkinson, I. and others (2013) 'Quantifying the benefit of early climate change mitigation in avoiding biodiversity loss', *Nature Climate Change*, vol 3, pp. 678–682.

Whigham, D. F. (2004) 'Ecology of woodland herbs in deciduous forests', *Annual Review of Ecology, Evolution, and Systematics*, vol 35, pp. 583–621.

Wiens, J. J., Ackerly, D. D., Allen, A. P., Anacker, B. L., Buckley, L. B. and others (2010) 'Niche conservatism as an emerging principle in ecology and conservation biology', *Ecology Letters*, vol 13, pp. 1310–1324.

Wing, S. L. and Currano, E. D. (2013) 'Plant response to a global greenhouse event 56 million years ago', *American Journal of Botany*, vol 100, pp. 1234–1254.

Wisz, M. S., Pottier, J., Kissling, W. D., Pellissier, L., Lenoir, J. and others (2013) 'The role of biotic interactions in shaping distributions and realised assemblages of species: implications for species distribution modelling', *Biological Reviews*, vol 88, pp. 15–30.

Zhu, K., Woodall, C. W. and Clark, J. S. (2012) 'Failure to migrate: lack of tree range expansion in response to climate change', *Global Change Biology*, vol 18, pp. 1042–1052.

Zhu, K., Woodall, C. W., Ghosh, S., Gelfand, A. E. and Clark, J. S. (2014) 'Dual impacts of climate change: forest migration and turnover through life history', *Global Change Biology*, vol 20, pp. 251–264.

37
FOREST CARBON BUDGETS AND CLIMATE CHANGE

Yadvinder Malhi, Sam Moore and Terhi Riutta

Key concepts in forest carbon budgets

The net carbon budget of a forest is the product of the balance between processes of carbon gain (photosynthesis, tree growth, carbon accumulation in soils), and processes of carbon loss (respiration of living biomass, tree mortality, microbial decomposition of litter, oxidation of soil carbon, fire, degradation and type of disturbance). These processes are influenced by a number of climatic and environmental variables, such as temperature, atmospheric carbon dioxide (CO_2) concentration, moisture availability and frequency of disturbance. It is these variables that create large differences in carbon allocation and storage patterns between forest types such as the low-latitude humid or dry tropical forests, mid-latitude temperate forests and the high-latitude cold boreal forests, requiring that the major forest biomes need to be treated separately.

Whether in the wet tropics or the boreal regions, old-growth forests are often assumed to be mostly in a state of equilibrium, such that over a period of several years their carbon balance would be neutral. However, increasing evidence from forest plot studies has challenged this view and it is now more commonly accepted that undisturbed areas of forest also sequester carbon (net carbon sink), either because the demographic processes of carbon accumulation can persist for centuries or even millennia, or because the ecosystems are responding to global environmental change (see below). This stored carbon in ecosystems is eventually returned to the atmosphere, either naturally, when individual trees or groups of trees die through natural senescence and/or external environmental disturbance and then decompose, or through human-induced reasons such as logging, burning or forest clearance.

To fully understand forest carbon budgets, a more complete understanding of the processes involved in carbon gain and carbon loss is useful (Figure 37.1).

The *Gross Primary Production (GPP)* is the total amount of carbon fixed from atmospheric CO_2 into carbohydrates through the process of photosynthesis by plants in an ecosystem, such as a stand of trees in a forest. Globally, total terrestrial GPP is estimated to be about 120 Pg C yr^{-1} (Beer *et al.* 2010). 1 Pg = 1 Gt = 10^{15} g.

The *Net Primary Production (NPP)* is the net rate of production of new organic matter (whether in canopy leaves, reproductive tissue, wood, fine roots, or in more subtle components such as volatile organic compounds [VOCs] or root exudates). It equals the difference between

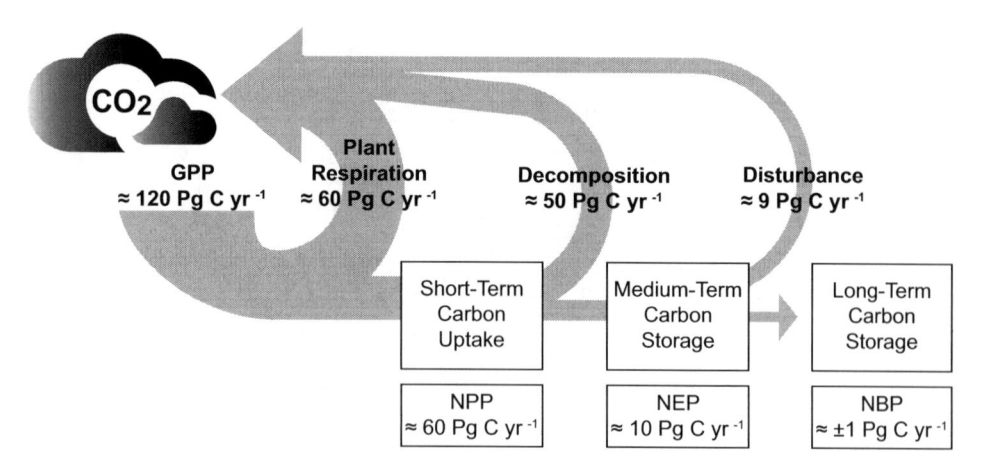

Figure 37.1 Gross primary production (taken from IPCC 2000 and adapted from IGBP 1998)

GPP and plant respiration (autotrophic respiration, R_a), which is carbon used to power the plants' own metabolisms, with CO_2 release as a waste product. Hence:

$$NPP = GPP - R_a.$$

Global NPP is often estimated to be about half of the GPP, about 60 Pg C yr^{-1} (Waring et al. 1998) though measurements from many old-growth boreal and tropical forests suggest the ratio is more often 0.3–0.4 (DeLucia et al. 2007). Hence global NPP is in the range 35–60 Pg C yr^{-1} and global plant respiration is about 60–85 Pg C yr^{-1}. Sometimes the more "solid" components of NPP (the canopy, reproductive, woody and fine root production) are distinguished from NPP and termed the *Biomass Production (BP)* (Vicca et al. 2012).

The *Net Ecosystem Production (NEP)* is the net accumulation of carbon by an ecosystem at a local spatial scale, following the loss of carbon through autotrophic and heterotrophic respiration (R_h). The sum of R_a and R_h is the *Total Ecosystem Respiration (R_{ECO})*.

$$NEP = NPP - R_h \text{ or } NEP = GPP - (R_a + R_h) \text{ or } NEP = GPP - R_{ECO}$$

Heterotrophic respiration is respiration by organisms that gain their carbon by consuming organic matter rather than producing it themselves, including fungi, bacteria and animals, predominantly through the decomposition of dead organic matter. NEP is defined to be negative if an ecosystem is losing carbon over time, e.g. through enhanced heterotrophic respiration following a mortality event or during a warming event.

Global terrestrial NEP is estimated to be about 10 Pg C yr^{-1}, implying that about 50 Pg C yr^{-1} is lost through heterotrophic respiration (IPCC 2000). NEP can be measured either by: (i) measuring changes in carbon stocks in vegetation and soil over time; or by (ii) measuring the flux of CO_2 into and out of the forest (the *Net Ecosystem Exchange, NEE*) with eddy covariance instrumentation or other micrometeorological techniques, just above the forest canopy.

If a forest stand is enmeshed with a spatial matrix of occasional local disturbance and gradual recovery, the NEP tends to be positive (a carbon sink) at local scale but at landscape scale this carbon accumulation is offset by disturbance events that are discrete in space and time, such as fires, pest outbreaks or storm blowdowns. To measure larger scale ecosystem carbon balance, we need to incorporate the spatial and temporal patterns of the disturbance regime. This concept is captured by the *Net Biome Production (NBP)*, the net carbon balance over larger spatial and temporal scales:

NBP = NEP − disturbance

NBP is appropriate for describing the net carbon balance of large areas (100–1,000 km^2) over longer periods of time (several years and longer). Current global terrestrial NBP is estimated to be about 2.3±0.5 Pg C yr^{-1} (see below), implying that about 8 Pg C yr^{-1} is lost through disturbance. Therefore, compared to the primary biosphere–atmosphere fluxes (such as GPP or respiration), global NBP is comparatively small.

Forest carbon fluxes and stocks vary significantly depending on different climates, productivity and disturbance regimes. However, the total flux of carbon in an ecosystem, or in a particular carbon pool of an ecosystem, gives relatively little insight into its importance as a long-term carbon store. The time of retention of carbon in a particular carbon pool of a forest ecosystem has a strong influence on that pool's importance as a carbon store. This concept is captured by the carbon *residence time*, defined as equilibrium carbon pool size divided by carbon input or carbon loss. For example, the annual rate of production of leaves may be much higher than that of wood in a forest, but the short residence times of leaves (around one year) compared to woody biomass (>100 years in many trees) means that wood is a much larger and important carbon store than leaf material.

As another example, arctic tundra and a tropical forest may store the same amount of carbon as one another, but the carbon turnover in the tropical forest may be 500 times more rapid than in the arctic tundra. Carbon isotope methods have been used to show that in the tropics, 85 per cent of the ^{14}C that entered ecosystems during the era of atmospheric nuclear testing in the 1950s has been converted to humus, whereas this proportion is only 50 per cent in temperate soils and approximately 0 per cent in boreal soils, further indicating a more rapid turnover of soil organic carbon in the tropics than at high latitudes (Trumbore 1993, Trumbore and Harden 1997).

In assessing the carbon sequestration potential of forest ecosystems now and in the future, we need to consider the different timescales over which the carbon gain is measured or estimated. These can vary from daily (solar radiation), to monthly (seasonal), to annual (length of growing season) and longer term timescales (forest life cycle). Over the longterm, the carbon sink capacity of any forest is determined by the size of the above and below ground woody productivity, rates of input of organic material into soils, and woody and soil carbon residence times. Thus, additional carbon can be stored in an ecosystem only if more carbon is kept for the same period of time or the same amounts of carbon are kept over longer periods of time. A reduced rate of disturbance could therefore enhance carbon storage. Increased growth, on the other hand, will not add to the long-term sink if disturbances increase in frequency.

Carbon stocks and flows of tropical, temperate and boreal forests

Nature and extent of intact and human-modified forests

Forests cover about 30 per cent (3,720 Mha) of the total global land area (Food and Agriculture Organisation 2010). Tropical forests are the most extensive forest biome, accounting for about 44 per cent of the forest area, followed by boreal (32 per cent) and temperate (24 per cent) forests (Table 37.1). The undisturbed, pre-agricultural Holocene global forest area has been estimated at 5,680 Mha, of which about 65 per cent remains today (Goldewijk 2001). Forests have been used and cleared by humans since the dawn of agriculture, but rapid and extensive deforestation is a recent phenomenon driven by technology. Until 300 years ago probably only 7 per cent of the global forest area had been lost (Goldewijk 2001), although in Europe, India and China a much higher proportion of forests had been cleared by that time (Williams 2000, Malhi *et al.* 2002 and references therein). Today, over 50 per cent of the original forest area in the temperate zone has been deforested or fragmented, nearly 25 per cent in the tropics, but only 4 per cent in the boreal zone (Wade *et al.* 2003). In the temperate zone, rates of deforestation accelerated during industrialisation and the colonisation of North America, peaking in the late nineteenth and early twentieth century (Goldewijk 2001), but the forest area has now stabilised and is even increasing in many temperate regions as marginal agricultural lands are abandoned (Food and Agriculture Organisation 2010) and forests are no longer a significant supply of biomass energy. The deforestation rate of tropical forests accelerated sharply in the late twentieth century and remains high, accounting for nearly all of contemporary net forest loss (Goldewijk 2001, DeLucia *et al.* 2007). Tropical deforestation rates are highest in South East Asia and in Latin America, although over the last decade there has been a strong slowdown in deforestation rates in Brazilian Amazonia, which previously accounted for over half of global net forest loss (Waring *et al.* 1998). Currently, the net loss of C from forest degradation and deforestation is approximately 1.1 ± 0.8 Pg C year^{-1} (Table 37.1). Boreal forests have faced considerably less land use pressure because of their cold climates and poor soils, with most of the original area remaining.

Many remaining forests are heavily human-modified, with only 36 per cent of the remaining global forest area in a relativly undisturbed state (Food and Agriculture Organisation 2010), even less than this (24 per cent) in large, >500 km^2 unfragmented patches (Potapov *et al.* 2008). In the temperate zone, only 6 per cent of the remaining forest area is relatively undisturbed (Hannah *et al.* 1995). Almost all northern hemisphere temperate forests were subject to management practices that reduced the forest C stocks (harvesting, litter raking, conversion to agricultural land) up to the mid-twentieth century (Goodale *et al.* 2002). In the boreal zone, 44 per cent of the forest area is relatively undisturbed, most of this in north-western Canada and Russia, although in Russia fragmentation is on the rise (Potapov *et al.* 2008, Bradshaw *et al.* 2009). Currently, temperate and boreal forests are used for commercial wood production, mostly using native species, and either planted or left to regenerate naturally after harvesting. The wood production is sustainable in terms of stable yields and biomass stocks, although not necessarily in terms of biodiversity and ecosystem functioning (Paillet *et al.* 2010). Approximately 30 per cent of the tropical forests area is relatively intact, most of this in Amazonia or the Congo basin (Potapov *et al.* 2008).

Carbon stocks in vegetation and soil

Globally, forest soils and vegetation contain about 1,200 Pg of carbon, which is about 50 per cent of all terrestrial carbon (IPCC 2000). This is partitioned between about 300 Pg C in vegetation live biomass and 900 Pg C in soils (Table 37.1). The forest C pool is larger than the atmospheric pool (830 Pg) and of similar size to the surface ocean pool (1,055 Pg) (Ciais *et al.* 2013). The total C stock is of similar size in the boreal (42 per cent of the global forest C stock) and tropical (40 per cent) zones, and considerably lower (18 per cent) in the temperate zone (Table 37.1). The small C stock in the temperate zone is due to both the small extant forest area and the depleted C density (stock per unit area) as a legacy of the past disturbances. However, some of the world's most carbon dense (high biomass) forests are old-growth forests located in the temperate zone in areas such as the Pacific Northwest, and southern hemisphere temperate zones in Chile, Australia and New Zealand (Keith *et al.* 2009, Pan *et al.* 2013).

The C pool in vegetation biomass (300 Pg C) is considerably larger in tropical forests (67 per cent of global forest vegetation C pool) than in temperate (17 per cent) or boreal (16 per cent) forests (Table 37.1). Most of the vegetation C, approximately 60–80 per cent in mature forests, is above ground (Mokany *et al.* 2006). For a given shoot biomass, the mean root:shoot ratio is relatively invariant across the forest biomes (Mokany *et al.* 2006). However, the root:shoot ratio decreases with increasing shoot biomass. Therefore, the root:shoot ratio in mature forests is highest in the boreal zone where the mean shoot biomass is lowest, and the ratio decreases towards the tropics (Mokany *et al.* 2006).

Globally, soils contain more C (in the range of 1,800 Pg C) than vegetation, atmosphere or surface oceans. Forest soils constitute approximately half of the soil C pool (about 900 Pg C). The estimates of the global soil C pools are however, uncertain, due to the high spatial variability and lack of reliable remote sensing techniques. Recent (since 1990) estimates of global soil C pool range from 991 to 2,200 Pg C (Scharlemann *et al.* 2014). Some of the largest sources of uncertainties include sampling depth, inconsistent methods, and scarcity of data from remote boreal regions, where organic soils with high C content are common (Jobbágy and Jackson 2000, Scharlemann *et al.* 2014). In forests, on average about 25–35 per cent of the carbon is contained in vegetation and 65–75 per cent in soil (Table 37.1). However, there are marked differences among biomes. In tropical forests, typically 42 per cent of the carbon is in vegetation and 58 per cent in the soil, whereas in boreal forests, 10 per cent of the carbon is in vegetation and 90 per cent in the soil. Temperate forests are intermediate. This latitudinal pattern is mainly driven by temperature, which limits the organic matter decomposition and nutrient cycling rates in colder climes more than it limits productivity, resulting in accumulation of soil organic matter. Approximately 30 per cent of the soil C is below the top 1 m depth (Jobbágy and Jackson 2000), although this layer is not included in most soil C estimates.

Productivity, carbon fluxes and residence time

Productivity of the forest biomes follows the latitudinal and temperature gradient: both GPP and NPP decrease from the tropics towards the colder regions, particularly so in conifer-dominated boreal forests (Figure 37.2, Table 37.1). This decline appears more driven by extension of the dormant season (winter) than by temperature effects on plant physiology in the growing season (Malhi 2010). However, as both autotrophic and heterotrophic respiration follow this same pattern of latitudinal decline, NEP in the three forest types is similar (Figure 37.2). NPP in most temperate forests is only slightly lower than in tropical forests, and they are a bigger C sink per unit area (Table 37.1, Figure 37.2). The current NEP in the temperate zone is probably due to the predominance of young age classes, as the forests are recovering from past disturbance, and due to the positive response to the high nitrogen deposition in large parts of the region (Magnani *et al.* 2007).

The NPP allocation to above and below ground varies with latitude: tropical forests allocate approximately 75 per cent of NPP to above ground components compared with 63 per cent in temperate forests and 54 per cent in boreal forests (Figure 37.2). The mean residence time of carbon in the vegetation components is similar across biomes, about 1–3 years in leaves and fine roots, and 30–100 years in wood (Malhi *et al.* 1999). The residence time of soil carbon is similar to that of vegetation in tropical and temperate forests, but much longer (a century) in the boreal zone (Malhi *et al.* 1999). It is likely, though, that there are large differences between soil layers,

Table 37.1 Forest area and forest C stocks and fluxes in tropical, temperate and boreal zones

	Tropical	Temperate	Boreal	Global[a]	Reference
Area (Mha)[b]	1,620	900	1,200	3,720	Food and Agriculture Organisation (2010)
C density, vegetation (Mg C ha^{-1})	124	57	40	81	Dixon et al. (1994); Saatchi et al. (2011); Thurner et al. (2014)
C density, soil (Mg C ha^{-1}), 0–1 m depth + 1–3 m depth	107 + 63	128 + 44	276 + 95	167 + 69	See footnote[c]
Biome C stock, vegetation (Pg C)	201	51	48	300	Calculated as vegetation C density × Area
Biome C stock, soil (Pg C), 0–1 m depth + 1–3 m depth	173 + 102	115 + 40	331 + 114	620 + 256	Calculated as soil C density × Area
NPP (Mg C ha^{-1} year^{-1})	8.4	7.4	2.7	6.3	Luyssaert et al. (2007)
NEP (Mg C ha^{-1} year^{-1})	0.6	0.9	0.4	0.6	Calculated as NBP/Area
Biome NPP (Pg C year^{-1})	13.6	6.7	3.2	23.5	Calculated as NPP × Area
NBP (Pg C year^{-1})	1.0±0.5[d]	0.8±0.1	0.5±0.1	2.3±0.5	Pan et al. (2011)
C loss from forest degradation and deforestation (Pg C year^{-1})	1.1±0.8	0	0	1.1±0.8	Ciais et al. (2013)
Biome C balance	−0.1±0.9	0.8±0.1	0.5±0.1	1.2±0.9	Calculated as NBP − C loss

[a] Sum or area-weighted mean.

[b] Definition of forest: Land spanning more than 0.5 ha with trees higher than 5 metres and canopy cover >10 per cent, or trees able to reach these thresholds *in situ*.

[c] Calculated by combining global soil organic carbon map (Hiederer and Köchy 2011, Panagos et al. 2012) and FAO maps of global land distribution by land cover type and occurrence of forest, which are part of the Food Insecurity, Poverty and Environment Global GIS database (FGGD) (http://www.fao.org/geonetwork/srv/en/main.home#landcover; http://www.fao.org/geonetwork/srv/en/metadata.show?id=14066). Estimates of the soil C stock at 1–3 m depth are based on Jobbágy and Jackson (2000).

[d] Estimate for low disturbance forests not subjected to degradation or clearance.

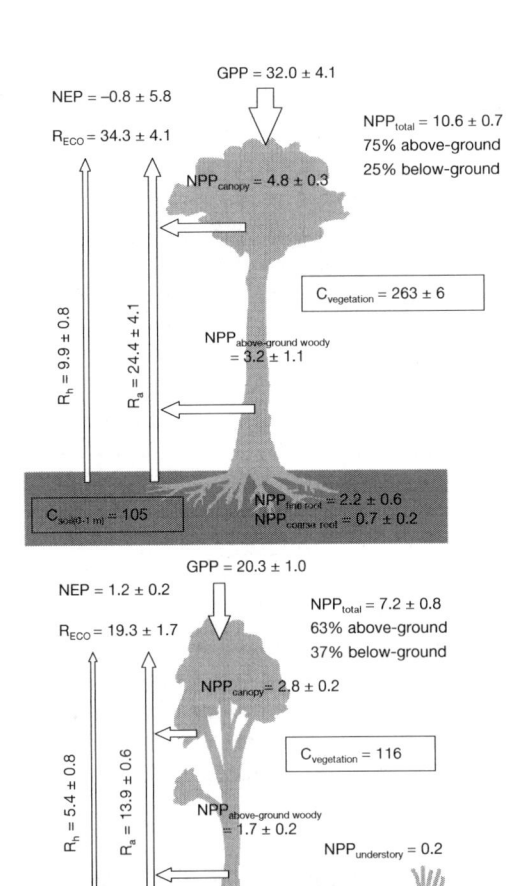

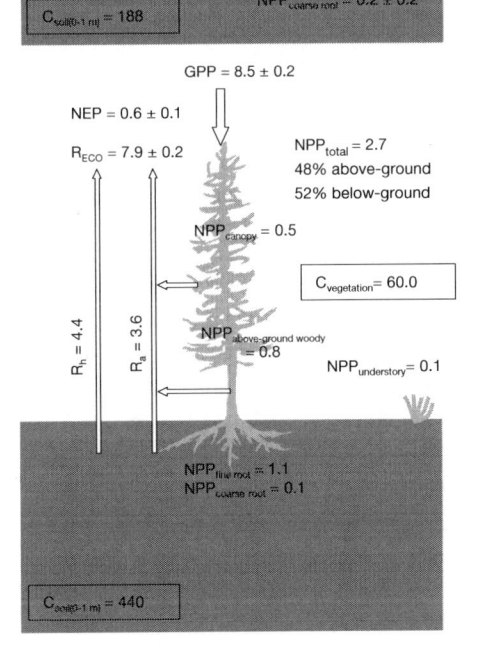

Figure 37.2 Example of vegetation and soil C pools (Mg C ha^{-1}) and C fluxes (Mg C ha^{-1} year^{-1}) in mature, closed canopy forests in tropical, temperate and boreal zones. The values are based on data from the individual sites, not biome means. Upper panel: Humid lowland tropical forest in Amazonia (Caxiuanã, Brazil), evergreen broadleaved forest with ~200 tree species/hectare (Aragão et al. 2009, Malhi et al. 2009, Malhi 2012). Middle panel: Temperate broadleaf deciduous ancient woodland in the UK (Wytham Woods), dominated by broadleaved *Acer pseudoplatanus*, *Fraxinus excelsior* and *Quercus robur* (Butt et al. 2009, Fenn et al. 2010, Thomas et al. 2011, Fenn et al. 2015). Lower panel: Evergreen, coniferous forest in Saskatchewan, Canada (BOREAS SOB site), dominated by *Picea mariana*, with a thick, peaty organic layer (Gower et al. 1997, Jarvis et al. 1997, Malhi et al. 1999, O'Connell et al. 2003). In the tropical site, the soil C stock was sampled to 1 m depth, but in the temperate and boreal site only to 0.3 m (139 Mg C ha^{-1}) and to 0.7 m (404 Mg C ha^{-1}) depth, respectively. The additional fraction of C down to 1 m depth was estimated based on Jobbágy and Jackson (2000)

with the inactive deep soil carbon having a residence time of more than a millennium and the labile top soil carbon being circulated with a much higher turnover rate.

Interannual variability in productivity and C fluxes is high in all biomes, driven by variations in weather. Extreme weather events, such as heat waves and droughts may switch these ecosystems from C sinks to C sources (Ciais *et al.* 2005, Gatti *et al.* 2014), indicating that the forest C dynamics are sensitive to changes in climate. Most of the interannual variability in the global terrestrial carbon cycle is driven by tropical vegetation, with warm El Niño years associated with carbon sources in the tropics (caused by a combination of higher temperatures and, in some regions, drought and increased fire activity). The interannual variability of the tropical terrestrial carbon cycle appears to have increased in recent years (Wang *et al.* 2014).

Boreal forests have an extreme seasonal pattern, with clear growing seasons and dormant seasons. The seasonality is governed by day length, which in the northernmost parts of the boreal zone ranges from 0 to 24 hours, and by temperature, ranging from sub-zero temperatures in the winter to the high 20s in the summer. Medium-term (five-day) mean air temperature in the spring is the most important trigger of the onset of photosynthesis (Suni *et al.* 2003). Because of the long days in the summer, the diurnal variation in carbon fluxes is smaller than in the lower latitudes. Temperate forests in the mid-latitudes vary in their degree of seasonality, both because of the variation in day length within this large zone and because the maritime regions (particularly on the western edge of continents where predominant winds come from the oceans) have far less pronounced variation in temperature than the continental regions. In maritime climates, evergreen coniferous trees can photosynthesise throughout the year, although the short day length during the winter limits their productivity. In the more continental regions with larger temperature variation between summer and winter, the seasonal pattern is similar to that in boreal forests, but the growing season is longer, and the day length during the growing season is shorter.

In tropical forests, seasonal variation in productivity and carbon dynamics is linked to rainfall patterns, and the direction and magnitude of the change in productivity between wet and dry seasons depend on the severity of the water stress. Regions with strong dry seasons, such as south-eastern Amazonia, show a distinct decline in productivity due to water limitations (Restrepo-Coupe *et al.* 2013). On the other hand, in regions where water does not become a limiting factor during the dry season, such as Equatorial Amazonia and Borneo, seasonal variation in productivity is modest and the difference in cloudiness (and therefore solar radiation) between seasons may be a more important factor behind seasonal patterns than the difference in moisture (Kho *et al.* 2013, Restrepo-Coupe *et al.* 2013).

Disturbance regime

Disturbance is the process that distinguishes NEP from Net Biome Production (Figure 37.1), and is a source of carbon to the atmosphere. The scale of natural disturbances varies in space and time, which can mean they are difficult to capture through localised forest census or eddy covariance approaches and are often better described through remote sensing approaches. Typical disturbances include windthrows, natural or human-induced crown and surface fires, insect outbreaks and droughts.

Disturbances can have a large effect on the carbon dynamics of the site, by reducing stock and altering productivity. Immediately after disturbance, productivity is low due to the diminished photosynthesising biomass, but it can recover quickly and NPP can even exceed pre-disturbance values. Disturbance typically leads to a release of carbon from the vegetation and soil C stocks, and that, coupled with a decrease in GPP, leads to an initial decrease in net carbon uptake, possibly turning the system into a carbon source. NPP and NEP are highest in forests that have started to recover from disturbance, and decline when the forest ages and canopy closes (Magnani et al. 2007).

In regions that lack regular large-scale disturbances such as cyclones or extensive wildfires, the effect of disturbances on the landscape-scale carbon balance is dominated by the small-scale disturbances, due to their high frequency. Espírito-Santo et al. (2014) showed that in the Amazon basin, small-scale (<0.1 ha) disturbances account for 88 per cent of the carbon lost due to natural disturbances, while intermediate- (0.1–5) and large- (>5 ha) scale disturbances account for 13 per cent and 0.2 per cent, respectively. Small gap disturbance rate, resulting from individual tree fall events, is similar in the three biomes, with a return time of approximately 100 years (Hiura et al. 1996, Masek et al. 2008, Espírito-Santo et al. 2014). Intermediate- and large-scale disturbances have return times ranging from a century to greater than 10,000 years, the largest disturbances being extremely infrequent (Hiura et al. 1996, Espírito-Santo et al. 2014).

Box 37.1 Montane forests

Approximately 2.4 per cent of the world's forests are montane forests (Schmitt et al. 2009). Forests in high altitudes have different characteristics from lowland forests. Temperature decreases towards higher altitudes, and the change in the forest C stocks and dynamics along an elevational gradient shares many features with the change along the latitudinal gradient from tropical to boreal zones, even though the spatial scales are very different, the former variation taking place within a hundred kilometres and the latter across thousands. A key difference between elevation and latitudinal gradients is that the metapopulation of potential species is fixed in elevation gradient; the accidents of biogeographical history and the time lag responses to past climate change can be more important in latitudinal gradients. Temperature is the dominant but not sole environmental variable that changes with elevation and variation in precipitation; cloud immersion and UV radiation can also be important. Vegetation C stock and productivity typically decrease with elevation (Girardin et al. 2014) while soil C stock in the organic layer (although not in the mineral layer) increases (Leuschner et al. 2007, Girardin et al. 2010). Productivity decreases, and allocation to below ground often increases along with elevation. As an example, in an altitudinal transect in the Peruvian Andes, at a lowland humid tropical forest plot at 194 m above sea level (a.s.l.) (mean annual temperature 26.4°C), total NPP is 13.6 Mg C ha^{-1} year^{-1} (Girardin et al. 2010). At the same latitude but at 3,025 m a.s.l. in a montane cloud forest (mean annual temperature 12.5°C), total NPP is 5.09 Mg C ha^{-1} year^{-1}. Soil carbon stocks can be very high in tropical montane forests (>100 Mg C ha^{-1}; Gibbon et al. 2010), but are often stored in shallow soils (<1 m depth) that are vulnerable to loss through erosion and warming. In a warming world, montane forests are likely to be sensitive to climate change, and may turn into net carbon sources as rates of decomposition increase.

Box 37.2 Peatland forests

Approximately 104 Mha, or 3 per cent, of the world's forests are on peat soils, about 55 per cent of them in the boreal zone and 45 per cent in the tropics (Zoltai and Martikainen 1996, Page et al. 2011, Pan et al. 2013). The areal estimates are, however, scarce and highly uncertain, and new tropical peatlands have recently been described in Amazonia (Lähteenoja *et al.* 2012) and in the Congo basin. Excessive moisture and suitable vegetation (high productivity and/or recalcitrant litter quality) are the prerequisites for peat formation. In such environments, litter and coarse woody debris production rate exceeds decomposition rate, which is impeded by anoxic conditions, and the partially decomposed organic matter accumulates as peat. The soil C content of these ecosystems is typically high, with values of about 1,810 Mg C ha^{-1} in the northern latitudes (Korhola *et al.* 1995, Borren *et al.* 2004, Akumu and McLaughlin 2013) and 2,010 Mg C ha^{-1} in the tropics (Page *et al.* 2011). Therefore, although peatland forests make up a small proportion of the total forest area, they constitute a significant fraction of the total forest soil C pool (at least 22 per cent or 197 Pg; the figures are likely to be even higher given the recent discovery of new forested tropical peatland areas). Vegetation C stock and productivity in peatland forests are similar to the forests on mineral soil in the same region (Bond-Lamberty and Gower 2008, Peregon *et al.* 2008, Hirano *et al.* 2012). In contrast to forests on mineral soil, which are small methane (CH_4) sinks, pristine forested peatlands are CH_4 sources to the atmosphere, emitting 0.1–0.2 Mg CH_4-C ha^{-1} year^{-1} (Minkkinen *et al.* 2002, Couwenberg *et al.* 2010). However, the CH_4 emissions from forested peatlands are approximately five times lower than emissions from low tree cover and open peatlands, and contribute only about 10 per cent to the total emissions from natural wetlands. The C balance of peatland ecosystems is highly sensitive to human disturbance and changes in climate (Page *et al.* 2002, Ise *et al.* 2008, Frolking *et al.* 2011, Moore *et al.* 2013), and because of their large soil C pool, they have the potential to turn into substantial C sources. However, these high carbon ecosystems, along with other wetlands, have only recently become a focus of coherent monitoring efforts (IPCC 2014).

The carbon balance of forests

Forests in the context of the global carbon budget

Since 1750, continuously increasing anthropogenic CO_2 emissions and land use change have perturbed the carbon cycle. Over the period 1750–2011, fossil fuel emissions and cement production have emitted 375±30 Pg C to the atmosphere, with a further 180±80 Pg C emitted by net land use change (mainly deforestation). Of these CO_2 emissions, only 43 per cent (240±10 Pg C) have remained as an observed net increase in atmospheric CO_2 concentrations, with 28 per cent (155±30 Pg C) absorbed by the oceans and 29 per cent (160±90 Pg C) by the terrestrial biosphere (Ciais *et al.* 2013). Considering the last decade (2002–2011), mean annual fossil fuel and cement production emissions (8.3±0.7 Pg C yr^{-1}) and net land use change emissions dominated by tropical deforestation (0.9±0.8 Pg C yr^{-1}) are apportioned between a 47 per cent (4.3±0.2 Pg C yr^{-1}) atmospheric increase, a 26 per cent (2.4±0.7 Pg C yr^{-1}) ocean sink and a 27 per cent (2.5±1.3 Pg C yr^{-1}) land biosphere sink (Ciais *et al.* 2013).

Hence there is a large sink of carbon in the land biosphere away from areas of deforestation and degradation. Forests almost certainly are the major contributor to this sink as the largest components of terrestrial biomass and productivity: whether in the high, mid or low latitudes, forests are particularly important as a carbon reservoir because trees hold much more carbon per unit area than other types of vegetation. If this land biosphere carbon sink were not present, and this excess carbon were instead allocated to ocean and atmospheric pools at the current fractions, atmospheric CO_2 would rise at a rate 37 per cent higher than currently observed; the global biosphere "brakes" the rate of atmospheric CO_2 rise by about 27 per cent. Forests are certain to be a large part of this "brake", and one of the major questions in forest carbon cycle ecology is to understand the nature, distribution and stability of this climate change brake.

Pan et al. (2011) presented a synthesis of forest inventory data and long-term carbon cycle studies to estimate a global forest carbon sink (over the period 2000–2007) of 2.30 ± 0.49 Pg C yr^{-1} (i.e. forests account for almost all of the biosphere carbon sink) partitioned between 22 per cent in boreal forests (0.50 ± 0.08 Pg C yr^{-1}), 34 per cent in temperate forests (0.78 ± 0.09 Pg C yr^{-1}) and 44 per cent in intact tropical forests (1.02 ± 0.47 Pg C yr^{-1}). There is a further sink in regrowing tropical forests, but this is closely entwined with gross tropical deforestation and incorporated in the net tropical land use change emission of 0.9 ± 0.8 Pg C yr^{-1}. However, this tropical regrowth sink could be large: Pan et al. (2011) estimate it to be 1.72 ± 0.54 Pg C yr^{-1}, resulting in a total tropical forest sink of 2.74 ± 0.72 Pg C yr^{-1}.

Causes of imbalance in forest carbon budgets

Next, we explore why forests appear to be the cause of such a large net carbon sink. In principle, mature forest stands are close to carbon balance (Net Ecosystem Productivity ~0), with the growth of individual trees being balanced by the death and decomposition of other stands, although there is evidence that some forests can continue to accumulate biomass for millennia. At a landscape scale, the continued growth of some stands is balanced by the degradation of other stands through natural disturbance, leading to expectations of a Net Biome Productivity close to zero for mature landscapes away from anthropogenic disturbance.

In practice, many forest stands are not in carbon balance. Following a disturbance event (whether natural or anthropogenic), forests are short-term carbon sources (for up to several years) as dead organic material decomposition dominates, but eventually become large carbon sinks as new biomass accumulation and recruitment of trees dominates over mortality and decomposition. Natural disturbances can be spatially stochastic with a characteristic size-frequency distribution, as in the case of tree blowdowns in Amazonia (Espírito-Santo et al. 2014) or fire occurrence in boreal forests. Such disturbance regimes mean that any particular stand monitored over time shows long-term carbon accumulation, followed by carbon release in a disturbance. Understanding the large-scale pattern of disturbance and recovery is essential for understanding the regional forest carbon balance.

In addition to stochastic disturbance events, there can be more spatially extensive past disturbance events than result in broadscale, spatially coherent regrowth and recovery and a large-scale carbon sink. Examples include extensive drought or fire caused by a large-scale weather event such as an El Niño, or forest recovery after extensive deforestation followed by agricultural abandonment, as is the case of much of eastern North America and parts of Europe, or simply a reduction in the intensity of human use of forests caused by population collapse or rural emigration. The latter has been speculated for some tropical regions such as Amazonia following the arrival of European colonists and diseases in the Americas, or West and Central Africa following the peak of the slave trade. Conversely, gradual intensification of human use and degradation of forests for fuelwood or timber can result in long-term carbon sources.

A further agent of change in the carbon balance of forests is long-term atmospheric change. This can take a number of forms. Shifts in local rainfall regimes can cause gradual loss or increase in biomass, although sometimes in unexpected directions. Fauset et al. (2012) demonstrated that long-term drying in Ghana caused an increase in forest biomass (i.e. a net biomass carbon sink) because of increasing dominance of fast-growing, high biomass species.

Increases in temperature could potentially increase growth rates in temperature-constrained regions such as high-latitude or montane forests, through direct effects on growth rate, through extension of the growing season, or indirectly through increasing litter decomposition and nutrient recycling (Jarvis and Linder 2000). However, warming can also cause soil organic matter decomposition rates to increase, resulting in net carbon emissions from soils and necromass that could offset or exceed the carbon accumulation in live biomass. Climate change could also cause a shift in disturbance regimes (e.g. the frequency of blowdowns, fires, insect outbreaks or droughts), leading to either net carbon emissions or uptake depending on whether disturbance intensity and frequency increases or decreases.

Increasing atmospheric carbon dioxide is expected to increase rates of photosynthesis, stimulating tree growth and generating a net carbon sink. Most global biosphere models assume such a CO_2 fertilisation response, and predict enhanced vegetation productivity in a high CO_2 world. Such a response may be greater in the tropics than in high latitudes because of greater sensitivity of photosynthesis to CO_2 at high temperatures (Hickler et al. 2008). However, a key uncertainty is whether plants are able to sufficiently utilise such an abundance of CO_2 by increasing availability and access of other limiting factors like nutrients such as nitrogen and phosphorus. A number of Free Air Carbon Dioxide Enrichment (FACE) experiments have been conducted on forests, all of them on temperate secondary forests or plantations. Experiments have shown that most C3 plants (all trees) respond to elevated concentrations of CO_2 with increased rates of photosynthesis, increased productivity and increased biomass (Norby et al. 2005). A range of studies have demonstrated 20–30 per cent increases in biomass in response to elevated CO_2 (Luo et al. 2006). However, it should be noted that this increase is not universally observed (Körner et al. 2005) and despite the initial increases in productivity observed in trees under elevated concentrations of CO_2, experiments at the ecosystem level and experiments longer than a few years suggest much reduced responses. Furthermore, productivity is not equivalent to carbon storage. If an increase in productivity is in tissues with a rapid turnover (e.g. fine roots, foliage), the enhanced growth may be respired within a year or two, leading to little or no gain in carbon storage. Of particular significance to forest ecosystems is that soil and litter carbon pools may also increase under elevated CO_2 (Jastrow et al. 2005). Higher concentrations of CO_2 enable plants to acquire the same amount of carbon with a smaller loss of water through their stomata. This increased water-use efficiency reduces the effects of drought. Higher levels of CO_2 may also alleviate other stresses of plants, such as high temperatures. The observation that productivity is increased relatively more in low productivity years suggests that the indirect effects of CO_2 in ameliorating stress may be more important than the direct effects of CO_2 on photosynthesis (Luo et al. 1999).

Finally, atmospheric pollution can either have positive or negative effects on atmospheric carbon balance. In many temperate regions, huge increases in nitrogen deposition (Galloway et al. 2008) are thought to have stimulated forest growth rates and caused a net carbon sink (Magnani et al. 2007). On the other hand, excessive nitrogen is also associated with acid rain, which may inhibit forest growth. Other pollutants such as ozone can also inhibit leaf photosynthesis and forest carbon uptake.

Future trajectories in forest carbon budgets

As the twenty-first century proceeds the forces of global change will continue to strongly influence the carbon budget and balance of tropical forests, and in return forest ecosystems will have a strong influence on atmospheric CO_2 concentrations and global climate.

Deforestation and degradation

Loss of forests through direct deforestation and logging and degradation of forests continue to be substantial sources of carbon, particularly in tropical regions (Malhi 2010). In recent years there have been hopeful signs with the large reduction of deforestation rates in the Brazilian Amazon (previously Brazil accounted for over half of global deforestation), but deforestation rates are increasing in other parts of Amazonia, and in many parts of Asia. Africa presents generally low rates of deforestation at the moment. A key threat to many tropical forest regions is the spread of large-scale plantations, pastoralism and agriculture. Oil palm is a particularly prominent threat, with the potential to expand rapidly in Amazonia and Africa in the same manner as it has done in Asia in recent decades.

Fire and other disturbances

Fire regime has a strong influence on long-term carbon balance. In many thinly populated tropical forests, boreal forests and semi-arid regions, increasing settlement and population density may increase fire return frequencies, decrease forest recovery between fire events and result in net emissions of carbon. On the other hand, in many long-settled and prosperous regions, such as North America and Australia, active fire suppression leads to increase in vegetation biomass, albeit this is vulnerable to extreme fire events because of the accumulated fuel loads. Climate change may also change the frequency and intensity of other forms of disturbance, such as extreme storm events, tropical cyclones and extreme drought or precipitation events, although the details and dynamics of such changes are hard to predict. The Anthropocene is characterised by multiple drivers of change acting in synchrony. The interaction of increasing drought frequency, forest degradation and fire pressure may lead to more sustained degradation and loss of forest, with net carbon emissions (Malhi *et al.* 2009, Brando *et al.* 2014).

Second-growth

Over the twentieth century, many temperate forest regions have passed through a "forest transition", with a net reduction of pressure on forests and increase in forest biomass and area, as marginal and uneconomic agricultural areas are abandoned, forests are no longer a source of wood fuel, and as economies move to a more industrial base and intensify agriculture. Over the twenty-first century, parts of the tropics are also likely to experience such a transition, and there is evidence it is already under way in countries such as Costa Rica. Such "second-growth" forests are a substantial carbon sink, and will become of increasing importance for regional ecosystem functioning, including carbon dynamics.

Atmospheric and climate change

The impacts of atmospheric change on the carbon balance of forests are one of the key uncertainties that will determine the CO_2 concentration in the atmosphere at the end of the century. Most ecophysiological models suggest that rising CO_2 will continue to increase photosynthesis rates, tree growth rates and carbon stocks (Ciais *et al.* 2013). However, this sink is likely to slow down and possibly reverse because of saturation of photosynthesis, feedbacks that increase tree mortality rates, or increased carbon emissions from warming soils. A state where large regions of the biosphere such as tropical or boreal forests enter a positive feedback mode, with warming causing release of CO_2 which causes further warming, could be regarded as a tipping point, a dangerous threshold that needs to be avoided. Identifying the nature and proximity of such a threshold remains a priority for forest carbon cycle science in the uncertain waters of the Anthropocene.

References

Akumu, C. E. and McLaughlin, J. W. (2013) Regional variation in peatland carbon stock assessments, northern Ontario, Canada. *Geoderma,* vols 209–210, pp. 161–167

Aragão, L. E. O. C., Malhi, Y., Metcalfe, D. B., Silva-Espejo, J. E., Jiménez, E. and others. (2009) Above- and below-ground net primary productivity across ten Amazonian forests on contrasting soils. *Biogeosciences,* vol 6, no 12, pp. 2759–2778

Beer, C., Reichstein, M., Tomelleri, E., Ciais, P., Jung, M. and others. (2010) Terrestrial gross carbon dioxide uptake: Global distribution and covariation with climate. *Science,* vol 329, no 5993, pp. 834–838

Bond-Lamberty, B. and Gower, S. T. (2008) Decomposition and fragmentation of coarse woody debris: Re-visiting a boreal black spruce chronosequence. *Ecosystems,* vol 11, no 6, pp. 831–840

Borren, W., Bleuten, W. and Lapshina, E. D. (2004) Holocene peat and carbon accumulation rates in the southern taiga of western Siberia. *Quaternary Research,* vol 61, no 1, pp. 42–51

Bradshaw, C. J. A., Warkentin, I. G. and Sodhi, N. S. (2009) Urgent preservation of boreal carbon stocks and biodiversity. *Trends in Ecology and Evolution,* vol 24, no 10, pp. 541–548

Brando, P. M., Balch, J. K., Nepstad, D. C., Morton, D. C., Putz, F. E. and others. (2014) Abrupt increases in Amazonian tree mortality due to drought–fire interactions. *Proceedings of the National Academy of Sciences,* vol 111, no 17, pp. 6347–6352

Butt, N., Campbell, G., Malhi, Y., Morecroft, M., Fenn, K. and Thomas, M. (2009) Initial results from establishment of a long-term broadleaf monitoring plot at Wytham Woods, Oxford, UK. University of Oxford Report. University of Oxford, Oxford, UK

Ciais, P., Reichstein, M., Viovy, N., Granier, A., Ogee, J. and others. (2005) Europe-wide reduction in primary productivity caused by the heat and drought in 2003. *Nature,* vol 437, no 7058, pp. 529–533

Ciais, P., Sabine, C., Bala, G., Bopp, L., Brovkin, V. and others. (2013) Carbon and other biochemical cycles. In T. F. Stocker, D. Qin, G.-K. Plattner, M. Tignor, S. K. Allen, and others (eds), *Climate Change 2013: The Physical Science Basis. Contribution of Working Group I to the Fifth Assessment Report of the Intergovernmental Panel on Climate Change,* Cambridge University Press, Cambridge, United Kingdom and New York, NY, USA

Couwenberg, J., Dommain, R. and Joosten, H. (2010) Greenhouse gas fluxes from tropical peatlands in south-east Asia. *Global Change Biology,* vol 16, no 6, pp. 1715–1732

DeLucia, E. H., Drake, J. E., Thomas, R. B. and Gonzalez-Meler, M. (2007) Forest carbon use efficiency: Is respiration a constant fraction of gross primary production? *Global Change Biology,* vol 13, no 6, pp. 1157–1167

Dixon, R. K., Brown, S., Houghton, R. A., Solomon, A. M., Trexler, M. C. and Wisniewski, J. (1994) Carbon pools and flux of global forest ecosystems. *Science,* vol 263, no 5144, pp. 185–190

Espírito-Santo, F. D. B., Gloor, M., Keller, M., Malhi, Y., Saatchi, S. and others. (2014) Size and frequency of natural forest disturbances and the Amazon forest carbon balance. *Nature Communications,* vol 5, no 3434, pp. 1–6

Fauset, S., Baker, T. R., Lewis, S. L., Feldpausch, T. R., Affum-Baffoe, K. and others. (2012) Drought-induced shifts in the floristic and functional composition of tropical forests in Ghana. *Ecology Letters,* vol 15, no 10, pp. 1120–1129

Fenn, K. M., Malhi, Y. and Morecroft, M. D. (2010) Soil CO2 efflux in a temperate deciduous forest: Environmental drivers and component contributions. *Soil Biology and Biochemistry,* vol 42, no 10, pp. 1685–1693

Fenn, K., Malhi, Y., Morecroft, M., Lloyd, C. and Thomas, M. (2015) The carbon cycle of a maritime ancient temperate broadleaved woodland at seasonal and annual scales. *Ecosystems,* vol 18, no 1, pp. 1–15

Food and Agriculture Organisation (2010) *Global Forest Resources Assessment 2010.* Food and Agriculture Organisation, Rome

Frolking, S., Talbot, J., Jones, M. C., Treat, C. C., Kauffman, J. B. and others. (2011) Peatlands in the Earth's 21st century climate system. *Environmental Reviews,* vol 19, pp. 371–396

Galloway, J. N., Townsend, A. R., Erisman, J. W., Bekunda, M., Cai, Z. and others. (2008) Transformation of the nitrogen cycle: Recent trends, questions, and potential solutions. *Science,* vol 320, no 5878, pp. 889–892

Gatti, L., Gloor, M., Miller, J., Doughty, C., Malhi, Y. and others. (2014) Drought sensitivity of Amazonian carbon balance revealed by atmospheric measurements. *Nature,* vol 506, no 7486, pp. 76–80

Gibbon, A., Silman, M. R., Malhi, Y., Fisher, J. B., Meir, P. and others. (2010) Ecosystem carbon storage across the grassland–forest transition in the high Andes of Manu National Park, Peru. *Ecosystems,* vol 13, no 7, pp. 1097–1111

Girardin, C. A. J., Malhi, Y., Aragão, L. E. O. C., Mamani, M., Huaraca Huasco, W. and others. (2010) Net primary productivity allocation and cycling of carbon along a tropical forest elevational transect in the Peruvian Andes. *Global Change Biology,* vol 16, no 12, pp. 3176–3192

Girardin, C. A. J., Silva-Espejo, J. E. S., Doughty, C. E., Huaraca Huasco, W., Metcalfe, D. B. and others. (2014) Productivity and carbon allocation in a tropical montane cloud forest in the Peruvian Andes. *Plant Ecology and Diversity,* vol 7, nos 1–2, pp. 107–123

Goldewijk, K. K. (2001) Estimating global land use change over the past 300 years: The HYDE Database. *Global Biogeochemical Cycles,* vol 15, no 2, pp. 417–433

Goodale, C. L., Apps, M. J., Birdsey, R. A., Field, C. B., Heath, L. S. and others. (2002) Forest carbon sinks in the Northern Hemisphere. *Ecological Applications,* vol 12, no 3, pp. 891–899

Gower, S. T., Vogel, J. G., Norman, J. M., Kucharik, C. J., Steele, S. J. and Stow, T. K. (1997) Carbon distribution and aboveground net primary production in aspen, jack pine, and black spruce stands in Saskatchewan and Manitoba, Canada. *Journal of Geophysical Research: Atmospheres,* vol 102, no D24, pp. 29029–29041

Hannah, L., Carr, J. L. and Landerani, A. (1995) Human disturbance and natural habitat – A biome level analysis of a global data set. *Biodiversity and Conservation,* vol 4, no 2, pp. 128–155

Hickler, T., Smith, B., Prentice, I. C., Mjöfors, K., Miller, P. and others. (2008) CO2 fertilization in temperate FACE experiments not representative of boreal and tropical forests. *Global Change Biology,* vol 14, no 7, pp. 1531–1542

Hiederer, R. and Köchy, M. (2011) *Global Soil Organic Carbon Estimates and the Harmonized World Soil Database.* Publications Office of the European Union, EUR 25225 EN

Hirano, T., Segah, H., Kusin, K., Limin, S., Takahashi, H. and Osaki, M. (2012) Effects of disturbances on the carbon balance of tropical peat swamp forests. *Global Change Biology,* vol 18, no 11, pp. 3410–3422

Hiura, T., Sano, J. and Konno, Y. (1996) Age structure and response to fine-scale disturbances of Abies sachalinensis, Picea jezoensis, Picea glehnii, and Betula ermanii growing under the influence of a dwarf bamboo understory in northern Japan. *Canadian Journal of Forest Research,* vol 26, no 2, pp. 289–297

IGBP (1998) The terrestrial carbon cycle: Implications for the Kyoto Protocol. *Science,* vol 280, no 5368, pp. 1393–1394

IPCC (2000) *Land Use, Land-use Change, and Forestry: A Special Report of the Intergovernmental Panel on Climate Change.* Cambridge University Press, UK

IPCC (2014) *2013 Supplement to the 2006 IPCC Guidelines for National Greenhouse Gas Inventories: Wetlands.* IPCC, Switzerland

Ise, T., Dunn, A. L., Wofsy, S. C. and Moorcroft, P. R. (2008) High sensitivity of peat decomposition to climate change through water-table feedback. *Nature Geoscience,* vol 1, no 11, pp. 763–766

Jarvis, P. and Linder, S. (2000) Constraints to growth of boreal forests. *Nature,* vol 405, no 6789, p. 904

Jarvis, P. G., Massheder, J. M., Hale, S. E., Moncrieff, J. B., Rayment, M. and Scott, S. L. (1997) Seasonal variation of carbon dioxide, water vapor, and energy exchanges of a boreal black spruce forest. *Journal of Geophysical Research: Atmospheres,* vol 102, no D24, pp. 28953–28966

Jastrow, J. D., Miller, R. M., Matamala, R., Norby, R. J., Boutton, T. W. and others. (2005) Elevated atmospheric carbon dioxide increases soil carbon. *Global Change Biology,* vol 11, no 12, pp. 2057–2064

Jobbágy, E. G. and Jackson, R. B. (2000) The vertical distribution of soil organic carbon and its relation to climate and vegetation. *Ecological Applications,* vol 10, no 2, pp. 423–436

Keith, H., Mackey, B. G. and Lindenmayer, D. B. (2009) Re-evaluation of forest biomass carbon stocks and lessons from the world's most carbon-dense forests. *Proceedings of the National Academy of Sciences of the United States of America,* vol 106, no 28, pp. 11635–11640

Kho, L. K., Malhi, Y. and Tan, S. K. S. (2013) Annual budget and seasonal variation of aboveground and belowground net primary productivity in a lowland dipterocarp forest in Borneo. *Journal of Geophysical Research-Biogeosciences,* vol 118, no 3, pp. 1282–1296

Korhola, A., Tolonen, K., Turunen, J. and Jungner, H. (1995) Estimating long-term carbon accumulation rates in boreal peatlands by radiocarbon dating. *Radiocarbon,* vol 37, no 2, pp. 575–584

Körner, C., Asshoff, R., Bignucolo, O., Hättenschwiler, S., Keel, S. G. and others. (2005) Carbon flux and growth in mature deciduous forest trees exposed to elevated CO2. *Science,* vol 309, no 5739, pp. 1360–1362

Lähteenoja, O., Reátegui, Y. R., Räsänen, M., Torres, D. D. C., Oinonen, M. and Page, S. (2012) The large Amazonian peatland carbon sink in the subsiding Pastaza-Marañón foreland basin, Peru. *Global Change Biology,* vol 18, pp. 164–178

Leuschner, C., Moser, G., Bertsch, C., Röderstein, M. and Hertel, D. (2007) Large altitudinal increase in tree root/shoot ratio in tropical mountain forests of Ecuador. *Basic and Applied Ecology,* vol 8, no 3, pp. 219–230

Luo, Y., Reynolds, J., Wang, Y. and Wolfe, D. (1999) A search for predictive understanding of plant responses to elevated [CO2]. *Global Change Biology,* vol 5, no 2, pp. 143–156

Luo, Y., Hui, D. and Zhang, D. (2006) Elevated CO2 stimulates net accumulations of carbon and nitrogen in land ecosystems: A meta-analysis. *Ecology,* vol 87, no 1, pp. 53–63

Luyssaert, S., Inglima, I., Jung, M., Richardson, A. D., Reichsteins, M. and others. (2007) CO2 balance of boreal, temperate, and tropical forests derived from a global database. *Global Change Biology,* vol 13, no 12, pp. 2509–2537

Magnani, F., Mencuccini, M., Borghetti, M., Berbigier, P., Berninger, F. and others. (2007) The human footprint in the carbon cycle of temperate and boreal forests. *Nature,* vol 447, no 7146, pp. 848–850

Malhi, Y. (2010) The carbon balance of tropical forest regions, 1990–2005. *Current Opinion in Environmental Sustainability,* vol 2, no 4, pp. 237–244

Malhi, Y. (2012) The productivity, metabolism and carbon cycle of tropical forest vegetation. *Journal of Ecology,* vol 100, no 1, pp. 65–75

Malhi, Y., Baldocchi, D. D. and Jarvis, P. G. (1999) The carbon balance of tropical, temperate and boreal forests. *Plant Cell and Environment,* vol 22, no 6, pp. 715–740

Malhi, Y., Meir, P. and Brown, S. (2002) Forests, carbon and global climate. *Philosophical Transactions of the Royal Society of London. Series A: Mathematical, Physical and Engineering Sciences,* vol 360, no 1797, pp. 1567–1591

Malhi, Y., Aragão, L. E. O. C., Metcalfe, D. B., Paiva, R., Quesada, C. A. and others. (2009) Comprehensive assessment of carbon productivity, allocation and storage in three Amazonian forests. *Global Change Biology,* vol 15, no 5, pp. 1255–1274

Masek, J. G., Huang, C., Wolfe, R., Cohen, W., Hall, F. and others. (2008) North American forest disturbance mapped from a decadal Landsat record. *Remote Sensing of Environment,* vol 112, no 6, pp. 2914–2926

Minkkinen, K., Korhonen, R., Savolainen, I. and Laine, J. (2002) Carbon balance and radiative forcing of Finnish peatlands 1900–2100 – The impact of forestry drainage. *Global Change Biology,* vol 8, no 8, pp. 785–799

Mokany, K., Raison, R. J. and Prokushkin, A. S. (2006) Critical analysis of root : shoot ratios in terrestrial biomes. *Global Change Biology,* vol 12, no 1, pp. 84–96

Moore, S., Evans, C. D., Page, S. E., Garnett, M. H., Jones, T. G. and others. (2013) Deep instability of deforested tropical peatlands revealed by fluvial organic carbon fluxes. *Nature,* vol 493, no 7434, pp. 660–663

Norby, R. J., DeLucia, E. H., Gielen, B., Calfapietra, C., Giardina, C. P. and others. (2005) Forest response to elevated CO2 is conserved across a broad range of productivity. *Proceedings of the National Academy of Sciences of the United States of America,* vol 102, no 50, pp. 18052–18056

O'Connell, K. E. B., Gower, S. T. and Norman, J. M. (2003) Net ecosystem production of two contrasting boreal black spruce forest communities. *Ecosystems,* vol 6, no 3, pp. 248–260

Page, S. E., Siegert, F., Rieley, J. O., Boehm, H.-D. V., Jaya, A. and Limin, S. (2002) The amount of carbon released from peat and forest fires in Indonesia during 1997. *Nature,* vol 420, no 6911, pp. 61–65

Page, S. E., Rieley, J. O. and Banks, C. J. (2011) Global and regional importance of the tropical peatland carbon pool. *Global Change Biology,* vol 17, no 2, pp. 798–818

Paillet, Y., Berges, L., Hjälten, J., Odor, P., Avon, C. and others. (2010) Biodiversity differences between managed and unmanaged forests: Meta-analysis of species richness in Europe. *Conservation Biology*, vol 24, no 1, pp. 101–112

Pan, Y., Birdsey, R. A., Fang, J., Houghton, R., Kauppi, P. E. and others. (2011) A large and persistent carbon sink in the world's forests. *Science*, vol 333, no 6045, pp. 988–993

Pan, Y., Birdsey, R. A., Phillips, O. L. and Jackson, R. B. (2013) The structure, distribution and biomass of the world's forests. *Annual Review of Ecology, Evolution, and Systematics*, vol 44, pp. 593–622

Panagos, P., Van Liedekerke, M., Jones, A. and Montanarella, L. (2012) European Soil Data Centre: Response to European policy support and public data requirements. *Land Use Policy*, vol 29, no 2, pp. 329–338

Peregon, A., Maksyutov, S., Kosykh, N. P. and Mironycheva-Tokareva, N. P. (2008) Map-based inventory of wetland biomass and net primary production in western Siberia. *Journal of Geophysical Research: Biogeosciences*, vol 113, no G1, p. G01007

Potapov, P., Yaroshenko, A., Turubanova, S., Dubinin, M., Laestadius, L. and others. (2008) Mapping the world's intact forest landscapes by remote sensing. *Ecology and Society*, vol 13, no 2, p. 51

Restrepo-Coupe, N., da Rocha, H. R., Hutyra, L. R., da Araujo, A. C., Borma, L. S. and others. (2013) What drives the seasonality of photosynthesis across the Amazon basin? A cross-site analysis of eddy flux tower measurements from the Brasil flux network. *Agricultural and Forest Meteorology*, vols 182–183, pp. 128–144

Saatchi, S. S., Harris, N. L., Brown, S., Lefsky, M., Mitchard, E. T. A. and others. (2011) Benchmark map of forest carbon stocks in tropical regions across three continents. *Proceedings of the National Academy of Sciences*, vol 108, no 24, pp. 9899–9904

Scharlemann, J. P. W., Tanner, E. V. J., Hiederer, R. and Kapos, V. (2014) Global soil carbon: Understanding and managing the largest terrestrial carbon pool. *Carbon Management*, vol 5, no 1, pp. 81–91

Schmitt, C. B., Burgess, N. D., Coad, L., Belokurov, A., Besançon, C. and others. (2009) Global analysis of the protection status of the world's forests. *Biological Conservation*, vol 142, no 10, pp. 2122–2130

Suni, T., Berninger, F., Vesala, T., Markkanen, T., Hari, P. and others. (2003) Air temperature triggers the recovery of evergreen boreal forest photosynthesis in spring. *Global Change Biology*, vol 9, no 10, pp. 1410–1426

Thomas, M. V., Malhi, Y., Fenn, K. M., Fisher, J. B., Morecroft, M. D. and others. (2011) Carbon dioxide fluxes over an ancient broadleaved deciduous woodland in southern England. *Biogeosciences*, vol 8, no 6, pp. 1595–1613

Thurner, M., Beer, C., Santoro, M., Carvalhais, N., Wutzler, T. and others. (2014) Carbon stock and density of northern boreal and temperate forests. *Global Ecology and Biogeography*, vol 23, no 3, pp. 297–310

Trumbore, S. E. (1993) Comparison of carbon dynamics in tropical and temperate soils using radiocarbon measurements. *Global Biogeochemical Cycles*, vol 7, no 2, pp. 275–290

Trumbore, S. E. and Harden, J. W. (1997) Accumulation and turnover of carbon in organic and mineral soils of the BOREAS northern study area. *Journal of Geophysical Research: Atmospheres*, vol 102, no D24, pp. 28817–28830

Vicca, S., Luyssaert, S., Peñuelas, J., Campioli, M., Chapin, F. S. and others. (2012) Fertile forests produce biomass more efficiently. *Ecology Letters*, vol 15, no 6, pp. 520–526

Wade, T. G., Riitters, K. H., Wickham, J. D. and Jones, K. B. (2003) Distribution and causes of global forest fragmentation. *Conservation Ecology*, vol 7, no 2, p. 7

Wang, X., Shilong, P., Philippe, C., Pierre, F., Ranga, B. M. and others. (2014) A two-fold increase of carbon cycle sensitivity to tropical temperature variations. *Nature*, vol 506, no 7487, pp. 212–215

Waring, R. H., Landsberg, J. J. and Williams, M. (1998) Net primary production of forests: A constant fraction of gross primary production? *Tree Physiology*, vol 18, no 2, pp. 129–134

Williams, M. (2000) Dark ages and dark areas: Global deforestation in the deep past. *Journal of Historical Geography*, vol 26, no 1, pp. 28–46

Zoltai, S. C. and Martikainen, P. J. (1996) Estimated extent of forested peatlands and their role in the global carbon cycle. In M. J. Apps and D. T. Price (eds), *Forest Ecosystems, Forest Management and the Global Carbon Cycle*. Springer. pp. 47–58

38
MODELLING CLIMATE IMPACTS ON FOREST ECOSYSTEMS

David R. Galbraith and Bradley O. Christoffersen

Introduction

The grand challenge of understanding how climate affects the functioning of Earth's forests has prompted the development of a variety of numerical modelling approaches, traversing a wide range of temporal and spatial scales. How will leaf photosynthesis respond to increasing temperature and atmospheric CO_2 concentrations? How does climate regulate interannual variability in the productivity of forest stands? What shifts can be expected in the geographical distribution of key timber species? What is the risk of forest die-back under drier and warmer climate scenarios? These are all questions for which numerical modelling approaches have helped to provide insight.

This chapter has four main aims: a) to provide the reader with a broad overview of the main modelling approaches used to simulate forest ecosystems; b) to introduce the reader to the notion of uncertainty in model predictions, using the case study of 'Amazon die-back' as an example; c) to outline how field observations can inform and improve ecosystem models; and d) to highlight some of the latest approaches holding great promise for modelling climate impacts on forests.

Overview of forest ecosystem models

The literature abounds with examples of modelling approaches applied to forest ecosystems, spanning a wide gradient of complexity. These different approaches can be sub-divided in many ways, e.g. *discrete* vs. *continuous* in time, *deterministic* vs. *stochastic*, *spatially heterogeneous* vs. *spatially homogeneous*. Perhaps one of the most useful conceptual divisions is that of *empirical* vs. *process-based models*. Empirical models are based on statistical relationships between different variables (e.g. regression) without any suggestion of causal mechanisms. Process-based models, on the other hand, aim to represent ecological processes in a mechanistic manner. In reality, these classifications represent the extreme ends of the spectrum and the majority of process-based models will include some element of empirical representation of relationships between different variables. Ultimately, forest models need to strike a balance between *realism* (the extent to which model structure mimics a real-forest system), *precision* (the degree to which model predictions are quantitatively expressed in specific mathematical expressions). It is generally not possible for a model to attain high performance in all three of these criteria – there are necessarily trade-offs involved (Levins 1966). Ultimately, the specific nature of the model used will depend on the objective and scale of the study – i.e. the tool needs to be the right one for the job.

In this section, we focus on some of the most common approaches used to simulate different aspects of forest ecology and functioning. Each of the approaches considered has traditionally been driven by distinctly different motivating research questions and different scientific communities. We focus on four modelling approaches (not necessarily mutually exclusive) occupying different positions along the empirical–mechanistic continuum: bioclimatic envelope models (BEMs), forest gap models, terrestrial biophysics and biogeochemistry models and dynamic global vegetation models (DGVMs).

BEMs

BEMs, also known as niche-based models, use associations between specific aspects of climate and the current ranges of species or ecosystems to predict their distribution under alternative climate scenarios. This approach can be dated back to the pioneering work of Köppen who mapped global climate zones based on vegetation distribution as far back as 1900 (Kottek *et al.* 2006), and Holdridge (1947), who developed a biome mapping scheme based on temperature, annual precipitation and potential evapotranspiration/precipitation ratio (potential evapotranspiration is the amount of evapotranspiration that would occur under no moisture limitation).

BEMs are commonly used to simulate species distributions under future climate scenarios (Sinclair *et al.* 2010) – such BEMs are known as species distribution models (SDMs). SDMs use statistical approaches such as general linear models or general additive models calibrated with data for current species distributions to make predictions of future species distributions under climate change. Thuiller (2003) compared four different statistical modelling approaches to simulate range shifts under climate change for 61 European tree species and found that the results depended strongly on the choice of statistical model employed. Bioclimatic envelope approaches are also used at larger scales to model biome distributions. For example, Zelazowski *et al.* (2011) used output from 17 coupled climate model projections to evaluate the potential distribution of humid tropical forests under two different climate scenarios. In this study, the authors first defined the climatological niche occupied by contemporary humid tropical forests by developing statistical models relating different precipitation and temperature variables to humid tropical forest distribution. They found that those variables most related to ecosystem water stress such as the maximum cumulative water deficit and precipitation of the driest month had the greatest explanatory power for the distribution of humid tropical forests. The variables that were found to be the best predictors of current forest distribution were then used to analyse the future climatological niche of tropical forests and model their expansion and contraction under the two climatic scenarios considered. Using this approach, the authors found considerable potential for forest expansion, especially in the Congo region, but also some potential for forest contraction in the Neotropics (Figure 38.3 in Zelazowski *et al.*, 2011). Using a similar approach, Malhi *et al.* (2009) analysed the output from 20 global climate models to investigate the likelihood of biome shift of Amazonian rainforests under twenty-first century climate change and found a tendency towards a shift in climate more suitable for seasonal forests.

Although BEMs have some clear advantages (i.e. they are simple to use and do not require much data to parameterise), they have been widely criticised based on the rather simplistic assumptions that underpin them. The major criticism is that BEMs are predicated on equilibrium climate–species or climate–biome relationships. Thus, they may not adequately capture the transient dynamics of ecosystems under anthropogenic climate change. BEMs also assume that the distributions of species and biomes are limited only by climate. In reality, feedbacks with factors such as competition and disturbance (e.g. fire) play an important role in species distribution and are not accounted for within BEMs. Finally, BEMs are calibrated based on actual climate–distribution relationships. Future climate scenarios may well take us into the domain of non-analogue climates, well beyond the calibration range of current models. Ultimately, they lack any mechanistic basis for the processes that result in changes in forest extent over time.

Forest succession (gap) models

The term 'gap model' is used to describe a diverse class of *individual-based* models developed to simulate changes in forest structure, composition and demography (the distribution of tree age and size within a forest) over time at fine scales. In the early gap models, forests were represented as a mosaic of discrete patches, typically 100–1,000 m^2, within which successional processes take place (Bugmann 2001). The size of these patches corresponds roughly to the size of a forest gap (a small clearing left in a forest due to a tree fall), hence the origin of the name used to describe this class of models. However, modern use of the term is broader than this, including also spatially explicit individual-based models. The other modelling approaches described in this chapter implicitly assume that the unique characteristics of individual trees are ultimately unimportant for ecosystem-level properties such as carbon storage. This is definitely not the case with gap models, where each individual tree is simulated as an independent entity. Gap models track the growth of individual trees over their life cycles and simulate their interactions with neighbouring trees. Indeed, in gap models community and ecosystem-level biomass, forest structure and taxonomic composition emerge from the integrated dynamics (recruitment, growth, mortality) of numerous individual trees. Community structure and dynamics, in turn, influence the local microenvironment (e.g. light availability) of individual trees, affecting growth at the level of the individual.

A large number of gap models have been developed so that the number and complexity of processes simulated varies greatly across models. However, a core set of processes is common to all gap models, including establishment, growth, light competition and mortality. The simulation of diameter growth rates typically involves the calculation of a species-specific optimal diameter increment which is modulated by environmental factors such as light and temperature. Vertical competition for light between individuals takes into consideration the height and projected leaf area of individuals, usually a function of allometric relationships with diameter. The simulation of mortality often includes two separate modes: age-related mortality and stress-related mortality. Age-related mortality depends on a species-specific aximum lifespan; the actual individuals that die over a given period of time are determined in a stochastic manner. Stress-related mortality is often linked to diameter growth efficiency. Tree establishment and regeneration is modulated by environmental conditions but also has a stochastic component. Traditional gap models, such as the JABOWA model of Botkin *et al.* (1972), did not explicitly simulate plant physiological processes although later models, such as the Hybrid model (Friend *et al.* 1993), included more mechanistic treatments of processes such as photosynthesis and nutrient cycling.

Gap models have been used in numerous applications across a diverse array of global forests. Commonly used gap models include FORCLIM (Bugmann and Solomon 1995), developed for simulation of European mountain forests, and the spatially explicit SORTIE model (Pacala *et al.* 1996), first developed and parameterised for hardwood forests in the north-eastern United

States but subsequently adapted and applied in tropical, temperate and boreal forests in various locations. One of the major uses of gap models has been to simulate changes in forest composition under climate change. Shuman et al. (2011) used FAREAST, a gap model parameterised for Russian boreal forests to simulate the effects of climate change on the composition of Siberian boreal forests. Under future warming scenarios, the model predicted that larch forests would become vulnerable to replacement by evergreen species. Gap models can also be used to simulate the impact of disturbances on ecosystems. For example, Uriarte and Papaik (2007) used the SORTIE model to evaluate the impact of hurricanes on the dynamics and structure of forests in New England.

Gap models are undoubtedly very useful tools for predicting the responses of particular species and ecosystems to rapidly changing climate. They allow for detailed predictions of compositional change that other modelling approaches are unable to provide. However, they lack generality and are also parameterisation-intensive. For example, the FAREAST model introduced in the previous paragraph simulates 58 dominant species in Eurasian boreal forests, each described by a set of 25 parameters reflecting their ability to compete for light and nutrients and their sensitivity to environmental conditions. While the detailed species-level silvicultural data needed for gap model parameterisation is available for some temperate and boreal sites, it is generally lacking for tropical forests.

Terrestrial biophysics and biogeochemistry models

While the individual-based framework employed by gap models provides important insights into forest dynamics at small scales, their use at continental and global scales is challenging given the high computational costs that would be required and the considerable amount of species-level parameterisation required. *Terrestrial biosphere models*, on the other hand, were developed for global application. These can be further classified into *soil–vegetation–atmosphere models* (SVATs), *terrestrial biogeochemistry models* (TBMs) and DGVMs (covered in the next section). Whereas the focal processes of gap models involve forest stand growth and dynamics, SVATs and TBMs focus on accounting for all major components of energy, water and carbon budgets in the system. Vegetation cover in SVATs and TBMs is prescribed and thus does not evolve over time as it does in gap models.

SVATs were developed initially by the atmospheric science community as *land surface models* (LSMs) to be embedded within AOGCMs (atmosphere–ocean general circulation models). Their initial focus was to provide process-based descriptions of *biophysical* processes such as the transfer of energy, water and momentum between the land surface and atmosphere. The Simple Biosphere Model (SiB) developed by Sellers et al. (1986) was one of the first SVATs. In its original configuration, the model simulated the transfer of radiation within the canopy, the partitioning of surface energy budget into sensible, latent (water vapour) and ground heat fluxes, as well as the dissipation of turbulent atmospheric momentum. A key component of the coupled energy and water budget was the treatment of canopy conductance (the degree to which stomata allow water to evaporate through leaf surfaces), which it simulated as a function of leaf area index (LAI), vegetation type and environmental conditions.

The rationale for the development of TBMs, on the other hand, was to adequately represent *biogeochemical* processes, particularly the carbon cycle. Their main goal is to simulate how carbon fluxes in forest ecosystems respond to changes in temperature, soil moisture, ambient CO_2 and light conditions. Commonly used TBMs include BIOME-BGC (Running and Hunt 1993), and SiB-CASA (Schaefer et al. 2008), developed by coupling the SiB SVAT with the CASA biogeochemistry model.

The simulation of photosynthesis as an enzyme-kinetic process (Farquhar et al. 1980) lies at the heart of TBMs. The Farquhar et al. equations condense the complexity of the photosynthesis reactions into a mechanistic framework whereby the rate of leaf photosynthesis is calculated as the minimum of two or three photosynthetic rates: the ability of the enzyme rubisco to fix carbon, the light-limited rate of electron transport and, optionally, the capacity of the leaf to export the products of photosynthesis (Collatz et al. 1991). In TBMs, the uptake of CO_2 for photosynthesis and loss of water via transpiration is coupled via stomatal conductance. The Collatz et al. (1991) model used by many TBMs achieves this coupling by employing a semi-empirical relationship between stomatal conductance, net leaf photosynthesis, relative humidity and ambient CO_2. To scale from leaf- to canopy-level photosynthesis, the simplest approach is the 'big leaf' approach whereby canopy-level fluxes such as gross primary productivity (GPP) can be derived from simulating leaf-level fluxes (e.g. photosynthetic rate) for one reference leaf. One common assumption for 'scaling-up' photosynthesis from leaf to canopy is that photosynthetically active radiation (PAR) and photosynthetic capacity vary proportionally through the forest canopy. Plant respiration is typically simulated as having separate *growth* and *maintenance* components. The former relates to the respiration of carbohydrates occurring during processes that result in the production of plant biomass, while the latter encompasses core metabolic functions such as the maintenance of ion gradients across membranes, which do not produce biomass. Most TBMs (and all DGVMs) simulate the biomass stocks of different plant tissues; hence the representation of *allocation* of NPP to these different tissues and the *turnover* of carbon in those pools is required. The simplest approaches involve assigning fixed carbon allocation coefficients and turnover rates to different pools, although approaches based on resource limitation are also commonly used. Finally, most TBMs also simulate the cycling of carbon through soil (e.g. decomposition, heterotrophic respiration and soil carbon storage).

One common use of TBMs with prescribed vegetation is to investigate the interannual variability of net ecosystem carbon fluxes, providing insight into how climate governs the source or sink state of ecosystems for carbon. For example, Keenan et al. (2012) evaluated the performance of 16 terrestrial biosphere models, the majority of which were TBMs, against forest–atmosphere CO_2 exchange data from 11 flux towers located in North American forests. The authors found that the models evaluated were capable of reproducing the mean magnitude of the observed annual flux variability but not its timing, systematically failing to reproduce observed variability in spring GPP. Thus, TBMs and SVATs provide useful information on the response of fluxes of carbon and water to climate but their assumption of equilibrium vegetation cover means that they are unable to simulate potential climate change-induced biome shifts.

DGVMs

DGVMs are currently the state-of-the-art models for simulating climate change impacts on ecosystems over large spatial scales (typically globally), and arguably are the class of vegetation models currently under the most intense development. Essentially, they integrate the research carried out in the BEM, gap model, SVAT and TBM communities into a coherent framework capable of simulating the effects of climate change on plant geography, vegetation dynamics, biophysics and biogeochemistry. Thus, DGVMs simulate a number of processes occurring on different timescales, with physiological and biophysical processes such as photosynthesis and evapotranspiration typically simulated on sub-daily timescales while ecological processes such as tree mortality or competition among plant functional types (PFTs) simulated on longer timescales (months to years). Among the most widely used DGVMs are JULES-TRIFFID

(Clark *et al.* 2011), the Lund–Potsdam–Jena (LPJ) model (Sitch *et al.* 2003) and its variants (Smith *et al.* 2001), and IBIS (Foley *et al.* 1996).

DGVMs use the concept of PFTs to represent the global functional diversity of plant life. The number of PFTs represented by DGVMs varies across models. For example, the original LPJ model had ten PFTs whereas JULES-TRIFFID only has five. JULES-TRIFFID includes only two forest PFTs (needleleaved trees and broadleaf trees) whereas LPJ includes eight forest PFTs spanning tropical, temperate and boreal biomes, broadleaf and needleleaf physiognomies and temperature and moisture-driven leaf phenologies. Each PFT is defined by a set of parameters which is invariant in time and space. These parameters ultimately determine the relative sensitivity of different PFTs to environmental change. The suite of parameters possessed by PFTs varies from model to model but typically includes parameters linked to physiological functioning, such as the temperature optimum of photosynthesis, as well as tree-level carbon cycling, such as turnover rates of different plant tissues.

The schemes used in DGVMs to simulate biophysical and biogeochemical processes are practically identical to those used in SVATs and TBMs. The main difference between DGVMs and SVATs and TBMs is the representation of dynamic rather than static vegetation cover, in which the fractional cover of PFTs evolves over time, usually as a function of their respective growth (NPP) rates. The underlying plant community structure, however, differs among DGVMs. The earliest (or *first generation*) DGVMs such as IBIS and TRIFFID contain no real representation of plant population dynamics. Instead, they consider only aggregate vegetation blocks which occupy a specific fraction of a grid cell (Figure 38.1, top left panel). In these

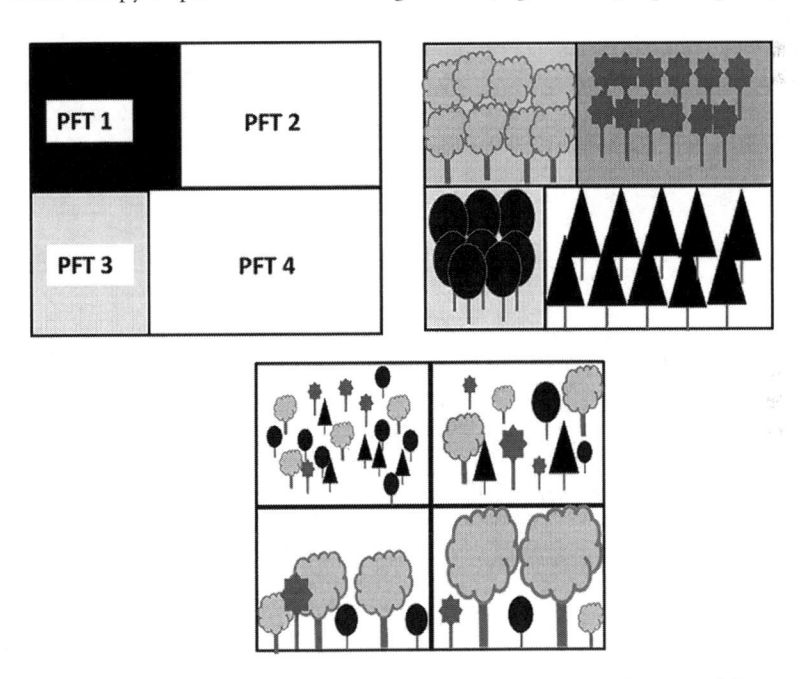

Figure 38.1 Schematic illustrating the representation of forest structure in different modelling approaches. Top left: area-based DGVM where trees are not simulated explicitly. In this approach the simulated area is divided into blocks of different sizes occupied by different plant functional types (PFTs). Top right: 'average individual' approach where PFT blocks are occupied by a simulated density of individuals with identical properties. Bottom: Gap model landscape mosaic where simulated patches contain trees of different species and different sizes at different stages of succession

models, vegetation biomass is calculated on an area basis without any consideration of ecosystem structure such as number of trees and their size distribution. In contrast, LPJ simulates varying densities of individual trees within each grid cell. However, all of these trees are identical and there is no size/age structure present (Figure 38.1, top right panel). In recent years, DGVMs have begun to include more ecologically realistic representations of vegetation dynamics. Such models are known as *second generation* DGVMs. For example, the Ecosystem Demography (ED) model (Moorcroft et al. 2001) explicitly represents forest age and size structure, classifying vegetation into discrete cohorts of a specific age/size, corresponding to a particular time since disturbance. Other examples of second generation DGVMs include LPJ-GUESS (Smith et al. 2001) and aDGVM (Scheiter and Higgins 2009).

DGVMs can be run in *offline* mode where they are forced with climate data either from historical datasets or future climate scenarios from general circulation models (GCMs). Offline simulations, however, do not allow for examination of the feedbacks of the land surface on the climate system. If appropriate *in situ* calibration data is available, offline site-level DGVM simulations can be run. These are often carried out for sites where there is good validation data for evaluation of DGVM simulations. More often, however, DGVMs are applied regionally or globally to simulate larger-scale patterns of productivity, biomass and biome distribution. An example of regional DGVM output is shown in Figure 38.2, which depicts changes (2010–2099) in above ground biomass over Amazonia simulated by JULES-TRIFFID when forced with climate data from the CCSM climate model under different land use and CO_2 fertilisation scenarios. DGVMs can also be run in *online* mode, where they are interactively coupled to climate models. The term 'Earth System Model' is often used to describe coupled land–atmosphere–ocean models that include interactions between components. Such coupled simulations have been used to assess the strength of land surface feedbacks on the carbon cycle. For example, Friedlingstein et al. (2006) documented a very large uncertainty in climate–carbon cycle feedbacks across coupled models, with the additional atmospheric CO_2 concentrations due to land surface feedbacks ranging from 20 parts per million volume (ppmv) to 200 ppmv up to 2100.

Uncertainty in model projections: the case study of Amazon die-back

Substantial uncertainty surrounds projections of climate change impacts on forests and other ecosystems. This uncertainty arises from several sources (Figure 38.3), which we will illustrate with reference to the modelled result of Amazonian rainforest 'die-back'. This result refers to the large-scale loss of rainforest under severe twenty-first century drying and warming in Amazonia and was first published by White et al. (1999) in simulations with the Hybrid vegetation model when forced with climate data from the HadCM3 climate model. The result has been replicated by several studies since then (e.g. Galbraith et al. 2010) but a number of other studies suggest large-scale biomass gains over Amazonia over the next 100 years rather than losses (e.g. Huntingford et al. 2013). The large differences in model projections illustrate the considerable uncertainty that surrounds the result.

It is helpful to partition the uncertainty surrounding predictions of Amazon die-back into three major sources associated with: a) future atmospheric *greenhouse gas (GHG) concentrations*; b) different *projections of future climate* by GCMs for given GHG concentration scenarios; and c) the *simulated response of the Amazon rainforest* to a given change in climate. Future GHG concentrations are uncertain because they ultimately depend on future use of fossil fuels by humanity. This in turn depends upon economic, technological and societal changes which cannot easily be foreseen. To account for this uncertainty, different scenarios proposed by the International Panel on Climate Change (IPCC) have been used which outline alternative

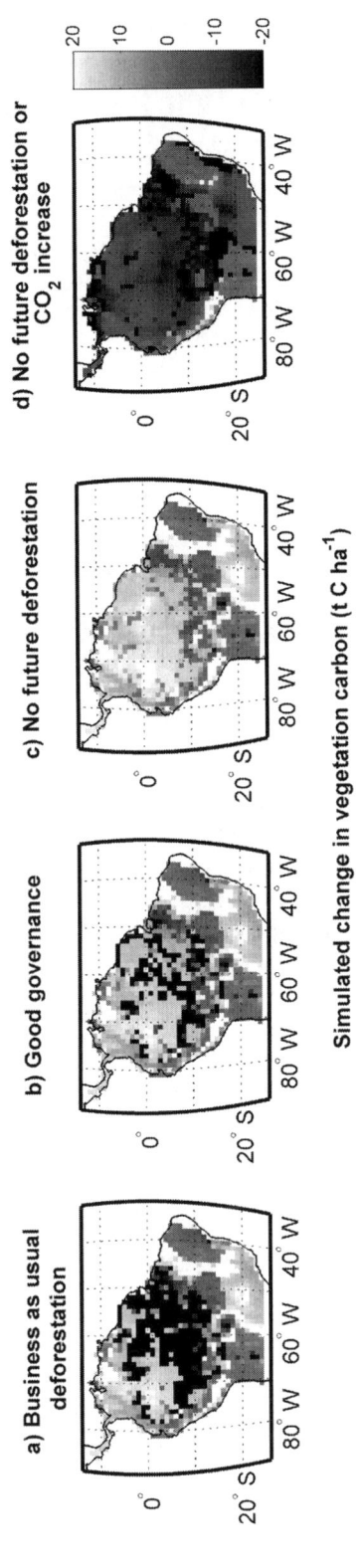

Figure 38.2 Change in vegetation carbon stocks (2009–2100) as simulated by the JULES-TRIFFID model forced with climate data from the CCSM climate model for four different land use and CO_2 scenarios: a) increasing CO_2 and business-as-usual land use; b) increasing CO_2 and good governance land use scenario; c) increasing CO_2 and no future deforestation scenario; d) no future deforestation and no future increases in atmospheric CO_2 concentration

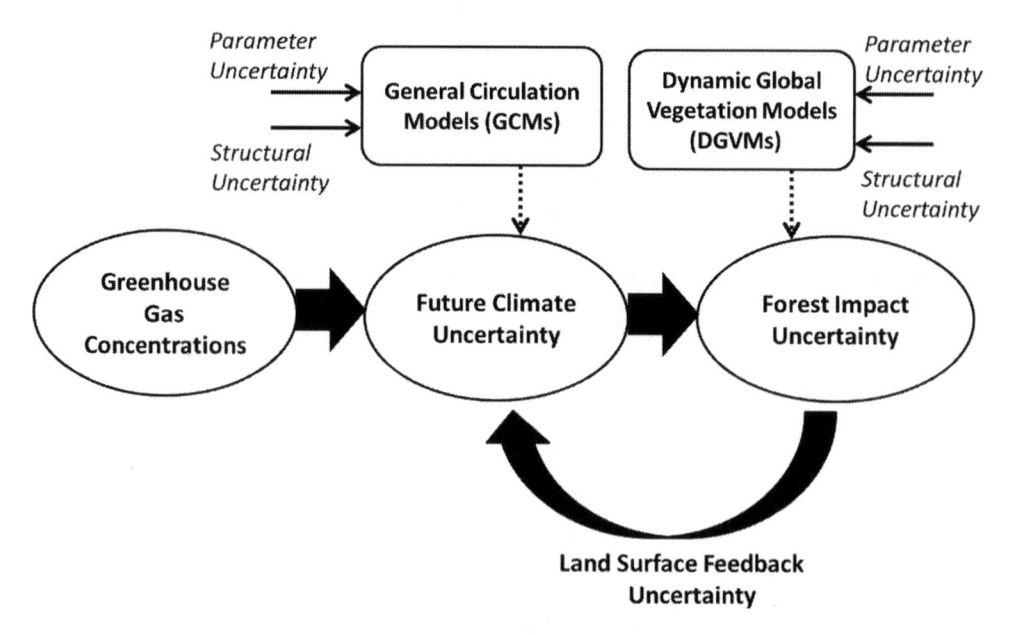

Figure 38.3 Summary of the major sources of uncertainty associated with the simulation of climate change impacts on forests

development and economic pathways which may ensue in the future. In the latest IPCC report, these are called Representative Concentration Pathways (RCPs). These pathways project future atmospheric CO_2 concentrations of between 421–936 ppmv in 2100.

While uncertainty in future GHG concentrations is mainly a result of uncertainty in socio-economics and policy, the uncertainty due to GCM predictions of climate and their resulting impacts on vegetation reflect scientific uncertainty in our understanding of climate and ecosystem processes. For a given GHG concentration scenario, different GCMs give widely varying predictions of future rainfall for Amazonia, although there is a general trend across most models towards decreases in dry season rainfall, at least in eastern Amazonia (Malhi *et al.* 2009). Whereas all GCMs predict increases in average temperatures over Amazonia this century, the range of predictions is quite broad (2–8°C for mid-range climate scenarios) (Malhi *et al.* 2009). Predictions of die-back have generally been most severe where vegetation models have been forced with HadCM3 climate data (White *et al.* 1999, Galbraith *et al.* 2010), as this GCM projects particularly large decreases in rainfall and increases in temperature over Amazonia. Loss of forest cover has been reported under other climate model predictions, although to a lesser degree.

Finally, Amazonian rainforests in some DGVMs are more sensitive to temperature increases and rainfall reductions than in other DGVMs. For example, Sitch *et al.* (2008) found that losses of vegetation carbon occurred in five DGVMs when forced with HadCM3 climate patterns and with CO_2 concentrations corresponding to the same emissions scenario, but that these varied widely in magnitude and geographical extent. Galbraith *et al.* (2010) further showed that the climatic factors causing Amazon die-back differed significantly among three DGVMs when forced with the same HadCM3 data. Temperature was a much more important cause of losses of Amazonian biomass than precipitation in the MOSES-TRIFFID and Hyland models, while in LPJ the contributions of both of these drivers to Amazon rainforest biomass losses were similar.

The uncertainty in climate and vegetation model responses can further be partitioned into *structural* and *parametric* uncertainty. Parametric uncertainty arises because the best values for model parameters are not well known. For example, vegetation models vary widely in the assumed turnover rates of woody biomass in tropical forest trees (Galbraith et al. 2013) or the assumed rooting depth of the vegetation. To understand the implications of this uncertainty, perturbed parameter ensemble simulations are often conducted where the values of the parameters are varied within realistic ranges. Structural uncertainty stems from uncertainty in the form of the mathematical representation of processes within models. An important example of this is the representation of the temperature dependency of plant respiration, where it has been shown that an exponential dependency of plant respiration on temperature (as in the TRIFFID model) results in considerably greater die-back than a formulation in which the respiration response to temperature mimics that of photosynthetic capacity, with increases up to a temperature optimum, beyond which a decline occurs (Huntingford et al. 2013). This is by no means an exhaustive list of the sources of uncertainty in model projections. In coupled simulations, uncertainty in the strength of feedbacks is also important (Friedlingstein et al. 2006), while there are further uncertainties arising from the specification of initial conditions and, in some cases, intrinsic uncertainties resulting from the stochastic nature of dynamical equations.

How data informs the modelling process

Field data plays an important role in the calibration, validation and development of models used in the simulation of climate impacts on forests. The level of calibration required depends on the nature of the model used. Gap models require species-level calibration while the PFTs used in DGVMs require parameter values which represent average values across large areas. A wide range of forest data sources have been used in the validation of vegetation models. This includes data from flux tower sites on forest exchange of water and carbon with the atmosphere, forest inventory data and regional datasets derived from remote sensing products. More recently, validation of ecosystem models has also been extended to include data from global change experiments such as large-scale drought experiments (Powell et al. 2013) and Free Air CO_2 Enrichment (FACE) experiments (De Kauwe et al. 2014). Continuous evaluation of model performance with available data is an essential aspect of the model development process as it can highlight important model deficiencies and key areas of improvement. This process is illustrated well by the study of Powell et al. (2013), who compared output from four DGVMs against observations from a manipulated drought experiment in Caxiuanã National Forest in the Brazilian Amazon. In this simulation, the models were run with *in situ* climate data, parameterised with *in situ* soil information and followed the rainfall exclusion conditions implemented in the experiment (incident rainfall was reduced by ~50 per cent). The study found that models agreed well with observed carbon flux measurements under control (non-drought) conditions but struggled to capture the impacts of the drought. This was particularly true for simulated changes in above ground biomass, which the models greatly underestimated compared to the observations. This reflects the lack of appropriate simulation of tree mortality processes in the models.

Frontiers in forest model development

Models that simulate climate impacts on forests are continually being refined and improved. This is especially true of global vegetation models, where there is a strong drive towards incorporation of additional processes that exert important controls on ecosystem functioning. A major focus of development has been the inclusion of interactive nitrogen dynamics and some groups have also made advances in including interactive phosphorus schemes. The inclusion of land use change and land management processes is another area of considerable activity in the global vegetation modelling community. Other development areas receiving considerable amounts of attention are fire modelling, ozone impacts on vegetation, emissions of volatile organic compounds and drought-induced mortality.

Future developments in global forest modelling are likely to include more sophisticated representations of biodiversity and ecological processes that thus far have virtually been ignored such as seed dispersal and trophic interactions. Already there are some exciting developments in place around these themes. Over the last couple of years, trait-based approaches have emerged to characterise forest tree functional diversity and have been used both in individual-based models (Fyllas *et al.* 2014) and global plant functional trade-off-based models (Pavlick *et al.* 2013). Scheiter *et al.* (2013) recently developed an individual-based model (aDGVM2) in which individual plants adopt a unique combination of traits that defines how they grow and compete with other plants, implemented via a scheme that allows for trait mutation and inheritance. This model explicitly simulates the dynamic evolution of functional trait distributions over time and in response to environmental stress. Finally, the first global ecosystem model, named the Madingley Model (Harfoot *et al.* 2014) has just been published. This new class of model simulates the fate of nearly all organisms, including herbivores, omnivores and carnivores, as well as plants and their interactions. Thus, this new generation of models will allow us to address a host of new, exciting questions about the current and future dynamics of forest systems.

References

Botkin DB, Janak JF, Wallis JR (1972) Some ecological consequences of a computer model of forest growth. *Journal of Ecology*, **60**, 849–872.

Bugmann H (2001) A review of forest gap models. *Climatic Change*, **51**, 259–305.

Bugmann HKM, Solomon AM (1995) The use of a European Forest Model in North America: a study of ecosystem response to climatic gradients. *Journal of Biogeography*, **22**, 477–484.

Clark DB, Mercado LM, Sitch S, Jones CD, Gedney N and others (2011) The Joint UK Land Environment Simulator (JULES), model description – Part 2: Carbon fluxes and vegetation dynamics. *Geoscientific Model Development*, **4**, 701–722.

Collatz GJ, Ball JT, Grivet C, Berry JA (1991) Physiological and environmental regulation of photosynthesis and transpiration – A model that includes a laminar boundary layer. *Agricultural and Forest Meteorology*, **54**, 107–136.

De Kauwe MG, Medlyn BE, Zaehle S, Walker AP, Dietze MC and others (2014) Where does the carbon go? A model-data intercomparison of vegetation carbon allocation and turnover processes at two temperate forest free-air CO_2 enrichment sites. *The New Phytologist*, **203**, 883–899.

Farquhar GD, Caemmerer SV, Berry JA (1980) A biochemical model of photosynthetic CO_2 assimilation in leaves of C3 species. *Planta*, **149**, 78–90.

Foley JA, Prentice IC, Ramankutty N, Levis S, Pollard D and others (1996) An integrated biosphere model of land surface processes, terrestrial carbon balance, and vegetation dynamics. *Global Biogeochemical Cycles*, **10**, 603–628.

Friedlingstein P, Cox P, Betts R, Bopp L, von Bloh W and others (2006) Climate-carbon cycle feedback analysis: Results from the (CMIP)-M-4 model intercomparison. *Journal of Climate*, **19**, 3337–3353.

Friend AD, Shugart HH, Running SW (1993) A physiology-based gap model of forest dynamics. *Ecology*, **74**, 792–797.

Fyllas NM, Gloor E, Mercado LM, Sitch S, Quesada CA and others (2014) Analysing forest productivity using a new individual and trait-based model (TFS v. 1.0). *Geoscientific Model Development*, **7**, 1251–1269.

Galbraith D, Levy PE, Sitch S, Huntingford C, Cox P and others (2010) Multiple mechanisms of Amazonian forest biomass losses in three dynamic global vegetation models under climate change. *New Phytologist*, **187**, 647–665.

Galbraith D, Malhi Y, Quesada B, Castanho AD, Doughty CE and others (2013) The residence time of woody biomass in tropical forests. *Plant Ecology and Diversity*, **6**, 139–157.

Harfoot MBJ, Newbold T, Tittensor DP, Emmet S, Hutton J and others (2014) Emergent global patterns of ecosystem structure and function from a mechanistic general ecosystem model. *Plos Biology*, **12**, e1001841. doi:10.1371/journal.pbio.1001841

Holdridge LR (1947) Determination of world plant formations from simple climatic data. *Science*, **105**, 367–368.

Huntingford C, Zelazowski P, Galbraith D, Mercado LM, Sitch S and others (2013) Simulated resilience of tropical rainforests to CO_2-induced climate change. *Nature Geoscience*, **6**, 268–273.

Keenan TF, Baker I, Barr A, Ciais P, Davis K and others (2012) Terrestrial biosphere model performance for inter-annual variability of land-atmosphere CO_2 exchange. *Global Change Biology*, **18**, 1971–1987.

Kottek M, Grieser J, Beck C, Rudolf B, Rubel F (2006) World map of the Koppen-Geiger climate classification updated. *Meteorologische Zeitschrift*, **15**, 259–263.

Levins R (1966) The strategy of model building in population biology. *American Scientist*, **54**, 11.

Malhi Y, Aragao L, Galbraith D, Huntingford C, Fisher R and others (2009) Exploring the likelihood and mechanism of a climate-change-induced dieback of the Amazon rainforest. *Proceedings of the National Academy of Sciences of the United States of America*, **106**, 20610–20615.

Moorcroft PR, Hurtt GC, Pacala SW (2001) A method for scaling vegetation dynamics: The ecosystem demography model (ED). *Ecological Monographs*, **71**, 557–585.

Pacala SW, Canham CD, Saponara J, Silander JA, Kobe RK, Ribbens E (1996) Forest models defined by field measurements: Estimation, error analysis and dynamics. *Ecological Monographs*, **66**, 1–43.

Pavlick R, Drewry DT, Bohn K, Reu B, Kleidon A (2013) The Jena Diversity-Dynamic Global Vegetation Model (JeDi-DGVM): A diverse approach to representing terrestrial biogeography and biogeochemistry based on plant functional trade-offs. *Biogeosciences*, **10**, 4137–4177.

Powell TL, Galbraith DR, Christoffersen BO, Harper A, Imbuzeiro H and others (2013) Confronting model predictions of carbon fluxes with measurements of Amazon forests subjected to experimental drought. *New Phytologist*, **200**, 350–364.

Running SW, Hunt ER (1993) Generalization of a forest ecosystem process model for other biomes, BIOME-BGC, and an application for global-scale models. Scaling processes between leaf and landscape levels. In: Ehleringer JR, Field CB (eds) *Scaling Physiological Processes: Leaf to Globe*. Academic Press, San Diego, US, 141–158.

Schaefer K, Collatz GJ, Tans P, Denning SA, Baker I and others (2008) Combined Simple Biosphere/Carnegie-Ames-Stanford Approach terrestrial carbon cycle model. *Journal of Geophysical Research-Biogeosciences*, **113**.

Scheiter S and Higgins SI (2009) Impacts of climite change on the vegetation of Africa: an adaptive dynamic vegetation modelling approach. *Global Change Biology*, **15**, 2224–2246.

Sellers PJ, Mintz Y, Sud YC, Dalcher A (1986) A simple biosphere model (SiB) for use within general circulation models. *Journal of the Atmospheric Sciences*, **43**, 505–531.

Shuman JK, Shugart HH, O'halloran TL (2011) Sensitivity of Siberian larch forests to climate change. *Global Change Biology*, **17**, 2370–2384.

Sinclair SJ, White MD, Newell GR (2010) How useful are species distribution models for managing biodiversity under future climates? *Ecology and Society*, **15**, 8.

Sitch S, Smith B, Prentice IC, Arneth A, Bondeau A and others (2003) Evaluation of ecosystem dynamics, plant geography and terrestrial carbon cycling in the LPJ dynamic global vegetation model. *Global Change Biology*, **9**, 161–185.

Sitch S, Huntingford C, Gedney N, Levy PE, Lomas M and others (2008) Evaluation of the terrestrial carbon cycle, future plant geography and climate-carbon cycle feedbacks using five Dynamic Global Vegetation Models (DGVMs). *Global Change Biology* **14**, 2015–2039.

Smith B, Prentice IC, Sykes MT (2001) Representation of vegetation dynamics in the modelling of terrestrial ecosystems: Comparing two contrasting approaches within European climate space. *Global Ecology and Biogeography*, **10**, 621–637.

Thuiller W (2003) BIOMOD – Optimizing predictions of species distributions and projecting potential future shifts under global change. *Global Change Biology* **9**, 1353–1362.

Uriarte M, Papaik M (2007) Hurricane impacts on dynamics, structure and carbon sequestration potential of forest ecosystems in Southern New England, USA. *Tellus Series a-Dynamic Meteorology and Oceanography*, **59**, 519–528.

White A, Cannell MGR, Friend AD (1999) Climate change impacts on ecosystems and the terrestrial carbon sink: A new assessment. *Global Environmental Change-Human and Policy Dimensions*, **9**, S21-S30.

Zelazowski P, Malhi Y, Huntingford C, Sitch S, Fisher JB (2011) Changes in the potential distribution of humid tropical forests on a warmer planet. *Philosophical Transactions of the Royal Society a-Mathematical Physical and Engineering Sciences*, **369**, 137–160: http://rsta.royalsocietypublishing.org/content/369/1934/137

PART VII

Human ecology

39
MULTIPLE ROLES OF NON-TIMBER FOREST PRODUCTS IN ECOLOGIES, ECONOMIES AND LIVELIHOODS

Charlie M. Shackleton

Introduction

During the colonial period, forests were valued by governments and conservators or forest officers mostly for their timber and watershed functions. Although millions of forest-dwelling or adjacent people made extensive uses of 'other' forest resources, these resources were not recognised by colonial foresters or administrators. Other than a few which became global commodities (such as rubber, Brazil nuts, rattan), the majority, at best, attracted interest only as curiosities by anthropologists and ethnographers of the era. As tropical forest deforestation accelerated in the latter half of the 1900s, a peculiar coalition of interests and processes examined the importance and potential of these other forest resources through new lenses, positing that they offered solutions or alternatives to seemingly disparate concerns of: (i) tropical forest deforestation; (ii) loss of culture and tradition amongst forest peoples; and (iii) underdevelopment and poor welfare in what were often marginalised and remote communities. These positions were crystallised into a strong call for action on the basis of economic analyses showing that the use of such products could rival the income from tropical timber, whilst simultaneously having far better outcomes in conserving forests and sustaining or improving livelihoods (Peters *et al.* 1989, De Beer and McDermott 1996). The result was a steadily growing understanding of the various roles these resources play, increasingly coherent terminology, and inclusion of these resources in management plans for forests and, in some instances, national or sub-national development policies.

Defining non-timber forest products

Because these other products grow in multiple habitats, are used by people for utilitarian and cultural purposes, and may be marketed, they have been examined from multiple disciplinary perspectives which resulted in an unsatisfying array of names, terms and acronyms. Fortunately, with time there has been convergence, and most commentators now refer to them as either non-timber forest products (NTFPs) or non-wood forest products (NWFPs). The latter is an artefact of the structural organisation of the Food and Agriculture Organisation (FAO), which had a Wood Division at the time, and is not used as much as the term NTFPs (Belcher 2003), which will be used here. The interdisciplinary nature of the subject also resulted in a range of definitions of the same terms, which through time have been narrowed down. Whilst specific words may differ between definitions, the key elements of the NTFP concept are widely accepted to include (De Beer and McDermott 1996, Belcher 2003, Shackleton *et al.* 2011): (i) largely wild (i.e. there can be some harvest from fields or cultivated populations; but the bulk of the species populations must be wild); (ii) biological products; (iii) from natural, semi-natural and transformed lands (despite the name, not just forests!); (iv) that are used or traded by local communities; and (v) the bulk of the benefits (cultural, material or economic) accrue to those local communities who make management decisions pertaining to the NTFPs.

The importance of NTFPs in ecosystem composition and dynamics

The majority of work on NTFPs has focussed on their contribution to provisioning services for human welfare, along with the argument that conserving forests to secure NTFP stocks will help protect natural systems from land transformation and hence conserve biodiversity. These arguments overlook that NTFPs also play roles in local ecologies and provide supporting and regulating services within forests and other ecosystems, which may at times or in certain situations result in trade-offs between their different roles. They are also components of biodiversity in their own right, and need not be relegated solely to being a surrogate reason for protecting other species. Indeed, they can constitute a very rich component of forest flora and fauna. For example, approximately 94 per cent of canopy tree species and 77 per cent of sub-canopy tree species encountered in South African indigenous forests have some recorded traditional or commercial use (Geldenhuys 1999). Similarly, Peters *et al.* (1989) reported that 42 per cent of all the individual trees in one hectare of tropical forest in Peru provided products that were marketed in the closest town, with the proportion being even higher if non-marketed NTFPs were included. These two examples show that NTFPs can constitute a significant proportion of forest biodiversity in terms of species and individuals.

Many NTFP species are community dominants by either stature or abundance, and thus play significant roles in shaping community composition, dynamics and nutrient cycles, as well as providing resources to other organisms. For example, riparian reedbeds are common sources of fibre for thatching and weaving by rural people around the world. It is also common for reedbed communities to be dominated by only a few species (such as *Phragmites*, *Papyrus*, *Juncus* or *Typha*) and therefore changes in the abundance of such reeds through fire, harvesting or flooding have marked effects on the abundance and performance of other reedbed species, including plants, invertebrates and birds (e.g. Valkama *et al.* 2008). Reedbeds are also extremely important in riparian nutrient dynamics. At a species level, the marula tree (*Sclerocarya birrea* subsp. *caffra*) can be used as an example (Hall *et al.* 2002). It is a widespread tree species in the semi-arid areas of sub-Saharan Africa, with multiple uses by local communities for its fruit (eaten raw or fermented into various liquors; the kernel is also extracted and eaten), bark for medicine, stones for decoration, kindling for fire, truncheons for live fencing and lightweight timber for carving. *Sclerocarya birrea* is also typically a large tree which plays a significant role in the ecological community as described by Hall *et al.* (2002). The fruit is a key food source to dozens of vertebrate and invertebrate species. The copious flowering makes it a significant resource for insect pollinators. The species is a preferred host to several mistletoe species (one of which, *Erianthimum dregei*, is also an NTFP to wood carvers) and is the only host to

Agelanthus crassifolius. The large canopy provides shade, which alters the understorey light, temperature and moisture regimes and promotes a markedly different sub-canopy herbaceous composition relative to adjacent areas. The overall productivity in these sub-canopy locations is also higher, which in turn alters natural fire intensities. Similar arguments could be made for many other dominant NTFPs, such as bamboos in Southeast Asia, vegetable ivory palms and Brazil nuts in Amazonia, or faunal species such as edible caterpillars. These are all NTFPs whilst also having significant, but largely unstudied, roles in community structure, dynamics or composition.

The ecology of NTFPs

Given that NTFPs span dozens of life forms and functional types (from fish and caterpillars to bulbs and roots through to leaves, fruits and resins), thousands of species and often many different parts of the same species it is impossible to derive a functional ecology of NTFPs (by functional ecology I mean the functions and roles of specific types of organisms in communities and landscapes and the organismal traits that characterise specific functional groups). But there is one characteristic that unites them all, namely, that they are harvested by humans. Therefore, the central question within autecological studies (i.e. the ecology of an individual species) of any NTFP species is 'how does it respond to harvesting at different levels (frequencies, intensities, seasons and techniques)?' This is typically part of a broader package of applied questions around what the ecologically sustainable yield for a specific NTFP is, how that yield varies temporally and spatially, and how harvesting interacts with other environmental pressures on a population (e.g. fire, herbivory, disease, drought, pollution) to undermine or facilitate individual or population persistence and allowable offtake. Interestingly, despite the millions of people involved in harvesting NTFPs for household consumption or trade, and the billions of dollars in value annually that this represents, we have very little autecological knowledge of most NTFP species, even those on high value markets.

Although it is impossible to derive generalisable biological similarities between all NTFP species, there are some emerging generalities that are potentially useful starting points in helping understand their nature and hence potential management approaches (Ticktin and Shackleton 2011), but one must always be alert to the exceptions:

- As biological organisms all NTFPs can be harvested at ecologically sustainable levels. Whether this can be achieved in practice with a satisfactory 'economic' return is a different question.
- Just because an NTFP is harvested, it does not doom the species to inevitable overharvesting and decline as many more conservation-minded observers fear. The meta-review by Stanley et al. (2012) showed that almost two-thirds of NTFP harvest systems they examined appeared to be ecologically sustainable. Whether or not ecological sustainability is achieved or maintained in the face of changing contexts depends upon the governance system, which itself is vested in the cultural, economic and political milieu.
- The supply of NTFPs is never static in time and place, and hence management plans and approaches must be adaptive.
- The impacts of harvesting of NTFPs are evident at various hierarchical scales above and below that of the population (from genetic to ecosystem) and should be examined at all scales for a full understanding and appropriate management (Ticktin 2004).
- The impact of harvesting on organism vital rates (i.e. rates of survival, growth and reproduction), and in turn, the sensitivity of the population growth rate to individual vital

rates, are key attributes in determining ecological persistence and growth (Ticktin and Shackleton 2011).
- It is harder and more complex to manage multipurpose use species and landscapes sustainably than it is for single use species (Herrero-Jáuregui et al. 2013).
- Species with large populations and fast growth rates offer better prospects for ecologically sustainable management than do species with small populations and low growth rates.
- Species with high and continuous recruitment have a better likelihood for ecologically sustainable management than do species with low and episodic recruitment (i.e. species with erratic and unpredictable periods of recruitment).

The roles of NTFPs in livelihoods

The importance of NTFPs in rural, and some urban, livelihoods in the developing world is undisputed. The nature and magnitude of the contribution varies spatially and temporally (Table 39.1), but there are no rural communities that do not make some use of one or more NTFP species to some degree. Understanding the spatial and temporal variation is important for management of NTFPs, natural habitats generally and also to optimise the opportunities for human benefits. To be able to do so requires disaggregation of the various roles or contributions that NTFPs make and to whom they are made (Shackleton and Shackleton 2004).

Table 39.1 provides some illustrative figures from a handful of studies. There are many such studies in the international literature. Meta-analyses are compounded by differences in the methods used and discrepancies in what NTFPs are included (or excluded), not to mention whether values are calculated as gross or net. Most early studies were based on one-off questionnaire surveys and did not take costs or seasonal differences into account. As the discipline has progressed so have the methods employed, with the Poverty and Environment Network of CIFOR seeking to inculcate a set of minimum requirements (seasonal or quarterly questionnaires; household and village scale; costs to be captured; analysis by quintiles or quartiles) (Angelsen and Lund 2011, Angelsen et al. 2014). Nonetheless, no single method is without its drawbacks, and questionnaire surveys typically underestimate values of commonly used NTFPs and overestimate the value of less used ones (Gram 2001, Menton et al. 2010). Therefore, several methods should be deployed to foster triangulation and combine quantitative rigour with qualitative depth. Thus, techniques such as household diaries, participant observation, market surveys, key-informant interviews and a host of Participatory Rural

Table 39.1 Illustrative recent examples of the range of NTFP contribution to rural household income (cash and non-cash combined)

Contribution to household income (%)	Location	Reference
64	Northeast Peru	L'Roe and Naughton-Treves (2014)
59	Cameroon	Angelsen et al. (2014)
44	Northwest Zambia	Kalaba et al. (2013)
39	Central Ethiopia	Mamo et al. (2007)
38	Central Ghana	Appiah et al. (2009)
34	Central India	Mitra and Mishra (2011)
32	Southern China	Hogarth et al. (2013)
15	Southern Malawi	Kamanga et al. (2009)
14	Southeast Bangladesh	Kar and Jacobson (2012)

Appraisal (PRA) tools can be used to great success, especially when implemented over extended periods to account for intra- and inter-annual changes in yields, prices and market demand, including in times of difficulty or crisis.

What Table 39.1 does not show is the inter- and intra-household variation in the importance of NTFPs. Findings over the last decade show that such variation can be substantial (Angelsen *et al.* 2014). Commonly reported patterns, with the inevitable exceptions of course, are that:

- NTFPs contribute significantly more to total household income of poorer households than richer ones.
- This often also has a gender dimension because most (albeit not all) female-headed households are usually poorer than male-headed ones in most developing countries.
- The absolute quantities consumed may not differ greatly between wealth quintiles, or may even be higher in richer ones, but this does not translate into higher contributions of NTFPs because richer households have greater total household incomes when all sources are considered (e.g. Kalaba *et al.* 2013).
- Richer households tend to dominate trade in high value NTFPs, whereas poor households lead trade in high volume, low value NTFPs with low capital requirements.
- However, specialisation in trade in low value products can result in poverty traps for some households as they typically have insufficient means or assets to move to other potentially more rewarding activities if returns decline.
- Local trade between households or in local markets can be substantial and provides an income equalising role as poorer households sell common NTFPs to richer ones, who may buy rather than collect their own (e.g. Shackleton and Shackleton 2004).
- The value attributed to NTFP collection, use, and at times, trade, by children has been grossly overlooked and requires more work.

NTFPs in household provisioning

The most widespread use of NTFPs is the frequent collection of locally accessible resources to meet household needs for food (fruits, seeds, nuts, leafy vegetables, mushrooms, honey, bushmeat), energy (firewood, charcoal, kindling), shelter/protection (wooden poles for housing and fences; thatch grass, reeds or palm fronds for roofing), medicines (bark, roots, leaves), fibres (lianas, palm fronds, rattans, grasses, bark), and agricultural implements (made from local woods). Individual households can use dozens of species during the course of a year, and within communities, hundreds of species are known and used. Shackleton and Shackleton (2004) reported the mean amounts used per household annually for some NTFPs across several sites in South Africa as approximately 58 kg of wild leafy vegetables, 5.2 tons of firewood, 104 kg of wild fruits, 180 poles for building.

NTFPs allow cash saving

A second advantage of NTFP use for household provisioning is that collection of locally available NTFPs allows households, especially the poorest ones, to allocate scarce cash for livelihood needs that cannot be collected from the wild, for example agricultural inputs such as seeds or fertilisers, school fees or books, buying of stock to trade, or some accumulation of savings against future misfortune, such as medical or transport expenses (Shackleton and Shackleton 2004). Thus, the use of NTFPs for food, energy, or any other purposes means the household does not have to purchase that same need. This represents a double benefit from the direct use of NTFPs.

NTFPs for cash income generation

Historically, most NTFPs were used for household provisioning. But there are few communities remaining that are totally divorced from the formal economy and monetisation of inter-household transfers. This presents opportunities for cash generation through the sale of NTFPs on local markets and further afield. Such trade may take the form of selling small amounts at *ad hoc* intervals to supplement other livelihood activities, through bulk sales of low value products on local markets, to value-added products on specialised or niche markets, nationally or internationally. The income earned and its relative contribution to total household income depends on the nature of the product, the nature of the market and the degree of engagement in trading by the household. Because many households only engage in NTFP trade as a supplementary livelihood activity, the absolute incomes earned per week or per month or per season can be low. This reduces the mean income earned per household within a community such that external commentators frequently claim that NTFP trade is not a viable rural development or poverty alleviation option. Therefore, it is better to examine the relative cash returns on the basis of per hour worked, which typically reveals that even low-scale NTFP trade is a viable use of time, netting 2–7 times better returns per hour than local wage labour (Shackleton *et al.* 2007). Thus, if there is sufficient market demand, this may, in some instances, be a better option for low-skilled workers than daily wage labour. For those engaged in NTFP trade on a more full-time basis, incomes can be well above national poverty lines, although it is highly dependent on the nature of the product (Shackleton *et al.* 2008, Cunningham 2011).

Local or wider trade in NTFPs provides other benefits beyond just the cash income. Several researchers have generated substantive lists of non-monetary benefits, such as building of social networks, development of business skills, being one's own boss, building of pride and confidence in making a living against difficult odds, being able to work from home and multitask with other household needs, contributing to maintenance of cultural processes and products based on traditional knowledge by offering cultural products in the public eye, and the like. These cannot be valued in monetary terms, but if deemed valuable by NTFP traders themselves, then they need to be acknowledged in development programmes and policies as constituents of well-being and human capital.

NTFPs as safety nets

The contribution of NTFPs as safety nets is receiving increasing attention, but is difficult to place a value upon. The safety net function is the role that NTFPs play as a fall-back option when households experience misfortune, such as drought, death or retrenchment of a breadwinner, a natural disaster and the like (Wunder *et al.* 2014). This safety net option can be mobilised via three means (Shackleton and Shackleton 2004). The first is for a household to increase the use of an NTFP that is already part of their provisioning needs (e.g. to consume more wild foods to replace cultivated or purchased foods). The second is to start using an NTFP that they typically have not used previously (e.g. use of firewood instead of commercial energies). The third is to engage in temporary sale of NTFPs to generate cash to tide them over until they can adapt to the impacts of the misfortune. Interestingly, however, once engaged in NTFP trade, many continue trading even after the effects of the misfortunate have passed or the household has adapted. Asking traders why they entered the NTFP trade often elicits a reply that it was in response to some emergency or misfortune for which cash was needed to cope and adjust. If NTFPs were not available as a safety net, what other coping strategies do rural households have? Several authors, such as McSweeney (2005), Hunter et al. (2011) and Paumgarten and Shackleton (2011) have explored this, but a great deal more is required to understand what strategies are chosen under different shocks and contexts. The large-scale inter-continental study of Wunder et al. (2014) revealed that 65 per cent of households reported at least one shock over the period of the last twelve months, many reported more than one shock event. Of these, 44 per cent used NTFPs as a safety net, although not necessarily as their primary choice. Most studies show that most affected households usually first turn to family and friends. However, if the shock is a covariate one, i.e. most or all households in the community are affected (e.g. via drought or a communicable disease epidemic), then this option is limited. Another coping strategy is to decrease consumption rates (of food or energy), which is not viable for extended periods. Alternatives may be to sell household assets or borrow cash from micro-lenders, both of which may offer an immediate coping strategy but often at the expense of longer term livelihood sustainability. Migration of some family members to seek employment, or a permanent move of the entire household, has also been reported. Along with these other coping strategies, NTFPs offer a variety of safety net options which are employed by millions of people around the globe every year.

NTFPs in local culture

It is methodologically complicated to place a value on the cultural importance of NTFPs to local communities (but it has been done), but that should not mean its importance should be overlooked by researchers or policy makers. It may be that the presence of culturally important species or sites underpins a specific community's interest in, or even identity with, a designated natural area and how it is managed or conserved. For example, many traditional communities have sacred or taboo areas where normal extraction, or land transformation activities, are limited or prohibited, with the result that these areas are important for biodiversity and the provision of other ecosystem services (e.g. Bhagwat et al. 2005). At a species level, many rural people make use of a range of species for culturally mandated spiritual needs and beliefs, such as the warding off of perceived evil influences or bad luck. The cultural values of NTFPs also play a role in cultural tourism opportunities if done in a respectful, responsible and consensual manner. These may include traditional ceremonies at which specific NTFPs are required, promotion of cultural foods and drinks, or the sale of cultural crafts and artefacts made from local NTFPs.

The value of NTFPs in international market chains

Many of the roles described above relate to local use of NTFPs, and hence they are frequently underappreciated by planners and developers. The bigger markets and export opportunities associated with the formal economy attract their attention. Many NTFPs have certainly 'made it' in such markets (argan and marula oil, aloe gels, active extracts for medicinals, rattans, mushrooms, narcotics, etc.), and there is significant potential for many more to do so. For example, the over-the-counter value of products containing *Prunus africana* bark is over US$200 million p.a, with it all sourced from wild populations. The export of Brazil nuts from Bolivia alone is worth over US$70 million p.a. Shea butter exports from Africa were estimated in 2007 at 150,000 tons of dry shea kernel earning about US$30 million p.a. and growing. Aloe gel exports from South Africa are worth over US$2 million p.a. Such high market values can, through time, prompt different scenarios depending on the governance systems in place. The first scenario is that high prices attract increasing numbers of market players, which may result in conflict over land and the resource and perhaps even overexploitation which jeopardises livelihoods and market supply in the medium- to long-term. The second scenario is increasing governance intervention, at either local or higher levels (national and international) to regulate or promote the value chain for sustainable outcomes, ecologically, economically and, increasingly, socially (e.g. fair trade products), as well as earn tax revenues from the trade. The third is increasing cultivation or domestication of the NTFP in plantation or field settings. This improves the potential surety of supply but drives up financial costs and typically concentrates the value in the hands of fewer operators, unless a conscious effort is made to engage smallholders. It also divorces the NTFP from the wild, undermining its usefulness as a vehicle for promoting habitat and species conservation.

A common feature of international value chains for NTFPs is that the local harvester or producer receives only a small proportion of the total value or cost paid by the final consumer. This varies per product and per country typically ranging from 10–20 per cent. Conscious efforts towards either value addition close to source, or equitable sharing, can increase this to 40–60 per cent, thereby better fulfilling the requirement (v) of the NTFP concept discussed in the section defining non-timber forest products.

Realising the multiple values of NTFPs

Perhaps one 'universal' truth about NTFPs is the frequent concern that they are not adequately recognised and hence incorporated into appropriate national and international policies pertaining to human welfare and development, forestry, land use planning and conservation. Several steps have been suggested by Shackleton and Pandey (2014) for improving their recognition and mainstreaming.

Promotion and adoption of multipurpose land uses

NTFPs include thousands of species and can be found in almost all habitats. Thus, there are multiple opportunities for use and perhaps management of NTFPs in parallel with other land use management objectives. For example, the simultaneous promotion of income streams from high value forest timber and from understorey floral greenery as practised in the southern Cape forests in South Africa (Durrheim and Vermeulen 1996). Or the management of forests for ecotourism and water yield along with harvesting of forest berries or mushrooms that occurs in many parts of Scandinavia (Eriksson *et al.* 2012). The extractive conservation areas in South America are prime examples of management agencies seeking to optimise the supply of many different products for local benefit as well as cash income.

Achievement of this requires a change in mindset and a focus out from viewing land management as a process of maximising the supply of a single good or service, to one where many goods and services are promoted at varying ratios in time and space. This not only benefits the supply of NTFPs and therefore probably the welfare of forest dependent people (depending on tenure, governance and revenue sharing mechanism), but also the biodiversity and ecological resilience and processes.

Inclusion of NTFPs in standardised forest inventories

The lack of inventory and research on key NTFPs is one of the main constraints to stimulating sustainable development initiatives through NTFP use (Cocksedge 2006, Vantomme 2003), even in countries with enabling NTFP policies, such as Nepal (Heinen and Shreshtha-Acharya 2011). If national or sub-national governments do not know what they have, how can they protect them or promote sustainable use strategies or facilitate markets? Incorporation of NTFP inventories into existing forest inventory systems would prevent duplication of effort and costs (Cocksedge 2006). Indeed, the inclusion of timber related NTFPs into such inventories will require very little additional effort (Guariguata *et al.* 2011). Other NTFP forms, such as fronds, bulbs, fibres and the like would require additional effort, costs and standardisation of inventory approaches. Guariguata *et al.* (2011) describe the difficulties of incorporating NTFP inventories into standard forest inventories because of the: (i) large diversity of NTFP species; (ii) different maturation and hence harvest rotation rates; and (iii) the variety of products other than just timber, from NTFP species. Whilst these are significant, they can be addressed through a systematic approach commencing with a subset of the more important NTFPs within a given region (Shackleton and Pandey 2014). Inventory tools already exist for almost every type of NTFP.

Investments of expertise, time and finances into understanding the autecology of key NTFP species with guidelines for ecologically sustainable yields in different habitats and regions

The sustainable harvesting limits are known for a woefully small number of NTFP species internationally (Ticktin and Shackleton 2011). The high aggregate value to local people and on national and international markets means that some of the value could be directed towards management and research regarding their autecology, sustainable yield and appropriate management options. Examples include multiple works on *Nardostachys* (e.g. Chauhan and Nautiyal 2005, Ghimire *et al.* 2005), and Brazil nuts (e.g. Myers *et al.* 2000, Wadt *et al.* 2005). Most developing countries have State agricultural and/or forest research agencies, but few (e.g. some States in India) have an equivalent for NTFP research. An added benefit would be that such research support would also generate a cohort of new scientists interested in and knowledgeable about the ecology and harvesting of NTFPs for human well-being, adding to national skills development and a broadening of resource management focus into the future.

Ensuring that land management or development plans include NTFPs in trade-off decisions

In many areas, the biggest threat to biodiversity, including NTFPs, is not overharvesting, but land transformation to plantations, agriculture or other purposes, such as dams, tourism estates, protected areas, roads and infrastructure (Millennium Ecosystem Assessment 2005). When wild or semi-wild lands are converted to intensive land uses or heavily logged, the myriad resources (including NTFPs) from that land previously used by communities are lost (Ntshona *et al.* 2010, Shanley and Luz 2003). This may force people to swap to other resources or species (thereby increasing the relative harvesting pressure), travel further afield to access the NTFPs from alternative places (with consequent increases in opportunity costs), or resort to buying the NTFP or a suitable substitute (which diminishes their disposable cash). It is therefore imperative that development and zonation plans recognise the need for and use of NTFPs and trade-offs are carefully assessed when land transformation decisions loom.

As commented by Shackleton and Pandey (2014) the mechanisms for dealing with trade-off decisions depend on local governance and management processes and frameworks in each country and at the local level. Where there are established and accepted environmental governance frameworks, tools such as environment impact assessments, strategic impact assessments or cost–benefit analysis can be used. However, trade-offs related to NTFPs are only possible via these mechanisms if there is an adequate inventory of stocks and demands for NTFPs, and thus their contribution to well-being. Where the use of such tools is less prevalent, incorporation of NTFPs into local land zonation or use decisions would need to happen at a more local level through illustration of how local people and the NTFP 'values' will be impacted under the range of competing land use options. For example, Schaafsma et al. (2014) spatially mapped stocks and flows of selected NTFPs in eastern Tanzania and demonstrated the loss of value to local communities (approximately US$42 million p.a.) if they were excluded from accessing resources in proposed new conservation areas. Van Beukering et al. (2003) valued the income streams for different land use options in and around Mount Leuser National Park in Indonesia and explicitly included important NTFPs used by local communities, and were able to show how the benefits to different stakeholders changed under various mixes of land use options. This valuation provided the basis for informed decision making regarding land uses.

Conclusion

This chapter has outlined the multiple and significant roles of NTFPs in forest ecology and human well-being. However, the gap between the two is currently large with ecology usually being the domain of foresters, ecologists and conservationists, whilst the understanding of local livelihoods is perceived as the realm of economists, development planners and anthropologists. If the real values of NTFPs are to be manifest for the benefit of natural habitats and people, a shift in mindsets is required. The first shift is for different disciplines to work together to provide an integrated understanding of NTFPs and the need for appropriate policies for their recognition, management and securing of benefit flows. The second shift is for all role players from local to national level to view land and forests as more than just timber trees and, more recently, carbon sinks.

References

Angelsen, A., Jagger, P., Babigumira, R., Belcher, B., Hogarth, N.J. and others (2014) Environmental income and rural livelihoods: a global comparative analysis. *World Development*, http://dx.doi.org/10.1016/j.worlddev.2014.03.006

Angelsen, A. and Lund, F. (2011) Designing the household questionnaire. In: Angelsen, A., Larsen, H.O. and Lund, F.F. (eds). *Measuring livelihoods and Environmental Dependence: Methods for Research and Fieldwork*. Earthscan, London. pp 107–126.

Appiah, M. B., Blay, D., Lawrence, E., Damnyag, L., Dwomoh, F.K. and others (2009) Dependence on forest resources and tropical deforestation in Ghana. *Environment, Development and Sustainability*, 11, 471–487.

Belcher, B.M. (2003) What isn't an NTFP? *International Forestry Review*, 5, 161–168.

Bhagwat, S.A., Kushalappa, C.G., Williams, P.H. and Brown, N.D. (2005) The role of informal protected areas in maintaining biodiversity in the Western Ghats of India. *Ecology and Society*, 10 (1) 8 online.

Chauhan, R.S. and Nautiyal, M.C. (2005) Commercial viability of cultivation of an endangered medicinal herb *Nardostachys jatamansi* at three different agroclimatic zones. *Current Science*, 89, 1481–1488.

Cocksedge, W. (2006) Incorporating non-timber forest products into sustainable forest management: an overview for forest managers. Royal Roads University, Victoria. (Available on-line http://cntr.royalroads.ca)

Cunningham, A.B. (2011) Non-timber products and markets: lessons for export-oriented enterprise development from Africa. In: Shackleton, S.E., Shackleton, C.M., and Shanley, P. (eds). *Non-Timber Forest Products in the Global Context*. Springer, Heidelberg, Germany. pp. 83–106.

de Beer, J.H. and McDermott, M.J. (1996) *The Economic Value of Non-Timber Forest Products in Southeast Asia* (2nd ed.) IUCN, Amsterdam.

Durrheim, G.P. and Vermeulen, B.J. (1996) Sustainable multiple use management of indigenous evergreen high forest in the southern Cape, South Africa. In Mushove, P.T., Shumba, E.M. and Matose, F. (eds) *Sustainable Management of Indigenous Forest in the Dry Tropics*, Harare, Zimbabwe: Zimbabwe Forestry Commission/SIDA, pp. 62–69.

Eriksson, L., Nordlund, A.M., Olsson, O. and Westin, K. (2012) Recreation in different forest settings: a scene preference study. *Forests*, 3, 923–943.

Geldenhuys, C.J. (1999) Requirements for improved and sustainable use of forest biodiversity: examples of multiple use forest in South Africa. In: Poker, J., Stein, I., Werder, U. (eds.), *Proceedings Forum Biodiversity Treasures in the World's Forests*. Alfred Toepfer Akademie, Germany, pp. 72–82.

Ghimire, S.K., McKey, D. and Aumeeruddy-Thomas, Y. (2005) Conservation of Himalayan medicinal plants: harvesting patterns and ecology of two threatened species, *Nardostachys grandiflora* DC. and *Neopicrorhiza scrophulariiflora* (Pennell) Hong. *Biological Conservation*, 124, 463–475.

Gram, S. (2001) Economic valuation of special forest products: an assessment of methodological shortcomings. *Ecological Economics*, 36, 109–117.

Guariguata, M.R., García-Fernández, C., Nasi, R., Sheil, D., Herrero-Jáuregui, C. and others (2011) Timber and non-timber forest product extraction and management in the tropics: towards compatibility? In: Shackleton, S.E., Shackleton, C.M. and Shanley, P. (eds), *Non-Timber Forest Products in the Global Context*. Springer, Heidelberg, Germany. pp. 171–188.

Hall, J.B., O'Brien, E.M. and Sinclair, F.L. (2002). *Sclerocarya birrea*: a monograph. *School of Agricultural and Forest Science publication no 19*, University of Wales, Bangor. 157.

Heinen, J.T. and Shreshtha-Acharya, R. (2011) The non-timber forest products sector in Nepal: emerging policy issues in plant conservation and utilisation for sustainable development. *Journal of Sustainable Forestry*, 30, 543–563.

Herrero-Jáuregui, C., Guariguata, M.R., Cárdenas, D., Vilanova, E., Robles, M. and others (2013) Assessing the extent of 'conflict of use' in multipurpose tropical forest trees: a regional view. *Journal of Environmental Management*, 130, 40–47.

Hogarth, N.J., Belcher, B., Campbell, B. and Stacey, N. (2013) The role of forest-related income in household economies and rural livelihoods in the border-region of Southern China. *World Development*, 43, 111–123.

Hunter, L.M., Twine, W.C. and Johnson, A. (2011) Adult mortality and natural resource use in rural South Africa: evidence from the Agincourt Health and Demographic Surveillance site. *Society and Natural Resources*, 24, 256–275.

Kalaba, F.K., Quinn, C.H. and Dougill, A.J. (2013) Contribution of forest provisioning ecosystem services to rural livelihoods in the Miombo woodlands of Zambia. *Population and Environment*, 35, 159–182.

Kamanga, P., Vedeld, P. and Sjaastad, E. (2009) Forest incomes and rural livelihoods in Chiradzulu District, Malawi. *Ecological Economics*, 68, 613–624.

Kar, S.P. and Jacobson, M.G. (2012) NTFP income contribution to household economy and related socio-economic factors: lessons from Bangladesh. *Forest Policy and Economics*, 14, 136–142.

L'Roe, J. and Naughton-Treves, L. (2014) Effects of a policy-induced income shock on forest-dependent households in the Peruvian Amazon. *Ecological Economics*, 97, 1–9.

McSweeney, K. (2005) Natural insurance, forest access, and compounded misfortune: forest resources in smallholder coping strategies before and after Hurricane Mitch, Northeastern Honduras. *World Development*, 33, 1453–1471.

Mamo, G., Sjaastad, E. and Vedeld, P. (2007) Economic dependence on forest resources: a case from Dendi District, Ethiopia. *Forest Policy and Economics*, 9, 916–927.

Menton, M.C., Lawrence, A., Merry, F. and Brown, N.D. (2010) Estimating natural resource harvests: conjectures? *Ecological Economics*, 69, 1330–1335.

Millennium Ecosystem Assessment (2005) *Ecosystems and Human Wellbeing: Synthesis*. Island Press, Washington DC.

Mitra, A. and Mishra, D.K. (2011) Environmental resource consumption pattern in rural Arunachal Pradesh. *Forest Policy and Economics*, 13, 166–170.

Myers, G.P., Newton, A.C. and Melgarejo, O. (2000) The influence of canopy gap size on natural regeneration of Brazil nut (*Bertholletia excelsa*) in Bolivia. *Forest Ecology and Management*, 127, 119–128.

Ntshona, Z., Kraai, M., Kepe, T. and Saliwa, P. (2010) From land rights to environmental entitlements: community discontent in the 'successful' Dwesa-Cwebe land claim in South Africa. *Development Southern Africa*, 27, 353–361.

Paumgarten, F. and Shackleton, C.M. (2011) The role of non-timber forest products in household coping strategies in South Africa: the influence of household wealth and gender. *Population and Environment*, 33, 108–131.

Peters, C.A., Gentry, A. and Mendelsohn, R. (1989) Valuation of an Amazonian rain forest. *Nature*, 339, 655–656.

Schaafsma, M., Morse Jones, S., Posen, P., Swetnam, R.D., Balmford, A. and others (2014) The importance of local forest benefits: valuation of non-timber forest products in the Eastern Arc mountains in Tanzania. *Global Environmental Change*, 24, 295–305.

Shackleton, C.M. and Shackleton, S.E. (2004) The importance of non-timber forest products in rural livelihood security and as safety-nets: evidence from South Africa. *South African Journal of Science*, 100, 658–664.

Shackleton, C.M. and Pandey, A.K. (2014) Positioning non-timber forest products on the development agenda. *Forest Policy and Economics*, 38, 1–7.

Shackleton, C.M., Shackleton, S.E., Buiten, E. and Bird, N. (2007) The importance of dry forests and woodlands in rural livelihoods and poverty alleviation in South Africa. *Forest Policy and Economics*, 9, 558–577.

Shackleton, C.M., Delang, C., Shackleton, S.E. and Shanley, P. (2011) Non-timber forest products: concept and definition. In: Shackleton, S.E., Shackleton, C.M., and Shanley, P. (eds). *Non-Timber Forest Products in the Global Context*. Springer, Heidelberg, Germany. pp. 3–21.

Shackleton, S.E., Campbell, B., Lotz-Sisitka, H. and Shackleton, C.M. (2008) Links between the local trade in natural products, livelihoods and poverty alleviation in a semi-arid region of South Africa. *World Development*, 36, 505–526.

Shanley, P. and Luz, L. (2003) The impacts of forest degradation on medicinal plant use and implications for health care in Eastern Amazonia. *Bioscience*, 53, 573–784.

Stanley, D., Voeks, R. and Short, L. (2012) Is non-timber forest product harvest sustainable in the less developed world? A systematic review of the recent economic and ecological literature. *Ethnobiology and Conservation*, 1 (9) online: www.ethniobioconservation.com

Ticktin, T. (2004) The ecological implications of harvesting non-timber forest products. *Journal of Applied Ecology*, 41, 11–21.

Ticktin, T. and Shackleton, C.M. (2011) Harvesting non-timber forest products sustainably: opportunities and challenges. In: Shackleton, S.E., Shackleton, C.M., and Shanley, P. (eds). *Non-Timber Forest Products in the Global Context*. Springer, Heidelberg, Germany. pp. 149–170.

Valkama, E., Lyytinena, S. and Koricheva, J. (2008) The impact of reed management on wildlife: a meta-analytical review of European studies. *Biological Conservation*, 141, 364–374.

van Beukering, P.J.H., Cesar, H.S.J. and Janssen, M.A. (2003) Economic valuation of the Leuser National Park on Sumatra, Indonesia. *Ecological Economics*, 44, 43–62.

Vantomme, P. (2003) Compiling statistics on non-wood forest products as policy and decision-making tools at the national level. *International Forestry Review*, 5, 156–160.

Wadt, L.H., Kainer, K.A. and Gomes-Silva, D.A. (2005) Population structure and nut yield of a *Bertholletia excela* stand in Southwestern Amazonia. *Forest Ecology and Management*, 211, 371–384.

Wunder, S., Börner, J., Shively, G. and Wyman, M. (2014) Safety nets, gap filling and forests: a global comparative perspective. *World Development*. DOI http://dx.doi.org/10.1016/j.worlddev.2014.03.005

40
AGRICULTURE IN THE FOREST
Ecology and rationale of shifting cultivation

Olivier Ducourtieux

Introduction

Agriculture is now one of the main types of land use in the world. According to the FAO, forests covered only 31 per cent of the land in the world in 2011, while land used for agriculture – i.e. arable land, permanent crops and pastures – counted for 38 per cent. However, in the history of humankind, agriculture is a recent invention (Mazoyer and Roudart, 2005). With the transition from the hunter-gatherer lifestyle to one of agriculture during the Neolithic Revolution, farmers quickly expanded their activity into forest areas. Here, they developed shifting cultivation, an agricultural practice with temporal rotations in forest ecosystems.

Shifting cultivation still exists today in tropical regions and is the livelihood basis for probably in the order of hundreds of millions (Mertz *et al.*, 2009). The range in the population estimate is wide, because of the intrinsic difficulties in counting smallholders spread out under the tree canopy at the margins of tropical forests, but also because of confusion that exists in differentiating shifting cultivators from other farmers involved in small-scale agriculture (Thrupp *et al.*, 1997). We will focus on rotational shifting cultivation, when agriculture takes place for a long time in the forest, but not on its conversion into permanent open fields by shifting and burning the trees. To back up the biological, ecological, technical, social or economical facets of shifting cultivation, the illustrations and data provided hereafter will be based on localized case studies, mostly from Southeast Asia. The resultant generalization may make cultivation appear more concrete and intelligible, but it must be remembered that shifting cultivation systems differ significantly across the globe.

Shifting cultivation principles and fallow ecology

The practices of shifting cultivators vary widely depending on difference between the cultures, regions and ecosystems where it is practiced. Key authors have built up definitions and typologies for shifting cultivation (Cairns, 2015; Palm *et al.*, 2005; Ramakrishnan, 1992), which help us to reduce the diversity to a minimal common point: roughly, shifting cultivation comprises a short phase of cultivation followed by a long phase of fallow development.

The cultivation phase

The cultivation phase begins with the selection of a plot of forest to fell. Shortly after the beginning of the dry season, farmers decide on the surface area to clear depending on many constraints: the surface area available, the needs of the household and the expected yields, the topography of the plot, the previous year's harvest and prospects of stock or shortages, how far away the new plot is from the village and its potential fertility – based on past crops and observations such as the soil texture or colour, the standing vegetation, i.e. tree species, density and sizes (Cairns, 2007). Arrangement of plots can vary with farmers proceeding at the village community level, with the household fields grouped together in a regulated plot allotment, with a mandatory rotation. Elsewhere, farmers may select the plot independently at the household level, with small family fields scattered over the forest. However, in both cases, the cultivation is a household activity.

After the plot selection, the villagers fell the forest. They begin by clearing the understory with machetes. Then, they attack the bigger trees with axes, large machetes, or even chain saws. To reduce the exhausting nature of the work, the trees are often cut above the large base and buttresses of the trunks, sometimes stumps taller than two meters remain (Schmidt-Vogt, 1999). Clear cutting is rare: some trees are left standing in the field, because they are too difficult to cut down, or because they are too valuable for the villagers, providing other resources: fruit or building materials. Farmers often pollard the standing trees to limit the shading over the crops (Rerkasem et al., 2009). The slashed vegetation is left drying in the field for the dry season, covering the soil.

After a few months (1–4) – depending on the local climate and the density of the fallen biomass, the farmers burn the dry biomass. To limit the risk of uncontrolled fire expanding into the forest, they first clean the border of the plot and then proceed by pushing the burning material against the wind (Ducourtieux, 2006). After the first run, the leftover small branches are grouped and burnt again, while the bigger ones are moved to mark the limits of the fields and are used as a source of firewood for the home[1]. With the first rains, the field is now ready for cultivation. Villagers sow their fields with a dominant source of staple food (carbohydrate) – either cereals such as rice (*Oriza sativa*), maize (*Zea mays*), millet (*Pennisetum glaucum*, *Setaria italica*, *Panicum miliaceum*, *Eleusine coracana*, etc.), sorghum (*Sorghum bicolor*), wheat (*Triticum* spp.), barley (*Hordeum vulgare*), quinoa (*Chenopodium quinoa*) and fonio (*Digitaria* spp.); or tubers such as yams (*Dioscorea* spp.), taro (*Colocasia esculenta*) and cassava (*Manihot esculenta*). Often 30 to 50 species of vegetables and herbs (e.g. cucurbits, crucifers, peppers, sunflower, groundnut, etc.) are grown in a complex system (Conklin, 1957; Rerkasem et al., 2009). The diversity of crops provides many services to the farmers:

- The different species covering the ground ensure: (i) the optimal use of sunlight; (ii) limiting the growth of weeds; and (iii) limited erosion. The varied roots explore the different soil horizons for the optimal use of water and nutrients (Nye and Greenland, 1960; Ramakrishnan, 1992).
- It provides diversified and balanced food for farmers' households (Hladik et al., 1993).
- It minimizes the impact of the farming risks, either from market hazards (price variation, evolution of demand, etc.), or from natural ones (drought or flooding, pests, etc.), which affect only part of the crops (Brookfield, 2001).

---

1 Despite the fact that shifting cultivators rely on firewood for cooking and heating, they do not need to source it by cutting down trees. The fuel source comes from the swidden.

On a plot, sowing is neither standardized nor random. The farmers decide how to carry out sowing based on their experience in using all the environment's resources, on a very precise scale, per square meter. For example, the sowing density will depend on the slope, with tubers being preferentially planted in large heaps of ashes and maize in the wettest part. After sowing, the farmers dedicate most of their time to weeding the swidden field, carried out quickly and regularly, to prevent weeds from putting a strain on the yields. For example, weeding done too late lets the weeds bloom and go to seed, complicating the control of weeds (Roder, 2001).

The harvest can last for a few months, beginning with short-cycle crops (e.g. maize); farmers continue to harvest perennial crops (e.g. bananas, cassava, fruit trees, etc.) well into the fallow period. For the staple cereal, average yields range generally from 0.5–2 tonnes ha^{-1} (rice, fonio) up to 4–5 tonnes ha^{-1} (maize) for a first year of cultivation after slashing. The results vary widely according to the species and the pedo-climatic conditions, but the level decreases quickly with successive years of cultivation[2]. Such yields appear low compared to the current ones from chemically fertilized agriculture (e.g. 5–12 tonnes ha^{-1}), but are similar (or higher) to manure-fertilized grain farming (Mazoyer and Roudart, 2005). Moreover, we must take into account the associated crops: when added to the other services they provide (see above), their value can reach the level of that of the staple crop (Ducourtieux, 2006).

During the cultivation phase, the stock of mineral nutrients as well as of organic matter is depleted, the soil becomes compacted; in parallel, the field is progressively invaded by weeds (notably by wind dispersal of seeds, i.e. anemochory) and pests (insects and nematodes). Such a combination explains the decreasing yields from the first crop cycle, while the workload for weeding redoubles (Ducourtieux, 2006). After a few years of cultivation with decreasing results, the farmers give up the cultivation of the field. The second phase of shifting cultivation begins – the fallow phase – with a development of spontaneous plants from herbaceous to forest ecosystems.

Renewal of fertility: rotation and on-site biomass accumulation

In agriculture, the fertility required by the cultivated plants is generally imported to the field, from a wide array of sources: chemical fertilizers, animal manure, alluvial deposits, etc. However, the renewal of fertility is based on lateral transfer and concentration of nutrients towards the cultivated field (Mazoyer and Roudart, 2005).

Shifting cultivation contrasts with this general rule: the soil fertility comes from the biomass accumulated on-site during the fallow period. In the fallow period, the organic matter is stored in the above ground vegetation, but also in the soil. The accumulated biomass is mobilized for the cultivation phase by its mineralization. The process is continuous and spontaneous in the litter as well as in the organic horizons of the soil. The desiccation following the clearing of the plot, and more importantly the burning of the slashed vegetation, amplifies and speeds up the mineralization. The burning affects the soil by: (i) increasing the amount of available nutrients (P, K); (ii) the alkalinization, which increases the bioavailability of phosphates; and (iii) the thermal disinfection, which cleans the top ground of pests (insects, nematodes, fungus, etc.) and dormant weed seed banks (Gerold *et al.*, 2004; Nye and Greenland, 1960; Ramakrishnan, 1992; Roder, 2001).

---

2 For example, rice yields are often divided by two for the second year, and the field is almost never farmed for more than three years. On the contrary, maize yields decrease too, but less sharply and a field is more often cultivated over four to six successive years in similar soil and climate conditions. The sensitivity to nematodes for aerobic rice roots explains the difference (Roder, 2001).

During the burning, most of the nitrogen is released into the atmosphere. However, the cultivated plants are supplied by the subsequent mineralization of humus, accelerated by both the burning and the exposure of the soil to solar light, and by the biological fixation of atmospheric nitrogen by legumes as associated crops (Szott et al., 1999). With the fast mineralization of the biomass and the crop build-up, the soil fertility decreases quickly during the cultivation phase. The labour productivity wanes, inviting the farmers to stop cultivating the field, which is then left for regrowth of spontaneous vegetation. A new fallow phase begins, with biomass slowly accumulating above and below ground: fertility renewal depends on rotation in shifting cultivation (see Figure 40.1).

Fertility is classically presented as proportional to the fallow duration (Ramakrishnan, 1992). The progressive build-up of biomass has been proven[3], but recent research works challenge the oversimplified interpretation of linear progression. The staple crop yield is not strictly proportional to the fallow length (Mertz et al., 2008; Roder, 2001; van Keer, 2003), but overall increases, with diminishing marginal returns.

The farmers have to find a trade-off between contradictory factors for the duration of the fallow phase. The first one is the available land at the household or village community level, depending mainly on land suitability (e.g. slopes, depth of soil) and on demography (see Figure 40.2).

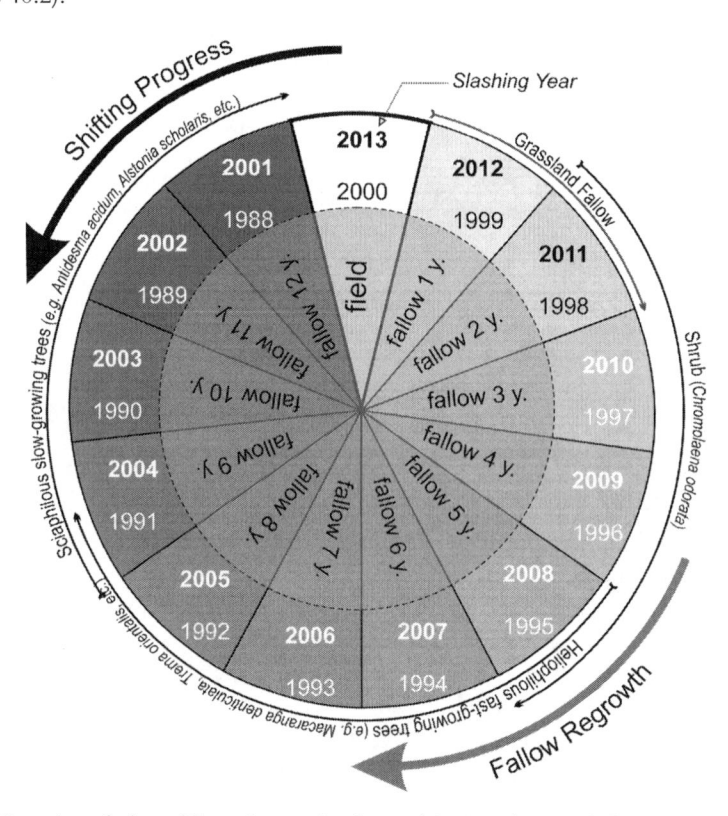

Figure 40.1 Rotation of a long fallow phase and a short cultivation phase in shifting cultivation
Example from Ban Samlang village (Phongsaly, northern Laos; own work)

---

3 For example, in northern Thailand, van Keer found a biomass over 20 tonnes ha^{-1} after a three-year fallow period, then 30 tonnes ha^{-1} after seven years, 70 tonnes ha^{-1} after ten years and 80 tonnes ha^{-1} after eighteen years (2003).

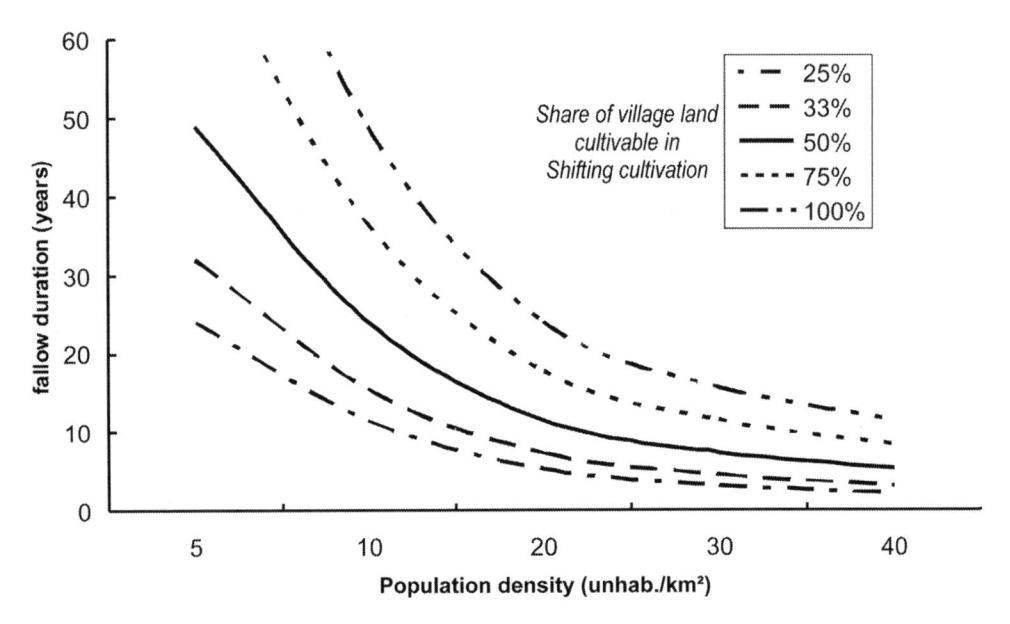

Figure 40.2 Population density and fallow duration (in accordance with the proportion of the land cultivable in the total area)

Based on a household of two adults/three children, 0.5ha slashed/cultivated per adult and per year; own work

With older fallows, the yields are higher and the labour required for weeding is reduced; that induces farmers to look for long rotations. However, with old fallows, they are subjected to an increased workload for slashing, for accessing the field and bringing back the crops[4]. When plenty of land is available, farmers may decide to limit the area of fallow in rotation and to allocate the remote land to community reserve: for example, the Phunoy people from Samlang village (Ducourtieux, 2006) decided that the difference in yield after a 12- or a 15-year fallow (see Figure 40.1) was not sufficient to compensate for the extra access time to the field. They, therefore, decided to follow a 12-year fallow; incidentally, all land further than three hours walking from the village is left out of the rotation, in reserve.

Ecology of the fallows

As soon as the farmers stop weeding the field, the spontaneous vegetation tends to develop, beginning with pioneer herbaceous species, whose seeds had arrived[5] since the clearing and that farmers struggled to control by weeding during the cultivation phase.

While the clearing and the subsequent cultivation phase are an allogenic succession (human disruption), the secondary succession after the plot is left to fallow corresponds to an autogenic one, with successive seral communities that vary widely according to regional biomes and local ecosystems (Gómez-Pompa *et al.*, 1991). In this chapter, the examples of species refer to a local ecological succession in Southeast Asia, around Phongsaly[6], northern Laos (see Figure 40.3).

4 For example, two labourers that comprise a household can cultivate one hectare; they have to carry back home 30 to 50 bags of 50kg each after harvesting.
5 Either via anemochory (wind dispersal) or zoochory (animal dispersal).
6 500–1,500m asl, 21°41′N/102°06′E.

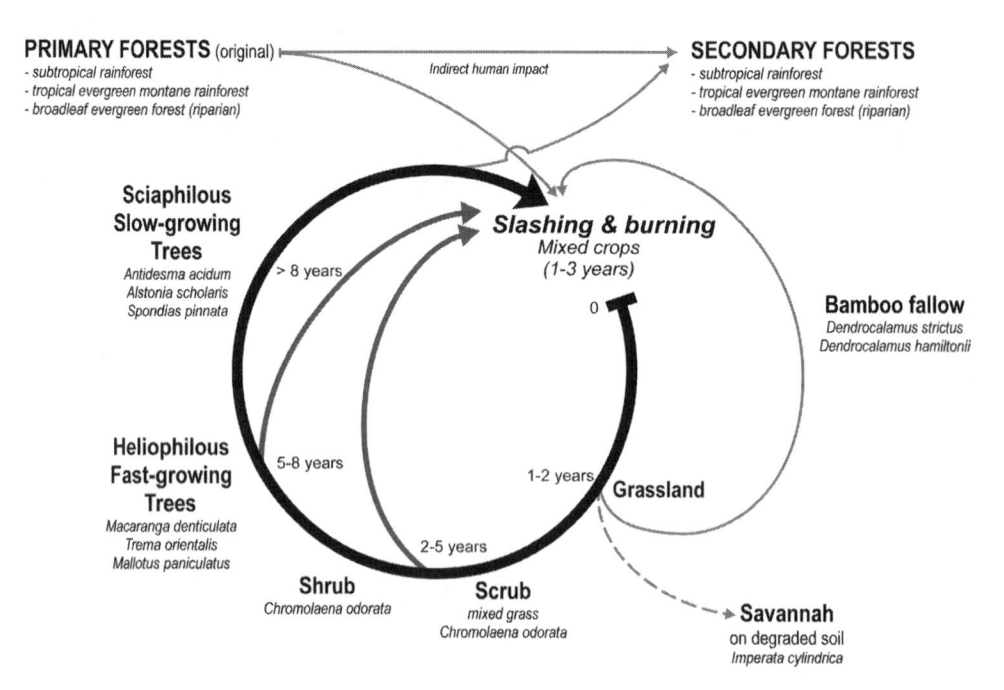

Figure 40.3 Secondary succession in a shifting cultivation fallow
Example from Phongsaly, northern Laos (own work)

The first seral community comprises pioneer herbaceous plants (e.g. *Pennisetum* spp., *Imperata cylindrica*, *Ageratum conyzoides*, *Miscanthus* spp.). Scrub species (e.g. *Chromolaena odorata*) quickly grow over the weeds and supplant them in a few years as the second seral community. After five years or more, the first trees appear (i.e. stand initiation), from heliophilous fast-growing species (e.g. *Trema orientalis*, *Macaranga denticulata*, *Mallotus paniculatus*). When the canopy closes, only shade-tolerant (sciaphilous) plants survive in the understory. The trees are engaged in a competition for light (stem exclusion), which modifies their habits and exposes them to rapid decay in windthrow, leaving room for the last seral community by stand-replacing of sciaphilous slow-growing trees (e.g. *Antidesma acidum*, *Alstonia scholaris*, *Spondias pinnata*). They comprise the old fallow, a secondary forest over ten years old.

Such a sequence is common in forest ecology (Bermingham *et al.*, 2005; Gómez-Pompa *et al.*, 1991), but secondary successions in shifting cultivation fallow differ from spontaneous ones because of four human-based specific and interactive factors (Cairns, 2007; de Jong *et al.*, 2001; Ramakrishnan, 1992; Rerkasem *et al.*, 2009; Schmidt-Vogt, 1999):

- The plot slashing without grubbing advantages the tree species able to resprout from the stump. The large scale of the clearing furthers heliophilous species and burning favours fire-tolerant ones for the first steps of the succession. For example, shifting cultivation has contributed to the wide extent of Siam Weed (*Chromolaena odorata*) in Southeast Asia since the end of the nineteenth century (Roder, 2001).
- Farmers use fallows for grazing their livestock, for hunting and for gathering spontaneous fruits, vegetables, building material (e.g. wood, rattan, weed straw) and medicinal plants, either for self-consumption or for sale.

- Farmers do not cut down all the trees when slashing the fallow. They maintain some of them, particularly useful species, such as palms[7], fruit trees (e.g. *Mangifera* spp., *Carica papaya*, *Musa* spp., etc.), resin trees (e.g. *Styrax benzoin*), etc. Cycle after cycle, the reproduction of those species is favoured and the plant density increased in the fallow.
- Farmers may plant perennial plants in the young fallow, either for harvesting[8] or for improving the fertility renewal[9].

All these human interventions affect the secondary successions and lead to very specific ecosystems. Under shifting cultivation, the landscape is a mosaic in 'leopard skin', the size of the spots of which[10] depends on the allotment system and on the size of the social and productive units (village/clan communities, nuclear or extended families). Old fallows are secondary forests, but they strongly differ from old-growth forests. Fallows are not left-alone spaces, but productive areas that farmers exploit and manage. At the extreme, the economic value of products collected from the fallow can reach a level that dissuades the farmers from slashing the fallow, which then becomes a permanent garden, for example in Indonesia with rubber agroforest in Kalimantan or dammar agroforest gardens in Java (Babin, 2004; Cairns, 2007; Michon, 2005). Thus, shifting cultivation appears as a variety of agroforestry (Cairns, 2015).

Forest dynamics and Environmental impact of shifting cultivation

Sustainability limits of shifting cultivation: forest degradation and demographic crisis

Agriculture emerged around 10,000 years ago in half a dozen Neolithic centres, which were mostly in open ecosystems: steppes of the Fertile Crescent, floodplains, peatlands, etc. (Mazoyer and Roudart, 2005)[11]. Quickly, expansion of agriculture faced forests, the dominant ecosystems, which had been in extension after the last Glacial Period (Williams, 2006). To conquer those forest frontiers, the farmers had to design new tools, new techniques and new farming systems which led to the appearance of shifting cultivation. For example, based on artefacts, archaeologists conclude that influence of agriculture – i.e. shifting cultivation – progressed at a pace of one kilometer per year in Neolithic Europe[12]. However, should we imagine a pioneer front, inexorably progressing from southeast to northwest, with forest ahead and open field behind? This is unlikely. Clearing a primary dense forest[13], either a temperate or a tropical one, requires ten times more work than a 15–20 year-old fallow. With a population limited to less than

---

7 *Arecaceae* spp. e.g. sugar palms (*Borassus flabellifer*, *Arenga pinnata* and *Caryota urens*), oil palm (*Elaeis* spp.), betel palm (*Areca catechu*) and coconut palm (*Cocos nucifera*).
8 For example (Cairns, 2007), fruit trees such as banana (*Musa* spp.); timber species such as teak (*Tectona grandis*); essential oil species such as cardamom (*Amomum* spp.), *Styrax tonkinensis* for benzoin; and other cash crops such as oil palm (*Elaeis* spp.), coffee tree (*coffea* spp.), rubber tree (*Hevea brasiliensis*), rattan (*Calameae*), paper mulberry (*Broussonetia papyrifera*), lacquer tree (*Toxicodendron vernicifluum*), or dammar tree (*Shorea* spp. and *Hopea* spp.), etc.
9 Legume trees, such as *Alnus nepalensis* in Himalaya, *Leucaena leucocephala* in Latin America and Southeast Asia, *Sesbania grandiflora* (edible flowers and fruits) in Southeast Asia, etc. (Cairns, 2007).
10 For example, with plot allotment regulated in a village of 30 nuclear-family households, the surface unit reaches 15–30ha; if the allotment is dispersed, the unit is limited to 0.5–1ha.
11 See notably Price T.D. and Gebauer A.B. (eds, 1995), *Last hunters, first farmers: New perspectives on the prehistoric transition to agriculture*, School of American Research, Santa Fe, NM, US.
12 See notably Harris D.R. (ed., 1996), *The origins and spread of agriculture and pastoralism in Eurasia*, Smithsonian Institution Press, Washington, DC.
13 'Primary forest' means here a forest untouched by agriculture, whether it was disturbed or not by hunters and gatherers.

100 million people at the dawn of our era, labour was the scarcest factor of production (Mazoyer and Roudart, 2005). Neolithic farmers were definitely more interested in living in sedentary settlements and in basing their livelihood on rotational shifting cultivation around their village, with long fallows, than in being nomadic along a pioneer deforestation front. Such a rationale still prevails nowadays, when population densities remain low and compatible with long fallows (see Figure 40.2).

The forest space included in the rotation tends to increase with the needs of a growing population[14]. To keep the yield levels – i.e. maintain the duration of the fallow – the villagers had to expand their territory in the rotational system. However, that was not a permanent solution: eventually, they faced impassable obstacles (e.g. large river, mountains) or borders of forests cultivated by other villages; they might be reluctant to increase to an unbearable level the access time to the fields; some might reject the pressure of community rules enforced by more and more distant village or clan authorities. Therefore, some people left the village to found a new community settlement further into the primary forest, while most of the households stayed. The inclusion of primary forests into shifting cultivation rotation combines these two processes of crown expansion and new settlement leap; it ends when the pristine forest resource is no longer available, or too far away. In a mature shifting cultivation system, the villages' agroforests are adjoining, leaving no room for expansion.

To face continual demographic growth, farmers must then rely on speeding up the rotation to increase the crop area in a limited accessible space. The early stages are hardly noticeable[15], but, with shorter and shorter fallows: (i) the biomass accumulation is more and more limited – the yields drop due to inadequate fertility (see Figure 40.4); (ii) the fallow time becomes insufficient for forest secondary successions. The forest is more and more degraded, and finally vanishes while people convert to permanent farming (Mazoyer and Roudart, 2005; Williams, 2006).

This classic demographic crisis (Ramakrishnan, 1992) explains how shifting cultivation contributed to the deforestation of the most fragile forests, close to a Neolithic centre. For example, the Mediterranean forest quickly disappeared, before our era[16], while the western European one was still wealthy at the time of the great medieval clearings ('assarting'), which were not for shifting cultivation, but for permanent agriculture, whose fertility renewal relied on lateral transfer from pastures to fields (Mazoyer and Roudart, 2005). Agriculture was no longer in the forest, but instead of the forest. Slashing and burning were just two of the techniques used in the pioneer front to convert forest into definitive open field, in a rationale that cannot be compared with the rotational shifting cultivation, which occurs nowadays only in tropical and equatorial forests.

Shifting cultivation and biodiversity: a complex nexus

With deforestation, another facet of the environmental impact of shifting cultivation is biodiversity. At the plot level, the issue is clear: the biodiversity level of an old-growth forest

---

14 At a demographic growth rate of 0.1 per cent per year, a population doubles in seven centuries; at 0.5 per cent per year, in 140 years; and at 2 per cent per year in 36 years.
15 For example, the year-to-year yield variability due to climate hazards conceals the average limited reduction when fallow time decreases from 20 to 18 or to 15 years (Mertz et al., 2008).
16 Shifting cultivation is not the sole culprit. The development of harbour cities and the expansion of trade (and wars) by wooden ships required a huge quantity of wood (Williams, 2006).

Agriculture in the forest

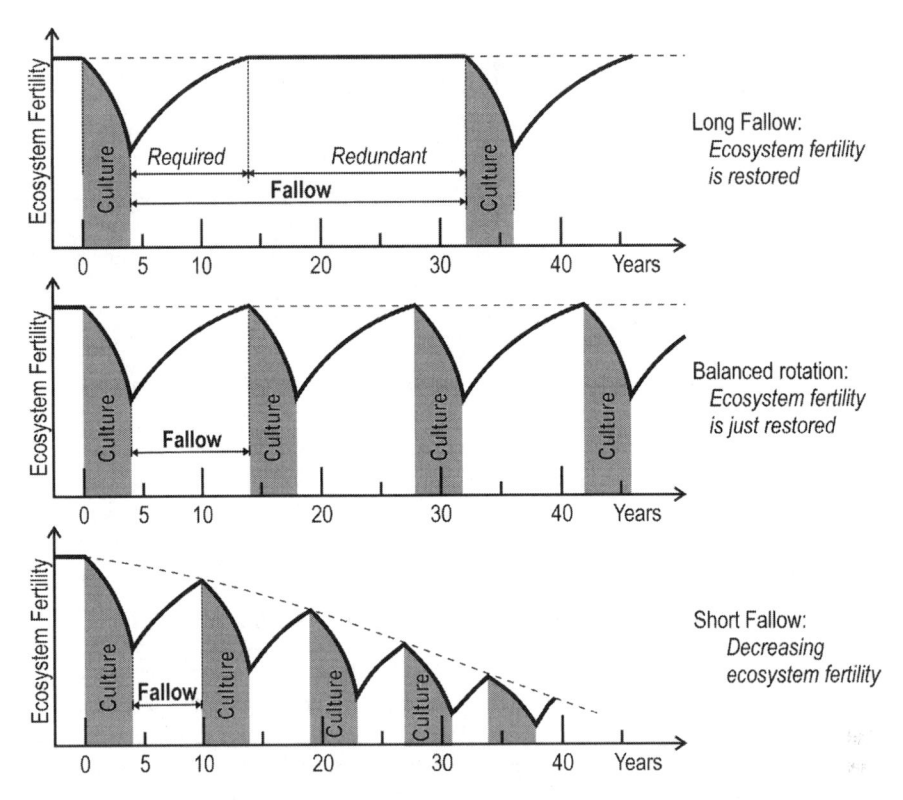

Figure 40.4 Shifting cultivation crisis with accelerating rotation
Based on Mazoyer and Roudart (2005, p. 115), from earlier works by Guillemin (1956) and Ruthenberg (1976)

(either a primary or late seral one) exceeds those of a secondary seral community[17] (Cairns, 2015; Gerold et al., 2004; Ramakrishnan, 1992). The drop can be huge when the shifting cultivation faces the demographic crisis, with poor grass and scrub fallows that dominate the landscape. However, the comparison becomes more complicated when biodiversity is compared at wider scales, either village or regional ones. The biodiversity level of an old-growth forest is nearly isotropic; it marginally increases with the size of the sample plot. In a shifting cultivation landscape, the association of species differs from place to place according to the time since the last cultivation phase. The biodiversity level is the total for the different stages of the fallow mosaic plus that of the cultivated field (Rerkasem et al., 2009). The number of species in the fallow varies with the number of seral communities comprising the mosaic of ecosystems (Gerold et al., 2004; Schroth et al., 2004). As those seral communities are successive, the species differ in the rotation (see. Figure 40.1 and Figure 40.3); the total number of species in a shifting cultivation landscape (i.e. a fallow matrix) may top those of an old-growth forest (Babin, 2004). However, the biodiversity of an ecosystem cannot be reduced to the number of species it comprises. Specialist species, localized and often rare, may be superseded by generalist species when old-growth forests are converted into secondary forests.

---

17 In West Kalimantan for example, a mature forest comprises 0.05 tree species per square meter, while a woody fallow peaks at 0.04 (Rerkasem et al., 2009).

Villagers can also manage the fallow to extract more value from the different ecosystems (Cairns, 2007), with ambivalent impact on biodiversity: favouring some valuable species, sometimes by planting them, may increase the total number of species per fallow plot; but if these species become numerous, they can supplant the spontaneous ones, thus reducing the biodiversity[18].

In the shifting cultivation field, the biodiversity combines inter- and intra-specific agrobiodiversity. Up to 30 or 50 different species can be planted in a swidden (Conklin, 1957) and, for each, farmers tend to manage a large set of varieties and cultivars to adapt the production to: (i) the micro-specificities of the plot; (ii) pest and climate hazards, reducing sensitivity to the hazards by mixing cultivars; (iii) the current household needs[19]; and (iv) the level of the former harvest[20] (Brookfield, 2001). The stocks for the different cultivars are managed at household level, but also at the community one with exchanges between households (Conklin, 1957; Rerkasem et al., 2009; Xu Jianchu and Mikesell, 2003). Kept away from the Green Revolution, shifting cultivation has suffered less from the erosion of genetic diversity for the main cereals (rice, maize, wheat). For example, more than half of the 90,000 rice germplasms stored at the IRRI world gene-bank (Los Baños, Philippines) come from the uplands of two countries, India and Laos (Xu Jianchu and Mikesell, 2003).

Shifting agriculture, carbon cycles and global climate change

As shifting cultivation implies burning down forest, it has been included very early in the causes of global climate change. In the first glaring publication about the greenhouse effect in 1978, Wong, an oceanographer, incriminated shifting agriculture for the massive net release of carbon into the atmosphere: 'the atmospheric input of carbon dioxide from burning forests is mainly due to the forest fires from the shifting cultivation in tropical regions'[21].

Although land preparation (slashing, then burning) causes a massive, but one-off emission of CO_2, the fallow phase is a significant carbon sink. For many researchers, the net carbon balance is close to zero during the complete cycle of shifting cultivation and cannot be assimilated with the high level of net release when forest is converted into permanent agricultural land (Achard et al., 2002; Bruun et al., 2009; Tinker et al., 1996; Ziegler et al., 2012). The CO_2 released during the spectacular burning of fallow is part of a short carbon cycle, in which the storage/release process takes a few years (Stocker et al., 2013). This is quite different from the gaseous emissions of chemical-based farming systems, which come from fossil carbon.

However, the long-term balance is not completely neutral, because: (i) the first conversion of the old-growth forest into shifting cultivation released large amounts of CO_2, as primary forests can store three times more carbon than fallow in long rotation[22]; and (ii) if the rotation quickens, the amount of carbon stored in the fallow tends to decrease.

---

18 In West Kalimantan for example, a rubber garden comprises less than 0.02 tree species per square meter, less than half of a woody fallow (Rerkasem et al., 2009).
19 For example, a nuclear household with a low number of active workers compared to the number of people to feed will favour staple food cultivar with high and secure yield, while a household with only a few young children to feed may choose to plant cultivars of lower yield potential, but with more recreative (e.g. flavour, use for alcoholic distillation or pastry) or market value.
20 Particularly important for limiting the staple food shortage period the year after a poor harvest.
21 Wong, C.S. (1978), 'Atmospheric input of carbon dioxide from burning wood', *Science*, vol. 200, no 4338, p. 197.
22 Palm put forward 300 tonnes ha^{-1} for a primary forest, and 90 tonnes ha^{-1} for a 25-year fallow in humid tropics (Palm et al., 2000).

Political and social dimensions of forest dynamics under shifting cultivation

Based on manual work and low yields, shifting cultivation provides only limited labour productivity[23], as well as land productivity[24] compared to input-intensive farming systems. Over the basic needs of the family, the surplus-labour is so limited that the accumulation of capital at the household level is minimal, as is the possibility of a social body to capture it. Shifting cultivation cannot support large and dense populations, in contrast with other agrarian systems (Mazoyer and Roudart, 2005). In history, societies that based their economy on shifting cultivation were not able to build strong states, contrary to hydraulic empires for example. It has caused dissymmetric relationships, balances of power and conflicts with neighbouring states based on open field agriculture, which have progressively taken control of the forest people's resources: land, forest, water and workforce. Shifting cultivation populations have been marginalized (Scott, 2009), forced to submit to disrupting levies in metal (gold or iron tributes), in kind (grains, animal, wood tributes) or even in labour (slavery, forced labour), despised and heavily criticized by dominant powers, based in the open field plains. Western European colonialism globalized such an approach.

Widespread condemnation of shifting cultivation

When the British and French colonial empires expanded during the nineteenth century, the explorers, military, executives and scientists discovered many instances of shifting cultivation in tropical regions. They built up, very quickly, a discourse advocating for the elimination of shifting agriculture. For example, Thorel wrote in 1868 while exploring Indochina, the pearl of the French empire[25]:

> The second mode of rice cultivation, which is practiced in forests, is a barbarian, transitory method that is destined to disappear with the progress of civilization. [...] It is practiced from Saigon to China, but more frequently in Cambodia and Laos, where civilization is still much more backward and where forests are more extensive.

The text encapsulates the colonizers' position towards shifting cultivation, which was judged an archaic, unproductive and degrading practice throughout the time the empires persisted. The condemnation took its origins in the very sources of Western European colonialism: racism, the civilizing mission and greed for land. At the time of independence, the new national powers, built on elites trained by the former colonizers, as well as the United Nations bodies[26], endorsed this critical position.

By the second half of the 1980s, a new concern rose up in the world's scientific community, civil society, media and political network: how to prevent the degradation of the environment.

---

23 For example, with 0.5ha per worker and a yield of 2 tonnes ha^{-1} (staple and mixed crops, in grain equivalent), the labour productivity reaches only 1 tonne/worker/year, compared to 3 tonnes/worker/year for irrigated rice farming systems, to 10 tonnes/worker/year in three-field rotation with heavy plough farming systems, or even to 500–1,000 tonnes/worker/year in mechanized and chemically fertilized agriculture (Mazoyer and Roudart, 2005).
24 For example, with a yield of 2 tonnes (grain equivalent) per cultivated hectare and 15 years of fallow, the land productivity reaches only 125kg per hectare of total land per year.
25 See Thorel C. (2001), *Agriculture and ethnobotany of the Mekong Basin*, White Lotus, Bangkok, p. 79 and p. 185.
26 Among others, the Forestry Division of FAO: in 1957, Unasylva published a special issue (vol. 11/1) entitled *Shifting Cultivation: An Appeal by FAO to Governments, Research Centers, Associations and Private Persons Who Are in a Position to Help*, which began with the forewords 'Shifting cultivation, in the humid tropical countries, is the greatest obstacle not only to the immediate increase of agricultural production, but also to the conservation of the production potential for the future, in the form of soils and forests.'

The FAO and the World Bank launched in 1985 the Tropical Forestry Action Plan, followed by many bilateral agencies and environmental NGOs. For the first time, it aimed at stopping tropical deforestation for environmental grounds and, of course, shifting cultivation was again presented as the main cause of deforestation. Even the General Assembly of the United Nations joined the movement at the 1992 Earth Summit in Rio: among the objectives of Agenda 21 is the recommendation to 'limit and aim to halt destructive shifting cultivation by addressing the underlying social and ecological causes' (Articles 11–13).

Based upon the ideas about shifting cultivation that: (i) it is not a sustainable type of agriculture (see Figure 40.4); (ii) it is too unproductive – thus remains a poverty trap for the forest ethnic minorities – to efficiently participate in national economic development; (iii) the externalities – deforestation, erosion, biodiversity loss – affect everybody, the State feels it legitimate to blame shifting cultivators, to ban their practices, to promote alternative agriculture and forest reserves. The State, as the guarantor of the general interest, assumes the right to judge shifting cultivators, who often pile up handicaps: cultural/language/lifestyle differences, lack of political representatives. The point of view, highly interested, from the cities and open field elites outclasses the interests of the forest minorities (Hecht and Cockburn, 2010; Xu Jianchu and Mikesell, 2003). From the cities, it is easy for media, politicians and many scientists to disgrace shifting cultivators by just showing a picture of fallows recently burnt, and presenting it as the demise of tropical forests.

Shifting cultivators: forest destroyers or forests gardeners

More and more scientists have disputed this logic by demonstrating that: (i) shifting cultivators have proved, on many occasions, their ability to adapt their agricultural practices and their livelihoods to a changing environment (Cramb, 2007; Hansen and Mertz, 2006; Michon, 2005); and (ii) the quick deforestation that occurred in the Tropics during the twentieth century resulted mainly from other causes and drivers than rotational shifting cultivation.

The tropical forest covered around 15 per cent of Earth's surface in the 1950s; this had decreased to 12 per cent by the mid-1970s and is at 6 per cent nowadays (Achard *et al.*, 2002; Rudel, 2005). Such a steady pace raises strong and legitimate concerns in terms of biodiversity loss and climate impact. If shifting cultivators are the sole or main perpetrators of the forest slaughter as widely accepted, the solution is easy: we just have to ban shifting cultivation and provide alternative livelihoods for those populations, with either the promotion of new farming systems or direct compensatory payments for environmental services (Gullison *et al.*, 2007; Palm *et al.*, 2005). However, the factors of tropical deforestation are not so simple. In the last decade, many researchers (Achard *et al.*, 2002; Boucher *et al.*, 2011; Lambin *et al.*, 2001; Rudel, 2005) have shown that the causes of the quick deforestation are multiple and shifting cultivation plays only a minor role (see Figure 40.5), behind forest conversion to permanent agriculture – by either smallholders or large estates; mainly for soybean, oil palm, hevea and sugarcane cropping, or ranching – logging, migration settlements and urban expansion, encroachments due to roads and other large-scale infrastructures (mines, hydropower dams, etc.).

When focusing on shifting cultivation, the environmental policies fail to reduce the progress of deforestation (Rudel, 2005) and they further marginalize the shifting cultivators (Fox *et al.*, 2009). Moreover, when a ban of shifting cultivation is promulgated and enforced, the fallow forests become access-free for other economic agents; the ultimate conversion to an immediate deforestation is favoured instead of a progressive, long-term degradation with shifting cultivation. The wrong political path to protecting tropical forests relies on some misunderstandings: (i) timescale confusion, with a demographic crisis being only a mid- or long-term prospect; (ii) a confusion of techniques, when slashing and burning may be the first steps for converting forest to

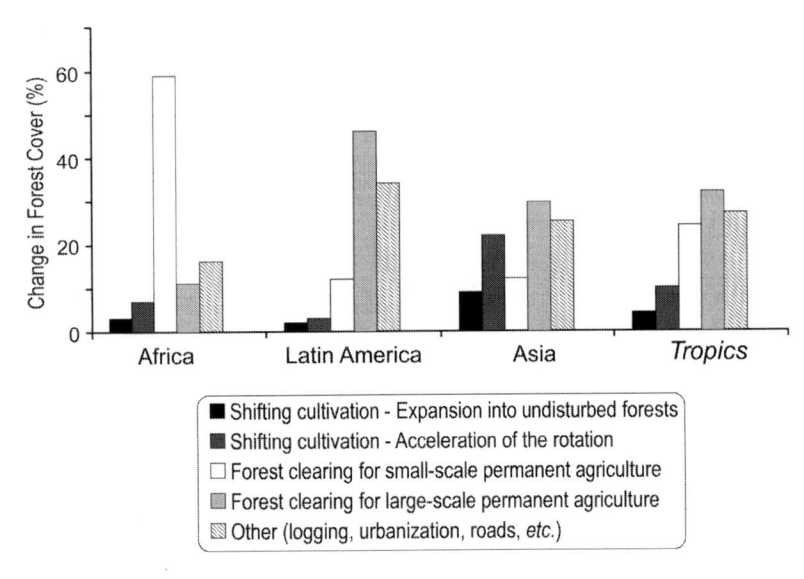

Figure 40.5 Factors of tropical forest cover decline
Based on FAO (2001), *Global Forest Resources Assessment 2000 – Main report*, FAO, Rome, (www.fao.org/documents/show_cdr.asp?url_file=/DOCREP/004/Y1997E/y1997e0t.htm)

open land, as well as bulldozers, for example, or steps in the rotation pattern for shifting cultivation in forests (Ziegler et al., 2011). Such misunderstandings lead to overestimations of the population involved in shifting cultivation (Mertz et al., 2009; Thrupp et al., 1997). They base inefficient policies that aim to protect the forest by restricting shifting cultivation, but that may trigger a demographic crisis when the land allowed for fallow is shrunk or when shifting cultivators' villages are resettled and consolidated on limited land (Lestrelin and Giordano, 2006). Moreover, these policies may serve the interests of powerful economic agents and lobbies, easing their access to forest resources and land (Hecht and Cockburn, 2010; Ziegler et al., 2009).

Conclusion

Shifting cultivation covers a tremendous diversity of farming systems, which originated from the first steps of agriculture. Nowadays, it perdures in tropical forest ecosystems. The originality of shifting cultivation consists in the principle that the soil fertility is renewed by the forest biomass accumulated *in situ* during the fallow period, resulting in a landscape mosaic of secondary plant formations, forests included. Without (fallow) forest, there is no shifting cultivation: the forest is intrinsic to shifting cultivation.

World maps of deforestation are superimposed on those of shifting cultivation (Hansen et al., 2013; Thrupp et al., 1997). Inferring that shifting cultivation is the main or sole cause of current tropical deforestation is a too-easy conclusion, based on confusion, prejudice and collusion of interests (Fox et al., 2009). Another interpretation could be advanced: the tropical forest lasts where human societies have the experience to protect it and the interest in doing so. The only heritage and asset of the shifting cultivators is the forest, whose degradation and depletion occur only long term because of the multiple and complex social regulations the forest farmers have developed to preserve the basis of their livelihood, a far remove from the tragedy of the commons (Cairns, 2007, 2015; Cramb, 2007; Forsyth and Walker, 2008; Hecht and Cockburn, 2010). The deforestation and shifting cultivation maps may be understood in a different way: tropical forests remain where shifting cultivators remain (Ziegler et al., 2011). The current eagerness to blame the forest farmers overlooks their stewardship management of forest resources, and masks the responsibility of other actors involved in definitive deforestation (Rudel, 2005; Williams, 2006): permanent agriculture, unsound forest exploitation, urbanization, or large-scale infrastructures. Focusing on eradicating shifting cultivation fails to prevent deforestation, and even may hasten its trend.

References

Achard F., Eva H. D., Stibig H.-J., Mayaux P., Gallego J. and others (2002), 'Determination of deforestation rates of the world's humid tropical forests', *Science*, vol 297, pp. 999–1002.

Babin D. (ed., 2004). *Beyond tropical deforestation: From tropical deforestation to forest cover dynamics and forest development*. UNESCO/Cirad, Paris.

Bermingham D. E., Dick C. and Moritz C. (eds, 2005), *Tropical rainforests: Past, present, and future*. University of Chicago Press, Chicago, IL, US.

Boucher D., Elias P., Lininger K., Roquemore S. and Saxon E. (2011), *The Root of the problem: What is driving tropical deforestation today?* Union of Concerned Scientists, Cambridge, US.

Brookfield H. C. (2001), *Exploring agrodiversity*. Columbia University Press, New York City, NY.

Bruun, T., de Neergaard, A., Lawrence, D., and Ziegler, A. (2009), 'Environmental consequences of the demise in swidden cultivation in Southeast Asia: Carbon storage and soil quality', *Human Ecology*, vol 37, no 3, pp. 375–388.

Cairns M. F. (ed., 2007). *Voices from the forest: Integrating indigenous knowledge into sustainable upland farming*. Earthscan, London.

Cairns M. F. (ed., 2015). *Shifting cultivation and environmental change: Indigenous people, agriculture and forest conservation*. Earthscan, London.

Conklin H. C. (1957), *Hanunóo agriculture: A report on an integral system of shifting cultivation in the Philippines*. FAO, Rome.

Cramb R. A. (2007), *Land and longhouse: Agrarian transformation in the uplands of Sarawak*. NIAS, Copenhagen.

De Jong W., Chokkalingam U. and Perera D. (2001), 'The evolution of swidden fallow secondary forests in Asia', *Journal of Tropical Forest Science*, vol 13, no 4, pp. 800–815.

Ducourtieux O. (2006), 'Is the diversity of shifting cultivation held in high enough esteem?', *Moussons*, vols 9–10, pp. 61–86.

Forsyth T. and Walker A. (2008), *Forest guardians, forest destroyers: The politics of environmental knowledge in Northern Thailand*. University of Washington Press, Washington, D.C.

Fox J. M., Fujita Y., Ngidang D., Peluso N., Potter L. and others (2009), 'Policies, political-economy, and swidden in Southeast Asia', *Human Ecology*, vol 37, no 3, pp. 305–322.

Gerold G., Fremerey M. and Guhardja E. (eds, 2004). *Land use, nature conservation and the stability of rainforest margins in Southeast Asia*. Springer, Berlin.

Gómez-Pompa A., Whitmore T. C. and Hadley M. (eds, 1991). *Rain forest regeneration and management*. Unesco, Paris.

Guillemin R. (1956), 'Evolution de l'agriculture autochtone dans les savanes de l'Oubangui', *Agronomie Tropicale*, vol 11, no 2, pp. 143–176.

Gullison R. E., Frumhoff P. C., Canadell J. G., Field C. B., Nepstad D. C. and others (2007), 'Tropical forests and climate policy', *Science*, vol 316, pp. 985–986.

Hansen M. C., Potapov P. V., Moore R., Hancher M., Turubanova S. A. and others (2013), 'High-resolution global maps of 21st-century forest cover change', *Science*, vol 342, pp. 850–853.

Hansen T. S. and Mertz O. (2006), 'Extinction or adaptation? Three decades of change in shifting cultivation in Sarawak, Malaysia', *Land Degradation and Development*, vol 17, pp. 135–148.

Hecht S. B. and Cockburn A. (2010), *The fate of the forest: Developers, destroyers, and defenders of the Amazon*. 2nd edition, University of Chicago Press, Chicago, IL, US.

Hladik C. M., Hladik A., Linares O. F., Pagezy H., Semple A. and Hadley M. (eds, 1993). *Tropical forests, people and food: Biocultural interactions and applications to development*. UNESCO, Paris.

Lambin E. F., Turner II B. L., Geist H. J., Agbola S. B., Angelsen A. and others (2001), 'The causes of land-use and land-cover change: Moving beyond the myths', *Global Environmental Change*, vol 11, no 4, pp. 261–269.

Lestrelin G. and Giordano M. (2006), 'Upland development policy, livelihood change and land degradation: interactions from a Laotian village', *Land Degradation and Development,* vol 18, no 1, pp. 55–76.

Mazoyer M. and Roudart L. (2005), *A history of world agriculture: From the Neolithic age to current crisis.* Monthly Review Press, New York City, NY.

Mertz O., Wadley R. L., Nielsen U., Bruun T. B., Colfer C. J. P. and others (2008), 'A fresh look at shifting cultivation: Fallow length an uncertain indicator of productivity', *Agricultural Systems,* vol 96, pp. 75–84.

Mertz O., Leisz S., Heinimann A., Rerkasem K., Thiha T. and others (2009), 'Who counts? Demography of swidden cultivators in Southeast Asia', *Human Ecology,* vol 37, no 3, pp. 281–289.

Michon G. (ed., 2005), *Domesticating forests: How farmers manage forest resources.* IRD/Cifor/Icraf, Bogor.

Nye P. H. and Greenland D. J. (1960), *The soil under shifting cultivation.* CABI, Wallingford, UK.

Palm C. A., Woomer P. L., Alegre J. C., Arévalo L., Castilla C. E. and others (2000), *Carbon sequestration and trace gas emissions in slash-and-burn and alternative land uses in the humid tropics,* Climate Change Working Group Final Report, Phase II, ICRAF, Nairobi.

Palm C. A., Vosti S. A., Sanchez P. A. and Ericksen P. J. (eds, 2005), *Slash-and-burn agriculture: The search for alternatives.* Columbia University Press, New York City, NY.

Ramakrishnan P. S. (1992), *Shifting agriculture and sustainable development: An interdisciplinary study from north-eastern India.* UNESCO, Paris.

Rerkasem K., Lawrence D., Padoch C., Schmidt-Vogt D., Ziegler A. and Bruun T. (2009), 'Consequences of swidden transitions for crop and fallow biodiversity in Southeast Asia', *Human Ecology,* vol 37, no 3, pp. 347–360.

Roder W. (ed., 2001). *Slash-and-burn rice systems in the hills of northern Lao PDR: Description, challenges, and opportunities.* Irri, Los Baños, Philippines.

Rudel T. K. (2005), *Tropical forests: Regional paths of destruction and regeneration in the late twentieth century.* Columbia University Press, New York City, NY.

Ruthenberg H. (1976), *Farming systems in the Tropics,* Oxford University Press, Oxford, UK.

Schmidt-Vogt D. (1999), *Swidden farming and fallow vegetation in northern Thailand.* Franz Steiner, Stuttgart, Germany.

Schroth G., Da Fonseca G. A. B., Harvey C. A., Gascon C., Vasconcelos H. L. and Izac A.-M. N. (eds, 2004), *Agroforestry and biodiversity conservation in tropical landscapes.* Island Press, Washington, DC.

Scott J.C. (2009), *The Art of not being governed: An anarchist history of upland Southeast Asia.* Yale University Press, New Haven, CT, US.

Stocker T. F., Qin D., Plattner G.-K., Tignor M., Allen S. K. and others (eds, 2013), *Climate change 2013: The physical science basis. Contribution of Working Group I to the Fifth Assessment Report of the Intergovernmental Panel on Climate Change,* Cambridge University Press, New York City, NY.

Szott L. T., Palm C. A. and Buresh R. J. (1999), 'Ecosystem fertility and fallow function in the humid and subhumid tropics', *Agroforestry Systems,* vol 47, nos 1–3, pp. 163–196.

Thrupp L. A., Hecht S. B. and Browder J. O. (1997), *The diversity and dynamics of shifting cultivation: Myths, realities, and policy implications.* World Resources Institute, Washington, DC.

Tinker, B. P., Ingram, J. S. I., and Struwe, S. (1996), 'Effects of slash-and-burn agriculture and deforestation on climate change', *Agriculture, Ecosystems and Environment,* vol 58 no 1–2, pp. 13–22.

Van Keer K. (2003), 'On-farm agronomic diagnosis of transitional upland rice swidden cropping systems in northern Thailand', PhD thesis, Faculteit Landbouwkundige en Toegepaste Biologische Wetenschappen, Katholieke Universiteit Leuven, Leuven, Belgium.

Williams M. (2006), *Deforesting the Earth: From prehistory to global crisis, an abridgment.* University of Chicago, Chicago, IL, US.

Xu Jianchu and Mikesell S. (eds, 2003), *Landscapes of diversity: Indigenous knowledge, sustainable livelihoods and resource governance in montane mainland Southeast Asia.* Yunnan Science and Technology Press, Kunming, China.

Ziegler A. D., Fox J. M. and Jianchu Xu (2009), 'The rubber juggernaut', *Science,* vol 324, pp. 1024–1025.

Ziegler A. D., Fox J. M., Webb E. L., Padoch C., Leisz S. J. and others (2011), 'Recognizing contemporary roles of swidden agriculture in transforming landscapes of southeast Asia', *Conservation Biology,* vol 25, no 4, pp. 846–848.

Ziegler A. D., Phelps J., Jia Qi Yuen, Webb E. L., Lawrence D., and others (2012), 'Carbon outcomes of major land-cover transitions in SE Asia: Great uncertainties and REDD+ policy implications', *Global Change Biology,* vol 18, no 10, pp. 3087–3099.

41
INDIGENOUS FOREST KNOWLEDGE

Hugo Asselin

Introduction

There are more than 370 million indigenous people in the world, in some 90 countries (United Nations, 2009). Article 33 of the United Nations Declaration on the Rights of Indigenous Peoples recognizes the principle of self-identification, whereby indigenous peoples themselves define their own identity based on various criteria, among which are culture, language and the occupation of ancestral lands (United Nations, 2009). Several indigenous peoples live in forested ecosystems and rely to various extents on ecosystem goods and services to meet their needs. In many countries, indigenous identities, cultures and practices are tightly linked to traditional lands. This intimate connection between cultures and ecosystems is reflected in the vast body of indigenous forest knowledge. Akin to the traditional ecological knowledge concept, it refers to "a cumulative body of knowledge, practice and belief evolving by adaptive processes and handed down through generations by cultural transmission, about the relationship of living beings (including humans) with one another and with their environment" (Berkes, 2012, p. 7).

Indigenous forest knowledge is still largely ignored in planning and management (Cheveau *et al.*, 2008), and indigenous peoples are generally at best considered as "just another stakeholder" (Stevenson and Webb, 2003). Mere modifications to the prevailing science- and technology-based forestry regimes to take into account indigenous needs and views will not suffice to achieve "indigenous forestry". A paradigm shift is required, to base forest planning and management on aboriginal values, practices and knowledge. Drawing on examples from various forest ecosystems around the world, this chapter reviews the properties of indigenous forest knowledge that make it complementary to scientific knowledge as well as key contributions that indigenous knowledge can make to forest management.

Properties of indigenous forest knowledge

Indigenous and scientific knowledge are not engaged in a "credibility contest" (Trosper and Parrotta, 2012), and they should instead be considered on equal footing. Scientific "validation" of indigenous knowledge is futile at best, and actually carries a risk of information distortion (Mazzocchi, 2006) and perpetuation of power imbalances (Castleden *et al.*, 2009). In fact, the intrinsic properties of indigenous knowledge make it complementary to scientific knowledge in several ways. Indigenous knowledge is cumulative, dynamic, long term, holistic, local, embedded, moral and spiritual (Menzies and Butler, 2006).

Indigenous forest knowledge is cumulative as it is the sum of empirical observations acquired through trial-and-error by passing generations. This is somewhat similar to scientific knowledge, also based on empirical evidence. The term "traditional", often used to qualify indigenous knowledge, should not be understood as meaning "static", "ancient" or "outdated". It rather implies that each generation modifies the accumulated body of knowledge, practices and beliefs according to its own experiences, in a dynamic and adaptive fashion. Indigenous forest knowledge is thus constantly being updated by modifying, adding or deleting information to take into account the evolving context and the apparition of innovations. For example, an indigenous hunter using a rifle and a GPS (global positioning system) is using contemporary tools but still can rely on traditional knowledge.

Being rooted in traditions and still mostly transmitted orally, indigenous forest knowledge provides a long-term perspective, which can prove particularly useful in areas where written archives or instrumental data are recent, discontinuous or unavailable. There is growing concern over the loss of knowledge and cultural erosion that proceeds at an accelerating pace. It is indeed estimated that up to 90 per cent of the world's languages will become extinct or threatened by extinction by the end of the twenty-first century (United Nations, 2009), largely because of acculturation. A study conducted with the Tsimane' people of Bolivia showed a 20 per cent decrease of traditional plant use reports between 2000 and 2009, irrespective of individuals' age, sex, schooling or Spanish fluency (Reyes-Garcia *et al.*, 2013). The decrease was more pronounced for men than women and for informants living in villages close to market towns than for those in remote settlements. Nevertheless, plant use reports in this study did not significantly differ according to decade of birth (from the 1920s to the 1980s), illustrating the effectiveness of intergenerational knowledge transfer.

Indigenous forest knowledge is holistic, meaning that it considers all elements of the environment (including humans) to be interconnected and influencing each other. For example, the Huu-ay-aht First Nation traditional forest management practices are based on *Hishuk Tsawak*, a worldview that means "everything is one, everything is connected" (Castleden *et al.*, 2009). This is in contrast with scientific knowledge, which is reductionist, meaning that it considers the different elements of the environment separately. In addition, scientific knowledge generally assumes that mind and matter are separate, and thus that humans are outside the environment and can control or manage it. Indigenous people rather consider mind and matter on the same level, and that it is not resources that need to be managed, but resource users. In Morocco, Berber communities have developed a forest management system, known as *agdal*, which is based on a holistic worldview (Genin and Simenel, 2011). Following customary laws, they use periodicity of cutting, harvesting quotas, species selection and functional zoning to shape forest landscapes into diversified patches in order to satisfy their material, social and cultural needs, while preserving biodiversity.

Indigenous forest knowledge is local and provides a level of detail that is hard to achieve by other means. For example, in southern India, *amla* trees (*Phyllanthus emblica* and *Phyllanthus indofischeri*) are infested by *Taxillus tomentosus*, a native but invasive mistletoe (Rist *et al.*, 2010). People from the local Soliga indigenous community get > 10 per cent of their cash income from selling the fruit of the *amla* tree. They identified almost three times more secondary host species to mistletoe than scientific surveys (35 vs. 12 species), they pointed to mammals as important mistletoe dispersers (whereas scientific information was restricted to bird dispersers) and they pointed to fire suppression as a possible cause for mistletoe spread in the area. Such information could be key to designing effective management strategies.

Although indigenous forest knowledge is local, most people use extensive territories. Just as scientific studies need several sampling sites over wide areas to be generalizable; it is possible to scale-up indigenous knowledge by interviewing several individuals, families, communities or peoples. Indigenous knowledge holders make it in fact possible to gather information over large areas (e.g., thousand km^2) and extended periods, something that would be almost impossible – or then very costly – using conventional scientific studies. One such case was documented in the Northern Territory of Australia, where limited and localized surveys suggested major declines of some mammal species, but with inconsistencies in the data. Interviews with indigenous knowledge holders made it possible to substantially expand the spatial and temporal scales of data acquisition (Ziembicki et al., 2013). Indigenous knowledge confirmed decline for several mammal species and pointed towards habitat modifications due to cessation of traditional surface burning practices as a probable cause.

Indigenous forest knowledge is context dependent, not only environmentally, but also culturally. Knowledge is indeed embedded within each culture's idiosyncrasies that shape how people acquire and use knowledge. Hence, knowledge cannot be readily transferred from one context to another or from one community to another. For example, land use maps revealed different patterns for two Algonquin communities (Quebec, Canada) (Table 41.1). While both communities selected dense forest sites near water, the age and species composition of the selected stands were different. Furthermore, while people from Pikogan extensively used the forest road network to access family hunting grounds, this variable did not significantly explain site location for Kitcisakik, as people from this community used diverse travel modes, including the use of all-terrain vehicles and canoeing on lakes and rivers.

Indigenous knowledge is also moral and spiritual, whereas scientific knowledge generally adopts a mechanistic viewpoint, assuming that phenomena have rational explanations and that events may be connected as cause and effect. Thus, indigenous practices are governed by ethical principles. For example, the Nawash Ojibway (Ontario, Canada) have developed principles of wild harvest (LaRiviere and Crawford, 2013). Certain activities are only allowed during specific seasons in order to respect the animals' reproductive cycle. People thank Mother Nature for the provision of food, they never take more than they need, share their harvest with other community members, and avoid wasting.

Table 41.1 Most relevant forest characteristics explaining the location of cultural sites on the ancestral lands of two Algonquin communities (Quebec, Canada) (source: Hugo Asselin)

Site characteristic	*Pikogan*	*Kitcisakik*
Water proximity	< 60 m	< 60 m
Road proximity	< 100 m	Not significant
Stand type	Deciduous and mixed stands	Mixed stands, and stands with *Pinus strobus* or *Thuja occidentalis*
Stand age	Mature even-aged and young uneven-aged	Old uneven-aged
Stand density	> 60 per cent	> 60 per cent

Key contributions of indigenous knowledge to forest management

There are several ways in which indigenous knowledge can contribute to forest management: by providing biological information and ecological insight, as a source of alternative management practices, by favouring biodiversity conservation, by contributing to environmental monitoring and assessment, as a support for social development, and as a source of environmental ethics principles (Berkes, 2012).

Indigenous forest knowledge can provide extensive biological and ecological information at levels of detail that would be hard to reach using quantitative scientific research methods. For example, in a study conducted within seven Merap and Punan villages in East Kalimantan (Indonesia), 52 informants were asked if, where, and when they had observed eight species of regional conservation interest: rafflesia (*Rafflesia* spp.), black orchid (*Coelogyne pandurata*), sun bear (*Helarctos malayanus*), tarsier (*Tarsius bancanus*), slow loris (*Nycticebus coucang*), proboscis monkey (*Nasalis larvatus*), clouded leopard (*Neofelis diardi/N. nebulosa*), and orangutan (*Pongo pygmaeus*). The study was able to gather important information on the elusive rafflesia and clouded leopard, and extended the known distributions of sun bear, tarsier, slow loris, and clouded leopard (Padmanaba *et al.*, 2013). It was completed within only six weeks, at a total cost estimated to have been one or two orders of magnitude lower than expert field surveys.

Indigenous forest knowledge can sometimes be limited to larger or more culturally salient species (Ziembicki *et al.*, 2013), such as cultural keystone species, i.e., species that shape in a major way the cultural identity of a people, as reflected in the fundamental roles they play in diet, materials, medicine, and/or spiritual practices (Garibaldi and Turner, 2004). An example of cultural keystone species is the eastern white pine (*Pinus strobus*), perceived as the "king of the forest" by the Kitcisakik Algonquin community of Quebec, Canada (Uprety *et al.*, 2013). White pine – *cigwâtik* in the local language – is mentioned in traditional stories and myths. It provides habitat for culturally important wildlife species, such as bald eagle (*Haliaeetus leucocephalus*), moose (*Alces alces*) and marten (*Martes americana*). White pine is also an important medicinal plant species, it is used as timber, and tall individuals towering above the canopy serve as landmarks (Figure 41.1). Shelterwood cuts and understorey plantations are often used to manage white pine stands. According to the Kitcisakik people, association with balsam fir (*Abies balsamea*) should however be avoided, as a legend says the two species are enemies.

Animal species are important to indigenous people, not only because they provide food and other goods (leather, fur and wool for clothing, bones for tools or art craft, etc.), but also for their ecological roles. For example, people in Korea have been practicing native beekeeping for at least 2,000 years (Park and Yeo-Chang, 2012). They possess significant knowledge on bee ecology that they use in traditional practices that contribute to sustainable forest management. There is indeed a positive correlation between the number of native bee colonies in a province and the total area of high-biodiversity zones within the area. A similar positive correlation between biodiversity and traditional management practices was reported for bird species in China (Shen *et al.*, 2012).

One classic example of how forest management practices can be influenced by indigenous knowledge is the management by the Menominee tribe of Wisconsin (USA) of their traditional forests (Trosper, 2007). By historically preferring selective cutting of the oldest trees over short-rotation clear-cutting (advised by federal authorities) they were able to produce high-quality wood for which they could get good prices on the market, while maintaining forest cover for cultural activities. The difference between Menominee and conventional forestry in terms of residual forest cover is striking and can easily be seen by typing "Menominee County, Wisconsin" in Google Earth™. Another example of

Figure 41.1 Two white pines (*Pinus strobus*) towering above the surrounding canopy and used as landmarks by the members of the Kitcisakik Algonquin community. White pine is a cultural keystone species to this community

Source: Pierre Cartier

indigenous forest management involves the monkey-puzzle tree (*Araucaria araucana*), a sacred species to the Mapuche Pewenche people of Chile (literally the people – *che* – of the monkey-puzzle tree – *pewen*). The species is listed as vulnerable by the International Union for Conservation of Nature, following decades of intense exploitation by the timber-extraction industry and substitution by plantations of exotic species (*Pinus radiata*, *Eucalyptus* spp.) or conversion to pastureland. Seeds of the monkey-puzzle tree are an important part of the diet of the Mapuche Pewenche people who have developed profound knowledge on the species' ecology, as well as techniques of *in-situ* and *ex-situ* conservation (Herrmann, 2006). When harvesting seeds, people make sure to leave some in the trees to favour regeneration. Some seeds fall on the ground during harvesting and not only are they left there, but they are trampled for better positioning in the soil, which also favours regeneration. Nurseries have been developed, where *Araucaria* seeds are planted between rows of willow shrubs (*Salix* spp.) which provide shelter.

Non-timber forest products are also used and managed by indigenous peoples. While edible and medicinal plants have received much attention, indigenous forest knowledge also extends to more cryptic or lesser-studied species. For example, the California Indian Tribes were reported to traditionally use at least 36 mushroom species as food, medicine, or for other uses such as to make a red pigment for war paints, to start or carry fire, to polish animal skin or to make bowls (Anderson and Lake, 2013). They have developed mushroom management practices involving prescribed burning, based on their precise knowledge of mushroom habitat, seasonality and ecological relationships.

Whereas inventories usually focus on tree species targeted for timber production, data on non-timber forest products should also be acquired. Indigenous people are prominent users of non-timber forest products and their knowledge could contribute to improve inventory comprehensiveness. For example, a pilot study was undertaken jointly by the United States Department of Agriculture Forest Service and the Great Lakes Indian Fish and Wildlife Commission to document indigenous knowledge of paper birch (*Betula papyrifera*) bark properties needed for traditional uses such as making canoes and food containers. The information was to be incorporated into an inventory field guide (Emery *et al.*, 2014).

Indigenous forest knowledge can also be used in the management of so-called "nuisance species". For example, beaver (*Castor canadensis*) can cause substantial damage to forest roads in North America by blocking culverts to stop water flow and create ponds where they build huts to raise their young (Figure 41.2). Forestry companies sometimes hire indigenous trappers to get rid of problematic beavers. However, problems usually arise in summer, when beaver meat is unpalatable and when fur quality is low. Hence, trappers are forced to leave the beavers in the forest, feeling guilty for such a waste. It would be more culturally-sound to put preventive trapping programmes into place. Trappers possess profound knowledge of habitat use by beavers and could identify in advance the areas that might be problematic in summer, and harvest the animals at a time of year when fur and meat are usable.

Certain traditional practices are essential to maintain some species in specific areas. For example, controlled fires set by indigenous people of southern British Columbia (Canada) have contributed to maintain Garry oak (*Quercus garryana*) in the landscape over most of the last 3,000 years, despite an unfavourable, wet and cool climate (Pellatt and Gedalof, 2014). Indigenous population decline since European colonization and the subsequent application of fire suppression policies have caused replacement of Garry oak by conifer species.

Ecological restoration involves management practices to assist the recovery of degraded, damaged or destroyed ecosystems. Uprety *et al.* (2012) reviewed the main contributions of indigenous knowledge to ecological restoration. The site-specific nature of indigenous knowledge makes it particularly applicable to restoration initiatives. Indigenous knowledge can be used to document the characteristics of reference ecosystems, select appropriate species and sites to maximize restoration success, provide expertise through traditional management practices, manage invasive species and conduct post-restoration monitoring.

While protected areas are a key element of biodiversity conservation strategies, they are somewhat problematic from an indigenous viewpoint. Considering themselves as integral parts of the ecosystems, indigenous people cannot contemplate "putting a dome" over an area and preventing themselves from using it. Nevertheless, forests with a certain level of protection are present in several cultures. So-called "sacred groves" (Bhagwat and Rutte, 2006) are managed following locally-defined rules that often prohibit logging or hunting, but that allow other uses, such as harvesting of medicinal plants, fodder or firewood. Some have been protected for a very long time and are biodiversity hotspots. Their importance is even higher as they are not part of the official protected areas network. Advances have been made in conservation practices, and protected areas increasingly accommodate a range of indigenous practices (including subsistence hunting and the collection of plants for traditional purposes), especially in North America (Gladu *et al.*, 2003).

Environmental monitoring and assessment are time consuming and costly. They are thus often limited to short and specific time frames in order to reduce operation costs. Indigenous people are usually present year-round on the land and will be the first ones to notice any change in the environment. They can take measurements on a continuous basis and maintain long-term monitoring, whereas forest managers usually focus on areas where logging is planned in the short term. Governments sometimes maintain permanent sampling plots, but these are often not numerous enough or cannot be sampled frequently enough. Community-based environmental monitoring can be used to facilitate mapping of indigenous knowledge,

Figure 41.2 Beaver (*Castor canadensis*) dam preventing stream water flow through a culvert. The resulting pond flooded the nearby forest road (to the right, not shown on photograph), causing major damage

Source: Claude-Michel Bouchard

observations, practices and land uses in a dynamic and adaptive way. Participatory GIS (geographic information systems) are used in different settings to inform decision makers. For example, the Sami people of Sweden and the Cree and Naskapi First Nations of eastern Canada collect field data on the risks and impacts of resource extraction industries on habitat use by reindeer/caribou (*Rangifer tarandus*) and indigenous livelihoods (Herrmann et al., 2014). The data collected by the Sami, Cree and Naskapi peoples are site-specific, but the accumulation of data over extended areas and during several consecutive years provides a landscape perspective allowing for a thorough understanding of human–environment relationships that would otherwise be difficult to grasp.

The Tinto Community Forest, in Cameroon, was used as a case study to evaluate if – and how – participatory GIS contributed to good governance of local resources (McCall and Minang, 2005). The results indicated that participatory mapping (with and without GIS) contributed positively to good governance, notably by improving dialogue, legitimizing and using local knowledge, redistributing access to and control over resources, and empowering local communities with new skills in geospatial analysis. However, challenges remained, mostly regarding equity, data ownership and information sharing, likely because the study was not controlled by the Tinto people and adopted an external viewpoint. Women were underrepresented in the decision-making process. This is particularly problematic because women are the main collectors of non-timber forest resources. Tinto people had easier access to paper maps than to digitized maps and were mostly excluded from the final, key stages of the decision-making process. Furthermore, small-scale maps were generally used, whereas large-scale maps would have been more appropriate to fully express detailed indigenous knowledge.

Effective decision-support tools are needed to foster integration of indigenous knowledge at all steps of the sustainable forest management process, from inventory to monitoring, through planning and management. As social acceptability of forestry operations largely depends on the cohabitation of different forest users seeking different ecosystem goods and services, maps are probably the most important tool to develop. A challenge in this regard is the reluctance of indigenous people to share culturally sensitive information such as precise locations of sites of interest (Berkes, 2012). Functional zoning offers a working compromise. The territory is divided into zones ascribed to one or more uses (e.g., wood harvesting, berry picking, hunting or spiritual activities). However, such a compartmented view of the land does not correspond to the holistic perception of indigenous people. More culturally appropriate mapping could, for example, involve the calculation of a sort of "indigenous suitability index", where characteristics of sites of interest would be compared with those of randomly selected locations to determine which site characteristics are avoided or sought after. Using this information, a map could be produced, displaying a colour gradation from zones of high to low indigenous interest, without having to disclose the actual locations of the sites of interest. Such maps would provide guidelines for forest management planning without giving away key information, thus granting indigenous people increased control over the decision-making process.

Concurrently with the rise of environmental certification, there has been increased interest in local frameworks of criteria and indicators (C&I) of sustainable forest management developed by indigenous communities. While certification encourages the adoption of best practices theoretically leading to sustainable forest management, C&I frameworks measure if – and to what extent – practices are sustainable. Hence, certification is about means and C&I are about ends. In addition, certification is mostly used by forest managers to display conformity with ecological, social and economic norms, whereas C&I frameworks are used by local communities in an empowerment perspective. A collaborative research project with the Kitcisakik Algonquin community (Quebec, Canada) enabled the documentation of the social representations of forest and forestry and to use the information to elaborate a local framework of 5 principles and 22 criteria of sustainable forest management (Saint-Arnaud et al., 2009). The framework includes cultural, ethical, ecological, educational and economic principles. The ethical and educational principles are unique to this framework and play a key role in bridging the apparently irreconcilable indigenous and industrial viewpoints.

Putting indigenous knowledge to work will contribute to increase social acceptability of forest management practices. Furthermore, adopting a holistic viewpoint and recognizing the intricate link between people and the land will necessarily lead to local-scale management stemming from a bottom-up decision-making process. Such "human-scale forestry" will be more acceptable if it allows more needs to be met through the provision of a wider range of ecosystem goods and services. Social development and community well-being and resilience will ensue, as indigenous peoples establish a clear link between healthy land and healthy people. This is illustrated by the *miyupimaatisiiun* concept shared by most indigenous people from the Algic language family in North America, which means "being alive well", and where health is not only defined in terms of individual physiology, but also in terms of social relations, cultural identity and link to the land (Adelson, 2000). In Ecuador and Bolivia, the similar concept of *buen vivir* (the "good life") has emerged as an alternative political paradigm to capitalism, emphasizing the harmonious relationship between human beings, society and nature (Altmann, 2013).

The notion of respect is central to several indigenous peoples' environmental ethics: respect for other beings – living and non-living – by taking only what is needed, not wasting and sharing. O'Flaherty et al. (2009) reported four lessons learned from the Pikangikum Ojibway elders (Ontario, Canada), that nicely summarize indigenous environmental ethics and complementarity with scientific knowledge. First, humbly accept that management practices can only partly meet the needs of other beings with whom indigenous people share the land, and who will be impacted by management activities. Second, adopt a precautionary approach and develop a monitoring system based on indigenous knowledge. Third, establish effective communication channels between indigenous people and other stakeholders. Fourth, consider indigenous and scientific knowledge on equal footing, each equally contributing to an integrated land management strategy.

Conclusion

Several properties of indigenous forest knowledge make it complementary to scientific knowledge. Perhaps the most striking difference between the two knowledge types is that indigenous knowledge is based on a holistic perception of the environment, implying that mind and matter are on the same plane. In other words, indigenous people consider themselves an integral part of forest ecosystems, rather than considering they can control or manage it from the outside. This is far from being trivial, especially considering the rising interest in ecosystem-based forest management:

> a management approach that aims to maintain healthy and resilient forest ecosystems by focusing on a reduction of differences between natural and managed landscapes to ensure long-term maintenance of ecosystem functions and thereby retain the social and economic benefits they provide to society.
>
> *(Gauthier* et al., *2009, p26)*

Considering that indigenous knowledge and practices have existed for millennia, that indigenous people are a key component of many forest ecosystems and that the idea of a pristine, untouched forest is often a myth (Denevan, 1992), ecosystem-based forest management is probably the best meeting ground for indigenous and scientific forest knowledge. Indeed, the various contributions of indigenous forest knowledge to forest management practices highlighted in this chapter suggest several ways to reduce the gap between natural and managed landscapes, as most indigenous practices can be considered drivers of the so-called "historic range of variability". Small steps have been made towards integration of indigenous knowledge and needs in forestry, showing for example that it does not severely impact profitability (e.g., Dhital et al., 2013). The next (big) step will need to be moving from simple integration of indigenous knowledge within a scientific paradigm, to full-fledged indigenous forestry.

References

Adelson, N. (2000) *"Being Alive Well": Health and the Politics of Cree Well-Being*, University of Toronto Press, Toronto, Ontario, Canada.
Altmann, P. (2013) 'Good life as a social movement proposal for natural resource use: The indigenous movement in Ecuador', *Consilience: The Journal of Sustainable Development*, vol 10, no 1, pp. 59–71.
Anderson, M. K. and Lake, F. K. (2013) 'California Indian ethnomycology and associated forest management', *Journal of Ethnobiology*, vol 33, no 1, pp. 33–85.
Berkes, F. (2012) *Sacred Ecology*, 3rd edition, Routledge, New York, NY, USA.
Bhagwat, S. A. and Rutte, C. (2006) 'Sacred groves: Potential for biodiversity management', *Frontiers in Ecology and the Environment*, vol 4, no 10, pp. 519–524.
Castleden, H., Garvin, T. and Huu-ay-aht First Nation. (2009) '"Hishuk Tsawak" (Everything is one/Connected): A Huu-ay-aht worldview for seeing forestry in British Columbia, Canada', *Society and Natural Resources*, vol 22, no 9, pp. 789–804.
Cheveau, M., Imbeau, L., Drapeau, P. and Bélanger, L. (2008) 'Current status and future directions of traditional ecological knowledge in forest management: A review', *Forestry Chronicle*, vol 84, no 2, pp. 231–243.
Denevan, W. M. (1992) 'The pristine myth: The landscape of the Americas in 1492', *Annals of the Association of American Geographers*, vol 82, no 3, 369–385.
Dhital, N., Raulier, F., Asselin, H., Imbeau, L., Valeria, O. and Bergeron, Y. (2013) 'Emulating boreal forest disturbances dynamics: Can we maintain timber supply, aboriginal land use, and woodland caribou habitat?', *Forestry Chronicle*, vol 89, no 1, pp. 54–65.
Emery, M. R., Wrobel, A., Hansen, M. H., Dockry, M., Moser, W. K. and others (2014) 'Using traditional ecological knowledge as a basis for targeted forest inventories: Paper birch (*Betula papyrifera*) in the US Great Lakes region', *Journal of Forestry*, vol 112, no 2, 207–214.
Garibaldi, A. and Turner, N. (2004) 'Cultural keystone species: Implications for ecological conservation and restoration', *Ecology and Society*, vol 9, no 3, art 1.
Gauthier, S., Vaillancourt, M.-A., Kneeshaw, D., Drapeau, P., De Grandpré, L. and others (2009) 'Forest ecosystem management: Origins and foundations', in S. Gauthier, M.-A. Vaillancourt, A. Leduc, L. De Grandpré, D. Kneeshaw and others (eds), *Ecosystem Management in the Boreal Forest*, Presses de l'Université du Québec, Québec, Canada.
Genin, D. and Simenel, R. (2011) 'Endogenous Berber forest management and the functional shaping of rural forests in Southern Morocco: Implications for shared forest management options', *Human Ecology*, vol 39, no 3, pp. 257–269.
Gladu, J. P., Brubacher, D., Cundiff, B., Baggio, A., Bell, A. and Gray, T. (2003) *Honouring the Promise: Aboriginal Values in Protected Areas in Canada*, National Aboriginal Forestry Association and Wildlands League, Ottawa and Toronto, Ontario, Canada.
Herrmann, T. M. (2006) 'Indigenous knowledge and management of *Araucaria araucana* forest in the Chilean Andes: Implications for native forest conservation', *Biodiversity and Conservation*, vol 15, no 2, pp. 647–662.
Herrmann, T. M., Sandström, P., Granqvist, K., D'Astous, N., Vannar, J. and others (2014) 'Effects of mining on reindeer/caribou populations and indigenous livelihoods: Community-based monitoring

by Sami reindeer herders in Sweden and First Nations in Canada', *The Polar Journal*, vol 4, no 1, pp. 28–51.

LaRiviere, C. M. and Crawford, S. S. (2013) 'Indigenous principles of wild harvest and management: An Ojibway community as a case study', *Human Ecology*, vol 41, no 6, pp. 947–960.

Mazzocchi, F. (2006) 'Western science and traditional knowledge', *EMBO Reports*, vol 7, no 5, pp. 463–466.

McCall, M. K. and Minang, P. A. (2005) 'Assessing participatory GIS for community-based natural resource management: Claiming community forests in Cameroon', *Geographical Journal*, vol 171, no 4, pp. 340–356.

Menzies, C. R. and Butler, C. (2006) 'Introduction. Understanding ecological knowledge', in C. R. Menzies (ed.), *Traditional Ecological Knowledge and Natural Resource Management*, University of Nebraska Press, Lincoln, NE, USA.

O'Flaherty, R. M., Davidson-Hunt, I. J. and Miller, A. M. (2009) 'Anishinaabe stewardship values for sustainable forest management of the Whitefeather forest, Pikangikum First Nation, Ontario', in M. G. Stevenson and D. C. Natcher (eds), *Changing the Culture of Forestry in Canada: Building Effective Institutions for Aboriginal Engagement in Sustainable Forest Management*, Canadian Circumpolar Institute Press, Edmonton, Alberta, Canada.

Padmanaba, M., Sheil, D., Basuki, I. and Nining, L. (2013) 'Accessing local knowledge to identify where species of conservation concern occur in a tropical forest landscape', *Environmental Management*, vol 52, no 2, pp. 348–359.

Park, M. S. and Yeo-Chang, Y. (2012) 'Traditional knowledge of Korean native beekeeping and sustainable forest management', *Forest Policy and Economics*, vol 15, pp. 37–45.

Pellatt, M. G. and Gedalof, Z. (2014) 'Environmental change in Garry oak (*Quercus garryana*) ecosystems: The evolution of an eco-cultural landscape', *Biodiversity and Conservation*, vol 23, no 8, pp. 2053–2067.

Reyes-Garcia, V., Guèze, M., Luz, A. C., Paneque-Galvez, J., Macia, M. J. and others (2013) 'Evidence of traditional knowledge loss among a contemporary indigenous society', *Evolution and Human Behaviour*, vol 34, no 4, pp. 249–257.

Rist, L., Shaanker, R. U., Milner-Gulland, E. J. and Ghazoul, J. (2010) 'The use of traditional ecological knowledge in forest management: An example from India', *Ecology and Society*, vol 15, no 1, art 3.

Saint-Arnaud, M., Asselin, H., Dubé, C., Croteau, Y. and Papatie, C. (2009) 'Developing criteria and indicators for aboriginal forestry: Mutual learning through collaborative research', in M. G. Stevenson and D. C. Natcher (eds) *Changing the Culture of Forestry in Canada: Building Effective Institutions for Aboriginal Engagement in Sustainable Forest Management*, Canadian Circumpolar Institute Press, Edmonton, Alberta, Canada.

Shen, X., Li, S., Chen, N., Li, S., McShea, W. J. and Lu, Z. (2012) 'Does science replace traditions? Correlates between traditional Tibetan culture and local bird diversity in Southwest China', *Biological Conservation*, vol 145, no 1, pp. 160–170.

Stevenson, M. and Webb, J. (2003) 'Just another stakeholder? First Nations and sustainable forest management in Canada's boreal forest', in P. J. Burton, C. Messier, D. W. Smith and W. L. Adamowicz (eds) *Towards Sustainable Management of the Boreal Forest*, NRC Research Press, Ottawa, Ontario, Canada.

Trosper, R. L. (2007) 'Indigenous influence on forest management on the Menominee Indian Reservation', *Forest Ecology and Management*, vol 249, no 1, pp. 134–139.

Trosper, R. L. and Parrotta, J. A. (2012) 'Introduction: The growing importance of traditional forest-related knowledge', in J. A. Parrotta and R. L. Trosper (eds) *Traditional Forest-Related Knowledge. Sustaining Communities, Ecosystems and Biocultural Diversity*, Springer, New York, NY, USA.

United Nations. (2009) *State of the World's Indigenous Peoples*, United Nations, Department of Economic and Social Affairs, Division for Social Policy and Development, Secretariat of the Permanent Forum on Indigenous Issues. United Nations Publication ST/ESA/328, New York, NY, USA.

Uprety, Y., Asselin, H., Bergeron, Y., Doyon, F. and Boucher, J.-F. (2012) 'Contribution of traditional knowledge to ecological restoration: Practices and applications', *Ecoscience*, vol 19, no 3, pp. 225–237.

Uprety, Y., Asselin, H. and Bergeron, Y. (2013) 'Cultural importance of white pine (*Pinus strobus* L.) to the Kitcisakik Algonquin community of western Quebec, Canada', *Canadian Journal of Forest Research*, vol 43, no 6, pp. 544–551.

Ziembicki, M. R., Woinarski, J. C. Z. and Mackey, B. (2013) 'Evaluating the status of species using Indigenous knowledge: Novel evidence for major native mammal declines in northern Australia', *Biological Conservation*, vol 157, pp. 78–92.

42
RECREATION IN FORESTS

Bruce Prideaux

Introduction

Throughout history forests have been used for a range of recreational activities including hunting, hiking and fishing. In the contemporary era the demand for traditional recreational activities has increased and new forms of recreation have emerged driven in part by the tourism sector. Many of these are commercial in nature, and along with traditional activities, have placed new pressures on forests. As the number of users grows but the available areas of forest shrink the need for new management strategies to maintain ecological integrity will increase. In many countries the public sector has responded by implementing policies to protect unique ecosystems and landscapes, reduce deforestation and regulate activities such as recreation and tourism. Governments are also beginning to realise that they must place a higher priority on long term ecological sustainability if forest based activities are to continue to prosper into the future. The objective of this chapter is to briefly outline major factors that may affect the provision of sustainable recreational activities in forests.

To ensure long term ecological sustainability of forests used for recreational purposes proactive visitor management strategies are required to guard against overuse or damage. In publicly owned forests management strategies may include zoning (Synge, 2004), voluntary codes of conduct, permits, licensing and in some cases permission to operate commercial activities. Where required there may also be a need to employ strategies that recognise indigenous prior use and the need to balance existing land use patterns such as agriculture and grazing with the need for ecosystem conservation. Not all forests are publicly owned and a growing number of privately owned forests (including parcels of private forest adjacent to or within the borders of publicly owned forests) and those under traditional ownership are now beginning to be used for tourism and recreation purposes. In many cases the requirements for the operation of businesses in private forests differ from those that apply to protected forests.

This chapter first overviews the growing range of recreational uses of forests followed by a discussion on management practices used to ensure that recreational and tourism activities are undertaken in a sustainable manner. A short case study based on the Wet Tropics World Heritage Rainforest (WTQWHA) in Queensland, Australia is used to illustrate how recreation including tourism can be undertaken in a manner that does not detract from the ecological values of forests. The discussion is framed around Figure 42.1 which illustrates the role of economic factors (demand and supply), the private and public sectors and sustainability issues in the management of forest resources for recreation.

Recreational and tourism activities

This chapter discusses two user groups (tourists and local) who may engage in forest recreation. While both groups use forest resources it is important to differentiate between them, particularly from the viewpoint of management and infrastructure requirements. Members of local communities may have a different use profile to tourists because of proximity, customary practices and activity types. For the purposes of this chapter however, tourists and local communities who use the forest for recreation are referred to as visitors unless the discussion dictates that a distinction be made between the two groups.

As Newsome et al. (2012) noted, while there is a significant body of work on the biophysical impact of recreation, including tourism, little attention has been given to visitor perceptions of the biological impacts of tourism and even fewer papers have attempted to combine biophysical impacts with social science. The perceptions of visitors including tourists are important and as Prideaux, Thompson and McNamara (2012) highlight, tourists do often show concern about their impact on forests and many are concerned that tourism poses a moderate threat to the ecological integrity and sustainability of forests.

In response to concerns about the impact of recreation on forest ecology a new specialised field of study, recreation ecology, has emerged (Monz et al., 2013). Defined as 'the study of the impacts of outdoor recreation and nature-based tourism activities in natural or semi-natural environments' (Monz et al., 2013: 441) recreation ecology is concerned with issues such as hiking, camping and other activities where visitor activities are concentrated. Trampling, for example, has emerged as a major concern (Hill and Pickering, 2009 – see Monz et al., 2013) and resulted in the emergence of a number of trampling protocols (e.g. Cole and Bayfield, 1993). As Monz et al. (2013) note, the relationship between increasing use and ecological change may be generalised as a curvilinear, asymptotic relationship where at some point ecological damage is so high that additional demand creates little additional damage.

One critique of current research directions is the propensity of many researchers to focus on specific issues such as trampling rather than taking a system wide approach to issues related to ecological sustainability. Only by employing a holistic system wide approach will managers be able to understand how to balance the demand for recreation with the need to develop management regimes that are able to achieve long term ecological sustainability. With this need in mind this chapter proposes the Forest Tourism and Recreation Model outlined in Figure 42.1.

Forest recreation

Recreation describes a range of activities that tourists and local communities participate in. While there is some disagreement over a precise definition of recreation, Bryan's (2000: 198) view that it can be described as a 'continuum from general interest and low involvement to specialised interest and high involvement' in a sport or activity setting that requires a specific set of equipment and skills, is a useful starting point. While both tourists and local community members (locals) engage in forest recreation, locals are likely to have different use patterns and levels of emotional attachment to forests compared to overnight tourists who generally live some distance from the forest. For example, in research undertaken in the WTQWHA Carmody and Prideaux (2008) found that members of the local community reported high levels of emotional attachment to the forest and visited the forest up to four times per year.

Traditional forest activities, some of which are consumptive, include hunting, fishing, hiking and foraging. In North America, sustainable hunting of fauna including deer and birds continues to be practiced. Hunting, both legal and illegal is also widespread in other parts of the world and is often very damaging to ecosystem health (Kahler et al., 2013). Many traditional forms of forest use such as hiking and camping have become more organised and even commercialised with purpose built hiking trails and organised camping sites provided by forest owners. The introduction of the automobile in the early twentieth century and the rapid growth of road construction that followed allowed far greater access to forests than had hitherto been the case. Rapid growth in automobile ownership has also generated a significant demand for recreation including off-road driving. Rapid urbanisation, growing disposable income, the introduction of annual leave and shorter working hours has also created the demand side conditions that have made forest recreation increasingly popular.

New types of activities have emerged including photography, wildlife viewing, and a growing number of new forest activities including mountain bike riding, bungee jumping, zip sailing, motor sports (including four wheel driving and cross country motorbike riding), climbing, canyoning and white water rafting. While mostly non-consumptive, many of these activities require the construction of specific infrastructure and need close monitoring to ensure that long term ecological sustainability is not compromised.

Growth in recreation has led to changes in the way in which forests are used and managed, including the provision of paved roads, parking areas, toilets, interpretation and walking trails. A far greater emphasis is now placed on interpretation, conservation and strategies to enhance visitors' understanding (Stewart et al., 1998). In high demand areas, visitor facilities may include commercial activities such as food and beverage operations, commercial tours, souvenir sales and accommodation.

The commercialisation of forest experiences that is a characteristic of increasing recreational use has created numerous opportunities for forest managers to supplement budgets by applying entrance fees and in some cases issuing leases to commercial operators. In many developing countries tourism, more so than local community demand, has created a new range of opportunities for local communities to become involved in both visitor related activities and in strategies that enhance long term sustainability. Recent research in the Amazon rainforest by Rodrigues and Prideaux (2014), found that the river people of the Mamori Lake area (located about 100 kilometres from Manaus, Brazil) who were actively engaged in tourism activities reported that they were able to reduce or even stop slash and burn agricultural practices and wildlife hunting. In a similar finding Hanisdah Saikim et al. (2011) reported that the creation of jobs servicing the tourism industry in the Tabin Wildlife Reserve, Sabah led to a significant reduction in illegal poaching.

Visitor management

A number of authors (Cheung, 2013) state that non-consumptive use of natural areas for recreation is often associated with negative environmental impacts. In response to these concerns a number of management and planning frameworks have been suggested including carrying capacity (Stankey, 1991), Limits to Acceptable Change (LAC) (Stankey et al., 1985), Visitor Activity Management Process (VAMP) (Graham et al., 1988), the Recreation Opportunity Spectrum (ROS) (Clark and Stankey, 1979) and Visitor Impact Management (VIM) (Graefe et al., 1990). As Mason (2005) observed, these approaches have provided a form of 'toolbox' able to be used by management authorities when developing visitor management strategies. Cheung (2013) provides a useful summary of these strategies and their application and observed that in many cases management models of this type have neglected the need to incorporate considerations of the impact of degradation of natural resources on visitor preferences. In his research into Hong Kong's country park system, Cheung (2013) observed that understanding visitors' views and preferences provides useful data to assist selection of the most appropriate management strategies to achieve ecological sustainability while maintaining visitor enjoyment.

Modelling recreation and tourism use for forests

As the preceding discussion has suggested there are a wide variety of activities that forests are used for. Efficient use requires not only a high level of environmental knowledge but also a detailed understanding of demand and supply forces and how they operate in forest recreation settings. Use patterns for specific forests are generally determined by ownership, accessibility to population centres, landscapes, flora and fauna, climatic conditions and the level of infrastructure that has been provided for visitors. Access is a particularly important factor for both local residents and tourists. Infrastructure is also an important element (Carmody and Prideaux, 2008) and may include commercial activities such as food and beverage services and accommodation, which may be located onsite or offsite.

Figure 42.1 illustrates the relationships between major factors that influence the demand for and supply of recreational opportunities in forests but does not attempt to show the complexity of these relationships. The level of visitor services provided in a particular forest will be determined by the ecological quality and visual amenity of the resource, the philosophy towards visitor use involving: the owners of the resource; the supply and quality of visitor experiences and services provided; the economic sustainability of businesses providing recreational services; and, in the long run, the need to maintain ecological sustainability.

Management structures

Aside from ecological considerations one of the most important relationships outlined in Figure 42.1 is the attitude of the owners of the forest resource which, as Figure 42.1 suggests, may be the public sector, private owners or traditional owners. In publicly owned forests the prevailing economic philosophy of the government is likely to be a major factor in how forests are managed. If governments have a neoliberal philosophical view the user pays principle may prevail (Laarman and Gregersen, 1996) and entrance fees are charged. Moreover, if greater economic benefit is able to be derived from other uses such as lumber, access to recreational users may be denied. Governments that have adopted social democracy ideals (Giddens, 1998) are more likely to view forests and protected areas as a free good that should be available to all members of the community, irrespective of their place in society or income (see Park et al., 2010). In the case of privately owned forests, decisions to invest in visitor infrastructure are usually based on potential economic benefit or in the case of traditional owners, on cultural factors including traditional hunting activities as well as tourism (see Mendoza-Ramos and Prideaux, 2013).

Tools available for managing the use of forests include zoning, participatory management, monitoring regimes and ecosystem protection. Zoning strategies in relation to visitors should aim to allow visitors access to key scenic sites and areas where recreation facilities can be provided (tracks and camp sites, for example) while ensuring minimal disturbance of flora and fauna and minimising risks associated with fire, invasive species and fragmentation. Strategies of this nature become particularly important where protected forests include human populations

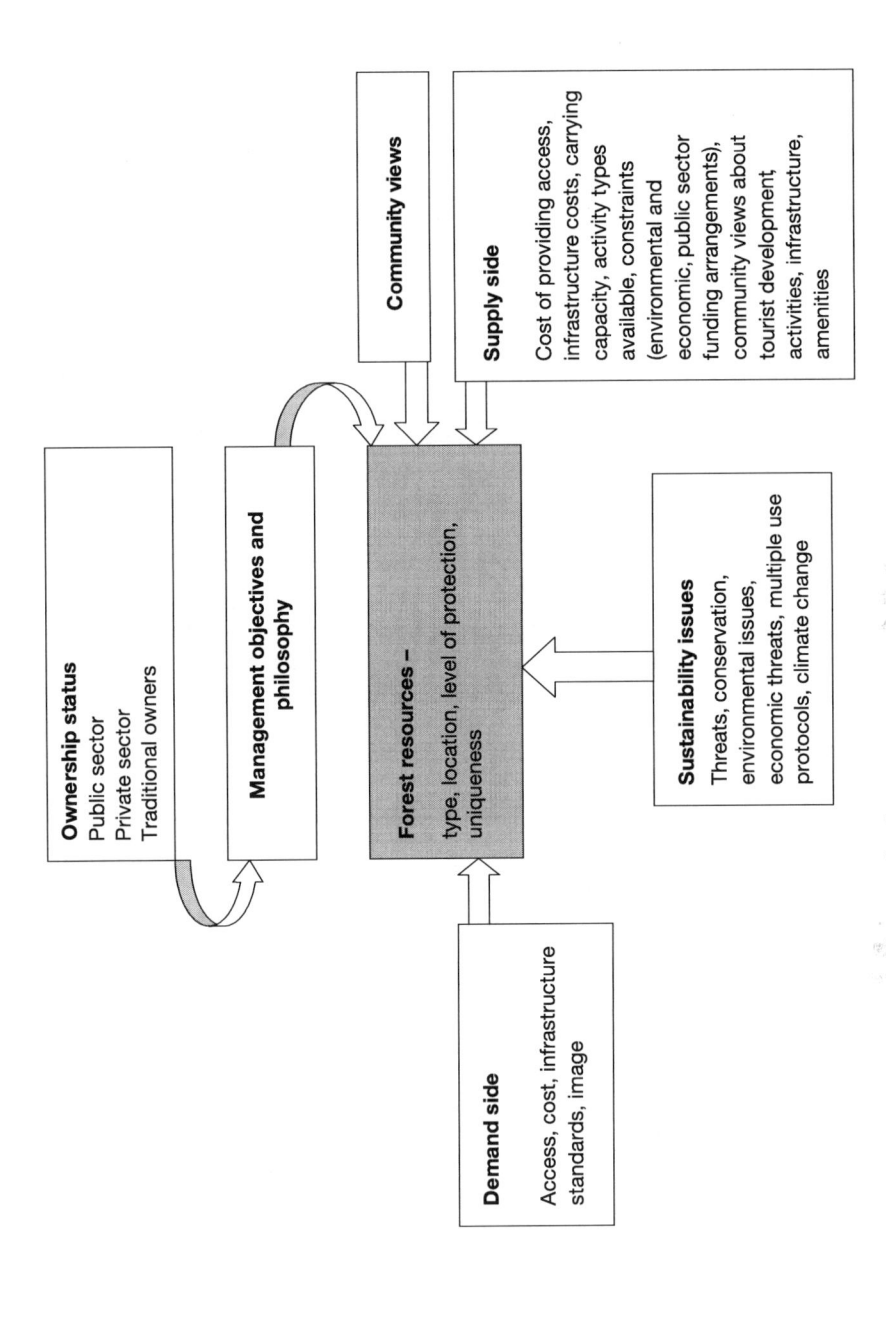

Figure 42.1 Forest Tourism and Recreation Model: Macro level
Adapted from Prideaux (2014)

and traditional land use practices such as farming. This is also important where the size of human settlements located on or near the borders of protected forests is increasing. In Uganda, for example, rapid population growth in areas adjacent to parks including Kibale Forest National Park, home to a large chimpanzee population, has created friction between local communities and park authorities as the local communities look to the forest for a range of resources such as traditional medicines, food, fibre and timber.

Participatory management offers one solution for ensuring that a balance is achieved between the needs of local communities and visitors and the need for ecological sustainability. As Borrini-Feyerabend et al. (2004) noted in relation to the Spanish Balearic Islands, the need to balance environmental protection with the needs of its human population was addressed through participatory management.

Monitoring of visitors and the natural environment is also an important management strategy designed to detect problems and implement strategies to prevent irreversible change (Hadwen et al., 2008). A number of researchers (Monz, 2002) have noted that unregulated or under-regulated visitation may have adverse environmental impacts. Problems caused by erosion of walking tracks, damage to flora and disturbance to fauna may be overcome with simple measures such as hardened tracks, barriers, interpretation and, in some cases, by ranger patrols. On a larger scale overcrowding may lead to significant problems and carrying capacity (see Buckley, 1999) may need to be determined and entry rationed if required. The use of entry fees is a simple yet effecting strategy to control crowding.

Supply side

The supply side of the model refers to the type of experiences that particular forests have to offer visitors and the quality and quantity of facilities that are provided. Two categories of experiences may be identified, natural experiences and built experiences. 'Natural experiences' refers to how the attributes of a particular forest may be used to provide visitor activities and experiences. Attributes include the type of forest (upland forest, rainforest, savannah and so on), type of protection (National Park, for example), ownership status, endemic flora and fauna, climatic characteristics, land forms (rivers, lakes and other geographic features), threats and vulnerabilities (from fire, for example) and accessibility. 'Built experiences' refers to a range of commercial and non-commercial infrastructure that may be provided for visitors. Facilities of this type may be located within the forest or nearby. Commercial facilities may include: guiding services; tour operators (wildlife and photographic, for example); food and beverage outlets; souvenirs and adventure equipment outlets; accommodation including hotels, hostels and camp grounds; tree top walks, cantilever platforms and observation towers; commercial forest interpretative centres; private zoos; mini theme parks; adventure activities including bungee jumping, white water rafting, abseiling; and themed transport such as train rides and cable cars. Figure 42.2 illustrates three commercial business operations found in the WTQWHA case study.

Non-commercial infrastructure (e.g. toilets, visitor interpretation and viewing platforms as illustrated in Figure 42.2) is generally provided at no cost, although sometimes a small fee may be levied. Some infrastructure may be provided by the private sector as illustrated in Figure 42.3.

The scope for providing commercial and non-commercial visitor infrastructure in a specific forest is generally determined by the ability of the forest to support recreation and the attractiveness of the forest to potential users. Actual demand will be a function of management philosophy, types of experiences, profitability of commercial operations and environmental

Figure 42.2 Typical non-commercial forest infrastructure: (A) interpretation, (B) a low energy composting toilet and (C) an elevated walkway

Photos courtesy of Bruce Prideaux

Figure 42.3 Commercial businesses operating in the WTQWHA. From left to right the businesses are a cableway, a scenic railway and a small theme park

Photos courtesy of Bruce Prideaux

impacts. The quantity and quality of infrastructure in public forests is determined by the resources allocated by government and any cost recovery strategies. In private forests willingness to pay by customers will determine the level of investment in infrastructure while in forests managed by traditional owners, cultural considerations and other traditional activities may be an important element in decisions on how the forest is used.

Demand side

Demand for forest recreation is, as Figure 42.1 suggests, a function of a range of direct and indirect factors. Visitor type is a key factor. Local residents, for example, are likely to use a nearby forest more frequently and for different reasons than tourists. In Rotorua, New Zealand, the Whakarewarewa Forest is used by local residents for daily recreation activities such as hiking, bike riding and jogging while tourists are much more likely to visit the area to view giant Californian Redwoods planted over a century ago. Direct factors include access, the cost of a particular forest experience, the quality of the natural and built experiences available in a particular forest, quality of infrastructure, and overall visitor appeal described as image. Collectively, direct factors will determine the level of interest a potential visitor will have in visiting a particular forest. Clearly, cost is also an important factor as well as the uniqueness of the activities available, as is access. In decisions made to visit a particular forest access has three components; the monetary cost of visiting the forest; the cost of time incurred in visiting the site; and, for tourists without independent transport, the frequency of public transport or tour coaches to the site. Thus while Uganda's Kibale Forest National Park, an important primate attraction, has significant appeal, the high access cost makes it less appealing than South Africa's Kruger National Park even though the latter lacks large primates. Infrastructure is another factor that may influence visitors and includes the quality and location of commercial infrastructure as well as interpretation and other signage.

Indirect factors describe the image that a particular forest has as a tourist attraction. Image is a function of the quality of the forest landscape and the commercial and non-commercial activities associated with it, the effectiveness of its promotion as an attraction, and its reputation as a place worth visiting. Yellowstone National Park in the USA and Kruger National Park in South Africa are examples of protected areas that have an iconic reputation that greatly assists in their promotion as tourism attractions.

Sustainability issues

Long term ecological sustainability is a critical issue for forest managers particularly where there is an emphasis on attracting visitors. Globally, forests are under pressure from growing human population, urbanisation, pollution, agriculture, resource extraction, recreation and tourism, and in the near future from climate change (Turton *et al.*, 2010). At the same time there is a growing recognition that remaining forests need protection and in many nations large areas of forest have been given protection as national parks or similar. For the purposes of this chapter discussion of issues related to sustainability will focus on recreation use.

One of the key elements in the attractiveness of a particular forest is the quality of the forest experience including visual appeal and standards of visitor infrastructure. If the quality of the experience declines, demand is likely to fall. Other factors, including crowding, poor maintenance of facilities, loss of competitiveness and damage to the ecosystem, are also important. While overcrowding and maintenance deficiencies are relatively easy to rectify through administrative actions such as reducing visitor numbers and enhanced maintenance, damage to the ecosystem is more serious and potentially long term. Damage from fire, invasive species, illegal consumptive activities, pollution and fragmentation are some examples of ecosystem damage.

In the long run it can be expected that climate change will create an entire new range of environmental risks including in and out migrations of flora and fauna, change in weather regimes as well as damage from increased fire events and in some cases from wind storm events. Problems of this nature may lead to a decline in the quality of the experience through loss of scenic values or in Alpine forests, loss of snow (Burki *et al.*, 2007). One climate change related problem that may affect forests is modification of local climates through increasing temperatures, changes in seasonal weather patterns and changes in precipitation. In addition to in and out migration of temperature sensitive flora and fauna there may also be local extinctions. Factors of this type may have a negative impact in some areas and a possible positive impact in others. From the visitors' perspective, not all climate change related changes in weather are negative. As Richardson and Loomis (2004) note, increased temperatures in the Rocky Mountains National Park (Canada) may attract additional visitors with estimated increases of 7–12 per cent predicted for the 2020s. In similar research Scott *et al.* (2007) concluded that visitor numbers were likely to increase in the Watson Lakes National Park, Canada.

Case study

To illustrate many of the issues raised in this chapter the Wet Tropics Queensland World Heritage Area (WTQWHA) located on the north-east coast of Australia is used as a case study. The WTQWHA is located in the tropics, is 894,000 hectares in size, includes lowland and upland forests and was inscribed on the World Heritage list in 1988 in recognition of its outstanding natural universal values. The WTQWHA also has significant cultural value for Rainforest Aboriginal people. While not included in the original listing there is an active movement to relist the area to include its cultural values (see http://www.wettropics.gov.au/rainforest-aboriginal-landscape accessed 10 February 2014).

Before it was declared a World Heritage Area (WHA) the forest was a patchwork of protected and unprotected national parks, forestry reserves and private property. Following World Heritage listing the timber industry was wound down and quickly replaced by tourism. The WHA is managed by an independent statutory body, the Wet Tropics Management Authority (WTMA) (http://www.wettropics.gov.au/wet-tropics-management-authority accessed 10 February 2014) overseen by a joint Federal State Ministerial Council. Day to day maintenance, including site infrastructure construction, pest and weed control, site hardening, enforcement and commercial activity permit approvals, is undertaken by Queensland Parks and Wild Life Service. Funding is primarily provided by the Australian Federal Government. To ensure that the WTQWHA's key environmental values are maintained a zoning system based on four zones has been established with Zone A, for example, protecting areas with a 'high level of integrity and remote from disturbances that are associated with modern technological society' (see http://www.wettropics.gov.au/zoning-system). This system has been successful with the WTQWHA able to handle large volumes of visitors will little adverse environmental impact.

The WTQWHA along with the nearby World Heritage listed Great Barrier Reef are the main natural attractions of the region which receives in the order of 2.35 million domestic and international visitors per year. In response to the demand for recreational facilities by the local community and the tourism industry the WTMA has constructed a range of visitor facilities including viewing platforms, walks, boardwalks (see Figure 42.2) and swimming sites. Over 100 scenic sites are maintained and are able to be accessed by a 600 kilometre road network including public highway. Most of the most popular sites have visitor infrastructure that includes interpretation (see Figure 42.2). Commercial operators have developed a range of businesses located, in some cases, within the WHA (see examples in Figure 42.3) or on/near the boundary of the WHA.

Although both the Queensland state and Federal governments have adopted neoliberal economic philosophies, entry to the WHA remains free except for tourists travelling on commercial tours. Because there are multiple unmonitored entrance points visitor numbers are based on irregular surveys, the latest of which was undertaken in 2004 which estimated that 4.4 million people visit the WHA. It was further estimated that 40 per cent of all visits were by residents of communities adjoining the area (Bentrupperbäumer, O'Farrell and Reser, 2004). In 2006 the estimated gross economic value of tourism directly generated by the WTQWHA (without flow on effects) was AU$426 million (Prideaux and Falco-Mammone, 2007).

From its inception the WTMA has adopted a range of strategies similar to those suggested by Cheung (2013) including a community outreach program, an active education program and regular consultation with a number of community groups and commercial operators. As part of its consultative mechanism the WTMA is advised by two statutory advisory committees and two liaison groups all of which report to the Authority and its board. The Community Consultative Committee represents a wide range of community groups and advises the Authority on views of the community about policies and programs. Members of the Scientific Advisory Committee represent a wide variety of science and social disciplines and advise the Authority on issues related to conservation, biodiversity and related issues. The Conservation Sector Liaison Group provides the conservation sector with an opportunity to channel information and opinions of the conservation sector to the Authority. The Wet Tropics Tourism Network representatives advise the Authority on a range of tourism related issues. The Authority also has a close working relationship with the Rainforest Aboriginal people with whom a Wet Tropics Regional Agreement outlining a mechanism for cooperative management was signed in 2005.

As Carmody and Prideaux (2008) observed, the strategies employed by the WTMA to build cooperative natural area management consultation meets or exceeds many of the criteria suggested by Dickman (2002) for community consultation and management support. Dickman (2002) suggested three possible approaches, interactive, proactive and opportunistic. The WTMA has adopted a suite of strategies that can be broadly categorised as proactive and which aim to achieve long term management of community values while maintaining ecological integrity. Aside from community groups and business groups the WTMA actively engages with the Rainforest Aboriginal people who are represented on the Authorities Board.

The management strategies adopted by the WTMA are designed to balance the need for long term ecological sustainability with the recreational needs of nearby communities and the provision of opportunities for the tourism industry has been very successful. Research by Carmody and Prideaux (2008) found that the local community had a very positive view of the opportunities for recreation in the WHA indicating that the Authority has been largely effective in balancing the need to maintain long run ecological sustainability with the recreation needs of the community and the tourism industry.

While management strategies have to date overcome many of the problems generated by recreational and tourism activity the long term is less clear particularly as problems associated with climate change begin to emerge. As Williams *et al.* (2008) note climate change impacts may lead to changes in assemblage composition, changes in ecosystem process and loss of genetic diversity which in turn can have a cascading effect that will change community composition and structure. It is not clear how the effects of climate change will impact on visitors. In part this is because future visitors will have a different set of destination and activity selection criteria, possibly using technologies that differ from those that exist today.

Conclusion

The objective of this chapter was to examine a range of factors that may affect the provision of recreational activities in forests. The Forest Tourism and Recreation Model (Figure 42.1) was suggested as one approach that may be adopted. As the chapter has noted the tourism sector is an increasingly important element in recreational use and for this reason the chapter examined recreation from two perspectives, the needs of local communities and tourists. Obviously no two forests are the same and the factors outlined in this chapter will apply to specific forests based on their location, the range of activities that may be undertaken, management strategies, accessibility and so forth. Some of the literature (Cole, 2004; Cunha, 2010) paints a gloomy picture of visitors generating irreversible effects on the natural environment but as the WTQWHA case study illustrates this need not be the case if management authorities have the vision and legislative and financial resources to balance visitor demands with ecological sustainability.

It is obvious that as the rate of global urbanisation accelerates in parallel with increases in global GDP the demand for recreation is likely to grow, particularly in forests that are located near to urban areas. Responding to this demand in a manner that does not compromise ecological sustainability will be a major challenge to forest managers and as the impacts of climate change escalate the challenges will grow in complexity.

There are many management models and strategies that may be employed to manage the demand for recreation. Those outlined in the chapter illustrate some of the many options available. The governance structures that guide the operation of the WTMA discussed earlier provide a useful benchmark for other forests and demonstrate how it is possible for a large protected forest to be managed to provide sustainable recreational opportunities for the local community as well as visitors in a manner that meets or exceeds the needs for conservation and ecological sustainability.

References

Bentrupperbäumer, J.M., O'Farrell, S. and Reser, J.P. (2004) *Visitor Monitoring System for the Wet Tropics World Heritage Area: Volume 2 Visitor Monitoring Process - from Pre-Destination to Post-Destination*. Cairns, Australia: Cooperative Research Centre for Tropical Rainforest Ecology and Management (Rainforest CRC).

Borrini-Feyerabend, G., Larrucea, J. and Synge, H. (2004) Participatory Management in the Minorca Biosphere Reserve Spain, in Synge, H. (ed.) *European Models of Good Practice in Protected Areas*, Cambridge, UK: IUCN, pp. 16–21.

Bryan, H. (2000) Recreation specialisation revisited, *Journal of Leisure Research*, 32.1 (First Quarter 2000): 18–21.

Buckley, R. (1999) An ecological perspective on carrying capacity. *Annals of Tourism Research*, 26: 705–708.

Burki, R., Elsasser, H., Abegg, B. and Koenig, U. (2007) Climate change and tourism in the Swiss Alps, in Hall, C. M. and Higham, J. (eds) *Tourism, Recreation and Climate Change*, Clevedon, UK: Channel View, pp. 155–164.

Carmody, J. and Prideaux, B. (2008) *Community Attitudes, Knowledge, Perceptions and Use of the Wet Tropics of Queensland World Heritage Area in 2007*. Report to the Marine and Tropical Sciences Research Facility. Cairns, Australia: Reef and Rainforest Research Centre (120pp).

Cheung, L. (2013) Improving visitor management approaches for the changing preferences and behaviours of country park visitors in Hong Kong, *Natural Resources Forum* 37: 231–241.

Clark, R. and Stankey, G. (1979) *The Recreation Opportunity Spectrum: A Framework for Planning, Management, and Research*, Portland, OR, US: US Department of Agriculture Forest Service, Pacific Northwest Forest and Range Experimentation Station.

Cole, D. (2004) Impacts of hiking and camping on soils and vegetation: A review, in Buckley, R. (ed.), *Environmental Impacts of Ecotourism*, Wallingford, UK: CABI Publishing.

Cole, D. and Bayfield, N. (1993) Recreational trampling on vegetation: standard experimental procedures, *Biology Conservation* 63: 209–215.

Cunha, A. (2010) Negative effects of tourism in a Brazilian Atlantic Forest National Park, *Journal of Nature Conservation*, 1894: 291–295.

Dickman, C. (2002) Community versus research based conservation: What are the paradigms? in D. Lunney, C. Dickman and S. Burgin (eds). *A Clash of Paradigms: Community and Research Based Conservation*. Australia: Royal Zoological Society of New South Wales.

Giddens, A. (1998) *The Third Way: The Renewal of Social Democracy*, Cambridge, UK: Polity Press.

Graefe, A., Kuss, F. and Vaske, J. (1990) *Visitor Impact Management: The Planning Framework*, Washington, DC, US: National Parks and Conservation Association.

Graham, R., Nilsen, P., Payne, R. (1988) Visitor management in Canadian national parks, *Tourism Management*, 9(1): 44–62.

Hadwen, W., Hill, W. and Pickering, M. (2008) Linking visitor impact research to visitor impact monitoring in protected areas, 7(1): 87–93 (DOI: 10.2167/joe193.0)

Hanisdah Saikim, F., Prideaux, B., Hamzah, Z. and Mohamed, M. (2011) *Using Tourism as a Mechanism to Reduce Poaching and Hunting: A Case Study of the Tidong Community, Sabah*, APTA Annual Conference, Seoul, Asia Pacific Tourism Association, June.

Hill, R. and Pickering, D. (2009) Differences in resistance of three subtropical vegetation types to experimental trampling, *Journal of Environmental Management*, 90: 1305–1312.

Kahler, J., Roloff, G. and Gore, M. (2013) Poaching risks in community-based natural resource management, *Conservation Biology*, 27(1): 177–186.

Laarman, J. G., and Gregersen, H. M. (1996) Pricing policy in nature-based tourism. *Tourism Management*, 17(4): 247–254.

Mason, P. (2005) Visitor management in protected areas: from 'hard' to 'soft' approaches, *Current Issues in Tourism*, 8(2–3): 181–194.

Mendoza-Ramos, A. and Prideaux, B. (2013) Indigenous ecotourism in the Mayan Rainforest of Palenque: Empowerment issues in sustainable development, *Journal of Sustainable Tourism*, http://dx.doi.org/10.1080/09669582.2013.828730.

Monz, C. (2002) The response of two Arctic tundra communities to human trampling disturbance, *Journal of Environmental Management*, 64: 207–217.

Monz, C., Pickering, C. and Hadwen, W. (2013) Recent advances in recreation ecology and the implications of different relationships between recreation use and ecological impacts, *Frontiers in Ecology and the Environment*, 11(8): 441–446.

Newsome, D., Moore, S. and Dowling, R. (2012) *Natural Area Tourism Ecology, Impact and Management*, 2nd Ed, Bristol, UK: Channel View.

Park, J., Ellis, G., Kim, S.S. and Prideaux, B. (2010) An investigation of perceptions of social equity and price acceptability judgments for campers in the U.S. national forest, *Tourism Management*, Volume 31, Issue 2: 202–212.

Prideaux, B. (2014) Factors governing the development of tourism in rainforest regions, in Prideaux, Bruce, (ed.) *Rainforest Tourism, Conservation and Management: Challenges for Sustainable Development*. New York, NY, US: Routledge, pp. 163–176.

Prideaux, B. and Falco-Mammone, F. (2007) *Economic Values of Tourism in the Wet Tropics World Heritage Area*. James Cook University, Cairns, Australia: Cooperative Research Centre for Tropical Rainforest Ecology and Management.

Prideaux, B., Thompson, M. and McNamara, K. (2012) The irony of tourism: Visitor reflections of their impacts on Australia's World Heritage rainforest, *Journal of Ecotourism*, 11(2): 102–117 DOI:10.1080/14724049.2012.683006

Richardson, R. and Loomis, J. (2004) Adaptive recreation planning and climate change: A contingent visitation approach. *Ecological Economics*, 50: 83–99.

Rodrigues, C. and Prideaux, B. (2014) Developing backpacker tourism in the Brazilian Amazon rainforest: Adding value to the forest, in Prideaux, B. (ed.), *Rainforest Tourism, Conservation and Management: Challenges for Sustainable Development*, New York, NY, US: Routledge, pp. 163–176.

Scott, D., Jones, B. and Konopek, J. (2007) Implications of climate and environmental change for nature-based tourism in the Canadian Rocky Mountains: A case study of Waterton Lakes National Park, *Tourism Management*, 28: 570–579.

Stankey, G. (1991) Conservation, recreation and tourism: The good, the bad and the ugly, in Diensbier, R. (ed.), *Proceedings of the 1990 Congress on Coastal and Marine Tourism: a Symposium and Workshop on Balancing Conservation and Economic Development*. Honolulu, HI, US: USDA.

Stankey, G., Cole, D., Lucas, R., Petersen, M. and Frissel, S. (1985) *The Limits of Acceptable Change (LAC) System for Wilderness Planning*, Ogden, UT, US: USDA.

Stewart, E., Hayward, P., Devlin, P. and Kirby, V. (1998) The 'place' of interpretation: A new approach to the evaluation of interpretation, *Tourism Management*, 19: 257–266.

Synge, H. (2004) *European Models of Good Practice in Protected Areas*, IUCN, Cambridge, UK.

Turton, S., Dickson, T., Hadwen, W., Jorgensen, B., Pham, T. and others (2010) Developing an approach for tourism climate change assessment: Evidence from four contrasting Australian case studies, *Journal of Sustainable Tourism*, vol 18, no 3: 429–447.

Williams, S., Shoo, L., Isaac, J., Hoffmann, A. and Langham, G. (2008) Towards an integrated framework for assessing the vulnerability of species to climate change, *PLoS Biology*, 6(12): 2621–2626.

43
IMPACTS OF HUNTING IN FORESTS

Rhett D. Harrison

Introduction

Humans have hunted forest animals since the hominid lineage first appeared in Africa approximately 6 million years ago (Cartmill 1993; Stanford 1999). Seeing the coordinated movement of chimpanzees hunting colobus monkeys, their excitement as they approach the kill, and the subsequent sharing of the meat – one cannot help but recognize parallels in humans. As large predators, early hominid populations presumably had an impact on the populations of other animals. Nonetheless, these were probably of much the same order as those of chimpanzees in the forests of Africa today. However, as human technology improved and populations increased, humans moved into the realm of a "super predator." Although not all in forests, the late Pleistocene megafaunal extinctions, which occurred across the globe, have been attributed to the arrival of new human populations and the development of improved hunting technology (Sandom et al. 2014). This process has continued into the modern era (Dirzo et al. 2014).

Without doubt, hunting is the most pervasive human activity to impact forests. Across the tropics, much more forest is affected by hunting than by deforestation, logging or other extractive activities (Chazdon 2014). Indeed it is doubtful if there are forests in any part of the world that are not exposed to some level of human hunting pressure. Nevertheless, depending on factors such as human population density, legal regulation and societal attitudes, the types of technology employed and the natural abundance of wildlife, hunting pressure and the impacts of hunting vary tremendously from place to place.

Hunting in temperate and boreal forests

Many temperate and boreal forests lost their large carnivores during historical times, largely as a consequence of hunting. Wolves, bears, and wild cats, such as lynx in Europe or cougars in the Americas, were exterminated from much of their former ranges. Often these animals were persecuted because of the perceived threat they represented to people or livestock. For example, in an attempt to protect sheep the Tasmanian government offered bounties for dead Tasmanian tigers (*Thylacinus cynocephalus*) (Paddle 2002). Between 1888 and 1909 it paid out 2,184 bounties, although many more Thylacines are believed to have been killed. The last wild one was shot in 1930 and the species became extinct with the death of the last individual in captivity at Hobart Zoo in 1936 (Paddle 2002). Similarly, in Scotland wolves were perceived to be such a threat to travelers and livestock that in 1427 King James I made it compulsory to hold three wolf hunts a year during the cubbing season, and the killing of the last wolf in Scotland, said to have been in 1743, is celebrated in folklore (Buczacki 2002).

An important feature of hunting practices in temperate forests is that for a long time, several centuries in many places, strict regulations have governed hunting (Bubeník 1989). Although specifics vary, usually there is a requirement to hold a license to hunt specific species or in a specific area, and there are controls concerning the hunting season and the numbers of animals that can be shot. Restricting hunting to game species and managing these populations has meant that modern hunting in temperate forests is rarely a threat to species survival. The abundance of wildlife in temperate forests is more dependent on habitat factors, including the quality and size of the forest habitat and the suitability of the agricultural matrix for wildlife (Lindenmayer et al. 2006). Importantly from a conservation perspective, hunting in protected areas is prohibited and effectively policed in most countries.

Consequences of hunting for ecosystem processes in temperate forests

Impacts of the loss of apex predators

Unsurprisingly, the selective removal of apex predators has had an enormous impact on the ecology of forests in temperate regions. In particular, in the absence of management populations of large herbivores, such as deer, wild sheep or wallabies, can attain pest proportions and cause enormous damage to forests by over-browsing seedlings and killing older trees through the peeling of bark during winter (Palmer and Truscott 2003; Gordon et al. 2004; Côté et al. 2004; Dexter et al. 2012). For example, in Scotland, despite cull quotas for landowners, red deer (*Cervus elephus*) populations have increased and, together with farmed sheep, are having a major impact in preventing the natural regeneration of native forests (Palmer and Truscott 2003; Clutton-Brock et al. 2004). In areas where deer and sheep have been excluded through fencing, forests have regenerated surprisingly quickly (Figure 43.1). Managing large herbivore populations, usually through a combination of fences and hunting, is a major concern in temperate forest management, both for conservation outcomes and because of the economic losses caused through damage to trees (Côté et al. 2004).

Another way in which the extirpation of apex predators may affect the ecology of forests is through a process known as mesopredator release (Crooks and Soulé 1999; Prugh et al. 2009). Populations of small to medium sized predators that were previously held in check by the apex predator increase, which in turn drives declines and extirpations of smaller prey species. For example, across North America while apex predator populations have declined since the nineteenth century, 60 percent of mesopredator species ranges have increased, although responses varied from an expansion in range of over 40 percent (coyotes) to contractions of nearly 100 percent (black-footed ferret) (Prugh et al. 2009). Mesopredator release has been implicated in declines of a wide variety of prey species, including scallops, corals, terrestrial invertebrates, sea turtles, lizards, birds and mammals (Brashares et al. 2010). Sometimes these mesopredators become major pests, as in the case of the olive baboon in West Africa, which increased in population after lions (Henschel et al. 2014) and leopards were hunted out in many areas and is now a serious agricultural pest (Brashares et al. 2010). Nevertheless, controlling outbreaks of mesopredators may have their own unintended consequences. In SE Australia attempts to reduce populations of introduced foxes resulted in a dramatic increase in the

Figure 43.1 Deer exclosure experiments, such as this one in Mountain Lake Biological Station (Virginia, USA), demonstrate the impact deer can have on forest regeneration. Behind the fence a dense carpet of seedlings of canopy species has established, whereas outside the fence (foreground) no seedlings are visible

Photo: Benjamin S. Ramage

population of wallabies, which through over-browsing are now preventing the natural regeneration of forests following fires (Dexter *et al.* 2013).

In North America and parts of Europe large predators are making a come-back (Deinet *et al.* 2013) and studies on the impacts of their recolonization indicate how fundamentally the ecology of forests must have been altered when these apex predators were hunted out. For example, large areas of Yellowstone National Park that were previously covered in over-grazed grasslands have returned to forest since wolves recolonized. The dramatic effect is thought to be driven more by the "fear of predation," which causes deer to avoid grazing close to cover that might conceal a predator, rather than through the direct effects of increased mortality through predation (Ripple *et al.* 2001).

Consequences of modern hunting practices

Hunting practices have affected and continue to affect the ecology of temperate forests in other, perhaps less obvious, ways. Since the Middle Ages, wealthy landowners have reserved hunting rights to forests throughout Europe (Henneberger 1996). On the one hand, the preservation of these hunting reserves succeeded in providing protected habitat for native biota in a way not dissimilar from modern nature reserves. On the other hand, vermin – that is small to medium sized predators and many raptors – were often persecuted with a view to increasing game populations (Reynolds and Tapper 2008). Since their protection under law in many areas, a number of these smaller predators and raptors have made a come-back, sometimes assisted through reintroductions (Deinet *et al.* 2013). Nevertheless, continued persecution from gamekeepers remains a problem (Smart *et al.* 2010; Packer and Birks 2002). Perhaps not strictly a hunting impact, but another way in which modern hunting may affect the ecology of forests arises through the practice of releasing thousands of game birds, in particular pheasants and ducks, for shooting. This is not a well-researched issue, but this massive augmentation of game bird populations may affect resident wildlife in a number of ways (Champagnon *et al.* 2009). Where birds are being released into a native population this is likely to have a profound impact on the genetics of the wild population. Released birds may also compete with other native species for food or nest sites and there may be a knock-on effect in increasing predator populations, which may in turn threaten the populations of rare native species. However, other aspects of land management for game birds and shooting, such as maintaining a more open canopy, may be beneficial to native wildlife (Draycott *et al.* 2007).

Hunting in tropical and subtropical forests

Although tropical and subtropical forests have been occupied by humans for as long as other parts of the planet, until recently human populations in many areas were low, largely because of vector borne diseases like malaria. Moreover, comparatively low levels of development meant that hunters often could not afford cartridges for shotguns or modes of transport, such as motorcycles and outboard motors, that today afford them access to almost any remaining piece of forest. In addition, large areas of tropical forest have been cleared and, through the activities of logging and mining companies, roads have been constructed through remaining forest blocks (Abernethy *et al.* 2013). Improved communications resulting from these developments have enabled hunters both to access more forest and to transport bushmeat out to urban markets (Laurance *et al.* 2009). In short, over most of the tropics and subtropics, hunting pressure on forest wildlife has increase dramatically over the past 2–3 decades as a consequence of increased human population densities, improved road access to forests and markets, more widespread use of shotguns and motorized transport, the emergence of professional hunters who sell bushmeat to urban populations, and the declining area of forest to support animal populations (Brashares *et al.* 2011; Wilkie *et al.* 2011; Robinson and Bennett 2000; Abernethy *et al.* 2013; Harrison 2011). The quantities of bushmeat being hunted from tropical forests are simply staggering: an estimated 6 million animals harvested annually from the Bornean states of Sarawak and Sabah during the early 1990s (or ~36 animals km^{-2}) (Bennett *et al.* 2000); an estimated 4 million tonnes of bushmeat exported annually from the Congo basin (Fa and Brown 2009); 120,000 Amur falcons traded through a single market in Nagaland, NE India in 2013 (Dalvi *et al.* 2013). Moreover, although nature reserves in some countries have been effective in enforcing hunting bans, this tends to be the exception rather than the rule. Enforcement and management of tropical reserves is highly variable (Laurance *et al.* 2012) and often reserves are as likely to be defaunated as elsewhere (Harrison 2011). The spread of the empty forest syndrome (Redford 1992) suggests that hunting may be the greatest threat to the persistence of wildlife throughout tropical and subtropical forests.

Although primary production in tropical forest is high, much of the plant biomass is in the form of wood or inaccessible and largely unpalatable leaves. Consequently, vertebrate biomass is low as compared to other ecosystems. To remain sustainable hunting pressure in tropical forests cannot exceed approximately one person km^{-2} (Robinson and Bennett 2004). Throughout much of the tropics hunting pressures greatly exceed this level, and have resulted in the extirpation of large numbers of species from remaining forests (Redford 1992; Robinson 1999). Moreover, one of the distinctive features of hunting in tropical forests, as compared to temperate areas, is that it is relatively non-selective (Robinson and Bennett 2000). Encounter rates with wildlife are low and so hunters tend to shoot anything they see above some threshold size, and as areas become increasingly over-hunted people tend to shoot smaller and smaller species. This multi-species opportunistic approach is an important reason behind the extirpation of so many species, because even as species become too rare to justify hunting on their own, they may be shot by hunters nominally out looking for other quarry (Branch et al. 2013).

In the mid-2000s, only about 35 percent, 9 percent and 1 percent of the Neotropical, Afrotropical and Indo-Malayan regions, respectively, supported intact megafaunal communities (> 20 kg bodyweight) (Morrison et al. 2007). Less well understood is that outside these areas, often all animals larger than about 1 kg may have been extirpated, barring a few hunting tolerant species such as pigs (Harrison 2011). Nevertheless, hunting pressure also varies among tropical forests. Perhaps the largest influence is that of culture. Some communities may hunt very little or not at all, while neighboring groups have over-hunted their forests. For example, in Peninsular Malaysia in some areas primates, such as gibbons or siamang, can be found in forests on the edge of suburban areas, but elsewhere these same species are missing from remote forests. In Kalimantan, the primary determinants of the likelihood that people killed orangutans was religion and the amount of forest remaining in the vicinity of the village (Davis et al. 2013). Across much of South Asia, wildlife coexists broadly with people (Ranganathan et al. 2008), but in northeast India, which is culturally and biogeographically part of SE Asia, even remote forests are empty (Datta et al. 2008; Dalvi et al. 2013). Sometimes people have taboos against hunting certain species. For example, in eastern Madagascar although people were found to eat a wide variety of bushmeat, including several protected species, there was a taboo against eating Indri (Jenkins et al. 2011). Hunting and bushmeat consumption is also determined by economic status. Across Africa, it was found that in villages near forest bushmeat was usually eaten by poorer people, but further from the forest it was more commonly eaten by wealthier classes as a symbol of status (Brashares et al. 2011). In parts of the tropics with substantial areas of native forest remaining, such as in the Amazon and Congo basins, hunting pressure may be negatively correlated with distance to the nearest point of access, usually a road or river, and areas more than 2–3 days from an access point may be rarely hunted (Peres and Lake 2003). Moreover, where hunting is less intense population declines among larger, hunting sensitive species may result in compensatory increases in the populations of smaller, more hunting tolerant species (Peres and Dolman 2000).

Often the increased hunting pressure is accompanied by other drivers of species loss, such as habitat loss and fragmentation, increased isolation and conversion of matrix habitats to agricultural plantations. Hence, it may be difficult to attribute species extirpations exclusively to hunting. However, one approach is to employ species-area models. Using parameters validated elsewhere, these models can be used to predict the number of species one would expect to lose through habitat effects alone and if there is a surplus of extirpated species these can be attributed to hunting (Sreekar et al. 2015) (Figure 43.2). Nevertheless, it should be pointed out that such methods are still conservative with respect to the impacts of hunting, because species-area models will over-estimate extirpations over shorter time periods. Unfortunately, researchers often ignore hunting or simply treat it as a consequence of habitat changes. For example, a recent review of 1,220 studies on the impacts of tropical forest fragmentation on mammals found that only five studies considered the separate effects of fragmentation and hunting (Kosydar et al. 2014). A failure to consider the impacts of hunting is undoubtedly inflating the impacts of these other drivers and distorting our understanding of the processes. From a conservation perspective, a better understanding of the separate effects might lead to radically different approaches to conserving tropical biodiversity. Over much of South Asia wildlife coexists with people across highly altered landscapes, comprising forest patches, agroforestry plantations and agricultural fields, largely because few people hunt

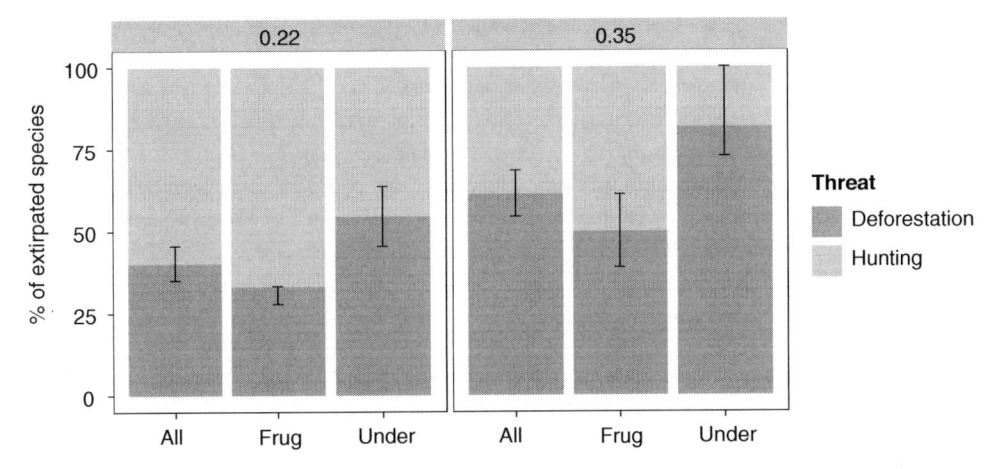

Figure 43.2 Separating the effects of deforestation and hunting on the extirpation of birds in a fragmented tropical landscape in Xishuangbanna, China (Sreekar *et al.* 2015). Proportion of total extirpations of all forest birds (All), forest frugivores (Frugivores) and forest understorey insectivores (Understorey) partitioned into species extirpated by deforestation only (dark grey), and the estimated additional species extirpated by hunting (light grey) using different area models (countryside [slope=0.22] and matrix calibrated [slope=0.35]). Bar heights show medians and error bars represent 95 percent confidence intervals

(Ranganathan *et al.* 2008). This suggests that in other tropical areas a bigger focus on controlling hunting might benefit conservation.

Consequences of hunting for ecosystem processes in tropical forests

Background

All organisms in a community, through their activities, affect ecosystem processes. Animals that are hunted may, for example, include predators, herbivores, browsers, seed dispersers and seed predators. Indeed, because hunting often simultaneously affects the populations of species that perform very different roles in the community, it can be difficult to make predictions about impacts of hunting on ecosystem processes (Dirzo *et al.* 2014; Wright 2003). For example, if the dispersers and predators of large seeds are both impacted by hunting, what can one say about the likely fate of tree species with large seeds? Moreover, the redundancy, or the degree to which the roles performed by a particular species may be replaced by other species, varies. However, because there are fewer large species in a community, the redundancy among such species is usually low. For example, if an Asian forest loses its tigers, the role this species plays in the forest is unlikely to be replaced by other predators, as they are either much smaller or have very different biologies (e.g. pythons). In contrast, among medium to small predators Asian forests may support 3–4 species of wild cat, wild dogs and several species of civets. As we have already learned, larger animals tend to be targeted by hunters and are often among the first species to be extirpated from forests. A further difficulty in understanding the likely impacts of hunting on a forest is that responses are likely to vary in a non-linear manner with differences in hunting pressure. For example, whereas at relatively low levels of hunting the largest carnivores might be extirpated possibly leading to an increase in the populations densities of large herbivores, as hunting pressure increases herbivore populations may also decline. Last, in attempting to understand the consequences of hunting on ecosystem processes in forests, we are ultimately interested in how the dynamics of the tree community may be affected. However, hunting impacts may only become evident after multiple tree-generations and as trees are long-lived species this could take several centuries or even millennia. In summary, it is difficult to generalize about the impacts of hunting, because hunted animals perform a variety of ecological functions, levels of redundancy among hunted species vary, ecological responses to differences in hunting pressure are likely to be non-linear, and the plant community responses may take decades to centuries to play out (Wright 2003; Stoner *et al.* 2007; Harrison *et al.* 2013).

Impacts of Pleistocene megafaunal extinctions

It is important to appreciate that forests are dynamic entities that have been profoundly impacted by people since humans first evolved and spread around the world. Increasing evidence suggests that many mature, and apparently pristine, tropical forests have regrown on land that was cultivated by people just 500 years previously (Chazdon 2014). Even more surprising is the suggestion that tropical forests today may still be adjusting to the disappearance of the Pleistocene megafauna. For example, several large seeded Neotropical trees may have been adapted to dispersal by gomphotheres, as there are no contemporary seed dispersers large enough to consume their fruit (Janzen and Martin 1982). Perhaps even more remarkable, a recent analysis suggests that through their feeding and defecating activities the Pleistocene megafauna may have been responsible for redistributing nutrients on a huge scale and that current tree distributions in the Amazon may still be adjusting to the reduced flux of nutrients, particularly phosphorous (Doughty *et al.* 2013). To think that tropical forests may still be responding to over-hunting that occurred more than 10,000 years ago is profoundly sobering.

Population declines of large carnivores

As mentioned earlier, in many temperate forests large carnivores were hunted out in historical times. Large areas of subtropical and tropical forest have lost their largest carnivores within the past few decades. If populations of large herbivore prey are controlled by top-down processes, as opposed to food availability, then removal of predators will result in an increase in herbivore populations. This appears to be what has happened in many temperate areas and drier tropical forests. Herbivore populations often attain pest levels, if they are not managed, and can alter natural restoration pathways, changing the composition of forests, or inhibit restoration altogether (Dexter *et al.* 2013; Clutton-Brock *et al.* 2004). However, the case that predators limit prey populations in humid tropical forests is more equivocal. Large deer and wild cattle are often naturally rare, whereas pigs have a high reproductive capacity and can increase populations rapidly when food resources are abundant. Importantly, there does not seem to be any evidence that large herbivore populations attain pest levels in rain forests following the extirpation of apex predators. Indeed a more commonly reported pattern in hunted forests is that populations of large herbivores are declining or have even been extirpated (Dirzo and Miranda 1991).

Population declines of large herbivores

The largest herbivores, such as elephants, rhinos and wild cattle, have relatively low intrinsic population growth and hence their populations are often negatively impacted even at low hunting pressure. Moreover, rhinos and elephants in particular are highly sought after.

Megafauna (> 20 kg) have been eliminated from over 90 percent of their former forest ranges over the past approximately 50 years (Morrison et al. 2007). In contrast, smaller deer and pigs have relatively high reproductive capacities and are rarely extirpated by hunting, despite being preferred for their meat. Nevertheless, hunting pressures are such that in many areas their populations are very low. Large herbivores are important browsers of tree seedlings, often causing mortality through the destruction of the terminal bud. In addition to feeding, pigs cut large numbers of seedlings to make nests (Ickes et al. 2012). In areas where hunting pressure is high, we would therefore expect seedling survival and seedling densities to increase. However, density-dependent seedling mortality factors, such as fungal diseases, are also understood to be prevalent in tropical forests (Comita et al. 2010), and if strong enough could compensate for reduced browsing pressure.

Herbivore exclosure experiments in both savanna woodlands in East Africa and moist forests in western Amazonia have shown that large herbivore exclusion leads to increased ground vegetation cover (Beck et al. 2013; Young et al. 2013). In a defaunated tropical rain forest in Mexico, Dirzo and Miranda (1991) demonstrated that levels of vertebrate herbivory were negligible and, in comparison with an intact site nearby, seedling abundance was much higher but diversity much lower. Thus, at the defaunated site release from vertebrate herbivory appeared to have enhanced the competitive dominance of a small set of species. Similarly, in a Bornean forest with very high hunting pressure sapling densities approximately doubled over a 15 year period, while there was a ~2 percent decrease in sapling diversity (Harrison et al. 2013).

Population declines of seed dispersers and seed predators

Seed dispersers are among the animal guilds most sensitive to hunting (McConkey et al. 2011). Even forests with relatively abundant wildlife may be missing large terrestrial herbivores – such as elephants – that are important dispersers for tree species with very large seeds, frugivorous primates and large frugivorous birds, such as hornbills. In part this reflects the relatively low intrinsic population growth of such species, but it is also probably a consequence of the fact that frugivores are relatively easy targets for hunters waiting under fruiting trees. For effective seed dispersal, seed dispersers must perform two related roles. First, they should remove seeds from the parent tree and second they need to deposit them at some distance away in a site suitable for germination and juvenile survival (Corlett 1998). Therefore, in forests impacted by hunting one would predict: (i) reduced removal of seeds; and (ii) reduced dispersal distances for animal dispersed species. In addition, larger bodied seed dispersers tend to consume larger seeds and deposit them at greater distances from the maternal plant. As larger animals tend to be more affected by hunting, we would therefore also predict that: (iii) the effects of hunting would be greater for plant species with larger seeds.

Perhaps the only general result to emerge from the literature on the impacts of hunting in tropical forests is a reduced removal of seeds and reduced recruitment of plants with animal dispersed seeds relative to species with abiotically dispersed seeds (e.g. wind) in hunted forests (Effiom et al. 2013; Vanthomme et al. 2010; Brodie et al. 2009; Wright et al. 2007; Holbrook and Loiselle 2009; Nunez-Iturri et al. 2008; Sethi and Howe 2009; Harrison et al. 2013; Terborgh et al. 2008; Nunez-Iturri and Howe 2007). In a heavily hunted forest in Borneo, it was also found that tree species with animal dispersed seeds were becoming more clustered over time, presumably as a consequence of reduced dispersal distances, and this effect was greater for species with larger seeds (Figure 43.3) (Harrison et al. 2013). However, the impact of seed size has been more equivocal. Studies from different forests have reported both increased and decreased recruitment of plant species with larger seeds. These results reflect an interaction between the impacts of

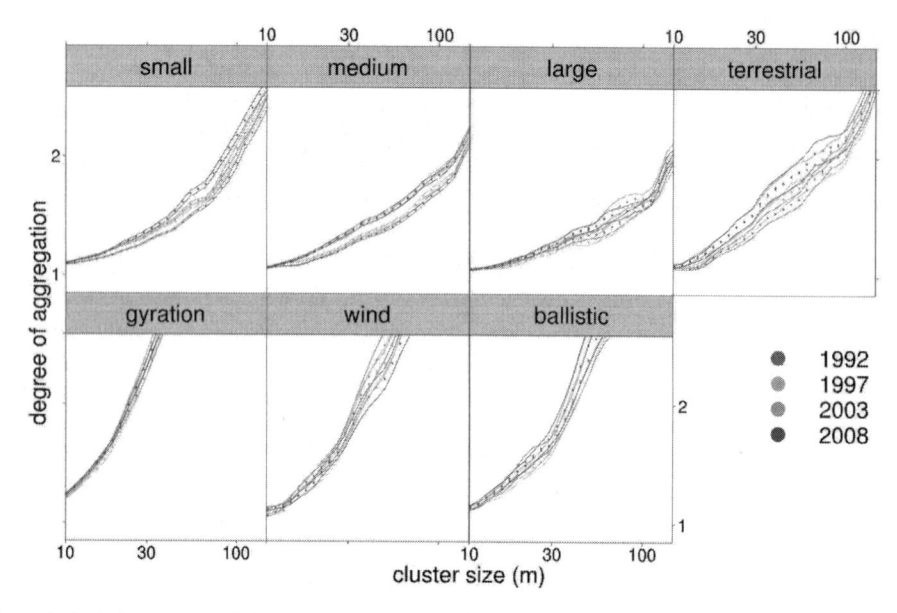

Figure 43.3 Defaunation and the loss of seed dispersers drives increased clustering of saplings over time among animal dispersed species in a Bornean rain forest (Harrison et al. 2013). Degree of sapling aggregation (y-axis) is plotted against cluster sizes (x-axis). Points represent means across species and lines are the boundaries of the 95 percent confidence interval. Top row: small (<20 mm long), medium (≥20 to <50 mm long), and large (≥50 mm long) seeds dispersed by canopy animals and seeds dispersed by terrestrial animals that initially fall from the canopy. Bottom row: tree species with gyration, wind and ballistically dispersed seeds. Tree species with animal-dispersed seeds (top row) show significantly increased clustering over time (curves move towards the left), while species with abiotically dispersed seeds (bottom row) do not

hunting on seed dispersers and seed predators (Beckman and Muller-Landau 2007). The dispersers of large seeds are often disproportionately affected by hunting, leading to reduced seed removal and when most seed mortality is caused by insects or fungal pathogens mortality of large seeds is expected to increase (Sethi and Howe 2009; Harrison et al. 2013; Terborgh et al. 2008; Nunez-Iturri et al. 2008; Vanthomme et al. 2010; Galetti et al. 2006). However, when important predators of big seeds, such as large rodents, are hunted then the survival of bigger seeds may increase in hunted forests (Dirzo et al. 2007; Wright et al. 2007).

Evolutionary consequences of hunting

By altering the selective environment, hunting is also expected to affect the evolutionary trajectories of species. In the Atlantic forests of Brazil, where some forests have been fragmented and defaunated for over a century, Galetti et al. (2013) found that seeds of the forest palm *Euterpe edulis* were smaller in defaunated sites that lacked seed dispersers with large gaps, than those in forest fragments that still had large seed dispersers. As seed size is fundamentally associated with plant life-history strategy this selection for smaller seeds is likely affecting many aspects of the biology of this species. We may also anticipate that release from browsing pressure of mammalian herbivores is altering the selective environment among seedlings. Increased growth rates may be favored over investing in defenses against herbivores.

Conclusions

Humans have hunted forest animals since the emergence of the hominid lineage in Africa approximately 6 million years ago. But since the late Pleistocene, technological advances have increased the impacts of human hunting way beyond that of a normal large predator. Forests today may still be adjusting to the impacts of the widespread, hunting induced, megafaunal extinctions that occurred in the late Pleistocene. In temperate forests apex predators were extirpated from much of their former ranges in historical times and the empty forest syndrome is a rapidly spreading phenomenon in the tropics today (Dirzo et al. 2014). Through the selective removal of larger animals, hunting is altering ecosystem processes and causing an extended erosion of biodiversity. Although some changes may be apparent within a few years, decades or centuries, the full consequences of modern hunting will only likely become evident after a much more substantial period of time. However, we may be able to reverse many of these changes by restoring the historical dynamics of forest ecosystems through reintroducing species (Seddon et al. 2014).

References

Abernethy, K.A. Coad, L., Taylor, G., Lee, M.E. and Maisels, F., 2013. Extent and ecological consequences of hunting in Central African rainforests in the twenty-first century. *Philosophical Transactions of the Royal Society B: Biological Sciences*, 368, p. 20120303. Available at: http://dx.doi.org/10.1098/rstb.2012.0303.

Beck, H., Snodgrassa, J.W. and Thebpanyab, P., 2013. Long-term exclosure of large terrestrial vertebrates: Implications of defaunation for seedling demographics in the Amazon rainforest. *Biological Conservation*, 163, pp. 115–121.

Beckman, N.G. and Muller-Landau, H.C., 2007. Differential effects of hunting on pre-dispersal seed predation and primary and secondary seed removal of two neotropical tree species. *Biotropica*, 39, pp. 328–339. Available at: file:///home/rhett/Documents/Library/PDF/Beckman-2007-Differential.

Bennett, E.L., Nyaoi, A.J. and Sompud, J., 2000. Saving Borneo's bacon: The sustainability of hunting in Sarawak and Sabah. In J.G. Robinson and E.L. Bennett, eds. *Hunting for Sustainability in Tropical Forests*. New York: Columbia University Press, pp. 305–324.

Branch, T.A, Lobo, A.S. and Purcell, S.W., 2013. Opportunistic exploitation: An overlooked pathway to extinction. *Trends in Ecology and Evolution*, 28(7), pp. 409–413. Available at: http://www.ncbi.nlm.nih.gov/pubmed/23562732 [Accessed July 14, 2014].

Brashares, J.S., Prugh, L.R., Stoner, C.J. and Epps, C.W., 2010. Ecological and conservation implications of mesopredator release. In J. Terborgh and J. A. Estes, eds. *Trophic Cascades: Predators, Prey, and the Changing Dynamics of Nature*. Washington, DC: Island Press, pp. 221–240.

Brashares, J.S., Golden, C.D., Weinbaum, K.Z., Barrett, C.B. and Okello, G.V., 2011. Economic and geographic drivers of wildlife consumption in rural Africa. *Proceedings of the National Academy of Sciences of the United States of America*, 108(34), pp. 13931–13936. Available at: http://www.pubmedcentral.nih.gov/articlerender.fcgi?artid=3161600&tool=pmcentrez&rendertype=abstract [Accessed August 13, 2014].

Brodie, J.F., Helmy, O.E., Brockelman, W.Y. and Maron, J.L., 2009. Bushmeat poaching reduces the seed dispersal and population growth rate of a mammal-dispersed tree. *Ecological Applications*, 19(4), pp. 854–863. Available at: http://www.ncbi.nlm.nih.gov/pubmed/19544729.

Bubeník, A.B., 1989. Sport hunting in continental Europe. In R.J. Hudson, K.R. Drew and L.M. Baskin, eds. *Wildlife Production Systems: Economic Utilisation of Wild Ungulates*. Cambridge, UK: Cambridge University Press, pp. 115–133.

Buczacki, S., 2002. *Fauna Britannica: Natural History, Myths and Legend, Folklore, Tales and Traditions*, London: Hamlyn.

Cartmill, M., 1993. *A View into Death in the Morning: Hunting and Nature through History*, Cambridge, Massachusetts, US: Harvard University Press.

Champagnon, J., Guillermain, M., Gauthier-Clerc, M., Lebreton, J.-D. and Elmberg, J., 2009. Consequences of massive bird releases for hunting purposes: Mallard *Anas platyrhynchos* in the Camargue, southern France. *Wildfowl*, 2, pp. 184–191.

Chazdon, R., 2014. *Second Growth: The Promise of Tropical Forest Regeneration in an Age of Deforestation*, Chicago, US: The University of Chicago Press.

Clutton-Brock, T.H., Coulson, T. and Milner, J.M., 2004. Red deer stocks in the Highlands of Scotland. *Nature*, 429, pp. 261–262.

Comita, L.S., Muller-Landau, H.C., Aguilar, S. and Hubbell, S.P., 2010. Asymmetric density dependence shapes species abundances in a tropical tree community. *Science*, 329(5989), pp. 330–332. Available at: http://www.ncbi.nlm.nih.gov/pubmed/20576853 [Accessed July 23, 2011].

Corlett, R.T., 1998. Frugivory and seed dispersal by vertebrates in the Oriental (Indomalayan) region. *Biological Reviews*, 73, pp. 413–448. Available at: file:///home/rhett/Documents/Library/PDF/Corlett1998–0002067200/Corlett1998.pdf.

Côté, S.D., Rooney, T.P., Tremblay, J.P., Dussault, C. and Waller, D.M., 2004. Ecological impacts of deer overabundance. *Annual Review of Ecology, Evolution, and Systematics*, 35(1), pp. 113–147. Available at: http://www.annualreviews.org/doi/abs/10.1146/annurev.ecolsys.35.021103.105725 [Accessed July 19, 2014].

Crooks, K.R. and Soulé, M.E., 1999. Mesopredator release and avifaunal extinctions in a fragmented system. *Nature*, 400, pp. 563–566.

Dalvi, S., Sreenivasan, R. and Price, T., 2013. Exploitation in Northeast India. *Science*, 339, p. 270.

Datta, A., Anand, M.O. and Naniwadekar, R., 2008. Empty forests: Large carnivore and prey abundance in Namdapha National Park, north-east India. *Biological Conservation*, 141, pp. 1429–1435. Available at: file:///home/rhett/Documents/Library/PDF/Datta-2008-Empty.

Davis, J.T., Mengersen, K., Abram, N.K., Ancrenaz, M., Wells, J.A. and Meijaard, E., 2013. It's not just conflict that motivates killing of orangutans. *PloS one*, 8(10), p. e75373.

Deinet, S., Ieronymidou, C., McRae, L., Burfield, I.J., Foppen, R.P., 2013. *Wildlife Comeback in Europe: The Recovery of Selected Mammal and Bird Species. Final Report*. London, UK: Zoological Society of London.

Dexter, N., Ramsey, D.S.L., MacGregor, C. and Lindenmayer, D., 2012. Predicting ecosystem wide impacts of wallaby management using a fuzzy cognitive map. *Ecosystems*, 15(8), pp. 1363–1379. Available at: http://link.springer.com/10.1007/s10021-012-9590-7 [Accessed August 11, 2014].

Dexter, N., Hudson, M., James, S., MacGregor, C. and Lindenmayer, D., 2013. Unintended consequences of invasive predator control in an Australian forest: Overabundant wallabies and vegetation change. *PloS one*, 8(8), p. e69087. Available at: http://www.pubmedcentral.nih.gov/articlerender.fcgi?artid=3749205&tool=pmcentrez&rendertype=abstract [Accessed August 17, 2014].

Dirzo, R. and Miranda, A., 1991. Altered patterns of herbivory and diversity in the forest understory: A case study of the possible consequences of defaunation. In P.W. Price, T.M. Lewinsohn, G.W. Fernandes, and W.W. Benson, eds. *Plant-Animal Interactions: Evolutionary Ecology in Tropical and Temperate Regions*. London: John Wiley & Sons, pp. 273–287.

Dirzo, R., Mendoza, E. and Ortiz, P., 2007. Size-related differential seed predation in a heavily defaunated neotropical rain forest. *Biotropica*, 39, pp. 355–362. Available at: file:///home/rhett/Documents/Library/PDF/Dirzo-2007-Size-related differe-1377046272/Dirzo-2007-Size-related differe.pdf.

Dirzo, R., Young, H.S., Galetti, M., Ceballos, G., Isaac, N.J.B. and Collen, B., 2014. Defaunation in the Anthropocene. *Science*, 345, pp. 401–406.

Doughty, C.E., Wolf, A. and Malhi, Y., 2013. The legacy of the Pleistocene megafauna extinctions on nutrient availability in Amazonia. *Nature Geoscience*, 6(9), pp. 761–764. Available at: http://www.nature.com/doifinder/10.1038/ngeo1895 [Accessed July 22, 2014].

Draycott, R.A.H., Hoodless, A.N. and Sage, R.B., 2007. Effects of pheasant management on vegetation and birds in lowland woodlands. *Journal of Applied Ecology*, 45(1), pp. 334–341. Available at: http://doi.wiley.com/10.1111/j.1365-2664.2007.01379.x [Accessed August 7, 2014].

Effiom, E.O., Nunez-Iturri, G., Smith, H.G., Ottosson, U. and Olsson, O., 2013. Bushmeat hunting changes regeneration of African rainforests. *Proceedings of the Royal Society B*, 280, p. 20130246.

Fa, J.E. and Brown, D., 2009. Impacts of hunting on mammals in African tropical moist forests: A review and synthesis. *Mammal Review*, 39(4), pp. 231–264.

Galetti, M., Donatti, C.I., Pires, A.S., Guimaraes, P.R. and Jordano, P., 2006. Seed survival and dispersal of an endemic Atlantic forest palm: The combined effects of defaunation and forest fragmentation. *Botanical Journal of the Linnean Society*, 151, pp. 141–149.

Galetti, M., Guevara, R., Cortes, M.C., Fadini, R., VonMatter, S. and others, 2013. Functional extinction of birds drives rapid evolutionary changes in seed size. *Science*, 340(May), pp. 1086–1090.

Gordon, I.J., Hester, A.J. and Festa-bianchet, M., 2004. The management of wild large herbivores to meet economic, conservation and environmental objectives. *Journal of Applied Ecology*, 41, pp. 1021–1031.

Harrison, R.D., 2011. Emptying the forest: Hunting and the extirpation of wildlife from tropical nature reserves. *BioScience*, 61(11), pp. 919–924. Available at: http://www.jstor.org/stable/info/10.1525/bio.2011.61.11.11 [Accessed November 18, 2011].

Harrison, R.D., Tan, S., Plotkin, J.B., Slik, F., Detto, M. and others, 2013. Consequences of defaunation for a tropical tree community. *Ecology Letters*, 16(5), pp. 687–694. Available at: http://www.ncbi.nlm.nih.gov/pubmed/23489437 [Accessed May 23, 2013].

Henneberger, J., 1996. Transformations in the concept of the park. *The Trumpeter*, 13(3), pp. 127–133. Available at: http://trumpeter.athabascau.ca/index.php/trumpet/article/viewArticle/252/369.

Henschel, P., Coad, L., Burton, C., Chataigner, B., Dunn, A. and others, 2014. The lion in West Africa is critically endangered. *PloS one*, 9(1), p. e83500. Available at: http://www.pubmedcentral.nih.gov/articlerender.fcgi?artid=3885426&tool=pmcentrez&rendertype=abstract [Accessed July 24, 2014].

Holbrook, K.M. and Loiselle, B.A., 2009. Dispersal in a Neotropical tree, *Virola flexuosa* (Myristicaceae): Does hunting of large vertebrates limit seed removal? *Ecology*, 90(6), pp. 1449–1455.

Ickes, K., Paciorek, C.J. and Thomas, S.C., 2012. Impacts of nest construction by native pigs (*Sus scrofa*) on lowland Malaysian rain forest saplings. *Ecology*, 86(6), pp. 1540–1547.

Janzen, D.H. and Martin, P.S., 1982. Neotropical anachronisms—the fruits the gomphotheres ate. *Science*, 215, pp. 19–27.

Jenkins, R.K.B., Keane, A., Rakotoarivelo, A.R., Rakotomboavonjy, V., Randrianandrianina, F.H. and others, 2011. Analysis of patterns of bushmeat consumption reveals extensive exploitation of protected species in eastern Madagascar. *PloS one*, 6(12), p. e27570. Available at: http://www.pubmedcentral.nih.gov/articlerender.fcgi?artid=3237412&tool=pmcentrez&rendertype=abstract [Accessed January 2, 2012].

Kosydar, A.J., Rumiz, D.I., Conquest, L.L. and Tewksbury, J.J., 2014. Effects of hunting and fragmentation on terrestrial mammals in the Chiquitano forests of Bolivia. *Tropical Conservation Science*, 7(2), pp. 288–307.

Laurance, W.F., Goosem, M. and Laurance, S.G.W., 2009. Impacts of roads and linear clearings on tropical forests. *Trends in Ecology and Evolution*, 24(12), pp. 659–669. Available at: http://www.ncbi.nlm.nih.gov/pubmed/19748151 [Accessed June 10, 2011].

Laurance, W.F., Useche, D.C., Rendeiro, J., Kalka, M., Bradshaw, C.J.A. and others, 2012. Averting biodiversity collapse in tropical forest protected areas. *Nature*, 489(7415), pp. 290–294. Available at: http://www.ncbi.nlm.nih.gov/pubmed/22832582 [Accessed July 24, 2012].

Lindenmayer, D.B., Franklin, J.F. and Fischer, J., 2006. General management principles and a checklist of strategies to guide forest biodiversity conservation. *Biological Conservation*, 131(3), pp. 433–445.

McConkey, K.R., Prasad, S., Corlett, R.T., Campos-Arceiz, A., Brodie, J.F. and others, 2011. Seed dispersal in changing landscapes. *Biological Conservation*, 146, pp. 1–13. Available at: http://linkinghub.elsevier.com/retrieve/pii/S0006320711004575 [Accessed January 5, 2012].

Morrison, J.C., Sechrest, W., Dinerstein, E., Wilcove, D.S. and Lamoreux, J.F., 2007. Persistence of large mammal faunas as indicators of global human impacts. *Journal of Mammology*, 88(6), pp. 1363–1380.

Nunez-Iturri, G. and Howe, H.F., 2007. Bushmeat and the fate of trees with seeds dispersed by large primates in a lowland rain forest in western Amazonia. *Biotropica*, 39, pp. 348–354. Available at: file:///home/rhett/Documents/Library/PDF/Nunez-Iturri-2007-Bushmeat and the fat-2585004544/Nunez-Iturri-2007-Bushmeat and the fat.pdf.

Nunez-Iturri, G., Olsson, O. and Howe, H.F., 2008. Hunting reduces recruitment of primate-dispersed trees in Amazonian Peru. *Biological Conservation*, 141, pp. 1536–1546. Available at: file:///home/rhett/Documents/Library/PDF/Nunez-Iturri-2008-Hunting.

Packer, J.J. and Birks, J.D.S., 2002. An assessment of British farmers' and gamekeepers' experiences, attitudes and practices in relation to the European Polecat *Mustela Putorius*. *Mammal Review*, 29(2), pp. 75–92.

Paddle, R., 2002. *The Last Tasmanian Tiger: The History and Extinction of the Thylacine*, Cambridge, UK: University of Cambridge.

Palmer, S.C. and Truscott, A.-M., 2003. Browsing by deer on naturally regenerating Scots pine (*Pinus sylvestris* L.) and its effects on sapling growth. *Forest Ecology and Management*, 182, pp. 31–47.

Peres, C.A. and Dolman, P.M., 2000. Density compensation in neotropical primate communities: Evidence from 56 hunted and nonhunted Amazonian forests of varying productivity. *Oecologia*, 122, pp. 175–189.

Peres, C.A. and Lake, I.R., 2003. Extent of nontimber resource extraction in tropical forests: Accessibility to game vertebrates by hunters in the Amazon basin. *Conservation Biology*, 17(2), pp. 521–535.

Prugh, L.R., Stoner, C.J., Epps, C.W., Bean, W.T., Ripple, W.J. and others, 2009. The rise of the mesopredator. *BioScience*, 59(9), pp. 779–791. Available at: http://www.jstor.org/stable/27735987 [Accessed July 9, 2014].

Ranganathan, J., Daniels, R.J.R., Chandran, M.D.S., Ehrlich, P.R. and Daily, G.C., 2008. Sustaining biodiversity in ancient tropical countryside. *Proceeding of the National Academy of Science*, 105, no. 46, pp. 17852–17854.

Redford, K.H., 1992. The empty forest. *BioScience*, 42(6), pp. 412–422.

Reynolds, J.C. and Tapper, S., 2008. Control of mammalian predators in game management and conservation. *Mammal Review*, 26(2–3), pp. 127–155.

Ripple, W.J., Larsen, E.J., Renkin, R.A. and Smith, D.W., 2001. Trophic cascades among wolves, elk and aspen on Yellowstone National Park's northern range. *Biological Conservation*, 102(3), pp. 227–234. Available at: http://linkinghub.elsevier.com/retrieve/pii/S0006320701001070.

Robinson, J.G., 1999. Losing the fat of the land. *ORYX*, 33, p. 1.

Robinson, J.G. and Bennett, E.L., 2000. *Hunting for Sustainability in Tropical Forests*, New York: Columbia University Press.

Robinson, J.G. and Bennett, E.L., 2004. Having your wildlife and eating it too: An analysis of hunting sustainability across tropical ecosystems. *Animal Conservation*, 7, pp. 397–408. Available at: file:///home/rhett/Documents/Library/PDF/Robinson-2004-Having your wildlife-3507757312/Robinson-2004-Having your wildlife.pdf.

Sandom, C., Faurby, S., Sandel, B. and Svenning, J.-C., 2014. Global late Quaternary megafauna extinctions linked to humans, not climate change. *Proceedings of the Royal Society B*, 281(June), p. 20133254. Available at: http://dx.doi.org/10.1098/rspb.2013.3254.

Seddon, P.J., Griffiths, C.J., Soorae, P.S. and Armstrong, D.P., 2014. Reversing defaunation: Restoring species in a changing world. *Science*, 345, pp. 406–412.

Sethi, P. and Howe, H.F., 2009. Recruitment of hornbill-dispersed trees in hunted and logged forests of the Indian Eastern Himalaya. *Conservation Biology*, 23(3), pp. 710–718.

Smart, J., Amar, A., Sim, I.M.W., Etheridge, B., Cameron, D. and others, 2010. Illegal killing slows population recovery of a re-introduced raptor of high conservation concern – The red kite *Milvus milvus*. *Biological Conservation*, 143(5), pp. 1278–1286. Available at: http://linkinghub.elsevier.com/retrieve/pii/S0006320710000911 [Accessed July 9, 2014].

Sreekar, R., Huang, G., Zhao, J.-B., Pasion, B.O., Yasuda, M. and others, 2015. The use of species-area relationships to partition the effects of hunting and deforestation on bird extirpations in a fragmented landscape. *Diversity and Distributions*, 21, pp. 441–450..

Stanford, C.B., 1999. *The Hunting Apes: Meat Eating and the Origins of Human Behavior*, Princeton, New Jersey, US: Princeton University Press.

Stoner, K.E., Vulinec, K., Wright, S.J. and Peres, C.A., 2007. Hunting and plant community dynamics in tropical forests: A synthesis and future directions. *Biotropica*, 39, pp. 385–392. Available at: file:///home/rhett/Documents/Library/PDF/Stoner-2007-Hunting and plant co-0588624128/Stoner-2007-Hunting and plant co.pdf.

Terborgh, J., Nunex-Iturri, G., Pitman, N.C.A., Valverde, F.H.C., Alvarez, P. and others, 2008. Tree recruitment in an empty forest. *Ecology*, 89, pp. 1757–1768. Available at: file:///home/rhett/Documents/Library/PDF/07–0479.1–0689195776/07–0479.1.pdf.

Vanthomme, H., Belle, B. and Forget, P., 2010. Bushmeat hunting alters recruitment of large-seeded plant species in Central Africa. *Biotropica*, 42(6), pp. 672–679.

Wilkie, D.S., Bennett, E.L., Peres, C.A. and Cunningham, A.A., 2011. The empty forest revisited. *Annals of the New York Academy of Sciences*, 1223, pp. 120–128. Available at: http://www.ncbi.nlm.nih.gov/pubmed/21449969 [Accessed July 16, 2011].

Wright, S.J., 2003. The myriad consequences of hunting for vertebrates and plants in tropical forests. *Perspectives in Plant Ecology Evolution and Systematics*, 6, pp. 73–86.

Wright, S.J., Hernandez, A. and Condit, R., 2007. The bushmeat harvest alters seedling banks by favoring lianas, large seeds, and seeds dispersed by bats, birds, and wind. *Biotropica*, 39, pp. 363–371. Available at: file:///home/rhett/Documents/Library/PDF/Wright-2007-The bushmeat harvest-2383788032/Wright-2007-The bushmeat harvest.pdf.

Young, H.S., McCauley, D.J., Helgen, K.M., Goheen, J.R., Otarola-Castillo, E. and others, 2013. Effects of mammalian herbivore declines on plant communities: Observations and experiments in an African savanna. *Journal of Ecology*, 101(4), pp. 1030–1041. Available at: http://www.pubmedcentral.nih.gov/articlerender.fcgi?artid=3758959&tool=pmcentrez&rendertype=abstract [Accessed August 8, 2014].

44
THE ECOLOGY OF URBAN FORESTS

Mark J. McDonnell and Dave Kendal

Forests in urban and suburban environments share many of the characteristics of other forests discussed in this volume, particularly their structure and function. For those involved in the modern science and practice of forest ecology, urban forests present unique challenges and opportunities due to the intensity and duration of human–forest interactions. Urban forests are composed of trees growing in parks, natural areas, along streets and in gardens within a matrix of buildings, roads and waterways, all of which exist primarily due to the actions of humans. The development of the discipline of forest science was driven by the increasing world demand for wood products in the eighteenth and nineteenth centuries, and most early foresters were focused on 'natural forests' or plantations and were uninterested in trees and forests in built environments (Miller 1988). Due mainly to the belief in the now outdated paradigm that humans were not components of ecosystems, this lack of interest in human settlements was also shared by the emerging, closely related scientific discipline of ecology (McDonnell 2011). As a consequence of the current rate and magnitude of the growth of urban areas around the globe, there is an increasing recognition of the need to create and maintain green, liveable, biodiversity rich, healthy and sustainable cities and towns (Forman 2014). Everyone involved in the science and practice of urban forest ecology and urban forestry has an opportunity to make significant contributions to the design, creation and management of sustainable human settlements in the future.

While foresters and ecologists were developing the core science of forest ecology in sparsely populated regions of the globe, another group of researchers and practitioners that included arborists, horticulturalists, nurserymen, landscapers, landscape architects and park and land managers (i.e. practitioners) developed a parallel, yet distinctively human focused science of urban forestry (Grey and Deneke 1986, Miller 1988). Unlike the traditional forest science focus of extracting wood products and preserving forests, practitioners and scientists working in urban environments engaged in growing, planting, managing and conserving trees in cities and towns primarily for the amenity they provided to humans (Grey 1996, Kowarik and Körner 2005, Dwyer *et al.* 2000, Konijnendijk *et al.* 2005, Konijnendijk *et al.* 2006, Carreiro *et al.* 2008).

Although arborists, city foresters and tree wardens were studying and managing trees and woodland in cities throughout the world in the late 1800s, the study of urban forest ecosystems is relatively new (Rowntree 1998, Konijnendijk *et al.* 2006), and urban forestry was not recognised as a discipline within the profession of forestry until the early 1970s. The recognised science of urban forest ecology emerged simultaneously in North America and Europe (Miller 1988, Rowntree 1998, Dwyer et al. 2000, Konijnendijk et al. 2005). Pioneers included James Schmid (1975) who explored vegetation patterns in residential gardens in Chicago and Aloys Bernatzky (1978) who brought a scientific approach to 'tree surgery'. Due to the inextricable connection between people and the science and practice of urban forestry, it was originally (and often still is) referred to as urban and community forestry, especially in the United States and Britain (Miller 1988, Konijnendijk et al. 2006). There is ongoing discussion in the literature as to the scope of the study and practice of urban forest ecology and urban forestry. The body of literature on urban forests that has developed over the past century shows that the discipline includes a diversity of topics such as the care and maintenance of street and park trees, the study and management of urban woodlands, the mapping and assessment of a city's entire tree canopy, green structures, infrastructure and networks, urban biodiversity conservation, ecological restoration, urban heat island effects, carbon flux, human health, aesthetics, human values, environmental justice and ecosystem services provided by urban trees and forests. Although there is no universally accepted definition of the study and practice of urban forest ecology and urban forestry, Konijnendijk et al. (2006) provide a comprehensive overview of the more popular published definitions.

Because urban forests possess many of the characteristics of non-urban forests and they occur in biomes throughout the world, much of the information presented in the previous chapters is applicable to anyone interested in the science and practice of urban forest ecology and urban forestry. What distinguishes the study of urban forest ecology is the emphasis on understanding and managing human–forest interactions in places where people live. The aim of this chapter is to provide: 1) an overview of the current scope of the science and practice of urban forest ecology and urban forestry; 2) a review of recent developments and emerging directions in the ecological study of urban forests; and 3) a current assessment of the study of people's relationships with urban forests.

Scope of the science and practice of urban forest ecology and science

The paramount books in the field urban forest ecology and forestry describe in detail a complex diversity of topics that delineate the discipline (Grey and Deneke 1986, Miller 1988, Bradley 1995, Bradshaw et al. 1995, Watson and Himelick 1997, Konijnendijk et al. 2005, Carreiro et al. 2008, Loeb 2011). In very simple terms, the practice and science of urban forest ecology and urban forestry is concerned with the stewardship of urban forest ecosystems, the interactions between forests and the environment, and the relationship between forests and the people that dwell in them (Figure 44.1). Stewardship refers to the creation, care, management and protection of urban forest ecosystems. The practitioners directly involved in urban forest stewardship typically include arborists, horticulturalists, nurserymen, landscapers, landscape architects, urban planners, park and land managers, policy makers and home gardeners. Indeed, in a perfect world everyone living in a city or town should be engaged in beneficial urban forest stewardship practices.

Over the past decade there has been a growing worldwide interest in creating and maintaining liveable, green, healthy, sustainable cities and towns (McDonnell et al. 2009). Ecosystem services are commonly used as a framework to explore how this can be achieved. Ecosystem services refer to the function(s) of urban forest ecosystems that provide a benefit for humans, especially ones that improve well-being (Daily 1997). The maintenance and enhancement of these services will ensure socially, economically and environmentally healthy cities and towns.

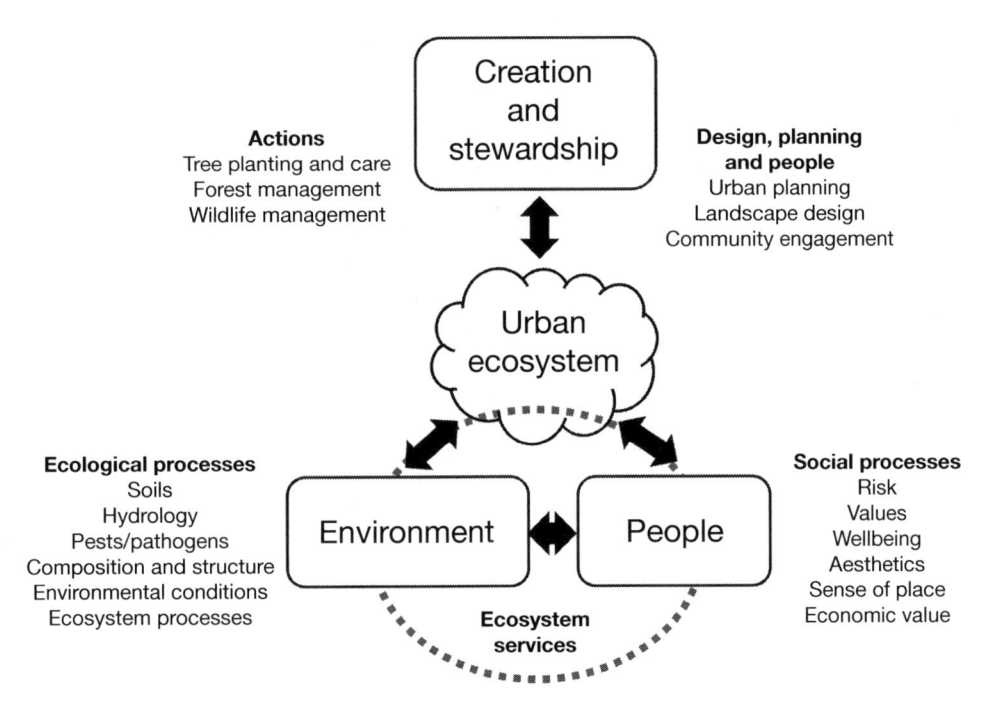

Figure 44.1 Components of the science and practice of urban forest ecology

The quantity and quality of ecosystem services provided by urban forest ecosystems are directly related to the interactions between forest structure and function, environmental conditions and the values and economic resources of city dwellers (de Groot *et al.* 2010).

Recent developments and emerging directions in the ecological study of urban forests

Here we identify five emerging themes in the study of urban forest ecosystems. The first is the use of new conceptual approaches in understanding urban forest ecology. Over the last 20 years the emergence of two new approaches to the study of urban ecosystems which we refer to as: 1) the ecology 'in' and 'of' cities; and 2) the ecology of urbanisation gradients have significantly improved research outcomes. Ecology 'in' cities studies are typically single discipline, small-scale and located within a city. In contrast, studies focused on the ecology 'of' cities are commonly interdisciplinary and multi-scale incorporating both the ecological and human dimensions of urban ecosystems (Grimm *et al.* 2000, Pickett *et al.* 2001). The use of the urban–rural gradient approach to the study of urban ecosystems (McDonnell and Pickett 1990) has proven a useful concept to ecologists and urban foresters around the world (McDonnell and Hahs 2008). In addition, the study of urbanisation gradients has fostered the growth of the comparative ecological study of cities and towns at regional and global scales (McDonnell *et al.* 2009).

Second, differences in ecological processes between urban and rural areas are being articulated. It is now fairly well established that urban environments and ecological processes vary across a gradient of urbanisation (McDonnell and Pickett 1990). In city centres, temperatures can be much higher due to the urban heat island effect, where impervious surfaces absorb heat from the sun and release this heat overnight, resulting in average temperatures commonly up to 5°C and sometimes up to 10°C warmer than nearby rural areas. Cities also tend to have changed hydrology and nutrient cycling (Grimm et al. 2008). For example, soil moisture levels are often lower than rural areas as water is captured by stormwater systems and piped into rivers, and nitrogen levels in soil can be higher due to nitrogen deposition (Grimm et al. 2008). Soils also tend to be compacted and polluted with heavy metals (Craul 1994).

The third emerging theme is the use of ecosystem services as a frame for understanding and measuring the important ecological functions of the urban forest. Urban trees may produce timber, firewood or food (particularly in cities in developing countries), but the main purpose of many urban trees is to improve the urban environment and provide amenity and other benefits to urban dwellers. As such, the ecosystem services that are important in urban areas are not the same as those that are important in non-urban forests. Ecosystem services are categorised as supporting (those that support other services, such as soil formation), regulating (e.g. disease regulation), provisioning (e.g. food, firewood) and cultural (e.g. spiritual, recreation) (Reid et al. 2005). Key services provided by urban ecosystems (Bolund and Hunhammar 1999), and urban forests in particular (Jim and Chen 2009), include:

- Regulating services
  - Microclimate regulation through shade and transpiration
  - Air filtering by particulate matter adhesion to leaves, and gaseous absorption
  - Storm-water regulation via rainfall interception
  - Carbon sequestration
- Provisioning services
  - Fruit
  - Timber
  - Firewood
- Cultural services
  - Aesthetics/beauty
  - Stress recovery/attention restoration
  - Cultural heritage
  - Recreational opportunities
- Supporting services
  - Photosynthesis

The fourth emerging theme is the role of urban forests in creating more sustainable cities. Increasing the ecosystem services provided by the urban forest within cities can reduce reliance on external inputs, such as food for human consumption or electricity for cooling, and lead to a more sustainable system (McPherson 1994). While much urban and social research focuses on the 'compact city' (high population density) as a more sustainable city model, research in ecology suggests that more complex forms are required to maximise ecosystem services and habitat provision for biodiversity, such as embedding higher density residential nodes in larger areas of green open space (Tratalos et al. 2007). A related theme is the growing interest in the role of resilience, the capacity of the system to absorb external shocks, in sustainability studies. Increasing temperatures due to climate change and the attacks by pests and diseases are threatening the urban forest in many cities around the globe. The resilience of the urban forest to these threats is critical to the ongoing provision of ecosystem services, and the maintenance of habitat for biodiversity. Current thinking is that resilience of the urban forest is increased by ensuring that no one species or cultivar dominates, and therefore the broader impact of the loss of a particular species is minimised (Kendal et al. 2014).

The last theme is the importance of the urban forest for broader biodiversity and conservation outcomes. While the ecosystem services frame is proving useful for understanding how the urban forest contributes to human well-being, it is not as well suited for understanding the role of the urban forest for the well-being of non-human animals. Instead, there is a growing body of evidence that the urban forest is an important resource for many animals (McDonnell et al. 2009). As the awareness of the importance of cities to broader ecological processes and conservation outcomes increases, so does the importance of the role of urban forest in conservation outcomes.

Mapping the urban forest

The mapping of urban forests in the 1980s was a significant advancement in the development of the practice and science of urban forest ecology (Rowntree 1998). A number of tools (e.g. i-Tree, ArborPro, TreeKeeper) are now commonly used to manage the urban forest and to quantify the provision of ecosystem services by the forest. Computerised tree inventories are now widely used to record information about trees under management. These systems typically record the species, location, size, health, hazard level and maintenance history information that allow maintenance work to be scheduled efficiently. Inventory data is also commonly used to calculate measures of taxonomic diversity of the forest for use in strategic planning (Kendal et al. 2014). Where inventory data is not available, trees can be sampled to estimate forest diversity and structure (Nowak et al. 2008).

A common rule of thumb to increase the diversity, and therefore the resilience, of the urban forest has called for management benchmarks to be set for the relative abundance of the most taxa at different taxonomic levels i.e. the maximum proportion of trees that are from any one species, genus or family. This widely applied rule has suggested benchmarks of 10 per cent at the species level, 20 per cent at the genus level, and 30 per cent at the family level (Santamour 1990). However, recent research suggests that these values may be too low at the genus and family levels, and that benchmarks should be lower in temperate climates, and higher in cold climates and in streetscapes (Kendal et al. 2014).

Remote sensing using satellite imagery or aerial photography can be used to measure the extent and health of the forest canopy (Dwyer et al. 2000). Recent advances in remote sensing include the use of LIDAR imagery to generate three-dimensional images of the forest, and thermal infrared imagery to detect localised heat effects. Relative canopy cover can vary greatly between cities, and is related to surrounding environmental conditions, for example, desert cities are likely to have lower urban forest cover than cities in forested areas (Dwyer et al. 2000). Being able to measure urban forest cover allows targets for increased cover (based on cover in comparable cities) to be set. Inventory and remote sensing information are often combined in Geographic Information Systems, where they can also be used in conjunction with other information on infrastructure and socio-economics to make planning and management decisions – for example, patterns of ecosystems services, diversity and tree cover – and can be used to identify areas of urban forest vulnerability within a city (Dobbs et al. 2014).

Policy and governance

With increasing sophistication in the tools available for urban forest managers, there has also been an increase in the sophistication of the policy instruments available to urban forest managers. Historically, municipal tree ordinances (or bylaws) have been used as the main policy instrument (Conway and Urbani 2007). Risk is a major driver of professional urban forest management, and ordinances are typically used to regulate the planting and maintenance of trees to ensure community safety, in addition to restricting removal of trees.

More recently, urban foresters have been using strategic policy documents containing broader scale (e.g. citywide) targets for urban forest characteristics such as canopy cover and diversity (using relative abundance of common taxa at different taxonomic levels). These strategic targets can then be implemented within more fine-scale and detailed landscape planning (e.g. at park, street or suburb levels). This strategic approach ensures that locally important issues, such as maintaining cultural heritage or protected native vegetation can be managed adequately within a broader policy framework. Simple benchmarking based on forest cover has been criticised for being simplistic and taking a two-dimensional approach to a multi-dimensional issue, and that more sophisticated criteria including governance, tree health and risk should be incorporated into management criteria (Kenney *et al.* 2011). Current trends suggest that future strategic planning will be able to incorporate multiple criteria into benchmarking and monitoring to ensure more holistic management outcomes.

Incorporating a social perspective into the science and practice of urban forest ecology

Preference and values

While there was some early psychological research on people's relationship with trees (Sommer 2001), it was not until late in the twentieth century that local people's perception of forests began to be incorporated into research, planning and management. Following on from work in the 1970s exploring how much people prefer different kinds of forested landscapes in rural areas, perception research began to be applied to urban forests in the USA. Many initial studies used an environmental psychology perspective, exploring the way people perceived the urban forest through studies of landscape preference (Schroeder 1982) and environmental values (Dwyer *et al.* 1991). This perspective distinguished between attitudes, which are the judgments people make about the world around them (e.g. preference, or how much they like something), and values which are the underlying things that are important to people, and which help to shape higher cognitive constructs such as attitudes (Ives and Kendal 2014).

Preference research has consistently shown that people generally like landscapes that include trees more than urban settings without vegetation (Kaplan and Kaplan 1989). People respond positively to characteristics of trees such as larger size, spreading or globular canopies (Sommer and Summit 1995) and coarse foliage (Williams 2002). Many of these findings have been explained with reference to evolutionary psychology, which argues that humans evolved in the African savannah landscapes where preference for good quality habitat improved survival. These theories suggest that all people respond to landscapes in similar ways. However, there has been criticism of this approach, and much recent work has explored *differences* in people's perception of landscape. Comparative preference studies have been conducted in Australia and Europe and have found many similarities, but some important differences in the way people perceive urban trees in different countries (Williams 2002, Tyrväinen *et al.* 2003, Schroeder *et al.* 2006). For example, street trees with spreading canopies were not as preferred in Australia (Williams 2002) and smaller trees were preferred in the UK (Schroeder *et al.* 2006).

A useful method of understanding the different ways that people perceived the urban forest has been explorations of values (Dwyer *et al.* 1991). A wide range of values related to urban trees have been identified in the USA including peace and tranquillity, religious symbolism, environmental improvement, and ecosystem services such as the provision of shade and air pollution removal (Dwyer *et al.* 1991, Lohr and Pearson-Mims 2004). In Europe, a similar suite of values related to aesthetics, spirituality, culture and nature have been identified (Tyrväinen *et al.* 2003, Tyrväinen *et al.* 2007).

Benefits

There are clear links between the way people value the urban forest and the benefits they derive from them, and there is a rapidly growing body of research focused on the benefits provided by the urban forest to people. There are two main perspectives used in this research. The first uses social research methods to explore how different kinds of people respond to vegetation in the city. This includes research in psychology which has shown that the forest provides a wide range of psychological benefits including stress relief and attention restoration, reduced aggression, improved productivity, improved personal well-being and place identity and attachment (Konijnendijk *et al.* 2005). Research in medicine and population health has focused on the positive physical health outcomes of urban forests. In a now classic study, Roger Ulrich found that patients staying in hospital rooms where trees were visible from their window have shorter stays, used fewer painkillers and had fewer negative evaluations by nursing staff (Ulrich 1984). More recent research has shown that more nearby green space is positively related to self-reported health and lower rates of morbidity for many types of disease (e.g. Mitchell and Popham 2007).

Social research methods have also been used to assess the economic importance of the urban forest. Numerous studies have shown that the urban forest can improve neighbourhood property values, commonly by 5 per cent, although some studies suggest increases of 10 per cent or even 20 per cent (Nowak and Dwyer 2007). While most of these studies have been conducted in North America, similar studies have been undertaken in other regions. For example, house prices have also been positively related to urban forest features in Hong Kong (Jim and Chen 2010). The urban forest also has a significant economic value in itself, including both the value of benefits provided and replacement cost. Using these methods, the value of the urban forest in the USA has been estimated to be in the trillions of dollars (Nowak and Dwyer 2007).

The second strand of benefits research uses biophysical research methods to model previously identified benefits generated by the forest in a particular city, based on data from remote sensing or tree inventories. These typically measure intermediate ecosystem services known to be related to human health and well-being. For example, inventories and satellite images can be used to model carbon sequestration and the space available for recreation which indirectly affects people's thermal well-being and levels of physical activity (e.g. Dobbs *et al.* 2014). This kind of research is particularly useful for land managers, as it allows general knowledge about the benefits of the urban forest to be mapped on to the particular forest being managed. While there has been much research focusing on urban forest benefits, there are also costs, disservices and other detriments associated with the urban forest that must be acknowledged in planning and management. For example, the London Plane Tree (*Platanus* x *acerifolia*) releases pollen which can cause asthma, and this disservice can be mapped and included in benefit calculations (e.g. Dobbs *et al.* 2014).

Environmental justice

A different strand of research has explored environmental justice in the urban forest. In North and South America, this research has focussed on the relationship between income (and population density) and urban forest cover (Iverson and Cook 2000), or plant diversity (e.g. Hope *et al.* 2003). This has been largely attributed to the capacity of wealthier residents with the 'economic wherewithal' to move to preferred areas with high-levels of tree cover and diversity, or to plant vegetation. However, research in geography and political ecology has highlighted that drivers of inequality are more likely to be related to top-down political factors, including the capacity of residents to influence public investment in the urban forest (e.g. street trees), or limited space availability and property controls that restrict some residents such as renters planting their own gardens (Heynen and Lindsey 2003). These studies show that a range of factors influence inequality in addition to income, including ethnicity and education level. This continues to be explored in other countries, where similar results have been found, although often with a greater emphasis on education levels than income (e.g. Kendal *et al.* 2012b).

Implications for management of the urban forests

Incorporating this understanding of the social environment in which the urban forest is located, and the effects of the urban forest on people, is critical to the holistic management of the forest. Increasingly, managers are planning urban forests around concepts of human benefits such as ecosystem services. Strategic planning tools are often coarse, such as benchmarking tree cover and species diversity. This narrow focus can lead to undesirable outcomes. For example, programs that promote increased tree canopy by giving trees to homeowners can increase inequality in urban forest coverage as disadvantaged people are unlikely to participate in these programs (Perkins *et al.* 2004).

Additional tools that allow strategic planning the equitable distribution of the forest, mapping the community's values and measuring public attitudes are required for the urban forest to satisfy the needs of the community. There are two important advances in urban forestry that are facilitating this strategic planning: 1) community consultation and community-based urban forest management are increasingly being used by urban forest managers to communicate strategies to the local community, and to incorporate community perspectives into urban forest planning; and 2) the growing use of Geographic Information Systems by urban forest managers is facilitating this by allowing information such as measures of socio-economic status to be cross-referenced with urban forest canopy cover and diversity.

While many managers will neither have the resources nor the capacity to undertake this kind of strategic planning, there are still important lessons from this existing body of research that can be incorporated into everyday planning and management activities:

1 *Different kinds of trees are valued and liked by different people.* For example, many cities around the world incorporate a significant proportion of native trees. People with high environmental values may consider these trees as important components of the urban forest. However, people with stronger cultural values may prefer exotic species (for a variety of reasons). Clearly then, both exotic trees and native trees are required in urban forests to satisfy the needs and values of the broader community.
2 *Attitudes and values of experts differ from those of the public.* Numerous studies have shown that the way the public perceive landscapes is different from landscape professionals. While it might be obvious to an arborist or tree planner which species perform best in the urban forest based on environmental tolerances or other professional considerations, there are many other criteria used by the public to judge trees.
3 *Disadvantaged people are also likely to be disadvantaged by planning and management decisions.* Unless disadvantage is explicitly considered in planning processes, inequality in the urban forest is likely to be reinforced. Disadvantaged people are less likely to participate in planning processes or everyday management (e.g. requesting a new tree). It is the

responsibility of urban forest planners and managers to consider the effect of decisions on inequality in the urban forest.

Future social science research directions

An extensive body of social research on the urban forest has developed over the last 40 years, yet there are still many areas of the relationship between people and the urban forest that are not well understood. We know little about what drives manager decision-making around the urban forest. In the public realm, decisions are being made every day that shape the composition and structure of the urban forest. In other areas of natural resource management there is a growing understanding of the factors shaping decision-making, yet we know almost nothing about the influences on decision-making in the urban forest. This also applies to the private realm. A large proportion of urban forest in many cities exists in the private realm, yet we know little about how private landowners select the species they plant and manage the urban forest they are responsible for. This is particularly important given the challenges of climate change, which is likely to see a shift in the composition of the urban forest (Kendal et al. 2012a).

We also know relatively little about cultural ecosystem services, which are currently poorly defined (Daniel et al. 2012). While we know much about the effect of the urban forest on human well-being, this knowledge is yet to be incorporated into the ecosystem services framework. Parallel with developments in broader natural resource management, there has been some interest in community participation in the planning and management of urban forests. In practice, this rarely extends beyond top-down communication from managers and planners to the community. More research is required to understand how the community can be effectively incorporated into urban forest planning and management processes.

Summary

The science and practice of urban forest ecology is now a holistic discipline, drawing on arboriculture, horticulture, economics, ecology, soil science, civil engineering, design, law, psychology, sociology, history and geography to plan and manage (mostly) trees in urban areas. Stewards of urban forests must understand how a particular species of tree will grow in a given location, how to manage that tree, what the costs of planting and maintaining the tree will be, but also what role the tree will have in the broader urban forest, its effects on nearby infrastructure, the role the tree will play in the ecological functioning of the system, and the benefits (and detriments) to the people both nearby and across the city. We propose that as the study and management of urban forests progresses, we will experience a blurring of the existing boundaries between researchers and practitioners focused on the ecological, economic and social aspects of urban forest ecosystems. In the future the science of urban forest ecology will evolve and emerge as a truly transdisciplinary science.

References

Bernatzky, A. (1978). *Tree Ecology and Preservation* (p. 357). Elsevier Scientific Pub. Co., New York.

Bolund, P. and Hunhammar, S. (1999). Ecosystem services in urban areas. *Ecological Economics*, 29(2), 293–301.

Bradley, G. A. (1995). *Urban Forest Landscapes: Integrating Multidisciplinary Perspectives*. University of Washington Press, US.

Bradshaw, A. D., Hunt, B., and Walmsley, T. (1995). *Trees in the Urban Landscape: Principles and Practice*. E & FN Spon, London.

Carreiro, M., Song, Y-C. and Wu, J. (2008). *Ecology, Planning and Management of Urban Forests*. Springer, New York.

Conway, T. T. and Urbani, L. (2007). Variations in municipal urban forestry policies: A case study of Toronto, Canada. *Urban Forestry and Urban Greening*, 6(3), 181–192.

Craul, P. J. (1994). Urban soils: An overview and their future. In G. W. Watson and D. Neely (Eds), *The landscape below Ground* (pp. 115–125). International Society of Arboriculture, Savoy, IL, US.

Daily, G. C. (1997). Introduction: What are ecosystem services? In: Daily, G. C. (Ed.), *Nature's Services: Societal Dependence on Natural Ecosystems*. Island Press, Washington, D.C., pp. 1–10.

Daniel, T. C., Muhar, A., Arnberger, A., Aznar, O., Boyd, J. W. and others (2012). Contributions of cultural services to the ecosystem services agenda. *Proceedings of the National Academy of Sciences*, 109(23), 8812–8819.

de Groot, R. S., Alkemade, R., Braat, L., Hein, L. and Willemen, L. (2010). Challenges in integrating the concept of ecosystem services and values in landscape planning, management and decision making. *Ecological Complexity* 7, 260–272.

Dobbs, C., Kendal, D. and Nitschke, C. R. (2014). Multiple ecosystem services and disservices of the urban forest establishing their connections with landscape structure and sociodemographics. *Ecological Indicators*, 43, 44–55.

Dwyer, J. F., Schroeder, H. and Gobster, P. (1991). The significance of urban trees and forests: Toward a deeper understanding of values. *Journal of Arboriculture*, 17(10), 276–284.

Dwyer, J. F., Nowak, D. J., Noble, M. H. and Sisinni, S. (2000). *Connecting People with Ecosystems in the 21st Century: An Assessment of our Nation's Urban Forests*. Gen. Tech. Rep. PNW-GTR-490. US Department of Agriculture, Forest Service, Pacific Northwest Research Station, Portland, OR, US.

Forman, R. T. T. (2014). *Urban Ecology: Science of Cities*. Cambridge University Press, UK.

Grey, G. W. (1996). *The Urban Forest: Comprehensive Management*. John Wiley & Sons, New York.

Grey, G. W. and Deneke, F. J. (1986). *Urban Forestry*. John Wiley & Sons, New York.

Grimm, N. B., Grove, J. M., Pickett, S. T. A. and Redman, C. A. (2000). Integrated approaches to long-term studies of urban ecological systems. *BioScience*, 50: 571–584.

Grimm, N., Faeth, S., Golubiewski, N., Redman, C., Wu, J. and others (2008). Global change and the ecology of cities. *Science*, 319(5864), 756–760.

Heynen, N. C. and Lindsey, G. (2003). Correlates of urban forest canopy cover: implications for local public works. *Public Works Management and Policy*, 8(1), 33–47.

Hope, D., Gries, C., Zhu, W., Fagan, W. F., Redman, C. L. and others (2003). Socioeconomics drive urban plant diversity. *Proceedings of the National Academy of Sciences of the United States of America*, 100(15), 8788–8792.

Iverson, L. R. and Cook, E. A. (2000). Urban forest cover of the Chicago region and its relation to household density and income. *Urban Ecosystems*, 4(2), 105–124.

Ives, C. D. and Kendal, D. (2014). The role of social values in the management of ecological systems. *Journal of Environmental Management*, 144, 67–72.

Jim, C. Y. and Chen, W. (2009). Ecosystem services and valuation of urban forests in China. *Cities*, 26(4), 187–194.

Jim, C. Y. and Chen, W. (2010). External effects of neighbourhood parks and landscape elements on high-rise residential value. *Land Use Policy*, 27(2), 662–670.

Kaplan, R. and Kaplan, S. (1989). *The Experience of Nature: A Psychological Perspective*. Cambridge University Press, New York.

Kendal, D., Williams, N. S. G. and Williams, K. J. H. (2012a). A cultivated environment: Exploring the global distribution of plants in gardens, parks and streetscapes. *Urban Ecosystems*, 15, 637–652.

Kendal, D., Williams, N. S. G. and Williams, K. J. H. (2012b). Drivers of diversity and tree cover in gardens, parks and streetscapes in an Australian city. *Urban Forestry and Urban Greening*, 11(3), 257–265.

Kendal, D., Dobbs, C. and Lohr, V. I. (2014). Global patterns of diversity in the urban forest: Is there evidence to support the 10/20/30 rule? *Urban Forestry and Urban Greening*, 13(3, 411–417.

Kenney, W. A., Wassenaer, P. J. E. and Van Satel, A. L. (2011). Criteria and indicators for strategic urban forest planning and management, *Arboriculture and Urban Forestry*, 37(3), 108–117.

Konijnendijk, C. C., Nilsson, K., Randrup, T. B. and Schipperijn, J. (2005). *Urban Forests and Trees*. Springer, Berlin.

Konijnendijk, C. C., Ricard, R. M., Kenney, A. and Randrup, T. B. (2006). Defining urban forestry – A comparative perspective of North America and Europe. *Urban Forestry and Urban Greening*, 4:93–103.

Kowarik, I. and Körner, S. (2005). *Wild Urban Woodlands*. Springer, Berlin.

Loeb, R. E. (2011). *Old-Growth Urban Forests*. Springer, New York.

Lohr, V. I. and Pearson-Mims, C. H. (2004). How urban residents rate and rank the benefits and problems associated with trees in cities. *Journal of Arboriculture*, 30(1), 28–35.

McDonnell, M. J. (2011). The history of urban ecology. In: Niemelä, J., J. H. Breuste, G.Guntenspergen, N. E. McIntyre, T. Elmqvist, and P. James (Eds) *Urban Ecology: Patterns, Processes, and Applications*. Pp. 5–13. Oxford University Press, Oxford, UK.

McDonnell, M. J. and Pickett, S. T. A. (1990). Ecosystem structure and function along urban-rural gradients : An Unexploited Opportunity for Ecology. *Ecology*, 71(4), 1232–1237.

McDonnell, M. J. and Hahs, A. K. (2008). The use of gradient analysis studies in advancing our understanding of the ecology of urbanising landscapes: Current status and future directions. *Landscape Ecology* 23, 1143–1155.

McDonnell, M. J., Hahs, A. K. and Breuste, J. (2009). *Ecology of Cities and Towns: A Comparative Approach*. Cambridge University Press, Cambridge, UK.

McPherson, G. (1994). Cooling urban heat islands with sustainable landscapes. In R. H. Platt, R. A. Rowntree, and P. C. Muick (Eds), *The Ecological City: Preserving and Restoring Urban Biodiversity*. The University of Massachusetts Press, US.

Miller, R. W. (1988). *Urban Forestry: Planning and Managing Urban Greenspaces*. Prentice Hall, Englewood Cliffs, N.J., US.

Mitchell, R. and Popham, F. (2007). Greenspace, urbanity and health: Relationships in England. *Journal of Epidemiology and Community Health*, 61(8), 681–683.

Nowak, D. J. and Dwyer, J. F. (2007). Understanding the benefits and costs of urban forest ecosystems. In J. E. Kuser (Ed.), *Urban and Community Forestry in the Northeast*. Springer, New York.

Nowak, D. J., Crane, D. E., Stevens, J. C., Hoehn, R. E., Walton, J. T. and Bond, J. (2008). A ground-based method of assessing urban forest structure and ecosystem services. *Arboriculture and Urban Forestry*, 34(6), 347–358.

Perkins, H., Heynen, N. and Wilson, J. (2004). Inequitable access to urban reforestation: The impact of urban political economy on housing tenure and urban forests. *Cities*, 21(4), 291–299.

Pickett, S. T. A., Cadenasso, M. L., Grove, J. M., Nilon, C.H., Pouyat R. V. and others (2001). Urban ecological systems: Linking terrestrial ecology, physical and socioeconomic components of metropolitan areas. *Annual Review of Ecology and Systematics*, 32, 127–157.

Reid, W. V, Mooney, H. A. and Cropper, A. (2005). *Millennium Ecosystem Assessment: Ecosystems and Human Well-Being—Synthesis Report*. Washington, DC.

Rowntree, R. (1998). Urban forest ecology: Conceptual points of departure. *Journal of Arboriculture* 24, 62–71.

Santamour, F. S. (1990). Trees for urban planting: Diversity, uniformity, and common sense. *Proceedings of the Seventh Conference of the Metropolitan Tree Improvement Alliance (METRIA)*, 57–65.

Schmid, J. A. (1975). *Urban Vegetation: A Review and Chicago Case Study*. University of Chicago, Chicago, US.

Schroeder, H. (1982). Preferred features of urban parks and forests. *Journal of Arboriculture*, 8(1), 317–322.

Schroeder, H., Flannigan, J., and Coles, R. (2006). Residents' attitude toward street trees in the UK and US communities. *Arboriculture and Urban Forestry*, 32(5), 236–246.

Sommer, R. (2001). The dendro-psychoses of J.O. Quantz. *Journal of Arboriculture*, 27(1), 40–43.

Sommer, R., and Summit, J. (1995). An exploratory study of preferred tree form. *Environment and Behavior*, 27(4), 540–557.

Tratalos, J., Fuller, R., Warren, P., Davies, R. and Gaston, K. (2007). Urban form, biodiversity potential and ecosystem services. *Landscape and Urban Planning*, 83(4), 308–317.

Tyrväinen, L., Silvennoinen, H. and Kolehmainen, O. (2003). Ecological and aesthetic values in urban forest management. *Urban Forestry and Urban Greening*, 1(3), 135–149.

Tyrväinen, L., Mäkinen, K. and Schipperijn, J. (2007). Tools for mapping social values of urban woodlands and other green areas. *Landscape and Urban Planning*, 79(1), 5–19.

Ulrich, R. S. (1984). View through a window may influence recovery from surgery. *Science*, 224(4647), 420.

Watson, G. W. and Himelick, E. B. (1997). *Principles and Practice of Planting Trees and Shrubs*. International Society of Arboriculture, Savoy, IL, US.

Williams, K. (2002). Exploring resident preferences for street trees in Melbourne, Australia. *Journal of Arboriculture*, 28(4), 161–170.

INDEX

abiotic conditions 424–5
aboriginal values 586
abundance *see* species abundance
acclimation 518
acorns 37–8
advance regeneration 81
aerial habitat 201–6
aerial seedbanks 120
aerosols 437
afforestation 346
Africa: hunting 610–11, 614, 617, 619; subtropical forests 46, 48–9, 51; tropical forests 57, 63–6
Africanized bees 452, 461
Afrosoricida 267, 270
Afrotheria 267, 269–70
agriculture 571–86; biodiversity 578–80; biomass accumulation 573–5; boreal forests 25–6; carbon cycles 580; climate change 580; condemnation issues 581–2; destroyers/gardeners 582–3; dynamics 577–80; environmental impacts 577–80; fallow ecology 571–7; fertility renewal 573–5; political dimensions 581–3; rational 571–86; restoration 397, 401, 405, 409; rotation 573–5; social dimensions 581–3; sustainability 577–8
air coolers/humidifiers 208–9
air pollution 177, 436–7
Alamgir, Mohammed 127–40
Alaska 7, 11–13, 149–50

Alberta 11–15
Aleutian Island chains 8, 11–13
alga 250, 252–5, 258
Algonquin communities 588–90, 593–4
Ali, Adam A. 473–87
alien species 275, 452–6
allocation 352–3, 361–5
Altai–Sayan region 17
altitudes 57, 60, 215–16, 303–4
Amazon die-back 544, 550–3
Amazon rainforest 599
AMO *see* Atlantic Multidecadal Oscillation
amphibians 297–8, 301–3
Amur leopards/tigers 26
Amur River 18–19
ancient shield geology 62
animals: biodiversity 296–8, 301–4; boreal forests 25–6; fire effects 117–20, 124; indigenous knowledge 589; lianas 187, 192–3; lichens 255–7; northern temperate forests 37–43; vascular epiphytes 204, 208, *see also* mammals
Anticosti Island 377
ants 206–7, 452
apex predator hunting 611–12
arboreal ants 207
arborists 623–4, 630
arbuscular mycorrhiza 310–21
Argentina 46, 51
artificial regeneration 80–1
Artiodactyla 267, 273–4

Index

Asia: birds 279–81, 286; boreal forests 7, 11–12, 16–20; cyclones/winds 127–8, 131, 136; northern temperate forests 30–1, 33, 36–43; subtropical forests 48–50, 52–3; tropical forests 57, 63–6
aspen 377
Asselin, Hugo 586–96
assisted migration 522–3
Atlantic, boreal forests 7–8, 11–12, 15–16
Atlantic Multidecadal Oscillation (AMO) 60
atmosphere 488, 537–40, 547–8
atmosphere–ocean general circulation models 547–8
Australia: indigenous knowledge 588; recreational activities 597–8, 602–7; subtropical forests 46, 48, 52; tropical cyclones 130–1, 133–6; tropical forests 57, 60, 62–5
autecology 567
auto-cyclic stand replacement 23–4
autotrophic green alga 250, 252–5, 258
autotrophic respiration 353–4
avian species *see* birds

bacteria 226–36
Baker-Fedorov hypothesis (BFH) 159
Baldwin, Ken 7–29
Banin, Lindsay F. 56–74
bark beetles 94, 98, 100–3, 106–7, 462
basal area change 173–9, 494
bats 264, 267, 271, 273
Battisti, Andrea 215–25
bauxite mining 407
beavers 452, 591–2
beech bark disease 230
bees 38, 452, 461, 589
below ground hydrology 340–1
below ground plot-based evidence 176–8
BEM *see* bioclimatic envelope models
Benzing, David H. 198–214
Berbers 587
Bering Strait 16–17
Bernier, Louis 226–38
Berninger, Frank 352–65
BFH *see* Baker-Fedorov hypothesis
biases, plot-based evidence 173–4
biochemical cycles 325, 327–8
bioclimatic envelope models (BEM) 545–6

biodiversity: agriculture cultivation 578–80; amphibians 297–8, 301–3; animals 296–8, 301–4; birds 296–8, 301–4; boreal forests 297, 299–300, 302–4; bryophytes 239–42; classification 295, 299–301; climate change 295, 299; conifers 295, 297–8, 300–1, 303–5; criteria 295–7; data 295, 297–8; deforestation/forest degradation 385–6, 388, 390–5; fragmentation 411–21; gap dynamics 145–7; global patterns 295–306; indicators 295–7; indigenous knowledge 591; insects 216–17; IUCN 297–301; logging 425–6; mammals 297–8, 301–2; managed forests 77, 88–9; non-timber forest products 560, 565–8; pathogens/pests 226, 233–6; photosynthesis 360; plants 295–305; *Red List of Threatened Species* 297–8; soil types 304; species data 295, 297–8; strong wind effects 128–9, 132–3, 137; succession 146–7; temperate forests 297, 300–5; tropical forests 64–6, 295, 297, 299–305; urban forests 627; vascular plants 301–4; vertebrates 297–8
biofuel production 246–7
biogeochemistry: boreal forests 328; changes over time 328–32; climate change 545, 547–9; cycling 325–38; demand 330–1; fungi 332–6; lichens 254–5; mycorrhizal fungi 332–6; nutrients 325–38; plant–soil interactions 332–6; retrogression 335–6; services 336; soil–plant interactions 332–6; stand development 328–30; succession 335–6; supply and demand 330–1; temperate forests 328; tropical forests 328
biogeography 30, 48–52, 64–6
biological features, pathogens/pests 227–9
biological invasions 452–69; birds 452, 462–3; conservation 452–69; earthworms 452, 456–8; future directions 466; insects 452, 458–62; mammals 452, 463, 465–6; management 452–69; nematodes 452, 456–8; plants 452–6; reptiles 452, 462–3; slugs/snails 452, 456–8
biomass: agriculture 573–5; boreal forest droughts 489–96; lianas 186, 192–3; production 528–9
biophysics, climate change 545, 547–9

biosphere feedbacks 488
biota, vascular epiphytes 206–8
biotic responses 334, 427, 444–5
birds 279–94; biodiversity 296–8, 301–4; biological invasions 452, 462–3; carnivores 282; climate change 287; conservation 287–8; deforestation 285–6; diseases 286–7; dispersal agents 284; diversity 279–80, 283; feeding 280–2, 284; fire effects 119–20; forest degradation 285–6; fragmentation 285–6; future prospects 287–8; infectious diseases 286–7; insect disturbances 99; insectivory 281–2, 285; invasive species 286–7; migratory stopover habitat 283; neotropics 279–80, 285; nesting sites 282–3; nutrients/nutrition 280–2, 284; over-exploitation 286; rainforests 279–81; refugia 283; seed dispersal 284; shelter 280–2; Southeast Asia 279–81, 286; subtropical forests 279–81, 286; sustenance 280–2, 284; thermal refugia 283; threats 285–7; tropical forests 279–87; urbanization 287; wintering sites 283
black spruce 500–12
Blarquez, Olivier 473–87
boreal forests 7–29; animals 25–6; autocyclic stands 23–4; biodiversity 297, 299–300, 302–4; biogeochemical cycling 328; carbon budgets 529–37, 539; climate change 26–7, 474–7, 480–4, 488–98, 500–12; climates 9–11, 25–7; cold temperature adaptation 20–1; conifers 9–10; cyclic disturbances 24–5; cyclones/winds 128, 131, 134; droughts 488–98; dynamic patterns 23–5; fire 21, 24, 474–7, 480–4; floristic subdivisions 11–20; gap dynamics 146; genetic diversity 154–5, 157, 159, 161–2, 164–5; harvesting 371, 375–7; human use 25–6; hunting 610–13; insects 21–2, 24, 95, 107; lichens 252–3, 256–7; major disturbances 23–5; mixture maintenance 24–5; natural disturbances 21–2; natural regeneration 371, 375–7; northern temperate forests 30–1, 33, 39; permafrost 22–3; photosynthesis 358; physiological strategies 20; plot-based evidence 175, 177–80; soil conditions 22–3, 25; species adaptation/survival 20–3; succession 24, 149–50; temperature adaptation 20–1; threats 25–7; transition zones 9; tree growth responses 500–12; vegetation 11–20; zone extent 7–9
boring beetles 94, 101–2
Brazil 46, 51, 66
Bredemeier, Michael 339–51
British Columbia 13–15
brushtail possum 452
bryophytes 239–49; biodiversity 239–42; biofuel production 246–7; biomes 239, 241–5; climate change 247; continental diversity 239–41; disturbances 245; fire 245; fragmentation 245–6; function variation 242–5; global diversity 239–41; habitat 245–7; harvesting 245–7; human impacts 245–6; landscape diversity 241; nutrient cycling 244–5; pulp production 246–7; retention capacity 242–5; timber production 245–7; water retention 242–3
budmoths 220–1
budworms 21, 94–107, 502–4
bushmeat 274, 563, 613–14
buttress creep/trees 174

calcium 325–30, 358–9
cambial activity 499–502, 505–8, 510–12
Cameroon 593
Canada: boreal forests 7, 11–15, 488–96, 500–12; climate change/tree growth 500–12; droughts 488–96; fire dynamics 116–17; indigenous knowledge 588–91, 593–4; insect disturbances 93–5, 97
Canary Islands 49–50
carbohydrates 361, 364–5
carbon: biogeochemical cycling 331–5; cycles 68–9, 77–8, 81, 87–9, 580; lianas 186, 192; plant movement/climate change 517–18; productivity/residence times 531–6; sequestration 333–4, 389–90; stocks, logging 427–9; storage 389–90
carbon balance/budgets: boreal forests 488–96, 529–37, 539; climate change 527–43; disturbance regimes 534–6, 539; droughts 488–96; human-modified forests 529–30; imbalance causes 537–8; intact-modified forests 529–30; key concepts

527–9; logging 427–9; photosynthesis 353–4; soil 530–1, 533, 535–6; temperate forests 529–34, 537–9; tropical forests 529–39; vegetation 530–1, 533, 535
carbon dioxide: carbon budgets 527–8, 536–40; climate change 517–18, 527–8, 536–40, 551–3; deforestation/forest degradation 385, 389–90, 394; fires 390; forest conversion 389; insects 222–4; lianas 194; plant movement 517–18; plot-based evidence 176–80; succession 149–51; vascular epiphytes 210–11
caribou 256–7
carnivores 264, 267, 272–3, 282, 610–12, 615–16
cash income 562–4
cash savings 563–4
catchment hydrology 345–6
caterpillars 98, 106
cats 264, see also tigers
causality, pollution 444–8
censusing 173
Central African tropical forests 57
Central American tropical forests 57
Central Siberian boreal forests 11–12, 18
change dynamics, plot-based evidence 173–80
changes over time, biogeochemistry 328–32
characteristics, lianas 185
characterization, fires 114–17
charcoal accumulation 477, 479–80
chemical warfare 208
Chilean subtropical forests 51
Chinese subtropical forests 46, 48–50, 52
Chiroptera 267, 273
Christoersen, Bradley O. 544–5
Cingulata 267, 269
classic management methods 77–81
classification: biodiversity 295, 299–301; climate change modelling 544–5; insects 217
clear-cutting 82–6, 423
Clements, Frederick E. 141, 144
climate: biogeochemical cycling 327; boreal forests 9–11, 25–7; insect disturbances 96–9, 106–7; lianas 186–91; lichens 258; northern temperate forests 30–1; pollution 445–7; subtropical forests 46–8; tropical forests 56–61, 64–6; vascular epiphytes 201–4, 208–12
climate change: agriculture cultivation 580; bioclimatic envelope models 545–6; biodiversity 295, 299; biogeochemistry 545, 547–9; biophysics 545, 547–9; birds 287; boreal forests 26–7, 474–7, 480–4, 488–98, 500–12; bryophytes 247; carbon budgets 527–43; deforestation 385–96, 539; development model frontiers 553–4; droughts 488–98, 501–2, 506–8, 511–12; field observations 544, 553; fires 473–87, 539; forest degradation 385–96, 539; fuel 474–6; future perspectives 473–87, 538–40; gap models 545–7; genetic diversity 164–6; global carbon budgets 536–7; growth 499–516; human activities 473–4, 481, 484; insects 106–7, 222–4, 502–4; invertebrate pests 235; modelling 521, 544–5; mycorrhizal symbiosis 318–19; niche-based models 545–6; North America 474–85; northern temperate forests 42–3; pests 235; plant functional types 548–50, 553; plant movement responses 517–26; pollution 445–7, 538; REDD+ 69, 385–96; restoration 397–8; species distribution models 545–6; subtropical forests 53, 385–90, 394; succession 149–51, 545–7; temperate forests 474, 481; temperature 500–12, 517–23, 527, 531–8; tree growth 499–516; tropical forests 60–1, 385–96; uncertainty 544, 550–3; vascular epiphytes 209–12; vegetation 475, 545, 547–54; winds 136–7
close-to-nature forestry 429–30
cloud cover 60
cloud forests 208–10
cloud water capture 209
coastal forests 135–6
Cohen, Shabtai 339–51
cold temperature adaptation 20–1
Coll, Lluís 371–84
Commander Islands 19–20
community dynamics, lichens 258–60
community interactions, fragmentation 417–18
community-level effects, pollution 440–1

competitive effects, lianas 185–6, 191–3
complexity, succession 148–9
compositional change, succession 141–4
composition/dynamics, non-timber products 560–1
conifers: biodiversity 295, 297–8, 300–1, 303–5; boreal forests 9–10; budworms 94–5, 97–100, 102–7; harvesting 371, 375–7, 380–1; managed forests 75; natural regeneration 371, 375–7, 380–1; northern temperate forests 35–6
Connell and Slatyer's models 144
conservation: biological invasions 452–69; birds 287–8; deforestation/degradation 385–96; fragmentation 411–21; genetic diversity 164; harvesting 371–84; indigenous knowledge 591; lichens 260–1; logging 422–35; managed forests 77, 88–9; management 411–21; natural regeneration 371–84; non-timber products 559–61, 565–8; northern temperate forests 39–43; pollution 436–51; regeneration 371–84; restoration 397–410; tropical forests 385–96; urban forests 627
continental diversity 239–41
conventional logging approaches 423–4
Cooke, Barry 93–113
coppicing 78–9, 83–7
Cordilleran foothills 13–15
Corlett, Richard T. 46–55, 264–78, 517–26
corridor fragmentation 418–19
Cowles, Henry Chandler 141–2
Coxson, Darwyn 250–63
criteria: biodiversity 295–7; indigenous knowledge 593–4
crown fires 116–20, 475–7
crown injuries 117–19
cultivation: agriculture 571–86; managed forests 76
culture: indigenous knowledge 586–94; non-timber products 565; urban forests 626
cyanobacterium 250, 253–5, 258
cyclic disturbances 24–5, 95–6
cycling regulation, ectomycorrhizal fungi 334–6
cycling within organisms 325–38
cyclones 127–37, see also tropical cyclones

Dasyuromorphia 267–8
data: biodiversity 295, 297–8; boreal forest droughts 489–91; climate change modelling 553
dating sediments 480
dead wood 260
decay processes 33–4
deciduous forests 280, 371, 378–9
decision-support tools 593
decomposition rates 244
DED *see* Dutch elm disease
defoliators 98–103, 106–7
deforestation 69, 385–96; birds 285–6; carbon budgets 539; climate change 385–96, 539; conservation 385–96; degradation 385–96; forest degradation 385–96; REDD+ 69, 385–96; subtropical forests 385–90, 394; tropical forests 385–96
degradation *see* forest degradation
DeGrandpre, Louis 93–113
demand, biogeochemical cycling 330–1
demand side management 601, 604
demography, agriculture cultivation 577–8
dendrochronology 500–4, 511–12
density, lianas 186–91
Dermoptera 267, 271
DGVM *see* dynamic global vegetation models
Didelphimorphia 267, 269
differentiation, genetic diversity 155–61
Diprotodontia 267–8
direct seeding 400–2, 407–8
diseases 42, 230–1, 286–7
dispersal agents 228, 284, 399–409
distribution: lianas 186–91; lichens 252–4; mammals 264–6; mycorrhizal symbiosis 313–15; plant movement/climate change 521–2
disturbances: bryophytes 245; carbon budgets 534–6, 539; harvesting 372–3; logging 429–30; mycorrhizal symbiosis 317–18; northern temperate forests 34–5; plot-based evidence 176, 178; restoration 397–400, 403, 405; temperate forests 187; tropical forests 57; wind dynamics 127–40, *see also individual disturbances*

Index

diversity: bats 264, 267, 271, 273; birds 279–80, 283; boreal forests 154–5, 157, 159, 161–2, 164–5; bryophytes 239–42; gap dynamics 145–7; global patterns 295–306; insects 216–17; lianas 185–94; lichens 250–4; mammals 264; managed forests 77, 88–9; photosynthesis 360; primates 264, 267, 271; rodents 264, 271–2; strong wind effects 128–9, 132–3, 137; succession 146–7; tropical forests 64–6, 154–5, 159, 161, 163–4; vascular epiphytes 200–1, *see also* biodiversity; genetic diversity
DNA, genetic diversity 155–6, 160–2
domestic livestock overgrazing 40
drift, genetic diversity 155–6, 159
droughts: boreal forests 488–98; climate change 488–98, 501–2, 506–8, 511–12; hydrological cycling 343–5; northern temperate forests 43; plant allocation 363–4; succession 150; tree growth 501–2, 506–8, 511–12; vascular epiphytes 200–1, 203–4, 209–10
dry biomass 352
dry Mediterranean ecosystems 371, 379–81
Ducourtieux, Olivier 571–86
dust pollution 437
Dutch elm disease (DED) 230, 234, 462
dynamic global vegetation models (DGVM) 521, 545, 547–53
dynamics: agriculture 577–80; boreal forests 23–5; fires 114–26; gap dynamics 141–53; gene flows 154–71; insects 93–113; lichens 258–60; non-timber products 560–1; plot-based evidence 172–82; succession 141–53; tree genetic diversity 154–71; winds 127–40

early blooming species 36–7
earthworms 235, 452, 456–8
East African tropical forests 57
East Asian subtropical forests 48–50
Eastern Canadian droughts 494–5
Eastern North American boreal forests 11–12, 15–16
ecological effects: cyclones/winds 127–37; lianas 185–97; logging 422–35; northern temperate forests 33–9; pathogens/pests 229–31; urban forests 624–7
Ecological Restoration 397–8, 404, 406–7
economics/economy: non-timber products 559–70; REDD+ 393–4; urban forests 624–5
ecosystem assembly 402–4
ecosystem services 219, 624–5
ectomycorrhiza 310–21, 332–6
edge effects 173, 415–16
Edwards, David P. 422–35
elephants 616–17
El Niño Southern Oscillation (ENSO), tropical forests 58–60
emissions 69, 385–96, *see also* climate change; pollution
energy: hydrological cycling 339–51; mycorrhizal symbiosis 309–24; primary production 352–65
ENSO *see* El Niño Southern Oscillation
environmental impacts: agriculture 577–80; fragmentation 417; indigenous knowledge 591–4; insects 221–4; photosynthesis 355–60; REDD+ 69, 385–96; urban forests 624–5
environmental justice 629–30
epidemics 230
epigenetic effects 165
epiphytism 198–214, 250–61, *see also* lichens
ericoid mycorrhiza 310–15, 318–19
Esseen, Per-Anders 250–63
Euarchontoglires 267, 270–2
eucalyptus forest 401, 407
Eulipotyphla 267, 272
Eurasia 94–5
Europe: boreal forests 7, 11–12, 16; cyclones/winds 127–8, 131, 133–6; insects 93–4, 98, 100, 102, 106–7; northern temperate forests 30–4, 36, 38–43; subtropical forests 49
evapotranspiration 341–2, 348–9
evergreen species 35–6
evolutionary consequences, hunting 618–19
expertise investment 567
exploitation, birds 286

fallow ecology 571–7
farming *see* agriculture

Faroe Islands 8
feeding regimes 217–19, 228–9, 280–2, 284
felling 572
Fenton, Nicole J. 239–49
fertility renewal 573–5
fertilization 67–8
field observations, climate change 544, 553
finance investments 567
fine roots 362
Finland 7
fires: agriculture cultivation 573–4; effects on animals 117–20, 124; boreal forests 21, 24, 474–7, 480–4; bryophytes 245; carbon budgets 539; carbon dioxide 390; characterization 114–17; climate change 473–87, 539; dynamics 114–26; flaming combustion 114–16, 119; fuel 474–6; harvesting/natural regeneration 376; human activities 115, 473–4, 481, 484; indigenous knowledge 591; insect disturbances 103; lightning 115–16, 232, 474–5; modelling 480–4; mycorrhizal symbiosis 317; North America 474–85; northern temperate forests 42; pathogens/invertebrate pests 230, 232; effects on plants 117–24; recolonization 120–4; role of 114–26; smouldering combustion 114–16, 119–23; temperate forests 115, 474, 481; tropical forests 115; winds 115
firewood 572
fishing 597, 599
flaming combustion 114–16, 119
flood control 346–8
floristic subdivisions 11–20
flow controls, water 345–6
forage lichens 256–7
forest conversion 39–40, 389–90
forest degradation 69, 385–96; agriculture cultivation 577–8; birds 285–6; carbon budgets 539; climate change 385–96, 539; conservation 385–96; deforestation 385–96; management 385–96; REDD+ 69, 385–96; subtropical forests 385–90, 394; tropical forests 385–96
forest fragmentation see fragmentation
forest functioning: bryophytes 242–5; fragmentation 417–18; lianas 187; lichens 254–8; mammals 264; mycorrhizal symbiosis 315–17; strong winds 128–9, 131
forest gap models 35, 37, 141–53, 545–7
forest inventories 65, 567, 590–1
forest productivity 499–512
forest recovery 134
forest structure see structural impacts
fragmentation: biodiversity 411–21; birds 285–6; bryophytes 245–6; community interactions 417–18; conservation 411–21; corridors 418–19; ecosystem functions 417–18; edge effects 415–16; environmental change 417; functioning impacts 417–18; habitat impacts 415–17, 419; impacts 412–18; isolation effects 416; landscape-scale 411–12; large-scale projects 412–13; management 411–21; matrix impacts 416; northern temperate forests 39–40, 42; plot-based evidence 173; species–area relationships 411–12; species-species effects 417; strong wind effects 130, 137; synergistic impacts 417
framework species method, restoration 404–6
Frelich, Lee E. 30–45, 141–53
fuel 474–6
fungi: biogeochemical cycling 332–6; clones/spores 120, 123; lichens 250, 253, 260; mycorrhizal 309–22, 332–6; pathogens 226–8, 230–5
future perspectives: biological invasions 466; birds 287–8; climate change 473–87, 538–40; fire 473–87; genetic diversity 166–7; plot-based evidence 178–80; pollution research 448–9; urban forest research 631

Galbraith, David R. 544–5
gap models 35, 37, 141–53, 545–7
gases see greenhouse gases
Gauthier, S. 594
gene flows 154–71
genetic change 518
genetic diversity: boreal forests 154–5, 157, 159, 161–2, 164–5; climate change 164–6; conservation 164; differentiation 155–61;

DNA 155–6, 160–2; drift 155–6; future perspectives 166–7; geographic variation 162–4; harvesting 164; migration 156; mutation 156, 165; phenotypes 154–5, 157, 162, 164–6; phylogeographic patterns 161–2; pollen/pollination 156–9, 164–5; population 154–61; quantitative trait variation 162–4; seeds 156–66; selection 157; sources 155–7; trees 154–71; tropical forests 154–5, 159, 161, 163–4
genetic drift 155–6, 159
genetic variation sources 155–7
geochemical cycles 326–8
geographic occurrence: genetic diversity 162–4; insects 223; vascular epiphytes 200–1
geography, tropical forests 57–64
geology, tropical forests 61–4
germination 121–2
Girardin, Martin P. 473–87
global biodiversity 239–41, 295–306
global carbon cycles 68–9, 77–8, 81, 87–9, 536–7
global changes: boreal forest droughts 495–6; managed forests 87–9; succession 149–51; vascular epiphytes 209–12
global climate models 480–1, 521, 580
global distribution: lianas 186–91; mammals 264–6
governance, urban forests 627–8
GPP *see* gross primary production
granivory 123
grazing animals 37–43
green alga 250, 252–5, 258
Greene, David F. 114–17
greenhouse experiments 509–10
greenhouse gases 437; boreal forests 26–7; carbon budgets 527–8, 536–40; deforestation/forest degradation 385–6, 389–94; plant movement 517–18; uncertainty 550–2
Greenland 8
Groot, Nikée E. 172–82
gross primary production (GPP) 352–7, 360, 365, 527–8, 531, 534–5
ground biomass 489–96
ground fires 116, 119, 123, 475
group selection, managed forests 83, 86

growth: boreal forest droughts 488–91, 494–6; climate change 499–516; inter-/intra-annual temporal scales 499–516; stem growth 364–5; tropical forests 67–8
gypsy moths 98, 102, 452, 461

habitat impacts 245–7, 415–17, 419
Halme, Panu 422–35
Harrison, Rhett D. 610–22
harvesting: agriculture cultivation 573; boreal forests 371, 375–7; bryophytes 245–7; conifers 371, 375–7, 380–1; conservation 371–84; deciduous ecosystems 371, 378–9; disturbances 372–3; genetic diversity 164; insects 106; managed forests 82–9; management 371–84; Mediterranean ecosystems 371, 379–81; mycorrhizal symbiosis 318; northern temperate forests 39–40; regeneration 371–84; temperate forests 371, 378–9
health/vitality, managed forests 77–9, 84–5, 87–9
Hély, Christelle 473–87
herbivores 275, 317, 458–62, 611, 615–17
heterotrophic fungus 250, 253, 260
heterotrophic respiration 353–4
high diversity plantings 404–6
hiking 597–9, 604
history: fire/climate change 473–87; logging 422–3; succession 141–2
hornworts 239
horticulturalists 623–4, 631
host interactions 222–3, 226–36
household provisioning 563–4
Huang, Jian-Guo 499–516
Hughes, Alice C. 46–55, 264–78
human activities/impacts: boreal forests 25–6; bryophytes 245–6; carbon budgets 529–30; fires 115, 473–4, 481, 484; insects 215; mammals 274–6; northern temperate forests 33
human ecology: agriculture 571–86; hunting impacts 610–22; indigenous knowledge 586–96; non-timber products 559–70; recreation 597–609; REDD+ 393–4; urban forests 623–33; well-being 393–4
hunting 610–22; boreal forests 610–13; consequences 611–13, 615–18; population

declines 616–19; predators 611–12, 617–18; recreational activities 597, 599–600; subtropical forests 613–19; temperate forests 610–13; tropical forests 613–19
hurricanes 127–37
Huu-ay-aht First Nation traditional forest management 587
hydrological cycling 339–51; below ground 340–1; catchment hydrology 345–6; drought 343–5; evapotranspiration 341–2, 348–9; flood control 346–8; leaf area index 343–4; mesoscale levels 342–8; trees 340; underground processes 340–1
Hylander, Kristoffer 239–49
hypersensitivity 209–10
Hyracoidea 267, 270

Iceland 8
imbalance causes, carbon budgets 537–8
immature forest bias 174
income generation 562–5
indicators: biodiversity 295–7; indigenous knowledge 593–4; lichens 260–1
indigenous forest knowledge 586–96; human ecology 586–96; inventories 590–1; key contributions 589–94; non-timber products 590–1; scientific knowledge 586–95
indirect logging effects 429
industrial pollution 436–51
industrial uses, boreal forests 25–6
infectious diseases 286–7
inflation 174
infrastructure, recreational activities 602–4
injury, plants by fire 117–19
inoculation 319–21
insectivores 264
insects 215–25; abundance 219–21; bark beetles 94, 98, 100–3, 106–7; biodiversity 216–17; birds 99, 281–2, 285; boreal forests 21–2, 24, 95, 107; budworms 94–5, 97–100, 102–7; Canada 93–5, 97; carbon dioxide 222–4; classification 217; climate 96–9, 106–7; climate change 106–7, 222–4, 502–4; defoliators 98–103, 106–7; disturbances 93–113; diversity 216–17; dynamics 93–113; environmental factors 221–4; Eurasia 94–5; Europe 93–4, 98, 100, 102, 106–7; feeding 217–19; fires 103; geographic range 223; human impacts 215; invasions 101–2, 452, 458–62; landscape effects 106, 108; lichens 255; management effects 103–6; nitrogen 222–4; North America 94–5, 97–8, 100–2, 106–7; northern temperate forests 37–8; nutrients 103, 222–4; outbreaks 95–102, 219–21, 223; pattern classification 95–102; pests 95–102; population dynamics 95–102, 219–21, 223; predators 96–8, 100–1, 107, 220–1, 223; specialization 217–18; tree growth 502–4; tropical forests 94–5; United States 94–5, 97–8, 100–2, 106–7
intact forest carbon budgets 529–30
inter-annual seasonality 58–60
inter-annual temporal scales 499–516
international market chains 565–6
International Union for Conservation of Nature (IUCN) 297–301
interpreting tree growth/climate change responses 511–12
inter-rotational methods 87
intolerant plant species interactions 35–7
intra-annual seasonality 58–60
intra-annual temporal scales 499–516
invasive species: birds 286–7; insects 101–2, 452, 458–62; northern temperate forests 40–3; pathogens/pests 233–5
inventories 65, 567, 590–1
invertebrates 226–38, 255
investments, non-timber products 567
isolation effects 416
IUCN *see* International Union for Conservation of Nature

Jacobs, Douglass F. 371–84
Japanese boreal forests 19–20
Jorgenson, Torre 7–29
JULES-TRIFFID model 548–53
Jürgen Bauhus 75–90
justice aspects 629–30

Kamchatka Peninsula 7–8, 11–13, 16–17, 19–20
Kendal, Dave 623–33

Kneeshaw, Daniel 93–113
Korean indigenous knowledge 589
Kozlov, Mikhail V. 436–51
Krestov, Pavel 7–29
Kuril 8, 13, 19–20

Labrador 7, 11–12, 15–16
Lagomorpha 267, 271
LAI *see* leaf area index
Lake Baikal 18–19
Lamb, David 397–410
land management 566–8
landscapes: bryophyte diversity 241; fragmentation 411–12; insect disturbances 106, 108; pollution 438–9; REDD+ 391–2; succession 143; urban forests 623–31
land surface models 547–8
land use changes/conversion 39–40, 211, 566–8
La Niña, tropical forests 60
large-scale fragmentation 412–13
latitude effects 188–9, 215–16
Laurasiatheria 267, 272–4
leaf area index (LAI) 343–4, 547–8
leopards 26, *see also* tigers
Lewis, Simon L. 56–74
lianas 185–97; abundance 186–94; anatomy 190–1; animal diversity 187, 192–3; biomass 186, 192–3; carbon accumulation 186, 192; carbon dioxide 194; characteristics 185; climate 186–91; competitive effects 185–6, 191–3; density 186–91; distribution 186–91; diversity 185–94; global distribution 186–91; latitudinal distribution 188–9; local tropical forests 187; negative effects 185–6, 191–3; neotropical forests 187, 193–4; nutrients 186, 190–1, 194; pan-tropical forests 187–8, 190; plant diversity 187, 192–3; positive effects 185–7, 191–3; precipitation 186–91, 194; productivity 193–4; rainfall 186–91, 194; roots 189–90; temperate forests 187–8; tree diversity 185–6, 190–4; tree performance 185–6; tropical forests 187–94; vascular epiphytes 198; vascular systems 188–91, 194

lichens 250–63; animals 255–7; autotrophic green alga 250, 252–5, 258; biogeochemical cycling 254–5; boreal forests 252–3, 256–7; climate 258; community dynamics 258–60; conservation 260–1; cyanobacterium 250, 253–5, 258; distribution 252–4; diversity 250–4; dynamics 258–60; forage feeders 256–7; functions 254–8; indicators 260–1; interactions 254–8; invertebrates 255; nutrient cycling 254–5; photobiont 250, 253–5, 258; temperate forests 253; tropical forests 254; vertebrates 255–6
light, photosynthesis 355–7
lightning, fires 115–16, 232, 474–5
limiting nutrient element changes 331–2
litter production 353
livelihoods, non-timber products 559–70
liverworts 239–42, 245
logging 422–35; abiotic conditions 424–5; biodiversity 425–6; biotic interactions 427; boreal forests 25–6; close-to-nature forestry 429–30; conservation 422–35; conventional approaches 423–4; cycles 424; forest structure 424–5; history 422–3; indirect effects 429; management 422–35; mitigating impacts 422–35; natural regeneration 376, 378; pathogens/pests 232; post-logging mitigation 432; reduced impact methods 431–2; retention forestry 430–1; significance 422–3; structural impacts 424–5
lowland tropical forests 279–80
Lund–Potsdam–Jena (LPJ) model 549–50, 552

McDonnell, Mark J. 623–33
MacLean, David A. 93–113
Macroscelidea 267, 269
Madagascar 46, 50–1, 57
magnesium 325–8, 334–6, 358
majestic forest bias 173
major disturbances 23–5, 34–5
Malaysian tropical forests 60
Malhi, Yadvinder 527–43
mammals 264–78; *Afrotheria* 267, 269–70; alien species 275; biodiversity 297–8, 301–2; biological invasions 452, 463,

465–6; carnivores 264, 267, 272–3; distribution 264–6; ecological roles 264, 267; *Euarchontoglires* 267, 270–2; global distribution 264–6; human impacts 274–6; importance 274; *Laurasiatheria* 267, 272–4; management issues 274–6; marsupial orders 268–9; monotremes 268; orders 264–74; overexploitation 274–5; phylogeny 264–5; reintroductions 275–6; restoration 275–6; role importance 274; *Xenarthra* 267, 269, see also animals

managed forests 75–90; biodiversity 77, 88–9; carbon cycles 77, 88–9; classic methods 77–81; conservation 77, 88–9; coppicing 78–9, 83–7; diversity 77, 88–9; global carbon cycles 77–8, 81, 87–9; global development 87–9; harvesting 82–9; health/vitality 77–9, 84–5, 87–9; inter-rotational methods 87; nutrient cycles 78, 81–4, 87; productive capacity 77–85, 88–9; pruning 81–2; regeneration 79–81, 87–8; shelterwood methods 78–81, 83, 85–7; silviculture 77–89; site preparation 82–7; socio-economic benefits 77–8, 87–9; stand level management 77–89; tending 81–2; thinning 81–2; water resources 77, 81–4, 87

management: biological invasions 452–69; forest degradation 385–96; fragmentation 411–21; harvesting 371–84; indigenous knowledge 586–96; insect disturbances 103–6; logging 422–35; mammals 274–6; pathogens/pests 227, 231–3; pollution 436–51; recreational activities 600–2; regeneration 371–84; restoration 397–410; tropical deforestation 385–96; tropical forests 57; urban forests 623–31

mangrove forests 135–6
Manitoba 13–15
mapping urban forests 627
Mapuche Pewenche peoples 590
market chains 565–6
Markewitz, Daniel 325–38
marsupials 268–9
Mediterranean ecosystems 371, 379–81, 506
Melanesian islands 57
Menominee tribe of Wisconsin 589–90
Merap villages 589

meristem activity 499–502, 505–8, 510–12
mesopredator release 275
mesoscale hydrological cycling 342–8
meta-networks, mycorrhizal symbiosis 316–17
Michaletz, Sean T. 114–17
Microbiotheria 267, 269
migration 156, 283, 522–3
mining 25–6, 407
Miocene forests 48–9
mitigating logging impacts 422–35
mixture maintenance 24–5
modelling: boreal forest droughts 490–6; climate change 521, 544–5; fires 480–4; plant movement 521; recreational activities 600–5
modern hunting practices 612–13
MODIS data 46, 266, 299–300, 355
moist forest geography 56–64
Mongolia 18–19
monitoring: restoration 407; tree growth/climate change 507–10; tropical forest inventory plots 65
Monotremata 267–8
monotremes 267–8
montane forests 535
Montgomery, Rebecca A. 30–45, 141–53
Moore, Sam 527–43
Morin, Hubert 499–516
Morocco 587
mortality 353, 488–96
mosses 120, 123, 239–41, 243–5, 247
moths 98, 234, 452, 461
multiple succession pathways 148–9
multiple values, non-timber products 559–68
mushroom production 321
mutation 156, 165, 206
mycobionts 250, 255
mycorrhizas 204, 309–24, 332–6; biogeochemical cycling 332–6; climate change 318–19; distribution 313–15; disturbance 317–18; functional roles 315–17; fungi 309–22, 332–6; symbiosis 309–24

national parks 604
natural disturbances 21–2, 317–18, 429–30

natural regeneration 371–81; boreal forests 371, 375–7; conifers 371, 375–7, 380–1; conservation 371–84; deciduous ecosystems 371, 378–9; harvesting 371–84; logging 376, 378; managed forests 79–81, 87–8; management 371–84; Mediterranean ecosystems 371, 379–81; restoration 398–400; seeds 371–81; temperate forests 371, 378–9
Nawash Ojibway 588
NBP *see* net biome production
'near-to-nature forestry' 429–30
nectar 210, 267, 273, 280–1
NEE *see* Net Ecosystem Exchange
nematodes 452, 456–8
neotropical forests 187, 193–4, 279–80, 285, 614
NEP *see* Net Ecosystem Production
nesting sites 282–3
net biome production (NBP) 353, 528–9
Net Ecosystem Exchange (NEE) 528–9
Net Ecosystem Production (NEP) 353, 528–9, 531–5, 537
Net Primary Production (NPP) 352–3, 357, 359, 361–2, 439–40, 527–8, 531–5
Newfoundland 7, 11–12, 15–16
New Guinea 57, 60
New Zealand 604
niche-based models 545–6
nitrogen: biogeochemical cycling 325, 327–36; ectomycorrhizal fungi 334–5; fixation 244, 254–5, 327, 329; insects 103, 222–4; mycorrhizal symbiosis 315; photosynthesis 357–8; vascular epiphytes 210–11
non-cash incomes 562–5
non-mycorrhizal plants 312–13, 315
non-native human-introduced taxa 452–69
non-timber forest products: autecology 567; biodiversity 560, 565–8; cash savings 563–4; conservation 559–61, 565–8; cultural importance 565; definitions 559–60; expertise investment 567; household provisioning 563–4; human ecology 559–70; importance of 560–1; income generation 562–5; indigenous knowledge 590–3; international market chains 565–6; inventories 567; investments 567; land management 566–8; multiple values 559–68; rural livelihoods 562–5; safety nets 564–5
non-wood forest products 560
North African subtropical forests 48
North America: bacteria 226–36; boreal forests 7, 11–16, 500–12; climate change 474–85, 500–12; fire 474–85; fungal pathogens 226–8, 230–5; harvesting/natural regeneration 375–7; insects 94–5, 97–8, 100–2, 106–7; pathogens/pests 226–38; subtropical forests 48–50; temperate forests 30–5, 38–41; tree growth 500–12; tropical cyclones 127–8, 131
Northeastern Siberian boreal forests 19
Northern Baikal Plateau 18–19
northern hemisphere subtropical forests 48–50
northern temperate forests 30–45; animals 37–43; biogeography 30; boreal forests 30–1, 33, 39; canopy structure dynamics 35; climate 30–1; climate change 42–3; conservation issues 39–43; decay processes 33–4; diseases 42; disturbance 34–5; ecological processes 33–9; fragmentation 39–40, 42; gap dynamics 35, 37; harvesting 39–40; human use 33; invasive species 40–3; land use conversion 39–40; major disturbance 34–5; nutrient cycling 33–4; overgrazing 40–1; pests 42; plant–animal interactions 37–43; plant species interactions 35–43; seasonality 30–1; site conditions 32; societal importance 33; soil conditions 32; subtropical forests 30–1, 39, 42; succession 34–5; tropical forests 39; zone extents 30
North Pacific 11–13
Norway 7, 11–12, 16, 135
NPP *see* Net Primary Production
nurse trees 404–5
nut dispersal 284
nutrients: biogeochemical cycling 325–38; birds 280–2, 284; bryophytes 244–5; insects 103, 222–4; lianas 186, 190–1, 194; lichens 254–5; managed forests 78, 81–4, 87; mycorrhizal symbiosis 309–24; northern temperate forests 33–4;

photosynthesis 357–9; plot-based evidence 176–80; strong winds 128–9, 131; vascular epiphytes 204, 207–8, 210–11
nutrition, birds 280–2, 284

Okhotsk–Kamchatka 19–20
'old-growth' forests 173–80
Oleksyn, Jacek 30–45
on-site biomass accumulation 573–5
Ontario 13–15
orchid mycorrhiza 310–14
organism distribution, lianas 186
organism-level effects, pollution 443
outbreaks, insects 95–102, 219–21, 223
over-exploitation 274–5, 286
overgrazing 40–1
ownership status 601
ozone 437–8

Pacific, boreal forests 8, 16–17, 19
paleoecological record 518
pan-tropical forests 187–8, 190
parametric uncertainty 552–3
parasitic plant pathogens 227–8
Paré, David 325–38
Parrotta, John A. 385–96
participatory management 602
patch habitats 415
pathogens: biodiversity 226, 233–6; biological features 227–9; climate change 235; ecological effects 229–31; forest management 231–3; host interactions 226–36; invasive species 233–5; mycorrhizal symbiosis 317; North America 226–38; traditional forest management 231–3
Paucituberculata 267, 269
peatland forests 535–6
Peh, Kelvin S.-H. 352–65
Peng, Changhui 488–98
perch trees 404–5
Perissodactyla 267, 273
permafrost 22–3
permanent sample plots (PSP) 489–91
pests: biodiversity 226, 233–6; biological features 227–9; climate change 235; ecological effects 229–31; forest management 231–3; host interactions 226–36; indigenous knowledge 591–2; insect disturbances 95–102; invasive species 233–5; North America 226–38; northern temperate forests 42; traditional forest management 231–3
PFT *see* plant functional types
Pharo, Emma J. 239–49
phenotypes 154–5, 157, 162, 164–6
Phillips, Oliver L. 56–74
Pholidota 267, 272
phorophytes 208
phosphorus 315–16, 325–36, 358
photobiont 250, 253–5, 258
photosynthesis: biodiversity 360; environmental effects 355–60; light 355–7; nitrogen 357–8; nutrients 357–9; primary production 352–65; respiration 353–4, 359–60; temperature 355–7; water 357
phylogeny 264–5, 314–15
phylogeographic patterns 161–2
physical geography, tropical forests 57–64
phytoplasmas 227, 229
Pickett's models 144–5
Pilosa 267, 269
pine beetles 95, 106, 220
plant–animal interactions 37–43
plant functional types (PFT) 521, 548–50, 553
planting seedlings 400, 402–8
plants: allocation 352–3, 361–5; biodiversity 295–305; biological invasions 452–6; drought 363–4; fine roots 362; fire effects 117–24; leaves 362; liana diversity 187, 192–3; movement responses 517–26; mycorrhizal symbiosis 309–22; pathogens 227–8; productivity 361–5; recolonization of fires 120–4; stem growth 364–5; strong winds 128–37; water 363–4; woody plant tissue 364–5
plant–soil interactions 332–6
plant–species interactions 35–43
plastic change 518
Pleistocene megafaunal extinctions 616
plot-based evidence for change dynamics 172–82; aboveground dynamics 175–6; basal area change 173–9; below-ground dynamics 176–8; biases 173–4; boreal

forests 175, 177–80; carbon dioxide 176–80; disturbance 176, 178; future perspectives 178–80; nutrient availability 176–80; pollution 177; precipitation 176–80; roots 176–8; soil 176–8; specific errors 173–4; succession 178; temperate forests 175–7, 179–80; temperature 176–80; tropical forests 175–7, 179–80
policy, urban forests 627–8
political dimensions, agriculture 581–3
pollen 156–9, 164–5
pollination 156–9, 164–5
pollinators, fragmentation 418
pollutants, definitions 436–7
polluters, definitions 436–7
pollution: biotic response variation 444–5; carbon budgets 538; causality of responses 444–8; climate/climate change 445–7, 538; community-level effects 440–1; conservation 436–51; definitions 436–7; ecosystem-level effects 439–40; extent 437–43; future research directions 448–9; landscape-level effects 438–9; management 436–51; organism-level effects 443; plot-based evidence for 177; population-level effects 441–3; restoration 447–8; severity 437–43; smelters 437–40, 442, 444–5, 447–8; species-level effects 441–3
polyphyletics 239–49
population: differentiation 155–61; genetic diversity 154–61; hunting impacts 616–19; insects 95–102, 219–21, 223; pollution 441–3
post-glacial fire history 481–4
post-logging mitigation 432
potassium 325–8, 332–6
potential cover 46–7
practicalities: restoration 404–9; urban forests 624–5
precipitation: lianas 186–91, 194; plot-based evidence 176–80; tropical forests 56–60; vascular epiphytes 188, 200, 203, 209–11
predators: fragmentation 417–18; hunting 611–12, 617–18; insects 96–8, 100–1, 107, 220–1, 223
predictive methods, fire/climate change 473–87

Prideaux, Bruce 597–609
primary production 352–65, 527–8, 531, 534–5
primates 264, 267, 271
pristine 'old-growth' forests 173–80
Proboscidea 264, 267, 270
processed based climate change modelling 544
production, tropical forests 67–8
productive capacity 77–85, 88–9
productivity: carbon fluxes 531–6; lianas 193–4; plant allocation 361–5; residence times 531–6
progressive fragmentation 173
properties, indigenous knowledge 586–8
provisioning services 626
pruning 81–2
PSP *see* permanent sample plots
pulp production 246–7
pulse recurrence patterns 95–6
Punan villages 589
pure forest stands 344–5
Pureswaran, Deepa 93–113
Pyttel, Patrick 75–90

quantitative trait variation 162–4
Québec 15–16, 377
Queensland, Australia 597–8, 602–7

radiation 60, 339
rainfall: lianas 186–91, 194; tropical forests 56–60; vascular epiphytes 188, 200, 203, 209–11
rainforests: birds 279–81; mycorrhizal symbiosis 313–14; recreational activities 597–8, 602–7; tropical forests 56, 60–1, 66–7
rapid climate change 517–26
recensusing 173
recolonization 120–4
reconstructing fire history/climate change 476–80
recreational activities: fishing 597, 599; hiking 597–9, 604; human ecology 597–609; hunting 597, 599–600; management 600–2; modelling 600–5; rainforests 597–8, 602–7; sustainability issues 597–607; tourism 597–607; visitor

management 599–604; Wet Tropics World Heritage Rainforest 597–8, 602–7
Red List of Threatened Species 297–8
Reduced Emissions from Deforestation and Degradation (REDD/REDD+) 69, 385–96
reduced impact logging 431–2
reforestation 320–1, 391–2
refugia 283, 521–2
regeneration/regrowth: conservation 371–84; fire induced 120–4; harvesting 371–84; managed forests 79–81, 87–8; management 371–84; process of 373–4; restoration 398–409; seeds 371–81
regulating urban forests 626–8
reindeer 256–7
reintroducing mammals 275–6
Rejmánek, Marcel 452–69
remote sensing 25, 627
reproduction 120, 123, 205–6
reptiles 452, 462–3
residence times 531–6
resilience/resistance strategies 165
respiration 353–4, 359–60
response causality, pollution 444–8
restoration: agriculture 397, 401, 405, 409; climate change 397–8; conservation 397–410; dispersal agents 399–409; disturbance 397–400, 403, 405; ecosystem assembly 402–4; establishment sequences 403; framework species method 404–6; high diversity plantings 404–6; mammals 275–6; management 397–410; methods 398–404; mining 407; monitoring 407; natural regeneration 398–400; nurse trees 404–5; perch trees 404–5; pollution 447–8; practicalities 404–9; REDD+ 391–2; regeneration 398–409; seed dispersal 399–409; seeding 400–2, 407–8; seedlings 400, 402–8; species abundance 403–4, 408; succession 397, 402–8
retention capacity 242–5
retention forestry 430–1
retrogression 335–6
rhinos 616–17
Riutta, Terhi 527–43
rodents 264, 271–2

roots: characteristics 176–8; injuries 117–19; lianas 189–90; pathogens/pests 227, 229, 231–2
Rossi, Sergio 499–516
rotation, agriculture 573–5
rural livelihoods 562–5
Russian boreal forests 7

"sacred groves" 591
safety nets 564–5
Sakhalin Island 19–20
Saucier, Jean-Pierre 7–29
Savannah River Site Corridor Experiment 412
Scandentia 267, 270
Schmitt, Christine B. 295–306
Schnitzer, Stefan A. 185–97
science: indigenous knowledge 586–95; urban forests 624–5, 628–31
Sea of Japan 19–20
seasonality 30–1, 56–60
sediment dating 480
seeding 400–2, 407–8
seedlings 400, 402–8
seeds: dispersal 284, 399–409, 617; fire induced recolonization 120–4; genetic diversity 156–66; natural regeneration 371–81; predator populations 617–18; strong wind effects 129, 131–3
Sekercioglu, Çagan H. 279–94
selection, genetic diversity 157
selective logging 423–4
selective reporting 174
Severe Tropical Cyclone Larry 135
severity, pollution 437–43
Shackleton, Charlie M. 559–70
shade 35–7, 208
shelter, birds 280–2
shelterwood methods 78–81, 83, 85–7, 423
shield geology 62
shifting cultivation *see* agriculture
SiB *see* Simple Biosphere Model
Siberian boreal forests 7, 11–12, 17–19
significance, logging 422–3
silviculture 77–89, 102–3, 227, 229, 231–2, 429–30
Simple Biosphere Model (SiB) 547–8
single tree selection 83, 86

site conditions 32
site preparation 82–7
slugs 452, 456–8
smelters 437–40, 442, 444–5, 447–8
Smith, Hazel K. 352–65
Smith, Sandy M. 226–38
smouldering combustion 114–16, 119–23
Snaddon, Jake L. 411–21
snails 452, 456–8
snakes 286, 452, 462–3
snow 502, 508, 511–12
social perspectives: agriculture 581–3; managed forests 77–8, 87–9; northern temperate forests 33; REDD+ 393–4; urban forests 624–5, 628–31
social science research 631
soil: animals 39; biodiversity 304; carbon budgets 530–1, 533, 535–6; characteristics 176–8; substitute creation 207–8
soil conditions: biogeochemical cycling 327; boreal forests 22–3, 25; managed forests 77; mycorrhizal symbiosis 318; northern temperate forests 32; strong wind effects 135–6; tropical forests 61–4, 67–8
soil–plant interactions 332–6
soil–vegetation–atmosphere models (SVAT) 547–9
solar radiation 60, 339
sources, genetic diversity/variation 155–7
South Africa 51, 604
South America 46, 48, 51, 57, 60, 62–3
South Asian tropical forests 57
Southeast Asia 48, 57, 279–81, 286, 575–6
southern hemisphere, subtropical forests 48, 50–2
sowing, agriculture cultivation 572–3
spatial scales, strong winds 128, 137
specialization, insects 217–18
species abundance: insects 219–21; lianas 186–94; plant movement 521; restoration 403–4, 408; vascular epiphytes 200–1
species adaptation/survival 20–3
species–area relationships 411–12
species composition 360
species distribution 216–17, 521–2, 545–6
species diversity 239–42, 295, 297–8, 360
species-level effects, pollution 441–3
species movement 150–1

species–species effects, fragmentation 417
specific errors 173–4
speed of climate change 518–19
spontaneous spread 452–69
SPOT-4 VEGETATION satellite 57
spruce budworm 102–3, 502–4
standardised forest inventories 567
stand-levels: biogeochemistry 328–30; bryophytes diversity 241–2; hydrological cycling 342; managed forests 77–89; regeneration 377; succession 142–3
stem growth 364–5
stem injuries 117–19
stochasity, succession 148–9
stopover habitats 283
Stratford, Jerey A. 279–94
strong winds: Australia 130–1, 133–6; climate change 136–7; diversity effects 128–9, 132–3, 137; dynamics 127–40; ecological effects 127–40; forest function 128–9, 131; forest recovery 134; forest structure 128–30, 135–7; fragmentation effects 130, 137; notable events 134–5; nutrient cycling 128–9, 131; plants 128–37; scales 128, 137; effects on seeds 129, 131–3; soil condition effects 135–6; structural effects 128–30, 135–7; succession 128–9, 131–2, 137; topographic conditions 135–6; trees 128–37; tropical forests 127–37; United States of America 131, 134–5, *see also* winds
structural impacts: bird communities 283; climate change 552–3; logging 424–5; strong winds 128–30, 135–7; succession 141–4; tropical forests 66–7
Sturtevant, Brian R. 93–113
subarctic forests 256–7
subtropical forests 46–55; biogeography 48–52; birds 279–81, 286; boreal forests 25; climate change 53, 385–90, 394; climates 46–8; cyclones/winds 127, 131; deforestation 385–90, 394; degradation 385–90, 394; hunting 613–19; lianas 187; mycorrhizal symbiosis 320–1; northern hemisphere 48–50; northern temperate forests 30–1, 39, 42; potential cover 46–7; REDD+ 385–90, 394; southern

hemisphere 48, 50–2; succession 146–7; tropical forests 46, 48–52; vegetation distribution 53; zone extents 46
suburban forests 623, 628
succession: agriculture cultivation 575–6; biodiversity 146–7; biogeochemistry 335–6; boreal forests 24, 149–50; climate change 149–51, 545–7; complexity 148–9; compositional change 141–4; diversity 146–7; drought 150; dynamics 141–53; global changes 149–51; history 141–2; mechanisms 144–5; multiple pathways 148–9; northern temperate forests 34–5; pathways 148–9; plot-based evidence 178; restoration 397, 402–8; species movement 150–1; stochasity 148–9; strong winds 128–9, 131–2, 137; structural changes 141–4; subtropical forests 146–7; temperature changes 149–51; tropical forests 146–7
sulfur 325
sulfur dioxide 437
supply and demand 330–1
supply side management 601–4
surface fires 116–19, 475–7
survival mechanisms 20–3
sustainability: agriculture 577–8; indigenous knowledge 593–4; recreational activities 597–607; urban forests 626
sustained insect disturbances 95–6
sustenance, birds 280–2, 284
SVAT *see* soil–vegetation–atmosphere models
Sweden 7, 135
symbiosis, mycorrhizal symbiosis 309–24
system-level changes, succession 141

Tardif, Jacques C. 473–87
Tedersoo, Leho 309–24
temperate forests: biodiversity 297, 300–5; biogeochemical cycling 328; birds 282; carbon budgets 529–34, 537–9; cyclones/winds 127–8, 131–4; fire 115, 474, 481; harvesting 371, 378–9; hunting 610–13; lianas 187–8; lichens 253; mycorrhizal symbiosis 313; natural regeneration 371, 378–9; northern regions 30–45;
photosynthesis 358; plot-based evidence 175–7, 179–80
temperature: boreal forests 20–1; carbon budgets 527, 531–8; climate change 500–12, 517–23, 527, 531–8; photosynthesis 355–7; plant movement 517–23; plot-based evidence 176–80; succession 149–51; tree growth 500–12; tropical forests 60; vascular epiphytes 209–10
temporal scales: bryophyte diversity 242; strong winds 128, 137; tree growth/climate change 499–516
tending managed forests 81–2
terrestrial biophysics 545, 547–9
terricolous lichens 252, 254, 256, 258, 260–1
thermal environments 222
thermal refugia 283
Thiffiault, Nelson 371–84
thinning managed forests 81–2
threats: birds 285–7; boreal forests 25–7
tigers 26, 610–11, 615
timber production 245–7
time investment 567
time-since-fire (TSF) maps 477–8
Tinto Community Forest 593
tolerant plant species interactions 35–7
topographic conditions 135–6
tornadoes 127–8, 132–3
tourism 597–607
traditional forest management 231–3
Transbaikalian floristic subdivision 18–19
transport, water 341–2
trapping 591–2
tree growth: boreal forest droughts 488–91, 494–6; climate change 499–516; inter-/intra-annual temporal scales 499–516; stem growth 364–5; tropical forests 67–8
tree rings 499–512
trees: community composition 186, 191–2; density 186; diversity 185–6, 190–4; gene flows 154–71; genetic diversity 154–71; hydrological cycling 340; mortality 488–96; performance 185–6; plant allocation 361–5; strong winds 128–37
Tremblay, Francine 154–71
TRIFFID model 548–53

Index

tropical cyclones 127–37; Asia 127–8, 131, 136; Australia 130–1, 133–6; boreal forests 128, 131, 134; Europe 127–8, 131, 133–6; North America 127–8, 131; sub-tropical forests 127, 131; temperate forests 127–8, 131–4

tropical forests 56–74; altitude 57, 60; biodiversity 64–6, 295, 297, 299–305; biogeochemistry 328; biogeography 64–6; birds 279–87; boreal forests 25; carbon budgets 529–39; carbon cycling 68–9; climate 56–61, 64–6; climate change 60–1, 385–96; cloud cover 60; conservation 385–96; cyclones/winds 127–37; deforestation/degradation 385–96; diversity 64–6, 154–5, 159, 161, 163–4; fire 115; genetic diversity 154–5, 159, 161, 163–4; geography 57–64; geology 61–4; global carbon cycling 68–9; growth 67–8; hunting 613–19; hurricanes 127–37; insect disturbances 94–5; inter-/intra-annual seasonality 58–60; inventory plots 65; lianas 187–94; lichens 254; moist forest geography 56–64; mycorrhizal symbiosis 313–14, 320–1; northern temperate forests 39; physical geography 57–64; plot-based evidence 175–7, 179–80; precipitation 56–60; production 67–8; radiation 60; rainfall 56–60; rainforests 56, 60–1, 66–7; REDD+ 385–96; seasonality 56–60; soil conditions 61–4, 67–8; strong winds 127–37; structure 66–7; subtropical forests 46, 48–52; succession 146–7; temperature 60; typhoons 127–37; water 67–8

TSF *see* time-since-fire maps
Tubulidentata 267, 270
Turner, Edgar C. 411–21
Turton, Stephen M. 127–40
typhoons 127–37

Uganda 604
uncertainty, climate change 544, 550–3
underground hydrological cycling 340–1
ungulate overgrazing 40–1
United Nations Framework Convention on Climate Change (UNFCCC) 385–6

United States of America (USA): cyclones/winds 131, 134–5; indigenous knowledge 589–90; insect disturbances 94–5, 97–8, 100–2, 106–7; recreational activities 604
uptake, water 341–2
Ural Mountains 7, 11–12, 16–17
urban forests: ecological processes 624–7; governance 627–8; human ecology 623–33; management 623–31; mapping 627; policy 627–8; practicalities 624–5; regulation 626–8; science 624–5, 628–31; social perspectives 624–5, 628–31
urbanization 287
urban livelihoods 562
USA *see* United States of America

value aspects, urban forests 628–9
vapor pressure deficits (VPD) 189–90
vascular epiphytes 198–214; abundance 200–1; aerial habitat 201–6; air coolers/humidifiers 208–9; animals 204, 208; ants 206–7; influences on biota 206–8; carbon dioxide 210–11; climate 201–4, 208–12; climate change 209–12; cloud forests 208–10; cloud water capture 209; diversity 200–1; drought 200–1, 203–4, 209–10; geographic occurrence 200–1; global change 209–12; hypersensitivity 209–10; land use changes 211; nitrogen 210–11; nutrients 204, 207–8, 210–11; phorophytes 208; precipitation 188, 200, 203, 209–11; rainfall 188, 200, 203, 209–11; reproduction 205–6; soil substitute creation 207–8; temperature 209–10; vines 198, 203
vascular plants: biodiversity 301–4; lianas 188–91, 194
vegetation: boreal forests 11–20; carbon budgets 530–1, 533, 535; classification 57; climate change 475, 545, 547–54; distribution 53, 56–7; fire 475; plot-based evidence 173
velocity of climate change 518–19
Verkhoyansk Range 19
vertebrates 255–6, 297–8
vines 198, 203
viroids 227
viruses 227, 229

visitor management 599–604
vitality, managed forests 77–9, 84–5, 87–9
voltinism 222
VPD *see* vapor pressure deficits

Waito, Justin 473–87
Wallander, Håkan 325–38
wasps 462
water: delivery 26–7; evapotranspiration 341–2, 348–9; flow controls 345–6; hydrological cycling 339–51; managed forests 77, 81–4, 87; photosynthesis 357; plant allocation 363–4; retention 242–3; transport 341–2; tropical forests 67–8; uptake 341–2
waterlogging 318
weeding 575–6
West–Central North America 7, 11–15
western Canada 494–5
Western Siberia 11–12, 17
Wet Tropics Queensland World Heritage Area Rainforest (WTQWHA) 597–8, 602–7
WHA *see* World Heritage Areas
whitebark pine 234
wild cats 610–11, 615
wild cattle 616–17
wildfires 232
wildlife restoration 402–9
Willcock, Simon 172–82
winds: Asia 127–8, 131, 136; boreal forests 128, 131, 134; climate change 136–7; dynamics 127–40; ecological effects 127–40; Europe 127–8, 131, 133–6; fire 115, 121; genetic diversity 157; mycorrhizal symbiosis 317–18; North America 127–8, 131; northern temperate forests 42; sub-tropical forests 127, 131; temperate forests 127–8, 131–4; tropical forests 127–37, *see also* strong winds
wintering sites 283
Wisconsin, Menominee tribe of 589–90
WKH *see* Woodland Key Habitats
Wog Wog Habitat Fragmentation Experiment 412
wolves 610–12
Woodcock, Paul 422–35
Woodland Key Habitats (WKH) 430
woody plant tissue 364–5
Work, Timothy 93–113
World Heritage Areas (WHA) 597–8, 602–7
World Wide Fund (WWF) 299–303
WTQHWA *see* Wet Tropics Queensland World Heritage Area Rainforest

Xenarthra 267, 269
xylem formation 499–512

Yakutia 16–17
Yukon Territory 11–15

zones: boreal forests 7–9; northern temperate forests 30; pollution 437–43; recreational activities 600–2; subtropical forests 46; tropical forests 56
Zvereva, Elena L. 436–51